THIRD EDITION

Molecular Biology and Biotechnology

A Guide for Teachers

THIRD EDITION

Molecular Biology and Biotechnology
A Guide for Teachers

Helen Kreuzer, Ph.D.
Pacific Northwest National Laboratory
Richland, Washington

Adrianne Massey, Ph.D.
A. Massey & Associates
Chapel Hill, North Carolina

ASM
PRESS
Washington, D.C.

Address editorial correspondence to ASM Press, 1752 N St. NW, Washington, DC
20036-2904, USA

Send orders to ASM Press, P.O. Box 605, Herndon, VA 20172, USA
Phone: 800-546-2416; 703-661-1593
Fax: 703-661-1501
E-mail: books@asmusa.org
Online: estore.asm.org

ISBN 978-1-55581-471-7

10 9 8 7 6 5 4 3 2 1

Cover figure: Genetic scientists with plasmids in the background
(copyright K. G. Murti/Visuals Unlimited)

To Bruce Alberts and Patrick Fitzgerald—without their insight, encouragement, and concern for science education, this book would not exist.

About the Authors

Helen Kreuzer

received her B.S. degree in chemistry from the University of Alabama and her Ph.D. degree from Duke University's Department of Microbiology and Immunology and the Duke University Program in Genetics. She worked as an educator for many years, both as a college professor and in the area of teacher professional development. She and Dr. Adrianne Massey first met one another when Dr. Kreuzer became coordinator of the North Carolina Biotechnology Center's nationally recognized teacher education program. There, they developed the molecular biology curriculum that was later published as the text *Recombinant DNA and Biotechnology: A Guide for Teachers* by the American Society for Microbiology. The present volume is the third edition of that curriculum. Dr. Kreuzer and Dr. Massey have also written a college-level textbook for general audiences, *Biology and Biotechnology: Science, Applications, and Issues.*

Dr. Kreuzer has developed and taught numerous molecular biology minicourses for practicing secondary school and college faculty members. She served on the faculties of Salem College and Elon University, where she taught introductory cell biology, genetics, biology for non-science majors, molecular biology, and evolutionary biology. In 2002, Dr. Kreuzer returned to research in the Department of Biology at the University of Utah. There, she learned to use stable isotope ratios as tools in forensic science, environmental research, and basic biology. She has become an expert on the use of stable isotopes in microbial forensics and serves on an advisory panel for the Federal Bureau of Investigation. Dr. Kreuzer joined the scientific staff of Pacific Northwest National Laboratory in September 2005.

Adrianne Massey

received her B.S. and M.S. degrees from the Biological Sciences Department at the University of Georgia and her Ph.D. in zoology from North Carolina State University. Her company furthers informed science and technology policy development by counseling governments, mediating consensus-building activities, helping nonscientists understand biology, and training scientists in communications. Prior to founding her company, she was Vice President for Education and Training at the North Carolina Biotechnology Center, where she directed the Center's public education and workforce-training programs.

Dr. Massey served as the science advisor for the PBS series *BREAKTHROUGH: Television's Journal of Science and Medicine,* was the original director of the North Carolina Environmental Technology Consortium, served on the biological sciences faculty at North Carolina State University, and developed interactive exhibits for science and technology museums. In addition to publishing scientific articles on her research in evolutionary ecology, she has been an invited contributor to a number of government reports on biotechnology. She has served on a number of federal and state advisory panels on science education and training, technology development, and public policy. Internationally, she has worked with a variety of governments in Africa and Asia on issues related to biotechnology research and policy, and she participated in the international negotiations of the Biosafety Protocol. At the request of the U.S. State Department, she has served as an advisor to government officials, universities, and citizens' groups in the United Kingdom, Italy, and India and to international organizations, such as the World Trade Organization and the U.N. Food and Agriculture Program.

Note to Readers

This book, *Molecular Biology and Biotechnology: A Guide for Teachers,* is the third edition of the book formerly known as *Recombinant DNA and Biotechnology: A Guide for Teachers.* It contains four parts: *Laying the Foundation* (Part I), *Classroom Activities* (Part II), *Societal Issues* (Part III), and *Appendixes* (Part IV). ASM Press also publishes a student version of this text, *Molecular Biology and Biotechnology: A Guide for Students.*

Parts I of the teacher guide and student guide are identical. However, Parts II, III, and IV of the teacher guide contain materials intended solely for teachers, as well as all of the corresponding pages found in the student guide. Thus, the page numbers for the teacher and student guides are different from Part II on. Pages from the student guide are clearly labeled as *"Student Activity"* in the teacher guide.

Contents

Preface

When we got together to discuss the third edition of the book formerly known as *Recombinant DNA and Biotechnology,* we reflected on the stunning progress in molecular biology and biotechnology that has occurred in the 10 years since ASM Press first published our book. As we were writing the first edition:

- Geneticists were still 5 years away from having the entire sequence of a single human chromosome—chromosome 22. By 2003, the Human Genome Project, which provided the sequence of all human chromosomes, had been completed. Now, scientists have the complete genome sequences for more than 500 organisms.

- The only media star named Dolly was a country music singer, not a sheep, and most biologists believed that a fully differentiated cell could not be forced back to its undifferentiated state. Scientists have now used somatic cell nuclear transfer to clone more than 13 mammalian species.

- No one was concerned about the ethics of using embryonic stem cells to treat diseases, for biologists were still 5 years away from perfecting techniques that would allow them to keep these cells alive in cell culture.

- We were all quite familiar with tRNA (transfer RNA), mRNA (messenger RNA), and rRNA (ribosomal RNA), but no one was talking about miRNA (micro-RNA), dsRNA (double-stranded RNA), siRNA (small interfering RNA), and shRNA (short hairpin RNA), all of which are involved in RNAi (RNA interference). The 1998 elucidation of RNAi as a mechanism for regulating gene expression was so significant that it has already garnered its discoverers the Nobel Prize for Physiology and Medicine.

We also could not resist reminiscing about the events that brought us to this most unexpected place of writing a third edition of a textbook.

You see, we never set out to write a first edition of this book. Our more modest goal had been to produce a three-ring binder of lesson plans and background material for North Carolina's science teachers. We both worked for the North Carolina Biotechnology Center, one of the first biotechnology institutions to appreciate the fundamental importance of public understanding and well-trained workers to biotechnology development. To that end, in 1987, the Center established the first statewide biotechnology training initiative with the goal of helping teachers incorporate modern biotechnology into the state-mandated science curriculum.

In 1994, with money provided by North Carolina's biotechnology companies and General Assembly, we published and distributed the three-ring binder, *Teaching Basic Biotechnology,* to the 700-plus teachers who had been trained through the Center's summer workshops. And that was that. Or so we thought.

Shortly after we released *Teaching Basic Biotechnology,* Bruce Alberts happened to be visiting North Carolina. Most people know Bruce as the past president of the National Academy of Sciences. What they do not know is that Bruce is the father of a high school chemistry teacher and, not surprisingly, a very vocal champion of science teachers and a tireless advocate for improving science education in primary and secondary schools. Bruce looked through our book and announced, with great certitude, that it needed to be published nationally. He was convinced that teachers across the country would benefit from the information we created for North Carolina teachers.

We greatly appreciated the compliment; Bruce had authored a textbook on cell biology that we both admired greatly. It was beautifully written and infused with the sort of awe for living organisms that had inspired us both to become biologists. But, if truth be told, we thought no one would be interested in publishing our book, and we also had no idea how to go about finding a publisher. In addition, our temperaments are not inclined toward self-marketing.

A colleague just happened to have an ancient business card from someone in the book-publishing business who, according to the card, was a book editor for the American Society for Microbiology (ASM) Press. We thought this fellow, Patrick Fitzgerald, might be able to point us toward a potential publisher. Had we known that Patrick had become the director of ASM Press, we never would have been so bold as to call him. True to Patrick's style, he answered his own phone, so it never occurred to us that we were speaking with the number-one guy. He did not say a word as we described the book and asked if he could tell us how to go about finding someone who might be interested in publishing it. After a painfully long (for us) silence, Patrick said, "Wait. Let me get this straight. This book is already written?" He received assurances from Stanley Falkow, a professor at Stanford University, and Marshall Bloom, a researcher at the National Institutes of Health, both of whom generously agreed to review a high school textbook by unknown authors, that our book was scientifically accurate, and Patrick decided that ASM Press would publish it.

Patrick's decision to publish *Recombinant DNA and Biotechnology* was quite brave, as ASM Press had never tried to enter the already crowded market of high school biology textbooks. ASM Press had a well-defined audience that it knew how to serve quite well: professional scientists, universities, and medical schools. Patrick took a risk with us, and we are forever grateful to him. Patrick also did not blink when we told him we wanted to dedicate a science textbook to Bear Bryant, the former head football coach at the University of Alabama. We knew we had a friend for life.

Bruce Alberts, who generously authored the foreword of our second book with ASM Press, *Biology and Biotechnology: Science, Applications, and Issues,* continues to support our efforts to produce accurate and engaging texts and works diligently to improve science education at all levels. Patrick has moved on to become Publisher for the Life Sciences at Columbia University Press. Our book is now published in three languages and used by teachers at secondary schools and universities across the United States and the European Union and in Brazil, Taiwan, India, Japan, Turkey, and even Antarctica! However, without Bruce and Patrick, *Recombinant DNA and Biotechnology* would never have existed, which is why we are dedicating this edition to them.

Those of you who are loyal users of this text will notice a number of significant changes in this edition, the most obvious being the book's new title, *Molecular Biology and Biotechnology.* The previous title no longer captured the breadth of the book's content.

Another conspicuous change is the inclusion of a CD with this edition. The CD contains electronic files of the text's graphics, worksheets, and templates required for certain activities. In addition, the CD has the appendixes from the second edition that are essential for conducting the wet laboratories:

- Recipes and instructions for making solutions and media
- Information on sterile technique and laboratory biosafety
- Descriptions of laboratory equipment needed for some activities, such as micropipettes and gel boxes, and instructions on how to use it

We have added a substantial amount of new material to the text. Part I, *Laying the Foundation,* now has a much-needed chapter on cell biology. Since we define biotechnology as using cells and biological molecules to solve problems or make useful products, the new chapter is long overdue.

In Part II, *Classroom Activities,* you will see a new section entitled *Genomics.* This section contains activities from earlier editions, as well as brand new activities, *Mapping a Disease Gene* and *Microarray Analysis of Genome Expression;* a new reading, *Personal Genomics,* which describes the new field of pharmacogenomics; and *Comparing Genomes,* which builds on *Analyzing Genetic Variation* from the second edition. Another addition to Part II is *Medical Sleuth: A Story of Genetics in Action,* which, like the rest of the material in that section, helps students make connections among DNA gene expression, protein function, and classical genetics.

We added a significant amount of information to Part III, *Societal Issues,* because we are increasingly impressed by the importance of helping students learn how to think rationally about technology. Sadly, the media's tendency to focus on fear and controversy has worsened over time. This predilection is exacerbated by a wealth of misinformation dispensed via the Internet. The teacher's role in countering misinformation and correcting sloppy thinking has become even more essential. We hope the critical-thinking tools and processes for rational analysis, as well as the factual information on gene flow in plants, animal cloning, genetic screening, and embryonic stem cells, will assist you in this important task.

Finally, to update educators on advances in biology and biotechnology, we have inserted new information throughout the text. For example, we have added a description of the breakthrough technology RNA interference, which promises to be a powerful research tool and will have many commercial applications. You will also find substantially more information on proteomics, microarray technology, and other tools for identifying and making productive use of nature's rich supply of genetic variation, such as random amplification of polymorphic DNA, total-community genomics, and radiation hybrid cell lines.

Some of you may be using our textbook for the first time and are unfamiliar with our goals in writing this book. First, we wanted to provide background information to help educators stay abreast of advances in biology. Many educators began teaching before the molecular biology explosion occurred. As a result, updating educators on scientific findings is a requisite first step in enabling them to bring new-found knowledge into their classrooms. The first part of this book, *Laying the Foundation,* provides this fundamental information on biotechnology and the science underlying its development.

However, without clear, easy-to-follow activities that respect the constraints under which most educators operate, successful transmission of new understandings from teachers to students is unlikely. Part II, *Classroom Activities,* provides wet- and dry-laboratory activities for teaching both the basic science of molecular genetics and the hands-on techniques of DNA- and protein-based technologies. The activities, which are appropriate for audiences ranging from middle school students to upper-level undergraduates, are presented in lesson plan format.

Our experience with educators across the country has taught us that both types of instructional materials—fundamental grounding in the science and activities in lesson plan format—are essential for teachers interested in incorporating biotechnology into biology courses.

Working with high school teachers has taught us another invaluable lesson about teaching biology at all educational levels. Some of the activities in Part II utilize paper models or other objects to mimic molecular phenomena or laboratory techniques. The value of these activities may not be apparent to some postsecondary biology instructors. In fact, the activities may seem like superfluous games intended to keep students busy.

We are sympathetic to that point of view because it used to be ours. Originally, we believed that precise words and clear pictures would allow almost anyone to grasp the concepts contained in this book. However, after leaving an academic research environment, teaching audiences of various ages and educational levels, and learning about teaching and learning from teachers, we realized that the ability to think abstractly and to learn through visual and verbal information is much rarer than we knew. Activities that seemed silly to us originally—mimicking protein synthesis by stringing beads together, acting out transcription and translation, or splicing genes by cutting and pasting pieces of construction paper—we now accept as crucial components of learning. For most people, irrespective of age and educational level, molecular events can be understood *only* through manipulating models or acting out cellular and molecular events. No matter how apt our words or lucid our graphics, we often do not see that telltale sign of understanding—the light that appears in a student's eyes—until we have them *do* something.

Over time, we have come to appreciate these "superfluous" activities for another reason. When students are working with new laboratory equipment and techniques, they are unable to focus on the biological principle the activity is intended to teach because they are preoccupied with learning the requisite hands-on skills. Using dry-laboratory activities to teach the principle prior to the wet laboratories that teach the technique and reinforce the principle greatly increases comprehension.

Because biotechnology has generated wide public debate about a number of important social issues, we would be remiss if we described only the scientific foundations and technological applications of biotechnology. All too often, discussions about the impacts of science and technology are merely emotional exchanges of opinions that may have nothing to do with facts. These unproductive debates contribute to rather than alleviate confusion. In Part III, *Societal Issues,* we offer information and tools for rationally analyzing and discussing biotechnology issues, including potential environmental impacts, bioethical ramifications, and various roles governments play in facilitating and impeding scientific research and technological change.

The final segment of the book, Part IV, provides appendixes that contain biosafety information and instructions on aseptic technique, and it is followed by a glossary.

In this book, we want not only to provide information and activities, but also to share our understanding and appreciation of biology, which comes from years spent in an academic environment. During that time, we were fortunate enough to be able to immerse ourselves in biology. With biology, it takes years of single-minded pursuit of knowledge for that knowledge to be transformed into understanding. With that understanding comes a very deep respect and reverence for the workings of the natural world. If we convey just a fraction of that reverence to our readers, then we will have succeeded.

Helen Kreuzer
and Adrianne Massey

Acknowledgments

We are indebted to all who have shared their time, insights, and talents with us during the production of all three editions of this textbook.

First and foremost, we thank teachers across the country, especially those in North Carolina. Their enthusiasm for learning new and difficult material, determination to introduce this exciting science to their classes, willingness to share so generously with their colleagues, and unflagging devotion to their students have inspired and reinforced us. We feel blessed to have had the opportunity to work with all of them, but those who deserve special thanks because of their help in the production and field-testing of the first and second editions are Sherri Andrews, Leslie Brinson, Beverly Cea, Nancy Evans, Marilyn Garner, Britt Hammond, Bobbie Hinson, Marlene Jacoby, Elizabeth Rue, Thea Sinclair, and Brian Wood.

A number of people made significant contributions to some of the classroom activities. Karyn Hede tested the new genomics activities and improved the writing in all three editions. Louisa Stark, Director of the Genetic Science Learning Center at the University of Utah, gave us permission to use the microarray activity they developed. Amy Clark helped us develop the protein and bioinformatics activities. Marlene Jacoby provided us with the ethical decision-making model she uses successfully with her classes. Thomas Martin assisted with photography, graphics, and computer applications.

The text has benefited from the skills of our copy editor, Elizabeth McGillicuddy, and the artwork of Patrick Lane. Their conscientious oversight and attention to detail improved the quality of the book immensely.

Finally, thanks to those at ASM Press who over the years have become like family to us: Susan Birch, our production editor for all three editions; Jennifer Adelman and Laura Ledbetter; and, especially, Jeff Holtmeier, director of ASM Press, for his encouragement, support, and eagerness to improve the quality of biotechnology education.

PART I

Laying the Foundation

*L*aying the Foundation *contains background information on both the science that is fundamental to biotechnology and the technologies themselves. We introduce the broad scope of biotechnology before explaining its scientific underpinnings—cell biology and genetics, especially molecular genetics. We then describe how scientists apply their understanding of cellular and molecular biology in answering scientific questions and developing new technologies to solve problems and make useful products.*

An Overview of Biotechnology

1

Introduction

When one reads or hears the word "biotechnology," the word "revolution" is often close behind. This combination of words is appropriate in many ways, for advancements in biotechnology have the potential to revolutionize major aspects of our lives and our relationship with the natural world.

In the field of human health, biotechnology will bring new ways to diagnose, treat, and prevent diseases. Every aspect of food production and processing, from the seed placed in the ground to the meals on our tables, will be affected. Biotechnology is often touted as an environmental savior of sorts, for it can provide new, cleaner, renewable energy sources; methods for detecting and cleaning up environmental contamination; and manufacturing processes that are more environmentally benign than those being used today.

Even though people are certain that biotechnology is important, many are not sure they know exactly what biotechnology is. Such confusion is understandable, for "biotechnology" is an ambiguous term. To make matters even worse, the word is used differently by different people. So what is biotechnology anyway?

Biotechnology definitions

Defining "biotechnology" is actually very easy. Break it into its root words, "bio" and "technology," and you have the following definition:

Biotechnology: the use of living organisms to solve problems or make useful products.

After reading that definition, you may question the appropriateness of the phrase "biotechnology revolution." Living organisms have always met our needs for sustenance and comfort by providing us with food, shelter, clothing, and fuel. Human use of living organisms is inseparable from modification of them. In a sense, Stone Age farmers started the biotechnology revolution over 10,000 years ago when they domesticated plants and animals, because genetic modification is inherent in the domestication process (Figure 1.1). Our ancestors extended their use of living organisms to microorganisms around 8,000 years ago, when they began to exploit bacteria, yeast, and other fungi to convert grapes into wine, milk into yogurt and cheese, and grains into raised breads through the process of **microbial fermentation.**

Human use and manipulation of microorganisms extend well beyond food fermentations. Virtually all antibiotics come from microbes, as do the vitamins added to breakfast cereals and the enzymes that convert cornstarch to high-fructose corn syrup. Farmers have used microbes since the 19th century to control insect crop pests and have inoculated the soil with nitrogen-fixing bacteria to improve crop yields. Microbes have been used extensively in sewage treatment for decades. Certain vaccines are based on the use of live, but weakened, viruses or bacteria.

If biotechnology is over 10,000 years old, what is it about today's biotechnology that sets it apart from ancient biotechnology and elevates it to the status of a revolution? Scientific understanding distinguishes modern from ancient biotechnology. With earlier biotechnology, when our ancestors used organisms and attempted to change them to better meet their needs, they did not understand the mechanics underlying the life process they wanted to control and improve. In fact, they did not even understand that microorganisms were responsible for the fermentations that turned grapes into wine and milk into cheese until 150 years ago—a mere 7,850 years after they began using microbes for that purpose.

The lack of understanding meant that exploitations and manipulations of plants, animals, and microorganisms were trial and error ventures. Over the

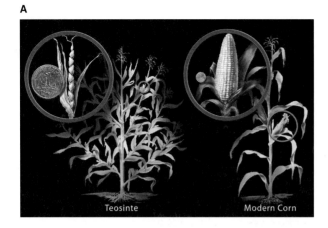

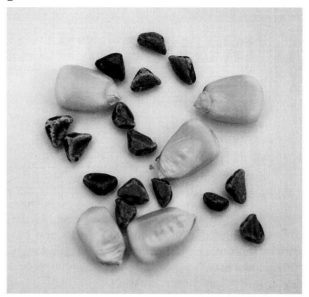

Figure 1.1 The wild ancestor of corn, teosinte, bears little resemblance to modern corn. (A) Ancient farmers in Central America used genetic modification through seed selection to convert teosinte, which had been a wild gathered plant and is only 3 to 4 in. long, into corn. (Image courtesy of Nicolle Rager Fuller, National Science Foundation.) (B) Seeds from modern corn and teosinte.

centuries, progress in the biological sciences provided insights into the inner workings of organisms, and with this understanding, people gradually became more proficient at using and improving them. During the 1960s and 1970s, knowledge of cellular and molecular biology reached the point where scientists could begin to use and manipulate organisms at those levels. Using and manipulating organisms for human advantage is not new. What is new is how people use and manipulate them. Now, they often use the cells and molecules of organisms in place of the whole organism. Now, they understand their manipulations at the most basic level, the molecular level. As a result, they are better able to predict what effect the manipulations will have, and they can direct the change they want with greater specificity.

Thus, "biotechnology" in the new sense of the word can be defined as follows:

"New" biotechnology: the use of cells and biological molecules to solve problems or make useful products.

Biological molecules are the large macromolecules unique to living organisms (Table 1.1). The biological molecules most often utilized in biotechnology today are **nucleic acids,** such as **DNA** and **RNA,** and **proteins.** As understanding of the roles other biological molecules play in cellular structure

and activities increases, researchers are finding, and no doubt will continue to find, new ways to use these molecules to our advantage.

For example, biologists have long thought of carbohydrates as comparatively simple biological molecules, important in energy storage and cell structure but lacking the complexity to carry out specialized functions that require molecular recognition and coordination. They have recently discovered that they underestimated the importance of carbohydrates in complex molecular interactions. For instance, carbohydrates stimulate immune responses by increasing the numbers and activities of white blood cells. They also act as road signs for the cells and chemicals involved in the inflammatory response. They are important for other physiological functions requiring cell-to-cell or molecule-to-molecule recognition, including antigen-antibody interactions. Both viruses and bacterial pathogens recognize and bind

Table 1.1 The four classes of biological molecules

Biological molecule	Examples
Lipids	Oils, steroid hormones, vitamin E
Proteins	Enzymes, collagen, hemoglobin
Carbohydrates	Glucose, starch, cellulose
Nucleic acids	DNA, RNA, ATP

to carbohydrates on the surfaces of their host cells. The improved understanding of carbohydrate functions and the capacity to manipulate carbohydrate molecules may lead to new types of health care products in the near future.

Biotechnology: A Collection of Technologies

Some of the confusion surrounding the word "biotechnology" could be eliminated by simply changing the singular noun to its plural form, "biotechnologies," because biotechnology is not a singular entity. Instead, biotechnology is a collection of technologies, all of which utilize cells and biological molecules. Some of the biotechnologies, such as bioprocessing technology and plant **tissue culture,** have been essential components of various industries, including pharmaceutical manufacturing, food and beverage production, and agriculture, for many decades. Other biotechnologies, such as genetic engineering and microarray technology, owe their existence to scientific discoveries made in the past few decades.

By developing technologies that use cells and biological molecules in place of whole multicellular organisms, researchers can capitalize on a critical aspect of life at the cellular and molecular level: the extraordinary specificity of the interactions. Because of this specificity, the tools and techniques of biotechnology are quite precise and, compared to earlier technologies, are tailored to operate in known, more predictable ways.

Any one of the biotechnologies can produce a wide variety of commercial products, and any one product may have diverse applications. For example, xanthan gum, currently produced by bacteria through bioprocessing technology, can be used by food processors to thicken salad dressings or by the petroleum industry to clean residual oil from oil wells. A genetically engineered bacterium could synthesize a protein-digesting enzyme that both dissolves blood clots and unstops drains.

Even though the investment community often refers to the "biotechnology industry," biotechnology is a set of enabling technologies used by a wide variety of industrial sectors. All of the biotechnologies described below can be used by the many industrial sectors listed in Table 1.2 to conduct basic and applied research, improve manufacturing processes, decrease costs, create new products, and improve existing products and services.

The Technologies and Their Uses

What are some of these technologies that use cells and biological molecules, and how are scientists using them?

MAb technology

Monoclonal-antibody (MAb) technology uses cells of the immune system that make proteins called **antibodies.** Your immune system is composed of a number of cell types that work together to locate and destroy substances that invade your body. One type of immune system cell, the **B lymphocyte,** responds to invaders by producing antibodies that bind to the foreign substance with extraordinary specificity. Scientists harness the ability of B lymphocytes to make these very specific antibodies. Because of their specificity, MAbs are powerful tools for detection, quantification, and localization, and measurements based on MAbs are fast, accurate, and extremely sensitive.

Diagnostic and Therapeutic Uses

The substances that MAbs detect, quantify, and localize are remarkably varied and are limited only by the substance's ability to trigger the production of antibodies. Home pregnancy kits use an MAb that binds to a hormone produced by the placenta. MAbs are currently being used to diagnose a number of infectious diseases, such as strep throat and gonorrhea. Because cancer cells differ biochemically from normal cells, medical researchers can make MAbs that detect cancers by binding selectively to tumor cells (Figure 1.2). In addition to diagnosing diseases in humans, MAbs are being used to detect plant and animal diseases, food contaminants, and environmental pollutants.

MAbs are also useful in the early stages of pharmaceutical research and development. Because of the specificity of MAbs, one molecule with therapeutic potential can be separated from a mixture of thousands of other molecules, which facilitates the discovery of new pharmaceuticals. Ultimately, physicians hope to have access to MAbs that can treat many types of cancer and other diseases. A MAb that specifically binds to tumors and is tagged with a radioisotope or toxin can deliver these tumor-killing agents directly to cancerous cells and bypass healthy cells. Physicians are currently using MAbs to treat autoimmune diseases, to prevent complications following heart bypass surgery, and to protect children at high risk from viral respiratory infections.

Table 1.2 Examples of industrial sectors affected by the biotechnologies

Human health care

Knowing the molecular basis of health and disease can lead to improved and novel methods for diagnosing, treating, and preventing diseases. Biotechnology products already on the market include detection tests for many infectious organisms, certain cancers, hormone levels, and genetic diseases; therapeutic compounds for rheumatic arthritis, diabetes, cystic fibrosis and other genetic diseases, multiple sclerosis, cardiovascular diseases, and many cancers; and vaccines for hepatitis B, meningitis, and whooping cough.

Agricultural production

The agricultural-production industry uses biotechnology to increase yields, decrease production costs, diagnose plant and animal disease, enhance pest resistance, improve the nutritional quality of animal feed, broaden the use of biological-control agents, and provide alternative uses for agricultural crops. Currently marketed products include insect- and disease-resistant crops, herbicide-tolerant crops, healthier oilseed crops, and crops that provide renewable sources of raw materials for soaps, detergents, and cosmetics.

Food and beverages

Food processing, brewing, and wine making have always relied on biotechnology to enhance the nutritional quality and processing characteristics of their starting materials—grains, fruits, and vegetables—as well as to improve the microorganisms that are essential to these industries. All fermented foods and beverages depend on the actions of microorganisms, which also serve as the sources of many food-processing aids, preservatives, texturing agents, flavorings, and nutritional additives, such as amino acids and vitamins. In addition, biotechnology-based diagnostic tests are improving food safety.

Enzyme industry

The enzyme industry and its products are essential to the operations of many of the other industrial sectors, such as food processing, textiles, and brewing. Microorganisms have been the essential manufacturing work force of this industry, and their impact will increase in the future as genetic engineering gives new manufacturing capabilities to standard production microorganisms and improves manufacturing-process efficiency and production economics.

Forestry/pulp and paper

Biotechnology is being used to create trees that are resistant to diseases and insects and to improve the efficiency with which trees convert solar energy to wood production. Extensive research is being conducted on microbes and their enzymes for pretreating and softening wood chips prior to pulping, removing pine pitch from pulp to improve the efficiency of paper making, enzymatically bleaching pulp rather than using chlorine, and deinking recycled paper.

Textiles

Many textiles, such as cotton, wool, and silk, occur naturally, while others are derived from natural substances, such as wood pulp. Biotechnology should have an indirect impact on the textile industry by improving the source materials, as well as a direct impact. Enzymes are currently used in natural-fiber preparation and value-added finishing of the final product, such as stonewashed denim jeans. Leather manufacturers use enzymes to remove hair and fat from skins and to make leather pliable. Genetically engineered microbes have produced textile dyes, such as indigo, and the protein found in spider silk.

Chemical manufacturing

Biotechnology can provide cleaner, more efficient ways of manufacturing chemicals than do current methods. Microbes have been used for decades to convert biological materials, such as corn, into feedstock chemicals. Public and private institutions are conducting research on increasing the use of plant biomass and microbial enzymes in chemical manufacturing, because both are likely to generate fewer toxic waste products.

Energy

Before fossil fuels can be used for energy production, sulfur must be removed, and biodesulfurization relies on microbes and their metabolic enzymes. Microbes have also been used to enhance oil recovery from in-ground crude oil formations for more than 30 years. In the future, as fossil fuels become depleted and oil prices increase, we will need to establish alternative energy sources, such as biomass-based fuels, like the ethanol that is currently added to gasoline. Advances in biotechnology are making the production of ethanol more attractive economically. Other potential areas of energy production include genetically engineered microbes to generate methane from agricultural or municipal wastes and photosynthetic microbes for hydrogen production. However, both will require a number of decades of research before they become economically viable.

Waste treatment

Microbes have always been essential for degrading organic wastes, whether the waste is generated by humans or agricultural and industrial operations. As the human population increases, supplies of potable water decrease, and the standard of living of people in developing countries improves, we will need to apply biotechnology to improving the efficiency of natural microbial degradation processes. In addition to utilizing microbes to break down wastes, we are also turning to microbes to help us clean up soils and water that have become contaminated with environmental pollutants.

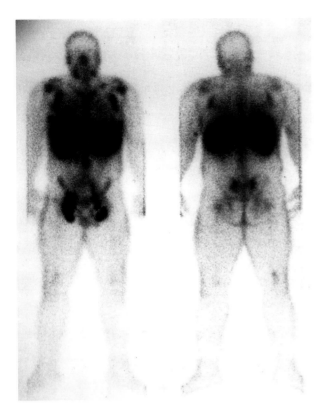

Figure 1.2 Radiolabeled antibodies confirm that the patient's cancer has spread to the lymph nodes, because radioactivity, indicated by the dark areas, is detected in the lymph nodes of the armpits, neck, and groin. A strong outline of the patient's body also verifies skin involvement in the cancer. The patient's liver and spleen are also darkened, because all antibodies normally collect in those organs. (Photograph courtesy of Jorge Carrasquillo, National Cancer Institute, National Institutes of Health.)

Bioprocessing technology

Bioprocessing technology, the oldest of the biotechnologies, uses living cells or components of their biochemical machinery to do what they normally do: synthesize molecules, change one molecule into another, break down molecules, and release energy. The living cells most frequently used are one-celled microorganisms, such as bacteria and yeasts, or mammalian cells; the cellular components most often used are proteins called **enzymes** that catalyze chemical reactions.

Enzymes are essential for life. They catalyze all cellular biochemical reactions, most of which would otherwise occur much too slowly to support life. Through enzyme-catalyzed reactions, organisms break down large organic molecules to obtain energy and to generate a supply of chemical building blocks for making new molecules. Companies are using the cells and their enzymes as **biocatalysts** to

commercially manufacture chemical compounds, generate energy, and break down chemical pollutants in the environment.

Fermentation and Mammalian Cell Culture

Companies commercially manufacture a wide variety of biotechnology products through large-scale fermentation and mammalian **cell culture,** two types of bioprocessing technologies that rely on the cellular enzymes that synthesize chemical substances. Because of the scale of the production systems, these technologies represent triumphs of both chemical engineering and molecular biology (Figure 1.3).

The oldest and most familiar bioprocessing technology is microbial fermentation. Originally, the microbial fermentation products people used were derived from the series of enzyme-catalyzed reactions that microbes use to break down glucose. In

Figure 1.3 Large-scale mammalian cell culture and microbial fermentation, carried out in bioreactors such as this, are biologically based manufacturing processes that utilize the biochemical machinery of cells to manufacture useful products. (Photograph courtesy of Diosynth RTP, Inc., a subsidiary of Organon.)

the process of metabolizing glucose to acquire energy, microbes synthesize useful by-products: carbon dioxide for leavening bread, ethanol for brewing wine and beer, lactic acid for making yogurt, and acetic acid (vinegar) for pickling foods (Figure 1.4A and B).

Companies extended their use of the rich biochemical machinery of microbes beyond the metabolic pathway for glucose breakdown once they discovered the wide array of useful products that certain microbial species produce, such as antibiotics and other pharmaceuticals, amino acids, hormones, vitamins, industrial solvents, pesticides, food-processing aids, pigments, enzymes, enzyme inhibitors, phar-

maceuticals, and even the botulinum toxin commonly known as Botox (Table 1.3). With the new capabilities provided by **genetic engineering,** researchers are providing various microorganisms with the capacity to synthesize novel molecules, including human insulin, essential blood proteins, and vaccines against human and animal diseases.

As experience with microbial production of recombinant proteins has grown, scientists have learned that some proteins must be produced by mammalian cells to be therapeutically effective. As a result, reliance on mammalian cell production systems has grown and is expected to increase further in the future.

Figure 1.4 (A) Useful metabolic products (in boldface type) of glucose breakdown provided by various microorganisms. (B) Chemicals currently produced by microbial fermentation of glucose and their industrial applications.

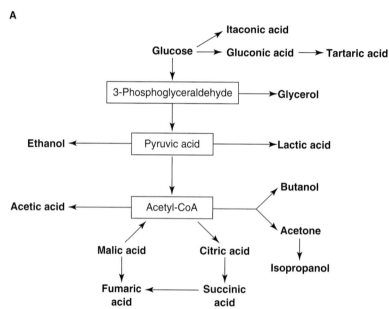

Organic chemical	Microbial sources	Industrial uses
Ethanol	*Saccharomyces*	Industrial solvent, fuel, beverages
Acetic acid	*Acetobacter*	Industrial solvent, rubber, plastics, food acidulant (vinegar)
Citric acid	*Aspergillus*	Food, pharmaceuticals, cosmetics, detergents
Gluconic acid	*Aspergillus*	Pharmaceuticals, food, detergent
Glycerol	*Saccharomyces*	Solvent, sweetener, printing, cosmetics, soaps, antifreeze
Isopropanol	*Clostridium*	Industrial solvent, cosmetic preparations, antifreeze, inks
Acetone	*Clostridium*	Industrial solvent, intermediate for many organic chemicals
Lactic acid	*Lactobacillus, Streptococcus*	Food acidulant, fruit juice, soft drinks, dyeing, leather treatment, pharmaceuticals, plastic
Butanol	*Clostridium*	Industrial solvent, intermediate for many organic chemicals
Fumaric acid	*Rhizopus*	Intermediate for synthetic resins, dyeing, acidulant, antioxidant
Succinic acid	*Rhizopus*	Manufacture of lacquers, dyes, and esters for perfumes
Malic acid	*Aspergillus*	Acidulant
Tartaric acid	*Acetobacter*	Acidulant, tanning, commercial esters for lacquers, printing
Itaconic acid	*Aspergillus*	Textiles, paper manufacture, paint

Table 1.3 Examples of the range of bioprocessing products and typical organisms used to manufacture them

Product	Typical organism or cell
Bulk organics	
Acetone/butanol	*Clostridium acetobutylicum*
Organic acids	
Citric acid	*Aspergillus niger*
Lactic acid	*Lactobacillus delbrueckii*
Amino acids	
Glutamic acid	*Corynebacterium glutanicum*
Lysine	*Brevibacterium flavum*
Microbial transformations	
Sorbitol to sorbose (vitamin C)	*Acetobacter suboxydans*
Steroids	*Rhizopus arrhizus*
Antibiotics	
Penicillins	*Penicillium chrysogenum*
Cephalosporins	*Cephalosporium acremonium*
Tetracyclines	*Streptomyces aureofaciens*
Extracellular polysaccharides	
Xanthan gum	*Xanthomonas campestris*
Dextran	*Leuconostoc mesenteroides*
Enzymes	
α-Amylase	*Bacillus amyloliquefaciens*
Pectinase	*Aspergillus niger*
Vitamins	
B_{12}	*Propionibacterium shermanii*
Riboflavin	*Eremothecium ashbyii*
Pigments	
β-Carotene	*Blakeslea trispora*
Vaccines	
Diphtheria	*Corynebacterium diphtheriae*
Poliomyelitis	Monkey kidney cells
Rubella	Hamster kidney cells
Hepatitis B	Recombinant yeast
Therapeutic proteins	
Insulin	Recombinant *Escherichia coli*
Growth hormone	Recombinant *Escherichia coli*
Erythropoietin	Recombinant mammal cells
Factor VIII-C	Recombinant mammal cells

Biodegradation

Intentionally using microbes to degrade unwanted substances also has a long history and has helped to lessen the impact human activities have had on the environment. For example, communities have long relied on complex populations of microbes in treating sewage. Microbes and the enzymes they use to break down organic molecules are helping us clean up new environmental problems: oil spills and toxic-waste sites. The use of microbial populations to clean up pollution is known as **bioremediation**. Probably the best-known example of bioremediation is the use of oil-eating bacteria to clean up oil spills, such as the *Exxon Valdez* spill in Alaska's Prince William Sound in 1989 and spills in Iraq after the 1991 Gulf War (Figure 1.5).

Environmental scientists are also exploring the possibility of using plants to treat contamination caused

Figure 1.5 *Exxon Valdez* oil spill, Alaska. Rocks on the beach before (right) and after (left) bioremediation are shown. (Photograph courtesy of the U.S. Environmental Protection Agency.)

Thanks to Microbes

Sometimes microbial pathways and enzymes are used, not for the end product, but for a specific step in a pathway that confounds chemists attempting to make a specific molecule. One of the most famous examples of using a microbe to carry out one step in a manufacturing process comes from the pharmaceutical industry. In the 1940s, pharmaceutical chemists discovered the anti-inflammatory properties of the steroid hormone cortisone. However, because they needed over 700 pounds of starting material (which had to be extracted from the bile or adrenal glands of slaughtered animals) to produce 1 gram of cortisone in a 37-step chemical process, they concluded that economical production of cortisone through chemistry was impossible.

The extreme inefficiency of the chemical synthesis process could be traced to one thing: the chemists needed to put an oxygen molecule on one specific carbon atom in a molecule that was a cortisone precursor. They found a microbe with an enzyme specific for this step, converting the 37-step process to one with only 11 steps and cutting the costs of production by 70%. They began commercial production of cortisone soon thereafter. Eventually, researchers found a number of microbes with

enzymes for other steps in the process, and the price of cortisone today is 400 times less than its original price.

In addition to saving lives and improving the quality of life of millions of people who have arthritis, asthma, and severe allergies, these microbes have changed history. Many historians believe that John F. Kennedy could never have been elected president if cortisone had not been commercially available. Kennedy took cortisone to treat two very debilitating diseases, chronic colitis and Addison's disease.

Steroids produced by microbes have changed history and our personal lives in other ways, as well. The most widely used steroids are the estrogens and progesterone found in birth control pills. Pharmaceutical chemists ran into the same sort of problems when they tried to synthesize sufficient quantities of these hormones at an economical price. Luckily, they found microbes capable of carrying out the specific reactions that were impeding commercial development. So, the widespread availability of birth control pills hinges on a specific enzyme that catalyzes one step in a metabolic pathway of a microbe.

by sources of pollution, such as contaminated wastewater from certain industrial manufacturing facilities. This practice, which is called **phytoremediation,** uses certain plants to remove, destroy, or sequester contaminants. Some plants secrete enzymes that break down the contaminants. Other plants act as sponges for pollutants, such as organic solvents, petroleum derivatives, and toxic metals like lead, zinc, cadmium, and mercury. Some cultivars of these plants can tolerate high levels of radioactive elements.

As mentioned above, when microbes break down molecules, they release energy contained in the molecule. Many people hope that in the future societies will be able to use sewage and agricultural refuse as renewable energy sources by exploiting microbes that degrade these organic compounds and, in the process, release energy.

Cell culture technology

Cell culture technology is the growing of cells in appropriate nutrients in laboratory containers or in large **bioreactors** in manufacturing facilities (Figure 1.3).

Plant Cell Culture

Plant cell culture is an essential aspect of plant biotechnology. The centrality of cell culture to plant biotechnology stems from a property unique to plant cells, their totipotency, or the potential to generate an entire multicellular plant from a single differentiated cell (Figure 1.6).

When scientists genetically engineer plants, they insert new genes into single cells. When a leaf cell is genetically engineered to contain a useful trait, such as resistance to insect pests, that cell must develop into a whole plant if it is to be useful to farmers. This regeneration is accomplished through cell and tissue culture.

Plant cell culture can also be used to produce valuable molecules naturally found in plants, such as food flavors and compounds that have therapeutic value (Table 1.4).

Animal Cell Culture

Plant cell culture is not the only type of cell culture being applied to agriculture. Using insect cell culture

Figure 1.6 Stages of plant cell culture from callus, a mass of undifferentiated plant tissue, to plantlet.

to grow viruses that infect insects may enable agricultural scientists to broaden the application of species-specific viruses in insect pest control. Mammalian cell culture is also being used in livestock breeding. Large numbers of bovine zygotes from genetically superior bulls and cows can be produced and cultured before being implanted into surrogate cows.

As described above, animal cell culture is used to produce therapeutic proteins and MAbs. The medical community also uses animal cell culture in the research laboratory to study such topics as the safety and efficacy of pharmaceutical compounds, the molecular mechanisms of viral infection and replication, the toxicity of compounds, and basic cell biochemistry.

ES Cell Culture

Recently, scientists have successfully cultured a unique type of human cell, embryonic stem (ES) cells. This feat deserves particular attention because of the potential medical benefits it may provide and the ethical issues it raises.

Your body has a number of tissue-specific cells that are permanently immature and not specialized to perform a specific function. When that tissue needs a supply of specialized cells, the unspecialized, immature cells, which are called **adult stem (AS) cells,** divide. Each AS cell gives rise to two cells: another stem cell that remains immature and a cell that becomes specialized. Unspecialized cells in the liver can differentiate into one of a number of specialized liver cells, such as cells that produce bile or epithelial cells that line the bile duct. Stem cells found in the bone marrow can produce all blood cell types, as well as muscle, cartilage, and bone cells. But under normal circumstances, liver cells do not differentiate into white blood cells, nor do bone marrow cells become specialized for bile production.

ES cells can give rise to virtually any type of cell. This complete developmental plasticity sets them apart from AS cells and provides great flexibility as potential therapeutics. For example, if researchers develop the skill necessary to control the differentiation of human ES cells, they may be able to produce replacement cells to treat diabetes, Parkinson's disease, and heart disease, among others. We discuss the laboratory methods for producing ES cells and the ethical issues ES cells raise in Part III, Societal Issues.

Table 1.4 Useful molecules naturally produced by plants

Industry	Molecular product	Source	Purpose
Pharmaceuticals	Digoxin	Foxglove	Heart disorders
	Codeine	Poppy	Sedative, pain
	Vinblastine	Periwinkle	Leukemia
	Atropine	Nightshade	Antispasmodic
Food	Vanilla	Orchid fruit	Flavor
	Cloves	Myrtle tree flower buds	Spice
	Cinnamon	Laurel tree bark	Spice
	Alginate	Brown seaweed	Emulsifier
	Carrageenan	Red seaweed	Texturizer
	Pectin	Citrus peel	Jelly, gelled candy
	Bromelin	Pineapple	Enzyme
Agriculture	Pyrethrins	Chrysanthemum	Insecticide
	Nicotine	Tobacco	Insecticide
Textiles	Indigo	Bean plant	Dye
	Starch	Corn kernel	Sizing
Personal care	Papain	Papaya	Contact lens cleaner
	Shikonin	*Lithnospermin* root	Cosmetic color
	Jasmine	Tea olive tree	Perfume

Biosensor technology

Biosensor technology represents the joining of molecular biology and microelectronics. A **biosensor** is a detecting device composed of a biological substance linked to a transducer (Figure 1.7). The biological substance might be a microbe, a single cell from a multicellular animal, or a cellular component, such as an enzyme or an antibody. Biosensors measure substances that occur at extremely low concentrations.

How do biosensors work? Biosensors generate digital electronic signals by exploiting the specificity of biological molecules. When molecules of the substance being measured collide with the biological detector, the transducer produces a tiny electrical current. This electrical signal is proportional to the concentration of the substance.

Biosensors are being developed for uses as varied as measuring the nutritional value, freshness, and safety of food; providing point-of-care analyses of blood gases, electrolyte concentrations, and blood-clotting capability in emergency rooms and intensive-care units; monitoring industrial processes in real time with immediate feedback for process control; locating and measuring pollutants; and detecting minute quantities of substances in blood. By coupling a glucose biosensor to an insulin infusion pump, the correct concentration of glucose in blood could be maintained at all times in diabetics.

Genetic engineering technology

Genetic engineering technology is often referred to as **recombinant DNA technology.** Recombinant DNA is made by joining or recombining genetic material from two different sources. In nature, genetic material is constantly recombining. Each of the following is just one of many ways nature joins genetic material from two sources:

- **Crossing over,** which occurs between **homologous** maternal and paternal chromosomes during gamete formation
- Fusion of egg and sperm during fertilization
- Exchange of genetic material by bacteria through **conjugation, transformation,** and **transduction**

In each of these examples of natural recombination, when genetic material from two different sources is combined, the result is increased genetic variation. The genetic variation that exists in nature has provided the raw material for evolutionary change driven either by natural selection or by artificial selection imposed by humans.

Using Existing Genetic Variation in Selective Breeding

Throughout 99% of human history, all of the plants and animals that people relied upon were wild. Around 8500 B.C., rather than gather wild plants and hunt wild animals, they began to save seeds and intentionally plant them and to corral certain animals. As soon as humans domesticated plants and animals, they began to alter their genetic makeup. Certain individuals in a population had traits, and therefore genes, that were valued, and our ancestors chose those individuals to serve as parents for the next generation. By selecting certain genetic variants from a population and excluding others, they intentionally directed the recombining of genetic material. As a result, they radically changed the genetic makeup of the organisms they domesticated.

For thousands of years, our ancestors altered the genetic makeup of plants and animals without understanding the biological basis of their manipulations. They knew very little about reproduction and nothing about genetics. However, they did know that offspring tended to look like their parents. Even though they did not understand how and why selective breeding worked, the minimal understanding they amassed through past experience was sufficient to make major genetic alterations in crops compared to wild plant ancestors (Figure 1.1).

This same pattern of genetically modifying the organisms that society relies upon also holds true for microorganisms. Long before people understood that microbes were responsible for the fermentation processes that converted grapes to wine and milk to yogurt, they selected the most desirable microbial cultures.

Figure 1.7 Schematic drawing of a simple biosensor.

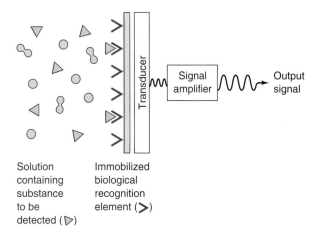

Solution containing substance to be detected (▷)

Immobilized biological recognition element (❭)

Transducer

Signal amplifier

Output signal

Thus, existing genetic variation has been a valuable natural resource that humans have exploited for centuries. The tools and knowledge to make selective breeding and microbial selection more predictable and more precise have continually evolved as scientific knowledge expanded. Genetic engineering is the next step in that continuum.

Generating Novel Genetic Variation with Genetic Engineering

The term recombinant DNA technology refers to the precise molecular techniques that join specific segments of DNA molecules from different sources. Scientists recombine DNA by using **restriction enzymes** (also called restriction endonucleases) designed to cut and join DNA in predictable ways. To ferry the DNA into the target organism, they usually use bacteria and viruses that transport DNA in nature, or simply their DNA molecules. Organisms that are provided with new genes from other organisms through molecular techniques are referred to as **transgenic** organisms.

Therefore, in addition to directing the recombining of genetic material through the intentional joining of eggs and sperm (or pollen in plants) in selective breeding, agricultural scientists can now recombine genetic material with greater precision by working at the molecular level.

Selective Breeding versus Genetic Engineering

Many scientists view genetic engineering as simply an extension of selective breeding, because both techniques join genetic material from different sources to create organisms that possess useful new traits. However, even though genetic engineering and selective breeding bear a fundamental resemblance to one another, they also differ in important ways (Table 1.5 and Figure 1.8).

In genetic engineering, scientists move single genes whose functions they know from one organism to another, while in selective breeding, sets of genes of unknown function are transferred. By increasing the precision and certainty of our genetic manipulations,

Table 1.5 Differences between selective breeding and genetic engineering

Parameter	Selective breeding	Genetic engineering
Level	Whole organism	Cell or molecule
Precision	Sets of genes	Single gene
Certainty	Genetic change poorly characterized	Gene well characterized
Taxonomic limitation	Usable only within and between species, sometimes genera	None

Figure 1.8 Schematic representation of the movement of a gene for a desired trait (colored circle) to a crop plant from its wild relative. Two methods of gene transfer are depicted: selective breeding and genetic engineering. Note that in selective breeding, sets of genes of unknown function are transferred from the wild relative to the crop plant. In genetic engineering, a single gene of known function is transferred.

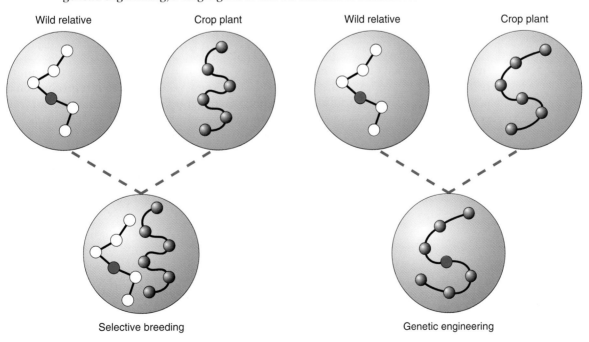

Wild relative   Crop plant   Wild relative   Crop plant

Selective breeding   Genetic engineering

the risk of producing organisms with unexpected traits decreases. The trial-and-error approach of selective breeding is circumvented.

In selective breeding, farmers and plant breeders have crossbred organisms in the same species, different species, and sometimes different genera—the next level of genetic difference—for centuries. For example, farmers began to artificially cross-pollinate plants in different species in the late 1700s. Sophisticated laboratory techniques have allowed modern plant breeders to create crops that would never exist naturally. Even so, plant breeders can force crop plants to crossbreed only with related plants. In genetic engineering, no taxonomic barriers exist. Genetic improvement of organisms is no longer restricted to the small pool of genetic variation within which plant breeders can manipulate breeding. A desirable gene from any organism may theoretically be placed in a second organism no matter how distantly related the two are. The potential to move genes between organisms provides great flexibility, because now society has access to all of nature's astounding genetic diversity. However, this unique capability also raises unique concerns, which we discuss in the last section of this book.

Case Study: *Bacillus thuringiensis*

An example of the flexibility genetic engineering provides is the research and development work that has been conducted on a variety of potential products that contain a gene from the bacterium *Bacillus thuringiensis,* or Bt for short. Bt is a naturally occurring organism, found in soils all over the world, that produces a protein that kills certain insects that ingest it, but Bt does not harm other organisms, such as fish, birds, or mammals. The degree of selectivity of the Bt protein is even more remarkable than being limited to insects. Each Bt strain is toxic only to a distinct group of insects. Certain Bt strains are toxic to the caterpillar stage of butterflies and moths, while others are toxic to beetles, and still others to mosquitoes. Each of these strains produces a slightly different version of the insecticidal protein.

Bt has been used for years as a biological control agent. The genes that code for the selectively toxic proteins in various Bt strains have been identified, and scientists have used recombinant DNA techniques to exploit the Bt that kills caterpillars in a variety of ways.

- Genetic engineers have modified the Bt organism itself by increasing the number of copies of the gene encoding the insecticidal protein. The modified strain of Bt can produce 10 times as much of the protein as the best nonmodified strains.

- The engineers have moved the Bt gene into a different bacterium that forms a protective capsule around the protein when the bacterium is killed. Within the capsule, the protein is protected against ultraviolet light and therefore persists in the field much longer than the Bt microbe does.

- Researchers have taken the Bt gene and moved it into bacteria that exist in symbiotic relationships with crop plants. For example, bacteria that naturally live inside cornstalks have been given the Bt gene to try to protect corn from caterpillars that bore into cornstalks. In a separate venture, a bacterium that lives on corn roots received a gene from a different type of Bt. This transgenic bacterium protected the plants against the corn rootworm, which is the immature stage of a beetle. In both cases, genetic engineers moved the gene into microbes whose lifestyles make them perfect candidates for pest control but that do not make pesticidal compounds.

- Finally, genetic engineers have moved the gene into a wide variety of plants, such as cotton, potato, and corn (Figure 1.9).

Protein engineering technology

Protein engineering technology is used, often in conjunction with genetic engineering, to improve existing proteins and to create proteins not found in nature. For example, medical researchers have used protein engineering to design novel proteins that can bind to and deactivate viruses and tumor-causing

Figure 1.9 (Left) Corn plant genetically engineered to produce the protein from Bt that kills caterpillars that ingest it. (Right) Damage by European corn borer caterpillars to cornstalks that do not contain the Bt gene. (Photograph courtesy of Syngenta, Inc., RTP, NC.)

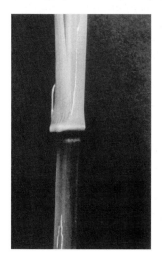

genes. As of now, research efforts are aimed primarily at modifying existing proteins, such as enzymes, antibodies, and cell receptors.

Why would researchers want to change existing proteins? The answer to this question varies with the protein. Research scientists alter cell receptor proteins to study how the less effective receptors change cell function. Modifying protein structure could also enhance the effectiveness of vaccines or decrease the allergenicity of proteins in food crops.

Enzyme Engineering

The most pervasive uses of protein engineering to date are applications that alter the catalytic properties of enzymes. Enzymes are beautifully designed for the role they play in nature as catalysts of the biochemical reactions on which living organisms depend. As such, enzymes function best under conditions that are compatible with life: neutral pH, mild temperature and pressure, and an aqueous (water-based) environment.

In certain manufacturing processes where catalysis by enzymes might prove useful, the conditions are too harsh for enzymes to function. Most enzymes literally fall apart at high temperatures, in very acidic or basic solutions, or when exposed to organic (non-water-based) solvents. Currently, researchers are modifying enzymes, both directly through chemical manipulations and indirectly by specifically mutating genes that code for enzymes, to increase their stability under harsh manufacturing conditions, to broaden their substrate specificities, and to improve their catalytic power.

Abzyme Engineering

Although most of the protein engineering work has been directed at changing the catalytic properties of existing enzymes, scientists have also invented a way to synthesize novel catalysts. Some researchers are synthesizing enzymes essentially from scratch, while others are creating antibodies with catalytic abilities, or **abzymes.** Antibodies resemble enzymes because both are proteins that bind to specific molecules. However, the similarity ends there. Antibodies bind for the sake of binding; enzymes bind to make reactions happen.

Scientists custom design abzymes to catalyze reactions for which there are no known enzymes. Some abzymes break down proteins, while others hydrolyze DNA. The most promising use of abzymes is as therapeutic agents capable of catalyzing specific proteins or activating antitumor chemotherapeutic compounds.

After inventing abzymes in the laboratory, researchers discovered that nature, once again, had beaten them to the punch. Antibodies with proteolytic (protein-degrading) and DNA-hydrolyzing activities occur naturally. The clearest example of the catalytic capabilities of some antibodies is hydrolysis of specific proteins in patients with autoimmune diseases, such as multiple sclerosis and autoimmune myocarditis. The roles of DNA-degrading abzymes are not quite clear yet. However, scientists have shown them to be powerful regulators of programmed cell death, or apoptosis, in certain cancers.

Cloning technology

Cloning technology allows scientists to generate a population of genetically identical molecules, cells, plants, or animals. Because cloning technology can be used to produce molecules, cells, plants, and some animals, its applications are extraordinarily broad.

Molecular, or gene, cloning, the process of creating genetically identical DNA molecules, provides the foundation of the molecular biology revolution and is a fundamental and essential tool of biotechnology research, development, and commercialization. Virtually all applications of recombinant DNA technology, from the Human Genome Project to pharmaceutical manufacturing to the production of transgenic crops, depend on molecular cloning.

In molecular cloning, the word clone refers to a gene or DNA fragment and also to the collection of cells or organisms, such as bacteria, containing the cloned piece of DNA. Because molecular cloning is such an essential tool of molecular biologists, in scientific circles, "to clone" has become synonymous with inserting a new piece of DNA into an existing DNA molecule.

Cellular cloning produces cell lines of identical cells and is also a fundamental tool of biotechnology research, development, and product manufacturing. All of the following applications depend on producing genetically identical copies of cells: the therapeutic and diagnostic uses of MAb technology described above, the regeneration of transgenic plants from single cells, pharmaceutical manufacturing based on mammalian cell culture, and generation of therapeutic cells and tissues, which is known as **therapeutic cloning.**

Animal cloning has helped animal breeders rapidly incorporate improvements into livestock herds for more than 2 decades and has been an important tool for scientific researchers since the 1950s. Animal cloning provides zoo conservationists with a tool for helping to save endangered species. In August 1998, for example, a rare breed of cow was cloned.

Although the 1997 debut of Dolly, the cloned sheep, brought animal cloning into the public consciousness, the production of an animal clone was not a new development. Dolly was considered a scientific breakthrough not because she was a clone, but because the source of the genetic material that was used to produce Dolly was an adult cell, not an embryonic one.

Recombinant DNA technology, in conjunction with animal cloning, is providing scientists with excellent animal models for studying genetic diseases, aging, and cancer and, in the future, will help them discover new drugs and evaluate other disease treatments, such as gene and cell therapy.

Antisense technology

Antisense technology is being used to block or decrease the production of certain proteins. Antisense technology is based on small, single-stranded nucleic acids (oligonucleotides) that prevent translation of the information encoded in DNA into a protein (Figure 1.10). Scientists have also discovered that using small pieces of double-stranded RNA has a similar effect. More details on the molecular aspects of antisense technology and the related **RNA interference** technology are provided in subsequent chapters.

The potential applicability of these closely related technologies is enormous. In any situation in which blocking a gene would be beneficial, antisense technology or RNA interference technology provides a valuable approach to the problem. An obvious example is a situation in which you want to block the production of a harmful protein product, such as the protein that causes cancer cells to proliferate or those involved in the inflammatory response. Currently, researchers are using this technology to slow food spoilage, control viral diseases, inhibit the inflammatory response, and treat asthma, cancers, and thalassemia, a hereditary form of anemia common in many parts of the world.

Metabolic Engineering

A less obvious but very exciting use of antisense technology is in **metabolic engineering.** In metabolic engineering, scientists use genetic modification techniques, such as antisense technology and genetic engineering, to shift existing biochemical pathways so that the production of certain molecules is favored over that of others and to provide organisms with novel metabolic capacities. To visualize how metabolic engineering might work, refer to Figure 1.4A, which depicts some of the products of glucose breakdown. If you wanted to maximize the production of isopropanol in a microbial fermentation process, which steps, and therefore enzymes, would you block with antisense technology?

Many compounds in nature that have great commercial applicability are not proteins. For example,

Figure 1.10 Schematic representation of antisense technology. In this example, the antisense oligonucleotide is an RNA molecule that blocks protein production by preventing the binding of messenger RNA (mRNA) to the ribosome.

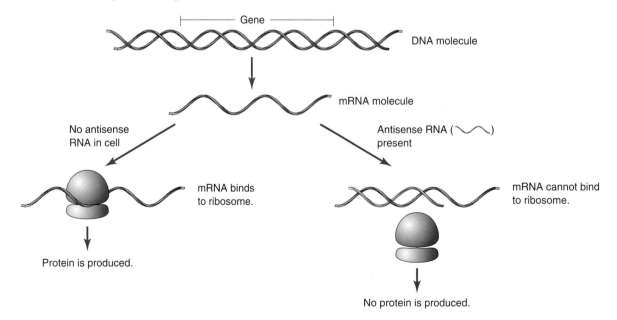

most compounds produced by plants to deter insect feeding could be useful as crop protectants but are not proteins. Metabolic engineering could be used to increase their production in crop plants. Antisense molecules would block the production of enzymes in certain pathways and reroute the plant's metabolism to favor the production of plant protectants.

On the other hand, researchers might want to decrease the production of a substance that is not a protein. An excellent example is cholesterol. Given the right antisense molecule, perhaps blocking key enzymatic steps in its synthesis would decrease cholesterol levels.

Microarray technology

Microarray technology is transforming laboratory research because it allows scientists to analyze tens of thousands of samples simultaneously. For example, thousands of different DNA, RNA, or protein molecules are placed on glass slides in a grid-like array to create **DNA chips, RNA chips,** and **protein chips.** Recent developments in microarray technology utilize customized beads in place of glass slides, but the principle remains the same.

DNA Microarrays

Researchers currently use microarray technology primarily for studies of gene structure and function. For analysis of genetic information, a DNA chip, also known as a **gene chip,** is created using a process similar to that used to manufacture microchips. However, instead of shining light through a series of masks to etch circuits into silicon, automated DNA chip makers use a series of masks to create a grid of DNA molecules. Each spot on the grid consists of identical pieces of DNA that are capable of bonding to complementary DNA and RNA molecules that are tagged with a fluorescent marker. Using a laser reader, a computer, and high-powered microscopes, scientists can determine the places where the fluorescently tagged DNA/RNA found a match with the chip-mounted DNA pieces.

In addition to playing important roles in research, DNA microarrays are already being used in clinics to detect mutations in disease-causing genes. In the future DNA chips will diagnose infectious diseases and tell doctors whether a pathogen is resistant to certain drugs, help crop scientists identify useful genes, and improve screening for microbes used in bioremediation.

Protein Microarrays

Gene sequence data provided by the Human Genome Project and other genome projects mean little until researchers determine what those genes do—which is where **protein microarrays** come in. While going from DNA arrays to protein arrays is a logical step, it is by no means simple to accomplish. The structures and functions of proteins are much more complicated than those of nucleic acids, and proteins are less stable than DNA. Each cell type contains thousands of different proteins, some of which are unique to that cell's job. In addition, a cell's protein profile varies with its health, age, and current and past environmental conditions.

In the future, protein microarrays will be used to discover protein **biomarkers** that indicate early stages of diseases, to assess the potential efficacy and toxicity of drugs before clinical trials, to measure differential protein production in different cell types and at various developmental stages, and to compare proteins produced by healthy and diseased cells.

Other Microarrays

Microarray technology was developed originally for genetic analysis, but the fundamental principle underlying this technology has inspired researchers to create many types of microarrays to answer scientific questions and discover new products (Figure 1.11).

Tissue microarrays, which allow the analysis of thousands of tissue samples on a single glass slide, are being used to detect protein profiles in healthy and diseased tissues and to validate potential drug targets. Brain tissue samples arrayed on slides with electrodes allow researchers to measure the electrical activity of nerve cells exposed to certain drugs.

Figure 1.11 Microarray technology. The "lab chip" devices shown here enable fast, automated analysis of DNA, RNA, proteins, and cells. These chips rely on the principles of microfluidics to manipulate tiny amounts of liquid within a miniaturized system. (Photograph courtesy of Agilent Technologies.)

Whole-cell microarrays circumvent the problem of protein stability in protein microarrays and permit a more accurate analysis of protein interactions within a cell. Small-molecule microarrays allow pharmaceutical companies to screen tens of thousands of potential drug candidates simultaneously.

Bioinformatics technology

Bioinformatics is the use and organization of information about biology. In biotechnology applications, bioinformatics technology exists at the interface of computer science, mathematics, and molecular biology.

The pace of discovery in molecular biology is breathtaking, due largely to the new research tools provided by biotechnology. One of the most formidable challenges facing biology researchers today is in informatics: how to make sense of the massive amount of data provided by biotechnology's powerful research tools and techniques. Scientists will simply drown in data unless they have informatics tools that allow them to collect, store, and retrieve information; manage data so that access is unhindered by location or the incompatibility of software and hardware; provide an integrated form of data analysis; and develop methods for visually representing molecular and cellular data.

Bioinformatics technology uses computational tools provided by the information technology revolution, such as statistical software, graphics simulation, and database management, for consistently organizing, accessing, processing, and integrating data from different sources. Bioinformatics consists, in general, of two branches. The first concerns the relatively simple problems of gathering, storing, and accessing data. The second branch focuses more on data integration, analysis, and modeling and is often referred to as **computational biology.**

As more and more information on genes, molecules, and cells is amassed, and as scientists need to integrate findings based on different experimental methods and different organisms, computers will become increasingly indispensable. **Systems biology** is the branch of biology that attempts to use biological data to create predictive models of cell processes, biochemical pathways, and, ultimately, whole organisms. Systems biologists attempt the Herculean task of elucidating the full complexity of the interactions that occur in biological systems. Only with iterative simulations generated by computers will they be able to develop a complete picture of the system they are studying. As an indicator of how essential

computers have become to biotechnology laboratories, the phrase "in silico" has joined in vivo and in vitro as a descriptor of experimental conditions.

Nanobiotechnology

Nanotechnology, which came into its own in 2000 with the birth of the National Nanotechnology Initiative, is the next stop in the miniaturization path that gave society microelectronics, microchips, and microcircuits. The word nanotechnology derives from nanometer, which is one-thousandth of a micrometer (micron), or the approximate size of a single molecule. Nanotechnology—the study, manipulation, and manufacture of ultrasmall structures and machines made of as few as one molecule—was made possible by the development of microscopic tools for imaging and manipulating single molecules and measuring the electromagnetic forces between them (Figure 1.12).

Nanobiotechnology, or bionanotechnology, joins the breakthroughs in nanotechnology to those in molecular biology, and practitioners in both fields benefit from one another. Nanotechnology researchers provide molecular biologists with tools for investigating the structures, functions, and interactions of biological molecules at the atomic level. Molecular biologists provide nanotechnologists with an understanding of and access to the nanostructures and nanomachines designed by 4 billion years of nature's engineering—the molecular components of a cell's machinery.

Using the tools provided by nanotechnologists, biologists can investigate the activities of individual molecules inside cells; visualize molecular interactions at the atomic level, such as those that occur between antibodies and antigens; measure and image the forces that bind molecules, such as the attraction of very specific molecules to cell membrane receptors; and investigate physiological function at the molecular level, such as interactions between actin and myosin (the molecular machines that drive muscle contraction).

Exploiting the extraordinary properties of biological molecules, nanotechnologists are able to accomplish many goals that are difficult or impossible to achieve by other means. For example, rather than build a framework for nanostructures by etching a piece of silicon, like a sculptor carving a piece of marble, nanotechnologists rely on the self-assembling properties of biological molecules to create nanostructures. DNA's ladder structure provides natural scaffolding for assembling nanostructures. In addition,

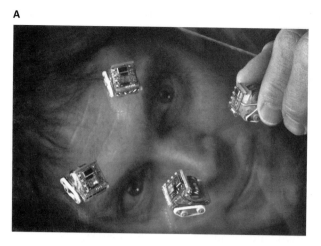

Figure 1.12 Nanotechnology. (A) Researchers at Sandia National Laboratories have built robots that are one-quarter of an inch long and weigh less than 1 oz. (B) Powered by three watch batteries, these autonomous, untethered robots may one day perform tasks carried out by large robots today, such as disabling land mines and detecting biological and chemical weapons. (Courtesy of Sandia National Laboratories; Randy Montoya, photographer.)

by using its highly specific bonding properties, DNA strands also serve as construction workers, creating a nanostructure by bringing together atoms bonded to its strands.

DNA has been used not only to build nanostructures, but also as an essential component of nanomachines. Most appropriately, DNA, the information storage molecule, may serve as the basis for the next generation of computers. The microchips that made the computer revolution possible by simultaneously getting smaller, cheaper, and faster are reaching size limitations imposed by the laws of physics. As microprocessors and microcircuits shrink to nanoprocessors and nanocircuits, the fundamental basis of computers may shift from electrons flowing through channels etched in silicon to DNA molecules mounted onto silicon chips. Such **biochips** are DNA-based processors that use DNA's extraordinary information storage capacity. They are conceptually very different from the DNA chips discussed above, even though the terms are sometimes used interchangeably. Biochips exploit the properties of DNA to solve computational problems; in essence, they use DNA to do math. Scientists have shown that computational problems a computer would take a century to solve can be solved by 1,000 DNA molecules in 4 months.

Biological molecules in addition to DNA are assisting in the continual quest to store and transmit more information in smaller spaces. For example, some researchers are using light-absorbing molecules, such as those found in the human eye's retina, to increase the storage capacity of CDs a thousandfold.

Some more imminent applications of bionanotechnology include increasing the speed and power of disease diagnostics, creating bionanostructures for getting pharmaceutical molecules into cells, and miniaturizing biosensors by integrating the biological and electronic components into a single minute component.

The Applications of Biotechnology

In the coming years, the diverse collection of technologies labeled "biotechnology" will yield an even broader array of products. Most of the commercial applications of biotechnology will be in three markets: human health care, agriculture, and environmental management (Figure 1.13).

Medical biotechnology

Biotechnology has already provided us with quicker and more accurate diagnostic tests, therapeutic compounds with fewer side effects, and safer vaccines.

Diagnostics

A physician's success in managing or curing a disease depends on diagnosing it accurately and early. Physicians can now detect many diseases and medical conditions more quickly and with greater accuracy because of the sensitivity of new diagnostic tools and techniques developed through biotechnology, such as MAbs, biosensors, DNA probes, DNA microarrays, **restriction fragment length**

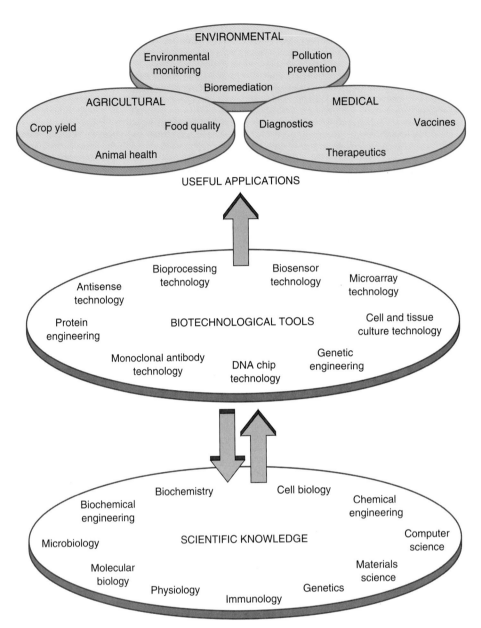

ENVIRONMENTAL
Environmental monitoring
Pollution prevention
Bioremediation

AGRICULTURAL
Crop yield
Food quality
Animal health

MEDICAL
Diagnostics
Vaccines
Therapeutics

USEFUL APPLICATIONS

BIOTECHNOLOGICAL TOOLS
Bioprocessing technology
Biosensor technology
Microarray technology
Antisense technology
Protein engineering
Cell and tissue culture technology
Monoclonal antibody technology
DNA chip technology
Genetic engineering

SCIENTIFIC KNOWLEDGE
Biochemistry
Cell biology
Chemical engineering
Biochemical engineering
Microbiology
Computer science
Molecular biology
Materials science
Physiology
Genetics
Immunology

Figure 1.13 Synthesis of scientific and technical knowledge from many academic disciplines has produced a set of enabling technologies—the biotechnologies. Any one technology will be applied to a number of industries to produce an even broader array of products.

polymorphisms, and the **polymerase chain reaction (PCR).** Most of these techniques are described in detail in subsequent chapters.

For example, the time required to diagnose infectious diseases, such as strep throat, gonorrhea, and chlamydia, has dropped from days to minutes. Early detection and intervention lessen the harmful effects of the infection on the patient and also have important implications for public health. Home pregnancy tests now allow women to determine if they are pregnant within days of conception.

Certain cancers are now diagnosed by simply taking a blood sample, thus eliminating the need for invasive and costly surgery. In some cases, the molecule that is used as the basis for diagnosis is secreted by precancerous cells, permitting intervention before cells turn cancerous and metastasize. Molecular footprints that are secreted by cells as the disease progresses from one stage to the next are known as **biomarkers.**

Through the Human Genome Project, scientists are also making remarkable progress in identifying and sequencing genes. These advances will greatly assist doctors in diagnosing hereditary diseases, such as type I diabetes, cystic fibrosis, early-onset Alzheimer's disease, and Parkinson's disease, that previously were detectable only after clinical symptoms appeared. In addition, genetic information may allow physicians to identify people with a propensity for diseases such as some cancers, asthma, emphysema, and osteoporosis. This knowledge gives patients an opportunity to prevent diseases by avoiding the triggers, such as smoking and diet.

In certain cases, biotechnology has also decreased the cost of disease diagnosis. A new blood test, developed through biotechnology, measures the amount of low-density lipoprotein, or "bad" cholesterol, in blood. Conventional methods require separate and expensive tests for total cholesterol, triglycerides, and high density lipoprotein cholesterol. Also, a patient must fast for 12 hours before the test. The new biotechnological test measures low-density lipoprotein in one test, and fasting is not necessary. We now use biotechnology-based tests to diagnose certain cancers, such as prostate and ovarian cancers, by taking a blood sample, eliminating the need for invasive and costly surgery.

Biotechnology-based diagnostics are also altering health care provision. Many of the new tests are portable, so physicians carry out tests and interpret results right at the patient's bedside. These point-of-care tests allow doctors to make decisions immediately rather than waiting hours or days for test results to come from a centralized hospital laboratory.

In addition, because many of the new diagnostic tests are based on color changes similar to a home pregnancy test, the results can be interpreted without expensive laboratory equipment, highly trained personnel, or costly facilities, making them more available to people in poorer rural communities and developing countries.

Therapeutics

Biotechnology will provide improved versions of today's therapeutic regimens, as well as treatments that would not be possible without these new techniques. Here are just a few examples of the novel therapeutic advances biotechnology now makes feasible.

Natural products as pharmaceuticals. Many plants produce compounds with human therapeutic value. For years, we have used a chemical derived from foxglove (digitalis) for treating heart conditions. A chemical extracted from yew trees is being used to treat breast and ovarian cancers (Figure 1.14). Scientists are also investigating ticks and leeches as sources for potential anticoagulant compounds and poison arrow frogs for painkillers.

Researchers have recently turned their attention to the extraordinarily diverse ecosystems found in the sea and have discovered compounds that heal wounds, destroy tumors, prevent inflammation, relieve pain, and kill microorganisms. As exciting as these developments are, the molecules with pharmaceutical potential could not be turned into commercial products without biotechnologies, such as

Figure 1.14 (A) The anticancer drug taxol occurs in the bark of the slow-growing Pacific yew tree, which reaches 5 ft in a number of decades. (B) It takes 30,000 pounds of bark to produce 1 kg of taxol. Between 2,000 and 4,000 trees are cut down to obtain that much bark. Plant cell culture provides another method for taxol production. (Photographs courtesy of National Institutes of Health; Mike Trumball, Hauser Northwest, photographer.)

A

B

cell culture, microbial fermentation, and animal cell culture. For example, researchers collected 2,400 kg of sponges to obtain 1 mg of a potential anticancer drug. Using these sponges as sources of a pharmaceutical would not be feasible economically; more important, it would be an ecological disaster. However, if the sponge cells can be maintained in cell culture, then the anticancer drug can be produced economically and without significant ecological impact. On the other hand, if the sponge cells are not amenable to cell culture, scientists might be able to identify the genes required to produce the compound and move them into organisms that do well under culture conditions.

Only by developing biotechnologies such as cell culture techniques, bioprocessing technologies, and recombinant DNA technologies can we maximize our use of the many natural compounds with therapeutic potential.

Endogenous therapeutic agents. The human body produces many of its own therapeutic compounds, and many of them are proteins. As proteins, they are prime candidates for possible production by genetically engineered bacteria. Such production would provide quantities that would allow us to better analyze their functions and would make their commercialization economically feasible. As we increase our understanding of these and other endogenous therapeutic agents, we will be able to capitalize more on the body's innate healing ability. Examples of endogenous compounds with therapeutic potential are the following.

- **Interleukin-2** activates **T-cell** responses (Figure 1.15).
- **Erythropoietin** regulates red blood cell production.
- **Tissue plasminogen activator** dissolves blood clots.

Biopolymers as medical devices. Nature has also provided us with substances that are useful medical devices. Some are superior to inorganic, manmade substances, because, being biological materials, they are more compatible with human tissues and are degraded and absorbed when their job is done. The manufacture of biological polymers is also more environmentally benign. Following are examples of naturally occurring biopolymers used as medical devices.

- The carbohydrate **hyaluronate** is a viscous, elastic, plastic-like, water-soluble substance that is used to treat arthritis, to prevent postsurgical scarring in cataract surgery, and for drug delivery.

Figure 1.15 Treating cancer with immune system molecules. Mice with lung cancer (left) were treated with activated killer T cells and interleukin-2. More than 250 tumor foci were reduced to fewer than 12 in mice receiving this treatment (right). (Photograph courtesy of Steven Rosenberg, National Cancer Institute, National Institutes of Health.)

- Adhesive protein polymers derived from living organisms are replacing sutures and staples in wound healing. They set quickly, produce strong bonds, and are absorbed.

- **Chitin,** a carbohydrate found in the exoskeletons of insects and crustaceans, combined with a natural fiber, polynosic, creates a material that limits bacterial and fungal growth.

Replacement therapies. Many disease states result from defective genes that cause the total lack or inadequate production of substances, usually proteins, the body normally produces. In the past, once medical researchers identified the missing or defective protein, physicians could treat the disease by giving patients the protein from other mammalian sources, if they could obtain large amounts of it. They gave diabetics insulin extracted from animal pancreatic tissue, collected at slaughterhouses; hemophiliacs relied on human blood transfusions to obtain the protein they lacked.

These life-saving tactics have some downsides, however. The animal protein is usually not identical to its human counterpart, so injecting it triggers an immune response in some people. In addition, acquiring the missing protein from extraneous sources carries with it the risk of contamination with pathogens or other harmful substances. Today, recombinant microbes and mammalian cells can manufacture human forms of the missing proteins under carefully controlled conditions.

Diseases now being treated by replacement proteins synthesized by recombinant microbes or mammalian cells include the following.

Hemophilia. Hemophiliacs lack certain proteins in the cascade that terminates in the formation of a blood clot. The missing protein in hemophilia A, **Factor VIII,** is now synthesized by recombinant mammalian cells, while the missing protein in hemophilia B, **Factor IX,** is synthesized by a recombinant microbe. Using pure forms of the missing proteins to treat hemophilia obviates the need for blood transfusions that can unknowingly transmit viruses to the recipient.

Type I diabetes. Type I diabetes results from an inadequate supply of **insulin,** a protein hormone that regulates blood glucose levels by affecting the cellular uptake of glucose. Biotechnology's first pharmaceutical product, human insulin, made with recombinant microbes, is now used to regulate blood glucose levels in most people with type I diabetes in the United States.

Emphysema. Some people who have never smoked tobacco begin to show signs of emphysema in their early 30s because they lack a functional form of the protein alpha 1-antitrypsin, which protects the lining of the lungs from a destructive enzyme produced by white blood cells.

Gaucher's disease. **Gaucher's disease** is caused by a deficiency of the enzyme that breaks down a particular type of lipid found in the membranes of red and white blood cells. Symptoms of Gaucher's disease include enlarged liver and spleen, painful bone lesions, and, in certain cases, damage to the nervous system.

Manufacturing pharmaceuticals by the use of transgenic plants and animals. Above, we described methods for producing large amounts of proteins by using microbial fermentation and mammalian cell culture. While these production systems have allowed pharmaceutical companies to produce sufficient amounts of therapeutic proteins, the costs of building and maintaining the manufacturing facilities are very high. Through genetic engineering, transgenic plants or milk from transgenic animals might become sources of pharmaceuticals.

Scientists have inserted the genes that code for therapeutic proteins into a variety of commonly grown crops, such as tobacco, corn, and soybeans. The proteins have maintained their stability for over 2 years in dried seeds of the transgenic plants. However, many people are concerned about using food crops to produce pharmaceutical compounds. In addition, some of the human proteins produced in plant systems are not therapeutically effective.

By using transgenic animals, such as goats and sheep, complex, therapeutically active human proteins can be expressed with great fidelity. Using goats rather than mammalian cell culture for production of a therapeutic protein could decrease the cost of production of 100 kg from $50 to 70 million to $16 to 25 million. However, obtaining purified proteins from milk is difficult, because milk contains fats, many proteins, minerals, sugars, and whole cells from mammary glands. Other concerns involve the ethics of using animals in this fashion.

Because both transgenic animal and plant production methods would require relatively small capital investments and minimal costs of production and maintenance, they may provide the only economically viable option for independent production of therapeutic proteins in developing countries.

Gene therapy. The ability to isolate and clone specific genes gives medical researchers the power to do far more than simply replace missing or dysfunctional proteins. Through gene therapy, physicians may be able to treat genetic disorders by giving patients functional genes in place of defective ones. Only certain inherited genetic diseases are amenable to correction via **replacement gene therapy.** Primary candidates for replacement gene therapy are hereditary diseases due to the lack of an enzyme, such as adenosine deaminase (ADA) deficiency, or the "bubble boy" disease. When ADA is not produced, intracellular toxins build up and kill cells of the immune system. Gene therapy experiments have been conducted on children who suffer from ADA deficiency (Figure 1.16) and cystic fibrosis, another inherited genetic disorder.

Hereditary disorders that can be traced to the production of a defective protein, such as **Huntington's disease,** are best treated with antisense technology, silencing the production of the defective protein.

Gene therapy has proved to be simpler in theory than it is in practice. The initial enthusiasm for the extraordinary promise of replacement gene therapy has been tempered with a large dose of reality. Enormous technical barriers must be overcome before replacement gene therapy can live up to its potential to cure inherited genetic disorders: getting replacement genes into the appropriate cells, inserting

A

B

Figure 1.16 Gene therapy. (A) Children with severe combined immune deficiency (SCID), which can have a number of causes, must spend their lives in germ-free environments. (Photograph courtesy of National Institute of Allergy and Infectious Diseases, National Institutes of Health.) (B) In 1991, two children with SCID received gene therapy to correct the genetic defect. (Photograph courtesy of Michael Blaese, National Institutes of Health.)

them into the proper site within the genomes of those cells, and getting them to function and respond to normal physiological signals.

Even though these technical impediments may have tempered the optimism for using replacement gene therapy for inherited genetic disorders, enthusiasm for the potential of more **transient gene therapy** as a tool for treating other diseases has grown. Leading scientists are investigating the use of briefly introduced genes as therapeutics for a variety of cancers, autoimmune diseases, acquired immune deficiency syndrome (AIDS), hemophilia, and other diseases (Table 1.6).

Cell therapy. Physicians may be able to correct certain diseases by providing patients with healthy versions of malfunctioning cells. The most familiar form of cell transplant therapy is the 20+-year practice of transplanting bone marrow cells into cancer patients. In some cases, the patient's own bone marrow cells have been removed, cultured, and reimplanted, because chemotherapy can be so devastating to bone marrow cells. If, however, the patient suffers from leukemia or another blood cell cancer, the transplanted bone marrow must come from a healthy donor with enough genetic similarity to the recipient for it to go unnoticed by the immune system.

For certain diseases, cell therapy, like gene therapy, comes closer to curing the disease than do repeated injections of missing proteins. For example, cell therapy may be a more favorable treatment for diabetics than daily insulin injections. Type I diabetes

is an autoimmune disease in which the body's immune system attacks the islet cells, the pancreatic cells that produce insulin. To treat type 1 diabetes, researchers have implanted insulin-producing cells

Table 1.6 Disorders for which gene therapy is being tested

Brain tumors
Hemophilia
Liver cancer
Prostate cancer
Hemoglobin disorders
Hyperlipidemia
Metabolic storage disorder
Human immunodeficiency virus infection
Colon cancer
Melanoma
Solid tumors
Graft vs. host in bone marrow transplants for leukemia
Point mutations in bone marrow cells
Head and neck cancer
Asthma
Focal muscle atrophy
Graft vs. host disease
Breast and ovarian cancer
Cardiovascular disease
Non-small-cell lung cancer
Liver cancer
Hypercholesterolemia
Infectious diseases
Neurodegenerative diseases

from organ donors into the subjects' livers. Eighty percent of the patients required no insulin injections 1 year after receiving pancreatic cells; after 2 years, 71% had no need for insulin injections.

Researchers have also achieved successful results with animals by giving them cardiac muscle cells to replace dead cells and neurons to correct neurological damage. Muscular dystrophy patients who received muscle cells have also shown some improvement.

Immunosuppressive therapies. Without a responsive immune system, we would face death as a real possibility every day. Sometimes, however, our vigilant immune system works against us. In organ transplant rejections and autoimmune diseases, suppressing our immune system would be in our best interest. Currently, physicians are testing the feasibility of using MAbs to accomplish this suppression. Here's how it works.

When exposed to foreign tissue in an organ transplant, the T cells of the immune system go to work to rid the body of this nonself component. By injecting the organ recipient with a MAb that binds to a protein found on the surfaces of T cells, the T cells are selectively incapacitated. Patients injected with this MAb show significantly less transplant rejection than do those given an immunosuppressive drug, such as cyclosporin. Because immunosuppressive drugs suppress all immune function, they leave organ transplant patients vulnerable to infection.

By using a specific MAb in place of an immunosuppressive drug, physicians selectively knock out only one aspect of the immune response, the T cells, leaving other aspects intact. Once again, because of the extraordinary specificity of this technique, they can zero in on the problem and perhaps cause fewer side effects.

In autoimmune diseases, such as rheumatoid arthritis and multiple sclerosis, our immune system turns against our own tissues, leading to the progressive degeneration of those tissues. MAbs are proving to be a successful therapy for slowing the progression of autoimmune diseases.

Cancer therapies. In addition to chemotherapeutic agents derived from plants and MAbs that selectively deliver toxins to tumors, biotechnology research tools have permitted progress in treating cancer on a variety of fronts. One is related to our new understanding of the genetic basis of cell growth and

differentiation. When genes involved in certain critical events of cell growth and development mutate, they become **oncogenes,** or tumor-producing genes. MAbs are being used to bind to and inactivate the proteins produced by these genes. Antisense technology may also provide us with a way to block the expression of these genes.

All of us normally have genes called **tumor suppressor genes.** When these genes function correctly, they suppress cell growth. When both copies of these genes become inactive or are absent, the genes then act as oncogenes. By introducing normal copies of the genes into tumor cells through gene therapy, the tumor may be made to regress.

A second front for treating cancer involves stimulating the immune systems of cancer patients through **cancer vaccines.** Unlike other vaccines, cancer vaccines are given after the patient has contracted the disease. The immune system has a difficult time distinguishing cancer cells from normal cells, because they are both "self." In addition, antigens from tumor cells stimulate a weak, ineffective attack by the immune system. Cancer vaccines help the immune system find and kill the tumor by intensifying the reactions between the immune system and the tumor antigen. Some cancer vaccines work by increasing the antigenicity of the tumor; others work by teaching the immune system to recognize the tumor as foreign (Figure 1.17).

Other candidate cancer therapies involve genetically engineering the tumor cells to make them more sensitive to drugs or to reduce their ability to evolve resistance to chemotherapeutic agents. Still other therapies use endogenous small proteins, such as the interleukins or **tumor necrosis factor,** to stimulate an immune system attack. Another interesting new approach is to use naturally occurring compounds, such as **endostatin** or other **antiangiogenesis** peptides, to starve tumors by cutting off their blood supply.

Regenerative Medicine

The human body has a remarkable capacity to repair and maintain itself. If necessary, your liver can regenerate up to 50% of itself, and by the time you finish reading this paragraph you will have manufactured 200 million brand new red blood cells. Medical scientists are excited about the prospect of using the body's natural healing processes, not simply to treat debilitating diseases, but to cure them. The body's toolbox for self-repair and maintenance contains many different proteins, such as growth factors, and

A 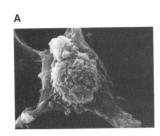 B 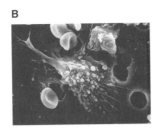 C 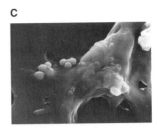 D

Figure 1.17 Some cancer vaccines teach the immune system to recognize tumors as foreign. (A) A metastatic cancer cell (note the long arms, or pseudopods, that permit certain cancer cells to move). (B) Immune system cells, the macrophages, recognize the cancer cell and begin to stick to it. (C) Macrophages inject toxins into the cancer cell, which begins losing its pseudopods. (D) The macrophages fuse with the cancer cell, which shrinks up and dies. (Scanning electron micrographs courtesy of Raouf Guirgus and Susan Arnold, National Cancer Institute, National Institutes of Health.)

various types of **stem cells** that have the capacity to cure diseases, repair injuries, and reverse age-related wear and tear.

Natural regenerative proteins. The human body produces minute amounts of a wide array of proteins, known as **growth factors,** that promote cell growth, stimulate cell division, and, in some cases, guide cell specialization into specific cell types. As proteins, they are prime candidates for large-scale production by recombinant cells and microbes, which makes their use as therapeutic agents economically feasible. Injections of these endogenous proteins can heal wounds and regenerate injured tissue.

- Epidermal growth factor stimulates skin cell division and could be used to encourage wound healing in burn victims.

- Fibroblast growth factor, which stimulates cell growth and division, has been effective in healing ulcers and broken bones and growing new blood vessels in patients with blocked coronary arteries.

- Transforming growth factor beta helps cells become specialized into different tissue types.

- Nerve growth factor encourages nerve cells to grow and could repair damage resulting from head and spinal cord injuries or degenerative neural diseases, such as Alzheimer's disease.

Stem cells. Most cells in the human body are fully differentiated, which means they have assumed a specific shape, size, and function. Some cells exist only to carry oxygen throughout the bloodstream, others to transmit nerve signals to the brain, and so forth. Stem cells are cells that have not yet become specialized.

After cells differentiate to form tissues and organs, some tissues and organs maintain a population of stem cells, known as AS cells, to replenish the supply of certain cells, such as red blood cells, and to replace tissue and organ cells that have died or been injured; other tissues have no resident stem cell populations.

AS cells are partially differentiated progenitor cells, and when an AS cell receives a cue to become fully differentiated, it first divides in two. One daughter cell specializes into a certain cell type, while the other remains undifferentiated, ensuring a continual supply of stem cells for regenerating that tissue type (Figure 1.18). All AS cells are **multipotent,** which means they can become various cell types within a certain class of cells, but different types of AS cells display varying degrees of plasticity regarding their potential fate. Bone marrow contains AS cells that can differentiate into any of the cell types found in blood, such as red blood cells and many types of white blood cells. Neural AS cells can become any of the specialized cells of the nervous system, such as neurons and the glial cells that support neurons. However, in the human body, bone marrow stem cells do not become neurons.

In the late 1990s, researchers reported that they had established lines of human ES cells by culturing some early, undifferentiated stem cells they had removed from **blastocysts**—the 5-day-old ball of approximately 150 cells that eventually develops into an embryo. This breakthrough opened up many avenues for regenerative medicine, because ES cells are **pluripotent,** which means they can become any kind of cell in the body. In addition to their total developmental plasticity, ES cells seem to be able to reproduce without limit.

The combined characteristics of developmental versatility and unlimited capacity for self-renewal make

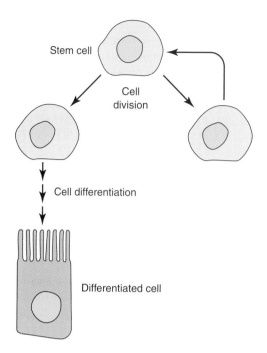

Stem cell

Cell division

Cell differentiation

Differentiated cell

Figure 1.18 To maintain a constant supply of stem cells while continuing to provide differentiated cells for renewing tissue, a single stem cell divides into two daughter cells. One daughter differentiates, and one daughter remains a stem cell.

both AS and ES cells excellent therapeutic tools with virtually unlimited potential. You are already aware that physicians use the AS cells found in bone marrow to treat various diseases. Now that researchers know how to maintain many different types of AS and ES cell lines in culture and are learning how to direct their development into specific cell types, their therapeutic potential might become reality. Although significant technical impediments exist and some very basic scientific questions remain unanswered, in the future, physicians may be able to use AS or ES cells to replace damaged or dead cells; reestablish function in stroke victims; cure diabetes; regenerate damaged heart muscle, spinal cord, or brain tissue; and treat diseases associated with aging, such as Alzheimer's disease.

The therapeutic potential of ES cells surpasses that of AS cells, because ES cells can become any type of cell. In addition, not all tissue types have a population of AS cells that can be mobilized for treating injuries and diseases. Encouraging results using ES cells to cure nervous system disorders, such as Parkinson's disease and spinal cord injuries, have been observed in laboratory animals. However, in spite of the encouraging results from animal research, using ES cells to treat human diseases is proceeding slowly in the United States due to ethical concerns. Some view therapies based on ES cells as unethical because 5-day-old embryos are the source of the ES cells. We

discuss the ethical and policy issues associated with ES cells in Part III, Societal Issues.

Tissue engineering. Tissue engineering, a marriage of cell biology and materials science, allows researchers to create semisynthetic tissues in the laboratory. These tissues consist of biodegradable scaffolding material plus living cells grown using cell culture techniques.

The most basic forms of tissue engineering use natural biological materials, such as collagen, for scaffolding. For example, one of the first products developed with this technology, two-layer skin, is made by infiltrating a collagen gel with **fibroblasts;** allowing them to grow, multiply, and become the dermis; and then adding a layer of **keratinocytes** to serve as the epidermis (Figure 1.19). The more sophisticated tissue engineering methods use implants composed of synthetic polymer scaffolding spiked with cells grown in the laboratory. The scaffolding is shaped to guide the structural development of the replacement tissue and then placed in the body where new tissue is needed. Adjacent cells, stimulated by the appropriate growth factors, invade the scaffolding, which is eventually degraded and absorbed.

Figure 1.19 Tissue engineering. A surgeon prepares to apply a sheet of artificially produced human skin to a patient. (Photograph copyright SIU/Visuals Unlimited.)

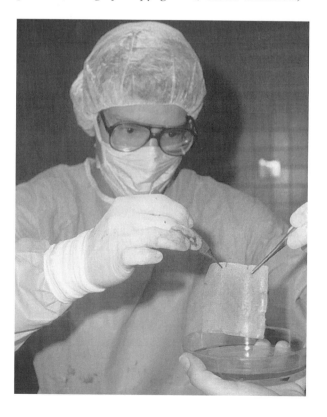

Vaccines

There is much truth to the adage "an ounce of prevention is worth a pound of cure." The best way to battle diseases is not to develop new therapeutics but to prevent the diseases. Through biotechnology, researchers are developing better preventative agents by improving on a practice that has been with us since the 19th century: vaccination.

Vaccine design and production. The vaccines that prevent smallpox and other diseases (polio, diphtheria, tetanus, and measles) are based on the use of either killed microorganisms or live but weakened microorganisms. When vaccinated with such a nonvirulent microbe, your body produces antibodies to that organism, but you don't get the disease. If you are exposed to that microbe again, your body has a ready supply of antibodies to defend itself. Vaccines are analogous to the "threat of war" that incites us to build up a supply of weapons. For the most part, vaccines cause no serious problems, but they do have side effects: allergic reactions, aches and pains, and fever. In a very few individuals, the vaccine has caused the disease it was intended to prevent.

A second problem with this method of vaccination is consistent production of virus-based vaccines. Growing large amounts of some human pathogenic viruses outside of the human body is not easy. Viruses are quasi-living organisms; they need the biochemical machinery of a living cell to reproduce (Figure 1.20).

Figure 1.20 To produce vaccines to viral diseases, the virus must be grown in living tissue. Typically, companies that manufacture vaccines use the embryos in chicken eggs. In this photograph, pockmarks (light areas) in chick embryonic tissue indicate colonies of the smallpox virus. (Photograph courtesy of John Noble, Centers for Disease Control and Prevention.)

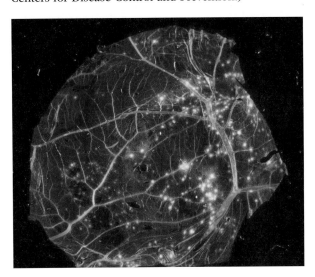

Finally, developing vaccinations for some deadly infectious diseases, such as human immunodeficiency virus infection/AIDS and malaria, is risky. Asking people to serve as volunteers to assess the safety and efficacy of these vaccines in early stages of development can also be ethically problematic. Because of these shortcomings, researchers have been unable to develop vaccines that successfully protect people from some of the deadliest diseases.

Using genetic engineering, scientists are improving existing vaccines and developing new ones. Usually, only one or a few proteins on the surface of the pathogen trigger the production of antibodies. By isolating the gene for the pathogen's cell surface protein(s) and inserting it into a yeast, or a bacterium such as *Escherichia coli,* bioprocess scientists can produce large quantities of the protein to serve as the vaccine. Because no live animals are needed for vaccine production, there are virtually no limitations on the amount of vaccine that can be produced. In addition, when the protein is injected, the body produces antibodies that can recognize the pathogen but suffers none of the adverse side effects that sometimes go hand in hand with vaccination. Using these new techniques of biotechnology, scientists have developed vaccines against diseases such as hepatitis B and meningitis.

DNA vaccines. Much to the surprise of many scientists, injecting naked DNA into muscles or skin cells also elicits an immune response. Researchers had assumed that DNA alone would not trigger an immune response of sufficient strength to impart protection against infectious diseases. However, in early gene therapy trials, the immune response against the therapeutic protein was too strong to make the gene therapy effective. While this result was disappointing to the gene therapy researchers, other scientists saw the positive side. The exciting discovery of DNA vaccines could lead to more advances in vaccine production, improved vaccines with fewer side effects, and more organisms for which we can develop effective vaccines.

How do DNA vaccines work? Researchers insert genes for one or more of the pathogen's proteins into a small, circular, noninfectious piece of DNA called a plasmid. After the plasmid is introduced into the host, the host cells synthesize the pathogen's protein(s) (Figure 1.21). Recognizing the protein as foreign, the immune system produces both antibodies and T cells specific for that antigen. DNA vaccines against AIDS, malaria, herpes, hepatitis B, and influenza are currently in clinical trials.

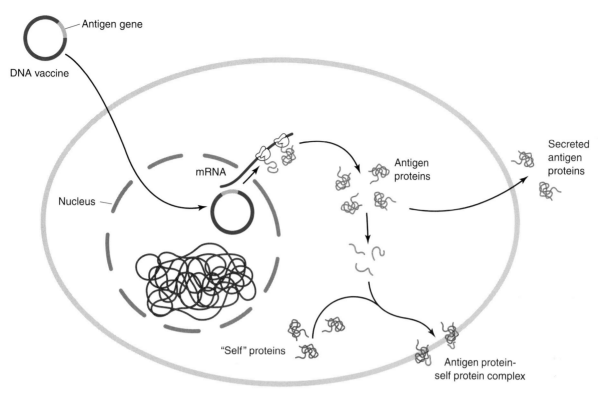

Figure 1.21 Plasmids altered to carry a gene for a protein (antigen) produced by a pathogen are injected into muscle cells. The gene encoding the antigen is transcribed in the nucleus into messenger RNA (mRNA), which moves to the ribosome, where it is translated into the antigen. The cell secretes some copies of the antigen into the bloodstream and chops others into small pieces. Proteins that identify every cell in the body as "self" carry the antigen pieces to the cell surface. In response, the immune system synthesizes T cells that will recognize the pathogen's antigen, while the secreted antigens trigger the production of antibodies by the B cells of the immune system.

Vaccine delivery systems. Whether the vaccine being developed is a live virus, a coat protein, or a piece of its DNA, the production of vaccines requires very elaborate and costly facilities and procedures, both of which are heavily regulated by the federal government. And then there is the issue of painful injections. How might the tools of biotechnology help to lessen these problems?

Earlier, we mentioned the prospect of using genetically engineered plants and animals for the production of therapeutic proteins. These transgenic organisms could also produce antigens for vaccines made of a pathogen's cell surface proteins.

Industrial and academic researchers are using biotechnology to develop edible vaccines. A company has genetically engineered goats to produce a malaria antigen in milk. University researchers have obtained positive results using human volunteers who consumed hepatitis vaccines in bananas and *E. coli* and cholera vaccines in potatoes. If edible vaccines become a reality, being vaccinated by drinking a glass of milk and eating a banana is much more appealing than a shot in the arm, isn't it? In addition, because these vaccines are genetically incorporated into food plants and need no refrigeration, sterilization equipment, or needles, they may be especially important for developing countries.

Medical Research Tools

One way that biotechnology will help us prevent diseases is less obvious than the ready example of vaccinations. For medical researchers, some of the most important outcomes of advances in biotechnology are not commercial products but the powerful research tools biotechnology provides.

The technologies of biotechnology discussed above are not simply techniques for producing vaccines, pharmaceuticals, and diagnostic kits. As research tools, these technologies have given us a much more valuable product: a deeper understanding of biological systems. They have provided us with new answers to old questions and have prompted us to ask questions we otherwise would never have dreamed of asking. Learning more about healthy biological processes and what goes wrong when they fail will enable researchers to develop even better diagnostics, therapeutics, and preventative agents.

One of the most powerful research tools provided by advances in biotechnology is targeted mutations, or **gene knockouts.** By disrupting specific genes, researchers gain valuable information about the role that the gene plays in healthy individuals. For example, scientists have used knockout mice that lack normal copies of DNA repair enzymes to investigate the roles these enzymes may play in cancer induction.

Agricultural biotechnology

The opportunities biotechnology will create for agriculture are as impressive and extensive as those for human health. We will witness progress in improving the quality, nutritional value, and yields of agricultural products and decreasing production costs. A related aspect of agriculture that will benefit from biotechnology is food processing.

Plant Agriculture

Because plants are genetically complex, plant agricultural biotechnology lagged behind medical advances in biotechnology. Another perhaps equally important factor that may explain the different rates of progress is the fact that over the years, animal research has received much more federal funding than plant research.

Nonetheless, scientists have made remarkable progress in plant biotechnology, largely because of improvements in two fundamental techniques of biotechnology: genetic engineering and plant cell and tissue culture. In the early 1980s, around the time companies were already using genetically engineered bacteria to produce human insulin, scientists discovered a way to genetically engineer plants by recombinant methods. To insert novel genes into plants, they exploited a genetic engineer that occurs in nature, *Agrobacterium tumefaciens.* A common soil bacterium, *A. tumefaciens* infects plants by injecting a portion of its DNA into plant cells (Figure 1.22). Since the mid-1980s, plant geneticists have used *A. tumefaciens* to genetically engineer many important agricultural crops and have developed a number of alternative methods for adding new genes to plants.

The traits agricultural scientists are incorporating into crops through genetic engineering are the same traits they have incorporated into crops

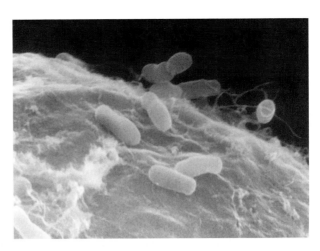

Figure 1.22 Scanning electron micrograph of *Agrobacterium tumefaciens* infecting a plant cell. (Photograph courtesy of Ann Matthysse, Biology Department, University of North Carolina, Chapel Hill.)

through selective breeding: improved nutritional content; resistance to diseases caused by bacteria, fungi, and viruses; better taste; the ability to withstand harsh environmental conditions, such as freezes and droughts; delayed ripening; and resistance to pests, such as insects, weeds, and nematodes.

Crop production and protection. In 2005, the 10th anniversary of the commercialization of transgenic crops, a farmer planted the billionth acre, the cumulative growth that has occurred in a decade. Since they first became available to farmers in 1996, the number of acres devoted to these crops has increased by double-digit growth rates every year. Globally, the number of acres of transgenic crops increased more than 50-fold in the first decade of commercial availability. This rate of adoption of a new farming technology is unprecedented in both American and global agriculture (Figure 1.23).

In its 2005 annual report on the global use of transgenic crops, the nonprofit organization International Service for the Acquisition of Agri-Biotech Applications (ISAAA) (http://www.isaaa.org) notes that in 2005, 8.5 million farmers in 21 countries planted 222 million acres of transgenic crops. All of these crops were genetically modified with recombinant DNA techniques for better pest management capabilities and included a number of insect-resistant, herbicide-tolerant, and disease-resistant crop varieties.

Seven of the 10 countries with the greatest number of acres of transgenic crops in 2005 were developing countries (Table 1.7). As you can see in Figure

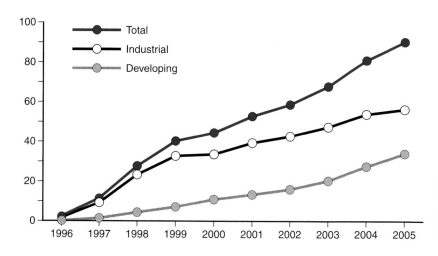

Figure 1.23 Global acreage of transgenic crops. Comparable data for adoption of transgenic crops by farmers in the United States can be found at http://www.ers.usda.gov/data/biotechcrops/alltables.xls.

1.23, the proportion of the global acreage of transgenic crops grown by developing countries has increased every year. Almost 40% of the 2005 acres were in developing countries, and the rate of increase from 2004 to 2005 was significantly greater for developing countries (23% growth rate) than industrialized countries (5% growth rate). Of the 8.5 million farmers who grew transgenic crops in 2005, 7.7 million were subsistence farmers in developing countries, primarily China, India, South Africa, and the Philippines, who had fewer than 10 acres of land.

Why has the rate of adoption of this new farming technology increased so dramatically, not only in the United States but also globally? According to ISAAA, "The continuing rapid rate of adoption of biotech crops reflects the substantial and consistent improvements in productivity, the environment, economics, health and social benefits realized by both large and small farmers . . . in both industrial and developing countries." More specifically, according to the most recent survey of the global impact of transgenic crops, conducted by Graham Brookes and Peter Barfoot, two economists from the United

Kingdom (http://www.pgeconomics.co.uk), the global net economic benefit for 2004 was $6.5 billion; the accumulated benefits from 1996 to 2004 were $27 billion, $15 billion of which accrued to developing countries. In addition to economic gains, Brookes and Barfoot also found a significant reduction in the global environmental impact of production agriculture from 1996 to 2004 associated with transgenic crops, due to a significant reduction in pesticide use (172 million kg of active ingredient); a 14% decrease in the environmental footprint associated with pesticide use (e.g., water use and soil erosion); and a reduction in greenhouse gas emissions that was equivalent to having 4.7 million cars off the road for a year.

Using biological methods to protect crops. Just as biotechnology is allowing us to make better use of the natural therapeutic compounds our bodies produce, it is also providing farmers with more opportunities to work with nature in plant agriculture.

Scientists have discovered that plants, like animals, have endogenous defense systems, the **hypersensitive response** and **systemic acquired resistance.** To exploit natural protection mechanisms, scientists are developing environmentally benign chemicals that can be used to trigger these two means of defense so that plants can better protect themselves against attack by insects and pathogens.

A truly extraordinary variety of alternatives to the chemical control of insects is available. All have this in common: they are biological solutions, based on understanding of the living organisms they seek to control.

Rachel Carson, Silent Spring, *1962*

To deter crop pests, we will also be able to rely more heavily on biological methods of pest control.

Table 1.7 Countries growing transgenic crops in 2005, ranked from most to fewest acres

1. United States	12. Romania
2. Argentina	13. Philippines
3. Brazil	14. Spain
4. Canada	15. Colombia
5. China	16. Iran
6. Paraguay	17. Honduras
7. India	18. Portugal
8. South Africa	19. Germany
9. Uruguay	20. France
10. Australia	21. Czech Republic
11. Mexico	

Biological control, or biocontrol, as it is often called, is the suppression of pests and diseases through the use of biological agents. For example, a virus may be used to control an insect pest, or a fungus may deter the growth of a weed (Figure 1.24).

Biological control has been used in agricultural systems in the United States since the 1800s, when farmers began using bacteria and a type of virus called a **baculovirus** to control pests, but the potential for biocontrol has been limited by the constraints that exist in nature. It is no simple task to find a specific microbe that kills a specific insect pest but not others, a certain fungus that causes diseases in weeds but not in crop plants, or an insect predator that preys only on "bad" and not "good" insects.

On the other hand, sometimes the unique characteristics of an organism or the ecological relationship between an organism and a crop would make the organism an excellent biocontrol agent, but it lacks the "ammunition." A perfect example of this situation is provided by the bacterium that lives in the stalks of corn plants that was mentioned above in the discussion of genetic engineering and Bt.

The flexibility that comes from genetic engineering removes those constraints. If scientists find a gene for insect resistance in an oak tree or a soil bacterium, they could incorporate it into crop plants. If a very rare mold that occurs only in showers in homes in Australia has a gene that codes for a protein that can kill weeds, scientists can isolate that gene, engineer it into bacteria, and have the bacteria manufacture the herbicidal protein using microbial-fermentation technology.

Figure 1.24 Parasitic wasps are effective biocontrol agents. Female wasps deposit eggs inside caterpillars, where larvae develop and feed. After the larvae emerge, they form white ovoid pupal cases attached to the caterpillar's body. (Image courtesy of Lacy L. Hyche, photographer.)

Exploiting cooperative relationships in nature. In addition to capitalizing on nature's negative interactions—predation and parasitism—to control pests, farmers might also use existing positive relationships that are important to plant growth. One example is the symbiosis between plants in the bean family and certain **nitrogen-fixing bacteria** (Figure 1.25). By providing the crop plant with a usable form of nitrogen, the bacteria encourage plant growth. Scientists are working to understand the genetic basis of this symbiotic relationship so that we can give nitrogen-fixing capabilities to crops other than legumes.

Figure 1.25 (A) Alfalfa nitrogen-fixing nodules. (Photograph copyright C. P. Vance/Visuals Unlimited.) (B) Alfalfa nodule cells containing actively nitrogen-fixing bacteroids. Magnification, ×640. (Photograph copyright C. P. Vance/Visuals Unlimited.)

A

B

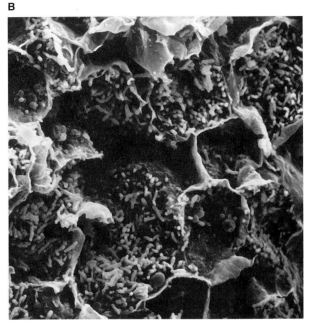

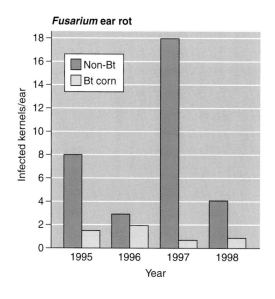

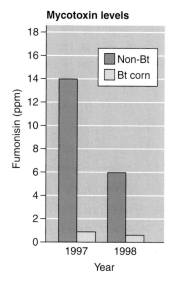

Figure 1.26 Because Bt corn provides better protection against insect damage, there are fewer holes in the corn plant that can be infected with fungal pathogens, such as *Fusarium*. University researchers in the United States, France, Italy, Spain, Turkey, and Argentina have documented lower levels of pathogen infection and mycotoxin contamination in Bt corn.

A second and less well-known example of a positive symbiotic relationship we might exploit for increasing crop production involves fungi and plant roots and is known as a **mycorrhiza.** In mycorrhizal associations, fungi extract nutrients from the soil and make them available to plants.

At present, plant genetic engineers (or any genetic engineers, for that matter) are limited to one- or two-gene traits, those controlled by one or two genes. Most of the agronomically valuable traits are multigenic; that is, they are controlled by many genes. One example is nitrogen fixation, which is controlled by at least 15 different genes. So, while the idea of having major crops obtain useful nitrogen from the air via a bacterium that lives on their roots is an appealing one, it is unrealistic in the near future.

Nutritional value of crops. The first generation of genetically engineered crops primarily benefited farmers. Although there are benefits to consumers in growing these crops, the benefits are largely invisible. For example, studies have shown that because Bt corn sustains relatively little insect damage, it is also infected significantly less often by fungi and molds that produce toxins that are fatal to livestock and harmful to humans (Figure 1.26).

The benefits of the next wave of agricultural biotechnology products to hit the market will be more obvious to consumers. Some of those benefits will involve improvements in food quality and safety, both of which we discuss later in this section. Biotechnology will also provide consumers with plant products that are designed specifically to be healthier and more nutritious.

Healthier cooking oils derived from biotechnology are currently being developed. Using genetic engineering, plant scientists have decreased the concentrations of saturated fatty acids in certain vegetable oils. They have also increased the conversion of linoleic acid to alpha-linoleic acid in these oils. Not only does this conversion decrease the saturated fatty acid concentration, it also increases the type of oil, found mainly in fish, that is associated with lowering cholesterol levels.

Another nutritional concern related to edible oils is the negative health effects produced when vegetable oils are hydrogenated to increase their heat stability for cooking or to solidify oils used in making margarine. Researchers have given soybean oil these same properties, not through hydrogenation, but by using genetic engineering to increase the amount of the naturally occurring fatty acid stearic acid, which provides greater heat stability.

Future nutritionally beneficial products may include grains engineered to have a complete protein profile; fruits, vegetables, and grains with improved vitamin and mineral contents; and potatoes with higher solid content, to reduce the amount of oil absorbed when they are fried.

Biotechnology also promises to improve the health benefits of functional foods. **Functional foods** are foods containing significant levels of biologically active components that impart health benefits or desirable physiological effects beyond our basic need for sufficient calories, the essential amino acids, vitamins, and minerals. Familiar examples of functional foods include compounds in garlic and onions that lower cholesterol and improve the immune response, antioxidants found in green tea, and the glucosinolates in broccoli and cabbage that stimulate anticancer enzymes. Scientists are using various biotechnologies to increase the availability of these types of compounds to consumers.

Animal Agriculture

The field of animal agriculture will continue to progress as biotechnology provides new ways to improve animal health and increase productivity. As is true of human health, improvements in animal health have come from advances in diagnostics, therapeutics, and vaccines. According to the U.S. Congress Office of Technology Assessment, animal diseases cost U.S. agriculture over $17 billion each year, so progress in detecting, curing, and preventing animal diseases will have considerable economic impact.

Animal health. Farmers are using MAbs on their livestock for early and more accurate diagnosis of certain diseases. The same principles discussed under human health apply here: because of their specificity, these diagnostics are faster, more accurate, and more sensitive than traditional diagnostics. Many of these diagnostics are also portable and allow veterinarians to diagnose livestock diseases on-site. Diagnosing a disease sooner and with greater accuracy means that appropriate therapy can be chosen and started sooner, thus decreasing the spread of the disease. Brucellosis, pseudorabies, scours, foot-and-mouth disease, and trichinosis are a few of the economically important infectious diseases veterinarians can now diagnose using detection technologies provided by biotechnology.

Another parallel exists in the type of therapies biotechnology is providing for both humans and animals. Genetic engineering is being used to produce sufficient quantities of endogenous therapeutic proteins found in animals. Cattle naturally make interferon and interleukin-2, proteins produced by the immune system to fight viruses. The genes for these proteins have been cloned and placed into bacteria for mass production. Injections of interferon and interleukin-2 can help decrease the incidence of shipping fever in cattle, a disease that costs the beef industry more than $250 million a year.

Great progress has also been made in disease prevention. Techniques being used to improve human vaccines are finding their way into veterinary medicine as well. Vaccines are now available for a number of diseases and parasites, including pseudorabies, foot-and-mouth disease, coccidiosis in poultry,

tapeworms, ticks, and tick-borne diseases. Scientists are also exploring edible vaccines for livestock.

A second route to disease prevention is to produce livestock that are resistant to diseases, just as agricultural scientists have created disease-resistant plants. Some breeds are naturally resistant to some bacterial diseases, such as mastitis, and this resistance has a genetic basis. If only one or a few genes are responsible for disease resistance, creating transgenic animals resistant to bacterial diseases may be possible.

Increasing productivity. Producers are always interested in improving the productivity of agricultural animals. Their goal is to obtain the same output (milk, eggs, meat, and wool) with less input (food), or increased output with the same input. Increasing muscle mass and decreasing fat in cattle and pigs has long been a goal of livestock breeders (Figure 1.27). These goals can be reached sooner by using genetic engineering to create transgenic animals with these traits. A number of industrial and academic researchers are investigating genetic engineering as a method for increasing the muscle mass in livestock.

Another method that is being used to increase the productivity of our livestock is a variation on the theme of selective breeding. Animal scientists first

Figure 1.27 A natural mutation of a gene involved in muscle growth leads to double muscling, which increases the productivity of beef cattle and improves the quality of their meat. (Photograph courtesy of Keith Weller, U.S. Department of Agriculture.)

choose those individuals that possess desirable traits. Then, instead of breeding the animals, they collect their gametes (eggs and sperm) and allow fertilization to occur in a laboratory dish. This in vitro fertilization is followed by embryo culture, a form of mammalian cell culture in which the fertilized egg develops into an embryo. At an early stage in development, the embryo is taken from the laboratory dish and implanted into a female of the same species but not necessarily of the same breed. This is known as **embryo transplantation.** Using this method, animal breeders can improve the genetic makeup of the herd more quickly than by simply relying on a single female who produces one calf per year. In the United States every year, surrogate mothers give birth to 100,000 calves produced from embryos of other, genetically superior cows (Figure 1.28).

Scientists have also tested ways to increase production efficiency by applying the same principle to some very different animals: cows, pigs, and fish. The principle involves increasing the level of endogenous growth hormone found in these animals. Protein synthesis in all animals is dependent upon growth hormone. As a result, growth hormone in cows, which is also known as bovine somatotropin (BST), is naturally important in growth and milk production. By slightly increasing a cow's blood growth hormone level with injections, farmers can increase the milk output or growth rate through stimulating protein synthesis.

Figure 1.28 Sylvia, the black and white cow, is the biological mother of all of the calves in the photograph. Embryos derived from Sylvia's eggs were implanted into the brown and white cows on the left, which served as surrogate mothers for Sylvia's calves. (Photograph courtesy of Colorado State University.)

One of the first products of biotechnology that became available to farmers was BST, or bovine growth hormone (BGH). Scientists isolated the gene for BST and inserted it into bacteria. Companies use these genetically engineered bacteria to produce BST. The BST is extracted from the fermentation tanks, purified, and injected into dairy cows, causing a 10 to 20% increase in milk production.

The safety of BST has been studied extensively by regulatory agencies in the United States, Canada, Europe, and many other parts of the world. Although all of these agencies have determined that BST is not active in humans and that meat and milk taken from cows injected with BST contain the same amount of BST as the milk of uninjected cows, some people remain concerned about the safety of BST. In addition, people who object to large-scale, industrialized agriculture see applications such as BST as furthering that trend. For both of these reasons, some consumers want to be able to purchase dairy products from cows that are not treated with BST. As a result, in certain grocery stores you may see milk labeled "rBGH free" ("r" for recombinant).

On the other hand, the product porcine somatotropin has applications that might be more attractive to some consumers. Injections of porcine somatotropin, the growth hormone in pigs that is the functional equivalent of BST, cause a dramatic increase in protein and a decrease in the amount of pork fat. Finally, scientists have introduced into salmon a gene that boosts the natural production of growth hormone. These farm-raised transgenic fish reach a marketable size sooner than fish with lower levels of fish growth hormone.

Food Processing

An aspect of food production that people often forget is food processing: what happens to the food when it leaves the farm. The food processing industry will be affected by a variety of developments in biotechnology. Some applications will involve the food product itself, but just as many, if not more, will be directed toward improving food additives.

Traditionally, microbes have been essential to the food processing industry, not only for the role they play in the production of fermented foods, but also as a rich source of food additives and enzymes used in food processing (Table 1.8). Through advances in genetic engineering and bioprocessing technology, their importance to the food industry will only increase in the future.

Table 1.8 Foods, food additives, and enzymes used in food processing for whose production microbial fermentation is essential

Foods

Cheese	Vinegar
Yogurt	Bologna
Buttermilk	Salami
Sour cream	Tofu
Soy sauce, tamari	Miso
Bread/baked goods	Sauerkraut
Wine, beer	Pickles
Cider	Olives

Food additives

Acidulants (citric and lactic acids)

Amino acids (glutamine, lysine)

Vitamins (β-carotene, riboflavin)

Flavor enhancers (monosodium glutamate)

Thickeners (xanthan gum)

Stabilizers (dextran)

Flavors (methyl salicylate-wintergreen, benzaldehyde-almond)

Enzymes

Proteinase (gouda and edam cheeses)

Peptidase (cheddar cheese)

Lipase (bleu cheese)

Amylase (high-fructose corn syrup)

Invertase (soft-centered candies)

Pectinase (fruit juices)

Product quality. Plant scientists are altering crops to have more desirable processing qualities. Many companies that produce soup, ketchup, and tomato paste now use tomatoes that were derived from a biotechnology technique, **somaclonal variant selection.** The new tomatoes contain 30% less water and are processed with greater efficiency. A 0.5% increase in the solid content is worth $35 million to the U.S. processed-tomato industry.

The first product of plant genetic engineering approved by the U.S. Food and Drug Administration was a tomato that is allowed to ripen on the vine instead of being picked while it is green. Calgene, the company that commercialized the Flavr Savr tomato, developed this tomato by biochemically separating the ripening process from the spoiling process. These processes usually go hand in hand, but in fact, some different enzymes, and therefore different genes, are involved. Using antisense technology, Calgene blocked a gene that codes for one of the enzymes involved in spoiling while leaving the ripening enzymatic pathways intact.

Food additives are substances used to increase nutritional value, retard spoilage, change consistency, and enhance flavor. The compounds food processors use as food additives are substances nature has provided and are usually of plant or microbial origin, such as xanthan gum and guar gum, which are produced by microbes.

Through genetic engineering, food processors will be able to produce many compounds that could serve as food additives but that now occur in scant supply or are found in microbes that are difficult to maintain in fermentation systems. We will also be able to capitalize on the extraordinary diversity of the microbial world and obtain new enzymes that will prove important in food processing.

Food safety. The most important advance in food processing will be in food safety. Food safety actually begins on the farm by decreasing the incidence of microbial contamination of plant and animal food products. As such, any of the biotechnology-based treatments for infectious diseases will enhance food safety. With genetic engineering, RNA interference, and antisense technology, plant scientists have tools that will allow them to decrease natural plant toxins and food allergens. In addition, MAbs, biosensors, and DNA probes are being developed that will be used to determine the presence of harmful bacteria, such as *Salmonella* species, *Clostridium botulinum,* and *E. coli* O157:H7, the strain of *E. coli* that has been responsible for a number of deaths in recent years. Once again, these tests will be quicker and more sensitive to low levels of contamination than were previous tests because of the increased specificity of molecular techniques. Biotechnology-based diagnostics have also been developed that will allow better detection of toxins produced by fungi that grow on crops. These same techniques will help food processors determine whether food products have inadvertently been contaminated with peanuts, a potent allergen for approximately 1% of the adult population in the United States.

Environmental biotechnology

Although many countries have become increasingly responsible about regulating the environmental impacts of human activities, societies continue to deplete the earth's nonrenewable resources; pollute the air, soil, and water; and generate massive amounts of wastes that do not degrade readily. The environmental problems caused by human activities are interrelated and interdependent, so one problem often amplifies another. Therefore, the only viable solutions to most environmental problems must encompass the whole system and be sustainable in the long term.

Natural systems and processes provide models of sustainability that are worth emulating. For example, nature's building materials, biological molecules, are both biodegradable and recyclable. Materials are used and reused again and again, and the concept of waste is meaningless. In addition, the processes are exceedingly efficient, so materials and energy are not squandered. Processes are tightly controlled, because a series of internal feedback loops provides real-time monitoring, instantaneous feedback, and immediate adjustments.

By using nature's design principles as a guide in developing new product and manufacturing processes, societies would:

• Conserve natural resources through efficient use of materials and energy
• Minimize environmental degradation by decreasing the amount of wastes and eliminating the use and production of hazardous or toxic substances

In other words, human activities would become more sustainable environmentally and economically if they were modeled on nature.

Rather than using nature's design principles and features as inspirations for new technological products and processes, why not do the simpler thing—steal them? Co-opt nature by using biological molecules as raw material and cellular metabolic machinery to carry out certain processes. Sound familiar? It should, because it is essentially the definition of modern biotechnology.

Many people hope that modern biotechnology will help solve some past environmental problems while causing fewer problems than previous physical or chemical technologies. This optimism is based on the use of biological, renewable resources in place of chemical, nonrenewable ones and the greater specificity, precision, and predictability that characterize biologically based technologies. As is true of medical and agricultural biotechnology, increased specificity and predictability should lead to industrial manufacturing technologies that generate fewer negative side effects and have fewer or less severe unintended consequences. In addition, just as biotechnology is providing society with new tools for diagnosing health problems and detecting harmful contaminants in food, it is yielding new methods of monitoring environmental conditions and detecting environmental pollutants.

Cleaning Up Pollution through Bioremediation

Using biotechnology to treat pollution problems is not a new idea. Communities have depended on complex populations of naturally occurring microbes for sewage treatment for many years. Microbes help purify water by breaking down solid organic wastes before the water is recycled. As is true of food fermentation, at first people had no idea they should be thanking microbes for breaking down human waste. Once they realized microbes were involved, engineers and biologists began developing various technologies for making waste degradation more efficient and effective; one avenue for improvement was genetic modification. Scientists began to intentionally modify the genetic makeup of microbes used in sewage treatment at least a century ago. Microbial populations, like all groups of living organisms, are composed of genetically variable individuals. Like early agriculturists improving crops and livestock, the waste treatment industry used selection of the very best strains to genetically modify microbes and improve their ability to break down human waste.

Solid organic wastes are not the only type of pollutants that need to be removed from water supplies. Some communities are finding that their water and soils are contaminated by chemical pollutants and are turning to microbes for help in removing these pollutants. The use of microbes to remove pollutants, or bioremediation, is most often employed to clean up oil spills, toxic-waste sites, and leakage from underground storage tanks (Figure 1.29).

Why microbes? Over the billions of years they have been on earth, microbial populations have adapted to every imaginable environment. No matter how harsh the habitat, some microbe has found a way to make a living there. Scientists have discovered microbes in hot springs that are the source of the geysers in Yellowstone, thousands of feet underwater in hydrothermal vents, in salty seas and lakes, and living off inorganic materials, such as copper sulfide in copper mines. Microbes that are able to live in such extreme environments are called **extremophiles,** and their enzymes are called **extremozymes** (Figure 1.30).

Life in unusual habitats makes for unique biochemical machinery, so the range of compounds microbes can degrade is enormous. Environmental engineers are capitalizing on this biochemical diversity to repair a number of pollution problems. Pulp and paper companies are using a fungus to clean up a noxious

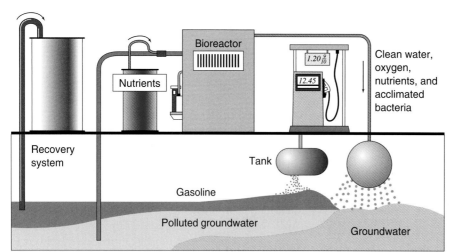

Figure 1.29 Gasoline from an underground storage tank seeps through the soil to the water table. After the leak is stopped, the free-floating gasoline is pumped out to a recovery tank, and polluted groundwater is pumped into a bioreactor with oxygen, nutrients, and hungry microbes. After the microbes eat the gasoline, the mixture of clean water, nutrients, and microbes is pumped back into the ground so that more of the pollutant can be degraded.

substance discharged during papermaking. Indigenous microbes that contribute to the natural cycling of metals, such as mercury, offer prospects for removing heavy metals from water. Other naturally occurring microbes that live on toxic-waste dumps are degrading wastes such as polychlorobiphenyls (PCBs) to harmless compounds.

Not all pollution problems that threaten human health result from human activities. During the hot summer months in Australia, aquatic cyanobacterial populations increase rapidly and secrete carcinogenic chemical compounds into the water supply. To counter these population blooms, copper sulfate, which kills the cyanobacteria but does not break down the toxins, is dumped into the water. As you

might expect, microbes that coexist with the cyanobacteria have evolved a better solution to the problem. Recently, scientists discovered a bacterium that produces three enzymes that sequentially break down the cyanobacterial toxins until they are harmless. Conveniently for scientists, the three genes that code for the enzymes are clustered together in a single functional unit, an operon.

In these examples, naturally occurring organisms are performing the cleaning duties. Using only naturally occurring microbes can be somewhat limiting, however. Microbes that degrade hazardous wastes in the clay soils of North Carolina may not work in the silty soils of the Mississippi delta. Genetic engineering gives environmental managers the flexibility they need to maximize the use of the biochemical capabilities of microbes and, at the same time, circumvent problems such as habitat specificity.

If useful enzymes are discovered in microbes that can survive only in certain habitats, the genes that code for the enzymes may be movable into microbes known to prosper in other habitats. For example, microbial ecologists have discovered a number of species of aquatic microbes that degrade some of the 200 naturally occurring halogenated hydrocarbons found in the ocean. Some of the worst soil pollutants belong to this class of chemical compounds. If microbiologists isolate the genes that code for the enzymes that degrade these compounds, they could then insert the genes into soil microorganisms. These genetically engineered microorganisms might then be able to clean up hazardous-waste sites.

The great majority of the biochemical potential that exists in microbial communities remains untapped, as less than 1% of the planet's microbes have been

Figure 1.30 Microbes living in extreme environments, such as hot pools in Yellowstone National Park, will provide industrial manufacturing companies with enzymes that retain their functions under extreme manufacturing conditions. (Photograph courtesy of Thomas A. Martin.)

cultured and characterized. Recently acquired information from large-scale microbial genomic studies allows better utilization of the wealth of genetic diversity in microbial populations. Because researchers may already know part of the sequence of the gene they are hoping to find, they are able to use DNA probes to fish, on a molecular level, for microbes with specific capabilities. Once caught, such organisms can be identified and cultured for their ability to carry out particular reactions. Before the days of recombinant DNA, any useful biochemical potential housed in the genome of these microbes would have remained inaccessible to scientists. The U.S. government, through the Department of Energy, initiated a large-scale program to identify useful microbes through its Genomes To Life program, which is modeled after the Human Genome Project and makes use of technologies that were developed to sequence the human genome.

As you might expect, many extremophiles cannot be cultured in the laboratory, nor do scientists know the identity of the gene(s) encoding the enzyme they hope to find. A few researchers have recently begun using a novel approach, known colloquially as total-community genomics, to discover useful genes. They collect samples from any natural environment, and then they use enzymes to break up the DNAs of all of the organisms in the sample. They make no attempt to separate the organisms in the sample or identify them, much less culture them. They are not interested in the organism that is the source of the gene, only the gene. The DNA fragments from the entire community of organisms are inserted into a bacterium that is easy to culture, such as *E. coli*. The host *E. coli* will express any unique genes it might have, and the proteins encoded by the novel genes are screened for desired activity.

Preventing Environmental Problems

If an ounce of prevention is worth a pound of cure in human health, surely the same can be said for environmental health. Rather than cleaning up environmental problems, wouldn't it be better not to create them? Modern biotechnology is opening up a number of avenues for preventing environmental problems.

Industrial sustainability through biotechnology. Industrial sustainability includes processes to discover useful products that meet the current consumer demand without compromising the resources and energy supply of future generations. How can industrial manufacturing achieve sustainability? The key words are "clean" and "efficient." Any change in production processes, practices, or products that makes production cleaner and more efficient per unit of production or consumption is a move toward sustainability. In practical terms, industrial sustainability means utilizing technologies and know-how to:

- Reduce material and energy inputs while maximizing renewable resources as inputs
- Minimize the generation of harmful pollutants or waste during product manufacture and use
- Produce recyclable or biodegradable products

Companies in a variety of industry sectors are practicing proactive environmental protection through industrial sustainability, and biotechnology is helping them achieve that objective.

Material and energy inputs. The economies of the industrialized world are based on fossil fuels, such as oil, coal, and natural gas, which are essentially long chains of carbon atoms hooked together. Fossil fuels provide the energy that drives the engines of economic growth and keeps people in the comfortable lifestyle to which they have become accustomed. They also provide the building-block molecules, which are also called feedstock chemicals, for thousands of consumer products. Society's dependence on fossil fuels leads to environmental, political, and economic problems.

- Fossil fuels are nonrenewable resources, as well as the major contributor, by far, to greenhouse gas emissions. Processes based on fossil fuels generate unwanted by-products, some of which are hazardous substances that pollute the air, soils, and water.

- Because most of the oil that U.S. citizens and industries rely on is located outside of our boundaries, the U.S. economy and our quality of life depend on a handful of oil-producing countries. America's heavy reliance on resources found in other countries represents a threat to national security and also contributes to the balance-of-trade deficit.

- The stability of the governments in countries with considerable amounts of petroleum affects the economic well-being of the United States. No matter how stable the U.S. government is, political turmoil in the oil-producing countries raises the price of petroleum and threatens access to that essential resource.

- As developing countries become industrialized, worldwide demand, and therefore competition, for this limited and nonrenewable resource will increase quite significantly.

All of these forces place considerable pressure on society to develop new, reliable, and affordable sources of energy and raw materials.

Life, like fossil fuels, is carbon based. Fossil fuels are nothing more than living organisms from past millennia. The fossilized carbon that industrialized societies depend on today is the carbon that was contained in the biological molecules of organisms when they died. The source of that carbon was the same then as it is today: carbon dioxide. Billions of years ago, photosynthetic microorganisms mastered the engineering feat of capturing solar energy and converting it into a usable form by hooking together the carbons found in carbon dioxide into glucose and the other biological molecules.

Green plants and other photosynthetic organisms continue to fix carbon in the very same way as those ancient organisms. The biological molecules they synthesize contain energy and molecular building blocks, just like the fossil fuels. Unlike fossil fuels, however, the energy and building-block molecules are renewable resources because photosynthetic organisms reproduce themselves. Increasing the use of renewable resources is a key pillar of industrial sustainability.

Using living biomass as a source of both energy and feedstock chemicals is definitely not a new or sophisticated concept. Biomass has always provided societies with energy directly, but biomass can also be used to create **biofuels.** With biofuels, the energy in biomass is not released immediately through burning but is stored in other organic molecules that will not rot. Fuel ethanol, or bioethanol; biogas; and biodiesel are examples of biofuels (Figure 1.31).

Figure 1.31 The bus uses diesel fuel derived from soybean oil. (Photograph courtesy of the National Renewable Energy Laboratory and the Nebraska Soybean Board.)

Biomass also used to be the source of feedstock chemicals, such as acetone, glycerol, and industrial alcohols, used by many industries. However, in the 20th century, technological improvements in oil refining decreased the cost of using petroleum as the starting material for producing chemical feedstocks and maximized the amounts of energy and feedstock chemicals that could be extracted from petroleum. Consequently, most processes that converted living biomass into feedstock molecules could not compete with comparable processes based on petroleum. This holds true today. Therefore, even though it has always made sense environmentally to shift from fossilized biomass to living biomass, economically that has not been the case.

Now, however, recognizing both the economic and political imperatives to lessen society's reliance on fossil fuels, the government has begun to establish policies and to commit more money to basic and applied research into alternative sources of energy and feedstock chemicals. The Genomes To Life program mentioned above is focused on biotechnology applications in energy production. Living biomass could serve as the raw material for both resources, and living organisms could provide the metabolic machinery to convert biomass into fuels and feedstocks. Many technical barriers will need to be overcome before biomass can make a significant contribution to the country's demand for energy and feedstock molecules. To solve some of these problems, government laboratories and industrial scientists have devoted significant resources to research projects that use biotechnologies, such as recombinant DNA technology, protein engineering, molecular evolution, and bioprocess engineering, to make biomass-derived energy and feedstocks economically competitive.

Process applications of biotechnology. Biotechnology's versatility as a tool for industrial sustainability is most apparent in the improved manufacturing processes being implemented by a wide variety of industries. The chemical, textiles, pharmaceutical, pulp and paper, food and feed, metal and minerals, and energy industries have all benefited from cleaner, more energy-efficient production made possible by incorporating biotechnology into their production processes. Almost always, these improvements are due to biocatalysts, which are living organisms and/or their enzymes (see the box on page 10). Because biocatalysts are more specific than chemical catalysts, they produce fewer unwanted byproducts. In addition, the environment benefits from industry's use of biocatalysts because they are water soluble and catalyze reactions most efficiently at low temperatures compared with nonbiological catalysts.

Many of these advantages can become limitations in certain industrial processes, such as paper making and textile processing, that require organic solvents, high or low temperatures, or extreme chemical conditions. Biocatalysts function in a narrow temperature range, and most fall apart at temperatures above 100°F or when subjected to chemical extremes. Because of the many environmental advantages offered by biocatalysts, however, industrial scientists are using biotechnology to attempt to circumvent the current limitations that restrict the use of biocatalysts in industrial manufacturing.

One limitation is economics; biocatalysts can be more expensive than chemical catalysts. To decrease the costs of biocatalysts, scientists are using genetic engineering to increase enzyme production in microorganisms that are commonly used in production or to give these cheap and easy-to-grow microbes the ability to make enzymes usually made by microbes that are too expensive to maintain or impossible to culture. Scientists are also applying the methods described above to discover novel biocatalysts that will function optimally at the relatively high levels of acidity, salinity, temperature, or pressures found in some industrial manufacturing processes.

A new extremozyme, or any other enzyme, that scientists may discover did not evolve to catalyze human industrial processes but to maximize its owner's survival and reproductive success. Thus, once a new enzyme is discovered, scientists may need to tweak its three-dimensional structure a bit to optimize its catalytic activity or broaden the reaction conditions under which it can function so that it is more compatible with existing industrial processes. Earlier, we described protein engineering, which makes slight changes in a protein's amino acid sequence to enhance the stability of its structure under conditions often found in industrial manufacturing.

Another method for improving natural biocatalysts is modeled after the evolutionary process, but on a molecular level. If an enzyme is newly discovered, scientists do not know which amino acid changes would create the best structure for the task at hand. In this situation, they use a number of different techniques to create a large number of genetic variants of the gene encoding the enzyme. They then screen the enzymes encoded by these genetic variants to identify slight variations of the enzyme that exhibit the greatest activity levels. This technique, known as directed protein evolution, optimizes the effectiveness and efficiency of the newly discovered enzyme.

Biodegradable products. As we described above, most chemical feedstocks are derived from petroleum. Many of the petrochemical feedstocks are small organic (carbon-containing) molecules, or monomers, that are hooked together to create polymers, such as polyethylene and plastic. Glucose, rather than petroleum, can be used as the starting material for producing these building-block monomers. In addition, biocatalysts, rather than chemical synthesis, can be responsible for synthesizing the monomers from glucose (Figure 1.32).

Almost all large chemical companies are building partnerships with biotechnology companies to develop enzymes that can break down plant starches and sugars, such as glucose, into useful building-block molecules that they can use in polymer synthesis. The first biomaterial came to market in 2004. Produced by a joint venture between Cargill, a grain commodity company, and Dow Chemical Company, this biomaterial is a polymer of lactic acid.

Figure 1.32 Biofine Corporation in Waltham, MA, has developed a process that converts plant starch, including cellulose, into small organic molecules that can serve as a monomer for synthesizing many chemicals. (Photograph courtesy of the Pacific Northwest National Laboratory.)

The Cargill-Dow process converts cornstarch to sugar and sugar to lactic acid. The lactic acid molecules are hooked together to create the polymer, polylactic acid, which is marketed under the name NatureWorks. The manufacturing process uses 20 to 50% less fossil fuels than manufacturing a comparable petrochemical polymer. NatureWorks is a biodegradable polymer that can be used as a starting material for manufacturing packaging materials and fibers for clothing, pillows, and comforters. The bedding company Pacific Coast Feather plans to replace 80% of its polyester products with polylactic acid over the next 5 years.

In addition, scientists are using recombinant DNA techniques to increase the availability of biodegradable products. Industrial scientists have genetically engineered plants to produce polyhydroxybutyrate, a naturally occurring bacterial polyester that could be used as a feedstock chemical for manufacturing biodegradable plastics. Cotton that is genetically engineered to contain a bacterial gene produces a polyester-like substance that is biodegradable and has the texture of cotton but is warmer. Other biopolymers with the potential to replace synthetic fabrics and fibers are under development in Japan and the United States. Finally, genetic engineering also provides the opportunity to produce abundant amounts of natural protein polymers, such as spider silk and adhesives from barnacles, through microbial fermentation processes.

In summary, no matter what stage of industrial production you choose—inputs, manufacturing process, or final product—modern biotechnology provides industry with tools, techniques, and know-how to move beyond command-and-control regulatory compliance to proactive pollution prevention and resource conservation strategies that are characteristic of industrial sustainability.

Monitoring the Environment
The techniques of biotechnology also provide novel methods for diagnosing environmental problems and assessing normal environmental conditions so that we can be more informed environmental stewards in the future.

Companies have developed methods for detecting harmful organic pollutants in the soil by using MAbs (Figure 1.33) and PCR, while scientists in government laboratories have produced antibody-based biosensors that detect explosives at old munitions sites. Not only are these methods cheaper and faster than the current laboratory methods, which require large and expensive instruments, but they are also

Figure 1.33 On-site monitoring of environmental pollutants with MAb technology. (Photograph courtesy of EnSys, Inc.)

portable. Rather than gathering soil samples and sending them to a laboratory for analysis, scientists can measure the level of contamination on-site and know the results immediately.

The remarkable ability of microbes to break down chemicals is proving useful not only in pollution remediation, but also in pollutant detection. A group of scientists at Los Alamos National Laboratory work with bacteria that metabolize a class of organic chemicals called phenols, many of which are considered pollutants by the Environmental Protection Agency. When the bacteria ingest phenolic compounds, the phenols attach to a receptor. The phenol-receptor complex then binds to DNA, triggering the activation of the genes involved in phenol metabolism. The Los Alamos scientists added to the bacteria a reporter gene that, when activated by a phenol-receptor complex, produces an easily detectable protein, thus indicating the presence of phenolic compounds in the environment. A variation on this theme involves linking fluorescent reporter genes to operons involved in the microbial breakdown of environmental pollutants.

Summary

Biotechnology is a set of very flexible and powerful tools that offer great potential for improving human health, increasing the quality and yield of our agricultural products, and improving our relationship with the environment. Some of the most important biotechnologies are MAb technology, cell culture, genetic engineering, bioprocessing technology, and protein engineering. The common thread that joins these technologies is the fact that they are based on the use of cells and biological molecules.

An essential advantage of biotechnology over other technologies is that it is based on biology. It can work with the biology of organisms in very specific, predictable ways to solve biological problems or make products. Of all the technologies developed so far, biotechnology has the potential to be most compatible with sustainable life on this planet.

Because biotechnology is in the earliest stages of development, societies are at a critical stage in deciding how best to utilize its power. They can use the tools of biotechnology to answer scientific questions, make new products, solve problems, and achieve goals deemed desirable by society. The issue before all of us is determining which questions, products, and problems are our highest priorities and deciding what sort of society we want in the new millennium.

Selected Readings

Anderson, W. F. 1995. Gene therapy. *Scientific American* 273:124–128. Somewhat dated, but still interesting, this article was written by a pioneer in gene therapy who was involved in the first replacement gene therapy trials. In 1997, *Scientific American* (276[5]) also produced a special edition on gene therapy.

Bains, William B. 2004. *Biotechnology from A to Z*. Oxford University Press, Oxford, England. A concise reference for some of the basic terms used in biotechnology and molecular biology.

Bloom, M., and J. Greenberg. 2006. *BSCS Biology: a Molecular Approach*. Glencoe/McGraw Hill Publishers, Blacklick, Ohio.

Chrispells, Martin, and D. E. Sandava. 2003. *Plants, Genes and Crop Biotechnology*. Jones and Bartlett, Boston, MA, and the American Society of Plant Biologists, Washington, DC. To date, the most comprehensive treatment of agricultural biotechnology, this book serves as an excellent introduction to agricultural ecology for students with little or no understanding of agriculture.

Clarke, Michael, and Michael Becker. 2006. Stem cells: the real culprit in cancer? *Scientific American* 295:48–52.

Cookson, Clive, et al. 2005. The future of stem cells. *Scientific American* 292:63–96. A series of articles by a number of authors on various aspects of the biology and politics of stem cells.

Ezzell, Carol. 2002. Proteins rule. *Scientific American* 286:26–33. An overview of proteomics.

Falkowski, Paul. 2002. The ocean's invisible forest. *Scientific American* 287:57–64. An overview of the wealth of microbial diversity in marine environments with a special emphasis on extremophiles.

Fischer, Alain, and Marina Cavazzana-Calvo. 2006. Whither gene therapy? *The Scientist* 20:36–42.

Glick, Bernard R., and Jack J. Pasternak. 2003. *Molecular Biotechnology*. ASM Press, Washington, DC. An excellent general reference on both the science and applications of biotechnology.

Gunter, Chris (ed.). 2004. Human genomics and medicine. *Nature* 429:440–475. A series of articles in *Nature Insights* focused on the transformation in disease diagnosis and treatment derived from an understanding of genomics.

Guttmacher, Alan E., and Francis Collins. 2002. Genomic medicine: a primer. *New England Journal of Medicine* 347:1512–1520.

James, Clive. 2005. *Global Review of Commercialized Transgenic Crops: 2005*. ISAAA, Ithaca, NY. Contains data on the global growth of plant agricultural biotechnology. ISAAA has published an excellent series of monographs on agricultural biotechnology and developing countries. For information on ordering this and other ISAAA publications on agricultural biotechnology in developing countries, visit their website (http://www.isaaa.org).

Kittredge, Claire. 2005. Gene therapy. *The Scientist* 19:14–19.

Kreuzer, Helen, and Adrianne Massey. 2005. *Biology and Biotechnology: Science, Applications, and Issues*. ASM Press, Washington, DC.

Lewin, Benjamin. 2006. *Genes IX*. Jones and Bartlett Publishers, Boston, MA. The title speaks for itself. Excellent general reference, as are Lewin's earlier editions, *Genes I* through *Genes VIII*.

Lysaght, Michael J., and Patrick Aebischer. 1999. Encapsulated cells as therapy. *Scientific American* 280:76–82.

Mooney, David J., and Antonios G. Mikos. 1999. Growing new organs. *Scientific American* 280:60–65.

Ng, Rick. 2004. *Drugs: from Discovery to Approval*. Wiley-Liss, Hoboken, NJ. An accessible and concise description of the drug development process that is appropriate for anyone interested in both the science and regulatory issues.

Nicholl, Desmond. 2002. *An Introduction to Genetic Engineering*. Cambridge University Press, Cambridge, England.

Organization for Economic Cooperation and Development. 1998. *Biotechnology for Clean Industrial Products and Processes: towards Industrial Sustainability*. OECD, Paris, France. A thorough treatment of biotechnology's potential to improve environmental effects of industrial processes. To order, visit the Organisation for Economic Co-Operation and Development website (http://www.oecd.org).

Parson, Anne B. 2004. *The Proteus Effect: Stem Cells and Their Promise for Medicine*. Joseph Henry Press, Washington, DC.

Rutledge, Colin, and Bjorn Kristiansen. 2001. *Basic Biotechnology*. Cambridge University Press, Cambridge, England. An excellent introduction to biotechnology for nonscientists, especially with respect to applied microbiology.

Smith, John E. 2006. *Biotechnology*. Cambridge University Press, Cambridge, England. Provides an excellent overview of the breadth of biotechnology applications and is especially useful for information on bioprocessing, cell culture, and industrial uses of microorganisms.

Thieman, William J., and M. A. Palladin. 2003. *Introduction to Biotechnology*. Benjamin Cummings, San Francisco, CA. This introductory textbook not only has information on molecular biology and its applications, it also provides more information on the biotechnology industry than most texts.

Cell Properties and Processes

2

Introduction

In chapter 1, we defined biotechnology as a set of tools for making useful products and solving problems. The fundamental problems confronting humans in the 21st century are essentially the same problems we have faced for centuries—growing and preserving our foods, staying healthy, and getting energy. For centuries, societies have used various technologies, including earlier biotechnologies, to solve these problems. Over time, the technological tools brought to bear on these problems have improved, especially during the last century, as scientific understanding buttressed technological progress. Modern biotechnology is the next step in a continuum of technological change that was initiated 10,000 years ago and began to accelerate significantly during the 20th century.

The common thread uniting the modern biotechnologies described in the previous chapter is their foundation: they are based on living cells and biological molecules. Using cells and biological molecules as the foundation of technologies allows technology developers to capitalize on innate attributes of life at that level. Therefore, learning some basic facts about the structure and function of cells and the biological molecules they contain is essential to understanding biotechnology's scientific foundations, applications, possible advantages over earlier technologies, and potential risks and benefits.

As you know, cells are the basic building blocks of all living things: single cells live and reproduce, but nothing less than a cell is alive, nor can it reproduce on its own. What you may not be conscious of is how truly remarkable cells are. How is it that a collection of lifeless molecules, each obeying the laws of physics and chemistry that govern inanimate matter, transcends the properties of the nonliving and emerges as life? Never has the expression "the total is greater than the sum of its parts" been more apropos.

The simplest living organisms, such as yeast and bacteria, consist of a single, self-sufficient cell that lives independently of other cells. In certain species, the one-celled organisms join together and form loosely organized cellular communities that carry out some simple functions, such as moving, as a unit. However, even in those species, the single cells are generalists and capable of existing independently. Cells can also be specialists and highly dependent on other cells. Creatures more familiar to us, such as plants, animals, and humans, consist of many different cell types—each of which performs a very specific job—that live in a highly integrated, mutually dependent cellular community.

In spite of their diversity, what is most striking about cells is their similarity. All cells:

- Are built according to the same basic design, using the same molecular construction materials
- Exhibit the same fundamental properties
- Operate using essentially the same processes, which are carried out in similar ways with similar molecular tools
- Abide by the same principles
- Speak essentially the same genetic language

This unity of life at the cellular level provides both the theoretical and operational foundations for modern biotechnology.

Universal Cell Properties

In 1665, a British scientist, Robert Hooke, sliced off a thin piece of cork and examined it under a microscope. He saw something that reminded him of a honeycomb—row upon row of tiny pores surrounded by walls. Hooke called the tiny pores cells because they reminded him of a monk's cell in a monastery (Figure 2.1). Hooke believed these cells had served as containers for the "noble juices" found in the cork tissue when it had been alive.

During the 18th and 19th centuries, dramatic technological improvements in microscopy allowed scientists to see cells in greater detail, and they discovered that every tissue of a plant or animal is made

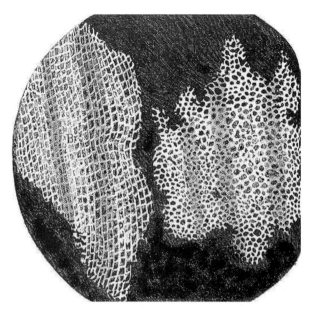

Figure 2.1 Hooke's drawing of cork cells.

of cells that share certain features. Each is enclosed by a structure, the **cell membrane,** that separates it from surrounding cells and the external environment. This physical barrier permits the internal makeup of cells to differ from their extracellular environment, and the internal makeup of cells stays remarkably constant even when there are dramatic changes in the external environment. The cell membrane, which is also known as the **plasma membrane,** also provides the interface between the cell and its environment. The cell takes up essential nutrients and rids itself of waste products by moving molecules across the plasma membrane.

Not only are cells separated from their environment by a membrane, but all cells also organize their genetic material into one or more **chromosomes** and localize the genetic material in a distinct region within the cell. Another striking feature of plant and animal cells observable with high-power microscopes is their extraordinary level of organization. In addition to having a cell membrane, they also use a series of internal membranes to create small compartments and to cordon off special structural elements known as **organelles.** Each organelle has a specific function (see the first sidebar). At the risk of repeating ourselves, we cannot help but ask, How can a sack of lifeless molecules maintain such an intricate level of organization?

We can trace the remarkable organization of cells to a few simple rules governing molecular interactions and certain structural and functional features of cells that derive from these principles of molecular interaction.

Molecular interactions

Below, we focus on two of the most important rules governing molecular interactions and describe how adherence to these rules leads inevitably to a certain degree of organization.

Interactions with Water

Cells live in and are designed for a watery environment. Single-celled organisms usually live in the water or in very moist environments. You may have read that your own body is 70% water—human bodies are typical of animal bodies in that regard. Animal body cells are bathed in watery body fluids, as are living plant cells. The internal environment of cells, too, is watery. The degree to which molecules interact comfortably with water determines in large measure how they are organized within a cell.

From your own experience, you know that only some chemical substances interact readily with water. Sugar dissolves in watery fluids; oil does not. But what does dissolving in water actually mean? The sugar granules dissolve as the sugar molecules interact with water molecules and become evenly dispersed throughout the watery fluid. The same is true for salt, vinegar, and lemon juice. Oil does not dissolve because the oil molecules do not disperse evenly throughout the water but congregate together and coalesce into a single unit (Figure 2.2A). If you stir the oil and water together, you can break the oil up into small droplets, but as soon as you stop stirring, the droplets will come back together into one large drop of oil.

Sugar and oil illustrate two fundamentally different types of interaction with water. Sugar is **hydrophilic** (hydro, water; philic, loving); it dissolves easily in water and disperses evenly throughout a watery liquid. Oil is **hydrophobic** (phobic, fearing); it does not dissolve in water. The interaction between water molecules and oil molecules is unfavorable, so water molecules keep to themselves and oil molecules keep to themselves. Therefore, the first rule governing molecular interactions is that hydrophilic molecules dissolve in water and hydrophobic molecules do not. As a result, when hydrophobic and hydrophilic molecules are mixed together, they spontaneously segregate themselves into two groups composed of the same type of molecules.

This tendency for hydrophilic molecules, like sugar, salt, and water, to congregate together and for hydrophobic molecules, like oils, to congregate together is an organizing force inside cells. Hydrophobic and hydrophilic molecules organize themselves. You can see in the oil/water example above that the oil and

Two Fundamental Types of Cells

This chapter focuses on the remarkable similarity of cells, whether the cell is a one-celled organism, such as a bacterium, or a highly specialized cell in a multicellular organism. All cells share certain basic features: they have molecular machinery for duplicating DNA and for breaking down and synthesizing molecules, they reproduce by dividing in two, they use the same molecular building blocks, and they are enclosed by a hydrophobic membrane that separates the cell from its surroundings.

However, only certain cells have the sort of internal compartmentalization described in this chapter. This distinction led biologists to divide cells into two different types, prokaryotic and eukaryotic cells. The organisms made of these two cell types are called prokaryotes and eukaryotes.

Prokaryotic cells are much simpler and significantly smaller than eukaryotic cells. The most obvious difference between prokaryotic and eukaryotic cells is the absence of internal membranes in prokaryotes. The lack of a nuclear membrane surrounding the genetic material is the origin of the name "prokaryote" (pro, before; karyon, kernel or nucleus). Because most prokaryotic cells are so small, they are able to function efficiently without subcellular compartments.

All prokaryotes are single-celled organisms, but not all single-celled organisms are prokaryotes. Some prokaryotes may live in colonies, but there are no true multicellular prokaryotes in which cells become specialized for certain tasks. The two major groups of prokaryotes are eubacteria and archaea. Eubacteria include the familiar bacteria, such as Escherichia coli, and blue-green algae

(sometimes called cyanobacteria). Archaea (also called archaebacteria) are a fascinating group of prokaryotic organisms that inhabit extreme environments. They are genetically and biochemically similar to one another but are as different from eubacteria as plants and animals are from each other. When we talk about prokaryotes or bacteria in this book, we are referring to eubacteria.

Because bacteria grow rapidly, are easy to grow in large quantities, and have much less genetic material than eukaryotic cells, they are a favorite tool of biotechnologists.

Eukaryotic cells are present in most organisms with which you are familiar, such as plants, animals, and fungi. While most plants, animals, and fungi are multicellular organisms that contain many specialized tissues and cell types, some eukaryotes are single-celled organisms, such as yeasts, green algae, or amoebae.

Like prokaryotes, eukaryotic cells are surrounded by a hydrophobic membrane, but unlike prokaryotes, they also have internal membranes. These internal membranes divide their cytoplasm into specialized compartments collectively known as organelles. Organelles perform specific functions within eukaryotic cells. For example, all eukaryotic cells contain organelles called mitochondria that play a crucial role in energy production. In plant cells, organelles called chloroplasts are responsible for capturing the energy in sunlight and converting it to chemical energy through the process known as photosynthesis.

(continued)

A. Prokaryotic cell

Genetic material   Plasma membrane   Cytoplasm

B. Eukaryotic animal cell

Nuclear membrane
Nucleus
Genetic material
Endoplasmic reticulum
Golgi apparatus
Cytoplasm
Mitochondrion
Plasma membrane

C. Eukaryotic plant cell

Chloroplast   Cell wall
Vacuole

Two Fundamental Types of Cells (continued)

As it turns out, the structural differences that are visible with sophisticated microscopy are mirrored in the invisible details of the molecular biology underlying many cell processes. Prokaryotes and eukaryotes carry out the same cell processes, but they often do so in very different ways.

*Trying to understand minute differences in the molecular mechanisms of cell processes in prokary-*otes and eukaryotes may seem like an arcane scientific exercise, but these differences are very important in biotechnology. Sometimes scientists exploit the differences and use them to accomplish a certain goal. For example, antibiotics kill bacteria without harming your own cells because they are targeted quite specifically to prokaryotic molecules and mechanisms.*

the water create two compartments inside their container: an oily, hydrophobic compartment and a watery, hydrophilic compartment. If you stir salt into the oil-water mixture, all of the salt dissolves in the water. When the water and oil separate, the water tastes salty, but the oil does not. If you add greenish olive oil to pale-yellow canola oil, the oils mix completely and form an intermediate shade. The color of the water stays the same because the chemicals that make olive oil green dissolve in oil, not water (Figure 2.2B and C).

We need to add two more pieces of information to the picture before moving on to the second principle underlying molecular interactions. In the examples above, we designated the entire molecule either hydrophobic or hydrophilic. While that is true for some molecules, like sugar, it does not have to be.

Some molecules have parts that are hydrophobic and parts that are hydrophilic. These molecules follow the same pattern as oil and sugar: the hydrophobic parts congregate (if possible) with other hydrophobic molecules (or parts of molecules), and the hydrophilic parts congregate (if possible) with other hydrophilic molecules or parts of molecules. Figure 2.3 illustrates some ways such molecules might arrange themselves.

You may be asking what makes a molecule hydrophobic or hydrophilic? We answer that question in detail in chapter 4. For now, we will simply say that atoms with unequal numbers of electrons and protons carry small positive or negative charges and are hydrophilic. Atoms with equal numbers of electrons and protons are electrically neutral (i.e., they carry no charge) and are hydrophobic.

Figure 2.2 Hydrophobic and hydrophilic molecules tend to congregate. (A) Oil, a hydrophobic substance, and water form two separate layers. (B) A hydrophilic substance, like salt, will dissolve in the water but not in the hydrophobic oil. (C) A hydrophobic substance will dissolve in the oil but not in the water.

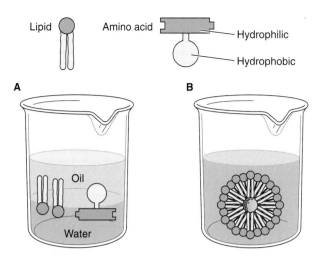

Lipid
Amino acid
Hydrophilic
Hydrophobic

A
B

Oil
Water

Figure 2.3 Some molecules have both hydrophobic and hydrophilic parts. (A) The most favorable arrangement for them is to keep the hydrophobic parts in hydrophobic environments and the hydrophilic parts in hydrophilic environments. (B) The molecules spontaneously form a sphere with the hydrophobic parts on the inside. You are looking at a cross section.

Molecular Binding Specificity

Molecules have distinct shapes, and their surfaces may have protrusions, nooks, pockets, and crannies (Figure 2.4). These variable shapes are key to organizing molecular interactions and, through them, cell processes.

In cells, molecules make things happen by fitting together. Scientists refer to molecules fitting together as binding and to the force between them that holds them together as a bond. When two molecules have shapes that fit together precisely, something happens to the molecules: a chemical reaction might occur or one of the molecules might be moved through the cell membrane, to give two examples. Because molecules have very specific shapes, even though a molecule may collide with thousands of other molecules in a cell, a chemical reaction occurs only when it binds to a molecule with the correct shape.

Fitting together depends not only on shape, but also on the chemical nature of the parts, especially whether the molecules are electrically neutral or have areas that are partially charged. For example, a hydrophobic pocket on one molecule would have to fit with a hydrophobic protrusion of the right size and shape. A correctly shaped hydrophilic protrusion would not be compatible with a hydrophobic pocket.

Remember the principle that molecular interactions require molecules with complementary shapes *and*

compatible chemistries. You will encounter many examples of molecules binding or not binding with each other in the sections and chapters to come. Virtually all healthy cell processes depend on the right molecules binding together, and conversely, diseases can be traced to mistakes in molecular binding.

Cellular membranes

Cellular organization can ultimately be traced to molecular interactions, but when biologists look at a cell under a microscope and are struck by how well organized this sack of molecules is, they are not seeing molecular interactions. The visible appearance or organization is due, in large part, to the existence of cellular membranes that contribute in various ways to the highly organized and tightly controlled activities observed through a microscope.

Figure 2.4 Three-dimensional structure of a molecule. In this style of representing a molecule, known as a space-filling model, individual atoms are shown as ball-like structures. (Image courtesy of Lawrence Berkeley National Laboratory.)

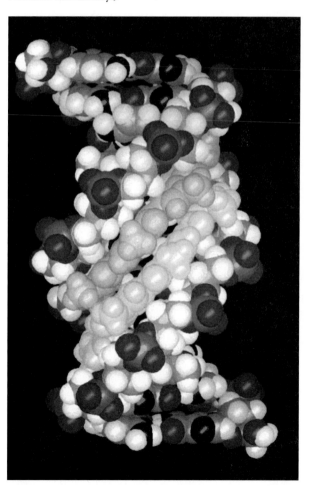

Structural Organization

First of all, cells maintain themselves as distinct units with a cellular membrane, the plasma membrane, that delineates the inside and outside of the cell. Eukaryotic cells also use a series of internal membranes to create small internal compartments and to cordon off essential structural elements within the cytoplasm, such as the cell nucleus, which houses the cell's genetic material. Like the plasma membrane, the internal membranes help keep things organized by determining which molecules enter and exit the structure or internal compartment they surround.

As you know, a key aspect of getting and staying organized is assigning certain items to specific locations. Similarly, the intracellular compartmentalization of certain molecules in specific locations helps cells keep some molecules in close physical proximity to one another and separates other molecules from each other. In addition, an extensive network of internal membranes creates interconnecting sacs, tubules, and channels. Two of these sets of membrane-bound intracellular spaces, the **endoplasmic reticulum** and the **Golgi apparatus,** sort and process the molecules that cells manufacture and efficiently transport the correct molecule to appropriate locations within the cell (Figure 2.5).

Figure 2.5 Micrograph of cell showing internal membranes. (Transmission electron micrograph courtesy of Douglas L. Schmucker, University of California—San Francisco.)

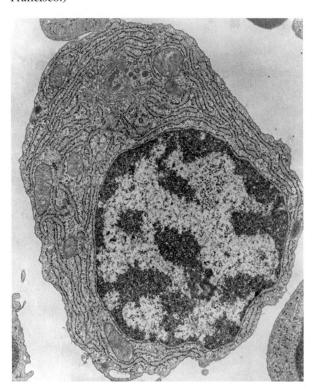

Membranes are obviously a crucial structural feature of cells. The structural organization provided by cell membranes is rooted in the rules governing interactions between hydrophilic and hydrophobic molecules. Cellular membranes are made of molecules with one hydrophobic end (a long tail) and one hydrophilic end. When placed in a watery environment, these bipolar molecules adhere to the molecular-interaction rule and spontaneously orient themselves so that the hydrophobic parts align with each other and avoid water, while the hydrophilic components face the watery environments inside and outside of the cell (Figure 2.6). This two-layer arrangement creates a hydrophobic barrier between the watery material inside the cell, the **cytoplasm,** and the watery external environment.

Internal Molecular Integrity

The contribution of cell membranes to cellular organization extends beyond the structural organization they provide. Both the plasma and internal membranes are actively involved in selecting which molecules are allowed to pass through the membrane. Membranes make these "decisions" using the two principles of molecular interaction described above: hydrophobic-hydrophilic interactions and precise binding of one molecule to another.

Only a few molecules can move easily through the hydrophobic barrier the cell membrane establishes between the cell and its environment. Those that can are small, electrically neutral molecules, such as carbon dioxide and oxygen, and small, very slightly charged molecules, such as water. However, many of the biological molecules essential to cell processes are large molecules with partially charged regions, such as glucose, or small molecules that are highly charged, such as calcium and potassium. Because many cell processes depend on these types of molecules, cells have developed mechanisms for getting the molecules across the hydrophobic barrier that cell membranes create.

Figure 2.6 A cell membrane is a two-molecule-thick layer with a hydrophobic core and hydrophilic faces. The cell membrane surrounds the cell like the membrane of a balloon. A cross section is shown.

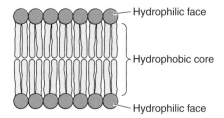

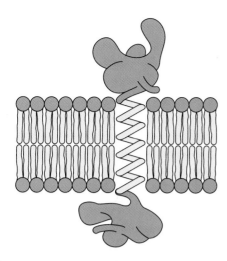

Figure 2.7 A protein embedded in a membrane. The portion of the protein inside the hydrophobic core of the membrane is also hydrophobic, while the external portions are hydrophilic.

In Figure 2.6, you can see that the cell membrane consists of two layers of molecules that have both a hydrophobic and a hydrophilic end. Embedded in that hydrophobic bilayer are various protein molecules with portions that are hydrophobic and portions that are hydrophilic (Figure 2.7). The wide variety of functions served by these proteins will be discussed in detail in later chapters. Irrespective of their functions, in every case, the membrane proteins bind to and interact with only one or a very few molecules. This extraordinary specificity, based on molecular shape and charge, allows cells to control very precisely the molecules that enter and leave the cell or move into or out of membrane-bound compartments within the cell.

Cell Chemistry

All biological phenomena are based on chemical interactions, so understanding the cellular processes described below and throughout this text requires a basic understanding of cell chemistry. The advances in microscopy that allowed scientists to observe and study cell structure coincided with advances in chemistry. Chemical analysis of cells revealed that different cells from different organisms had remarkably similar chemistries. In addition, scientists discovered that all cells contain four types of molecules—biological molecules—found only in living things: carbohydrates, lipids, proteins, and nucleic acids. Before we describe the biological molecules in more detail, however, we must first provide a very quick review of basic chemistry, all of which will be familiar to you.

Atoms, ions, and molecules

The most common constituents of biological molecules are four elements: carbon, oxygen, hydrogen, and nitrogen. An **element** is a substance that cannot be broken down further; it is a pure ingredient. For example, gold is gold and contains nothing else. Gold is an element. Water, on the other hand, is composed of hydrogen and oxygen (H_2O), and table salt is made up of sodium and chlorine. Table salt and water are not elements, but each contains two different elements.

The smallest piece of an element that is still recognizable as that element is an **atom.** An atom of a given element is made up of a specific number of protons and electrons, as well as some neutrons, which we will ignore in this discussion (Table 2.1). Protons carry a positive charge, and electrons carry a negative charge. In general, an atom isolated from other atoms carries no charge because the number of protons equals the number of electrons. If an atom gains or loses electrons, it is a charged particle, which is called an **ion.**

In addition to losing and gaining electrons, atoms can form bonds with each other by sharing electrons. An atom of a certain element can form a specific number of bonds. For example, an atom of hydrogen forms one bond, carbon atoms four, and oxygen two. Therefore, hydrogen can bond to one other atom, which might be another hydrogen atom or an atom from a different element, such as oxygen, and carbon can bond to four other atoms. The term **molecule** refers to two or more atoms bonded together. Water and table salt (sodium chloride) are molecules, as is the oxygen you breathe, which consists of two atoms of oxygen bonded to each other (O_2). Molecules are hydrophobic, hydrophilic, or some of each because of the characteristics of the bonds between their atoms, which in turn are a function of the identities of the atoms participating in the bond.

Table 2.1 Elements and ions

Element or ion	No. of protons (+ charge)	No. of electrons (− charge)	Charge
Element			
Hydrogen	1	1	0
Carbon	6	6	0
Nitrogen	7	7	0
Oxygen	8	8	0
Ion			
Calcium	20	18	+2
Sodium	11	10	+1
Potassium	19	18	+1
Chlorine[a]	17	18	−1

[a]In its ionic form, chlorine is referred to as a chloride ion.

Table 2.2 Chemical symbols for biologically important elements

Element	Symbol[a]
Hydrogen	H
Carbon	C
Nitrogen	N
Oxygen	O
Sulfur	S
Phosphorus	P
Sodium	Na
Potassium	K
Chlorine	Cl
Calcium	Ca

[a]The symbols for sodium (Na) and potassium (K) come from their Latin names, natrium and kalium.

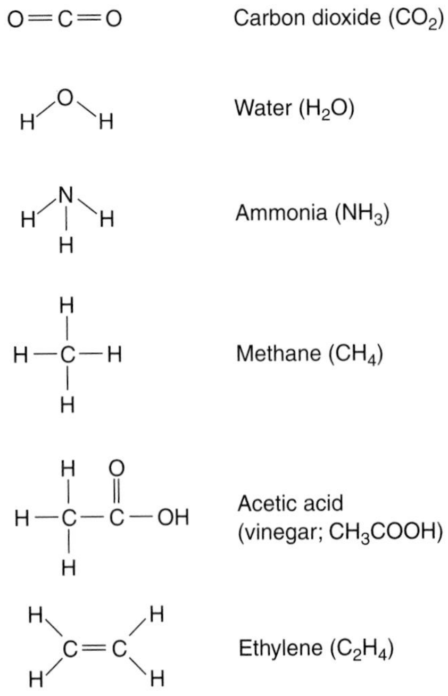

Chemical bonds contain energy, and when bonds are broken, energy is released. The converse is also true. When organisms synthesize large molecules by joining smaller molecules together, they need to be able to draw from an energy source to drive that reaction.

Drawing Molecules

Throughout this book, we will be depicting biological molecules in figures so that you can see important features or look at the changes molecules undergo during biological processes. Sometimes, when the individual atoms in the molecule are not particularly important to the information we are trying to convey, we will portray molecules as blobs. However, at other times we will want to emphasize, or at least acknowledge, the individual atoms that comprise a molecule. To show the atoms in a molecule, we will draw the molecules as ball-and-stick figures, usually two dimensional. In these figures, we will use the

Figure 2.8 Examples of small molecules drawn in ball-and-stick style.

chemical symbol of the element (which conveniently enough happens to be the first letter of the names of hydrogen, carbon, nitrogen, and oxygen) with lines representing the bonds it is forming. Table 2.2 lists the chemical abbreviations of elements you will encounter in this book, and some examples of small molecules, drawn in the ball-and-stick form, are shown in Figure 2.8.

When larger molecules need to be represented, drawing all the individual carbon and hydrogen molecules can get very cluttered and can interfere with

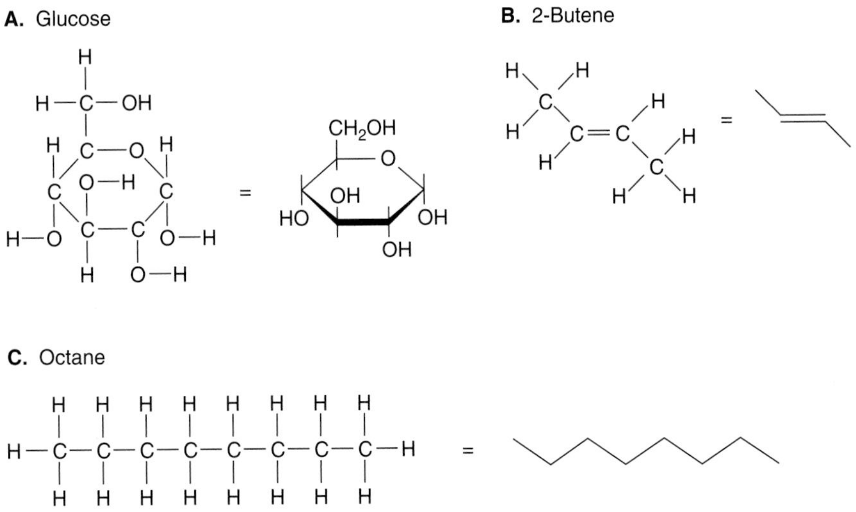

A. Glucose

B. 2-Butene

C. Octane

Figure 2.9 Skeleton representations of molecules. In representing biological molecules, unlabeled corners are assumed to be carbon atoms bound to hydrogen atoms and are not shown. Note that carbon atoms in a molecule always have four bonds.

Table 2.3 Subunits of biological molecules

Class of molecule	Examples	Smallest repeating unit
Lipid[a]	Fats, oils	Fatty acids
Carbohydrate	Starch, cellulose	Simple sugars
Nucleic acid	DNA, RNA	Nucleotide
Protein	Enzymes, antibodies	Amino acids

[a]Lipids are a heterogeneous group of molecules that are grouped together because they are insoluble in water (hydrophobic) and soluble in nonaqueous solvents, such as benzene. Lipids exhibit a great variety of chemical structures, sizes, and complexities, so they are not as tidy a class to describe as the other biological molecules. The "repeating unit" listed for lipids applies only to fats and oils and not to lipids such as cholesterol or vitamin E. The most useful fact about the biochemistry of lipids is that they are mostly hydrocarbons, which are chains of carbon molecules bound primarily to hydrogen atoms.

getting the point across. In these situations, we may use an abbreviated system in which carbons are understood to be at the corners of the skeleton and hydrogen atoms are understood to be filling the rest of the carbon's possible bonds. If another type of atom (say, oxygen) is present, it will be shown specifically (Figure 2.9).

Now that we have explained the various ways that we are going to present molecules visually, let's look at the four classes of biological molecules, the molecular stars of this text. As you read the descriptions of these large molecules below, you will notice that many of them are made of small repeating units that serve as molecular building blocks (Table 2.3). This is an important concept, because when organisms eat, they break the big molecules down into the building blocks, which can be recycled and strung together to make new large molecules. Breaking large molecules into building-block molecules releases energy stored in the bonds linking them. In addition, the building blocks can be degraded further. When the bonds in the biological molecule are broken, cells obtain the energy they need to stay alive.

Lipids

Lipids are a diverse collection of biological molecules that share the property of not being soluble (capable of dissolving) in water. Therefore, they are hydrophobic molecules, and they include fats, oils, cholesterol, and certain vitamins.

Lipid Structure

Fats and oils have similar molecular makeups. They are made of glycerol, a three-carbon molecule, combined with different numbers and types of **fatty acids.** Fatty acids are long chains of carbon and hydrogen atoms that terminate in a **carboxyl group** (Figure 2.10). The carboxyl groups react with glycerol to create a fat. If the fat is a liquid at room temperature, it is called an oil. No doubt you have heard of fatty acids and know that there are saturated and unsaturated fatty acids. A saturated fatty acid contains all of the hydrogens it can possibly hold, as shown in Figure 2.10. An unsaturated fatty acid, in contrast, does not. Instead, it has a double bond between at least two of its carbon atoms.

Saturated fatty acids form straight chains because of the geometry of the bonds between the carbon and

Figure 2.10 (A) Fats and oils are composed of a glycerol molecule joined to long-chain fatty acids. (B) The same saturated fatty acid represented in both ball-and-stick and skeleton forms. (C) One molecule of glycerol bonded to three saturated fatty acids. Will this fat be solid or liquid at room temperature?

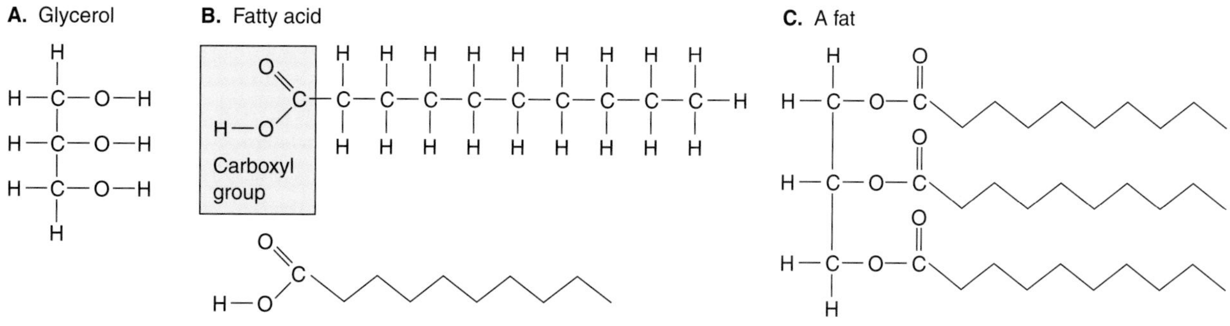

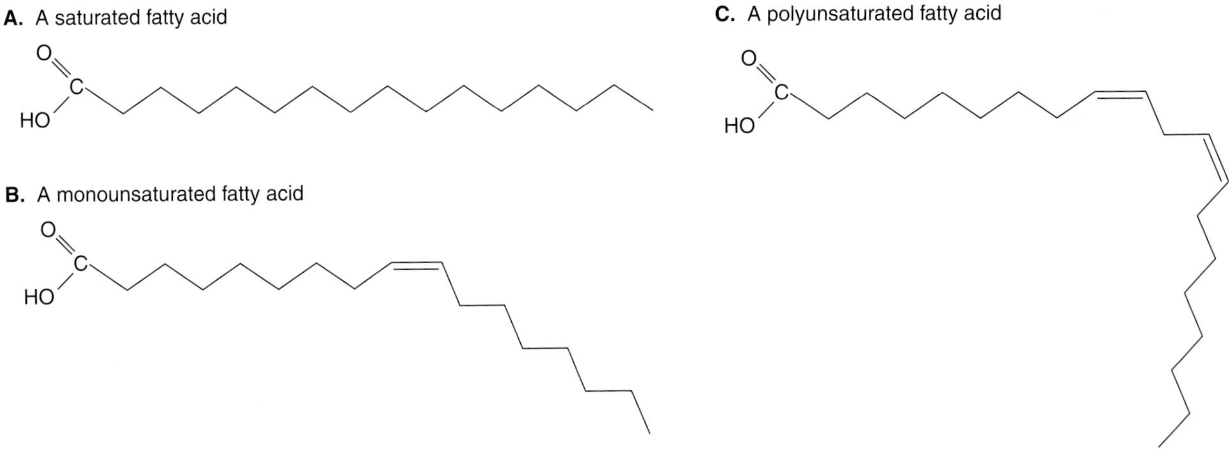

A. A saturated fatty acid

C. A polyunsaturated fatty acid

B. A monounsaturated fatty acid

Figure 2.11 One saturated and two unsaturated fatty acids.

hydrogen atoms. In the fat shown in Figure 2.10, the straight fatty acid chains pack neatly side by side. Because of this physical arrangement, saturated fats are solid at room temperature. Now, look at the unsaturated fatty acid in Figure 2.11 and note how the double bond changes the geometry of the fatty acid. Because of the bend(s) in the chain, unsaturated fatty acids cannot pack together neatly like saturated fats. The greater the number of double bonds, the less able the fatty acids are to align themselves and the more likely the unsaturated fat is to be liquid at lower temperatures.

Other lipids, such as cholesterol, have a very different structure. Cholesterol, as well as testosterone, estrogen, and cortisone, belongs to a class of lipids known as sterols. Sterols also contain large numbers of carbon atoms bonded primarily to hydrogen, but the carbons do not exist as long chains but in rings that are joined to one another (Figure 2.12). In spite of this change in structure, sterols share the defining

Figure 2.12 Cholesterol, a sterol lipid, is composed almost entirely of carbon and hydrogen atoms.

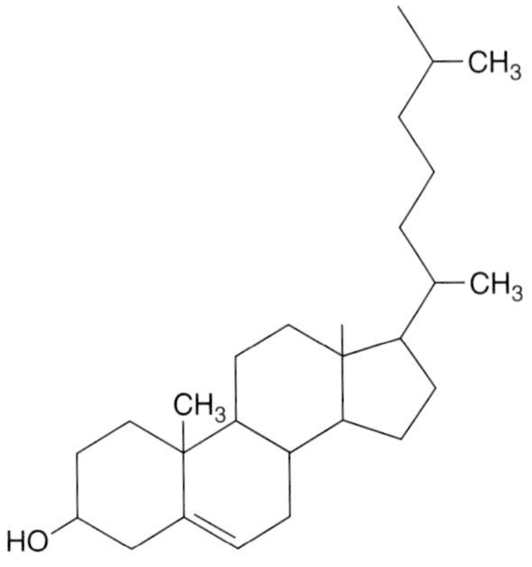

trait that makes fats, oils, and sterols lipids: that is, they are all hydrophobic molecules that are insoluble in water but soluble in other lipids.

Lipid Function

We have already discussed one of the important biological functions lipids serve: they are the primary structural components of cell membranes. Membrane lipids are almost always depicted schematically, as they are in Figure 2.3, as a circle attached to two long tails. The two tails represent two fatty acids, and the circle represents glycerol. You may be asking what is bonded to the third carbon in the glycerol, and the answer is a phosphate molecule. Phosphate molecules play very important roles in a number of biological molecules, so you will be hearing about them often. A molecule of phosphate contains an atom of phosphorus bonded to four oxygen atoms, three of which are also bonded to hydrogen atoms. When the phosphate molecule forms a bond with glycerol, two hydrogens are lost from the phosphate. The new grouping of atoms, known as a **phosphate group,** carries a charge and therefore is hydrophilic. The new molecule created by joining the two-chained lipid to a phosphate molecule is known as a **phospholipid** (Figure 2.13).

Another important function of lipids is energy storage. Animals use excess energy to manufacture lipids and then store those lipids until the energy is needed. Some plants, too, store energy for new plants (think of sunflower seeds, almonds, and walnuts, all high in oil). Lipids offer organisms a way to pack a lot of energy into a small space. The flip side of this, as any dieter knows, is that you have to expend lots of energy to burn up a single pound of unwanted fat. We will return to the topic of lipids as a source of energy later in this chapter.

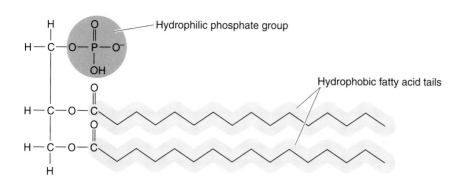
Hydrophilic phosphate group

Hydrophobic fatty acid tails

Figure 2.13 A phospholipid. Phospholipids are major components of cellular membranes. The charged phosphate group is hydrophilic, while the long fatty acid tails are hydrophobic.

Cholesterol is also an important molecular component of cellular membranes in animal cells. It helps to make membranes more oil-like and less fat-like; in other words, fluid. That fluidity is essential to cell function. So, even though cell membranes are always drawn as a static arrangement of two layers of lipids with embedded proteins, they are much more dynamic than that. In addition to being a membrane structural material, cholesterol is the starting molecule for synthesizing a key component of bile, the substance secreted by the liver that aids in fat digestion, and the steroid hormones, which include estrogen and testosterone (Figure 2.14). Plant cellular membranes, or any part of the plant for that matter, do not contain cholesterol, which is why food companies label vegetable oils "cholesterol free." Plant cells contain other sterols, however.

Carbohydrates

Carbohydrates are defined by their overall chemical formula; they contain a ratio of one carbon atom to two hydrogen atoms to one oxygen atom (CH_2O). Familiar examples of carbohydrates are sugar, starch, and cellulose.

Carbohydrate Structure

The smallest carbohydrates are the simple sugars, such as glucose, fructose, and ribose, whose chemical names are monosaccharides (Figure 2.15). Two simple sugars can be hooked together to form other carbohydrates, the disaccharides, such as sucrose (table sugar), which consists of a molecule of glucose bonded to a molecule of fructose, or milk sugar, lactose, which is one molecule of glucose bonded to one molecule of a different sugar, galactose. In addition, a large number of these simple sugar molecules (monosaccharides) can be hooked together to form large-chain molecules, or polysaccharides. For example, plants and animals link glucose molecules together in large chains called starch or glycogen, respectively (Figure 2.16). Carbohydrates are hydrophilic, like table sugar, but the large ones are so big that they do not dissolve in water.

Like lipids, carbohydrates play important roles in structure and energy storage. Carbohydrates are the primary components of the cell walls of bacteria and plants, and the hard exoskeleton of insects is made of another carbohydrate, chitin. The simple sugars provide a readily available source of energy for all cells, and polysaccharides provide a mechanism for storing sugar molecules until they are needed. When plants or animals need energy, they break glucose units off of their storage polysaccharides, which are termed starch and glycogen, respectively. Even though polysaccharides and lipids are both energy storage molecules, they represent two levels of

Figure 2.14 Steroid hormones. Note the similarity of the structure of each molecule to cholesterol, which provides the starting material for synthesizing these hormones. Also note the extraordinary similarity between the molecular structures of estrogen and testosterone.

A. Testosterone

B. Estrogen

C. Cortisol

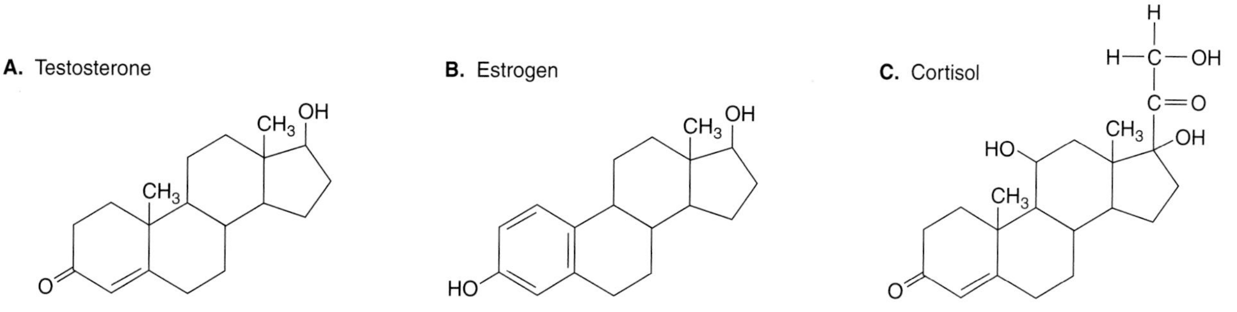

A. Simple sugars

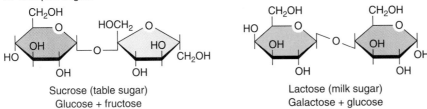

Glucose   Fructose   Galactose

B. Complex sugars

Sucrose (table sugar)
Glucose + fructose

Lactose (milk sugar)
Galactose + glucose

Figure 2.15 Simple sugars, such as glucose and fructose, can combine to make more complex sugars.

storage. Glycogen acts as a short-term storage molecule, and most people have a glycogen supply that could provide cells with energy for about 1 or 2 days. Fats are for long-term energy storage, and a normal-size person has enough fat to power cells for 4 to 6 weeks.

Finally, in addition to serving as sources of energy and structural molecules, carbohydrates are often found connected to other molecules on the outsides of cells. In these positions, carbohydrates play essential roles in cell-cell recognition, cell signaling, and adhesion of cells to each other.

Proteins

Thus far, you have learned that lipids and carbohydrates play roles as structural molecules, energy storage molecules, and recognition molecules. What has been missing up to this point is action: molecules that create other molecules, break them down, and make things happen. These molecules are proteins. We will introduce you to proteins only briefly in this section, because they receive a great deal of attention throughout the remainder of the text.

Protein Structure

Proteins are chains of building-block molecules called **amino acids.** All amino acids share a portion of their structure—a hydrophilic backbone. They differ from each other in which of 20 different side chains they contain (Figure 2.17). These side chains are highly variable in shape and chemical nature; some are hydrophilic, and others are hydrophobic. Some side chains are small, because they contain very few atoms; others take up more space.

Like polysaccharides, cells make proteins by stringing together a few or many hundreds of their building-block molecules. Unlike polysaccharides, which consist of a very few different monosaccharides, and sometimes only one, repeated over and over, proteins are made of 20 different building blocks in variable orders and proportions. Having 20 types of building blocks to choose from allows cells great variability in the proteins they synthesize, especially because a protein can contain thousands of amino acids. For a simplistic analogy, think of a chain made of various numbers of 20 different pop-beads randomly selected from a large pile of pop-beads. But a protein is more than a straight chain of

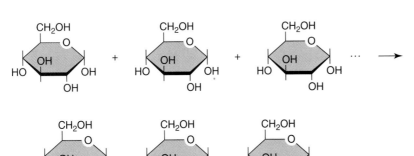

Figure 2.16 Plants store glucose as starch by linking glucose molecules together in a long chain.

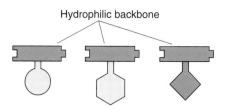

Hydrophilic backbone

Figure 2.17 All amino acid molecules have a hydrophilic part that is chemically identical to those of other amino acids, and they can be linked together to form a chain of amino acids. While the chain of amino acids has a hydrophilic backbone, each of the 20 amino acids has a different side chain that can be either hydrophobic (colored) or hydrophilic.

pop-beads. Imagine that the chain is folded and coiled into a specific three-dimensional shape (Figure 2.18). As you will see again and again, having that shape enables a protein to do its job. If its shape changes even slightly, the protein can lose its ability to function.

Protein Function

Some proteins, like collagen, are key structural elements in organisms; however, the real importance of proteins is as drivers of cellular processes. Proteins carry out a remarkable variety of functions as enzymes, signaling molecules, receptors, transporters, hormones, and antibodies. It would be reasonable to say that proteins carry out nearly every function necessary for life.

Irrespective of the diversity of tasks assigned to proteins, their mode of operation is constant. To carry out its function, a protein depends on its ability to fit with and bind to other molecules, which sometimes, but not always, are other proteins. When proteins bind, quite specifically and precisely, to other molecules, they cause something to happen. The

"something" might be breaking a chemical bond, forming a chemical bond, rearrangement of a part of the protein itself, or a combination of these changes. The ability to bind with specificity to a certain molecule is determined by the protein's three-dimensional structure, which in turn is determined by the sequence of amino acids. In chapter 4, you will learn about protein structure and function in great detail.

Nucleic acids

The best-known example of a nucleic acid is DNA, the genetic material. Nucleic acids, like proteins, are also a primary focus of this text, so our introduction to this class of biological molecule will be brief here.

Nucleic Acid Structure

Nucleic acids, like proteins and polysaccharides, consist of chains of building-block molecules. For nucleic acids, the building blocks are **nucleotides.** A nucleotide is composed of three separate subunits and, in that respect, is more complicated chemically than either an amino acid or a monosaccharide. The three subunits are a simple sugar, a phosphate molecule, and one of five possible nitrogenous bases (Figure 2.19).

The two nucleic acids, DNA and RNA, are composed of long chains of nucleotides; each contains four different nitrogenous bases. For DNA, those bases are thymine, adenine, cytosine, and guanine; in RNA, they are uracil, adenine, cytosine, and guanine. So, while a protein is made of 20 different building-block molecules, each nucleic acid is made of only four possible building blocks arranged randomly in a sequence. Other differences between the chemistries of DNA and RNA are minor, and we discuss them in chapter 4.

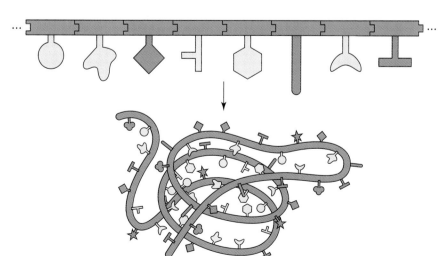

Figure 2.18 A protein is a chain of amino acids that folds into a specific three-dimensional shape. The colored side chains are hydrophilic.

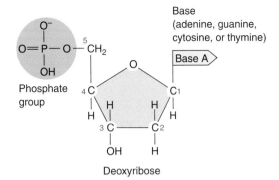

Figure 2.19 A nucleotide, the building block of DNA, is composed of deoxyribose sugar, a phosphate group, and one of four bases.

Nucleic Acid Function

Both DNA and RNA are information molecules. Hereditary information is encoded in the linear sequence of nucleotides in a DNA molecule. Expression of the hereditary information involves converting the information in the DNA sequence into a sequence of amino acids in a protein. RNA molecules mediate the translation of DNA into proteins.

Isolated nucleotides also have important roles in the biological system. For example, the nucleotide adenine bound to the sugar ribose and one phosphate group, **adenosine monophosphate,** or AMP, is important in cellular responses to signals. If two phosphate groups are added to AMP, a molecule of **adenosine triphosphate** (ATP) is formed. ATP, the energy molecule, is discussed in more detail below.

Cellular Processes

In addition to sharing many structural features and being composed of the same molecular building materials, all cells carry out a number of essential processes. These processes and the goals that they serve comprise the defining traits of life. The activities that distinguish living organisms from nonliving matter are extremely complicated but are remarkably similar in all organisms, from bacteria to humans. As a result, in spite of the complexity of the cell processes described below, if scientists understand a process for one type of cell, they understand it, except for minor details, for all cells.

Essential cell functions

Cells Grow and Reproduce

Over time, cells increase in size, and this growth requires energy and building materials. Conveniently, cells get both of these supplies from the same source—the large molecules that make up food: proteins, carbohydrates, fats, and nucleic acids. Cells break down large, complex food molecules to get the smaller, simpler building-block molecules, and in the process, the energy in food is released. Cells then recombine the simple molecular raw materials into the large molecules they need in order to grow, build and rebuild their structures, and carry out activities (Figure 2.20). This process of breaking down large molecules for building blocks and energy and then channeling both materials and energy into

Figure 2.20 Cell metabolism. The biological molecules in food are broken down for energy and building-block molecules and then reassembled according to the needs of the cell.

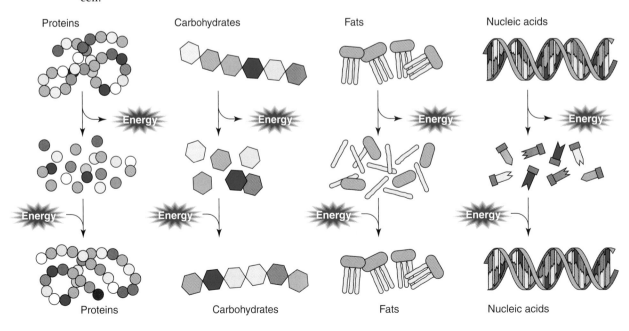

molecular manufacturing is known as **metabolism.** We discuss cellular metabolism in more detail later in this chapter.

The quintessential feature of living entities is that they are self-reproducing. Just as new animals and plants are generated by the reproduction of animals and plants, new cells are generated by the reproduction of cells. All cells grow to a certain size and then reproduce by dividing in two. Once separated, the two daughter cells grow to a certain size, and then they, like the parental cell that gave rise to them, also divide.

The cycles of growth and division require careful management. Each daughter cell must receive a full complement of the cell parts and molecules it needs to be a complete, living cell. During the growth phase, the cell manufactures molecules and cell structures. When the cell senses that its supply of molecules and structures is sufficient for two fully functional cells, it copies its genetic material and divides in two.

Receiving an entire correct copy of the genetic material from the parental cell is essential for survival. Therefore, cell division must be timed so that it follows DNA replication, and all cells must have a mechanism in place to ensure that each daughter cell contains a complete genome. In bacterial cells, the DNA is connected to the cell membrane, and after it replicates, the DNA of the daughter cell attaches to a different part of the bacterial membrane than the parental strand. After DNA replication, the cell continues to grow, increasing the distance between the two points of attachment. When it reaches a certain size, the cell divides between the points on the membrane where the parental and replicated DNA strands are attached, thus ensuring that each cell has a complete genome. In eukaryotic cells, a much more complex process, known as **mitosis,** ensures that a complete complement of chromosomes is distributed to each daughter cell. The cyclical process of growth and division is known as the **cell cycle** (Figure 2.21A).

The cell cycle of eukaryotes consists of four distinct phases. The "birth" of a new daughter cell from a parental cell through cell division is followed by a phase (G_1) in which the cell grows but does not replicate its DNA. Following G_1 phase, the cell synthesizes DNA (S phase) in preparation for cell division. DNA replication in S phase is followed by another growth phase (G_2), in which the cell continues to manufacture molecules, including those that will be needed for mitosis. During the final

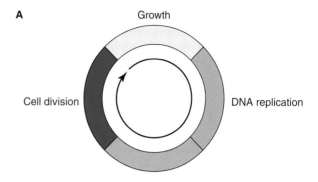

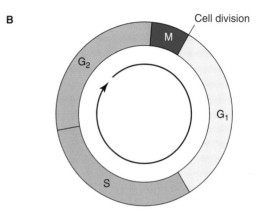

Figure 2.21 The cell cycle. (A) Cells go through cyclic periods of growth, DNA replication, and cell division as they reproduce. (B) The relative amounts of time a cell spends in each phase of the cycle. After mitosis (M), followed by cell division, the cell enters a growth phase (G_1). After the cell reaches a certain size, DNA replicates (S) in preparation for cell division and enters a second growth phase, G_2.

stage, mitosis, or M phase, DNA that was replicated during S phase is distributed equally between two nuclei. After nuclear division occurs, the cells divide. The relative lengths of the four phases of the eukaryotic cell cycle are illustrated in Figure 2.21B.

Cells Maintain Their Internal Environments

When the first cell biologists viewed cells with a microscope, one of the most obvious features was the cytoplasm's greater concentration than the external environment. In other words, even though cells contain the same molecules that are readily found in their external environments, the total numbers of nonwater molecules, such as salt and sugar, are greater internally, and the number of water molecules relative to nonwater molecules is greater externally.

If you compare the chemical makeup of a cell to the environment surrounding it, you will find that not only do the numbers of molecules differ, but also their identities. For example, it is not simply that your cells contain more sodium and potassium molecules per water molecule in their cytoplasm, i.e., 10 molecules of sodium plus 10 molecules of potassium internally compared to 1 molecule of each in the external environment. Your cells are bathed in a fluid that contains 10 times more sodium than the fluid inside your cells; the cytoplasm of your cells contains 20 to 40 times more potassium than your extracellular body fluids. In addition, cells manufacture and keep inside themselves many unique kinds of molecules not found in their immediate, external environment.

Why does this matter? Cells must work hard to maintain the concentration differences between the cytoplasm and the external environment. In the absence of this energy expenditure, water molecules would move back and forth across the cell membrane through **osmosis** until the intracellular and extracellular concentrations achieved osmotic balance, in which external and internal concentrations are equal. A cell in osmotic balance more often than not is a dead cell.

Cells also must be quite selective about which molecules enter and leave the cell. Hydrophobic molecules and some very small charged molecules, such as water, can cross the cell membrane easily. However, given that life is water based, the hydrophobic nature of the cell membrane effectively excludes most molecules with which it comes in contact. As you learned above, many important biological molecules, such as glucose and amino acids, are charged molecules. As a result, they cannot readily traverse the hydrophobic cell membrane, nor can ions such as potassium. These molecules enter and leave the cell with the assistance of **transport proteins** embedded in the cell membrane, like the protein in Figure 2.7.

Cells have a number of different types of membrane transport proteins. **Channel proteins** allow the movement of small, charged molecules, such as the ions potassium and sodium, in and out of the cell. Even though water can cross the cell membrane, a water-specific channel protein increases the speed with which water molecules move through the membrane. Another type of transport protein, a **carrier protein,** escorts molecules such as glucose across the membrane (Figure 2.22). Different carrier proteins fulfill escort duties for different charged molecules, and as you might expect, these transport proteins have shapes that allow them to bind specifically to the molecule they escort. Finally, membranes have proteins known as pumps that use energy to move charged molecules from regions of lower to regions of higher concentration. An example is the sodium-potassium pump, which moves potassium from the external environment to the cytoplasm and pumps sodium out of the cell.

Maintaining unequal internal and external concentrations of ions is essential to healthy cell function. In the absence of a concentration difference, nerve cells would not fire and muscles would not be able to contract. The molecular basis of cystic fibrosis is a malfunctioning membrane protein that is responsible for the movement of chloride ions from the cell's cytoplasm to the external environment.

Cells Respond to Their External Environments

As we alluded to in the above paragraph, cells respond to their environments. When the concentration of salt or water in the external environment changes, cells make changes to keep from becoming too salty or too watery inside. Cells respond to many molecules in addition to salt and water.

Figure 2.22 The glucose carrier protein binds to glucose and permits it to pass through the cell membrane. Transport is with the concentration gradient.

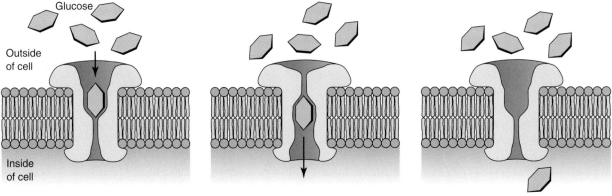

One-celled organisms, like bacteria, move toward food molecules and away from toxins and predators; under harsh environmental conditions, many enclose themselves in tough coverings and become inactive. A plant cell responds to an insect eating it by releasing protective, insect-killing chemicals. The presence of food causes stomach cells to release digestive juices.

To be capable of these and a variety of other kinds of responses, cells must be able both to sense a change in their environment and then to change themselves. Single-celled organisms sense and respond directly to the outside environment; cells in multicellular organisms often respond to each other. The ability of cells to sense environmental conditions and make appropriate changes is crucial to survival, because it allows them to protect themselves and to identify and access essential resources, such as food.

The variety of stimuli that cells detect includes chemical changes, electrical impulses, osmotic concentration, temperature, pressure, light, and sound. The categories of molecular responses that can be induced by these stimuli include enzyme activation, enzyme suppression, protein synthesis, suppression of protein synthesis, release of stored protein, and changes in cell permeability by altering channel proteins. As you review that list, we hope you notice that at the molecular level, the phrase "the cell responds" translates into some sort of change in the activity of one or more proteins. These changes in protein activity ultimately lead to a cellular change, such as moving, dividing, manufacturing molecules, or even dying.

Not only are proteins the mediators of the cell's response, they are also responsible for receiving the signal. Many receptor proteins are located in the cell membranes, but some are also found within the cell. If the signal molecule cannot cross the cell membrane, then it exerts its effect by binding to a membrane receptor. If the signal molecule is one that crosses the cell membrane naturally, is allowed to pass through a channel protein, or is escorted across by a carrier protein, then it is able to bond to an intracellular receptor protein. Whether the signal bonds to membrane receptors or intracellular receptors, the effect is the same: binding of the signal molecule changes the shape of the receptor protein, and this in turn alters its function. When the signal is no longer bound to the receptor, the receptor protein is unable to carry out the function permitted by the bonded signal molecule.

Cells Communicate with Each Other

The chemical signals that cells sense and respond to are often sent by other cells. Even though one-celled organisms are self-sufficient, they need to communicate with each other, if for no other reason than to find another cell for mating. The two sexes of yeast cells and other one-celled organisms use chemical signals to find one another and fuse. One-celled organisms also communicate for reasons other than mating. Certain disease-causing bacteria signal to each other to determine when their population size is sufficient to launch an attack. The signaling molecules secreted by the bacteria trigger molecular changes that lead to the production of the proteins that permit bacteria to infect cells. When too few of its fellow bacteria are present, a bacterium does not waste energy synthesizing these molecules. This phenomenon, known as quorum sensing, is reminiscent of the coordinated behavioral approaches of predators that hunt in packs, such as lions and wolves.

Cells in multicelled organisms communicate among themselves to coordinate their activities. External and internal conditions are continually assessed by cells, and that information is transferred to other cells so that they can act on that information. For example, when we encounter a snake in the woods, specialized receptor cells in our eyes pick up that signal from the external environment and stimulate the optic nerve cells; they then convey the information to the brain. The brain cells process the information and translate it into "snake." Certain nerves then convey the information from the brain to the adrenal gland, stimulating the release of adrenaline, a hormone, into the bloodstream. Adrenaline travels to the liver, heart, and blood vessels, where it stimulates specialized cells in each organ to release glucose into the blood, increase the heart rate, and raise the blood pressure. While nerves in one branch of the nervous system are conveying the snake message to the adrenal gland, nerve cells in a second branch are transmitting the message to muscle cells, giving them the signal to contract. For most of us, the message triggers the jump and run muscles in our legs. For a small minority of people, most of whom are field biologists or adolescent boys, the message stimulates the muscles that allow them to bend over and grab the snake.

In other cases, the cell-to-cell communication in multicellular organisms is completely internal, relying on cells that are incapable of sensing changes in the external environment but are extremely sensitive to activities of the other cells in the organism. For

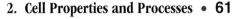

example, when the adrenaline causes liver cells to release glucose into the bloodstream, as in the situation discussed above, cells in a different organ, the pancreas, sense the increase in glucose. This sudden rise in glucose causes these pancreatic cells to secrete the hormone insulin, a protein that decreases blood glucose levels by changing the activities of muscle, fat, and liver cells. While the pancreas is attending to the sudden increase in blood glucose levels, cells in the cardiovascular system are also making appropriate adjustments by responding to chemical signals that were released in response to the changes in heart rate and blood pressure.

Cell signaling, which is the sending and receiving of intercellular messages, is a communication system in which both the signal and the receptor are molecules. The signaling cell produces a specific molecule, which can be a biological molecule, one of its building-block molecules, or a nonbiological molecule, such as salt or carbon dioxide. The target cell receives the message because it has receptor proteins that bind very specifically to the signaling molecule. Signaling cells may excrete signaling molecules, which then travel to the target cell, or the signaling molecule may be a component of the signaling cell's membrane. As for the target cell, the receptor protein may be embedded in the membrane or inside the target cell. The binding of the signaling molecule to the target cell receptor triggers the first step in a series of molecular events. In other words, the extracellular molecular signal is converted into intracellular signals that ultimately change target cell function.

Common principles governing cell processes

In addition to sharing a basic set of processes, cells carry out these processes in very similar ways. What are some of the key overarching principles underlying and uniting all cell processes?

Cell Processes Require Energy

It is clear from the discussion above that cells are very active and highly organized. They can do many different things simultaneously: break down molecules, sense, respond, communicate, and grow, to name a few. In addition, they build their own intricate internal structures from many types of simple building-block molecules. They constantly ensure that their internal chemical compositions differ from the external environment.

Carrying out activities, building and rebuilding molecules and structures, and maintaining an internal

environment that differs from the cell's surroundings all require energy. Cells must obtain that energy from external sources, because one of the laws of physics that cells abide by is that they cannot create energy de novo; they can, however, transform energy from one type to another. For all living cells, the immediate source of energy they rely upon is chemical-bond energy; for the great majority of organisms on the planet, the ultimate source of the energy stored in chemical bonds is light energy from the sun.

Virtually all plants and a few microorganisms are able to convert light energy into chemical-bond energy. Through the process of **photosynthesis** (Figure 2.23), green plants, algae, and some bacteria capture sunlight and transform it into the chemical-bond energy contained in the molecule glucose. More specifically, photosynthesis utilizes the sun's energy to drive a series of chemical reactions that combine carbon dioxide with hydrogen from water to create glucose molecules. The glucose molecules synthesized by photosynthetic organisms provide the requisite energy, not only for themselves, but also for other organisms that are unable to utilize light energy and must obtain energy from breaking down food molecules.

Because chemical bonds contain energy, when bonds are broken, energy is released. A good example of energy release through bond breaking is the burning of gasoline. You have to put in a little energy (from a match, perhaps) to get the reaction started, but then a huge amount of energy is released in the forms of heat and light. What is happening at the level of the gasoline molecules? Bonds are breaking. Gasoline is a mixture of hydrocarbons. Hydrocarbons are substances made from hydrogen (the hydro) and carbon. Bonds between carbon and hydrogen contain a great deal of energy. When a hydrocarbon burns, the bonds between hydrogen and carbon are broken, releasing energy.

Believe it or not, the same kind of process takes place in your cells. Bonds between carbon and hydrogen, and carbon and carbon, as well as others, are broken. In the body, too, the most energy-rich bonds are those between carbon and hydrogen.

Figure 2.23 Photosynthesis. Plants use the energy in sunlight to power the synthesis of glucose from carbon dioxide and water.

$$6\ CO_2 + 6\ H_2O + energy \longrightarrow C_6H_{12}O_6 + 6\ O_2$$

Carbon dioxide + water + energy $\longrightarrow$ Glucose + oxygen

Guess which molecules contain the most carbon-hydrogen bonds—lipids, especially fats and oils. That is why fats have more calories per gram than other types of food. When your body breaks down a fat, all that lovely energy in those carbon-hydrogen bonds is released.

It may be difficult to relate the burning of gasoline to the chemical breakdown of fats, since you obviously do not have a fire going on inside. The difference lies in how the process takes place, not in the end products, which are identical whether gasoline or a fat is being degraded: carbon dioxide and water. When gasoline burns, the reaction requires heat to get it started, and then the heat from the burning gasoline keeps the fire going. In your body, the fat-burning reactions can occur at a much lower temperature because enzymes are catalyzing the reactions. In the previous section, we spoke rather vaguely about enzymes causing "an action to occur." When fats are being broken down, the action occurring is the breaking of bonds. Certain enzymes break specific chemical bonds within fat molecules, and other enzymes, which we will discuss later, make chemical bonds between molecules to create fats, which can then be stored for later use as a source of chemical-bond energy.

Fats are not the only molecules that provide energy for running cell processes. Carbohydrates, simple sugars, other lipids, and amino acids also contain chemical bonds between atoms, and specific enzymes can break those bonds and release the energy stored within them. As molecules of lipids and glucose and other food molecules are broken down, the energy in those bonds is not used immediately to run cell processes. Instead, that energy is harnessed and is used to synthesize a different molecule, ATP.

ATP is the universal energy molecule that provides an immediate fuel source for activities in all cell types, both prokaryotic and eukaryotic. As its name implies, ATP consists of a molecule of adenosine (the nitrogenous base adenine bonded to a sugar) and three phosphate groups. One of the phosphate groups is bonded to adenosine, and the rest are bonded to each other (Figure 2.24). The bond between the second and third phosphate groups contains a large amount of energy and is readily broken. ATP is similar to a supply of quick cash that can be used immediately to drive, or "pay for," one of the cell processes described above. The other food molecules, such as carbohydrates and lipids, are more akin to checks that need to be cashed or money in a savings account; ATP is the energy equivalent of money in your pocket.

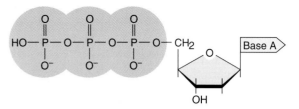

Figure 2.24 A molecule of ATP.

Cell Processes Consist of Chemical Reactions

Cells are tiny chemical factories that perform thousands of chemical reactions simultaneously. Through these chemical reactions, molecules inside the cell join together, break in two, and exchange atoms with each other. The reactions that occur and the molecules and atoms that take part in them are not fundamentally unique; they abide by the laws of physics and chemistry that apply equally to all matter, both animate and inanimate.

Above, we described, in very general terms, a set of chemical reactions that cells use in order to acquire the energy they need to stay alive. Large food molecules, such as carbohydrates and lipids, are broken down into smaller building-block molecules, such as simple sugars and fatty acids. The building-block molecules are degraded further into smaller molecules, such as carbon dioxide and water. Breaking down large food molecules into their component molecules and, ultimately, into even smaller molecules provides an obvious example of a set of chemical reactions that constitute a cell process. However, almost every aspect of the cell processes described above can ultimately be reduced to a set of chemical reactions.

Thinking about life as nothing more than a set of chemical reactions can be difficult, because living cells and the processes that keep them alive are so remarkable. Cells have the ability to sense and respond to environmental conditions in a way that seems purposeful. They also can increase in size, repair themselves, and copy themselves. So, even though molecules in the cell are composed of atoms found in nonliving matter and the chemical reactions between these molecules conform to the same laws that govern all chemical reactions, the chemistry that occurs in a test tube in your chemistry class is not identical to the chemical reactions that are happening in every cell of your body. What distinguishes lab chemistry from life chemistry?

One difference is the nature of the molecules involved in the chemical reactions. As we explained above, cells contain special molecules, the biological molecules, not found in inanimate matter. While some chemical reactions that occur in living organisms involve nonbiological molecules and are identical to reactions that would occur in a test tube, most cellular reactions involve biological molecules. Another difference is simply the number of chemical reactions that occur simultaneously, because every cell contains thousands upon thousands of molecules, and most are engaged in reactions with each other constantly.

Perhaps the most important difference between the chemical reactions that occur in a cell and those in a test tube is the rate at which the reactions occur. The great majority of the chemical reactions that occur in cells would occur very, very slowly if we poured the reactant molecules into a beaker. To drive these reactions, cells manufacture enzymes, which are protein catalysts. All catalysts speed up chemical reactions without being changed in the process. Chemical catalysts, such as platinum, are not especially specific; they can catalyze a number of different chemical reactions. In biochemical reactions, each enzyme is responsible for making a specific chemical reaction occur (Figure 2.25). The molecules enzymes operate on are called **substrates,** and the molecules they produce by operating on a substrate are referred to as their products. Even though an enzyme can operate on only a single substrate, any one substrate can be acted on by more than one enzyme. For example, if enzyme A binds to substrate W, it will lead to product W1; however, if enzyme B binds to substrate W, it will lead to product W2.

Recall that the function of a protein depends on its three-dimensional shape. To be capable of catalyzing a chemical reaction, an enzyme must fit correctly with its substrate molecule. The portion of the enzyme that binds with the substrate is known as its **active site.** Most enzymes are capable of binding to only one substrate and generating only one product from that substrate. The product has a different shape from the substrate and usually does not fit

with the enzyme; therefore, the enzyme releases the product once the reaction has occurred. If a different enzyme binds to the substrate, it leads to a different product. Therefore, by manufacturing some enzymes but not others, cells can control the chemical reactions that occur and those that do not.

Other differences between lab chemistry and life chemistry are related to the last two principles that characterize all cell processes.

Cell Processes Occur in a Series of Small Steps

As we said above, most cell processes can be reduced to a set of chemical reactions. The set of chemical reactions that constitute a certain cell process do not occur independently of each other but are linked into a sequence of consecutive reactions. Each small step in a cell process is usually carried out by a specific protein, such as an enzyme, that is assigned to that step and no other. Structuring cellular processes as a stepwise series of small, sequential events, each requiring a specific protein, provides a cell with many points at which to control a certain process. In general, stepwise cell processes are categorized as cascades and pathways.

A cascade is a series of events, resembling a chain reaction, that leads from a starting condition to an ending condition. An example of a cascade is the sequence of events in a target cell that is triggered by a signaling molecule. The signaling molecule binds to a specific protein receptor, which activates the protein by changing its shape. The shape change leads to a functional change that triggers a change in a second protein, and so forth. Ultimately the chain reaction of protein changes culminates in a significant cellular change, such as initiating the synthesis of a new protein (Figure 2.26).

The synthesis of that protein occurs via the second, more familiar category of a stepwise process mediated by proteins, a pathway. A pathway consist of a

Figure 2.25 An enzyme catalyzes one specific chemical reaction, during which molecules fit together precisely.

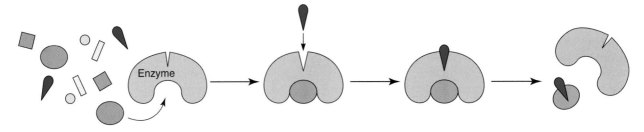

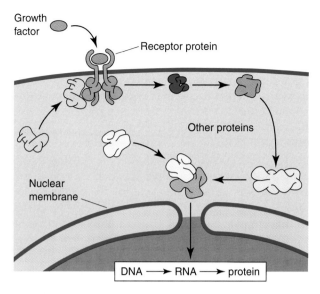

Figure 2.26 In this cascade of molecular interactions, a growth factor binds to a protein receptor in a cell membrane. The binding changes the shape of the receptor, which in turn leads to other intracelluar changes involving the proteins. Ultimately, the chain reaction leads to protein synthesis as DNA is transcribed to RNA.

Growth factor

Receptor protein

Other proteins

Nuclear membrane

DNA ⟶ RNA ⟶ protein

series of steps that chemically alter a starting material and convert it to a final end product of the pathway. Each step in a pathway is catalyzed by a specific enzyme, and each enzyme makes small changes in its substrate. Thus, little by little, the starting material at the beginning of the pathway is chemically altered through bond breakage and formation; the accumulation of small changes can lead to an ultimate end product that is very different from the starting material.

In the generalized pathway depicted in Figure 2.27, the first enzyme binds to its substrate, the starting material of the pathway, and an action occurs, resulting in the product of that enzyme-catalyzed reaction. The molecular product of reaction 1 provides the substrate for the second step. The second enzyme binds to that molecule, and another action occurs, resulting in the product of reaction number 2

and the substrate for the final step, which leads to the pathway's final end product. You can think of a pathway rather like an assembly line of robots, in which each robot has a shape that allows it to attach to its target and only its target.

Pathways do not exist in isolation from each other. Sometimes the end product of one pathway is the starting product of a different pathway, so the first pathway feeds into the second. Therefore, the operation of the second pathway depends on the operation of the first. In Figure 2.27, the materials produced by enzymes A and B on the way to the end product are called intermediates. Sometimes the end product of one pathway is an intermediate in another pathway, linking the two pathways into a network. In addition, pathways are very often branched, because there is more than one possible fate for an intermediate in the pathway. Another way to look at branching pathways is that two pathways share many steps, but at a certain intermediate, they diverge. Finally, some pathways converge with each other like two or more roads merging into one.

Cells break down and manufacture biological molecules in sequential sets of reactions known as metabolic pathways. Each step in a metabolic pathway is catalyzed by a specific enzyme. Biologists use the phrase catabolic pathway, or **catabolism,** for the breakdown, energy-yielding half of metabolism and anabolic pathway, or **anabolism,** for the synthesis, energy-requiring half.

Catabolic pathways. In the human body, energy and building materials are obtained from breaking down, or catabolizing, carbohydrates, lipids, and proteins. The first stage of catabolism, in which the large molecules are broken into their component parts—carbohydrates into sugars, proteins into amino acids, and fats into fatty acids and glycerol—occurs in the digestive tract. These small building-block molecules are absorbed into the bloodstream through the walls of the small intestine and are carried to the

Figure 2.27 Cellular processes are carried out by a sequence of enzymes, each of which promotes a specific reaction with a specific substrate, like a robot in an assembly line.

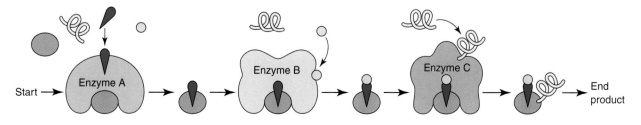

Start

Enzyme A

Enzyme B

Enzyme C

End product

body's cells. Within the cells, these molecules are broken down further through different metabolic pathways. For the purposes of our discussion, we will focus on the catabolism of only one building-block molecule: glucose.

Glucose is the major energy molecule for cells from almost all organisms, both prokaryotes and eukaryotes. Cells that need a quick shot of energy take up glucose, or another sugar from the blood, and break it down immediately. In animals, if the supply of glucose exceeds the body's demand for quick energy from glucose, enzymes hook the glucose units together into a glycogen molecule, which is stored until the demand for energy from glucose increases again.

Cells break down glucose to release the energy in its bonds via two sequential pathways. The first is a linear pathway, and the second is circular (Figure 2.28). The first pathway is found in bacteria, as well as

plants and animals, attesting to its evolutionary ancientness. In this nine-step pathway, known as **glycolysis,** glucose, which contains six carbon atoms, is converted into two molecules of the three-carbon compound pyruvic acid. No oxygen is required to convert glucose to pyruvic acid, so even microorganisms that live in the absence of oxygen can use this pathway. If oxygen is present, pyruvic acid is broken into two molecules: carbon dioxide and acetyl-coenzyme A (CoA). Acetyl-CoA is the starting material for the second pathway, which is called the tricarboxylic acid (TCA) cycle, or the Krebs cycle, after the biochemist who teased apart all of the steps in this circular pathway. After a number of turns of the TCA cycle, the six carbons in the original glucose molecule that provided the starting material for glycolysis are converted to six molecules of carbon dioxide. The great majority of the energy released from breaking down glucose to carbon dioxide is generated in the TCA cycle. When energy is released from breaking the chemical bonds in glucose, it is captured and, through a different pathway, is stored as ATP.

Figure 2.29 illustrates a metabolic branch point, as described above. If oxygen is present, the intermediate at this branch point, pyruvate, will be broken down through a series of chemicals to carbon dioxide and water. If not, your cells will convert it to lactic acid.

Fatty acids absorbed into the bloodstream can also be broken down by cells that need energy, such as

Figure 2.28 Schematic representation of glucose breakdown through glycolysis and, if oxygen is present, the TCA cycle.

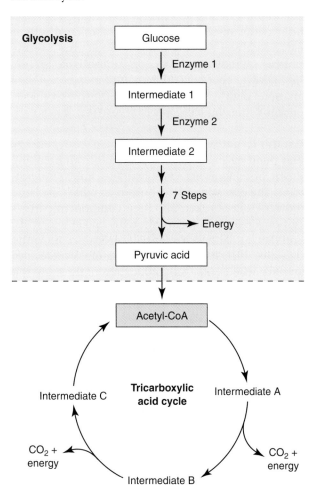

Figure 2.29 A branching metabolic pathway in glucose breakdown.

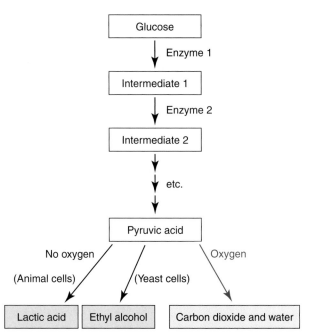

Exercise Burn and a Metabolic Branch Point

Have you ever exercised until you felt your muscles burn? The burning is a direct result of a branch point in the metabolic pathway for breaking down glucose.

When you exercise, your muscles need a lot of energy. Exercising muscle cells breaks down lots of glucose—muscles even have their own glycogen stores for just such situations. Breaking down glucose to carbon dioxide and water requires oxygen. When your muscles are working hard, you are breathing harder, and your heart is beating faster to supply your muscles with extra oxygen to power glucose breakdown to carbon dioxide.

Sometimes, though, you cannot get enough oxygen to your muscles. You might be putting extraordinary demands on isolated muscles, as when you lift weights, or demanding output from your entire body that exceeds your capacity to supply oxygen via breathing and blood circulation, as when you sprint for a while. When your muscles cannot get enough oxygen, the normal breakdown of glucose to carbon dioxide and water cannot happen.

Look at Figure 2.28. When oxygen is present, pyruvic acid is converted into acetyl-CoA and on to carbon dioxide and water. In Figure 2.29, you can see that this step is actually a metabolic branch point. If oxygen is not present, your muscle cells do something different with the pyruvic acid. They have another enzyme that converts it into a three-carbon compound called lactic acid. When lactic acid builds up in your muscles, you feel that burning sensation. The lactic acid burn is a sign that your muscles are working anaerobically (without oxygen). In other words, your breathing and heart rate are not keeping up with your muscles' demand for oxygen.

You've probably heard of aerobic exercise. The word "aerobic" means "with oxygen." Aerobic exercise by definition means your breathing can keep up with your muscles' need for oxygen, so you can sustain the exercise for longer periods of time. Because you are breathing harder and your heart is beating faster than usual during aerobic exercise, sustained exercise of this kind can strengthen your heart.

exercising muscle cells, or, if not needed by cells, they can be converted to fat and stored as fat droplets in fat cells. When cells absorb amino acids from the bloodstream, they usually link them together to build proteins or they break them down into their constituent atoms and remodel them into other amino acids. If necessary, cells can also use amino acids for energy, but that is not their primary fate. Cells would rather not break down amino acids for energy because proteins are responsible for making cells work. The protein inventory of a cell changes from one minute to the next, so cells always need a ready supply of all 20 amino acids for protein manufacturing.

How are molecules other than glucose broken down to get energy? All of them are broken down into molecules that are intermediates in the glucose breakdown pathway just described. Various sugars, such as fructose, enter the pathway during the early stages of glycolysis. Fatty acids are broken down via a multistep pathway to acetyl-CoA, which enters the TCA cycle. By 20 different, but converging, pathways, all of which first require the removal of nitrogen, amino acids are converted into pyruvic acid, acetyl-CoA, or one of the other intermediates in the TCA

cycle. Like cars merging onto an interstate highway, sooner or later the breakdown pathways for glucose, fatty acids, and amino acids merge into a common pathway.

Anabolic pathways. Cells synthesize many different molecules for their own purposes. There are hundreds of enzymes inside a cell, and every one is a protein the cell synthesized. The cell membrane is composed of lipids synthesized by the cell. Muscle and liver cells synthesize the energy storage carbohydrate glycogen. Plant cells synthesize and store starch. Cells synthesize large molecules by joining together the component building blocks, and they also synthesize the building blocks from scratch using just atoms.

The synthesis of large biological molecules and their building blocks costs energy. Cells power the energetically costly anabolic reactions by coupling them to energy-yielding catabolic reactions, rather like burning gasoline (an energy-yielding reaction) to heat water (an energy-costing process). If an organism's energy intake and output are balanced, its cells break down sugars and fatty acids to form ATP and then use that ATP to synthesize their own proteins, lipids, carbohydrates, nucleic acids, and other components, as well as to power movement and other processes. If the organism takes in more energy than it needs, it uses excess ATP to drive the synthesis of long-term energy storage molecules, such as starch, glycogen, and fats.

Many of the intermediates of glucose breakdown are used as starting materials for making amino acids, nucleotides, fatty acids, glycerol, and other small molecules, such as steroids. Therefore, sugar molecules, like glucose, can be used not only for energy but also to make amino acids. Amino acids and fatty acids can also be broken down and reassembled into glucose molecules. The carbon and hydrogen atoms in a fat may end up in DNA, after having been part of a glycogen molecule. If there is more energy available than the body needs, the energy and carbon atoms in glucose can be converted into a fatty acid for storage.

Metabolic problems. Because a unique enzyme catalyzes each metabolic step, if the protein at a single step malfunctions, it can derail the entire pathway. For example, if the second enzyme (B) in the pathway illustrated in Figure 2.27 is missing, then its particular operation will never be carried out. The third enzyme (C) cannot bind to the product of the first step, because it does not have the correct shape, so the product of the first step in the process will build up. A genetically determined enzyme defect leading to interruption of a metabolic pathway at a specific point is known as an inborn error of metabolism. The defective enzyme can be the catalyst for a catabolic or anabolic reaction. How does an inborn error manifest itself? It depends.

The cell's scheme for breaking down and synthesizing carbohydrates, fats, proteins, and their component building blocks consists of hundreds of steps, many of which link separate pathways into a network. Just like any network, sometimes the consequences of a malfunction go unnoticed because other parts of the network compensate for the error. Other times, the error causes a network failure, and the system crashes. The same is true for errors in genes that encode metabolic enzymes. Sometimes the missing or malfunctioning enzyme is so crucial to survival that a fertilized egg with this mutation dies very early in development, so people with these types of devastating mutations do not exist. Other times, mutations for defective or missing metabolic enzymes pass from parents to children, sometimes with serious negative consequences and other times with little or no impact on the child's health.

When it comes to assessing the potential impact of an inborn error, we can be sure that two things will happen when a metabolic enzyme is missing completely or seriously malfunctioning.

1. The substrate of the broken enzyme will not be converted into a product.
2. Without this product to serve as the substrate for the next enzyme in the pathway, subsequent products in that pathway cannot be produced.

Does either or both of these biochemical effects lead to serious medical consequences? Yes and no. If either the normal end product of the pathway or any of the intermediates between the broken enzyme and the end product is essential to the survival of the cell, then the cell dies. If the substrate of the broken enzyme cannot be turned into the product, it may accumulate in the cell. Is accumulation of the substrate harmful? It depends. Accumulation of certain molecules can lead to cell death, while accumulating other molecules may alter cell function without having serious medical consequences. On the other hand, accumulation may not occur. If the substrate of the malfunctioning enzyme is at a branch in the pathway or is the molecular link to another pathway, then the substrate of the broken enzyme will be shunted to the alternate pathway. But this may create a new problem. Does the increased availability of this intermediate lead to an excess of the end product of the second pathway? Sometimes, having too much of the end product of the alternate pathway might pose a problem.

Cells Regulate Their Processes

Cells carry out their life processes in an orderly, regulated way. They control which process occurs, when it occurs, and where it occurs. For example, cell reproduction requires DNA duplication and cell division into two equal halves, each with enough parts and molecules to enable self-sufficiency. DNA duplication and cell division must be regulated to ensure the following.

- They occur only after the cell reaches a certain size and has enough parts for two free-living cells.
- DNA duplication happens before cell division happens.
- The ratio of DNA duplication events to cell division events is 1 to 1.

Having a mechanism to regulate cell division is essential to survival. In multicellular organisms, when the mechanisms that regulate cell growth and division fail, the result is cancer—uncontrolled cell reproduction.

Cells regulate all of their processes in order to ensure they meet their needs, which change from one moment to the next, as efficiently and effectively as possible. As we mentioned above, the stepwise nature of pathways provides many possibilities for regulation, since any step can be controlled. Cells have a number of tactics for regulating their activities, and all involve the interaction of molecules, especially proteins, with each other. One way to modulate a pathway is to alter the activities of enzymes in the pathway. A second mechanism involves not synthesizing an enzyme in a certain pathway.

Altering enzyme activity. Cells have a variety of strategies for altering the activities of various enzymes. The first is very straightforward. The rate of an enzyme-catalyzed reaction depends, in part, upon the intracellular concentration of the substrate. If the number of substrate molecules is low, the enzyme will not encounter them often. Therefore, to ensure that cells function as efficiently as possible, certain reactions occur in specific compartments in the cell. This increases the probability that the enzyme and substrate will interact. In addition, the enzymes that catalyze sequential reactions in a pathway are often lined up sequentially along an intracellular membrane.

A second strategy, feedback inhibition, is based upon the end products of a pathway and not the substrates. Feedback inhibition is a common mechanism for controlling processes, even those that are not biological. The form of feedback inhibition you are probably most familiar with is the operation of a thermostat that turns off the furnace when the room temperature reaches a predetermined endpoint and then turns the furnace back on when the room gets too cold.

In biological systems, a substance produced in a pathway, typically the final end product, inhibits the activity of an enzyme at the beginning of the pathway. The inhibitor accomplishes this by binding to the enzyme in such a way that it alters the enzyme's activity. A number of the enzymes involved in breaking down molecules to meet immediate energy demands or storing them for future needs are regulated by the amount of available ATP. The addition and removal of a phosphate group from these enzymes shift their levels of activity from low to high, as appropriate.

A second form of feedback inhibition regulation involves allosteric proteins. Allosteric proteins have two binding sites: the active site and the regulatory site.

In metabolic pathways, the end product binds to the regulatory site of one of the enzymes in the pathway, usually the enzyme that catalyzes the first step. When the end product binds to the regulatory site, the shape of the enzyme changes and alters the active site in a way that prohibits bonding of the enzyme with its substrate. This action disables the pathway, and the inhibiting molecule (the end product) is no longer produced. When the concentration of end product molecules decreases, the first enzyme in the pathway regains its binding capability, and the metabolic pathway is reactivated. Using the analogy of the thermostat-controlled heating system, the heat acts as the inhibitor and is analogous to the pathway end product; the furnace serves in the role of the enzyme.

One of the simplest examples of feedback inhibition regulation is the pathway for synthesizing the amino acid isoleucine. The starting product in the isoleucine synthesis pathway is another amino acid, threonine. Isoleucine feeds back and inhibits the enzyme that catalyzes the first step in the pathway. When a surplus amount of isoleucine is available, it blocks its own synthesis. When the amount of isoleucine decreases, the enzyme is released from inhibition and isoleucine synthesis resumes (Figure 2.30).

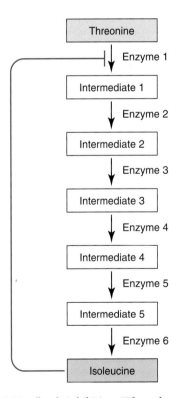

Figure 2.30 Feedback inhibition. When the amino acid isoleucine is plentiful, it blocks its own biosynthesis by inhibiting the first enzyme in the pathway that converts threonine, another amino acid, to isoleucine.

Using feedback inhibition, cells not only modulate the production of an end product, they also shunt activity down certain routes in branched pathways. In the amino acid synthesis pathway in Figure 2.31, the end product of the first branch, methionine, inhibits the first enzyme in that branch, while the end product of the second branch, threonine, inhibits the first enzyme in that branch. Therefore, if a cell has a sufficient amount of methionine, feedback inhibition allows it to block that metabolic route and shunt the intermediates into the pathway to synthesize threonine.

Altering enzyme synthesis. Cells can also regulate their metabolic pathways by altering which enzymes are produced. For example, most bacteria have genes encoding enzymes capable of breaking down a variety of sugars they might encounter. However, synthesizing these enzymes would be a waste of the cell's energy if the sugars were unavailable. Most bacteria can sense which sugars are available in their environment, and they use that information to synthesize the necessary enzyme(s) for breaking down that sugar. Conversely, most bacteria encode enzymes capable of synthesizing all of the amino acids bacteria need to make their proteins. However, synthesizing these enzymes would be wasteful if the needed amino acids were already present in the environment, so most bacteria produce a specific amino-acid-synthesizing enzyme only if that amino acid is not available to them from their environment.

Another example of synthesizing an enzyme only if necessary involves cell-to-cell communication in bacteria. In quorum sensing, bacteria do not produce the molecules that permit them to infect cells until a sufficient number of bacteria are present. No doubt the bacteria accomplish this by not manufacturing one or more of the enzymes that are necessary for production of the molecules, known as virulence factors.

The molecular mechanisms that underlie this and other, similar examples of enzyme production regulation are discussed in detail in chapter 4.

Summary

Biotechnology is based on the use of living cells and their component parts. Therefore, knowing something about the structure and function of cells and the molecules they contain is essential to understanding biotechnology's scientific foundations, potential applications, and possible limitations.

Cells, which are the basic building blocks of all living things, share certain structural and functional features. Whether the cells are eukaryotic or prokaryotic, they contain the same types of molecules, operate by chemical processes that follow the same rules, and speak similar genetic languages.

Cells are remarkable for the level of organization they maintain. A few simple rules governing molecular interactions contribute to cellular organization, primarily the ways molecules interact with water and with each other. Part of the organization is structural; in most cells, intracellular compartmentalization keeps certain molecules in specific locations. In eukaryotic cells, the structural organization reaches a higher level of complexity due to a set of internal membranes that define certain spaces.

All cells contain the same types of biological molecules that perform essentially the same types of roles. Lipids are the primary structural elements of cellular membranes and are energy storage molecules. Carbohydrates are also important energy sources. Some, such as cellulose, also have important structural roles. Proteins, which are chains of up to 20 different amino acids, are the molecules that

Figure 2.31 Feedback inhibition at metabolic branch points provides more specific control.

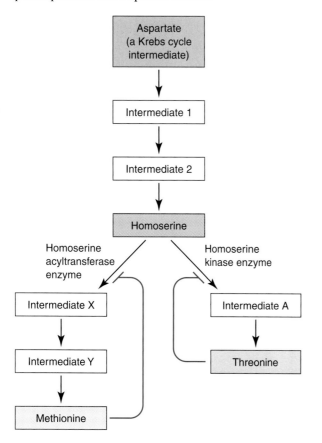

make things happen in cells and provide the molecular basis of life's functions: as enzymes, they create and break down molecules; as receptors, they allow cells to respond appropriately to signals from the external environment and other cells; as antibodies, they protect organisms from invaders. Proteins, such as collagen, also play important structural roles. Finally, nucleic acids, such as DNA and RNA, are information molecules, while another nucleic acid, ATP, is the universal energy molecule.

All living cells carry out a number of essential processes that are the defining traits of life. Cells grow and reproduce, maintain their internal environments, respond to the external environment, and communicate with each other. Cellular processes can be reduced to a series of chemical reactions, most of which are catalyzed by enzymes. To carry out all of these processes, cells require a constant supply of energy. Finally, cells also regulate their processes to ensure they are carried out in an orderly and efficient way.

Selected Readings

Alberts, Bruce, Dennis Bray, et al. 2003. *Essential Cell Biology: an Introduction to the Molecular Biology of the Cell.* Garland Publishing, Inc., New York, NY. This excellent textbook explains cell biology, from molecules to cell structure and function, through clear, compelling writing, helpful drawings, and beautiful electron micrographs. A more advanced version of the text by many of the same authors, *The Molecular Biology of the Cell,* contains many more electron micrographs.

American Society for Cell Biology. *CBE—Life Sciences Education* (http://www.lifescied.org). A free, online quarterly journal, launched in spring 2002 as *Cell Biology Education—A Journal of Life Science Education.*

Cooper, Geoffrey, and Robert Hausman. 2004. *The Cell: a Molecular Approach.* American Society for Microbiology, Washington, DC. Another excellent cell biology textbook with beautiful illustrations and thorough explanations. This book also includes a number of sidebars describing some of the scientists who made the most important discoveries in cell biology.

DeDuve, Christian. 1984. *A Guided Tour of the Living Cell.* Scientific American Library. W. H. Freeman and Co., New York, NY. A beautifully illustrated two-volume set by one of the pioneers in modern cell biology. The author, a Nobel laureate, takes readers on a tour of the cell. Another book by DeDuve, *Blueprint for a Cell: the Nature and Origin of Life,* is more advanced but equally engaging.

Goodsell, David. 1996. *Our Molecular Nature: the Body's Motors, Machines and Messages.* Copernicus, New York, NY. This book, which is essentially a hall of fame of certain molecules, does an excellent job of explaining molecular processes and players in a very compelling way.

Thomas, Lewis. 1974. *The Lives of a Cell: Notes of a Biology Watcher.* Viking Press, Inc., New York, NY. Thomas, a physician, is a writer who appeals to scientists and nonscientists alike. This series of essays, and those in another of his books, *The Medusa and the Snail,* convey the magic most biologists feel when they consider cell function.

Varmus, Harold, and Robert Weinberg. 1993. *Genes and the Biology of Cancer.* Scientific American Library. W. H. Freeman and Co., New York, NY. An excellent and accessible description of cancer at the cellular and molecular levels.

Wilson, John. 2002. *Molecular Biology of the Cell: a Problems Approach.* Garland Publishing, New York, NY. A companion to the book by Bruce Alberts et al.

Genes, Genetics, and Geneticists   **3**

The Centrality of Genes

The remainder of this book focuses primarily on genes: what they are, what they do, how they do it, and how researchers are using them. However, before we delve into the workings of genes on a molecular level, we would like to step back and look at these extraordinary particles from a broader perspective.

Everyone recognizes that students must learn about genes to understand genetics and heredity, the transmission of inherited traits from one generation to the next. What you may not realize is that an understanding of genes is central to an understanding of all of biology. Each of the diverse branches of biology can be seen as a way of investigating questions about genetics.

- Developmental biology is, in large part, the study of gene regulation: what turns genes on and off.
- Physiology and its companion science, morphology, focus on the structural and functional manifestations of gene expression.
- Ecologists study the genetic adaptation of organisms to their environments.
- Taxonomy is the study of the genetic differences and similarities within and between species.
- Evolutionary biologists investigate changes in gene frequencies in populations over time.

You can take any biological phenomenon of interest and explore it from the angle of the gene. First, using your mind's eye as a microscope, put the phenomenon under maximum magnification and ask about the molecular details of its genetic basis. Lower the magnification to the physiological level and ask how the function of this particular gene integrates with the functions of other genes in producing a living organism.

Now, look at the whole organism. How much of what you see has a genetic basis? How much can be explained by environmental influences? How do the two interact?

Now, take a giant step back and use your mind's eye like a telescope to explore the same gene over time and space. How has that gene's structure changed over time? Has the change in structure brought about a concomitant change in function? How much of that change has been meaningful in an evolutionary sense? How does the gene vary within the population? Between populations within a single species? Between species?

To an outsider, biology appears to be a fragmented field of study with many loosely related subdisciplines, and textbooks that dissect the branches of biology into separate, apparently unrelated chapters do little to correct this misconception. Yet all of the various threads of biological thought can be woven into a coherent whole with genetics as the unifying concept. Keep returning to the central landmark that genes provide, and you will be better able to integrate the field of biology into a coherent whole.

Genetics is central to biology, the study of life, because genes are central to life. The fundamental characteristic of living organisms is reproducibility. Living organisms produce more living organisms like themselves. For organisms other than those that reproduce by simple division, reproduction involves both the transmission of hereditary information from parent to offspring and a developmental process that transforms that information into a living organism. Both of these hallmarks of living organisms reflect the functioning of genes: faithful replication and information transmission. If life depends on genes, it only makes sense that the study of life, biology, would have genetics at its core.

A Brief History of Genetics

In the beginning

The relationship between genes, reproduction, and development seems obvious now, but it was not always so. In fact, for thousands of years—99% of human history—people were totally ignorant of the biological basis of reproduction. Sperm and eggs are invisible to the naked eye, and people could understand only those phenomena they could see. The invisible details of reproduction were left to their imaginations, and imagine they did!

Some truly great scientists developed theories of reproduction that seem laughable now. Hippocrates, Aristotle, and, much later, even Charles Darwin believed that offspring resulted from the blending of male and female genital secretions that contained "seeds" that had been shed by each body part: eye seeds, liver seeds, and heart seeds. During reproduction, these seed-filled fluids coagulated and were transformed into an organism as each organ seed developed.

It seems to me that among the things we commonly see there are wonders so incomprehensible that they surpass all the perplexity of miracles. What a wonderful thing it is that this drop of seed from which we are produced bears in itself the impressions not only of the bodily form, but also the thoughts and inclinations of our fathers! Where can that drop of fluid contain that infinite variety of forms?

Michel de Montaigne, 1570

The invention of the microscope in the 1600s permitted observation of human sperm and eggs but did not quell the development of imaginative theories of reproduction. Many scientists believed that tiny humans resided in each sperm cell and that these minuscule humans took root and grew in the female's uterus. In this scenario, the male provided all of the genetic material, and the female provided a nurturing womb for safe development. Some scientists even claimed that, using microscopes, they could see little chickens and horses in sperm taken from those animals.

Other scientists were skeptical, however. In fact, one scientist ridiculed this theory of reproduction by saying he had seen a tiny human enclosed in a sperm take off its even tinier coat!

Even though our ancestors were ignorant of the precise details of reproduction at the microscopic level, they could easily see consistent hereditary patterns in the natural world. Life begets life, and offspring look like their parents.

As early as 10,000 years ago, the first agriculturists put these observations to work by systematically crossbreeding the plants and animals they had begun to domesticate. As soon as they domesticated living organisms, they began to genetically manipulate them through this selective breeding. Our agrarian ancestors managed to genetically modify crops and livestock by trial and error, relying only on minimal understanding acquired through observation. Their knowledge was incorrect and incomplete but nonetheless sufficient to enable them to domesticate and genetically improve virtually all of the crops used today for food and fiber.

As successful as the early agriculturists were, their attempts at selective breeding were still hit-or-miss ventures. Sometimes the "magic" worked, and sometimes it didn't. Such uncertainty is not very satisfying to scientists, who want to explore and understand things from the inside out. The mysteries of heredity provided a fertile field for such exploration because inheritance patterns often seem unpredictable. Traits disappear in one generation and reappear in another. Some observable traits have a genetic basis and therefore can be subjected to successful selection; others do not. Some traits are controlled by one gene, while others are controlled by hundreds of genes.

For centuries, most people invoked supernatural powers to understand the apparently inexplicable nature of biological inheritance. A few, however, persisted in trying to understand. Through their efforts, scientists have come to understand so much about genes that now, not only do they know what genes are made of and what genes do, but they can also manipulate them quite specifically. In other words, now scientists know genes inside out.

How did biologists come to learn so much about genes? Who were the pioneers in the field of genetics? What questions did they ask, and how did they answer them? Placing the current understanding of genetics in a historical perspective is a valuable exercise. Tracing the development of this field not only informs readers about the workings of biological systems at a basic level, it also provides insights into how science advances. Sometimes the next critical discovery in a field is obvious, and scientists race each other in an attempt to be the one who captures the prize. At other times, the most important discoveries are ignored because the timing is not right. The field of genetics was born at such a time. No one could hear what the "Father of Genetics" was saying because he was years ahead of his time.

Discovery: the discrete nature of genes

One evening in February 1865, a monk named Gregor Mendel presented a lecture to the local scientific society in Brno, Czechoslovakia. He described in great detail the results of 8 years of data on cross-breeding thousands of peas. Those attending the lecture were not impressed, but the results he shared that night were some of the most important findings in the history of biology. Scientists are fortunate that his paper, like all papers of that society, was published in a journal kept in libraries across Europe. Thirty-five years later, his work was rediscovered, and only then did scientists understand its importance. (A translation of Mendel's paper is available at http://www.mendelweb.org.)

Every biology textbook discusses Mendel's work in detail. All biology teachers and students know of Mendel's work and could probably describe the methods and results in their sleep. Yellow peas, green peas, round peas, wrinkled peas. Do the crosses; see the results. Now do more crosses; count more offspring. But what was so significant about Mendel's work? Why do biologists speak of him in reverential tones?

Stated simply, Mendel changed the way we view the natural world. Although he never used the word "gene," his work revealed the inherent nature of genes and gave birth to a new branch of biology: genetics, the study of heredity. He demonstrated that the hereditary substance that passes from one generation to the next is organized as discrete packets of information. Heredity does not involve blending together fluidlike contributions. Instead, heredity depends upon combining discrete particles from both parents (Figure 3.1). When egg and sperm fuse during fertilization, the maternal and paternal hereditary particles do not become joined together but retain their distinct identities.

What is so important about having hereditary information packaged as discrete particles? Bundling up genetic information into separable chunks provides

a constant means of generating genetically variable offspring, even in the absence of **mutation.** Without genetic variation, life on earth could not have evolved. Now, that's important.

Because genes are discrete particles, the maternal and paternal genes for a given trait **(alleles)** separate from each other during the production of gametes. To understand the importance of this separation in producing genetic variation, contrast the two models of heredity: fluid blending and discrete particle.

The Fluid-Blending Model of Inheritance

Imagine pouring together milk (maternal genes) and chocolate syrup (paternal genes) to make chocolate milk (offspring). Now, take the chocolate milk and divide it into four equal parts (gametes) and pour two of them together (fertilization). You still have chocolate milk. No matter how many times you repeated this dividing and mixing, crossing offspring gametes to produce the next generation, you would still end up with chocolate milk offspring and only chocolate milk offspring. You would never again have a glass of pure chocolate or a glass of pure milk. In other words, the offspring would always be genetically identical.

Since genetic variation is the key to the evolution of life on earth, such a system of genetic blending would be an evolutionary dead end. For evolution to proceed, genetically variable offspring must be generated constantly. When used in conjunction with sexual reproduction, the discrete nature of genes provides this variation.

The Discrete-Particle Model of Inheritance

To demonstrate the discrete-particle model, use separate objects—jelly beans, M&Ms, marbles, or index cards—as genes, and mimic sexual reproduction: produce gametes through meiosis, and join them in fertilization. Now, mimic a second round of reproduction with the gametes of the offspring. For example, if you use two red (maternal) and two green (paternal) jelly bean genes, the first-generation offspring will all have a single genotype: red/green. Cross those individuals, and you will have three genotypes: pure red, red/green, and pure green—genetically variable offspring.

Mutation was not a factor in creating these genetically variable offspring. The variation results from reassorting existing alleles through meiosis and fertilization, the constituent elements of sexual reproduction.

Figure 3.1 Graphical representation of the different models of inheritance: fluid blending and discrete particle.

Fluid blending   Discrete particle

Mendelian principles. This separation of the maternal and paternal genes for the same trait (male and female alleles) during gamete formation is known as Mendel's Principle of Segregation (Figure 3.2).

A second important Mendelian principle also depends upon having hereditary information organized as discrete particles. This second principle, the Principle of Independent Assortment, involves separation of male and female genes for different traits (Figure 3.3).

Its essence is this: on average, any gamete has an equal number of maternal and paternal alleles. On a purely mechanical level, this means that during meiosis, the paternal chromosomes do not line up on one side of the divide and the maternal chromosomes on the other. Maternal and paternal chromosomes do not stick together as groups during gamete production. In humans, each new gamete has

23 chromosomes. On average, 11 or 12 are of maternal origin, and 11 or 12 are paternal. (Of course, crossing over between homologous chromosomes makes the situation less tidy than just described when one looks at the specific maternal and paternal genes on chromosomes.)

Dominant and recessive alleles. The discrete nature of the gene permitted Mendel to observe and describe the concept of dominant and recessive characteristics. Think back to the chocolate milk-versus-jelly bean analogy. For the chocolate × milk cross, all offspring in all generations are chocolate milk. For jelly beans, assume that one color is dominant and the other is recessive. The first cross, red × green, tells us very little, because all of the offspring look the same and have the same genetic makeup. The revealing cross is the next one, in which red/green offspring are crossed with each other. This cross results in both red and green offspring, as well as three genotypes. A trait that disappeared in the first generation reappears in pure form in the second. The traits do not blend into each other. One trait (**dominant allele**) simply overpowers the other (**recessive allele**) in the first-generation offspring.

Genotypes and phenotypes. Because Mendel chose traits that exhibited clear dominance relationships, he was able to elucidate another concept fundamental to our understanding of inheritance: the relationship between genotype and phenotype. The outward appearance of an organism (**phenotype**) may or may not directly reflect the genes that are present (**genotype**). When you want to understand the details of an organism's genetic makeup, sometimes the phenotype is helpful, and other times it can be very misleading.

Genetics, statistics, and probability. Another ancillary but very important feature of the discrete nature of genes involves not biological inheritance, but the ease with which genetic traits can be scientifically investigated. Because genes are discrete entities (like coins, dice, or cards), the observed results of crosses can be scientifically analyzed by using the laws of statistics and probability (as with coins, dice, and cards). Using observable differences, scientists can quantify patterns of inheritance and infer the genetic basis of traits without knowing the details of the molecular biology of the gene of interest.

It would be difficult to overstate the importance of Mendel's findings to our understanding of the workings of biological systems, from DNA to evolution. We could not learn all we know about the natural world until we could appreciate and understand his

Figure 3.2 The Principle of Segregation. Mendel observed that experimental crosses between pea plants with round or wrinkled seeds yielded plants with round seeds in the first generation and a 3:1 mix of plants with round and wrinkled seeds in the second generation. From this result and observations of other characteristics, he inferred that maternal and paternal hereditary information is packaged as discrete particles and that maternal and paternal particles separate from each other during gamete formation.

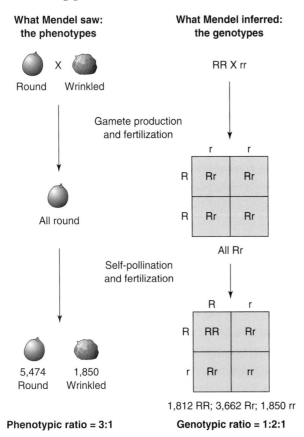

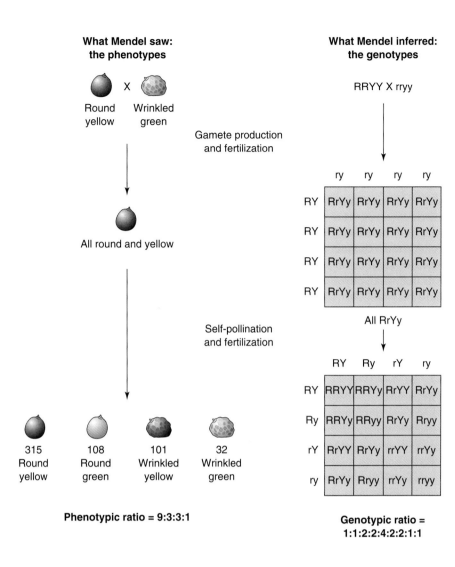

What Mendel saw: the phenotypes

Round yellow X Wrinkled green

Gamete production and fertilization

All round and yellow

Self-pollination and fertilization

| 315 Round yellow | 108 Round green | 101 Wrinkled yellow | 32 Wrinkled green |

Phenotypic ratio = 9:3:3:1

What Mendel inferred: the genotypes

RRYY X rryy

	ry	ry	ry	ry
RY	RrYy	RrYy	RrYy	RrYy
RY	RrYy	RrYy	RrYy	RrYy
RY	RrYy	RrYy	RrYy	RrYy
RY	RrYy	RrYy	RrYy	RrYy

All RrYy

	RY	Ry	rY	ry
RY	RRYY	RRYy	RrYY	RrYy
Ry	RRYy	RRyy	RrYy	Rryy
rY	RrYY	RrYy	rrYY	rrYy
ry	RrYy	Rryy	rrYy	rryy

Genotypic ratio = 1:1:2:2:4:2:2:1:1

Figure 3.3 The Principle of Independent Assortment. By observing the hereditary patterns of two separate traits, seed shape and color, Mendel inferred that during gamete formation, the alleles for one trait segregate independently of the alleles for a trait located on a different chromosome. What would Mendel have seen if the genes for seed color and shape were adjacent on a chromosome?

remarkable work on the particulate nature of inheritance. He could see patterns where everyone else saw disarray. From these patterns, he made inferences and established the fundamental concepts of genetics before anyone knew genes existed.

And yet, no one was impressed the night he presented his results or for 35 years afterward. Mendel revolutionized biology 16 years after he died. Science works like that more often than you might guess.

Discovery: the chromosomal nature of inheritance

Once Mendel established that the hereditary material is organized into packets of information that separate from each other during gamete formation, scientists had to localize these particles within the cell.

Improvements in microscopy permitted the next set of discoveries, because scientists could actually see, on a cellular level, the phenomena Mendel had

described. Using microscopy, biologists began to uncover the underlying mechanism of the principles Mendel had inferred from observing phenotypic variation. In other words, technological advances enabled the next generation of scientists to answer the next question: Where are these discrete particles located?

In 1902, 2 years after Mendel's work was rediscovered, Walter Sutton observed chromosomes behaving in ways that resembled the segregation of hereditary material Mendel had postulated. Studying meiosis in grasshoppers, Sutton, a graduate student at Columbia University, observed that chromosomes occur in morphologically similar pairs and that the two members of a chromosome pair separate from each other during gamete formation. He used this cytological evidence in conjunction with Mendel's findings to hypothesize that the hereditary material is associated with chromosomes. When he discovered this critical relationship, he ran excitedly to his major professor and mentor, E. B. Smith, who was one of the foremost biologists of his time. This is how Smith described what happened.

> *I well remember when in the early spring of 1902, Sutton first brought his main conclusion to my attention. I also clearly recall that at the time I did not fully comprehend his conception or realize its entire weight.*

Smith may not have understood what Sutton discovered, but Sutton was right on target. His theory was confirmed and elaborated by a remarkable group of biologists, also at Columbia University. In a 5-year period, they uncovered the major details of the chromosomal basis of heredity in Thomas Hunt Morgan's laboratory, an infamous room known to biologists as The Fly Room.

Thomas Hunt Morgan once joked that God made the fruit fly just for him. With its relatively simple genetic makeup of only eight chromosomes (four pairs) and a very short generation time (12 to 14 days), the fruit fly, *Drosophila melanogaster,* is an excellent experimental animal for studying genetics. Much of our basic understanding of classical genetics comes from fruit fly studies; the great bulk of that early work came from Morgan's laboratory.

In the early 1900s, Morgan began to use *Drosophila* to address problems in evolution. He was not particularly interested in studying the cellular mechanics of inheritance. Instead, as an experimental scientist, he wanted to test existing theories about the role of mutation in evolution. Over and over, he bombarded fruit flies with X rays or exposed them to chemicals in an attempt to mutate an easily observable trait.

One day, he finally got lucky. In the midst of thousands of red-eyed flies was a mutant at last: a white-eyed male. This single mutant mated with a red-eyed female, left its mutated gene, and in that act provided the raw material for a series of experiments that provided the foundation for classical genetics.

By all accounts, Morgan was quite an interesting character. On the one hand, he was informal, humorous, and a stereotypical absent-minded scientist. He was often mistaken for a beggar because his clothes were so ragged. His laboratory was just as disorderly as his clothes. Many desks and tables were crammed into a tiny room, leaving little room to move. Hundreds of half-pint milk bottles filled with rotting bananas and thousands of flies were everywhere. An equal number of flies lived free in the room, hovering around ubiquitous stalks of bananas in various stages of decay. He recorded data on scraps of paper and the backs of envelopes, which promptly got buried under the piles of paper on his desk.

Despite his appearance, he was from one of the finest families in Kentucky and an heir to the fortune of industrialist J. P. Morgan. He was known for giving money to graduate students who had a hard time making ends meet. He was also a brilliant investigator and an inspiring leader. He led an extraordinary group of young students to key findings that served as the footings on which to build our understanding of genetics. In a very few years, Morgan and his students, Alfred Sturtevant, Calvin Bridges, and Herman Muller:

- Proved that genes are chromosomally located
- Introduced the concept of **sex-linked inheritance**
- Introduced and elaborated the concept of **genetic linkage**
- Originated the idea of **gene mapping**
- Constructed the first genetic map of a chromosome
- Demonstrated crossing over between homologous chromosomes

Discovery: the chemical nature of genes

Now that scientists knew where genes were located, they began to ask questions about the molecular makeup of genes.

In 1869, a German chemist, Frederick Miescher, had isolated a novel substance from the nuclei of white blood cells. He gave it the name "nuclein." Unlike proteins, nuclein had a high concentration of phosphorus. His colleagues tried to convince Miescher that nuclein was just another protein and phosphorus was a contaminant, but Miescher persisted in his belief that nuclein was another type of molecule. Eventually, chemical analysis revealed that chromosomes are made of both protein and Miescher's nuclein, which we now call deoxyribonucleic acid (DNA).

Of these two substances, which carries hereditary information? Almost everyone thought that proteins must be the hereditary material. Chemically, proteins are much more complicated than DNA, and everyone knew that the molecule of heredity had to be able to contain an extraordinary amount of information. It did not seem possible that a simple molecule like DNA could be the hereditary material.

Of course, biologists now know that they were wrong. The heredity material is not protein but DNA. But how did geneticists discover this?

The Transforming Factor

The first chapter of the story occurred in 1928 in the laboratory of a man interested not in heredity or chromosomes but in vaccines. Frederick Griffith was trying to understand the differences between the strains of *Diplococcus pneumoniae* that cause pneumonia (virulent strains) and the strains that are harmless (nonvirulent strains). The virulent forms had polysaccharide capsules that gave them a smooth appearance. The nonvirulent bacteria had no capsules and were rough.

Griffith hoped that either heat-killed virulent strains or live nonvirulent strains could be used as a vaccine. He mixed heat-killed virulent and live nonvirulent strains together in the hope of making an effective vaccine. He injected the mixture into mice, and they died. How could dead bacteria be virulent? He was able to retrieve virulent bacteria from blood samples taken from the dead mice. Something had been transferred from the dead virulent bacteria to the live nonvirulent bacteria. Griffith called this substance the "transforming factor." Scientists now know the transforming factor was DNA, but Griffith did not know that (Figure 3.4). The Griffith story also illustrates how science progresses. Often, findings in one branch of science shed light on a different realm.

In 1943, O. T. Avery and his colleagues at the Rockefeller Institute purified the transforming factor and announced that it was DNA. Many scientists doubted this announcement and persisted in believing the hereditary material was made of protein. They doubted that DNA, with only four subunits, could carry enough information to turn a fertilized egg into a human being. Proteins, which have 20 subunits, can carry much more information. In a way, it seems incredible now that scientists doubted that four subunits could provide sufficient information when every day people transfer and process tremendous amounts of information via computer languages that have only two subunits—0 and 1.

DNA or Protein?

The final answer to the DNA-versus-protein debate was provided by definitive experiments conducted in 1952 by Alfred Hershey and Martha Chase. Because of the work done on the atomic bomb in World War II, scientists had access to the radioactive substances that have proved invaluable in scientific investigation. These isotopes were crucial to the Hershey-Chase series of experiments on the molecular nature of the hereditary material.

By the 1950s, researchers had begun to use experimental organisms that were even simpler than

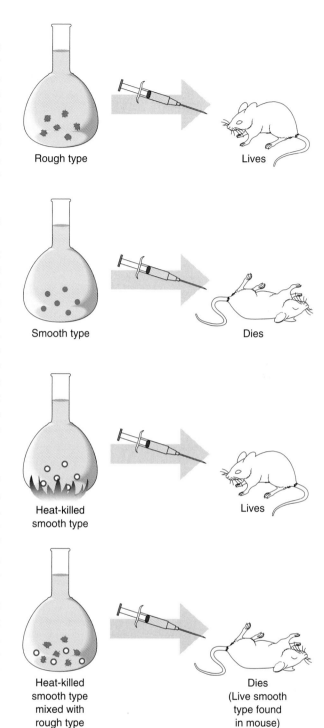

Figure 3.4 Discovery of the "transforming factor" by Griffith. The strain of pneumonia-causing bacteria (smooth) is virulent because of a gene that codes for a protective outer covering. A different strain (rough) does not cause pneumonia because it lacks the gene for the protective covering. When nonvirulent bacteria are mixed with heat-killed virulent bacteria, the nonvirulent bacteria are transformed into virulent bacteria. The gene for the protective outer layer moves from the dead smooth bacteria to the living nonvirulent strain.

bacteria, the **bacteriophages.** Bacteriophages are viruses that infect bacteria. Viruses consist of a coat protein with a small amount of genetic material, usually made of DNA, inside. They infect bacteria by injecting their genetic material into the bacterium while leaving the coat protein outside. Hershey and Chase exploited this bit of viral molecular biology to settle once and for all the question of the molecular nature of the genetic material.

Proteins and DNA differ chemically in many ways, but one simple chemical difference was critical to the Hershey-Chase experiment. DNA has phosphorus but not sulfur, and proteins have sulfur but not phosphorus. Hershey and Chase grew viruses labeled with the radioactive isotopes of sulfur and phosphorus, ^{35}S and ^{32}P. These viruses were used to infect *Escherichia coli* cells. The solutions containing the infected bacteria and the viruses were agitated in a blender and centrifuged to separate the viral coats from the infected bacteria. The supernatant containing the viral coats was rich in ^{35}S, while the pellet of infected bacterial cells contained ^{32}P. These infected *E. coli* cells produced phage progeny that contained ^{32}P and no ^{35}S (Figure 3.5).

Thus, the DNA-versus-protein debate was finally resolved. Because the phages injected their DNA into the bacteria while leaving their protein coats outside, Hershey and Chase concluded that DNA was the genetic material.

Again, note the recurring themes of scientific progress.

- Technological advances promote scientific discoveries.
- Findings in one branch of science shed light on others.

Discovery: the structure of DNA

As soon as the DNA-versus-protein debate was settled, another question was immediately born: What was the structure of the DNA molecule? Whatever the structure, scientists knew DNA's function depended on it. They also knew that whoever described the structure would be guaranteed an important place in history, and so, the race was on.

Before scientists proved conclusively that DNA is the genetic material, they actually knew a great deal about its biochemistry. They knew all of its components (phosphate, deoxyribose, and the four nitrogenous bases) and their molecular structures. By 1930, they knew that each subunit (nucleotide) of DNA consists of one phosphate, one deoxyribose, and one nitrogenous base. Finally, in 1952, they discovered that the phosphate and deoxyribose moieties are linked together to form a chain and that the nitrogenous base is attached to this chain.

Figure 3.5 The experiments of Alfred Hershey and Martha Chase. One group of viruses containing protein labeled with the radioactive isotope ^{35}S and a second group of viruses containing DNA labeled with the radioactive isotope ^{32}P infected bacterial cells by injecting their genetic material into the cell. Hershey and Chase separated the viral coats from the bacterial cells and found ^{35}S in the viral coats and ^{32}P within the bacterial cells. Viral progeny that resulted from the infection also contained radioactive ^{32}P. Hershey and Chase thus concluded that the genetic material of the virus was DNA and not protein.

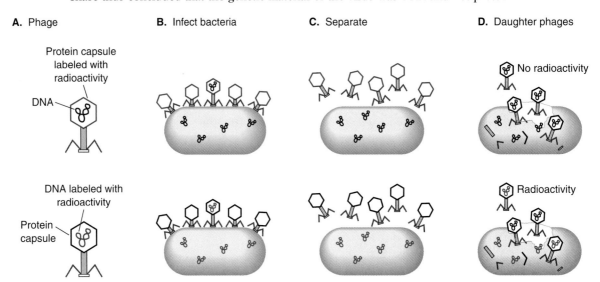

So what else was there to know? The three-dimensional structure. Biologists could not deduce the three-dimensional structure from the information available, and they knew that the key to explaining DNA's function resided in its three-dimensional structure. Findings from a number of branches of science coalesced and provided sufficient information to discover this next piece of the puzzle.

- The physicists Maurice Wilkins and Rosalind Franklin provided data from X-ray diffraction analysis of DNA crystals.

- The chemist Erwin Chargaff discovered that DNA molecules always have equal amounts of adenine and thymine and equal amounts of cytosine and guanine.

- Another chemist, Linus Pauling, described the rules governing the formation of chemical bonds and published details of a novel chemical structure that certain proteins could assume, the alpha helix.

Watson, Crick, and the Double Helix

Francis Crick, a graduate student in physics at Cambridge, and James Watson, an American geneticist at Cambridge on a postdoctoral fellowship, both young and brash, were determined to be the ones who unearthed DNA's structure, and they did. Their success stemmed not only from their determination, but also from a mutual distaste for sloppy thinking; their blunt, open, passionate style of communication; and a bit of luck.

They set out to discover DNA's three-dimensional structure by building a model. In 1953, when they happened upon the structure—a helix composed of two strands running in opposite directions with internal pairing between the bases, adenine, thymine, cytosine, and guanine—they knew immediately that they had it! They could easily see one DNA function, information replication, in its structure. Each strand could be used as a template for making another strand.

In their brief, understated report in the journal *Nature* in which they described DNA's structure, Watson and Crick state:

> *This structure has novel features which are of biological interest . . . It has not escaped our notice that the specific pairing we have postulated immediately suggests a possible copying mechanism for the genetic material.*

In his book *The Eighth Day of Creation,* Horace Freeland Judson beautifully describes their discovery in this way:

> *That morning, Watson and Crick knew, although still in mind only, the entire structure: It had emerged from the shadows of billions of years, absolute and simple, and was seen and understood for the first time.*

Ever since Watson and Crick discovered the structure of DNA, scientist after scientist has described DNA's structure as "beautiful." In reply, nonscientists usually ask why we see beauty in a molecule. DNA's beauty lies in its structural simplicity. It is a simple, straightforward molecule: simple in structure but astoundingly complex in its function.

The elegance of DNA has inspired awe in some nonscientists as well. A quote from Salvador Dali eloquently describes the feeling many scientists share with the painter regarding DNA:

> *And now the announcement of Watson and Crick about DNA. This, for me, is the real proof of the existence of God.*

In another report in *Nature* 2 months later, Watson and Crick provided insights into the second DNA function: the elaboration of the information it contains into a living organism:

> *It therefore seems likely that the precise sequence of the bases is the code that carries the genetic information.*

Thus, Watson and Crick were able to deduce how DNA carries out its two functions, replication of the information it contains and elaboration of that information into an organism, by simply looking at its structure. A fundamental principle of architecture, "form follows function," is also a fundamental truth in biology.

Theory: the central dogma

In 1957, Francis Crick proposed what was to become known as the central dogma describing the process of information translation:

$$DNA \rightarrow RNA \rightarrow Protein$$

Information flows from DNA through RNA to proteins. The information in DNA is encoded in the linear sequences of nucleotides, just as the information on this page is encoded in linear sequences of letters. The information in DNA is translated into the language of proteins, which is composed of different "characters": amino acids in place of nucleotides.

This translation is facilitated by a middleman, RNA, which, like DNA, is a nucleic acid composed of a sequence of nucleotides.

Translating the information in DNA into a protein is somewhat like copying pages from a book and taking those sheets of paper to someone who can translate it into Japanese.

Further studies supported the central dogma and led to the entrenched and very satisfying view of the gene as a stable, continuous segment of DNA whose linear nucleotide sequence corresponds to a linear sequence of amino acids. It was not long before this tidy picture of gene structure and function had to be revised, however. The more one learns about biological systems, the more one realizes how untidy they can be. If you look long enough, you can find exceptions to any rule or pattern in biology. Biological systems are too complex to always follow rules. This lack of predictability can be a source of both joy and frustration for biologists.

Revising the Central Dogma

As biologists have learned more about gene structure and function, they have fine-tuned and elaborated the original concept of DNA translation. First, they found that the flow of information from DNA to RNA is not one-way. Information can flow from RNA to DNA. They also discovered that not all DNA is translated into proteins. Often, a portion of a gene's nucleotide sequence does not contain information that prescribes an amino acid sequence, and this section is excised before the DNA's information is translated into proteins. The translated sequence of nucleotides in a stretch of DNA is known as an **exon;** the excised sequences are called **introns** (see chapter 4).

Scientific Models: Gateways or Barriers?

If you think the resistance of the scientific community to Mendel's discoveries was unique to that time, think again. The story of another brilliant geneticist, Barbara McClintock, parallels Mendel's so closely, it is eerie. McClintock is the scientist who discovered "jumping genes," or transposons, which we discuss in detail in chapter 13.

Like Mendel, she studied an organism so familiar it was almost dull—corn. Like Mendel, she was blessed with great powers of observation, an extraordinary intellect, and an uncommon amount of patience and persistence. For over 30 years, she carried out the slow, laborious work of cross-pollinating Indian corn plants and, months later, observing the results of her experimental crosses. By simply using her uncanny ability to find patterns in the color of corn kernels, she developed a theory of inheritance that called into question the fundamental nature of genes and eventually revolutionized our understanding of genetics.

McClintock proposed the existence of jumping genes to the scientific community in the 1950s. Unlike Mendel's audience, McClintock's was impressed—but not positively. In short, some thought a brilliant geneticist, who was an elected member of the prestigious National Academy of Sciences, had gotten a bit confused, to say the least. What she proposed was heresy to the existing view of genes and chromosomes, and so they concluded she was out of touch with reality, not that their model was wrong.

Often scientists forget that the models they develop are supposed to be useful tools for understanding and interpreting the workings of the system under study. Instead, they become wedded to the model and are unable to see and hear information that does not support it. Their model acts like a filter letting expected information pass through to their brains but blocking the information they do not expect.

This surely was the case during the 20 years when McClintock tried to explain to her colleagues what she saw in those corn kernels and was met with complete resistance. Like Mendel, she saw patterns where everyone else saw disarray. She open-mindedly asked what the patterns she saw revealed about the behavior of genes and chromosomes. She did not aim for a specific answer to support an existing theory. Instead, she sought true and deep understanding.

Other scientists were blind to the patterns and deaf to her reasoning. When they could not understand her message, they chose to ignore it because it did not fit with their preconceived notions of what was "supposed" to happen. Eventually, after other researchers discovered evidence of transposons in bacteria, yeasts, and fruit flies, her fellow scientists could finally hear what she had to say. In 1983, McClintock received the Nobel Prize for her work on transposons.

They also found sequences of nucleotides that repeat themselves over and over and over, like a phonograph record that is stuck in a certain place. Because each person's pattern of repeats is unique, these sequences are important in DNA typing. Scientists assumed such sequences were irrelevant to biological functions and assigned them to the category that is colloquially referred to as "junk" DNA. Now, however, there is evidence implicating these DNA repeats in some cancers and also in a number of neuro-muscular disorders.

Biologists then found that some nucleotide sequences in a gene do not stay put but hop around. They are known as transposable elements, transposons, or "jumping genes" (see the first sidebar).

Perhaps most surprising and intriguing, biologists found that the perception of RNA as a passive middleman relaying information between the "real" molecules, DNA and proteins, is not appropriate. RNA, like an enzyme, has the ability to catalyze chemical reactions. This discovery may help explain the origin of life on earth and, in particular, the evolution of nucleic acids and proteins.

Who knows what scientists will find next?

Understanding Genes

When biologists discuss genes, they sometimes sound as if they are talking about a supernatural force, because:

- Genes occupy a central place in the study of all biological sciences
- The structure of the DNA molecule is so simple and yet so powerful
- The mechanism of gene action is so awe inspiring
- Genetic variation is crucial to the evolution of life on earth

There is no doubt that genes are potent particles. If information is power, then genes are the powerhouses of the planet, for they contain the information of life. They are the repositories of the information that makes us who we are and that binds us to our species and separates us from other species. Genes forecast the future and reflect the past. Your genes contain information on your genealogical relationships and ethnic background and a chapter on the evolution of life on earth.

Yes, genes are powerful, but they are not omnipotent. Sometimes it is easy to forget that, especially in today's society, which often uses "it's in their DNA"

as the catch-all explanation for any natural tendency, not only of individuals but also of groups—the culture of corporations, the behavior of politicians, and characteristics of people living in certain countries.

In the public's mind, an aura of unambiguous predestination surrounds genes. However, genes represent potentialities, not inevitabilities. A gene is not the final arbiter of a trait but only one of many factors that determine whether that gene's potential becomes realized as the phenotype commonly associated with it. Virtually every observable trait represents the cumulative effect of the actions of many genes, some of which negate the others. Environmental factors throughout an organism's life influence phenotypic expression of genetic information, as well. Consequently, genotypes are not predictable from seeing phenotypes, and phenotypes are rarely predictable from knowing the genotype.

Figure 3.6 graphically illustrates the complex relationship among genes and observable traits. The path from a gene to its primary protein product to a visible trait is indirect and complex; many genetic factors other than the single gene influence the phenotypic expression of that gene. Overlay environmental influences that may affect each gene's production of its encoded protein on these complex genetic relationships, and you can easily see why the relationship between genotypes and phenotypes is rarely simple.

Having an accurate understanding of gene function, and its limits, can contribute to mental well-being,

Figure 3.6 The diagram illustrates the complex relationships among genes, the proteins they encode, and observable traits. The path from gene to trait is neither straightforward nor linear. One gene affects many traits, and one trait is the end result of the actions of many genes. Gene products produce feedback and alter the activities of other genes.

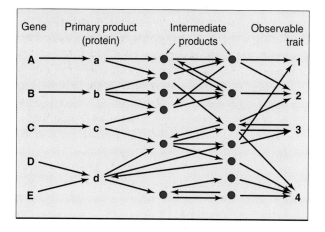

because many of the fears and concerns people express about the risks of biotechnology can be traced, in part, to misunderstanding the nature of genes. Appreciating the complicated path through which genetic information becomes realized as a phenotypic trait can help people make informed decisions about the safety of eating genetically modified foods and allow them to assess statements about the environmental risks of biotechnology. In addition, understanding the relationship between genes and traits can contribute to physical well-being. In the future, you will hear a great deal about scientists discovering genes associated with certain diseases. You will need to decide whether you want to be tested to determine if you have the normal form of that gene. Then, if you request the test, you will need to understand the test results to determine your propensity, or lack of one, to inherit, transmit, or develop certain diseases. If you use a faulty model of gene function to interpret the test results, you may well decide you are destined to develop a disease you will never have or, on the other hand, are *not* at risk for a disease that you will develop.

Misunderstandings about genes

The primary misconception that pervades the public's understanding of genetics is an implicit belief in a one-to-one relationship between a gene and an observable phenotypic trait that is simple, direct, and inevitable. In other words, most nonscientists believe that:

- One gene is responsible for one phenotypic trait
- A phenotypic trait can be traced to the activity of a single gene
- A certain genotype invariably prescribes a corresponding phenotype

Most genes affect many different phenotypic traits, not a single trait, a phenomenon known as **pleiotropy.** When you picture the many interrelated biochemical pathways described in the previous chapter, it should come as no surprise that one gene can have many effects. A single enzyme responsible for catalyzing a reaction that produces a molecule with a role in a number of pathways has an impact on all of those pathways. Pleiotropic effects result not only from many different biochemical pathways sharing the same molecules, but also from the interdependence of cells and organs in multicellular organisms. When you stop to think about it, pleiotropy should be the rule rather than the exception.

The second common misconception is that a phenotypic trait can always be traced to a certain genotype. In fact, a number of different genotypes can lead to the same observable phenotype. The most straightforward example of this occurs in the one-gene, two-allele traits of Mendel's peas. The homozygous dominant or heterozygous genotypes both produced the same phenotypes—yellow seeds, round seeds, purple flowers, and so forth—because one allele was completely dominant to the second. However, the same phenotype can also be traced to different genetic loci, not simply different alleles at the same locus. Under these circumstances, the phenomenon is known as **genetic heterogeneity.** Once again, this relationship of different genotypes to the same phenotype should come as no surprise. If different enzymes in a single pathway are mutated, the pathway's end product will not be produced. Expression of phenotypic traits requiring the presence of that end product will be compromised, no matter which step in the pathway is malfunctioning.

An example of a genetically heterogeneous trait in humans is retinitis pigmentosa, a disease characterized by degeneration of the retina that follows a very specific course: reduced night and peripheral vision, followed by rod, then cone, degeneration and blood vessel constriction. Scientists have tracked this disease to mutations at more than 15 separate loci, some of which are autosomal and some sex linked. Some of the new, mutated alleles are dominant to the normal allele, while others are recessive.

A final misconception we address here is that individuals with the same genotype always display the same phenotype. Individuals with the same genotype at a particular locus can have different phenotypes. Even after scientists have determined the phenotypes that typically derive from specific genotypes, for certain traits they cannot predict the phenotype based on the genotype. This uncoupling of genotype from phenotype may assume two forms, which geneticists characterize as genotype **penetrance** and genotype **expressivity.**

Penetrance is the percentage of the population with a certain genotype exhibiting the trait associated with that genotype. If 100% of the organisms with a certain genotype at a locus display the expected phenotype, the gene is completely penetrant. From what you know of Mendel's work, you would describe the allele encoding round seeds as completely penetrant. When some organisms have the appropriate genotype for expressing a trait but do not display the trait, the trait exhibits reduced or incomplete penetrance.

Think of penetrance as the probability that an individual with a mutation in a certain gene will display the mutant phenotype, assuming the mutant allele occurs in a combination (heterozygous or homozygous) that would permit expression. For example, because cystic fibrosis is a recessive trait, carriers—individuals having only one recessive allele of a recessive disorder—are not expected to have the disease. Because they are heterozygous for a recessive mutation, they are excluded from calculations of penetrance of the cystic fibrosis mutation.

Geneticists have now identified hundreds of different mutations in the gene that encodes the chloride channel protein associated with cystic fibrosis. All people who display clinical symptoms of cystic fibrosis have a mutation in that gene, but not all people with a mutation in that gene have clinical indications of the mutation. Therefore, the degree of penetrance of the cystic fibrosis mutation varies with the nature of the mutation (Figure 3.7). Certain mutations in the gene encoding the chloride channel protein are completely penetrant, because all who are homozygous recessive for that mutation exhibit some of the symptoms associated with cystic fibrosis. Other mutations

Figure 3.7 Gene changes in cystic fibrosis. Different mutations in the gene that encodes the chloride channel protein have different phenotypic effects.

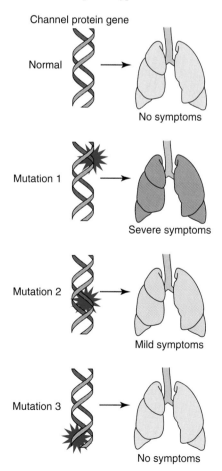

seem to be completely impenetrant; people with two alleles for these mutations have lived to be 60 years old without developing any symptoms. They happened to discover they have two mutant alleles at the chloride channel protein locus only because physicians now have access to tests for identifying carriers of certain mutations at that locus.

Expressivity refers to the range of phenotypes associated with a certain genotype and is related to the degree to which a gene is expressed in an individual. If a trait takes different forms in different individuals having the same genotype, it exhibits variable expressivity. For example, a dominant mutation at a single locus causes the white part of the human eye, known as the sclera, to turn blue. The allele's expressivity varies, ranging from a barely detectable light blue to a blue that is so dark it is almost black. In addition, because the blue-sclera allele is dominant to the normal allele, we expect any individual with a copy of the mutated allele to have blue sclera. However, only 90% of the people with a gene for blue sclera display the expected phenotype because the blue-sclera allele is also incompletely penetrant.

Cystic fibrosis is another trait that also exhibits variable expressivity. Some children develop severe symptoms early in life, others later, and the severity of the symptoms also varies. For years physicians have recognized this variation, and now molecular evidence showing that the gene has many mutant allelic forms may explain at least some of the variation in phenotypic expression. Interestingly, the severity of the mutation on the genetic level is sometimes not mirrored in the severity of clinical symptoms. People with mutations that lead to the total absence of the chloride channel protein often have few, if any, cystic fibrosis symptoms; those with a mutation that leads to a loss of a single amino acid suffer from severe cystic fibrosis and usually die before they are 25 years old.

Language can also create and reinforce misconceptions. In a typical conversation, you might hear someone say she "inherited musical ability from her father" or he "inherited sickle-cell anemia from both of his parents." However, traits are not inherited; genes are inherited. Traits are the ultimate manifestation of a many-stepped process affected by both genetic and environmental factors. Rather than saying a child with sickle-cell anemia has "inherited sickle-cell anemia from his parents," the correct (and nonmisleading) description of sickle-cell anemia inheritance would be: "He inherited a mutated form of the gene that encodes the hemoglobin protein from both of his parents. This mutation leads to the production of hemoglobin molecules that, under the environmental influence of low oxygen levels, stick together, or polymerize. The polymerization causes red blood cells to sickle and burst, leading to anemia." (See Figure 4.40.)

Equally misleading are media reports of scientists discovering "*the* gene for cystic fibrosis" or "*the* schizophrenia gene." There is no such thing as "*the* cystic fibrosis gene" or "*the* schizophrenia gene." There are genes that encode proteins that are associated with these diseases. Unfortunately, these misstatements, which are used by scientists and nonscientists alike, further entrench the widespread misconception of a one-to-one, predetermined relationship between having a gene and exhibiting a certain trait.

Rather than using an imprecise phrase like "the cystic fibrosis gene," a more accurate description is: "Cystic fibrosis is caused by a gene mutation that encodes a malfunctioning chloride channel protein in epithelial tissue, including the epithelial tissue lining the respiratory and gastrointestinal systems. This leads to the production of thick mucus that blocks pulmonary airways, accumulates in the lungs, and clogs secretory ducts of the pancreas, causing the set of respiratory and digestive problems associated with cystic fibrosis."

Here are a few additional misstatements commonly made about genes and their more accurate, albeit lengthier, explanations.

Genes make proteins. Genes do not make proteins. Genes contain the information for making proteins. Each nucleotide sequence in a gene specifies the amino acid sequence in a protein. Proteins are made by other proteins (enzymes) that use the gene as a manufacturing guide. Proteins cannot manufacture proteins without this guide, but the guide alone is not sufficient for protein production.

Genes are on chromosomes. Genes are not "on" chromosomes. A chromosome is composed of a very long molecule of DNA that is bound to proteins named histones. A gene is a given length of DNA that is translated into a chain of amino acids. The genes plus the bound proteins are the chromosome.

Genes replicate themselves. Genes do not replicate themselves. Proteins (enzymes) replicate genes using the gene as a guide.

DNA is like a blueprint. This is a very useful metaphor, up to a point. Then, like all metaphors, it becomes more constraining than useful. DNA is like a blueprint because it contains information for building proteins, just as architectural blueprints contain information for building a structure. But DNA is much more than that. DNA not only has information for building the structure of an organism, it also contains information for making enzymes, hormones, receptor molecules, transport proteins, and antibodies. In other words, to be like DNA, an architectural blueprint would have to contain instructions for:

- Making the tools and machines used to erect the building
- Fabricating the electrical wires, polyvinylchloride pipes, drywall, and insulation
- Securing the money to fund the construction
- Designing the precise details of heating and cooling units, security system, and telecommunications networks
- Creating the workers, supervisors, attorneys, bankers, and real estate agents
- Populating the building and directing the tenants' activities
- Even dismantling the building and constructing another
- And much, much more

That would be some blueprint!

Genes determine who we are. Genes do not determine who we are. Genes influence who we are. Envision all of the factors that could influence the process of constructing a building, from the architectural blueprint to the appearance of the final product. A few immediately come to mind: choice of materials, worker competence, and available funds. Because of these variables, nonidentical buildings could easily be constructed from the same blueprint. Now, recall that much more information is included in a DNA "blueprint," providing many more opportunities for extraneous factors to shape the final product.

Understanding Evolution

In addition to being the repositories of information for synthesizing proteins, genes are the units of evolutionary change. Thus, a complete understanding of genes and genetics comes not only from learning about the molecular structure and function of genes, but also from understanding and appreciating the role they play in evolution. After all, the study of evolution is the study of genetic variation over time and space.

Evolution is simply the change in the frequencies of certain genes in a species' gene pool. More specifically, it is the change in the relative proportions of alternative forms of a gene (alleles) in a population.

Evolution depends on two interrelated processes:

- Creation of genetic variation
- Selection of certain of these variants at the expense of others

The agents that create genetic variation are mutation and recombination. A number of evolutionary forces can change the frequencies of certain alleles in a population, but **natural selection** is the primary agent that changes gene frequencies, by selecting certain variants but not others. Because the creation of genetic variation and the selection of certain variants are essential to evolution, each process must be explained in more detail in order to understand evolution. The other forces that change the genetic makeup of populations are described in the second sidebar.

Creating genetic variation

As you might expect, because genetic variation drives evolution, nature has many mechanisms for generating genetic variation. All of these mechanisms are commonly grouped into two broad categories: mutation and recombination.

Types of Mutations

Biologists use the term "mutation" to refer to the changes in the genetic information a cell carries. Two classes of mutations, based on the amount of genetic information (DNA) that is changed by the mutation, are usually recognized:

1. Changes in large segments of DNA molecules are termed **macromutations, macrolesions,** or **chromosomal aberrations.** Examples of macromutations include:
   - Changes in the total amount of genetic information because of the loss (deletion) or addition (duplication) of a gene(s) or even entire chromosomes
   - Changes in the positions of the genes relative to one another without any change in the total amount of genetic information (inversions and translocations) (Figure 3.8)

Changing the Genetic Makeup of Populations

The genetic makeup of a population and, therefore, the amount of genetic variation in a population can be affected by these factors:

- *The rate of mutation*
- *Natural selection*
- *Gene flow*
- *Genetic drift*
- *Nonrandom mating*

Some increase the amount of genetic variation in populations; others decrease it. Some affect the frequencies of genotypes without changing the allele frequencies; others alter the frequencies of genes or alleles. The impacts these factors have on the genetic makeup of populations are described briefly here.

Natural selection

Differential survival and reproductive success of certain genotypes relative to others. However, survival is important only in the service of greater reproductive success. Natural selection is generally viewed as the most significant evolutionary force.

Mutation

Spontaneous changes in the genetic material that can create new alleles and, ultimately, new genes and through this change allele and gene frequencies. The direct impact of mutation on gene *frequencies differs conceptually from the role mutation plays in generating the genetic diversity that is subsequently acted on by evolutionary forces, especially natural selection. The degree to which mutation affects gene frequencies directly depends upon the spontaneous rate of mutation, which is generally too low to shift gene frequencies significantly. Even though environmental factors can affect the rate of mutation, the nature of the mutation is independent of the environment. Mutations occur randomly, with no regard to the benefit a new trait might have in a certain environment. The environment determines which of those randomly generated mutations is advantageous. In other words, mutations are not goal oriented.*

Gene flow

The movement of alleles or genes from one population to another that occurs when individuals change populations, assuming they ultimately reproduce, or in plants, when pollen moves from one population or species to another. Gene flow tends to decrease the genetic differences between populations because they share more genes after gene flow has occurred. We discuss this phenomenon at length in agricultural applications of biotechnology.

(continued)

Changing the Genetic Makeup of Populations (continued)

Nonrandom mating

Recall that sexual reproduction does not create new alleles but only juggles the distribution of alleles, just as dealing new hands of cards does not change the number of aces in a deck. Modern evolutionary biologists have mathematically demonstrated that the frequencies of both the alleles and genotypes (percentage of homozygotes and heterozygotes) in a population will stay constant as long as mating is random. In other words, even though sexual reproduction is very important to evolution because it generates genetic diversity, it alone will not change gene or genotype frequencies in a randomly mating population. This phenomenon is known as the Hardy-Weinberg equilibrium.

In nature, however, mating is often not random. Inbreeding is common, especially in plants, where some species self-pollinate, and scientists have observed numerous examples of assortative mating in animals. In positive assortative mating, those organisms that are alike in some trait selectively breed with each other; in negative assortative mating, dissimilar organisms prefer each other. Nonrandom mating can change the frequencies of genotypes in a population, but it alone does not lead to changes in allele frequencies and therefore does not cause evolution. It does, however, change the phenotypic variation that evolutionary forces act on. For example, both inbreeding and positive assortative mating increase the number of individuals who are homozygous for certain traits without changing
the relative frequencies of alleles in the gene pool. By altering the frequency of genotypes and increasing the proportion of organisms expressing the homozygous trait, nonrandom mating alters the relative distributions of phenotypes. The factor that determines whether these changes in genotypes and phenotypes lead to changes in gene frequencies is natural selection, including sexual selection.*

Genetic drift

*Random changes in the allele or gene frequencies in a population that result from pure chance. Genetic drift typically has a significant effect on gene frequencies when populations are founded by a small number of individuals (the **founder effect**) or when a large population has its numbers significantly reduced, which is known as a population bottleneck. As is true of all events governed by the laws of probability, the smaller the sample, the more likely a sampling error will occur. In other words, you are more likely to get a 1:1 ratio of heads to tails with 100 coin tosses than with 10 tosses. When a small number of individuals starts a new population or emerges from a population bottleneck, certain alleles may not be present in the new population solely by chance. In addition, a gene may have only one allele in the new population even though it had many alleles in the parent population. Whenever a gene is represented by only one allele in a population, evolutionary biologists say the allele has become fixed, because 100% of the individuals in that population have that allele.*

Figure 3.8 Diagrams of the four types of chromosomal aberrations, or macromutations. The letters denote genetic loci.

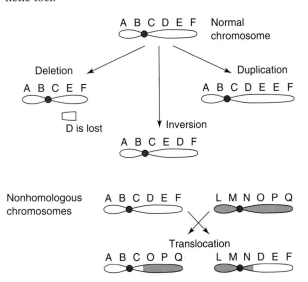

2. Changes in the sequence of nucleotides of single genes are termed **micromutations, microlesions,** or **point mutations,** because a relatively small amount of DNA is changed.

Effects of Mutations

What is the effect of a mutation on an organism? It depends on what has changed. Some mutations are beneficial, others are devastating, and many are neutral. Whether a mutation is harmless or devastating does not necessarily depend on whether it is a macro- or micromutation. Sometimes changing a single nucleotide has horrible effects on the phenotype, and yet, a complete doubling of genetic information has been very important in the evolution of many plants. In general, however, as you would expect, the larger the amount of genetic information that is changed, the more likely it is that the mutation will be harmful.

More information on the effects of mutations on protein structure and function and the resulting phenotypes will be provided in chapter 4. Because the topic at hand is evolution, we will focus here on mutations as sources of genetic variation that is then fed into the evolutionary process.

From the point of view of the evolutionary biologist, the two types of mutation vary greatly because they affect the evolutionary process in different ways. Micromutations, or point mutations, create new forms of genes, or as an evolutionary biologist would say, they can add new alleles to the existing gene pools if they are not harmful.

The effects of macromutations on the evolutionary process cannot be described in a single sentence because they vary in both type and significance according to the exact change that has occurred.

Deletions and duplications. When the total amount of genetic information changes, the evolutionary result may be profound, as in the immediate speciation that has sometimes followed chromosome number doubling, or it may be totally insignificant. For example, when a chromosome is lost or gained, the individual is often sterile, as is the case with XO and XXY humans. While this condition is devastating to the individual, the mutation is irrelevant from an evolutionary perspective because the genetic change does not become incorporated into the gene pool.

Another type of macromutation, gene duplication, can be both directly and indirectly important in evolution. If the gene duplication results in increased production of a protein that contributes to the survival of the individual, it will be favored by natural selection. The indirect benefits of gene duplication result from a different scenario. If a single gene is represented by a number of copies, some of these duplicate genes can mutate into new allelic forms without loss of the original gene's function.

Inversions and translocations. Macromutations that involve changing the positions of genetic loci relative to one another also vary in the effects they have on the evolutionary process. When a section of a chromosome breaks, rotates 180°, and then reinserts itself into the same chromosome (an inversion) or when nonhomologous chromosomes break and exchange pieces of the chromosomes (translocation), the result is usually partial sterility at best. Most gametes produced by individuals heterozygous for translocations do not contain a full complement of the genes involved in the translocation (Figure 3.9).

Consequently, macromutations that rearrange the positions of genes on chromosomes are not a common source of genetic variation that is "usable" for driving evolutionary change. However, in certain cases, translocations and inversions have become established components of a species' gene pool. Because they deter appropriate pairing and crossing over in meiosis (see below), the genes in the inverted or translocated segment become tightly linked and are transmitted as a single unit. Sometimes, these linked sets of genes are important evolutionarily, because they represent specific gene combinations that are highly adaptive and confer a selective advantage.

Transposable Elements

A special form of mutation that occurs in both prokaryotes and eukaryotes involves transposable elements, or "jumping genes." These transposable elements, which are also called transposons, are short segments in the DNA that include a gene for a protein called a transposase. This protein causes the short piece of DNA to "jump" from its original location into a new, often random location somewhere else in the genetic material of the cell. Sometimes the jumping involves forming a new copy of the transposon, and sometimes not (Figure 3.10).

The insertion of a transposon at a new location can pose a problem: the transposon might land in the middle of an important gene, alter the nucleotide sequence within the gene, and prevent the synthesis of the protein. In 1992, scientists linked the occurrence of a genetic disease to a transposition event. Neither parent of the child with the genetic defect was a carrier of the gene for the disease. A detailed examination of the child's defective gene showed that a transposon had inserted into it, causing a mutation. To date, scientists have found transposons in plants, animals, yeasts, and bacteria.

Recombination

The second mechanism for generating genetic variation is recombination, the joining of genetic information from two sources to produce new genetic combinations. Recombination differs from mutation in that specific genes are not changed but simply reassorted. Exactly how the genetic material from two sources gets combined varies with the mode of reproduction of the organism.

Sexual reproduction and recombination. For almost all organisms that reproduce sexually, reproduction involves two processes:

- Production of gametes that differ genetically from the individual producing them

Laying the Foundation

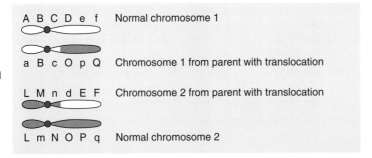

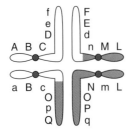

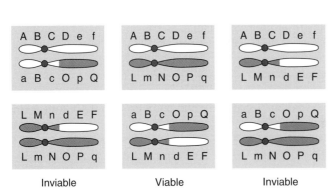

Figure 3.9 Diagrams of the disruptive effect of a translocation on chromosome pairing in meiosis and the resulting gametes. The parental genotype possesses one normal version of each of the two chromosomes involved in the translocation event. The letters indicate genetic loci. Lowercase and uppercase letters denote different alleles for the same trait. Four of the six possible gametes derived from the parental genotype are nonviable because they do not have a complete complement of genes.

- Fusion of these gametes to create an individual that differs genetically from both parents

Recombination occurs during both processes. Let's start with the most straightforward example of recombination: fertilization.

Fertilization is the fusion of two gametes (eggs and sperm or pollen) into a single zygote. When the gametes, which contain genetic information, fuse, the resulting zygote's genetic information is a new combination of genetic information from the two parents. Thus, the offspring that develops from the zygote contains genetic information from two sources and is genetically novel: it is not genetically identical to either parent. All offspring of sexual reproduction, including you, are genetic recombinants.

Recombination also occurs during the production of these gametes through meiosis. Increased genetic variation in populations is the result of two associated meiotic events: the independent assortment and

segregation of nonhomologous chromosomes and the crossing over that occurs between homologous chromosomes.

As a result of independent assortment and segregation of nonhomologous chromosomes in meiosis, a gamete differs genetically from the cell that gave rise to it by having one-half as much genetic material and a random assortment of the maternal and paternal genetic material. Thus, each gamete contains genetic material from two sources (maternal and paternal) in new combinations. The number of possible combinations of nonhomologous chromosomes that could be generated in the production of each human gamete is 2^{23}, or over 8,000,000.

Independent assortment and segregation alone are a rich source of genetically variable gametes, but another recombination event occurs in meiosis. During the pairing of homologous chromosomes, genetic material from the paternal chromosome is exchanged with genetic material from the maternal

A. Transposons that move by **replicative transposition** are duplicated in the process of jumping, a copy-and-paste mechanism.

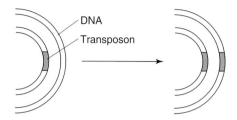

B. Nonreplicative transposons move in a cut-and-paste manner. No copy is made.

Figure 3.10 Replicative and nonreplicative transposons.

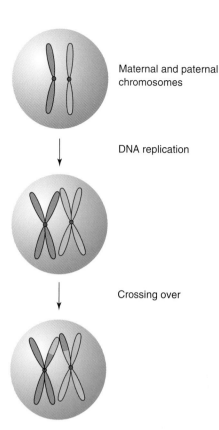

Figure 3.11 Crossing over between homologous chromosomes occurs during gamete production. As a result, the gametes differ from each other and from the parental genotype.

chromosome through the process of crossing over (Figure 3.11). The resulting chromosome is a recombinant: a hybrid containing genes of both maternal and paternal origin, or a novel combination of genetic material from two sources. A graphical summary of the relationship between sexual reproduction and recombination is provided in Figure 3.12.

Asexual reproduction and recombination. Many organisms do not reproduce sexually. In single-parent, or asexual, reproduction, a copy of the genome of the parent is passed along in its entirety to the offspring (Figure 3.13). The offspring is therefore genetically identical to the parent organism and is often called a clone of the parent. The group of organisms resulting from repeated reproduction is called a clonal population, because every member of the population is genetically identical to every other member.

A typical example of asexual reproduction is the division of a bacterium to produce two bacterial cells. These two cells then divide, the four progeny divide, and so on until a large clonal population is generated. Bacteria are not the only organisms that reproduce asexually. Yeasts and some other fungi, protozoans, and algae can reproduce by simple fission or by sexual reproduction. A few species of animals reproduce asexually through parthenogenesis (in which females

produce offspring without being fertilized by males). Many plants can propagate themselves asexually; humans take advantage of this to duplicate desirable plants by rooting leaves and cuttings.

You can see from comparing Figures 3.12 and 3.13 that populations of asexually reproducing organisms would be far less genetically diverse than sexually reproducing populations if they had no other mechanism for generating genetically variable individuals. In fact, though, bacteria have three natural mechanisms by which genetic materials from two sources can be combined: conjugation, transformation, and transduction. Each of these processes is not only important in providing genetic variation for the evolution of bacteria but also plays an important role in human interactions with microbes. Later in this book you will learn the importance of these three natural processes to research and technology and also some of their medically important consequences.

Conjugation (Figure 3.14) is a process by which one bacterium transmits a copy of some of its DNA

Maternal and paternal chromosomes

DNA replication

Crossing over

DNA

Transposon

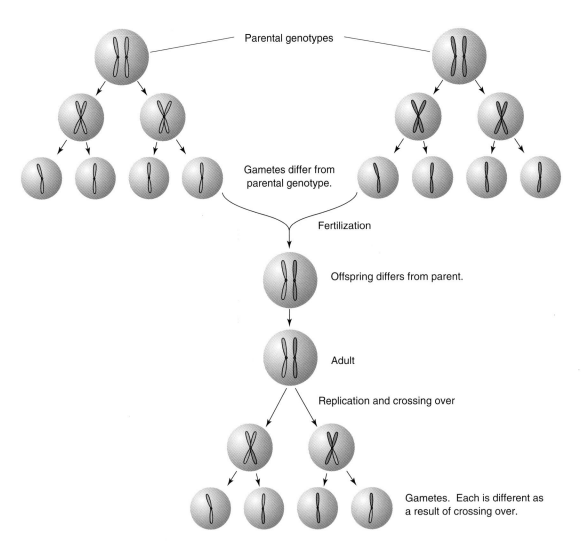

Figure 3.12 Sexual reproduction, genetic variation, and recombination. Genetic variation is created during three stages of sexual reproduction. When gametes are produced, they differ genetically from the parental genotype because they have half the amount of DNA and a random assortment of paternal and maternal chromosomes. During fertilization, genetic material from two sources is combined, creating an offspring that differs genetically from both parents. Finally, during gamete production, crossing over between maternal and paternal (homologous) chromosomes occurs, creating "within-chromosome" genetic variation.

Figure 3.13 Asexual reproduction. Offspring are genetically identical to the parent. Genetic variation is created through mutation.

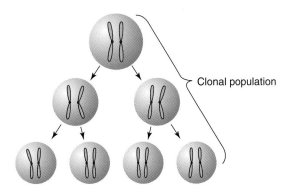

directly to another bacterium. Conjugation is often called bacterial sex, which in a strict biological sense is the mixing of genetic information. In conjugation, the genetic mixing occurs in the recipient cell and does not, in and of itself, result in the production of offspring. Therefore, bacterial sex is not the same thing as sexual reproduction. After genetic exchange, a bacterium may undergo asexual reproduction as usual, and its new genetic information would be passed to its offspring.

In transformation (Figure 3.15), a new combination of genetic material is created when cells take up DNA from the medium around them. In nature, some

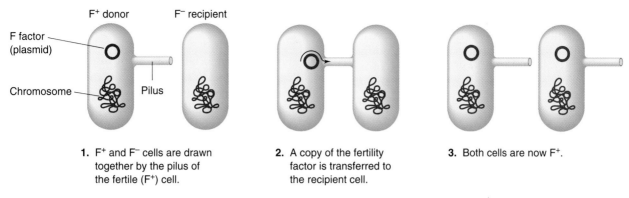

1. F⁺ and F⁻ cells are drawn together by the pilus of the fertile (F⁺) cell.

2. A copy of the fertility factor is transferred to the recipient cell.

3. Both cells are now F⁺.

Figure 3.14 Bacterial conjugation. Genetic material is exchanged between F^+ and F^- cells.

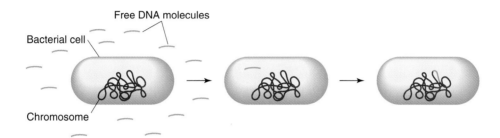

Figure 3.15 Transformation. A cell takes up free DNA from its environment, integrates it into its chromosome, and expresses the encoded products.

types of bacteria are easily transformed, while others are not. Scientists have learned to manipulate many kinds of cells to render them more susceptible to transformation in the laboratory.

In transduction (Figure 3.16), a virus intermediary carries DNA from one bacterial cell to a second cell. During a typical virus infection, viral nucleic acid is replicated in the host cell, and viral coat pro-teins are made. Near the end of the infection cycle, the viral nucleic acid is packaged into the new virus coats, and new virus particles are re-leased. For transduction to occur, some of the bac-terial DNA must be packaged into a virus particle by mistake. This bacterial DNA is then injected into another bacterium and combines with the resident DNA. No infection occurs, since no virus genome was injected.

Figure 3.16 Transduction. A virus serves as a conveyer of genetic material from one or-ganism to another. In the example here, the organism is a bacterium, and the virus is a bacteriophage.

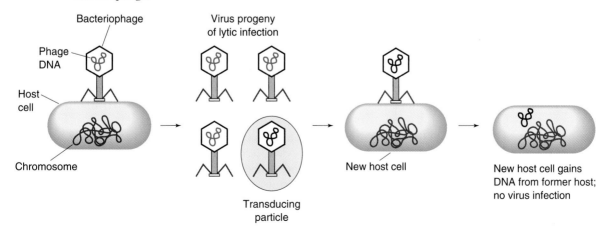

Selecting certain genetic variants

After genetically variable individuals are created by any of the methods described above, these individuals are exposed to the indifferent force of natural selection. Natural selection acts on individuals, but individuals do not evolve. Populations evolve. How does this happen?

What Is "Survival of the Fittest"?

Natural selection is a measure of reproductive success, but not necessarily of survival. Populations contain individuals that vary in their abilities to both survive and reproduce. This variation provides the raw material for evolutionary change. The fittest organisms are those that produce the most offspring that survive to reproductive maturity. If, however, an organism excels at skills required for survival but leaves relatively few offspring, it is not "fit."

On the other hand, traits that have questionable or even negative survival value persist in a population if they are beneficial in obtaining mates. Think of the peacock's tail. It is difficult to imagine how a 5-foot-long tail of brightly colored, iridescent feathers could have positive survival value, but the peahens think those tails are gorgeous, so males with long, brightly colored tails are the "fittest."

Natural selection favors (selects for) those organisms that produce the most offspring and rejects (selects against) those that produce fewer offspring. The genes of the organisms that natural selection has favored are passed to the next generation in greater numbers than the genes of organisms selected against. In this way, gene frequencies change over time, and this change is the process of biological evolution.

Observable Variation and Natural Selection

While populations are characterized by variation, only some of the individual variation that exists has a genetic basis. Some of the variation we observe in populations is exclusively phenotypic variation caused by environmental factors. Such variation may contribute to differences in survival and reproductive success and thus could be subject to natural selection, but it is not the raw material for evolution. Traits that contribute to an increase in fitness will evolve only if they have a genetic basis.

Thus, only some of the variation that exists in nature is significant in an evolutionary sense. Some is not significant now and may or may not be significant in the future, depending on its degree of heritability and the selective pressures that act on that variation.

Natural Selection as a Honing Device

Some people view evolution as somewhat sacred. They have faith that through natural selection, evolution has honed all traits in all organisms so that there is a perfect fit between an organism and its environment. That opinion is appealing but naive. Most traits are present simply because they are not harmful—at that time in that environment. Given another environment or another time, they might be.

Even though evolution is a change in gene frequencies, natural selection acts on phenotypic traits. At any one time, only some of the traits of an organism are the focus of natural selection. Others are irrelevant to selection and just happen to get passed on with the favored genes. At that time, operating in that environment, natural selection is not "interested" in those traits. As a result, it is not molding a fit between those traits and the environment. Next year, however, those neutral traits might become the focus of natural selection, or a useful attribute one year may be a handicap, or linked to a handicap, the next.

For example, imagine that a mutation produces a color pattern that provides a lizard with better camouflage. If having that trait means the lizard lives longer and has more offspring, that gene will be passed to the next generation in greater numbers. It will drag with it other, neutral genes that may have nothing to do with leaving more offspring. On the other hand, the new gene and the resulting color pattern might make the lizard more susceptible to predation. In that case, the gene would be selected against, as would the lizard's neutral genes and even other genes with positive survival value.

Now, imagine a situation in which the color pattern makes male lizards more susceptible to predation but is also the color pattern that female lizards prefer when choosing mates. In nature, we often have selective pressures working in opposition to each other.

In addition, as described above, some of an organism's traits may not depend entirely, if at all, on its genetic makeup. Only traits that have a genetic basis are subject to natural selection. Therefore, if a trait is caused largely by environmental factors, it could not have been "honed" to perfection by natural selection.

Many people fail to realize that disorder, uncertainty, and, therefore, flexibility are common in biological systems, particularly at and above the level of the organism. That understanding may seem irrelevant to biotechnology, but in fact, it is critical for accurately analyzing the environmental issues of biotechnology.

Variation, chance, and change

The pivotal concepts of evolutionary biology are contained in the abbreviated description of evolution: variation, chance, and change. A true understanding of genes, biology, and even biotechnology is possible only if you comprehend and appreciate the role these factors play in driving evolution and shaping life on earth. Therefore, they deserve a little more attention.

Variation

Genetic variation is the grist of the evolutionary mill. The great biologist Theodosius Dobzhansky, who was instrumental in elucidating the genetic basis of evolution, said that "nothing in biology makes sense except in the light of evolution." Therefore, understanding the nature of genetic variation is essential to understanding all of biology, not just evolution.

Evolutionary biologists have always studied genetic variation, even if they did not realize it. Charles Darwin, who did not know about genes and hereditary mechanisms, focused on phenotypic variation between species in an attempt to explain the diversity of species on earth. He studied polymorphisms (different types) visible to the naked eye: birds' beaks, iguanas' feeding behavior, and the shape of tortoises' shells. Later evolutionary biologists, understanding the relationship between genes and proteins, used electrophoresis to look at protein variation as a reflection of genetic variation. Today's evolutionary biologists are able to study genetic variation by looking directly at DNA variation through restriction fragment length polymorphism and nucleotide sequence analysis. Irrespective of the method used, when they ask questions about the amount and type of genetic variation, biologists are constantly changing focal lengths:

- How much variation is there within a population?
- How much variation is there between populations (within a species)?
- How much variation is there between species?

They also ask the flip sides of these questions: How much genetic similarity is there at each of these levels?

In studying biology, always keep in mind that both similarity and dissimilarity are informative and therefore important. All species are both genetically alike and genetically different from one another. Both genetic unity and genetic diversity characterize all of the earth's species. Within a species, the same is true: all individuals of a species (or a population) are genetically alike and genetically diverse.

Because a continuum of genetic variation and similarities exists within a species and extends outside of a species to related species, determining where one species stops and another begins can be difficult and, at times, arbitrary. Where one chooses to demarcate a species boundary is flexible and, more often than you might expect, subject to lively debate. A large and dynamic branch of biology, taxonomy, is devoted to the question, "Where should we draw the line?"

Because of that same genetic continuum, no one gene makes a fish a fish and an oak tree an oak tree any more than one gene that I have and you do not makes me more human than you.

Chance

Many people envision nature prior to human intervention as ideal and ordered: organisms perfectly adapted to unadulterated environments living in balance with each other through time. These people operate on the tacit assumption that Mother Nature would keep a tidy house with a "place for everything and everything in its place" if humans would stop introducing mess and disorder. While this idyllic view of nature may be appealing, it bears no resemblance to reality. The planet and its species are constantly changing independent of any human presence.

Life on earth is like a game of chance governed by constantly changing rules. An organism's opponent, natural selection, is also the game's unpredictable referee, rewriting the rulebook on every play. Natural selection is a measure of reproductive success but not necessarily survival.

For biological organisms, every round of reproduction is like a new round of cards. Chance determines which genes are combined. These new genotypes give rise to new phenotypes. The path from genotype to phenotype is susceptible to environmental influences, many of which are random occurrences. These new phenotypes then enter the game uncertain of how they will fare, because some of the rules have changed since the last round. Certain phenotypes might be improvements—in that environment at that time. These phenotypes will be selected for, causing the proportion of certain cards in nature's genetic deck to shift before the next hand is drawn. That same phenotype in a different environment at another time might be the kiss of death.

Given such an uncertain scenario, evolution has selected for organisms that hedge their bets through genetic diversity. A species persists from one year to the next by producing genetically variable offspring,

some of which manage to survive for another round. Each offspring represents a different combination of genetic cards; and in each game of cards, there's always at least one winning hand.

Nevertheless, the greater the diversity within a species, the less perfectly adapted to its environment that species is. Therefore, organisms attempt to balance diversity—a measure of adaptability—and adaptedness. For any species, a constant and very dynamic tension exists between these two properties.

Change

The idea that environmental factors lead to evolutionary change by exerting selective pressure on organisms is familiar. These environmental forces acting on species are constantly changing, and a species' survival depends upon its ability to respond with evolutionary changes. When you think of environmental factors as selective agents, you probably envision aspects of the physical environment, like drought. However, other living organisms, the biological environment, make up the most important component of a species' environment.

Thus, if one species responds to physical environmental changes with evolutionary changes, its evolutionary change becomes an environmental change for a second species.

Also keep in mind that the relationship between organisms and their environment is not linear but circular. The activities of organisms continually alter the earth's physical environment, often to the detriment of other species. Changes in the physical environment may then place new selective pressures on other organisms. Organism-driven environmental changes exert effects on individuals in the same species, as well. Individuals in the same species compete for access to resources. Some win at others' expense.

Constant flux characterizes the natural world. Species are continually evolving—changing genetically—in response to the physical environment and to each other. It really is a jungle out there.

Understanding the details of evolutionary change may seem unrelated to biotechnology, but it is not. You will see that our underlying assumptions about evolution and the resulting state of the natural world drive some of the concerns people have

about certain applications of biotechnology. These same assumptions encourage people to assign more power to biotechnology than it is due and to assume that the tasks in front of us are easier than they really are.

Selected Readings

Allen, Garland E. 1978. *Thomas Hunt Morgan: the Man and His Science.* Princeton University Press, Princeton, NJ.

Crick, Francis. 1988. *What Mad Pursuit: a Personal View of Scientific Discovery.* Basic Books, New York, NY.

Futuyma, Douglas. 2005. *Evolution.* Sinauer, Sunderland, MA.

Gibbs, W. W. 2003. The unseen genome: gems among the junk. *Scientific American* 289:46–53.

Gould, Stephen J. 2002. *The Structure of Evolutionary Thinking.* Belknap Press, Cambridge, MA.

Henig, Robin M. 2001. *The Monk in the Garden: the Lost and Found Genius of Gregor Mendel.* Mariner Books, Boston, MA.

Judson, Horace F. 1979. *The Eighth Day of Creation: Makers of the Revolution in Biology.* Simon & Schuster, New York, NY.

Keller, Evelyn Fox. 1983. *A Feeling for the Organism: the Life and Work of Barbara McClintock.* W. H. Freeman & Co., New York, NY.

Keller, Evelyn Fox. 2000. *The Century of the Gene.* Harvard University Press, Cambridge, MA.

Lagerkvist, Ulf. 1998. *DNA Pioneers and Their Legacy.* Yale University Press, New Haven, CT.

Mawer, Simon. 2006. *Gregor Mendel: Planting the Seeds of Genetics.* Harry N. Abrams, New York, NY.

Mayr, Ernst. 2002. *What Evolution Is.* Basic Books, New York, NY.

McConkey, Edwin H. 2004. *How the Human Genome Works.* Jones and Bartlett Publishers, Sudbury, MA.

Morange, Michel. 2002. *The Misunderstood Gene.* Harvard University Press, Cambridge, MA.

Watson, James. 1968. *The Double Helix.* Atheneum, New York, NY.

Watson, James, and Andrew Berry. 2005. *DNA: Life's Code.* Cambridge University Press, Cambridge, England.

Watson, James, and Francis Crick. 1953. Molecular structure of nucleic acids: a structure for deoxyribonucleic acid. *Nature* 171:737.

Watson, James, and Francis Crick. 1953. Genetical implications of the structure of deoxyribonucleic acid. *Nature* 171:964.

Wilson, Edward O. 2005. *From So Simple a Beginning: Darwin's Four Great Books.* W. W. Norton and Co., New York, NY.

An Overview of Molecular Biology 4

Gene Structure and Function

A hallmark of living systems is that they reproduce themselves. For many years, one of the greatest mysteries of science was the puzzle of how the tiniest seed or fertilized egg could contain all the information needed for the development of an entire organism. Classical geneticists deduced that individual traits were determined by invisible information units they called genes. They presumed that each cell duplicated its genes before dividing so that each daughter cell could receive a complete set. They also presumed that the genes present in sperm and egg cells would carry the genetic information to the next generation. Then, somehow, those genes would direct the development of a new individual.

It was clear that the genetic material must be capable of two extremely important functions. First, it must be in a form that can be copied very accurately so that correct information is transmitted from cell to cell and generation to generation. Second, its information must somehow be translated into a living organism. What molecule could possibly fulfill these two complex and critical requirements? Scientists naturally assumed that the genetic material must be a very complicated molecule or molecules.

Today, we know that DNA is the genetic material of all life on earth. However, at first, that discovery was so astonishing that many scientists refused to believe it. Why? Because DNA is such a simple molecule. How could such a simple molecule be responsible for the extraordinary diversity of life forms on this planet? The determination of the structure of DNA and the cracking of its genetic code are among the most important scientific achievements of the 20th century. The elegant simplicity of DNA structure and the beautiful efficiency with which that structure provides for the two essential functions of the molecule can inspire both the poet and the engineer within each of us.

Structure of DNA

The genetic material of all life on this planet is made of only six components. These components are a sugar molecule (deoxyribose), a phosphate group, and four different nitrogen-containing bases: adenine, guanine, cytosine, and thymine. The essential building block of the DNA molecule is called a nucleotide or, more precisely, a **deoxynucleotide.** A deoxynucleotide consists of a deoxyribose molecule with a phosphate attached at one place and one of the four bases attached at another (Figure 4.1). The carbon atoms of the deoxyribose sugar portion of a nucleotide are always numbered in the same way. The base is always attached to carbon 1, and the phosphate group is always attached to carbon 5.

In a DNA molecule, thousands or millions of these nucleotides are strung together in a chain by connecting the phosphate group on the number 5 carbon of one deoxyribose molecule to the number 3 carbon of a second deoxyribose molecule (a free water molecule is created in this process). Figure 4.2 shows an example in which three nucleotides are connected. The linkages formed between the deoxyribose molecules via the phosphate bridge are called **phosphodiester bonds.** Because the nucleotides are held together by bonds between their sugar and phosphate entities, DNA is often said to have a sugar-phosphate backbone. Notice that the ends of the molecule in Figure 4.2 are labeled 5′ and 3′ for the carbon atoms that would form the next links in the chain at either end.

The sugar-phosphate backbone of DNA is an important structural element, but all of the information is contained in the four bases. The key to the transmission of genetic information lies in a characteristic of these bases: that adenine and thymine together form a stable chemical pair and cytosine and guanine form a second stable pair. The pairs are formed through weak chemical interactions called **hydrogen bonds.** These two pairs, adenine-thymine and cytosine-guanine, are called complementary base

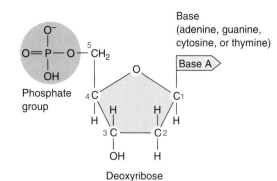

Figure 4.1 The nucleotide. Carbon atoms of the deoxyribose sugar portion are numbered according to chemical convention.

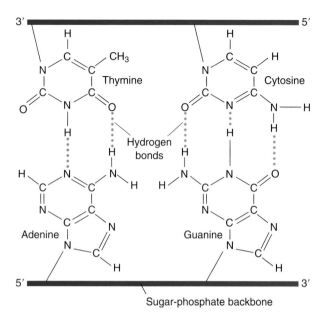

Figure 4.3 Complementary base pairs in DNA.

pairs (Figure 4.3). In a DNA molecule, two sugar-phosphate backbones lie side by side, one arranged from the 5′ end to the 3′ end, and the opposite strand arranged from the 3′ end to the 5′ end. The bases attached to one strand are paired with their partner bases attached to the opposite strand (Figure 4.3). Thus, the order of the specific bases on one strand is perfectly reflected in the order of the complementary bases on the other strand. Knowing the sequence of bases on one strand allows us to deduce the base sequence on the complementary strand. (See the activity *Constructing a Paper DNA Model*.)

DNA is often drawn as a flat molecule because that shape is easy both to draw and to look at. In reality, each of the two sugar-phosphate backbones is

Figure 4.2 A trinucleotide.

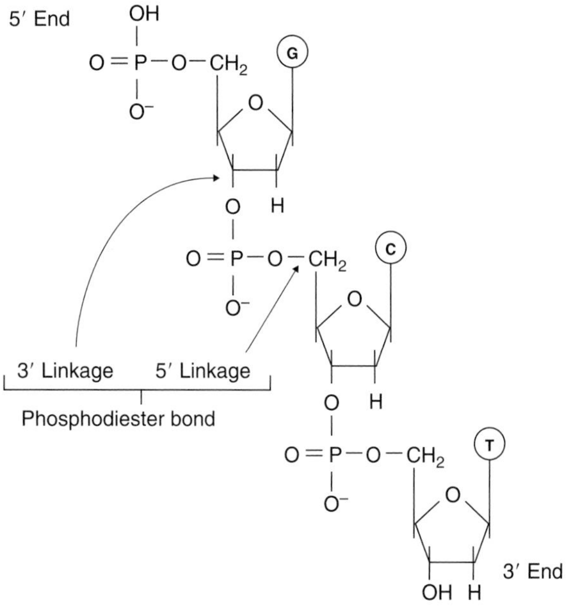

wrapped around the other in a conformation called a double helix. The base pairs are on the inside of the helix, like the rungs of a ladder. For ease of representation and viewing, DNA can be presented by using a model in which each backbone is represented as a thin ribbon and the base pairs are represented schematically (Figure 4.4). In fact, since the nucleotide sequence of one strand specifies the nucleotide sequence of the other strand, a DNA molecule or region of a molecule is very often represented by the sequence of only one strand, always written in the 5′-to-3′ direction.

DNA function: faithful replication

The structure of DNA immediately suggests how DNA carries out the first critical function of genetic material: faithful replication. You can see that either of the two strands of DNA can be used as a template, or pattern, to reproduce the opposite strand by using the rules of complementary base pairing. When a cell is ready to replicate its genetic material, the two opposite strands are gradually "unzipped" to expose the individual bases. Each exposed strand is used as a template to form two new strands (Figure 4.5A). The result is two daughter DNA molecules, each composed of one parental strand and one newly synthesized strand and each identical to the parent (Figure 4.5B).

How does DNA replication occur in a cell? DNA is duplicated by enzymes, the protein workhorses of cells. Special cellular enzymes work together to

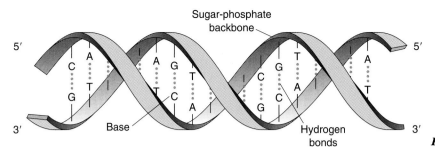

Figure 4.4 Ribbon model of DNA.

unzip the DNA double helix, capture free nucleotides, pair the correct new nucleotide with the template base, and make the new bonds of the growing sugar-phosphate backbone. The central player of this protein team is called **DNA polymerase.** It is the enzyme that actually makes the correct base pairs and forms the new phosphodiester bonds. Finally,

some of the DNA replication enzymes "proofread" the new DNA strand, checking for errors in base pairing and correcting any errors they find. These diligent enzymes ensure that very few errors occur during DNA replication so that genetic information is transmitted correctly. (See the activity *DNA Replication.*)

Figure 4.5 DNA replication. (A) Base pairing between an incoming nucleotide and the template strand of DNA guides the formation of a new daughter strand with a complementary base sequence. (B) In each round of DNA replication, each of the two DNA strands is used as a template for the synthesis of a new complementary strand, resulting in two daughter molecules, each with one "new" and one "old" strand.

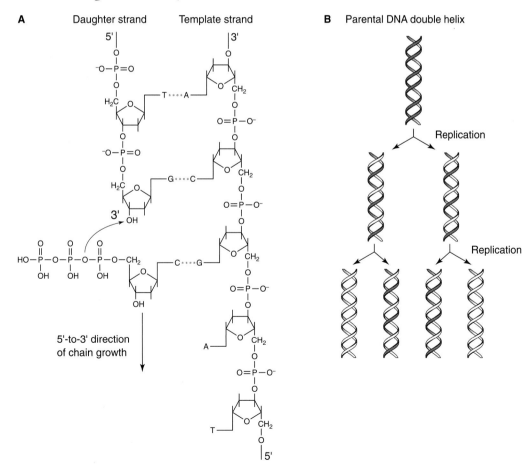

DNA function: information transmission

Although the structure of DNA immediately suggests how the molecule can fulfill the requirement for faithful transmission through duplication, it is not so obvious how such a simple molecule can determine the development of creatures as complex and varied as a blue whale or a rose. To understand how the structure of DNA elegantly fulfills this requirement, it is necessary to think about what makes a whale a whale or a rose a rose.

What does make a whale a whale? Its proteins. You read in chapter 2 that proteins are the biological molecules that carry out the actions of cell processes. Transport proteins carry oxygen, nutrients, hormones, and other important molecules throughout the whale's body and around its cells. Protein receptors embedded in cell surfaces bind with great specificity to its hormones, enabling the whale to grow and develop properly. Proteins of the whale's immune system defend it from infection. Enzymes digest the whale's foods, synthesize fats for its cell membranes and for energy storage in blubber, replicate its DNA for transmission to baby whales, and carry out all the other metabolic tasks necessary for whale cell life. Structural proteins, like collagen and keratin, form the bricks and mortar of its skin, muscles, organs, and tissues. Thus, proteins provide some of the structure and carry out essentially all the functions of the whale. The same is true for a rose, a fruit fly, a bacterium, a human being, and every other life form on earth.

It is not simply the nature of an organism's proteins that give the organism a particular identity. After all, the proteins in different animals can be pretty similar—muscle proteins, hemoglobin, enzymes to replicate DNA, and so on. The proteins of closely related animals can be very much alike (for an example of how evolutionary biologists use comparisons of similar proteins in different animals, see *Genotype and genome comparisons shed new light on evolutionary studies* in chapter 5 and the activity *Constructing an Amylase Evolutionary Tree*). Even organisms as different as animals and plants have some types of proteins in common.

Much of the diversity in nature is due to the organization of proteins within an organism, particularly structural proteins and those that synthesize additional structural body components. In some ways, the organization of similar proteins into different body plans is analogous to using a pile of bricks in construction. You could assemble them into the wall of a building or into a barbecue pit, or even a sidewalk. That analogy is an oversimplification, of course, since

proteins do differ between organisms, and the more distantly related the organisms, the more they differ. The organization of protein synthesis during the development of an organism leads to its unique form, whether it is a whale, a rose, a chrysanthemum, or a human. An organism does what it does and looks like what it is because of the nature of its proteins and how they are organized during development.

Back to questions. What determines the organization of protein synthesis during the development of an organism? Can you guess the answer? More proteins. When these proteins fail to function properly, the results can be dramatic, such as flies that grow legs where they should have antennae or fly larvae with two tail ends and no head. But the bottom line is that, in the end, proteins organize the body, compose or synthesize its structural components, and carry out its metabolic activities.

Therefore, for DNA to dictate the development of organisms, the information in the DNA must somehow be converted into proteins, the "stuff" of organisms. How does this conversion occur? To answer this question, we must know a bit more about the makeup of proteins.

Proteins

As you learned in chapter 2, a protein is a chain of amino acids, small organic molecules made mostly of carbon, hydrogen, oxygen, and nitrogen. Proteins are made from a pool of 20 different amino acids by connecting a few or thousands of these amino acids in various orders to form chains. In chapter 2, we used the analogy of making a chain of pop-beads from an assortment of 20 different colored beads—but a protein is more than a straight chain of beads. The chain of amino acids is folded and coiled into a specific three-dimensional shape (Figure 4.6). A protein's function is made possible by its unique three-dimensional structure.

For an example, let's consider a single receptor protein embedded in the outer membrane of a cell. Our imaginary receptor protein looks like an irregularly shaped glob with interesting nooks and crannies. The shapes of the nooks and crannies are absolutely critical to the protein's function: one cranny on the outside of the cell membrane is the place where a growth hormone molecule must fit exactly to signal the cell to grow. Other nooks and crannies on the inside are sites that fit precisely to other molecules for communication with the rest of the cell. Through interaction at these sites, the receptor protein can tell the cell that the hormone signal has

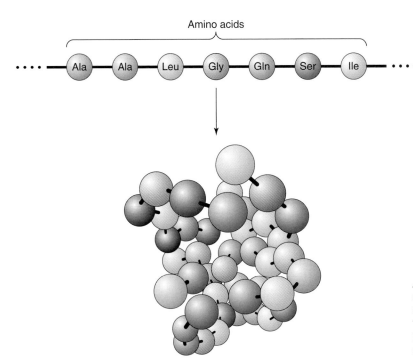

Amino acids

Ala Ala Leu Gly Gln Ser Ile

Figure 4.6 A protein is a chain of amino acids (represented by beads) that folds into a specific three-dimensional shape. The three-letter abbreviations on the beads are standard for specific amino acids (see Figure 4.23).

arrived. (For an example of a receptor protein and what happens when it fails to function, see the activity *An Adventure in Dog Hair, Part II: Yellow Labs.*)

Our imaginary receptor illustrates a crucial point about protein function. *A protein depends on its ability to bind to other molecules (sometimes other proteins) to carry out its functions. That ability is determined by its three-dimensional structure.*

What determines the three-dimensional structure of a protein? When amino acids are assembled into a protein chain, that chain immediately folds back upon itself to assume the most "comfortable," or energetically stable, shape. The most energetically stable shape is determined by the interactions of the individual amino acids that make up the protein. Therefore, the identities of the component amino acids and the order in which they occur in the chain govern the final three-dimensional structure of the protein. The order of the amino acids in the chain is thus extremely important to the function of the protein. As you can imagine, the possibilities for constructing different and unique protein chains are almost limitless. (Imagine how many unique chains of pop-beads you could make using 20 different colors of beads.) This variety is fortunate, considering the many and varied functions that proteins must perform. In fact, the forms and functions of proteins are so central to molecular biology that the second portion of this chapter is devoted to them.

For now, however, let us return to the question of how DNA controls the development of an organism. We have seen that an organism is the sum of its proteins. We have also seen that a protein's function depends upon its three-dimensional structure. Furthermore, its structure depends upon the sequence of amino acids in the protein chain. DNA determines the characteristics of an organism because it determines the amino acid sequences of all the proteins in that organism.

How does DNA determine an amino acid sequence? DNA contains a **genetic code** for amino acids in which each amino acid is represented by a sequence of three DNA bases (Table 4.1). These triplets of bases are called **codons.** The order of the codons in a DNA sequence is reflected in the order of the amino acids assembled in a protein chain (Figure 4.7). The complete stretch of DNA needed to determine the amino acid sequence of a single protein is a **gene,** the unit of heredity defined by the classical geneticists. The complete set of genes in an organism is called its **genome.**

Protein synthesis

The process by which proteins are produced from the genetic code has several steps. DNA is essentially a passive repository of information, rather like a blueprint. The "action" of making a protein occurs at special sites in the cell called **ribosomes.** Therefore, the first step in protein synthesis is to relay the

Table 4.1 The genetic code

First base in DNA triplet	A	A	A	A	A	A	A	A	G	G	G	G	G	C	C	C	C	C	T	T	T	T	T	T	T	T
Second base in DNA triplet	A	A	G	G	C	T	T	T	A	A	G	C	T	A	A	G	C	T	A	A	G	G	G	C	T	T
Choices for third base in DNA triplet	A,G	C,T	A,G	C,T	A,C,G,T	A	G	C,T	A,G	C,T	A,C,G,T	A,C,G,T	A,C,G,T	A,G	C,T	A,C,G,T	A,C,G,T	A,C,G,T	A,G	C,T	A	G	C,T	A,C,G,T	A,G	C,T
Amino acid encoded	Lysine	Asparagine	Arginine	Serine	Threonine	Isoleucine	Methionine	Isoleucine	Glutamic acid	Aspartic acid	Glycine	Alanine	Valine	Glutamine	Histidine	Arginine	Proline	Leucine	Stop	Tyrosine	Stop	Tryptophan	Cysteine	Serine	Leucine	Phenylalanine

information from the DNA to the ribosomes. To accomplish this step, cellular enzymes synthesize a working copy of a gene to carry its genetic code to the ribosomes. This working copy is called **messenger RNA (mRNA)** (RNA is a close molecular relative of DNA). mRNA carries the genetic code for a protein to the ribosomes. In the second step of protein synthesis, the codons in the mRNA must be matched to the correct amino acids. This step is carried out by a second type of RNA called **transfer RNA (tRNA).** Finally, the amino acids must be linked together to make a protein chain. The ribosome (which is made of proteins and RNA) performs this function. When the protein chain is complete, a genetic "stop sign" tells the ribosome to release the new protein into the cell.

RNA

Protein synthesis therefore requires a second type of nucleic acid molecule: RNA. Like DNA, RNA is made up of nucleotides composed of a sugar, a phosphate, and one of four different organic bases (Figure 4.1). However, there are three important differences between DNA and RNA, two of them chemical and one of them structural. The chemical differences are that (i) instead of the sugar deoxyribose, RNA contains the sugar **ribose** (hence the name ribonucleic acid), and (ii) instead of the base thymine, RNA contains the base **uracil** (Figure 4.8). The sugar-phosphate backbone of RNA is linked together like the DNA backbone, and the bases are attached to the number 1 carbon, as in DNA. The important structural difference is that, although RNA bases can also form complementary pairs, RNA is usually composed of only a single strand of sugar-phosphate backbone and bases. It does not have the base-paired double-helix structure of DNA (Figure 4.9), although it is capable of pairing with other single strands of DNA or RNA. As we shall see below, the single-stranded structure of RNA is ideally suited to its task of transferring information.

Synthesis of mRNA

The first step of protein synthesis is to make mRNA. This process resembles DNA replication in many ways. First, the DNA double helix must be unzipped

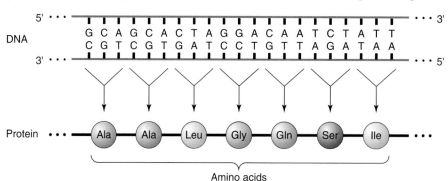

Figure 4.7 The base sequence of DNA determines the amino acid sequence of proteins.

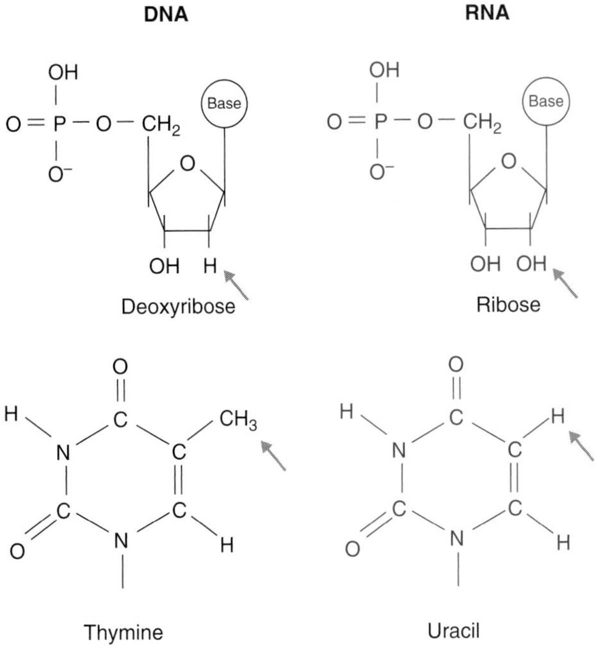

DNA   **RNA**

Deoxyribose   Ribose

Thymine   Uracil

Figure 4.8 Chemical differences between DNA and RNA.

to reveal the information-containing bases. Then, complementary nucleotides (which contain the sugar ribose, since we are making RNA) are paired with the exposed bases. During the synthesis of RNA, the base uracil substitutes for thymine and pairs with adenine. Phosphodiester bonds are made between the nucleotides, and the new mRNA contains a base sequence that is exactly complementary to the template DNA strand. The process of using a DNA template to create a complementary mRNA molecule is called **transcription** (Figure 4.10). tRNA and ribosomal RNA molecules are also encoded in DNA and synthesized by transcription, but unlike mRNA, they are not translated into protein.

There are two major differences between transcription and DNA replication (compare Figures 4.5 and

Figure 4.9 A single-stranded RNA molecule.

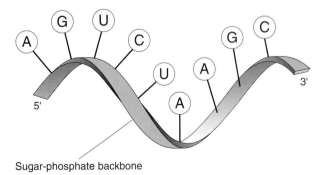

Sugar-phosphate backbone

4.10). In DNA replication, both strands are used as templates to generate two new strands for two new helices. In RNA synthesis, only one DNA strand is used as a template, and only a single RNA strand is made. The second difference is that the new mRNA molecule is released from the DNA template as it is made. The DNA double helix "zips back up" as the mRNA is released. Newly synthesized DNA remains part of a new DNA helix, paired with its parent strand.

How is mRNA actually synthesized in the cell? By now you know at least part of the answer: enzymes. The RNA-synthesizing enzyme **(RNA polymerase)** has an interesting task. Not only must it select the correct complementary nucleotides and link them together (as does its counterpart, DNA polymerase), but it must also decide where a gene is. A DNA helix can contain thousands or millions of base pairs. The RNA polymerase must determine exactly where to start and stop synthesizing RNA so that it will transcribe a complete gene.

How does RNA polymerase know where to start making mRNA? The answer is that special genetic "traffic signals" are built into the DNA base sequence. One very important traffic signal is called a **promoter.** A promoter is a special sequence of DNA bases that tells the RNA polymerase to start synthesizing RNA. As you might imagine, other signals tell RNA polymerase to stop synthesizing RNA and leave the DNA template. These signals are called **terminators.**

Using mRNA To Make a Protein

After transcription is complete, the mRNA moves to the ribosome, the site of protein synthesis. The ribosome recognizes the mRNA and holds it in proper alignment for its codons to be read correctly. Look back at Figure 4.7 and think about this for a minute. The task at hand is to translate the DNA base code in the mRNA into amino acids, as shown in the figure. You can translate a DNA base code into amino acids by looking at the genetic code table (Table 4.1). How could a cell do this with molecules? What if you had a molecule whose one end fit exactly to one and only one codon and whose other end was connected to the correct amino acid? If you had one of these molecules for each of the codons in Table 4.1, you would have a molecular "key" for translating codons into amino acids. This is what cells do—your cells, bacterial cells, lizard cells, and the mold cells on the old sandwich in your refrigerator. The molecules that make up the molecular key are another type of RNA: tRNA molecules.

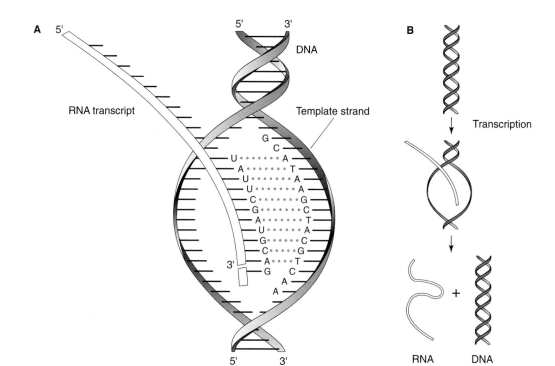

Figure 4.10 Transcription. (A) Base pairing between an incoming ribonucleotide and the DNA template guides the formation of a complementary mRNA molecule. The DNA template closes behind the RNA synthesis site, releasing the new RNA molecule. (B) In transcription, a single DNA strand is used as a template. The RNA transcript is released, leaving the DNA molecule intact.

tRNA molecules are folded in on themselves to resemble cloverleafs. At the tip of one of the lobes is a sequence of three bases called an **anticodon** (Figure 4.11). This anticodon pairs exactly with one of the codons on the single-stranded mRNA, using

Figure 4.11 A tRNA molecule. Complementary base pairing between different portions of the tRNA molecule maintains its shape.

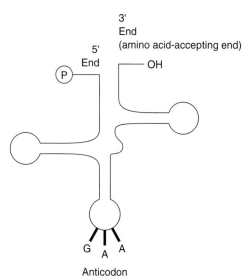

the rules of complementary base pairing. At the other end of the tRNA molecule is an amino acid. Each different tRNA is connected to the right amino acid for its anticodon. How does that happen? Enzymes, again. Every cell contains a host of exquisitely specific enzymes (called aminoacyl synthetases) whose job it is to recognize individual tRNAs and attach the correct amino acid to them, so that each type of tRNA has a specific amino acid attached to it. The result is that, when the anticodon on the tRNA pairs with its codon on the mRNA, the correct amino acid is brought to the ribosome.

The ribosome holds the mRNA molecule so that the tRNAs pair with their complementary codons one at a time, in order. As the tRNAs bring in the correct amino acids, the ribosomes link the amino acids into a growing protein chain. Once an amino acid has been linked to the chain, the tRNA molecule is separated and released from the mRNA-ribosome complex (Figure 4.12). This process, in which the mRNA base sequence is translated into a protein amino acid sequence, is called (amazingly enough) **translation.**

Protein synthesis is a complicated process involving a variety of interactions between enzymes and RNA

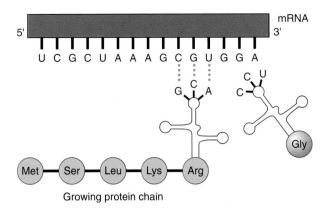

Figure 4.12 Translation. Complementary base pairing between the anticodons of incoming tRNA molecules and the codons of the mRNA guides the formation of the amino acid chain.

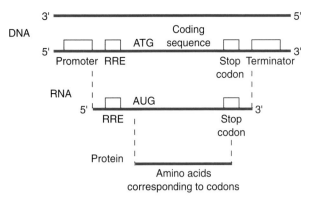

Figure 4.13 Major genetic traffic signals in bacteria. These signals tell RNA polymerase where to begin and end transcription, enable the ribosome to recognize mRNA, and direct the ribosome to start and stop protein synthesis. RRE, ribosome recognition element.

molecules. How does a ribosome distinguish an mRNA molecule from other RNA molecules in the cell? The answer is that every mRNA molecule contains more than just the codons needed to make a protein. It also contains traffic signals for the ribosome.

Foremost among these regulatory elements in bacteria is the signal for ribosomal recognition. In bacteria, this element commonly has the sequence 5′-GAGG-3′ or 5′-AGGA-3′ located 8 to 13 nucleotides upstream of the initiation codon. The ribosome recognizes the element and binds to the mRNA there. How does binding occur? The bacterial ribosomal RNA contains a base sequence complementary to the ribosomal recognition element on the mRNA, and the ribosomal RNA and mRNA associate through base pairing at that site.

The ribosome recognition element is followed by an **initiation codon** (usually AUG, which encodes methionine), where protein synthesis actually begins. At the end of the coding region, there is a stop signal (a **stop codon**) (Table 4.1). There are no tRNAs that pair with stop codons. Instead, proteins called termination factors bind to stop codons and cause protein synthesis to terminate. Since the recognition sequence, initiation codon, and stop codon are present in the base sequence of the mRNA, they must also be encoded in the original DNA template that was transcribed to make the mRNA.

A summary of the major genetic "traffic signals" is presented in Figure 4.13. The DNA molecule is a marvelous storehouse of complicated information. It contains not only the blueprints for the amino acid sequences of every protein an organism makes, but also the traffic signals that direct the cell to interpret the information properly. As we shall see below, DNA even contains signals that allow a cell to regulate the synthesis of its proteins to meet the needs of its environment.

Differences in the Molecular Biology of Prokaryotes and Eukaryotes

Although all cells on earth use the same genetic code and synthesize proteins via mRNA and tRNA, the genetic traffic signals that direct the processing of the coded information differ in prokaryotes and eukaryotes. For example, prokaryotic and eukaryotic promoters and RNA polymerases are different. A typical bacterial promoter consists of the sequence 5′-TTGACA separated by 17 bases from the sequence TATATT-3′ (some variation is permitted). Eukaryotic promoters are more variable. One component of a eukaryotic promoter is the sequence TATA located about 30 bases upstream of where transcription begins in yeasts and 60 bases upstream in mammals. This component is often called the TATA box. Usually, two more promoter components are found somewhat further upstream of the TATA box. These are the sequence CCAAT and a GC-rich sequence.

A more important difference is that, in contrast to prokaryotic RNA polymerase, eukaryotic RNA polymerase alone is not efficient at starting transcription from a promoter. Eukaryotic promoters are associated with a variety of additional DNA sequences where various transcription factors bind and stimulate transcription (see *Transcriptional Regulation through Activation* below).

The recognition of mRNA by ribosomes in eukaryotes is different from the prokaryotic system described above and appears to be more variable. In the simplest model, eukaryotic ribosomes are thought to bind at the 5′ end of the mRNA and to search down the mRNA for the first initiation codon to begin translation. Experiments in some laboratories suggest that this simple model does not explain all situations. In both prokaryotes and eukaryotes, translation begins at the initiation codon AUG. In *Escherichia coli,* initiation of protein synthesis also requires three proteins called **initiation factors.** In eukaryotes, many more protein factors are required. Both prokaryotic and eukaryotic genes contain stop codons as a signal to the ribosome to release the mRNA.

A more fundamental difference between prokaryotic and eukaryotic genes is that prokaryotic genes usually exist as a continuous sequence on a DNA molecule, and several are often transcribed from the same promoter. In contrast, eukaryotic genes are usually transcribed one at a time and are often found in many small pieces separated by stretches of noncoding DNA called introns.

When eukaryotic cells transcribe one of these split genes, a very long precursor RNA that includes all of the introns and the coding regions (exons) is synthesized first. Next, eukaryotic cells edit the RNA. In a process called **splicing,** the intron sequences are selectively cut out, and the exons are pieced together to make a functional mRNA (Figure 4.14). As you might imagine, the mRNA contains coded signals at the beginning and end of each intron that direct the cell to remove the intron. After the introns have been removed and the exons have been spliced together, the functional mRNA moves to the ribosome for translation. Splicing does not occur in eubacteria, and bacteria lack the enzymes necessary for splicing eukaryotic RNA. (See the activity *From Genes to Proteins.*)

These differences in promoters, ribosome recognition sequences, and splicing not only have fascinated researchers, but also have made life interesting for biotechnologists seeking to transfer information from eukaryotes to bacteria. As you might guess, simply transferring a gene from a mammal to a bacterium usually does not result in the production of a functional protein. In addition to the genetic information, biotechnologists must also provide the correct processing signals to the new host cell and "presplice" the gene if it contains introns. Many standard procedures have been developed to simplify these processes (see chapter 5).

Regulating gene expression

Cells must regulate the synthesis of their proteins in order to respond to environmental conditions. For example, most bacteria have genes encoding enzymes capable of breaking down quite a variety of sugars for energy. However, synthesizing these enzymes would be a waste of the cell's energy if the sugars were not available to the cell, so most bacteria synthesize an enzyme that breaks down a particular sugar only if that sugar is present in its environment.

Conversely, most bacteria encode enzymes capable of synthesizing all of the amino acids the bacteria need to make their proteins. However, synthesizing these enzymes would be wasteful if the needed amino acids were already present in the environment, so most bacteria produce the amino acid-synthesizing enzymes only if amino acids are not

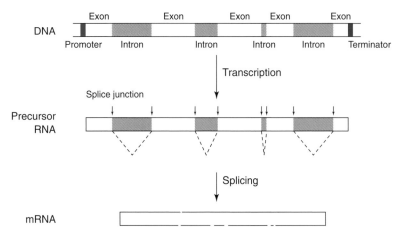

Figure 4.14 Splicing of precursor RNA to create mRNA.

available to them. Finally, many organisms, even bacteria, undergo some form of change during their life cycles. Plants and animals develop from a single fertilized egg. Some bacteria respond to adverse environments by forming spores. These structural changes also require controlled gene expression.

How is gene regulation achieved? There are many mechanisms. The best-understood ones involve regulation of transcription, the synthesis of mRNA. The rate of degradation of specific mRNAs can also be controlled. The translation rate of mRNA molecules is often regulated. Any step between DNA and protein is a potential regulatory target. However, the most common model for gene regulation involves regulation of transcription.

Transcriptional Regulation through Repression

In addition to promoters, many genes contain sites to which regulatory proteins can bind. These regulatory sites are often very near the promoter. In the most typical scenario in prokaryotes, the binding of the regulatory protein prevents transcription of the gene either by blocking access to the promoter or by preventing progression of RNA polymerase along the gene. Regulatory proteins that bind to DNA and prevent transcription are called **repressors.**

The sugar lactose can be used by the bacterium *E. coli* (and many other bacteria) as an energy source. *E. coli*'s lactose utilization genes are lined up in a row along its chromosome and are transcribed from a single promoter into one long mRNA. Several proteins are translated from this long message. The collection of lactose utilization genes is called the *lac* operon.

When lactose is present in the environment, *E. coli* synthesizes lactose utilization proteins from the *lac* operon. The lactose utilization proteins allow *E. coli* to derive energy from the sugar. When no lactose is present, these proteins are not synthesized. How does *E. coli* achieve this appropriate regulation of its *lac* genes? The following description is somewhat simplified but gives the basic idea.

E. coli synthesizes a *lac* repressor protein (Figure 4.15). In the absence of the sugar lactose, this protein binds to the *E. coli* chromosome at a specific sequence of bases (the operator) near the promoter of the *lac* genes and prevents transcription of the genes. Consequently, the bacterium does not waste energy making lactose-utilizing enzymes when there is no lactose in the cell.

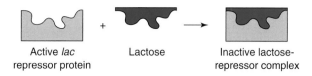

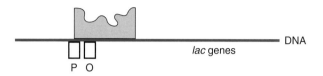

A. No lactose in cells: active repressor prevents transcription.

B. Lactose in cells: inactive lactose-repressor complex allows transcription.

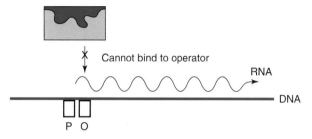

Figure 4.15 Transcriptional regulation of the *lac* operon. P is the promoter; O is the operator.

If lactose is present, however, the bacterium needs enzymes for using it. The *lac* regulation system allows the cell to respond beautifully to this new need. When lactose enters the cell, it interacts with a special site on the *lac* repressor protein. This interaction renders the repressor unable to bind to its site on the *E. coli* DNA, presumably by changing the shape of the protein. The repressor releases the DNA, leaving the gene free to be transcribed. The cell then can make the lactose-using enzymes and take advantage of the new energy source (Figure 4.15).

The *lac* genes of *E. coli* are "turned on" when the appropriate sugar is present and are otherwise "turned off." Sometimes, however, it is better for a cell to have genes normally turned on and to turn them off only under special circumstances. An example of this type of regulation is found in the transcription of the tryptophan-synthesizing enzymes of *E. coli*.

Tryptophan is an amino acid that, like other amino acids, is essential for protein synthesis. *E. coli* has genes that encode enzymes for synthesizing tryptophan from scratch (the *trp* genes). Like the *lac* genes, the *trp* genes are also lined up on the chromosome and transcribed from one promoter. Since *E. coli* constantly needs tryptophan for making new proteins, the *trp* genes are normally turned on. Occasionally, however, a lucky *E. coli* cell might find itself in an environment where tryptophan is plentiful. In this case, the bacterium conserves energy by stopping the synthesis of the *trp* proteins. Stopping expression of the *trp* genes in response to environmental changes is also achieved through a repressor protein (the following description is again somewhat simplified).

E. coli synthesizes a *trp* repressor protein, but the shape of the *trp* repressor does not allow it to bind to the chromosome (Figure 4.16). In its native state, the *trp* repressor cannot attach to its site near the promoter of the *trp* genes and therefore does not interfere with their transcription.

If the concentration of tryptophan inside the cell rises, however, the excess tryptophan interacts directly with the inactive *trp* repressor. Tryptophan binds to a special site on the protein, and its binding changes the shape of the repressor. The altered

Figure 4.16 Transcriptional regulation of the *trp* operon. P is the promoter; O is the operator.

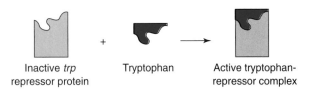

Inactive *trp* repressor protein   Tryptophan   Active tryptophan-repressor complex

A. Low tryptophan concentration: inactive repressor allows transcription.

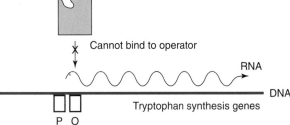

B. High tryptophan concentration: active repressor complex binds to operator and prevents transcription.

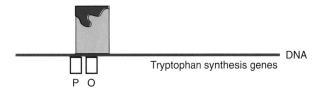

shape of the protein allows it to bind to the regulatory site near the promoter of the *trp* genes. When the active repressor binds, transcription of the *trp* genes is turned off. Thus, the presence of excess tryptophan in the cell leads to a shutoff of the tryptophan-synthesizing enzymes.

Transcriptional Regulation through Activation

The *lac* and *trp* operons provide excellent examples of transcriptional regulation in prokaryotes. Bacterial cells are easy to culture and manipulate, and transcription and gene regulation were first studied in them. When techniques for studying these processes in eukaryotic cells became available, scientists expected to find similar mechanisms at work. However, many years of work have suggested that the typical regulation mechanism in eukaryotic cells involves activation rather than repression. In these cells, most promoters are apparently not recognized efficiently by RNA polymerase alone. Instead, eukaryotic RNA polymerase requires helper proteins called **transcriptional activators,** or transcription factors, to help it bind to a promoter and begin transcription. Some of these activators associate with the RNA polymerase enzyme and do not bind to DNA themselves. Other activators bind to special base sequences in DNA and then interact with RNA polymerase to help it bind to a promoter (Figure 4.17).

The DNA-binding sites of these activators are often called **enhancers** because their presence enhances transcription from the associated promoter as long as the appropriate activator protein is present. The activities of genes can be regulated by the availability

Figure 4.17 Activator proteins are needed for transcription in eukaryotic cells.

A. No activators present: RNA polymerase cannot bind to promoter.

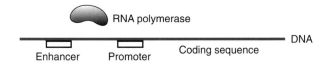

B. Activators present: RNA polymerase can bind to promoter and synthesize RNA.

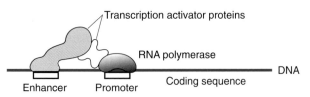

of the necessary transcriptional activators. For example, genes encoding the antibody proteins of the immune system have a specific enhancer, the immunoglobulin enhancer, associated with their promoters. This enhancer works only to enhance transcription in B lymphocytes (the antibody-producing cells of the immune system), because only B lymphocytes contain the transcriptional activator protein that binds the immunoglobulin enhancer. As a result, only your B lymphocytes make antibody proteins, even though all the cell nuclei in your body contain the antibody genes.

Another example of gene regulation through enhancers is the response of genes to steroid hormones, such as estrogen (not all hormones are steroid hormones; other hormone types exert their effects on cells through different mechanisms). Estrogen and other steroid hormones pass through the cell membrane and bind to specific receptor proteins inside the cell. When the hormone binds, the receptor changes its shape in a way that allows it to move to the nucleus, where the receptor-hormone complex acts as a transcriptional activator by binding specific enhancers. Thus, specific genes are turned on in response to the presence of the hormone.

Alteration of Promoter Recognition

Alteration of promoter recognition is a form of transcriptional regulation that occurs when a drastic change in gene expression is needed—a new set of genes must be transcribed and/or a currently transcribed set must be turned off. For example, when the bacterial virus T4 invades a host *E. coli* cell, the *E. coli* RNA polymerase begins to transcribe a few of the T4 genes that have *E. coli* style promoters. One of these genes encodes a protein that binds to the host RNA polymerase. The modified RNA polymerase can subsequently recognize only the virus's other promoters, which have a base sequence different from that of the normal host promoters. In this way, the virus switches transcription from the host genes to its genes.

Another example of this type of regulation is found in the bacterium *Bacillus subtilis*. This organism forms durable, dormant spores in response to adverse environmental conditions. Spore formation requires the expression of a number of genes that are not active during the normal life cycle of the bacterium. In addition, normal gene activity all but ceases during the spore stage. To achieve this gross change in gene expression, *B. subtilis* synthesizes a protein that binds to its RNA polymerase and causes it to recognize only the special promoters controlling sporulation genes.

RNA Interference

For many years, it was believed that regulation of gene expression occurred through the interaction of nucleic acid molecules and proteins, as described in the preceding examples. In the past few years, however, a previously unrecognized form of gene regulation mediated by short, double-stranded RNA (dsRNA) molecules has been discovered and characterized. This phenomenon has been named RNA interference, or RNAi. In a nutshell, the presence of short dsRNA molecules triggers enzymes to degrade mRNA molecules whose base sequence matches that of the dsRNA. The mechanism of RNAi is described in more detail in the activity *Antisense and RNA Interference.*

The discovery of RNAi ignited a blaze of interest in the cellular production of small, noncoding RNA molecules. Ironically, researchers had been purifying small RNAs for years without realizing that they could be important. Many procedures for isolating nucleic acids from cells yield both DNA and RNA molecules, including a mixture of small RNAs that researchers dismissed as trash, since they were too small to encode proteins. Now, scientists are taking a careful look at these formerly ignored micro-RNAs. They are produced in bacterial, plant, and animal cells, and there is evidence that they play roles in processes as diverse as plants' defenses against viral infection and normal development in animals, and in cancer. Understanding the normal cellular role of micro-RNAs is a focus of interest of much current research, and you may be hearing news stories about RNAi as discoveries are made.

Repression of Translation

Regulation of gene expression can be exerted through control of the rate of translation of an mRNA. A good example of translational repression can be found in the synthesis of the ribosomal proteins of *E. coli*. The ribosomes of *E. coli* are made up of large ribosomal RNA molecules and several proteins that bind to specific regions of the RNA. The genes for the ribosomal proteins are lined up in operons. A single mRNA is transcribed from each operon and translated into several proteins. As the proteins are translated, they find free ribosomal RNA in the cytoplasm and bind to their recognition sites. When all available ribosomal RNA is complexed with proteins, one of the proteins from that operon instead binds to the translation initiation region of the operon for the ribosomal proteins' mRNA. The binding of that protein to this mRNA prevents further translation. For each one of the ribosomal protein operons, one of the encoded proteins acts as a repressor of translation. Through this mechanism, a balance between the amount of available ribosomal RNA and ribosomal proteins is achieved.

These examples of simple gene regulation mechanisms are typical. We want to point out that repression occurs in eukaryotic cells and activation occurs in prokaryotes, too. Many steps in protein synthesis other than transcription can also be regulated. However, a central theme of gene regulation is that it involves interactions between proteins and other molecules: additional proteins, small molecules, RNA, and DNA. These interactions are dependent (as are all protein functions) upon the three-dimensional structures of the proteins. At the end of the second part of this chapter (*Protein Structure and Function*), we will look closely at the structures of two DNA-binding regulatory proteins and how they interact with DNA.

Genomic organization

The genes contained in DNA must be accessible to enzymes at the proper times for replication and transcription. Considering the extreme length of the DNA molecule in most organisms, this is a staggering requirement. Storage of DNA in cells is therefore not a haphazard affair, with the DNA "just lying around." DNA storage is a highly organized, complex phenomenon that is not well understood. Many scientists today are working to understand how cells manage their DNA information libraries.

Chromosomes

The physical packing of DNA inside cells presents a problem because of the extreme length of the DNA molecule relative to the size of the cell itself. If the DNA of *E. coli* were stretched out, it would be 1,000 times longer than the *E. coli* cell (Figure 4.18). The problem faced by eukaryotic cells is even more amazing: the DNA of a single human cell would stretch 2 m, yet the cell itself has a diameter of only 1/50 of a millimeter. It is clear that cellular DNA must be very highly folded. In cells, DNA is folded and packed in association with proteins. The DNA-protein material is called **chromatin,** and the packed structure is called a chromosome.

In eukaryotes, the genome is usually divided among several different linear chromosomes located in an organized nucleus. The ends of linear chromosomes are capped by structures called **telomeres,** which are segments of DNA consisting of short repeated sequences assembled into an unusual formation that includes a loop at the very end of the chromosome. Telomeres are not replicated by DNA polymerase. Instead, they are maintained by an enzyme called telomerase, which adds nucleotides to the ends of chromosomes, synthesizing the repeated sequence.

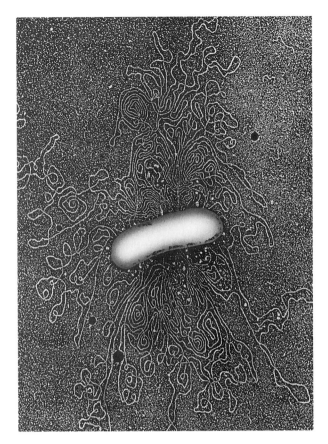

Figure 4.18 *E. coli* osmotically shocked to release DNA. (Photograph copyright K. G. Murti/Visuals Unlimited.)

Telomeres are essential for protecting the ends of linear chromosomes and may be linked to cell aging and death. Cells taken from higher eukaryotes and grown in culture have limited life spans, undergoing an average number of cell divisions before they die. The number of divisions they undergo is correlated with the life span of the organism from which the cells were taken and the age of the individual organism. As the cells undergo repeated divisions, the average length of their telomeres decreases. It is thought that this gradual loss of protection for the ends of the chromosomes may eventually lead to cell death. Interestingly, the telomeres in cancer cells, which are immortal in culture, do not shrink over time. It appears that the perpetuation of telomeres is linked to immortality, at least for cancer cells.

Chromosomal DNA is packed in an extremely ordered manner, with multiple layers of folding and coiling of the DNA molecules around the chromosomal proteins. However, even though the DNA is tightly packed, all of the DNA encoding essential proteins is available for transcription at the proper time. How the cell organizes the complex problem of storage and accessibility is not understood. Some scientists think that the large amount of noncoding DNA present in higher organisms (see below) may play an important role in DNA organization.

In most bacteria, DNA is present as one long circular molecule that is associated with special proteins in a single chromosome. These proteins are thought to hold the DNA in a condensed form so that it fits neatly into the cell. The circular chromosome is present in the cytoplasm, attached to the cell membrane.

Plasmids

Some cells, particularly bacteria, contain small rings of DNA outside of their chromosomes. These small circular pieces of DNA are called **plasmids.** Plasmids are replicated by the cell's enzymes and inherited by progeny bacteria. They typically contain a few thousand base pairs of DNA and encode a few proteins. None of these proteins are normally essential to the survival of the bacterium. Plasmids often contain genes for drug resistance or for poisons to kill rival bacteria and, in disease-causing bacteria, for proteins that contribute to the disease process. Because they are small and convenient to work with, plasmids are a favorite tool of biotechnologists. (See the activity *Recombinant Paper Plasmids* and section C, *Transfer of Genetic Information.*)

Mitochondrial and Chloroplast DNAs

Mitochondria and chloroplasts each contain their own circular DNA molecules, which are also referred to as genomes. These organelles carry out the essential functions of ATP synthesis and photosynthesis, respectively. Their circular genomes encode some, but not all, of the proteins needed for their essential functions. They also encode ribosomal and tRNAs, which are used to carry out translation inside the organelles. However, in both cases, many of the proteins needed for the organelle's function are encoded in nuclear DNA, synthesized in the cytoplasm, and transported into the organelle.

Inheritance of traits encoded by mitochondrial or chloroplast genes is different from the inheritance of other traits in eukaryotes. When egg and sperm unite to form the zygote, essentially all the cytoplasm, and thus the mitochondria and chloroplasts, are provided by the egg. Therefore, with possible rare exceptions, mitochondrial and chloroplast genes are inherited solely from the mother. Variegated leaves are a trait encoded in chloroplast

DNA. Defects in human mitochondrial DNA are associated with several different genetic diseases. The pattern of maternal inheritance of chloroplast and mitochondrial genomes also gives researchers a tool for following family trees (see the reading *Mitochondrial DNA* in chapter 27).

Virus Genomes

Viruses are not cells. They consist simply of genetic material enclosed in a capsule generally made of protein. Viral genetic material can be either DNA or RNA. Viruses require a host cell to make copies of themselves. When a virus infects a cell, it introduces its genetic material into that cell. The cell's enzymes and ribosomes transcribe and translate the viral genetic material and eventually reproduce virus particles. Outside of a host cell, viruses are biochemically dormant and are not even considered living.

It is easy to imagine a DNA virus genome substituting for the cell's own genes, but how does an RNA genome work? An RNA virus introduces its RNA genome into its host cell. When the viral RNA enters the cell, the host cell translates it as it would any other mRNA. Some RNA viruses encode in their RNA a message for a special enzyme that uses RNA as a template to synthesize new RNA, in rather the same way that DNA polymerase replicates DNA. While the viral enzyme is synthesizing more viral RNA, the host cell translates the RNAs to make viral proteins. Other RNA viruses encode an enzyme that uses RNA as a template to synthesize DNA. This enzyme is called **reverse transcriptase.** Once the viral RNA genome has been copied into DNA, the cell machinery of the host transcribes it for the virus.

Viruses are extremely specific about the host cells they can infect. Not only do viruses usually infect only one type of organism (such as humans or cats), but they are limited to certain cell types within the organism. For example, a "cold virus" infects only the cells of the upper respiratory tract, while a "stomach virus" infects only the cells of the digestive tract. This specificity explains why a pet dog does not get its master's cold. However, a few viruses, such as rabies, normally infect several different host species, and once in a while, a virus acquires the ability (possibly through mutation) to infect a new host.

In addition to viruses that infect humans or other animals, there are plant viruses specific to certain plant tissues and bacterial viruses specific to certain bacterial strains. Bacterial viruses have played an important role in the development of molecular biology and have been given a special name: bacteriophages. Electron micrographs of several viruses are shown in Figure 4.19.

Whether viruses infect plant cells, animal cells, or bacterial cells, the mechanisms by which they rec- ognize their own hosts are similar. Proteins that make up the outer capsule of the virus recognize and bind to a specific molecule, often a protein, on the surface of the host cell. This specific molecular recognition is what determines the extremely limited range of host cells a virus can infect.

Figure 4.19 Electron micrographs of various viruses. (A) Bacteriophage lambda (magnification, ×275,000). (Photograph copyright K. G. Murti/Visuals Unlimited.) (B) Purified bacteriophage T4. (Photograph courtesy of F. P. Booy; reprinted from J. D. Karam et al., ed., *Molecular Biology of Bacteriophage T4,* ASM Press, Washington, DC, 1994.) (C) Tobacco mosaic virus (magnification, ×144,000). (Photograph copyright K. G. Murti/ Visuals Unlimited.) (D) Vesicular stomatitis virus (rabies group) (magnification, ×100,000). (Photograph copyright K. G. Murti/Visuals Unlimited.)

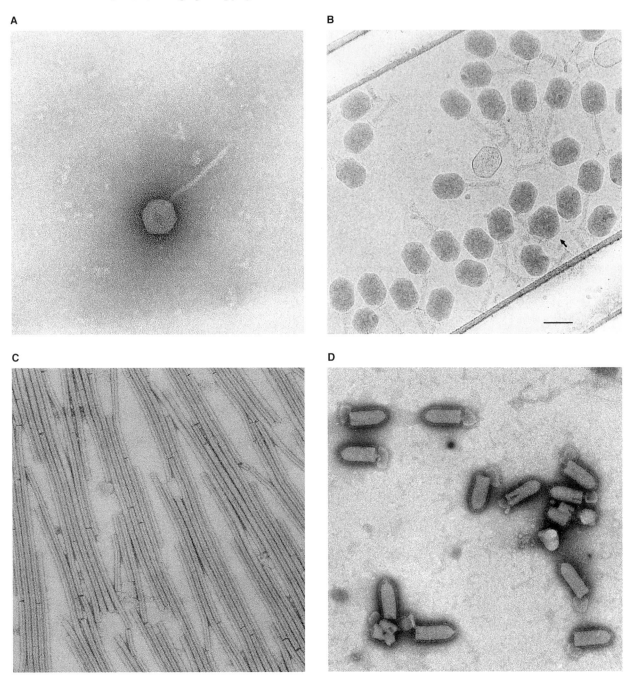

A

B

C

D

Viruses as Tools in Biotechnology

Viruses have proved to be useful tools for biotechnologists because of some of the properties mentioned above. Viruses are essentially containers of genetic material. They deliver that material into very specific cells. Scientists have developed ways of packaging new genetic material into several different types of viruses. The virus particles then inject that material into their host cells. Delivery of genetic material via viruses is an important method of gene transfer. (See the activity *Transduction of an Antibiotic Resistance Gene*.)

Noncoding DNA

How much DNA do organisms have? Naturally, it depends on the organism. The smallest viruses may have only a few thousand base pairs. Since the "average" protein requires 1,200 bases of coding sequence, these viruses encode only a few proteins. A bacterium such as *E. coli* contains about 4 million base pairs in its genome, while the human genome is composed of about 3 billion base pairs. Lest we grow smug about the complexity of the human genome, it bears noting that the genome of the mud puppy (an amphibian similar to newts and salamanders) is estimated to contain about 50 billion base pairs, and that of the lily contains 250 billion.

Does all of this DNA encode proteins? In bacteria and viruses, most of it appears to. Eukaryotes are a different story. In fact, only a small fraction of the DNA of multicellular eukaryotes (such as amphibians, mammals, and plants) is coding sequence. For example, the human genome contains about 22,000 genes with an average length of about 3,000 base pairs each. Simple multiplication predicts that about 66,000,000 base pairs, or only about 2%, of the 3-billion-base-pair human genome encodes proteins.

What is all that extra DNA in eukaryotes? A large portion of it is repeated sequences that are present throughout the genome, sometimes in large clusters and sometimes interspersed throughout the genome (Color plate 1). Some of the repeated sequences appear to be transposons or parts of transposons. No one knows what the purpose of any of this DNA is or even if it has a purpose. It has sometimes been called "junk DNA," though it seems presumptuous to label it junk before we understand it. (Think of the "junk RNA" that has now been discovered to play potentially important roles in gene regulation, as described above under *RNA Interference*.) A more neutral name for it is **noncoding DNA**. In general, the genomes of bacteria and viruses are compact arrays of genes along the DNA. However, your own genome and the genomes of plants or animals contain vast stretches of repeated DNA sequences, occasionally broken up by genes.

Genomics and bioinformatics

In 1990, an international consortium of scientists and governments set out to determine the complete base sequence of the human genome (for more information, see the reading *The Human Genome Project* in chapter 29). It was understood at that time that in order to complete the task, improvements in methods for DNA sequence determination had to be made, and therefore, improving sequencing technology was one of the project's goals. It was also recognized that software tools for handling the enormous amounts of information that would be generated by the genome project had to be developed. Improvements in sequencing technology during the 1990s led to the completion of a working draft of the sequence of the human genome in 2000, ahead of schedule. The software development that paralleled the sequencing effort has led to a suite of powerful computer analysis tools available free to anyone with an Internet connection. The development and use of software for the manipulation and analysis of DNA and protein sequence information has become a subdiscipline of biology in itself: **bioinformatics** (see chapter 34 for more information).

The improvements in DNA-sequencing technology have made the sequencing of a complete genome a much less daunting, though still formidable, task. Since the publication of the first complete genome sequence, that of the bacterium *Haemophilus influenzae*, in 1995, literally dozens of genome sequences have been determined. Among the sequenced organisms are numerous bacteria, yeast, dog, chicken, mosquito, honey bee, nematode, mouse, rat, chimpanzee, rice, paramecium, and zebrafish.

The determination of complete genome sequences, along with advances in software, allows scientists to make comparisons they were not previously equipped to make. It is now possible to compare the base sequences of chromosomes from one species to those of the next and determine in what ways their genetic organizations and contents differ. For example, comparison of the human and chimpanzee genomes shows that 29% of the encoded proteins are absolutely identical. The parts of the genomes that can be directly compared are 99% identical, and taking into account deletions, insertions, and rearrangements, the two genomes are still 96% identical. Thus, to understand the genetic differences that underlie the biological differences between humans and chimps, researchers can focus their efforts on those portions of the genomes that do differ (see the activity *Comparing Genomes* for more information). The characterization and comparisons of not only the structure, but also the patterns of gene expression in entire genomes, or large portions thereof, has come to be called **genomics.**

Mutations

Any change in a DNA sequence is called a mutation. It is a fact of life that mutations happen. They result from normal cellular processes and unavoidable environmental hazards. During DNA replication, the DNA polymerase enzyme occasionally makes errors that escape the proofreading that occurs during DNA replication. Environmental factors, such as ultraviolet light or mutagenic chemicals, damage DNA regularly. Most of this damage is corrected by DNA repair enzymes, which are proteins that recognize and repair abnormalities in DNA, but occasionally the repair enzymes miss something. When DNA polymerase attempts to use a damaged base as a template during DNA replication, it often cannot read the base properly and so inserts an incorrect base into the new strand, thus creating a mutation. Genetic events, such as transposition (see *Transposable Elements* in chapter 3), result in insertion or loss of segments of DNA, also changing the original sequence. Errors in cell division or recombination can lead to a rearrangement of the segments of chromosomes or even a change in the number of chromosomes.

What is the effect of a mutation? It depends. It depends on where the mutation is in the DNA, exactly what it is, and, often, what environment the organism inhabits. For example, a sequence change in one of the many noncoding regions would probably not have any effect on the organism. Similarly, a mutation in an intron sequence in a eukaryote might not have any effect unless it involved a processing signal for the splicing enzymes. Even changes in coding sequences may not have any effect on a protein. Many amino acids are encoded by more than one codon. For example, the codons TTT and TTC each encode the amino acid phenylalanine (Table 4.1). A mutation that changed TTT to TTC would not have any effect on the protein. Mutations with no effect on a protein are often called "silent" mutations.

In addition, many amino acid changes may not alter the function of a given protein in a significant way. Recalling the example of the receptor protein in the cell membrane, imagine that a mutation occurred in a region of the protein apart from the specific hormone recognition area. As long as the change did not distort the overall shape of the protein or impair its interaction with the cell, it might not affect the function of the protein.

Although many mutations are harmless, others can be devastating. Mutations in genetic traffic signals, such as promoters and ribosome recognition sequences, can completely shut down the synthesis of a protein. Sometimes a single base change in a coding region can result in an amino acid substitution that severely harms or destroys the protein's ability to perform its function. Recall again our example of the receptor protein embedded in the cell membrane. If anything distorted the shape of the receptor protein, the hormone might no longer fit and/or the receptor might no longer be able to communicate with the rest of the cell. The cell would lose its ability to respond to the body's signals for growth. Thus, a change in the protein's shape could have disastrous consequences for the organism. Similar examples could be given for proteins of the immune system that must recognize specific invaders, for transport proteins that carry specific nutrients, and so on.

Although most mutations are harmful or neutral, they can also be beneficial. Changes in the amino acid sequence of a protein might make it more resistant to heat, which could be an advantage if the organism's environment is becoming warmer. An alteration in the shape of another protein might allow it to bind to and break down a different type of sugar, which could be an advantage if the organism's environment contained that sugar. If a change in a protein's function is not immediately fatal, it might actually help the organism under the right environmental conditions.

In this discussion, we have been focusing on the effects of mutations in terms of how specific nucleic acid changes affect protein structure and function. The effect of a mutation on the organism will be a consequence of how any changes in protein function interact with the required cellular activities of the organism. This is a complex issue that depends on the role of the protein in question, whether the cell can compensate for changes in its activity, and, often, the nature of the specific environment the organism finds itself in, as discussed in chapter 2.

One consequence of mutations and recombination is that the chromosomes of all sexually reproducing individuals differ in many ways. Although all humans have similar chromosomes and produce the same sets of essential proteins, the exact DNA sequences in those chromosomes vary. One obvious source of variation is the same variation that causes us to look different: one individual has genes encoding blue eyes, and another has genes encoding brown eyes. A less obvious source of variation is the accumulation of changes in noncoding regions of DNA. Great variety can be present in these regions, particularly in the number of repeated sequences present, with no outwardly observable effects. In fact, it is extremely unlikely that any two individuals (except identical twins) would share the same sequences in all of their noncoding DNA. It is this variety that makes possible the new procedures of DNA fingerprinting. (See *Comparing Genotypes and Genomes* in chapter 5 and chapters 27 and 28.)

At the end of this chapter, we will look at specific examples of a harmful and a helpful mutation: the nucleotide changes, the amino acid changes, the effects on protein structure, the effects on the proteins' functions, and the effects of the changes in protein behavior on the phenotype of the organism. To put all of these things together, we first need to look at the link between genes and phenotype: proteins.

Protein Structure and Function

Genes are important because they supply information that directs the synthesis of proteins. It is the proteins that confer a phenotype on the cell: its biochemical capabilities, its shape, its communication channels, and so on. The function of a protein is determined by its three-dimensional shape, which is determined by the nature and sequence of its amino acids. The relationship between amino acid sequence and three-dimensional structure has been called the "second half of the genetic code," because it is the three-dimensional structure that leads to function and phenotype. Unfortunately, the relationship between the amino acid sequence and the three-dimensional structure is not simple. Although scientists have made great strides in developing computer programs for predicting the three-dimensional structure of a protein from its amino acid sequence, these programs are not yet perfect. Protein structure still has to be determined in the laboratory.

Over the past several years, the three-dimensional structures of hundreds of proteins have been painstakingly determined using the techniques of X-ray crystallography and nuclear magnetic resonance (see chapter 5). These structures have greatly increased our understanding of how amino acid chains fold up to be energetically stable proteins. Here is a summary of the basic principles.

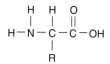

Figure 4.20 General structure of an amino acid. R signifies one of the 20 different side chains shown in Figure 4.23.

Amino acids and peptide bonds

The 20 acids that are normally found in proteins have the general chemical structure shown in Figure 4.20. The R in the figure can be any of 20 different so-called side chains. These different side chains give the 20 amino acids their separate identities.

When amino acids are joined to make a protein chain, the OH group on one end of the amino acid reacts with the NH_2 group of another amino acid. A water molecule (H_2O) is lost, forming what is called a **peptide bond** (Figure 4.21A). A protein consists of many amino acids joined together via peptide bonds. Like a DNA strand, a protein backbone has a direction, too. One end has a free NH_2 group, and the other has a free COOH group. These ends are called the N terminus and the C terminus, respectively. The overall effect is that a protein has a uniform peptide backbone with various amino acid side chains (Figure 4.21B). The identities and order of the side chains in a protein are called the **primary structure** of the protein. Primary structure is a direct consequence of the DNA base sequence in the gene encoding that protein.

You have probably already figured out that if all proteins have a uniform peptide backbone, then the nature of the amino acid side chains must be the factor that governs how an individual protein folds into a three-dimensional structure. That is true, and fortunately, one particular property of side chains is the most important for influencing the three-dimensional structure. This property has to do with how well the individual side chains interact with water molecules. To understand this property, let's start with a quick review of covalent bonds.

Polar and nonpolar covalent bonds

The atoms in a protein molecule are held together by covalent chemical bonds, which consist of electrons shared by two atomic nuclei. If the electrons are shared equally by the nuclei, then their negative charge is distributed evenly over the area of the bond and balanced by the positive charges in the nuclei. However, nuclei of certain elements attract

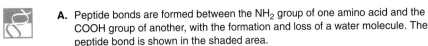
A. Peptide bonds are formed between the NH$_2$ group of one amino acid and the COOH group of another, with the formation and loss of a water molecule. The peptide bond is shown in the shaded area.

$$
\begin{array}{c}
\text{H} \quad \text{H} \quad \text{O} \\
| \quad\;\; | \quad\;\; || \\
\text{H}-\text{N}-\text{C}-\text{C}-\text{OH} \quad + \quad \text{H}-\text{N}-\text{C}-\text{C}-\text{OH} \\
| \qquad\qquad\qquad\qquad\qquad | \\
R_1 \qquad\qquad\qquad\qquad\qquad R_2
\end{array}
$$

$$\downarrow$$

$$
\begin{array}{c}
\text{H} \quad \text{H} \quad \text{O} \quad \text{H} \quad \text{H} \quad \text{O} \\
| \quad\;\; | \quad\;\; || \quad\;\; | \quad\;\; | \quad\;\; || \\
\text{H}-\text{N}-\text{C}-\text{C}-\text{N}-\text{C}-\text{C}-\text{OH} \quad + \quad \text{H}_2\text{O} \\
| \qquad\qquad\qquad | \\
R_1 \qquad\qquad\;\; R_2
\end{array}
$$

B. A protein is a polypeptide backbone with various amino acid side chains (R_n).

N terminus   C terminus

$$
\begin{array}{c}
\text{H} \quad \text{H} \quad \text{O} \quad \text{H} \quad \text{H} \quad \text{O} \quad \text{H} \quad \text{H} \quad \text{O} \quad \text{H} \quad \text{H} \quad \text{O} \quad \text{H} \quad \text{H} \quad \text{O} \\
\text{H}-\text{N}-\text{C}-\text{C}-\text{N}-\text{C}-\text{C}-\text{N}-\text{C}-\text{C}-\text{N}-\text{C}-\text{C}-\text{N}-\text{C}-\text{C}-\text{OH} \\
R_1 \quad\;\; R_2 \quad\;\; R_3 \quad\;\; R_4 \quad\;\; R_5
\end{array}
$$

Figure 4.21 To form proteins, amino acids are joined by peptide bonds.

electrons more strongly than other nuclei do (the ability to attract electrons is called **electronegativity).** If a strongly electronegative element forms a covalent bond with a less electronegative element, the electrons tend to be found near the strongly electronegative nucleus. You can think of the electronegative nucleus as pulling the bond electrons away from the less electronegative nucleus. This uneven distribution of electrons creates a partial negative charge around the electronegative nucleus and a partial positive charge at the other nucleus. Chemical bonds with this type of uneven charge distribution are called polar bonds because they have positive and negative poles (Figure 4.22).

Figure 4.22 A polar covalent bond. Although the oxygen and hydrogen nuclei are sharing two electrons, the highly electronegative oxygen nucleus tends to draw them away from the weakly electronegative hydrogen nucleus. As a result, the oxygen end of the bond acquires a partial negative charge, while the hydrogen end is partially positive.

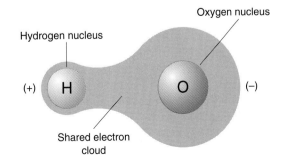

Amino acids (and proteins) are made primarily of carbon, hydrogen, oxygen, and nitrogen. Of these four elements, oxygen and nitrogen are strongly electronegative, and carbon and hydrogen are not. Thus, bonds between oxygen and hydrogen, oxygen and carbon, nitrogen and hydrogen, and nitrogen and carbon are polar. However, many of the bonds in a protein molecule are between carbon and hydrogen and are not polar. Whether a larger chemical group like the side chain of an amino acid is polar or not depends on its constituent chemical bonds.

The chemical structures of the side chains of all 20 normal amino acids are shown in Figure 4.23. The structures are grouped according to whether the side chains are fully charged (ionized) at physiological pH, are polar (partially charged), or are nonpolar. A fourth group, the aromatic amino acids, are called that because, like other aromatic compounds, they have a ring structure in their side chains. Of this group, phenylalanine is nonpolar, and the other two are somewhat polar.

Polarity and Stability

The polarity of the amino acid side chains is fundamentally important to protein structure, because polarity determines how stably a side chain interacts with other elements in the protein and in the cellular environment. A compound that has an overall polar character or is actually electrostatically charged is energetically stable when it associates with other compounds with complementary charges or partial

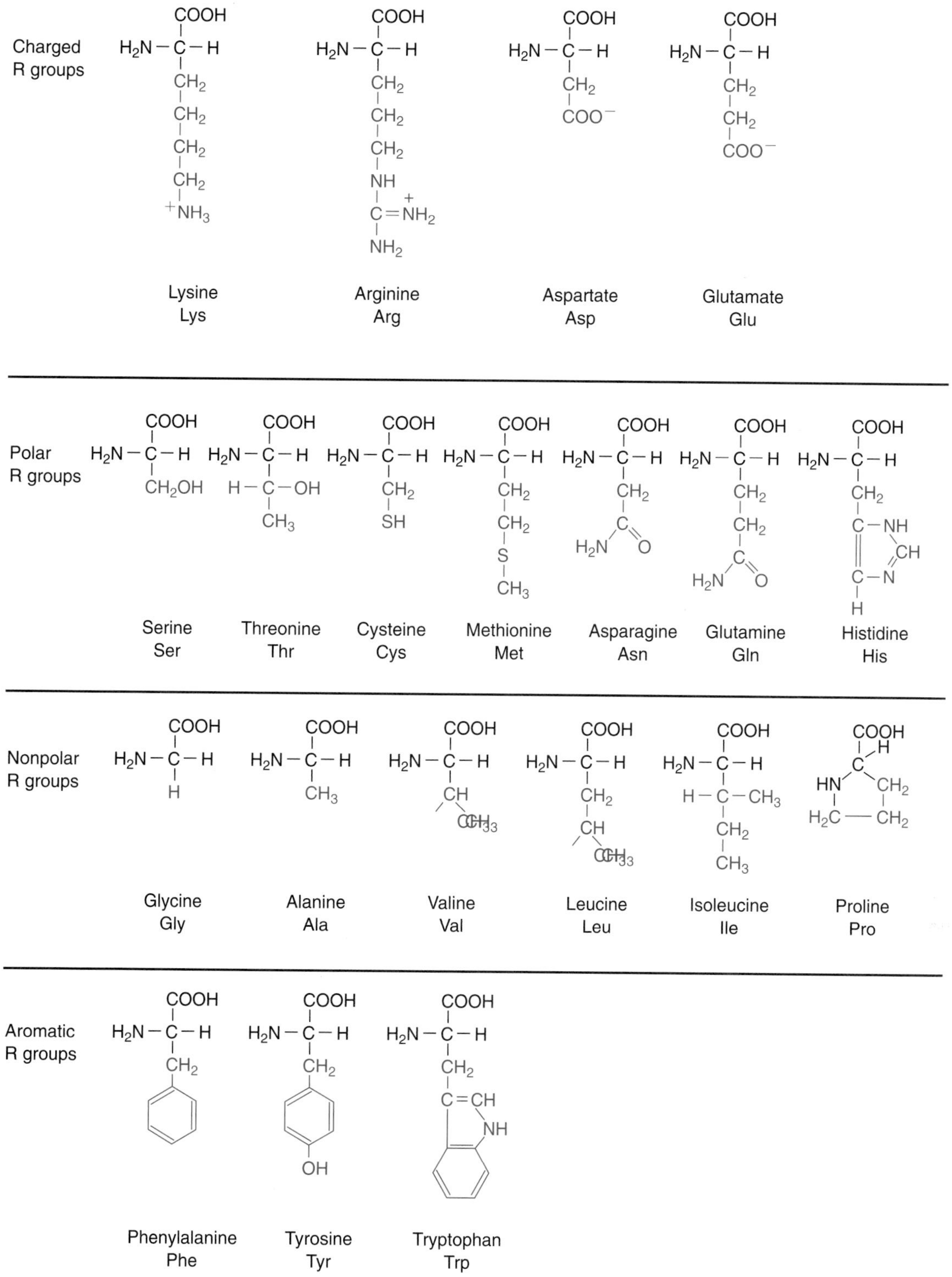

Figure 4.23 The amino acids commonly found in proteins. The three-letter abbreviation for each amino acid is shown beneath its full name.

charges. The complementary charges neutralize each other. Nonpolar molecules do not associate with charged or polar molecules in an energetically favorable manner. Instead, they associate comfortably with other nonpolar molecules.

We stated above that the single property of amino acids that most determines the three-dimensional protein structure is their ability to interact stably with water. Water is so important because the intracellular environment is water based, as are other body fluids. How stably an amino acid side chain interacts with water therefore determines how "comfortable" it is when exposed to the intracellular fluid.

Here is the bottom line about water. Water consists of two polar oxygen-hydrogen bonds and is a very polar molecule (Figure 4.24). It thus associates comfortably with other polar or charged molecules, and these molecules are therefore called hydrophilic. Since nonpolar molecules do not associate comfortably with water, they are called hydrophobic. In chapter 2, we discussed hydrophilic and hydrophobic molecules in a general way, using sugar, salt, and oil as examples. Sugars are polar molecules; they contain many OH bonds in which electrons are unequally shared. Salts are composed of charged ions. In water, the partial charges on sugar molecules and the full charges of the salt ions are neutralized through association with, and simultaneously neutralize, the complementary partial charges of water molecules. Hence, sugar and salt molecules can dissolve, or become evenly dispersed throughout water.

In contrast, oil, like other hydrophobic molecules, is composed of electrically neutral bonds (in the case of oil, CH). Oil molecules do not neutralize the partial charges of water molecules; water molecules in the presence of oil must stay together and self-neutralize. Hence, water and oil segregate and form separate compartments in solution.

Hydrophobic and hydrophilic amino acid side chains follow the same rules of molecular interactions as do other substances. Hydrophobic amino acid side chains (the nonpolar ones in Figure 4.23) do not associate stably with the intracellular fluid. Hydrophilic

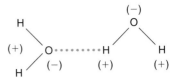

Figure 4.25 A hydrogen bond (dotted line) is a weak electrostatic attraction between opposite partial charges.

amino acid side chains (the charged and polar ones in Figure 4.23) do associate stably, because their charges or partial charges can be neutralized by complementary partial charges of polar water molecules.

Hydrogen Bonds

One type of neutralization that is particularly important in considering protein structure is the hydrogen bond, the same kind of bond found between base pairs in DNA (Figure 4.3). Hydrogen bonds are not covalent bonds. They are much weaker and form when two highly electronegative nuclei "share" a hydrogen atom that is formally bonded to only one of them. The partial positive charge on the bonded hydrogen neutralizes the partial negative charge on the second electronegative nucleus (Figure 4.25). Hydrogen bonds form only between the three most electronegative elements: oxygen, nitrogen, and fluorine. Of these three elements, only oxygen and nitrogen are common in biological systems, so you can forget about fluorine when thinking about protein structure. In proteins, the most important groups involved in hydrogen bonding are N, NH, O, OH, and CO groups. Water can form hydrogen bonds with all of them (Figure 4.26). Look at the hydrogen bonds in DNA in Figure 4.3, and you will see the same groups.

Figure 4.26 Common hydrogen bonds in biological systems.

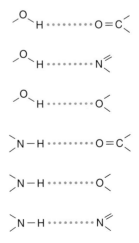

Figure 4.24 Water is a very polar molecule. The strongly electronegative oxygen nucleus hogs the electrons it shares with the hydrogen nuclei.

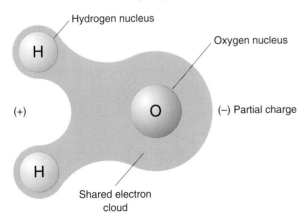

The fundamental consideration of protein structure

Now that we have looked at the features of covalent bonds that are important for understanding protein structure, let's see what it all boils down to. A protein molecule is a long peptide backbone with a mixture of charged, polar, and nonpolar amino acid side chains. Cytoplasm is a watery environment, so the charged and polar side chains will be stabilized through interactions with water molecules there. However, the hydrophobic amino acid side chains do not associate stably with water; they are more stable when clustered together away from water.

It appears that the basic rule underlying protein structure is, as much as possible, to fold up hydrophobic amino acid side chains together in the interior of the protein, creating a water-free hydrophobic environment. Hydrophilic side chains, meanwhile, are stable when exposed to the cytoplasm on the surface of the protein molecule. This is not to say that you would never find a hydrophilic amino acid on the interior of a protein or a hydrophobic one on the surface, but in general, the rule holds good. A protein is therefore said to have a hydrophobic core. The three-dimensional structure of each individual protein can be thought of as a solution to the problem of creating a stable hydrophobic core, given that protein's primary structure. However, every protein structure must solve one common problem. As you might guess, solving that problem forms another theme in protein structure.

Taking Care of the Hydrophilic Backbone

There is one major problem in folding a protein to create a hydrophobic core: the backbone. Look at Figure 4.21. The peptide backbone is full of NH and CO bonds, and both kinds of bonds are highly polar. On the surface of a protein, these partially charged bonds can be readily neutralized through hydrogen bonding with water. However, for a protein structure to be stable, the partial charges of the peptide backbone must also be neutralized inside the protein core, where there is no water. The solution to this problem is a major factor in determining protein structure.

The fundamental solution to the problem of the peptide backbone in the hydrophobic interior is for the backbone to neutralize its own partial charges. The NH groups can form hydrogen bonds with the CO groups (Figure 4.26), neutralizing both. Since every amino acid contributes one NH group and one CO group to the backbone, this solution is very convenient. However, because of geometric constraints, the NH and CO groups from the same amino acid are not in position to form hydrogen bonds with one another. Instead, the peptide backbone must be carefully arranged so that the NH and CO groups along it are in position to form hydrogen bonds with complementary groups elsewhere along the backbone. Two basic arrangements work well, and these two arrangements form major components of protein structure.

The first self-neutralization arrangement for the peptide backbone is for the backbone to form a helical coil, as if it were winding around a pole. The amino acid side chains point outward, away from the imaginary pole. The NH and CO groups along the backbone form hydrogen bonds with complementary groups above or below them on the pole, as shown in Figure 4.27. This arrangement is called an **alpha helix.**

In the second self-neutralization arrangement, stretches of the peptide backbone lie side by side, so that a CO group on one backbone can form a hydrogen bond with an NH group on the adjacent backbone (Figure 4.28). The amino acid side chains point alternately above and below the plane of the backbone. This arrangement is called a **beta sheet,** and the individual stretches of backbone involved in the sheet are called beta strands. Beta sheets are usually not flat, but twisted.

Figure 4.27 The alpha helix. C_α indicates the carbon atoms with side chains, which are not shown.

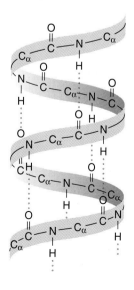

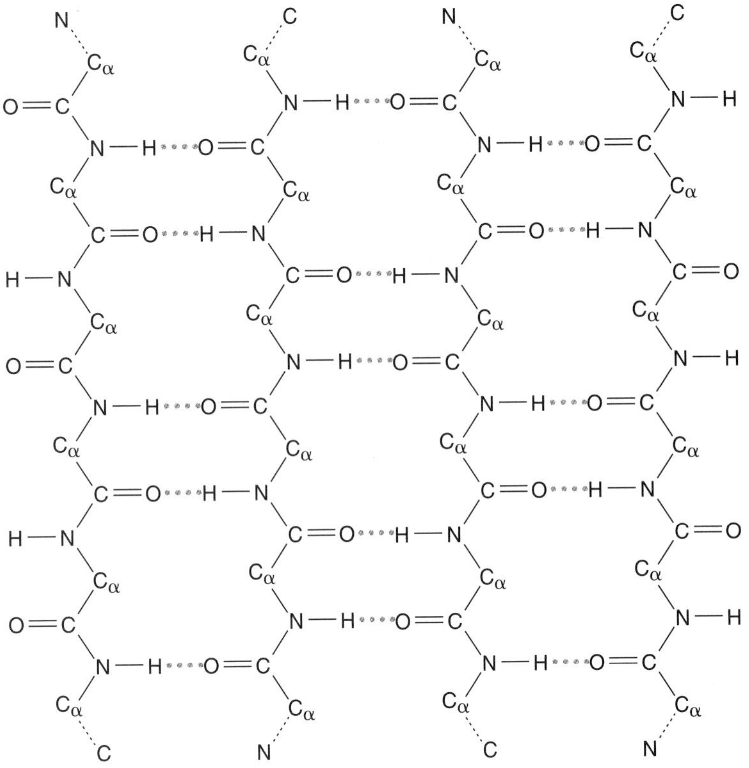

Figure 4.28 A beta sheet. C_α indicates the carbon atoms with side chains, which are not shown.

Secondary structure

Within a protein molecule, particular stretches of the amino acid chain may assume an alpha-helix or beta-sheet conformation. Certain amino acid sequences favor the formation of each one, although we cannot predict these structures from the primary structure with perfect accuracy. The regions of alpha helices and beta sheets within a protein are referred to as **secondary structure.**

Thus, within the hydrophobic core of a protein, some segments of the backbone may be found in the alpha-helix conformation while other segments may be arranged as beta sheets. (Some proteins are formed entirely from alpha helices; others, entirely from beta strands.) These secondary structures are often connected to one another via stretches of amino acids on the surface of the protein, where the partially charged backbone does not need to assume a particular secondary structure because it is neutralized by water in the cellular environment. The active sites of enzymes often involve these unorganized loops of amino acids, probably because the loops are freer to change conformation to bind a substrate.

Drawing Proteins

If you look at a picture of a protein in which all the atoms are shown, you cannot tell much (Figure 4.29).

Figure 4.29 In this drawing of the replication termination protein of *E. coli,* each sphere represents an atom. Even though this protein is small, its structure is complex. (Drawing courtesy of Stephen White, in whose laboratory the structure was determined.)

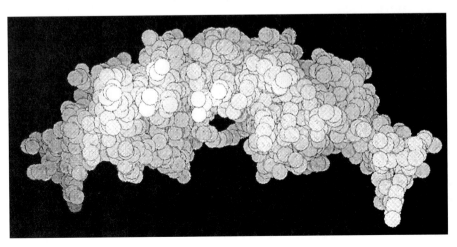

The picture contains too much information, too many atoms to tell what the big picture is. For purposes of understanding the overall structural plan, it is most helpful to look at just the configuration of the backbone. One popular way of showing this configuration is to draw the backbone as a ribbon, with alpha helices coiled and beta strands indicated with arrowheads. The arrows point toward the C terminus of the amino acid chain. Free loops are uncoiled regions of ribbon without arrowheads. Examples of some protein structures drawn in this manner are shown in Figure 4.30.

Look closely at the drawings in Figure 4.30. Plastocyanin (panel A) is composed of beta strands connected by loops. The center part of flavodoxin (panel B) is a twisted beta sheet. The strands of the beta sheet are connected by regions of alpha helix. The center of triose phosphate isomerase (panel C) also consists of beta strands, but they are arranged somewhat differently. Try using your finger to follow the peptide backbones of all three structures from N to C. You can see that adjacent strands in a beta sheet do not have to come from contiguous stretches of the backbone but may be segments that are widely removed in the primary structure.

You can also see from the flavodoxin and triose phosphate isomerase structures that alpha helices are apparently at the surfaces of these proteins. This arrangement is fairly common because of a handy property of alpha helices. Since the amino acid side chains stick out from the center of the imaginary barber pole, one side of the pole can have mostly hydrophilic side chains while the other has mostly hydrophobic ones. With such an arrangement, the helix can sit comfortably at the protein surface, its hydrophobic side buried in the protein core and its hydrophilic side exposed to the cellular fluid. Neat, isn't it?

Structural Motifs

Simple combinations of a few secondary-structure elements occur frequently in proteins. Protein structure scientists call them motifs. You can think of a motif as a little module of protein structure; many proteins are put together by assembling combinations of these modular motifs. Of course, that statement makes it sound as if the structure is independent of the primary amino acid sequence, which it is not. Still, if you simply compare lots of protein structures, it does seem as though various structural motifs form a sort of "tool kit" for assembling larger proteins. A couple of examples of structural motifs are the helix-loop-helix, the beta turn, the beta-alpha-beta, and the beta barrel. These motifs are shown in Figure 4.31. There are many others.

A. Plastocyanin

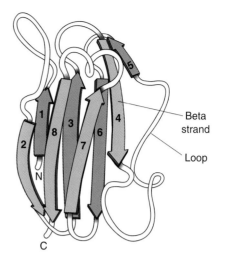

B. Flavodoxin

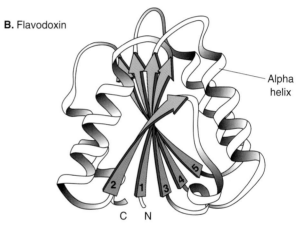

C. Triose phosphate isomerase

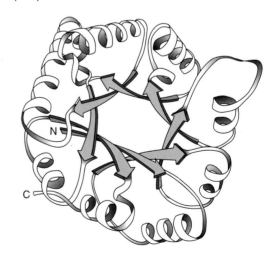

Figure 4.30 Ribbon drawings of protein structures. The beta strands in panels A and B are numbered in order from the N terminus to the C terminus of the amino acid chain. (Drawings courtesy of Jane Richardson.)

A. Helix-loop-helix

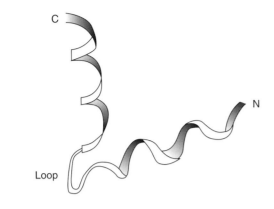

C

N

Loop

B. Beta turn

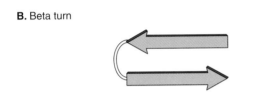

C. Beta-alpha-beta

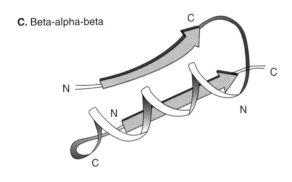

C

N

C

N

N

C

D. Beta barrel

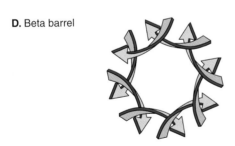

Figure 4.31 Some protein structure motifs. (Panels A and C are from C. Branden and J. Tooze, *Introduction to Protein Structure,* Garland Publishing, Inc., New York, NY, 1991; panels B and D are from A. Lehninger, D. Nelson, and M. Cox, *Principles of Biochemistry,* 2nd ed., Worth Publishers, Inc., New York, NY, 1993.)

Look at Figure 4.30. You can find three of the structural motifs in Figure 4.31 within the proteins of Figure 4.30. Plastocyanin has many beta-turn motifs. Both flavodoxin and triose phosphate isomerase contain many beta-alpha-beta motifs (trace the

backbones; you will discover there is a helix between each pair of beta strands). The center of triose phosphate isomerase is a beta barrel. Some structural motifs are associated with functional activities, though most are not. For example, one variety of helix-loop-helix (also called helix-turn-helix) is a DNA-binding motif that occurs in a variety of DNA-binding proteins. When these proteins bind their target DNA sequences, the helices of this motif sit next to the DNA, and the side chains on one of the helices reach into the major groove and make hydrogen bonds with the edges of specific bases there. (For an example, see *Structure and function* below.)

Domains

Several secondary-structure motifs usually combine to form a stable, compact, three-dimensional structure called a domain. Small proteins may consist of a single domain (such as those pictured in Figure 4.30); larger proteins may fold into several separate domains. The structures of individual domains within a protein and the way multiple domains fit together are called the **tertiary structure** of the protein.

Domains appear to be fundamental units of protein structure and function. Domains are usually formed from continuous stretches of amino acids and therefore are translated from continuous regions of mRNA. In multifunctional proteins, it is not uncommon to find that the protein folds into several domains and that each domain is associated with one function. Sometimes it is possible to separate the domains of a protein (through brief enzymatic digestion), and we sometimes find that the separate domains retain their individual functions.

An example of a bifunctional protein with two domains is the repressor protein of the bacterial virus lambda. The lambda repressor folds into two domains; the first 92 amino acids fold into the N-terminal domain, and amino acids 132 through 236 fold into the C-terminal domain. The other 40 amino acids connect the two domains (Figure 4.32A). Each domain of the lambda repressor has a separate function.

The two functions of the lambda repressor are to bind to the correct operator DNA and to form dimers by binding to a second molecule of itself. A dimeric protein is a stable association of two copies of the same polypeptide chain (see *Quaternary structure* below). Each of the chains is called a monomer. In the lambda repressor, the folded C-terminal domain of one monomer binds to the

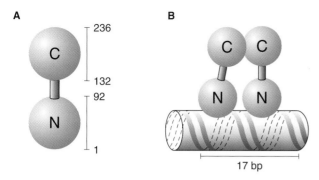

Figure 4.32 Domain structure of the bacteriophage lambda repressor protein. (A) The N-terminal domain consists of amino acids 1 through 92, and the C terminal domain consists of residues 132 through 236. (B) The repressor forms dimers through the interaction of the C-terminal domains. The N-terminal domains bind to a specific DNA sequence. (From M. Ptashne, *A Genetic Switch,* 3rd ed., Cold Spring Harbor Laboratory Press, Cold Spring Harbor, NY, 2004.)

C-terminal domain of a second repressor monomer to create the dimer. The N-terminal domains bind to DNA via a helix-turn-helix motif (Figure 4.32B). The N- and C-terminal domains can be separated by digesting the 40-amino-acid linker region, which is more vulnerable to digestion because it is not folded tightly into a tertiary structure. After separation, the N-terminal domain still binds its DNA recognition sequence but cannot dimerize, and the C-terminal domain can dimerize but cannot bind DNA. Not all domains retain a function when they are separated, but many do.

Quaternary structure

Many functional proteins consist of a single amino acid chain, but many contain more than one polypeptide chain. These chains can be multiple copies of the same chain, as in the dimeric lambda repressor described above. They can also be assemblies of different polypeptides; *E. coli* RNA polymerase contains five different chains encoded by five different genes. The identities and number of the polypeptide chains and how they fit together in the final protein are called the protein's **quaternary structure.**

Do polypeptide chains automatically fold into the correct secondary and tertiary structures as they are synthesized and then associate with the correct additional polypeptides into quaternary structures? Many of them do. However, the folding of some proteins is assisted by other proteins that have been named chaperone proteins. Some molecular chaperones seem to act by increasing the rate of final folding; others actually guide the folding itself, as well as the assembly of multiple polypeptide chains into complex quaternary structures.

Before we go on with our consideration of protein structure, stop and think about what protein folding might mean to a biotechnologist who wishes to move the gene(s) for a protein into a completely different type of cell and then obtain a functional protein. Examples of this kind of operation would include moving a eukaryotic gene or genes into a prokaryotic system or moving an animal gene or genes into a plant system. Substituting the correct DNA traffic signals so that the new cell can transcribe and translate the DNA may be only a small part of the battle. Getting the amino acid chain(s) to fold correctly may be much more challenging. Moving a gene from one system to another does not guarantee the production of an active protein.

Stability of protein structure

Three-dimensional protein structures are held together largely through the relatively weak chemical interactions of hydrogen bonds and the favorable interactions of hydrophobic side chains in the interior. Anything that disrupts these weak interactions—heat, extremes of pH, organic solvents, or detergents—can alter the folding of the protein. The most extreme form of alteration is the complete unfolding of the amino acid chain, a process called **denaturation.** Denaturation of a protein is often irreversible. You can observe denaturation by frying an egg. The egg white protein albumin is soluble in its native state. As you heat it, the albumin denatures and coagulates, forming a white solid. Cooling the cooked egg does not reverse the process.

Proteins vary in stability. One way to quantify stability is to measure the temperature at which a given protein denatures. This is sometimes called the melting temperature (T_m) of the protein. Some proteins require much higher temperatures to unfold than others do. For example, the enzymes of organisms that inhabit hot springs and ocean thermal vents are stable at very high temperatures. No one single thing makes these enzymes more thermostable. It appears that many different aspects of their primary and tertiary structures contribute to their heat resistance.

One feature of protein structure, however, makes a significant contribution to stability. This feature exploits the special properties of one amino acid, cysteine. In an oxidative environment, two properly positioned cysteine residues can react with each other to make a disulfide bridge (Figure 4.33A). The intracellular environment is not oxidative, so disulfide bridges are rare inside cells, but many extracellular proteins contain them. Disulfide bridges anchor regions of the protein in a specific configuration, stabilizing the structure (Figure 4.33B). Disulfide bridges can form between distant portions of the same domain or between different polypeptide chains within a quaternary structure.

Similar domains are found in different proteins

We said previously that domains appear to be fundamental units of protein structure and function. Scientists believe this statement because similar domains appear in different proteins, sometimes many different proteins. Some of the domains appear to be mostly structural; others are connected with specific functions. For example, a DNA-binding domain called the homeodomain (rhymes with Romeodomain) is found in a large number of transcriptional activator proteins that interact with a specific type of enhancer sequence. (See *Transcriptional Regulation through Activation* above.) These proteins activate different sets of genes and are found in a diverse set of organisms, including worms, fruit flies, and humans. Even so, the proteins all bind to DNA via their homeodomains. The amino acid sequences of the homeodomains are very similar in these proteins.

Another example of a functional domain found in many different proteins is the domain that binds the enzyme cofactor nicotinamide adenine dinucleotide (NAD). A number of enzymes use NAD as a cofactor in oxidation-reduction reactions. Each of these enzymes has a domain that binds NAD, and the structure of that domain is practically identical from enzyme to enzyme. In fact, many of these enzymes have two domains, the common NAD-binding domain and a second, unique domain containing the active site at which the substrate binds. Surprisingly, the amino acid sequences of NAD-binding domains vary, but the structure is almost perfectly conserved from enzyme to enzyme.

A repeated domain with no known biochemical activity is the kringle. It consists of about 85 amino acid residues folded into a shape that reminded some scientists of a certain Danish pastry, the kringle; thus, it was named. Kringle domains are found in a variety of proteins.

Modular Proteins

You may be wondering now if it is possible to put together proteins by connecting domains like modules or tinker toys. The answer is yes. Although not all proteins are made this way, many are, with perhaps a few unique regions thrown in. Figure 4.34 shows some modular protein structures. These modular structures suggest that many genes did not evolve from scratch. Rather, it looks as if many genes

Figure 4.33 Disulfide bridges stabilize protein structure. (Panel B is from C. Branden and J. Tooze, *Introduction to Protein Structure,* Garland Publishing, Inc., New York, NY, 1991.)

A. Two cysteine side chains can form a disulfide bridge.

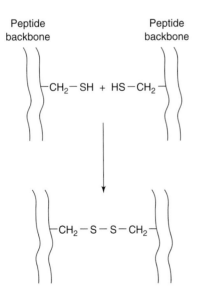

B. A disulfide bridge stabilizes the structure of this domain of an immunoglobulin protein (antibody).

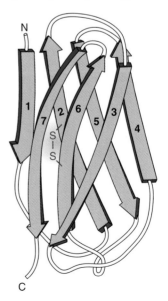

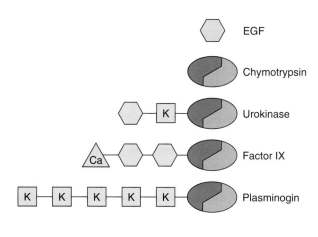

 Domains that are homologous to EGF, which is a small polypeptide chain of 53 amino acids.

 Serine proteinase domains that are homologous to chymotrypsin, which has about 245 amino acids arranged in two domains.

 Kringle domains that have a characteristic pattern of three internal disulfide bridges within a region of about 85 amino acid residues.

 Calcium-binding domain

Figure 4.34 Domain structures of some modular proteins. Epidermal growth factor (EGF) is a protein that signals several cell types to divide. The other four proteins are protein-cleaving enzymes with a variety of physiological roles. (From C. Branden and J. Tooze, *Introduction to Protein Structure,* Garland Publishing, Inc., New York, NY, 1991.)

were patched together from pieces or copies of pre-existing genes, resulting in proteins that contain domains common to many other proteins. It would be esthetically pleasing if protein domains were encoded by exons, with introns providing the bridges between them. For some domains, such as the epidermal growth factor domain shown in Figure 4.34, this seems to be true. The domain is encoded by a single exon. Unfortunately, the gene segments encoding more domains are distributed among several exons, with no apparent pattern to the exon-intron structure. Thus, the modular structures of proteins are not always related in a logically obvious (to us at this time) way to the modules of coding sequence in eukaryotic genes.

The modular structures of some proteins have implications for biotechnology. Nature has produced proteins with many different functions by joining similar domains in different ways over the course of evolution. Using recombinant DNA technology (described in chapter 5), we can now swap domains, too, by swapping portions of genes. For example, the

portion of the lambda repressor gene encoding the first 92 amino acids (the DNA-binding domain) can be replaced with a similar domain from a different repressor protein. The hybrid protein works as a repressor, and it recognizes the DNA-binding site of the second protein.

Structure and function

Protein structure is an interesting topic in and of itself, but it is so important because protein function depends on it. Let's look at a few specific examples: a structural protein, two DNA-binding proteins, and an enzyme.

Keratin: A Structural Protein

The **keratins** are a family of similar proteins that make up hair, wool, feathers, nails, claws, scales, hooves, and horns and are part of the skin. Keratin fibers also form part of the cytoskeleton. To fulfill their function, these proteins must be very strong. In addition, they must not be soluble in water (it would be quite unhandy if your hooves dissolved), even though most proteins are. Let's look at how the structure of keratin makes strong, water-insoluble fibers possible.

The amino acid chain of keratin folds into one long alpha helix. Almost all of its amino acids—alanine, isoleucine, valine, methionine, and phenylalanine—are hydrophobic, and they extend outward from the helical backbone. The presence of all these hydrophobic side chains everywhere violates the general rule that hydrophobic side chains must be buried in the protein's core. As you probably expect, this rule violation is important: it means that keratin is not energetically comfortable surrounded by water molecules and therefore is not soluble in water. Imagine, the reason your fingernails do not dissolve when you wash them is all those hydrophobic side chains.

Instead of associating with the watery environment inside cells, keratin molecules associate with each other in large groups. Their structure, a single long helix, lends itself to forming fibers. First, two alpha helices of keratin wind around each other. Their hydrophobic surface side chains interact favorably in this conformation. These two-chain coils lie end to end and side by side with from one to many other coils, forming fibers (Figure 4.35).

These fibers are not only held together by favorable hydrophobic interactions between side chains, but are also stabilized by disulfide bridges between the coils. The bridges make the fibers strong and rigid.

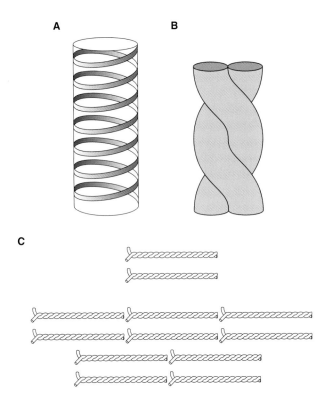

A   B

C

Figure 4.35 Keratin, a structural protein. (A) A single keratin molecule forms a long alpha helix. (B) Two keratin alpha helices then wrap around each other. (From J. D. Watson et al., *Molecular Biology of the Gene,* 4th ed., vol. 1, Benjamin/Cummings, Menlo Park, CA, 1987. Reprinted by permission of Addison Wesley Longman Publishers, Inc.) (C) Two-chain coils lie end to end and side by side, forming fibers. (From A. Lehninger, D. Nelson, and M. Cox, *Principles of Biochemistry,* 2nd ed., Worth Publishers, Inc., New York, NY, 1993.)

Different keratin proteins have different amounts of cysteine to make the bridges: the harder the final structure (hooves versus hair, for example), the more disulfide bridges are present. In the toughest keratins, such as tortoise shells, up to 18% of the amino acids are cysteines involved in disulfide bridges.

If you have ever had a permanent wave in your hair, you have manipulated the disulfide bridges of your keratin hair fibers. Recall that the disulfide bridges form only in the right kind of chemical environment (an oxidizing environment; see above). With the right chemicals, an oxidizing environment can be changed to its opposite (a reducing environment), causing the disulfide bridges to break apart.

When you get a permanent, the hair stylist first wraps your hair around small rods. Next, a smelly solution is applied to your hair. The smelly solution contains a reducing agent, a chemical that changes the oxidizing environment in your hair and breaks the disulfide bridges holding the keratin fibers side by side (the reducing agent is the component of the solution with the strong odor, too). While this is going on, the hair stylist has also arranged for your hair to be warm, either by placing you under a hair dryer or by putting a plastic bag over your hair. The moist heat breaks some hydrogen bonds that keep the keratin alpha helices stiff, allowing them to relax a little. The net effect is that the keratin helices move a little with respect to one another while they are wound around the rods.

After your hair has incubated sufficiently in the warm reducing environment, the stylist rinses out the reducing agent and applies a neutralizing solution. This solution restores the oxidizing environment, allowing disulfide bridges to re-form. But here's the catch. Your hair has been relaxed around the curling rods, and many of the cysteine SH groups will form disulfide bridges with new cysteine SH groups. These brand-new SH bonds hold the hair fibers in the conformation they were in around the curling rods: a permanent wave (Figure 4.36). Rinsing and cooling your hair allow the keratin helix hydrogen bonds to reestablish themselves, returning your hair to normal, except that new disulfide bridges now hold it in a wavy shape.

Lambda and *trp* Repressors: DNA-Binding Proteins

The bacteriophage lambda repressor protein binds to a specific operator DNA sequence in its bacteriophage genome and prevents RNA polymerase from transcribing certain genes, as do the bacterial *lac* and *trp* repressor proteins. As described above, a single lambda repressor polypeptide chain folds into two domains (Figure 4.32). The C-terminal domains of two polypeptides bind to one another, creating a dimeric protein.

The N-terminal domains of the lambda repressor bind to operator DNA sequence via a helix-turn-helix motif. The lambda operator DNA sequence is symmetrical; each N-terminal domain of repressor interacts with identical bases. Figure 4.37A shows how the helix-turn-helix motifs of the two N-terminal domains sit on the operator DNA. The amino acid side chains of helix 3 (part of the helix-turn-helix motif) contact specific bases within the DNA binding sequence (Figure 4.37B). When the protein binds to DNA, helix 3 sits along the DNA molecule so that these specific contacts can occur. Figure 4.37C shows one of these amino acid-base contacts in detail.

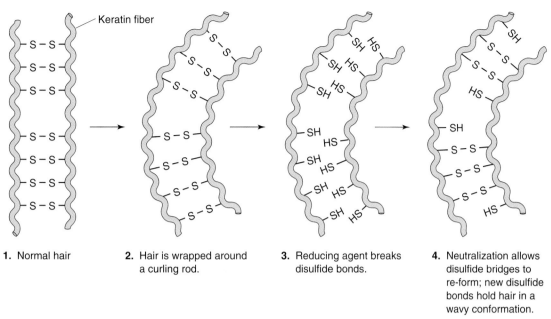

1. **Normal hair**

2. **Hair is wrapped around a curling rod.**

3. **Reducing agent breaks disulfide bonds.**

4. **Neutralization allows disulfide bridges to re-form; new disulfide bonds hold hair in a wavy conformation.**

Figure 4.36 The biochemistry of a permanent hair wave.

From this example, you can see that both the overall structure of the lambda repressor and its specific amino acid sequence are important to its function. The helix-turn-helix motif is oriented within the protein so that helix 3 can sit alongside the DNA molecule. The specific amino acids within helix 3 must contact specific bases for the binding to work, so the protein binds only to its recognition sequence. (For more information, see the works by Ptashne in *Selected Readings*.)

Now, let's revisit a protein we met earlier in this chapter, the repressor of the tryptophan operon, to see how a DNA-binding protein itself can be regulated by a second molecule. Recall that the *trp* repressor protein can bind both to the amino acid tryptophan and to a specific DNA base sequence, the *trp* operator sequence, but that the repressor binds to the operator DNA only when it is also binding to tryptophan (Figure 4.16).

Like the lambda repressor, the *trp* repressor is a dimeric protein that binds to DNA via a helix-turn-helix motif. The amino acid sequence of the DNA-binding helix is different, so it binds a different sequence of bases. Unlike the lambda repressor, the *trp* repressor's structure does not position the DNA-binding helix correctly for interacting with DNA.

Each of the *trp* repressor monomers has the helix-turn-helix motif; however, the two monomer chains interact in such a way that the DNA-binding helices are folded in toward the main body of the protein and are not positioned to fit alongside the DNA molecule (rather like the folded-in claws of a crab). Here is where tryptophan comes in. The repressor protein interacts with two molecules of tryptophan (one per monomer). The tryptophan molecules fit into the structure like wedges between the DNA-binding helices and the body of the protein, forcing the DNA-binding helices to swing out (imagine the crab's claws unfolding outward from its body) (Figure 4.38). Now the helices are positioned to bind their DNA recognition sequences. The tryptophan-repressor complex binds the *trp* operator and blocks further transcription of the *trp* operon.

Chymotrypsin: An Enzyme

Since enzymes are proteins that cause a chemical reaction, any discussion of their structure and function has to include the chemical reaction they promote. Chymotrypsin is a digestive enzyme that breaks down other proteins; it is called a proteinase (or protease [pronounced PRO-tee-ace]). Proteinases cleave the peptide bonds in protein backbones.

Chymotrypsin belongs to the family of proteinases called serine proteinases. They are called serine proteinases because a serine side chain participates in the actual cleavage of the substrate molecule (serine is pictured in Figure 4.23). The digestive enzyme trypsin is also a serine proteinase; the domain structures of some other serine proteinases are shown in Figure 4.34.

A. The orientation of helix-turn-helix DNA-binding regions on operator DNA

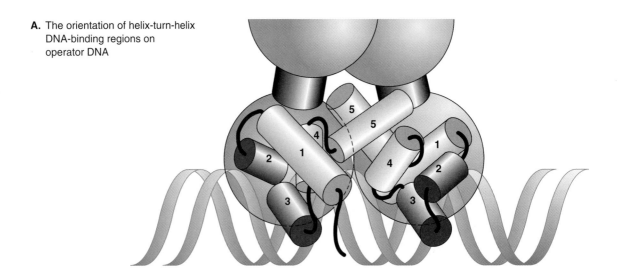

B. The amino acid sequence of helix 3. The specific bases in the operator sequence that are contacted by amino acid side chains are indicated by arrows.

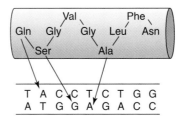

C. The contact between the first glutamine side chain in helix 3 and its target base

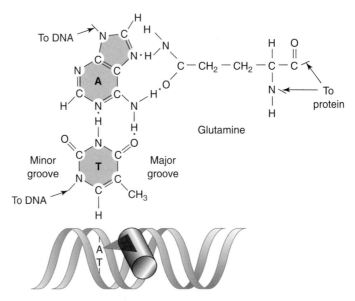

Figure 4.37 Binding of lambda repressor protein to DNA. (From M. Ptashne, *A Genetic Switch,* 3rd ed., Cold Spring Harbor Laboratory Press, Cold Spring Harbor, NY, 2004.)

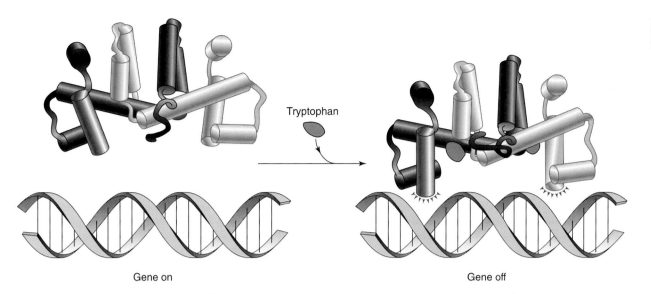

Tryptophan

Gene on   Gene off

Figure 4.38 Binding of the amino acid tryptophan to the *trp* repressor protein changes the conformation of the repressor so that it can bind to DNA. (Reprinted by permission from *Nature* 327:591–597, 1987.)

Chymotrypsin cleaves one peptide bond at a time, cutting a single polypeptide chain into two shorter ones (Figure 4.39A; see Figure 4.21 to review peptide bonds and polypeptides). The cleavage reaction occurs in two steps (Figure 4.39B). First, the peptide bond is cut, freeing one half of the original molecule but leaving the other half covalently bonded to the catalytic serine side chain. In the second step of the reaction, the bound substrate chain is released, restoring the enzyme to its original form. Both of these steps require a basic amino acid side chain in the proper position to hold onto a hydrogen. In chymotrypsin, the basic side chain is a histidine (Figure 4.23). A close-up of the first step of the reaction is shown in Figure 4.39C.

The structure of the serine proteinases positions the catalytic serine and the basic side chain so that the cleavage reaction can occur. A schematic of the structure of chymotrypsin is shown in Figure 4.39D. The protein has two domains, and the serine and histidine are each in a different one. They are located within loop regions where the domains fit together. Other amino acids in the loop help bind the polypeptide substrate in the proper position for cleavage, and some specific ones, which vary from proteinase to proteinase, confer substrate specificity. Chymotrypsin, for example, prefers to cleave peptide bonds adjacent to aromatic amino acid side chains, while trypsin prefers positively charged side chains. All these features of serine proteinase activity are understood in detail. We decided it was beyond the scope of this book to describe them all, and instead, we refer you to the excellent book by

Branden and Tooze that is listed in *Selected Readings,* where you will find a more complete discussion.

Effects of mutations on protein structure and function

Earlier in this chapter, we defined a mutation as any change in a DNA sequence and stated that the effects of a mutation depended on what, if any, effect it had on the production or functions of proteins. In this discussion, we focus on some specific effects of amino acid changes on proteins.

From the examples given previously, you have probably realized that the function of a protein really depends both on its three-dimensional structure and, often, on specific amino acids within that structure. For example, there are many helix-turn-helix DNA-binding proteins. For them to bind DNA, the helix-turn-helix structure has to be maintained. However, the base sequence these proteins bind to depends on the identities of the amino acids within the DNA-binding helix. In addition, many proteins have multiple jobs to do, and these jobs are usually carried out by different regions of the protein, often different domains (recall the lambda repressor protein described above: one domain binds DNA, and the other forms dimers). Thus, the effect of any amino acid change depends on what, if anything, that change does to critical protein structures, whether that particular amino acid is specifically involved in a function, and how that function or structure relates to the rest of what the protein does.

A. Overall reaction catalyzed by the enzyme. A and B represent the rest of the protein molecule on either side of the peptide bond to be cleaved (see Fig. 4.21).

$$A{-}\overset{\overset{\displaystyle O}{\|}}{C}{-}\overset{\overset{\displaystyle H}{|}}{N}{-}B + H_2O \longrightarrow A{-}\overset{\overset{\displaystyle O}{\|}}{C}{-}OH + H{-}\overset{\overset{\displaystyle H}{|}}{N}{-}B$$

B. The overall reaction proceeds in two steps. E—OH represents the chymotrypsin enzyme with —OH on the catalytic serine side chain (see Fig. 4.23).

Step 1. $$E{-}OH + A{-}\overset{\overset{\displaystyle O}{\|}}{C}{-}\overset{\overset{\displaystyle H}{|}}{N}{-}B \longrightarrow E{-}O{-}\overset{\overset{\displaystyle O}{\|}}{C}{-}A + H{-}\overset{\overset{\displaystyle H}{|}}{N}{-}B$$

Step 2. $$E{-}O{-}\overset{\overset{\displaystyle O}{\|}}{C}{-}A + H_2O \longrightarrow E{-}OH + HO{-}\overset{\overset{\displaystyle O}{\|}}{C}{-}A$$

C. Close-up of step 1, showing the role of the catalytic serine side chain and its histidine helper.

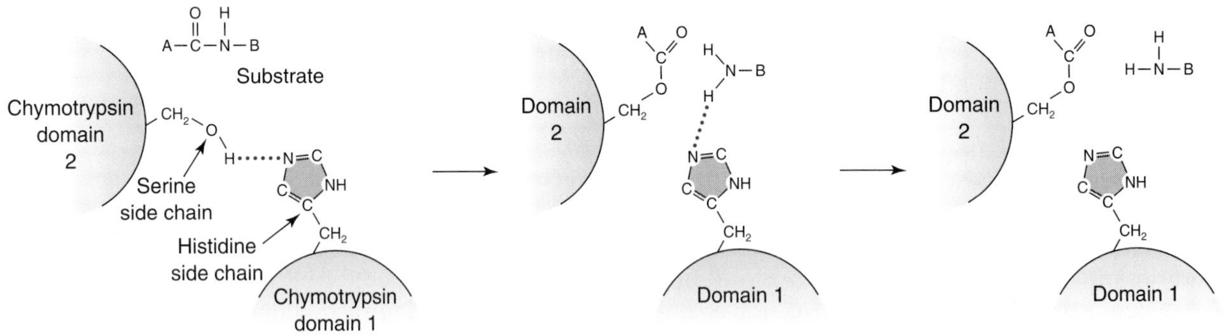

D. The domain structure of chymotrypsin, showing the positions of the serine (S195) and histidine (H57) side chains.

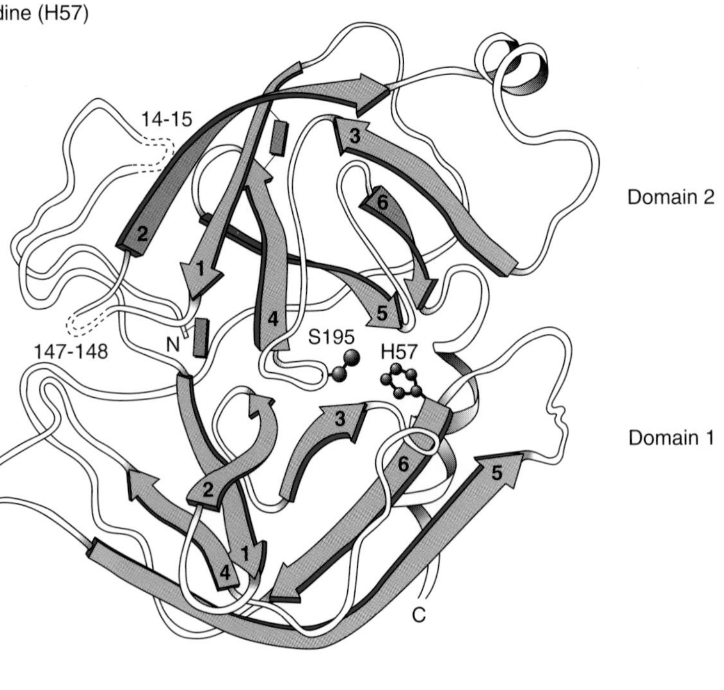

The accumulating data about DNA and protein sequences indicate that protein structures and functions can often be maintained through many amino acid changes. The keratin proteins are a good illustration. Their specific amino acid sequences vary, their cysteine contents vary, yet they are all recognizable, though different, versions of keratin. The genetic mutations that resulted in different versions of the protein did not destroy structure or function. Another example of this kind is hemoglobin, which has more than 300 known genetic variants in the human population alone. Most of these variants are single amino acid changes with only minor structural and functional effects. Protein structure, it seems, is reasonably robust in the face of many amino acid changes.

A Harmful Mutation

A specific example of a harmful mutation is found in the disease sickle-cell anemia. This disease is the result of 1 of the more than 300 variations in hemoglobin. In sickle-cell anemia, a single A-to-T mutation changes the sixth codon from GAG to GTG. This mutation changes the sixth amino acid in the 146-amino-acid protein from glutamic acid to valine. In the three-dimensional structure of hemoglobin, the sixth amino acid sits on the surface of the protein. Glutamic acid (the normal amino acid) is hydrophilic, so its side chain is stable when exposed to the intracellular fluid. Valine, however, is hydrophobic (Figure 4.23). Its side chain is more energetically stable when interacting with other hydrophobic molecules.

The problem with the valine side chain is not simply that it is hydrophobic. It does not really change the three-dimensional structure of hemoglobin for it to be on the surface, and the mutant hemoglobin can still bind oxygen. The problem is how the hydrophobic surface valine affects the way hemoglobin molecules interact with one another.

As it happens, the hydrophobic valine side chain on the surface just fits into a hydrophobic pocket that is exposed on the hemoglobin molecule when it is not bound to oxygen. The surface valine is not positioned to fit into its own pocket, but it can fit into the pocket on a second molecule. When the mutant hemoglobin molecules give up oxygen in the capillaries, the deoxygenated molecules fit together in a lock-and-key fashion. The surface valine fits into the

A. Normal hemoglobin

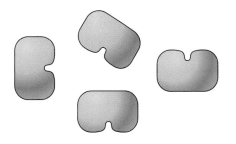

B. Sickle cell hemoglobin

Val-6

Figure 4.40 Representation of sickle-cell hemoglobin aggregation. (A) Normal hemoglobin molecules do not stick together. (B) The hydrophobic patch on the surface of sickle-cell hemoglobin caused by the glutamate-to-valine substitution at position 6 (Val-6) fits neatly into a hydrophobic pocket on a second molecule. Thus, sickle-cell hemoglobin molecules can polymerize in a head-to-tail fashion.

pocket of another molecule, which itself has a surface valine to fit into another molecule, and so on (Figure 4.40). The mutant hemoglobin molecules aggregate into long fibers, changing the red blood cells' shape to a sickle form and interfering with circulation of the cells through the capillaries. The impaired circulation gives rise to a number of deadly problems in the afflicted individual.

Thus, the fatal disease sickle-cell anemia is a consequence of hydrophobic interactions and altered protein structure. If the hydrophobic valine did not happen to fit into the hydrophobic pocket on the deoxygenated hemoglobin molecule, you would not get polymerization of the mutant hemoglobin, sickling of cells, and impaired circulation.

A Beneficial Mutation

An example of a beneficial mutation is the inherited condition benign erythrocytosis. Individuals with this condition have highly elevated red blood cell levels. Far from being ill, these individuals have greatly enhanced stamina. One such person, the Finnish

Figure 4.39 Mechanism of action of the proteinase chymotrypsin, an example of a serine protease. (Panel D is from C. Branden and J. Tooze, *Introduction to Protein Structure,* Garland Publishing, Inc., New York, NY, 1991.)

athlete Eero Mäntyranta, won three gold medals for cross-country skiing in the 1964 Winter Olympics. Scientists recently determined the molecular basis of benign erythrocytosis.

Red blood cells arise from progenitor cells called stem cells that are found in bone marrow. Stem cells are stimulated to mature into red blood cells by the hormone erythropoetin. The hormone communicates with the stem cell through a 550-amino-acid receptor protein embedded in the stem cell's outer membrane. The N-terminal portion of the protein lies outside the cell, forming a docking site for the hormone. When erythropoetin binds, the C-terminal portion of the receptor, positioned inside the cell, transmits the maturation signal. The C-terminal end of the receptor also contains a docking site for a cellular protein that prevents transmission of the maturation signal. Docking of this cellular protein thus acts as a molecular brake on red blood cell production (Figure 4.41A). The Finnish athlete and other members of his family carry a mutant version of the receptor gene. A G-to-A mutation changes codon 481 from TGG, for tryptophan, to TAG, a stop codon. This single base change causes the athlete's ribosomes to stop synthesis of the receptor protein 70 amino acids early.

Losing the 70 C-terminal amino acids does not disrupt the extracellular hormone-binding domain of the protein, its cell membrane-spanning domain, or the intracellular region responsible for transmitting the maturation signal. However, loss of those 70 amino acids removes the docking site for the braking protein. Thus, Eero Mäntyranta's red blood cell production has no molecular brakes, and he and other mutation-bearing family members have higher-than-normal levels of red blood cells (Figure 4.41B). Their blood can carry more oxygen than normal, so they have enhanced stamina. It is possible that their elevated red blood cell levels would be detrimental under some circumstances. High concentrations of red blood cells can cause the blood to thicken, leading to the possibility of heart attacks and strokes, particularly during hard exercise, and possibly death. Some reports state that people "afflicted" with benign erythrocytosis have normal life spans, while other reports claim that their average life spans are reduced.

Interestingly, the hormone erythropoetin is also a drug of abuse in sports. You may have heard it called "EPO." Athletes who cheat with EPO take the hormone to boost levels of red blood cells in their bodies and thereby achieve added stamina. In the late 1990s, several athletes were disqualified from the Tour de France bicycle race because of EPO use. This form of cheating can be detected because athletes who are using EPO have abnormally high levels of it in their systems, and those levels can be measured. Eero Mäntyranta did not have extra EPO in his body, but because his EPO receptor lacked molecular brakes, his body responded in a manner similar to the way most people would respond to added EPO, by producing additional red blood cells.

A Disease of Protein Structure?

In the early 1990s, an outbreak of so-called mad cow disease struck the British cattle industry. Symptoms of mad cow disease include erratic behavior and resemble symptoms of the sheep disease scrapie. In fact, the British mad cow disease appeared to have been transmitted to cattle through feed made in part from the remains of scrapie-infected sheep. Scrapie and mad cow disease are two of a group of diseases that affect the brain, including the human diseases Creutzfeldt-Jacob disease and kuru.

Using animal models, scientists were able to isolate an infectious agent that transmits scrapie. The agent passed through filters that would exclude bacteria, so it was assumed to be a virus. However, no one could detect either DNA or RNA in the scrapie agent, and it was unaffected by enzymes that destroy nucleic acids. However, the agent was neutralized by enzymes that degrade proteins. Researchers named the mysterious agent a prion. How could an agent that apparently contained no nucleic acid transmit a disease? Many scientists assumed the scrapie researchers must be in error.

Now we know that prion diseases are diseases of protein structure. Healthy individuals manufacture the prion protein, and it assumes its normal configuration. Under some circumstances, however, the prion protein can assume an alternative configuration. Once it is in this alternative configuration, it does not return to normal, and when the altered form of the protein comes into contact with the normal form of the protein, the normal form changes to the altered form. As the altered form becomes more and more prevalent, disease begins. Thus, the aberrant form of the protein can act as a disease agent.

Predicting three-dimensional protein structure

Predicting the three-dimensional structure of a protein from its amino acid sequence is a major unsolved problem in structural biology. Structural

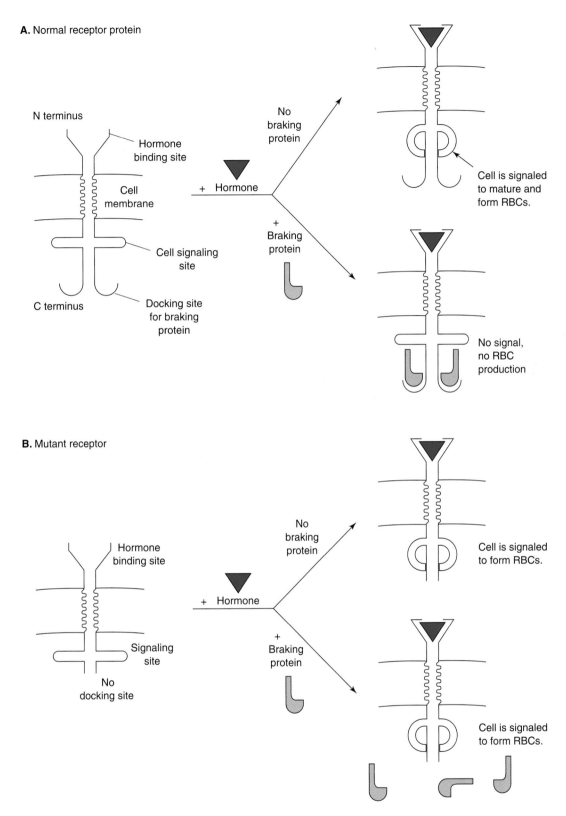

A. Normal receptor protein

N terminus

Hormone binding site

Cell membrane

Cell signaling site

C terminus

Docking site for braking protein

+ Hormone

No braking protein

+ Braking protein

Cell is signaled to mature and form RBCs.

No signal, no RBC production

B. Mutant receptor

Hormone binding site

Signaling site

No docking site

+ Hormone

No braking protein

+ Braking protein

Cell is signaled to form RBCs.

Cell is signaled to form RBCs.

Figure 4.41 Schematic representation of how the loss of 70 C-terminal amino acids from a receptor protein results in increased red blood cell (RBC) production.

biologists have searched and are searching for clues in solved protein structures, looking for amino acid patterns that correlate with specific structures. Structure prediction computer programs based upon these statistical studies have been developed, and they are useful but not perfect. If you are trying to predict a three-dimensional structure for a protein whose gene you have recently discovered, you are most likely to be successful if someone else has already discovered a similar protein and determined its structure.

DNA and protein sequence information is stored in large public databases, and computer software is available to compare these sequences (see chapter 34). Proteins with homologous amino acid sequences (sequences that match fairly well) have similar three-dimensional structures and generally have similar functions. Therefore, the first step in determining the three-dimensional structure of a protein is to compare its amino acid sequence with those of all proteins in the public databases to see if the structure of a homologous protein has already been solved. If you are lucky and find one whose structure has been determined, you have a good model for your protein. This model can then be used as a basis for identifying where the active sites of your protein are and can guide further experiments. If you do not find a homologous protein with a solved structure, the best you can do is to use the imperfect computer algorithms for structure prediction.

Sometimes you can get clues about what a protein does from amino acid sequence comparisons. For example, scientists identified the gene that is defective in cystic fibrosis (CF) patients in 1989. At the time, they did not know exactly what the protein product of the gene did, although they knew from the symptoms of CF that the protein's function was related to the movement of salt across cell membranes. (This movement is defective in CF patients, causing some cellular secretions to contain abnormal levels of salt and water.) When scientists compared the predicted amino acid sequence of the CF gene product to the amino acid sequences of known proteins, they found that part of the CF protein looked exactly like helical protein domains that span cell membranes. Thus, their amino acid sequence comparison strongly suggested that the CF protein sits in the cell membrane, a reasonable position for a protein involved in transporting things in and out of cells.

It is important to note that we are discussing *predicting* protein structure from the amino acid sequence, not *determining* that structure. The determination of a protein's structure is a scientific specialty all its own, requiring expensive instrumentation and highly specialized training. Most molecular biologists who study the functions of genes and proteins do not have the training or the instruments to determine a protein's structure. Even if they did, the methods available at this time do not work on all proteins. Therefore, predicting protein structures from DNA (and therefore amino acid) sequences is part of trying to figure out what a protein does and how it does it or what part of a protein is doing a specific thing.

Testing Structure-Function Predictions

Molecular biologists have come up with ways to test predictions about what parts of a protein are involved in specific functions even when they do not know what the structure is. How can they do this? In a way, they imitate nature. They introduce mutations into proteins (by manipulating the DNA sequence of the gene) and determine what the changes do to the protein's function.

For example, if you are studying a newly discovered DNA-binding protein and find that part of its amino acid sequence is consistent with a helix-turn-helix motif, you might suspect that the putative helix-turn-helix is the portion of the protein that binds DNA. Being a scientist, you would like to test your hypothesis.

First, you would have to clone the gene for your protein so that you could work with it (see chapter 5). Then, you could introduce specific base changes into the gene to cause specific amino acid changes in the protein. To test your hypothesis, you could change specific amino acids in the region you think might be a DNA-binding helix. If you found that your mutations caused the protein to bind to different specific DNA sequences, you would have good evidence that the region of the protein you altered was the DNA-binding region. Would you have proved that the structure was actually a helix-turn-helix? No, but your evidence that that particular area of the protein is directly involved in DNA binding would support the structure prediction.

Molecular biologists use approaches like the mutation experiment just described, often guided by predictions about a protein's structure, to try to learn which regions of proteins are involved in performing specific functions. In the end, though, the only way to be completely certain of a protein's structure is to use one of the instrumental methods described in chapter 5 for determining it, a long and exacting process.

Proteomics

Just as genomics is the study of the organization and expression of all the genes within an organism, **proteomics** is the study of the entire protein content of a given organism, cell, or tissue. The **proteome** of a cell can be defined as the sum of all the proteins expressed in a given cell at a given time. If you think about it, proteomics is a very important discipline: the cells of your cardiac muscle, neurons, skin, and intestinal lining all contain the same DNA. It is the expression of that genetic content, or their proteomes, that makes them different.

Proteomic studies generally consist of comparing the proteomes of different cell types, or of the same cells under different conditions. Since cells typically express thousands of proteins at once, meaningful proteomic studies are technically challenging. Interpreting protein expression data has only become possible for complex organisms since genomic sequence data have become available. The genome sequence provides information about the proteins that the cell *could* make; the expression data allow one to identify which proteins *are* being synthesized.

One way in which these studies are being conducted involves a chemical separation technique called mass spectrometry, combined with powerful bioinformatic software. Mass spectrometers use powerful magnets to separate charged versions of molecules according to their molecular masses, permitting exquisitely accurate determinations of those masses. From genomic data, scientists can predict the amino acid sequences of all possible proteins within the cell and the digestion products of those proteins when treated with specific proteinase enzymes. They then extract total cellular protein, digest the preparation with a proteinase, and separate the products in the mass spectrometer. The masses of the protein fragments are matched to the predicted masses, and the expressed proteins are identified.

Molecular biology, recombinant DNA technology, and scientific knowledge

Here, at the end of this chapter, is one final thought. You have just waded through a lot of information about basic molecular processes involving DNA and proteins. We started with the structures of DNA and genes and moved to protein synthesis, then to protein structure and function, and finally back to connections between changes in the DNA sequence and changes in protein function. Along the way, we tried to emphasize that all the cellular operations involving DNA are performed by enzymes. The next chapter is a description of how scientists have learned to use some of these enzymes to manipulate DNA and, thereby, proteins. The whole business of using cellular enzymes to manipulate DNA is often called recombinant DNA technology.

Recombinant DNA technology as such seems to make the news in connection with controversial biotechnology products, gene therapy, or scary science fiction, but thousands of scientists around the world are using recombinant DNA technology every day simply to learn more about how life works. Think about the preceding discussion of predicting protein structure and probing protein function. These efforts involve computerized comparisons of DNA sequences from hundreds of genes. They involve identifying and cloning the gene for the hypothetical DNA-binding protein. They involve altering the sequence of the cloned gene and producing protein from the altered DNA sequence to see what the effects of the amino acid changes were. In real life, accomplishing these steps would all involve recombinant DNA technology.

One of the wonderful things (at least for a scientist) about the field of molecular biology is that the technologies it has spawned have made it possible to learn even more basic biology. It seems that the more we learn, the more neat technologies we develop. The new technologies let us design new kinds of experiments for learning even more—almost a chain reaction of knowledge. This textbook itself provides an illustration (however insignificant) of the growth of molecular biological knowledge. The first edition of this book was published in 1996. At that time, the human genome had not been sequenced, nor had the genome of any other animal. The terms genomics and proteomics did not appear in the first edition. Neither did RNAi; all these areas of experimentation have developed since then. As you read chapter 5 and beyond, notice how the molecular technologies are applied to many scientific questions, as well as to medical and other applications in day-to-day human life.

Summary

All of earth's living creatures use DNA as their genetic material. The elegant structure of this molecule enables it to perform its two functions: replication and transmission of information. DNA controls the form of an organism by specifying the amino acid sequences of all the proteins in that organism. The proteins, in turn, form many of the organism's building blocks and carry out nearly all of its metabolic processes.

Protein synthesis in all known organisms uses the same genetic code and the same general process. The universality of this most basic life process demonstrates the interrelatedness of all life on the planet. The cells of all creatures synthesize mRNA. The codons of all their mRNAs carry the same meaning. The ribosomes of the simplest bacterium are capable of reading the genetic code and synthesizing the protein from a human brain cell.

All forms of earthly life arose as a result of genetic variation followed by natural selection. Mutation and recombination can result in the formation or addition of new genes, the destruction or deletion of old ones, major and minor changes in protein structures, and changes in gene regulation.

Now that scientists understand the inner workings of heredity and genetic variation, they have begun to experiment with those processes. People are beginning to harness the natural mechanisms of genetic variation and protein synthesis to manipulate the genetic contents of organisms and direct their gene expression. Are these experiments fundamentally different from natural evolution? No, except in the sense that humans are at the controls and they are occurring on a time scale that is immeasurably faster than nature's time scale.

The realization that humans now have the power to manipulate the genes of living organisms is exhilarating, awe inspiring, and potentially disturbing. It offers the hope of treatment for previously incurable diseases. It holds out promise for an increased and better food supply. At the same time, our newfound power raises a multitude of hard questions. It has now become the responsibility of all of us to understand this power of genetic manipulation. Only through understanding can we hope to use and regulate this power wisely.

Selected Readings

Bayley, H. 1997. Building doors into cells. *Scientific American* 277:62. Examples of protein engineering projects.

Branden, C., and J. Tooze. 1991. *Introduction to Protein Structure.* Garland Publishing, Inc., New York, NY. Well-illustrated, readable book about protein structure.

Cuerces-Amabile, C., and M. Chicurel. 1993. Horizontal gene transfer. *American Scientist* 81:332–341. Transfer of genetic information outside of parent to offspring.

Doolittle, R., and P. Bork. 1993. Evolutionarily mobile modules in proteins. *Scientific American* 269:50–56. Protein domain structure.

Gerstein, M., and M. Levitt. 1998. Simulating water and the molecules of life. *Scientific American* 279:100.

Gibbs, W. 2003. The unseen genome: gems among the junk. *Scientific American* 289:26–33.

Hall, S. 1995. Protein images update natural history. *Science* 267:620–624. How protein structure determinations and new computer models of protein structure are changing the science of biology.

Holtzman, D. 1991. A "jumping gene" caught in the act. *Science* 254:1728–1729. Two cases of human hemophilia A apparently caused by the movement of a transposon.

Kreuzer, H., and A. Massey. 2006. *Biology and Biotechnology: Science, Applications, and Issues.* ASM Press, Washington, DC. A discussion of biotechnology in the context of the biology of cells. Approximately half the book is devoted to an overview of cell biology. The other half presents detailed discussions of applications of biotechnology in agriculture, medicine, food, and nutrition, with emphasis on the role of public policy in the development of science.

McGinnis, W., and M. Kuziora. 1994. The molecular architects of body design. *Scientific American* 270:58–66. Genes, proteins, and development.

Micklos, D., and G. Freyer. 2003. *DNA Science,* 2nd ed. Cold Spring Harbor Laboratory Press, Cold Spring Harbor, NY.

Moxon, E. R., and C. Wills. 1999. DNA microsatellites: agents of evolution? *Scientific American* 280:94.

Prusiner, S. 1995. The prion diseases. *Scientific American* 272:48.

Ptashne, M. 1989. How gene activators work. *Scientific American* 260:40–47.

Ptashne, M. 2004. *A Genetic Switch,* 3rd ed. Cold Spring Harbor Laboratory Press, Cold Spring Harbor, NY. Excellent short book describing in molecular detail how bacteriophage lambda switches from lysogenic to lytic growth. Protein structure, protein-protein interactions, protein-DNA interactions, and gene regulation are combined in one well-understood biological system.

Rennie, J. 1993. DNA's new twists. *Scientific American* 266:122–132. Current thinking about DNA structure.

Rhodes, D., and A. Klug. 1993. Zinc fingers. *Scientific American* 268:56–65. Protein structure and gene regulation; zinc fingers are one structural motif used for DNA binding.

Richards, F. 1991. The protein folding problem. *Scientific American* 264:54–63. An overview of the attempt to understand what determines three-dimensional protein structure.

Roush, W. 1995. An "off switch" for red blood cells. *Science* 268:27–28. Description of the molecular biology of the stamina-enhancing mutation causing benign erythrocytosis.

Stix, G. 2004. Hitting the genetic off switch. *Scientific American* 291:98–101.

Tjian, R. 1995. Molecular machines that control genes. *Scientific American* 272:54–61. Description of the complex assembly of proteins needed for transcription of eukaryotic genes.

Wallace, D. 1997. Mitochondrial DNA in aging and disease. *Scientific American* 277:40.

Zamore, D. Z. 2002. Ancient pathways programmed by small RNAs. *Science* 296:1265–1269.

Recombinant DNA Technology

If you pay any attention at all to the news, you cannot avoid stories about biotechnology: sequencing a genome, identifying a gene, analyzing ancient DNA, fingerprinting DNA, cloning something, genetic engineering, etc. Have you ever wondered, "How do they do that?"

The technologies underlying the news stories are actually natural extensions of the scientific knowledge gained as researchers explored the details of how cells function. As scientists studied basic cellular processes, they often isolated cellular components, such as enzymes or DNA, from whole cells and studied their structures and activities outside the cell. When a process is studied in isolation, it can often be manipulated in ways that are impossible if it is still taking place inside a cell. In the pursuit of knowledge, then, scientists isolated and studied the activities of enzymes that manipulate DNA in many ways: for example, by copying it (DNA replication) and joining pieces of it (DNA replication, repair, and recombination). They also discovered natural processes in which DNA is transferred from one cell to another (transformation, transduction, and conjugation). As researchers discovered enzymes and processes, they experimented with using them for their own scientific purposes.

A by-product of the quest for knowledge was that scientists acquired a versatile tool kit for manipulating and analyzing DNA and proteins. Biological knowledge and biotechnology know-how increased hand in glove. Modern biotechnologies harness enzymes and procedures that scientists copied or modified from nature. In this chapter, we will look at some of the tools scientists use in carrying out research you read about in headlines. The chapter is not meant to be an exhaustive list (it's not) or a detailed how-to manual, but rather, a conceptual guide to basic procedures that are used over and over again in biotechnology. We will look at some of the enzymes and other fundamental tools for working with DNA and cells that are used to manipulate and analyze DNA, including determining its sequence; to

clone DNA; and to analyze proteins. You may find that many of the procedures are familiar to you, because they involve principles and enzymes you have already read about in this book.

In the next chapter, we will look at how these fundamental techniques are combined to achieve goals, such as finding genes, creating transgenic plants, or constructing a DNA fingerprint.

Manipulating and Analyzing DNA

The structure of DNA was published in 1953. At that time, scientists were not sure that DNA was the genetic material, much less aware of how its encoded information was translated into proteins, yet within 20 years, they had constructed the first artificial recombinant DNA molecules and used them to transfer traits in bacteria. Since then, advances in techniques for manipulating and analyzing DNA have continued to accumulate. Some of them have now become so routine that high school students often perform them as laboratory exercises.

This veritable explosion of DNA technologies is in part the result of the intense research focus on understanding the function of DNA that began in the 1950s. The other contributing factor is the fact that many of the tools used to manipulate and analyze DNA are natural enzymes and processes. Once scientists discovered and characterized them, they could exploit them.

Making recombinant DNA

One of the challenges facing scientists who wanted to study genes was that, in nature, many genes are found on one DNA molecule. Scientists had no way to separate a specific segment of DNA containing a single gene from the rest of a chromosome to study it in isolation. The discovery of enzymes that cut DNA at specific base sequences revolutionized biological science. For the first time, researchers could obtain reproducible, small fragments of DNA for study. By using other naturally occurring enzymes, they could join fragments together. Within just a few years of the discovery, researchers had joined together pieces of DNA and transferred the resulting molecules into bacteria. The discovery of DNA-cutting enzymes has been credited with making the DNA biotechnologies possible.

Cutting DNA

In the 1960s, scientists studying how certain bacteria resist infection by bacteriophages observed that in these bacteria, the DNA of the infecting virus was cut into segments. Pursuing their observation, they discovered that the phage-resistant bacteria contained enzymes that cut the bacteriophage DNA. Named **restriction endonucleases,** or restriction enzymes, they recognize specific base sequences in a DNA molecule and cut the DNA at or near the recognition sequence in a consistent way (Figure 5.1). Restriction enzymes can be isolated from the bacteria that produce them. When added to a test tube containing DNA, the enzyme will cut the DNA molecule at every occurrence of its recognition sequence.

The most commonly used restriction enzymes recognize palindromes, sequences in which both strands read the same in the 5′-to-3′ direction. An example of a palindromic sequence is 5′-GAATTC-3′. Remember that the complement of this sequence would read from 3′ to 5′ (left to right) as follows: 3′-CTTAAG-5′. Now, read the complementary sequence from 5′ to 3′. It is identical to the other strand.

The structure of a palindromic recognition sequence fits with the structure and function of the restriction endonuclease protein. The protein is composed of two identical subunits. Together, they

slide along the DNA helix, and when they reach the recognition sequence, each subunit binds to the exact same pattern of DNA bases, one per strand. Next, each subunit cuts the palindrome in exactly the same place (in the above example, between the G and the first A).

Over 100 different restriction endonucleases have been identified and isolated from many different bacteria. Their names indicate the organism from which they were purified (EcoRI from *Escherichia coli,* HindIII from *Haemophilus influenzae,* and so on). (See chapter 11 for more information.) Because they bind to and cut at specific DNA base sequences, they always cut a given DNA molecule in exactly the same way, producing a set of pieces referred to as **restriction fragments.** Scientists use restriction enzymes like precise DNA scissors, to cut DNA molecules into reproducible pieces.

Separating Mixtures of DNA Fragments

To cut a DNA molecule with a restriction enzyme, a solution containing the DNA is typically put into a small test tube and a solution containing the restriction enzyme is added. Time is allowed for the enzyme to find its recognition sequences and cleave the DNA molecule. Afterwards, scientists separate the mixture of fragments produced by the enzyme. The standard method used to separate DNA fragments is electrophoresis through gels made of agarose or polyacrylamide.

Agarose is a polysaccharide (as are agar and pectin) that dissolves in boiling water and then gels as it cools, like Jell-O. To perform agarose gel electrophoresis of DNA, a slab of gelled agarose is prepared, a solution containing DNA fragments is introduced into small pits in the slab called sample

Figure 5.1 Restriction endonucleases recognize and cut specific sites in a DNA molecule. The arrows indicate the cleavage sites of one such endonuclease.

DNA
```
...   A A T C G C T A G G C C A T T G C A A T G G C T A G G C C T A C G T T C A G G C C A A T C G A   ...
...   T T A G C G A T C C G G T A A C G T T A C C G A T C C G G A T G C A A G T C C G G T T A G C T   ...
```

Incubate DNA molecule
with restriction endonuclease.

```
...  A A T C G C T A G G          C C A T T G C A A T G G C T A G G          C C T A C G T T C A G G          C C A A T C G A  ...
...  T T A G C G A T C C          G G T A A C G T T A C C G A T C C          G G A T G C A A G T C C          G G T T A G C T  ...
```

Restriction fragments

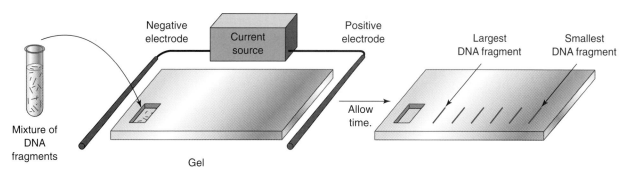

Figure 5.2 Gel electrophoresis separates DNA fragments by size.

wells, and then an electric current is applied across the gel. Since DNA is highly negatively charged (because of the phosphate groups along its backbones), it is attracted to the positive electrode. To get to the positive electrode, however, the DNA must migrate through the agarose gel.

Smaller DNA fragments can migrate through an agarose gel faster than large fragments. In fact, the rate of migration of linear DNA fragments through agarose is inversely proportional to the $\log_{10}$ of their molecular weights. What it boils down to is that if you apply a mixture of DNA fragments to an agarose gel, start current flowing, wait a little, and then look at the fragments, you will find that the fragments are spread out like runners in a race, with the smallest one closest to the positive electrode, the next smallest following it, and so on (Figure 5.2). Because of the mathematical relationship mentioned above, it is possible to calculate the exact size of a given fragment by comparing how far it migrated to the migration of molecules of known size. After gel electrophoresis, the DNA fragments in the gel are usually stained to render them visible (Figure 5.3). DNA fragments can be isolated and purified from agarose gels, or they can be tested for the presence of specific base sequences (described below). RNA molecules can be separated by electrophoresis, as well. (See chapters 12 and 13 for more information.)

Electrophoresis through polyacrylamide gels works in essentially the same way, except that polyacrylamide makes a tighter mesh than does agarose and so is better for separating smaller DNA molecules and DNA molecules that differ only slightly in size (for example, by 1 nucleotide).

Pasting DNA

Sometimes scientists want to put together fragments of DNA from different sources. This procedure is typically necessary if your goal is to insert a new gene into a microorganism, plant, or animal or to clone a gene as a prelude to characterizing it. We will

discuss some of these applications later. The tool used to join pieces of DNA together is another natural enzyme: **DNA ligase.**

DNA ligases join pieces of DNA together by forming new phosphodiester bonds between the pieces (Figure 5.4), a process called ligation. Every cell synthesizes ligase to seal gaps in DNA left by replication, repair, or recombination, and DNA ligases have been purified from many kinds of cells.

When two pieces of DNA from different sources are joined together, the result is said to be **recombinant DNA.** The word recombinant means "composed of parts that originally came from two or more sources." You, yourself, are a genetic recombinant in the sense that half of your chromosomes originally came from your mother's egg and the other half from your father's sperm. You could make a recombinant bicycle by taking apart two bicycles and assembling one from parts taken from each. But in the usual

Figure 5.3 A stained agarose gel showing separated DNA fragments. The outlines of the sample wells are visible at the top of the gel. The lanes at the far right and left contain a mixture of fragments of known size.

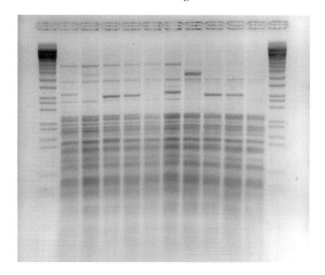

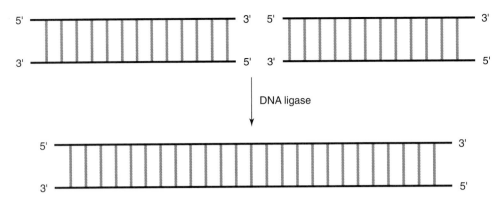

Figure 5.4 DNA ligase joins DNA fragments by forming bonds between the 3′ and 5′ ends of the two backbones.

sense, recombinant DNA refers to DNA molecules composed of fragments originally taken from different sources and joined together in a test tube. (See chapter 14 for more information.)

Detecting specific base sequences

Another thing that scientists frequently want to know is whether a specific base sequence is present in a DNA molecule or a DNA-containing sample. Think about what you have learned about the rules of base pairing in DNA. If you wanted to know whether the base sequence 5′-GGATGCGTCC-3′ was present in your sample, one way to find out would be to see whether any DNA in your sample could pair with a strand of DNA having the sequence 3′-CCTACGCAGG-5′. If it could, then the sequence of interest must be present in the sample. This example illustrates hybridization analysis, the means by which specific DNA sequences are usually detected.

Hybridization Analysis

DNA hybridization (also called annealing or renaturation) is the term used to describe the process in which two single DNA strands with complementary base sequences stick together to form a correctly base-paired double-stranded molecule (Figure 5.5). Hybridization occurs spontaneously: if two complementary single DNA strands are mixed together and left alone, they will hybridize. The time it takes for hybridization to occur is directly related to the lengths of the DNA sequences involved; as one might expect, short complementary sequences can line up correctly and form base pairs much faster than long sequences can. Hybridization also works with RNA molecules; complementary RNAs can hybridize to form a double-stranded RNA molecule, and a complementary RNA can hybridize to DNA.

To conduct a hybridization analysis, the double-stranded sample DNA must be separated into single strands. This process, called denaturation, is often carried out by heating the sample to 95°C. The heat overcomes the force of the hydrogen bonds holding the base pairs together, and the two strands separate, exposing their unpaired bases.

Once the strands are separated, their bases are free to form pairs with the test sequence. Scientists add something called a **probe** to the denatured sample DNA. A probe is a piece of single-stranded DNA containing the test sequence of bases. If the sample DNA contains the base sequence complementary to the probe DNA sequence, the probe will form base pairs with the sample DNA, or hybridize to it (Figure 5.6). After time has been allowed for hybridization to take place, the sample is washed to remove unbound probe and then tested to determine whether any hybridized probe is present.

Probes are chemically modified, or labeled, in some way so that their presence can be detected. For

Figure 5.5 Hybridization is the formation of base pairs between two complementary single-stranded nucleic acid molecules. The molecules can be the same or different lengths.

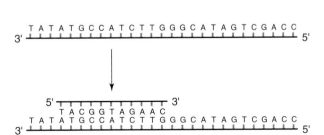

140 • **Laying the Foundation**

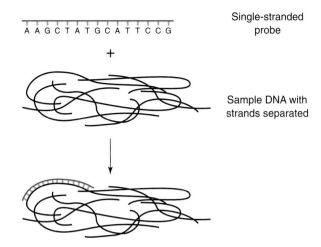

AAGCTATGCATTCCG

Single-stranded probe

+

Sample DNA with strands separated

If sample DNA contains the base sequence complementary to the probe sequence, the probe will form base pairs with the sample DNA, or hybridize to the sample, and physically stick to it.

Figure 5.6 Hybridization analysis. A single-stranded probe is added to denatured sample DNA.

example, a probe can be labeled with radioactive phosphorus. If it forms base pairs with the sample DNA, the radioactivity can be detected. Probes can also be labeled with fluorescent dyes or compounds that react with enzymes to make a colored product. (See chapter 16 for more information.)

Chemical Synthesis of Single-Stranded DNA

Where can you get a probe for a hybridization analysis? You can isolate naturally occurring DNA, label it, and heat it to denature it, or you can pick up the phone and call in an order for a DNA molecule of a particular sequence (within certain limitations) to any of a number of companies.

Just as biologists studying DNA figured out its role in the cell, chemists studying DNA figured out how to synthesize it in the laboratory. This process has been automated, and computer-controlled DNA synthesis machines loaded with bottles of starting reagents can churn out single-stranded molecules up to about 100 bases long. The user provides the desired base sequence. These relatively short single-stranded molecules are often called **oligonucleotides** (oligo, several). Labels such as fluorescent dyes can be added during synthesis, making the use of synthetic oligonucleotides as probes quite convenient.

The term "synthetic DNA" or "synthetic oligonucleotide" sounds as if the DNA was somehow different from naturally occurring DNA. It is not.

"Synthetic" refers to the fact that it was chemically synthesized and not to any differences between it and DNA isolated from an organism. If synthetic DNA is inserted into the genome of an organism, it functions just like any other DNA.

Locating a Specific DNA Base Sequence in a Specific Restriction Fragment

Often it is important to know which DNA fragment in a mixture contains a sequence of interest. For example, your goal might be to clone a particular gene from a given organism. Pursuant to that goal, you plan to isolate DNA from a sample of that organism's cells, cut it into fragments with a restriction enzyme, separate the fragments by electrophoresis, and then isolate the fragment that contains your gene. Obviously, you need to be able to identify which of all the restriction fragments in the mixture contain the gene of interest. Gel electrophoresis and staining alone cannot answer this question, because all DNA fragments look alike when stained. It is possible, however, to combine restriction digestion and hybridization analysis.

Hybridization does not work well on DNA fragments embedded in an agarose gel, so if your goal is to separate restriction fragments and then test them, you would perform a procedure called blotting, which is very similar to blotting ink writing. If you write something with a fountain pen and cover the writing carefully with a sheet of blotting paper, the pattern of your writing will be exactly transferred to the blotter. DNA blotting works in the same manner. DNA fragments are separated by agarose gel electrophoresis, and then the gel itself is covered with a membrane and blotting paper. The DNA fragments in the gel transfer to the membrane in exactly the same arrangement they had in the gel. Once the DNA fragments are on the membrane, they can be hybridized to probes to test for the presence of specific sequences (Figure 5.7). Blotting and hybrization analysis can be performed with RNA as well as DNA. (See chapter 16 for more information.)

Copying DNA

Yet another challenge for scientists working with DNA can be obtaining enough DNA to analyze. For example, you might have a very tiny sample that you know contains a gene or other DNA sequence you want to study, but the amount of DNA in your sample is too small to be useful. Again, nature provides a solution. DNA is copied inside cells every time a cell divides. Scientists take advantage of the DNA replication process to make copies of DNA outside of cells, too.

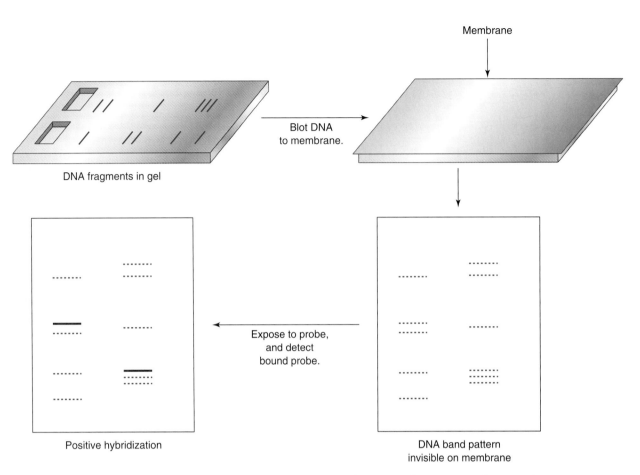

Membrane

Blot DNA
to membrane.

DNA fragments in gel

Expose to probe,
and detect
bound probe.

Positive hybridization

DNA band pattern
invisible on membrane

***Figure 5.*7** A given DNA sequence can be localized to a specific restriction fragment by blotting DNA fragments from an electrophoresis gel to a membrane and conducting hybridization analysis on the membrane.

DNA Polymerases

As you have read in previous chapters, DNA polymerases are the cellular enzymes that copy DNA. Using one strand of a parent DNA molecule as a template, they synthesize a complementary strand of DNA by adding nucleotides to the $3'$ end of the growing strand (Figure 4.5). DNA polymerase can be purified from a wide variety of cells and used in test tubes to copy existing DNA molecules.

Within the cell, DNA replication starts with the process of initiation, in which DNA strands are separated and an RNA primer is synthesized. (See chapter 7 for more information.) In a test tube, DNA polymerases also require a single-stranded template with a primer base paired to it. Scientists supply this structure by denaturing the parent DNA as for hybridization analysis (for example, by heating) and then adding a short, single-stranded DNA molecule that hybridizes to the template and serves as a primer. The first base of the newly synthesized DNA is attached via a phosphodiester bond to the

$3'$ end of the primer and is complementary to the base on the template strand. Synthesis proceeds as more bases are added to the primer (Figure 5.8). (See chapter 7 for more information.)

The single-stranded primer DNA is the same as a probe in hybridization analysis. It may or may not be labeled for detection. It is called a primer instead of a probe because its purpose is different: to serve as a starting point for DNA synthesis rather than simply to indicate the presence of a DNA sequence by hybridizing to it. Synthetic oligonucleotides are commonly used as primers for DNA polymerase reactions.

Making DNA from an RNA Template

Reverse transcriptases read an RNA sequence and synthesize a **complementary DNA** (often abbreviated as **cDNA**) sequence (Figure 5.9). These enzymes are made by RNA viruses that convert their RNA genomes into DNA when they infect a host. Reverse transcriptases allow scientists to synthesize a

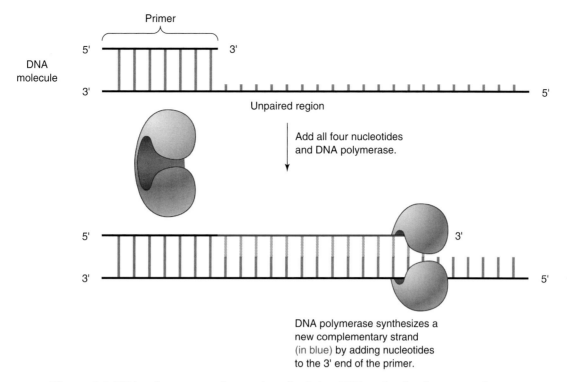

Figure 5.8 DNA polymerase makes copies of existing DNA molecules (known as the template). The DNA synthesis reaction requires a primer hybridized to single-stranded template DNA, the enzyme, and all four nucleotides. The primer becomes part of the new DNA molecule.

DNA gene from a messenger RNA (mRNA) molecule. This ability is useful for dealing with eukaryotic genes, since the original genes are often split into many small pieces separated by introns in the chromosome (see chapter 4). The mRNA from these

genes has undergone splicing in the eukaryotic cell, and the introns are gone, leaving only the coding sequences. Reverse transcriptase can convert this mRNA into a "prespliced" gene consisting only of protein-coding sequences.

Figure 5.9 Reverse transcriptase uses RNA as a template and synthesizes a cDNA copy. Through additional reactions, the RNA can be removed and replaced with a second DNA strand.

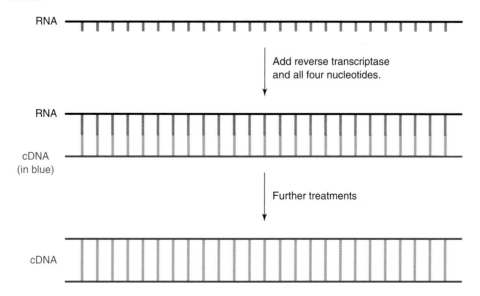

Why would someone want a prespliced gene? Suppose you are interested in moving a eukaryotic gene into a bacterium so that the bacterial cell can produce the protein for you. Since bacteria possess no equipment for splicing, they must be given a prespliced version of a gene if they are to express the correct protein product. Making cDNA from mRNA is important for expressing eukaryotic genes in prokaryotes, as in the production of human insulin in bacteria. Most diabetics now use human insulin made in bacterial cells instead of animal insulin.

Amplifying DNA

In the 1980s, a scientist driving late at night had a flash of inspiration about how he might use DNA polymerase and primers to create a chain reaction of DNA synthesis, making millions of copies of a target DNA segment. Back in his laboratory, he tested his idea and found that it worked. His invention, the polymerase chain reaction (PCR), is now used in many different kinds of applications, from disease diagnosis to DNA typing and studying ancient DNA.

PCR requires two primers that hybridize to opposite strands of the template DNA molecule at the boundaries of the segment to be copied. To perform PCR, you would denature the template DNA and allow it to hybridize to both primers, nucleotides, and DNA polymerase, and allow synthesis to begin. DNA polymerase then copies both of the strands, starting at the two primers.

After a few minutes, the reaction mixture is heated to 95°C to separate the product DNA strands from the templates. As the reaction mixture cools, primers hybridize to the newly synthesized products (as well as the original templates) and synthesis begins again. This cycle is repeated many times, and each cycle doubles the number of DNA molecules, resulting in the synthesis of millions or even billions of copies of the DNA segment that stretches from one primer sequence to the other (Figure 5.10). The process is also called DNA amplification.

The 95°C heating step may have caught your attention, since this temperature would be high enough to denature most enzymes and destroy their activity. For PCR, scientists use DNA polymerase enzymes purified from bacteria that live at very high temperatures, such as in hot springs and thermal vents in the ocean floor. Since their natural environments are

Figure 5.10 PCR produces many copies of a DNA segment lying between and including the sequences at which two single-stranded primers hybridize to the template DNA molecule. The primers are usually synthetic oligonucleotides.

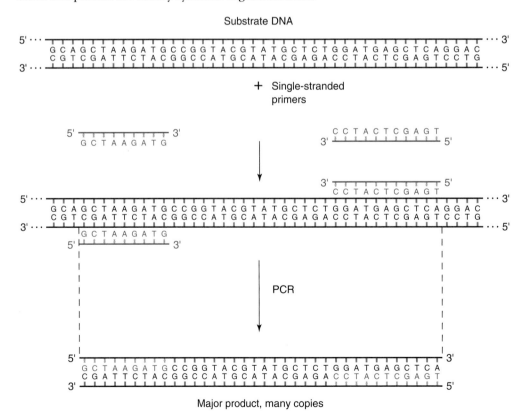

Major product, many copies

very hot, their DNA polymerases function well at high temperatures. These enzymes can be added to a PCR mixture and will survive repeated heating cycles. Because of this, the PCR has been automated. The reaction mixture is assembled and put into a machine that takes it through the repeated temperature cycles. (See chapter 17 for more information.)

PCR as a Detection Method

PCR can be used to amplify scarce DNA so that it can be studied or cloned, but it can also be used as a detection method. The reason it can be used as a detection method is that the primers for PCR must hybridize to the template DNA for amplification to occur. If you added primers to a sample that did not contain the specific template DNA, you would not get the PCR product. In a way, PCR is like hybridization analysis. In hybridization analysis, the hybridized probe is detected directly because of its label. In PCR, hybridization of the primers is detected by the synthesis of the DNA product (Figure 5.11).

Figure 5.11 Testing for the presence of a DNA sequence using PCR.

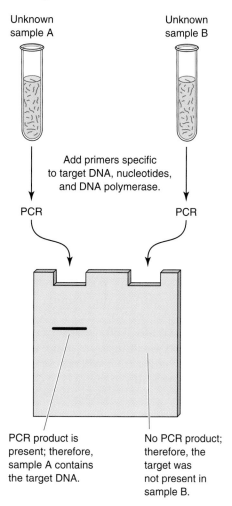

PCR is in fact a very good detection method. It is more sensitive than hybridization analysis precisely because it amplifies the target DNA. Regular hybridization analysis is limited by one's ability to detect the hybridized probe. If a sample contained only one copy of the DNA sequence of interest, only one molecule of probe could hybridize to it. In contrast, PCR can theoretically produce slightly over 1 billion copies of a DNA segment from a single starting molecule in 30 rounds of amplification. In fact, PCR has produced a detectable signal from a single copy of a target DNA sequence, far more sensitive than any method for direct detection of hybridized probe.

Sequencing DNA

The sequence of bases in a gene determines the sequence of amino acids in a protein. The genetic code was worked out in the 1960s, and the first method for determining the base sequence of DNA was also published during that decade. Initially, DNA-sequencing methods were completely chemical approaches and were slow and labor-intensive. As scientists began to work with DNA polymerase enzymes outside the cell, they developed ways to use the enzyme to reveal the sequence of a DNA molecule. Scientists knew that the DNA polymerase enzyme reflects the sequence of a template as it adds complementary bases to a growing new strand of DNA. They figured out how to detect the order in which bases were added, essentially looking over the shoulder of the enzyme as it synthesized DNA.

DNA Sequencing with Terminators

Enzymatic DNA sequencing employs DNA polymerase and synthetic oligonucleotides as primers. The nucleotides supplied to the DNA polymerase include nucleotide derivatives called terminators. As their name implies, terminators block further base addition by the polymerase enzyme. Following the sequencing reaction, scientists separate the partially replicated products and detect the order in which the bases are added to the growing strand. The order in which DNA polymerase adds new bases to the primer reveals the base sequence of the template.

DNA sequencing can be performed by a single individual in a laboratory, and when the procedure was first developed, that is how it was done. Researchers (often students) carried out the reactions, poured large polyacrylamide gels to separate the products, and read the DNA sequence themselves. After the invention of PCR, the sequencing process and the PCR were merged. DNA sequencing reactions can now be performed using the automated-PCR approach. The results are recorded by an instrument that detects the labeled reaction products, and the base sequence is fed directly into a computer file, where it can be analyzed with bioinformatics software (see chapter 34). DNA sequencing is now offered as a service by many biotechnology companies and centralized facilities at universities. You send them the DNA molecule (cloned into an appropriate vector [see below]) that you wish to have sequenced, and the facility returns to you a computer file with the sequence, typically charging a few dollars per base pair. (See chapter 18 for more information on DNA sequencing.)

Cloning

The term **cloning** means the production of identical copies of something. In biology, it specifically means genetically identical copies. Asexual reproduction is a natural cloning process, since it generates offspring that are identical to the parent. In fact, an asexually reproducing population is often called a clone, or a clonal population.

Cloning DNA

When applied to DNA, cloning is usually understood to mean the transferring of a piece of DNA into a cell in such a way that the DNA will be replicated and maintained along with the rest of the cell's DNA. As the cell reproduces, new identical copies of the DNA are produced. Although there are many ways to go about cloning DNA, it usually involves making a recombinant molecule using the fragment of interest and a carrier molecule referred to as a **vector**. A vector facilitates the transfer of DNA into the new host cell and its maintenance within that cell.

Cloning Vectors

Many kinds of cloning vectors are available, tailored to different kinds of host cells and applications. The most common kind of vector for cloning DNA into bacterial cells is a plasmid. Plasmids are small circular DNA molecules that contain an origin of replication. They are found within the cytoplasm of the bacterial cell and are replicated by the bacterial DNA polymerase. Many types of naturally occurring plasmids have been discovered, and they can be transferred between bacteria via transformation (see chapter 19) or conjugation (see chapter 20). Scientists typically use laboratory versions of the transformation process to introduce plasmids into host bacteria. Laboratory transformation involves mixing host cells and DNA together and then manipulating the environment of the host cells in such a way that they take up the DNA molecules.

Viruses are also commonly used as vectors, particularly in eukaryotic cells. When a virus is employed as a vector, its natural ability to infect a cell and introduce nucleic acid is exploited, but the natural viral DNA is replaced with the DNA to be cloned. (See chapter 21 for more information.) Other types of cloning vectors have themselves been constructed from naturally occurring elements: for example, artificial yeast chromosomes containing replication origins, centromeres, and telomeres have been assembled to serve as cloning vectors for yeast cells.

One particularly useful natural plasmid is the Ti plasmid from the plant pathogen *Agrobacterium tumefaciens*. After the bacterium infects a plant, the Ti plasmid transfers itself from the bacterial cell into the invaded plant and inserts part of itself, the so-called T-DNA, into the plant DNA. These genes encode proteins that promote tumor growth in the plant. Biotechnologists take advantage of the Ti plasmid by taking out the tumor genes and inserting genes of interest into the plasmid. The altered Ti plasmid is reintroduced into *A. tumefaciens,* the plasmid-containing bacterium is inoculated into a plant, and the altered Ti plasmid then transfers the new genes to the plant's DNA. (See chapter 22 for more information.) Unfortunately, only certain plants are susceptible to infection by *A. tumefaciens.*

Just as there are many kinds of cloning vectors, there are many ways to introduce DNA into a host cell. DNA can be injected directly into animal cells with a fine glass needle **(microinjection)** or blasted into plant cells on DNA-coated pellets with a "gene gun." Again, the technology used depends upon the host cell, the vector, and the application.

Marker Genes

Regardless of the vector and transfer method used, once the recombinant DNA has been introduced into a batch of host cells, the next task is to identify those cells that took up and are maintaining the recombinant DNA (the transformed cells). DNA transfer procedures are inherently inefficient, so only a fraction, sometimes a very small fraction, of the host cells exposed to recombinant DNA will take it up. Scientists often use **marker genes** to reveal host cells that have successfully received the recombinant DNA.

A marker gene can be any gene that, when expressed, confers a detectable phenotype upon its host. The most common marker genes encode resistance to antibiotics. Cells that take up the recombinant DNA can grow in the presence of the antibiotic, while the rest of the host cells cannot. By exposing a mixture of transformed and nontransformed cells to the antibiotic, scientists can kill the nontransformed cells and select the transformed ones. Other kinds of marker genes encode enzymes that cause cells to change color in the presence of specific substrates, relieve nutritional deficiencies by supplying an enzyme the host cells lack, or even make host cells glow in the dark.

Marker genes are part of the vector, and the phenotype conferred by a marker indicates the presence of the vector DNA in the host cell. To determine whether the fragment of interest is present in the vector DNA, additional analysis, such as hybridization to a probe matching the sequence of interest, PCR with appropriate primers, or purification and sequencing of the recombinant plasmid, is usually performed. (See chapter 14 for more information.)

An Illustration of Cloning

Figure 5.12 illustrates the cloning of DNA in recombinant bacterial plasmids. First, DNA from the donor organism is isolated and cut into segments with a restriction enzyme (or a combination of enzymes). The selected plasmid vector, carrying an antibiotic resistance gene, is cut with the same enzyme(s). The restriction fragments and the cut vector are mixed with DNA ligase, which forms phosphodiester bonds between the fragments and the vector. Alternatively, large quantities of the fragment to be cloned can be generated by performing PCR, using the source DNA as a template. The PCR product is then purified and ligated into a vector.

After time has been allowed for the ligase enzyme to join fragments of DNA, the mixture is introduced into a host cell via transformation. Only some of the host cells will take up and express recombinant plasmids. Following the DNA transfer procedure, the transformation mixture is spread on a solid growth medium to isolate individual cells from one another. The growth medium contains the antibiotic for which the plasmid encodes resistance. Under these conditions, only transformed cells can grow. Each individual transformed cell multiplies, generating a clonal population.

On a solid growth medium, the population takes the form of a little pile of cells called a colony. Colonies look like dots on the growth medium, and each dot is a genetically identical population containing one type of plasmid. The final step in the process of cloning DNA is to isolate plasmid DNA from several colonies, analyze it with restriction enzymes or by sequencing, and determine whether the plasmids

from these colonies contain the recombinant plasmid with the cloned fragment you wanted.

Cloning a particular fragment of DNA gives easy access to it. Once you have a given fragment of DNA in a recombinant plasmid in a bacterial cell, it is easy to produce it in quantity. You simply introduce a few of the bacterial cells into a container of sterile growth medium with antibiotic; allow them to multiply into a large, genetically identical population; harvest the cells; and isolate the plasmid DNA. The techniques described above for manipulating and analyzing DNA, such as restriction analysis and sequencing, require that you start with many identical copies of a given DNA molecule. Cloning a particular piece of DNA is essentially a prerequisite for manipulating it or studying it in detail.

DNA libraries

A common application of DNA cloning is to produce what is called a **DNA library,** or genetic library. To make a library, the entire genome of an organism is digested with restriction enzymes or otherwise fragmented, and the entire batch of products is mixed with vector DNA and ligated. The resulting mixture of recombinant plasmids is transformed into a host organism. The pool of transformed cells containing the collection of recombinant molecules is called a DNA library of that organism (Figure 5.13A). The idea is that the recombinant molecules in the library will contain fragments covering the whole genome of the organism.

A DNA library can be used like a resource. For example, if a scientist knows what gene he wishes to study, he can look for it in the library and pull out a cloned version. DNA libraries also serve as starting points for sequencing the entire genome of an organism. The inserted fragments of many, many recombinant molecules are sequenced, and the regions of overlap are identified. The various sequences are aligned by identifying overlaps until the entire genome is covered.

cDNA Libraries

A genomic library should contain clones of all the DNA in an organism, whether the DNA contained genes or was noncoding, and whether the genes were being expressed or not. If you were interested in studying only genes that were being expressed in a particular tissue or organism, you could make what is called a **cDNA library.** cDNA libraries are based on mRNA from the starting tissue, so they represent only those genes being actively transcribed. The first step in making one of these libraries is to isolate

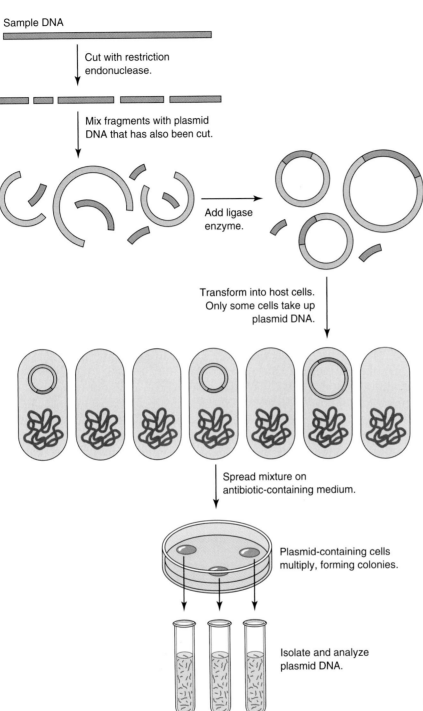

Sample DNA

Cut with restriction
endonuclease.

Mix fragments with plasmid
DNA that has also been cut.

Add ligase
enzyme.

Transform into host cells.
Only some cells take up
plasmid DNA.

Spread mixture on
antibiotic-containing medium.

Plasmid-containing cells
multiply, forming colonies.

Isolate and analyze
plasmid DNA.

Figure 5.12 Cloning of DNA in recombinant bacterial plasmids.

mRNA from the organism or tissue of interest. Then,
reverse transcriptase is used to generate cDNA copies
of all the mRNA molecules. These cDNA copies are
ligated to vector DNA molecules en masse, and the
resulting mixture of recombinant molecules is transformed into host cells, generating a cDNA library. The
cDNA library contains DNA with the coding sequences of all the proteins the organism was producing when the mRNA was isolated (Figure 5.13B).

Cloning complex organisms

When the term cloning is applied to plants and animals, it means the production of genetically identical organisms through asexual reproduction. If you
have ever taken a cutting of a plant and allowed it
to root, you have cloned something. The cutting
turned into a new plant that was genetically identical to the original plant.

A. Genomic DNA library. The insertions in the recombinant plasmids represent the entire DNA content of the organism.

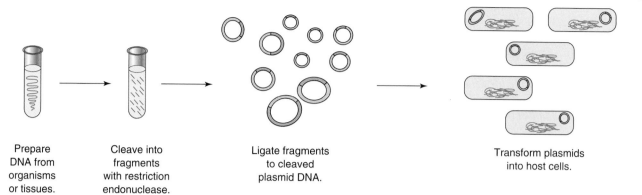

Prepare DNA from organisms or tissues. → Cleave into fragments with restriction endonuclease. → Ligate fragments to cleaved plasmid DNA. → Transform plasmids into host cells.

B. cDNA library. The insertions in the recombinant plasmids represent genes that were being expressed in the sample.

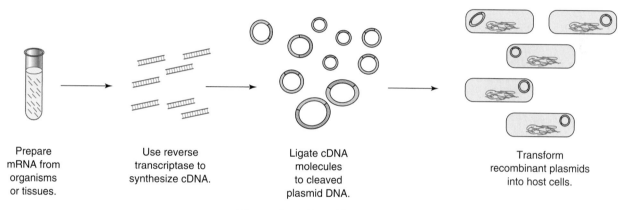

Prepare mRNA from organisms or tissues. → Use reverse transcriptase to synthesize cDNA. → Ligate cDNA molecules to cleaved plasmid DNA. → Transform recombinant plasmids into host cells.

Figure 5.13 DNA libraries.

The production of identical twins is a natural cloning process. Identical twins are formed when a very early embryo splits in two and each portion forms a baby. The two babies are genetically identical clones. Animal embryos formed by artificial insemination can be purposefully split to create identical siblings in a process called twinning.

The kind of cloning that tends to make the news, though, occurs when an adult animal is used as a source of DNA to produce a genetically identical offspring. In general, the way this is done is to remove the nuclei of an adult cell and an egg cell. The adult nucleus is implanted in the egg in place of its nucleus, and the egg with its new nucleus is implanted in the uterus of an appropriate female. If all goes well, the egg develops normally and the young animal is born.

It sounds simple, but it is not. A fertilized egg is totipotent—it has the potential to develop into all the body's tissues. Adult cells, however, are not. Scientists attempting to clone animals manipulate the donor cell to make the DNA more like that of a fertilized egg. Usually, something goes wrong. The success rates of animal-cloning procedures are very low.

Note that when the process succeeds, you get a baby animal whose nuclear genes are identical to those of the animal that donated the nucleus. In science fiction movies, "cloning" usually generates an identical, adult copy of the individual being cloned, and frequently the clone has the same memories as the original. This movie image is a far cry from reality. Real cloning generates a baby, and that baby would grow up with its own experiences and memories.

How identical might we expect an adult animal and its clone to be (age difference aside)? One clue comes from looking at identical twins. Even they, who are genetically identical and developed in the same uterine environment at the same time, are not perfectly identical. If identical twins, developing in the same uterus at the same time, are not identical, how much more different might clones that developed in the uteri of different mothers be? In addition, a small portion of eukaryotic DNA is found outside the nucleus. Mitochondria and chloroplasts contain a small amount of DNA, encoding several proteins (see chapter 27). When an animal clone is generated by nuclear transplantation, the mitochondria of the cloned offspring come from the egg donor, who could be unrelated to the nuclear donor. Taking these differences into account, we would expect a clone generated by nuclear transplant to be less like the donor parent than identical twins are like one another.

Analyzing Proteins

Proteins are the molecules through which genetic information is expressed, the mediators between genotype and phenotype. Learning what role a gene plays in a biological process requires learning what role is played by the protein the gene encodes. Likewise, manipulating the genes of an organism to change its characteristics is really about manipulating some aspect of the expression of proteins in that organism. Analyzing the expression of proteins is therefore an important component of research into the biology of cells, as well as biotechnology applications.

The proteins in a sample of cells or tissues can readily be extracted with simple chemical treatments. Like mixtures of DNA molecules, mixtures of proteins can be separated by gel electrophoresis (usually through polyacrylamide gels). The details of protein separation are different from the details of DNA separation, but the basic idea and approach are the same. Following electrophoresis, the entire gel can be soaked in a protein stain to reveal the pattern of all the proteins, or the proteins can be blotted to a membrane and tested for the presence of a particular one (see chapter 32).

Specific proteins can be detected by antibodies

Specific DNA sequences can be detected by hybridization tests with labeled probes. To detect individual proteins, we need something that will bind to them or react with them with similar specificity. Nature has provided a tool in the form of antibodies.

Antibodies are proteins produced by the B lymphocytes (B cells) of the immune system. Their function is to bind to foreign substances called antigens and tag them for elimination by other components of the immune system. The immune system has the capacity to produce millions of different antibodies. Each mature B cell produces only one kind, which binds to a specific chemical structure. The immune system can produce several different antibodies that bind to the same antigen; each antibody would recognize a different aspect of the antigen's three-dimensional shape and chemical nature and bind to it.

Monoclonal Antibodies

Scientists recognized that the most useful protein detection reagents would be pure antibodies of a single type. However, mature B lymphocytes are terminally differentiated cells and do not divide in culture, so it is not possible to culture them as a source of a single pure antibody. In the mid-1970s, a technique was developed that combined the antibody-producing ability of B cells with the ability to divide indefinitely in culture. The complex technology involved the forced fusion of normal B lymphocytes with cancerous ones (myeloma cells), which could divide indefinitely in culture. The result of the forced fusion and subsequent selection procedures was a series of cell lines, each of which could reproduce in culture and synthesized large quantities of one specific antibody molecule. These antibodies are called **monoclonal antibodies,** and they are a key tool in many biotechnology procedures.

To make monoclonal antibodies to a specific protein or other antigen, researchers inoculate a mouse with that substance. The mouse's immune system responds to the presence of the foreign antigen by making antibodies to it, just as your immune system responds to a foreign invader, like a virus. After the mouse has mounted its immune response, it is killed. B lymphocytes from its spleen are fused with myeloma cells, and the desired antibody-producing cells are identified.

Using Antibodies To Detect Proteins

To determine whether a specific protein is present in a sample, the sample is exposed to an antibody to that protein, washed, and then tested for the presence of bound antibody. The process is very analogous to a hybridization analysis for a DNA sequence. Antibodies, like DNA probes, can be labeled by the attachment of various chemicals—colored substances, fluorescent dyes, or radioactive elements.

Home pregnancy tests use antibodies to detect a pregnancy hormone called human chorionic gonadotropin, or HCG, that is secreted in the urine of pregnant women. Although the details of various test kits vary, in general, an absorbent wick is dipped into a urine sample. Antibodies to HCG labeled with a colored dye are distributed throughout the tip of the wick. Because the molecules are spread out, the color is not readily visible. However, if HCG is present in the sample, the antibodies will bind to it and be swept along as the urine migrates up the wick. As more urine travels up the wick and more HCG molecules encounter and are bound to the labeled antibodies, an increasingly visible line of color forms. In these test kits, the presence of a colored line after a given amount of time indicates pregnancy (Figure 5.14).

Rapid tests for strep throat work in a similar way, except that the antibodies are to *Streptococcus pyogenes* (also known as group A streptococcus), the bacterium that causes the disease, and the sample is a throat swab. If *S. pyogenes* is present in the throat swab, a colored product will appear. Confirming a diagnosis of strep throat used to require culturing of the bacterium, which could take up to 2 days. The antibody-based strep test works in 5 to 10 minutes.

Three-dimensional protein structure analysis

The function of a protein depends on its three-dimensional structure, and understanding that structure often provides insight into how a protein works.

Chemists and biochemists use the sophisticated techniques of X-ray crystallography and nuclear magnetic resonance (NMR) spectroscopy to determine protein structure. These methods were originally used to determine the structures of relatively simple molecules. For example, X-ray diffraction data obtained from DNA crystals by Rosalind Franklin were used by Watson and Crick to determine the three-dimensional structure of DNA. These techniques have been refined and augmented for application to complex molecules, such as proteins. Both methods are difficult and have significant limitations.

X-ray crystallography, the more widely used technique, requires the use of extremely pure protein crystals. Crystals are very regular, packed arrays of molecules, and since proteins can have quite irregular shapes, most are not easy to crystallize, and many cannot be crystallized at all. Even for those that can be crystallized, growing protein crystals large enough for crystallography studies can take months or years. Experimenters have even sent solutions into space aboard the Russian space station *Mir* and the U.S. space shuttle because larger, more perfect crystals can be grown in the absence of gravity. If a suitable crystal can be obtained, a beam of X rays is directed into it. The regular array of protein molecules within the crystal diffracts the X rays in a pattern that is recorded on film and used to deduce the arrangement of atoms within the molecules.

NMR exploits the magnetic properties of certain atomic nuclei (usually hydrogen in protein NMR) to

Figure 5.14 A home pregnancy test uses antibodies to detect the pregnancy hormone, HCG. (A) The absorbent wick contains dispersed antibody molecules labeled with a colored indicator. (B) To perform the test, the wick is dipped in urine, which migrates up the absorbent material. (C) If the urine contains HCG, the dispersed colored antibody molecules will bind to it and be swept along with the urine. (D) The indicator window sits over a line of molecular traps for the antibody. If HCG-antibody complexes are carried up the wick in the urine, they will be trapped there, forming a colored line under the window. The colored line indicates the presence of HCG and is a positive result.

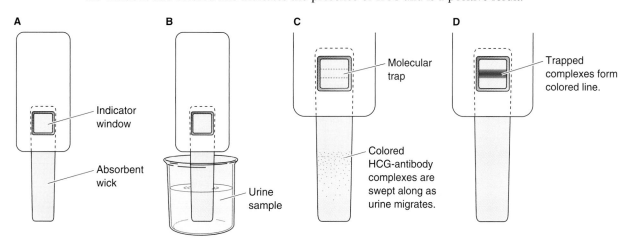

determine molecular structure. Protein NMR uses highly concentrated, pure solutions of protein, avoiding the need for crystals. This technique is limited to very small proteins that remain soluble at the required concentrations. NMR is used in many applications other than determining protein structure, including medical imaging.

The structures of the proteins analyzed to date dramatically illustrate the relationship of form and function. Proteins have grooves, pockets, and even pincer-like structures for binding to DNA or other molecules. Some proteins assume a different shape when bound to their targets. Knowing the structure of a protein often allows us to see how the protein performs its function. It also helps molecular biologists think about how other proteins might work. Finally, knowing the structure of a protein helps biotechnologists design ways to improve the protein for particular applications.

Proteomics

As scientists' ability to characterize complete genomes has improved, focus on characterizing the complete set of proteins expressed in an organism, tissue, or cell has increased, giving rise to a new scientific field: proteomics. Just as the genome is the complete set of an organism's genes, the proteome is the complete set of expressed proteins. Determining the identities of all the proteins expressed in a sample is a substantial challenge. A genome sequence provides a starting point: the sequence of every gene lays out the set of proteins (often tens of thousands) that could be present. Characterizing the array of proteins that are in fact present in a given sample requires powerful separation technology combined with analyses that permit the identification of individual proteins.

If an individual protein can be isolated, it can be fragmented by enzymatic digestion, and the precise masses of the fragments can be determined by sophisticated instruments called mass spectrometers. The precise masses are a function of the amino acids of which the protein fragments are composed. Computerized comparison of the masses generated to the predicted masses of all possible proteins encoded by the genome allows the protein to be identified. This approach works, but it is too time-consuming to be practical for large-scale proteomic analysis.

One approach being taken in proteomics is to digest the mixture of proteins extracted from a sample and determine precise masses for all of them. A "virtual digest" of all possible proteins in the genome reveals protein fragments with unique amino acid compositions and therefore unique masses. A computer can compare the masses of the fragments generated in the sample to the set of predicted unique masses to determine if any of those proteins are being expressed.

Applying the Technologies

We have just described a collection of biological tools and procedures that can be combined in various ways to answer questions, solve problems, and produce products. Now, we discuss ways in which these tools are used to accomplish specific goals, such as finding genes, analyzing genotypes, generating DNA fingerprints, and genetically engineering both plants and animals. As you read, you will see that the same kinds of undertakings can be used in a wide variety of fields, from basic biological research to forensic science, clinical medicine, agriculture, and beyond.

We will focus on three kinds of endeavors: finding genes, analyzing genotypes and genomes, and genetic engineering. These endeavors overlap. For example, if you find a mouse gene you believe has a particular function, the best way to test your hypothesis is to mutate that gene and see how the mouse is affected. Mutating the gene involves genetic engineering. Testing an engineered mouse to see if its genes have been altered in the way you intended involves analyzing its genotype. The techniques form more of a supportive web than a linear sequence.

Finding Genes

One of the first steps toward understanding the molecular basis of a particular trait is to find the genes that contribute to its expression. Identifying the genes is the beginning point for learning what proteins interact to produce a particular phenotype, how they interact, and how their expression is regulated. The most fundamental tool for finding genes involved in a specific trait is a mutant organism that expresses some kind of change in that trait. That organism's DNA can be analyzed in search of the causative mutation (which may sound easy but usually is not).

Mutant organisms are invaluable tools for finding specific genes

The notion of using mutant organisms to find genes may sound odd, since the term "mutant" is so often associated with movies in which mutants are menacing creatures greatly altered from their original state. In science, the term **mutant** is very specific: an organism with an alteration in its genotype.

Typically, scientists use the term to refer to organisms with genetic alterations that cause observable phenotypic alterations. Mutants are extraordinarily useful in scientific research, because each different one can provide information about genes and proteins involved in producing a trait. Scientists studying complex processes in an organism often collect as many different mutants with alterations in that process as they can, because each different mutant offers a different window into the process.

Microorganisms

Let us imagine that we would like to find the gene for a specific enzyme involved in the biosynthesis of the amino acid histidine in *Escherichia coli*. What we need is an *E. coli* strain that lacks this enzyme because of a mutation in a histidine biosynthesis gene. To find such a mutant, we could expose a culture of *E. coli* to a mutagen and screen the cells for individuals that had lost the ability to grow on histidine-free culture medium. The mutant we want cannot make histidine and therefore would not be able to live and reproduce unless its medium was supplemented with that amino acid. We could spread our mutagenized *E. coli* cells out on a histidine-containing medium, allow colonies to form, transfer some cells from each one to a histidine-free medium, and look for cells that could not grow there. If we identify one that cannot grow on the histidine-free medium, we can go back to the colony on the histidine-containing medium and retrieve cells from that colony for further testing (Figure 5.15). Once we have such a mutant (or a collection of several mutants), we can begin to hunt for the gene.

To find the gene that is defective in our mutant, we might next make a library of genes from a nonmutant *E. coli*. We could grow this *E. coli* on a histidine-free growth medium, so that it would have to synthesize its own histidine. We could isolate mRNA from the *E. coli* and use it to make a cDNA library

(see chapter 14). The plasmids in this library would contain inserted DNA fragments corresponding to every gene expressed in the normal *E. coli,* including the histidine biosynthesis genes. We transform a batch of the mutant, histidine-requiring *E. coli* bacteria with this mixture of plasmids and look for a transformed *E. coli* that has acquired the ability to make histidine. We should be able to find it easily because it can now grow on histidine-free medium (Figure 5.16).

The assumption is that any bacterium now able to make histidine must have received the missing gene on its new plasmid. We isolate the plasmid from that bacterium, determine the DNA sequence of the *E. coli* gene it contains, and with luck, we will have found the histidine biosynthesis gene we were looking for.

One very important model microorganism is yeast. Yeast is single celled and can be grown in culture, much like *E. coli*. It can be transformed with plasmids, mutated, and in general manipulated much like bacteria. However, yeast is a eukaryote. As a eukaryote, it is more closely related to organisms like fruit flies, mice, and humans than is the prokaryotic *E. coli*. Yeast provides an easy-to-use system that can be readily manipulated genetically and thus is an excellent system for finding genes that may have close relatives in animals.

Drosophila

The genetic method outlined above is simple and efficient but is applicable only under limited circumstances. Obviously, if you are looking for an animal gene, you cannot collect a batch of mutant animals, transform them with plasmids, and look for restoration of a normal phenotype. A few model systems, however, have well-developed mechanisms for making mutations and finding the genes associated with them.

Figure 5.15 Isolating *E. coli* mutants that cannot biosynthesize histidine (His⁻).

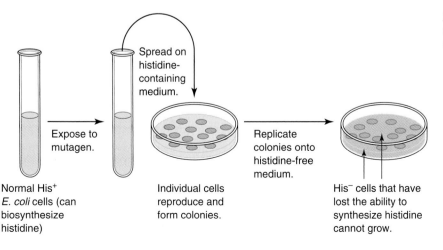

Normal His⁺ *E. coli* cells (can biosynthesize histidine)

Expose to mutagen.

Spread on histidine-containing medium.

Individual cells reproduce and form colonies.

Replicate colonies onto histidine-free medium.

His⁻ cells that have lost the ability to synthesize histidine cannot grow.

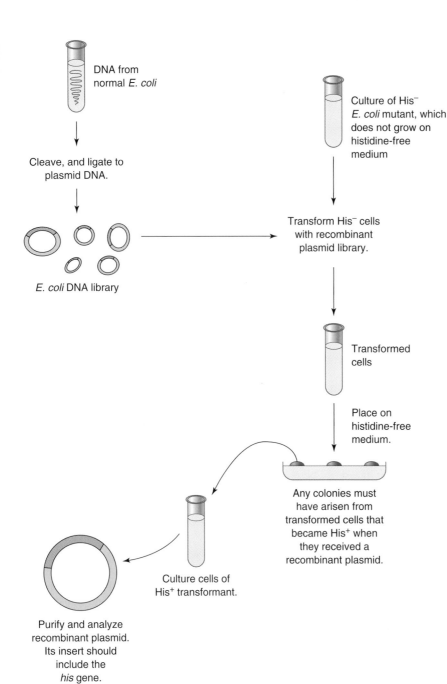

DNA from
normal *E. coli*

Cleave, and ligate to
plasmid DNA.

E. coli DNA library

Culture of His⁻
E. coli mutant, which
does not grow on
histidine-free
medium

Transform His⁻ cells
with recombinant
plasmid library.

Transformed
cells

Place on
histidine-free
medium.

Any colonies must
have arisen from
transformed cells that
became His⁺ when
they received a
recombinant plasmid.

Culture cells of
His⁺ transformant.

Purify and analyze
recombinant plasmid.
Its insert should
include the
his gene.

Figure 5.16 Using genetics to
find an *E. coli* gene for histidine
biosynthesis.

One of the best is *Drosophila*. A number of years ago, scientists comparing wild and laboratory strains of *Drosophila* discovered that many wild strains had acquired a transposon missing in the laboratory strains. If the transposon was introduced into a transposon-free laboratory fly, the transposon DNA would jump into random locations in the fly genome, causing mutations. This phenomenon has been developed into a system for creating mutant flies. If a scientist finds a transposon-created mutant with a phenotype of interest, he can then take advantage of the transposon again. The DNA sequence of the transposon is known, and so hybridization analysis can be used to find DNA segments where the transposon is located. Since the mutation of interest was caused by the transposon jumping into the middle of a gene, when the transposon is located, the gene is also presumably located.

Drosophila can also be mutagenized by feeding mutagenic chemicals to adults or by exposing them to radiation. The mutagens cause mutations during DNA replication associated with gamete production, and mutations appear in the progeny of the mutagenized flies. It is not as easy to find the site of a mutation created this way as it is to find the insertion site of a transposon.

Animals

Convenient mutagenesis systems like the *Drosophila* transposon do not exist for mammals. Instead, scientists must study the mutants nature offers them. You might be wondering where or how to find mutant animals, but in fact, many of them are available. Remember that to a scientist, a mutant is simply an organism with a genetic alteration, usually manifesting itself as a phenotypic difference. Humans have bred and collected mutant animals for centuries, perhaps because they liked the way they looked, as in the yellow, obese mice of Chinese mouse fanciers, or because the variant trait was useful, as in the short legs of dachshunds and basset hounds, which served hunters' purposes. Now, scientists are studying the genes of the obese mice to learn about weight regulation and analyzing genes of purebred dogs (see Color plate 7) to find the basis of their characteristic traits, including both anatomy and behavior.

Depending on how you think of it, you could also consider chocolate and yellow Labrador retrievers to be mutants, since each color variant is missing the activity of a protein. In fact, analysis of the genes of coat color variants has led to an understanding of the biological basis of coat color (see chapters 23 and 24).

In the end, whether you call a yellow Lab a mutant or just a phenotypic variant does not matter. The point is that any organism with an observably different phenotype that has a genetic basis can be a guide to finding the genes underlying that phenotype. We have identified many genes for human inherited diseases by studying individuals from families with those diseases.

To find a gene starting from a phenotypic variant, scientists need DNA samples from a large number of related individuals who carry the trait in their family. If they are looking for a human gene, they must find families that fit this description. If they are looking for an animal gene (say, in a mouse), they can breed the mice and make the family. By observing the progeny of breeding crosses or, in the case of humans, by studying the family's medical history, they can deduce which members have the trait and, if the trait is recessive, which members are carriers. (See chapter 29 for more information.)

Genetic Markers

DNA samples from several individuals are then characterized at many different locations. This can be done by digesting the DNA with many different restriction enzymes and comparing fragment patterns, by amplifying the DNA with many different sets of PCR primers and comparing the sizes of products, or by looking at specific nucleotides which are known to vary in individuals. The goal is to find specific variations between the patterns that are inherited by individuals in the family tree with the trait. A variation can be a unique restriction fragment, a single nucleotide polymorphism, or a unique PCR product. Such a distinctive fragment or pattern is called a **marker.** If a marker is found, it is assumed that the region of the chromosome containing the marker is very close to the disease gene.

After a marker for a trait has been identified, a new phase of work begins. Long stretches of DNA (often hundreds of thousands of base pairs) around the marker must be sequenced and searched for sequence patterns that look like genes (with promoters and coding regions). When potential genes are found, their sequences from the normal and mutant individuals must be compared. If the sequences of a putative gene from individuals with the abnormal trait are always different from the sequences of the same putative gene from individuals with the normal trait, then the gene is probably the culprit. The human gene BRCA-1, which is implicated in hereditary breast cancer, was found in this way.

This entire process can take years and is not foolproof, since a marker can be inherited with a trait purely by chance and not because it is close to the gene underlying that trait. Once a candidate for a particular gene has been identified, further study is required to confirm that the identification is correct.

Related organisms usually have similar genes

Finding a gene is usually much easier if you or someone else has already found the gene in a related organism. Given that closely related organisms have very similar DNA sequences, you could use the known gene sequence to make probes and search for similar sequences in your organism of interest. The study of model organisms and the sequencing of their genomes are paying off handsomely in this arena.

It simply is not practical to make tens of thousands of mutant mice to search for specific phenotypes, but it is quite practical to do so in yeast or *Drosophila*. Once a gene is identified in one of these easily manipulated model organisms, that gene sequence can often be used to locate the gene in other organisms. Now that the complete base sequences of the mouse and human genomes have been determined and are available in computer databases, it is not even always necessary to do physical experiments. Scientists can type base sequences into computerized search engines and ask whether any sequence resembling that one exists in the mouse or human genome (see chapter 34).

The fact that a similarity exists, of course, does not prove that the correct gene has been found, just as finding a marker that is inherited with a disease does not prove that a nearby gene is the culprit. Further research is needed to absolutely prove the gene's identity. This used to be a very difficult task in animals, but the genetic engineering technique of making so-called knockout mice (by inactivating a specific gene in a mouse [see below]) has given scientists a direct way to test whether a specific gene is involved in a specific phenotype. Finding genes for specific traits still is not an easy task by any means, but biotechnology has given us a whole new set of tools for tackling the project, and we are finding and analyzing genes today at a rate that would have seemed incredible just 20 years ago.

Gene sequences or markers can be used for genetic testing

Once a gene has been found and sequenced, the sequence of that gene in an individual can be determined. One application of this kind of genetic testing is to find out whether an individual has a form of a gene associated with an inherited disease. An inherited disease allele that is especially easy to test for is the one associated with sickle-cell anemia (SSA). In SSA, a mutant form of hemoglobin causes red blood cells to assume a sickle shape in low oxygen concentrations, impairing circulation and oxygen delivery (see chapter 4). The mutation that causes SSA alters a restriction site within the hemoglobin (called the β-globin) gene. To test for the presence of the mutation, laboratory technicians use PCR to amplify the portion of the genome where the mutation can occur, then digest the product DNA with the restriction enzyme and look at the pattern (Figure 5.17).

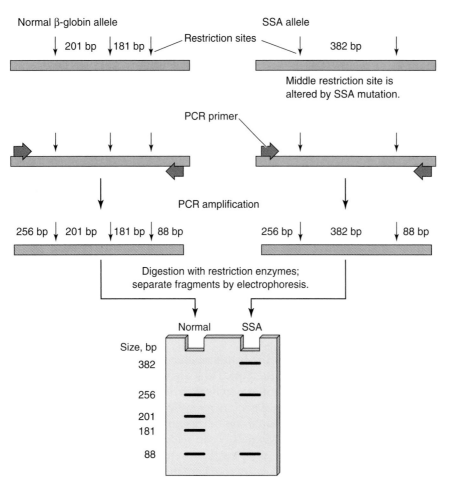

Figure 5.17 Genetic testing for SSA. bp, base pairs.

Comparing Genotypes and Genomes

There are many reasons, both academic and practical, to assess the degree of genetic difference between organisms. If we can compare the genotypes or genomes of different species, we can ask how closely related those species might be and draw conclusions about the course of evolution. Shifting our focus to the level of individuals, we can compare genotypes and draw conclusions about whether the individuals are related or whether two samples came from the same individual. In looking at the genotype of a single individual, we can ask about that individual's species, gender, inherited diseases, and more.

Before the advent of DNA technologies, it was impossible to analyze genotypes directly. Rather, comparisons had to be based on phenotypic characteristics, that is, some tangible manifestation of the information contained in the DNA molecule. We now know that much of the variation between the DNAs of individuals is found in their noncoding DNA and thus is phenotypically invisible. Further, some identical phenotypes can be encoded by different genotypes (for example, people with blood group genotypes AA and AO both express the phenotype of blood group A, despite the genetic difference). Also, adaptation to similar environmental circumstances can cause genetically dissimilar organisms to display similar phenotypic characteristics. Now that genotypes can be analyzed directly, variation that would previously have been invisible can be detected and measured.

Genotypes can be compared in several different ways

There are several ways to go about comparing genotypes. One way is to compare whole genomes through hybridization studies. In this approach, genomic DNAs from two species are prepared and then hybridized to each other. The more similar the nucleotide sequences of the two species, the more extensive the regions of base-paired duplex they will form. To measure the extent of the duplex DNA, the hybrid molecules are heated until the strands separate again. The temperature at which the strands separate (the melting temperature) is a function of the extent of base pairing. The more perfectly matched the two genomes, the higher the temperature required to separate the strands of the hybrid molecules. The melting temperature thus gives a gross measurement of DNA sequence similarity. This kind of assessment has been used to compare the similarity of the DNAs of different species of organisms.

At the opposite end of the spectrum in terms of the level of detail of the comparison is DNA sequence analysis. This approach gives detailed and accurate information but is impractical to apply to large stretches of large numbers of samples, because it is very time-consuming and relatively expensive. DNA sequence comparisons usually focus on small, specific regions. To give meaningful data, these target regions have to be areas of significant variation at the scale of interest. For example, to compare species that are distantly related, scientists select genes that all the species have in common, which means the genes cannot evolve so rapidly that they have become unrecognizable when one looks from yeast to rice to elephants (see *DNA and Protein Sequence Comparisons* below).

If the goal was to compare genotypes of individuals within a single species, however, scientists would need to choose DNA that did evolve rapidly so that differences would be found. Comparisons of most human genes would show very few, if any, differences between individuals. For this reason, sequencing studies of individuals of the same or closely related species often focus on highly variable regions of the nuclear genome or on mitochondrial DNA. Mitochondrial genomes undergo mutation at a much higher rate than the nuclear genome; thus, mitochondrial DNA sequences display much more variation between individuals than do most nuclear sequences. (For more information, see chapter 27.)

Genotype and genome comparisons shed new light on evolutionary studies

Evolution is a unifying theme in biology, and molecular biology has provided powerful new tools for studying it. Evolutionary biologists seek to discover how individual species arose from earlier forms and to understand the mechanisms of the process of evolution itself. Biotechnology has literally revolutionized both endeavors.

The essential concept of evolution is that, given a variable population of organisms, environmental conditions will favor the survival and reproduction of some and not other individuals. The genes of the favored individuals will be passed along to offspring more frequently than the genes of less-favored individuals, and over time, the genetic content of the population will shift toward the genotypes of the more successful forms. If two populations of the same organism find themselves in different environments or facing different selection pressures, over time they will become more and more genetically different. Thus, the amount of genetic difference between two species of organisms reflects the amount of time they have been diverging from one another.

DNA and Protein Sequence Comparisons

Formerly, biologists who sought to describe how modern species arose had to rely on morphological, ecological, and behavioral comparisons between various modern species and between modern and fossil forms to deduce degrees of kinship. DNA and protein sequence comparisons have given these scientists an entirely new set of data to consider. To use molecular data in evolutionary studies, biologists first assemble protein amino acid sequence or DNA base sequence data from a specific protein or region of the genome in the group of organisms under study. They then measure the degree of difference in the sequences. On the assumption that changes accumulate slowly and relatively steadily, they construct various "trees" that show how the different sequences could have been generated from a common ancestor or the order in which different groups of organisms may have evolved from one another. An evolutionary tree based on the amino acid sequence of the protein cytochrome *c* is shown in Figure 5.18. (For more information, see chapter 33.)

Using molecular techniques to probe evolutionary relationships between species is a large and active area of current research. For example, molecular evolutionary analysis found that the kiwi, a flightless New Zealand bird, is more closely related to the flightless birds of Australia than to the moas, the other major group of New Zealand flightless birds. DNA and protein sequence comparisons, as well as hybridization studies, found that humans and chimpanzees are more similar to one another than to any other species. These studies suggest that ancestors of modern humans and chimps probably diverged a mere 5 million years ago.

Ancient DNA

An interesting twist from the field of molecular evolution is the discovery that we can extract DNA fragments from appropriate ancient samples. DNA is rapidly degraded to small fragments after an organism dies, and only special combinations of circumstances will preserve soft tissue cells with DNA. However, samples preserved in bogs, including 17-million-year-old magnolia leaves, have yielded enough DNA for comparisons with modern species,

as have amber-encased insects. It is also possible to recover DNA fragments from bones and teeth, which are more commonly preserved than is soft tissue. PCR is particularly useful in studying ancient DNA because it can amplify the small amounts typically present in specimens.

Early attempts to amplify ancient DNA by PCR were often "learning experiences" for the scientists involved. What they learned was that it is extremely easy to amplify modern DNA contaminants of the ancient sample. In fact, many attempts to amplify ancient DNA resulted in the amplification of the scientists' own DNA. Through these experiences, laboratory personnel learned that extremely stringent protection against contamination is essential in these kinds of experiments.

In one headline-making study of ancient DNA, a group of scientists painstakingly amplified fragments of mitochondrial DNA from a Neanderthal human fossil. Neanderthals were a race of humans who lived in the Near East and Europe from about 125,000 to 30,000 years ago. Their anatomy was somewhat different from modern humans, and their average brain capacity was slightly larger than ours (Figure 5.19). They cared for one another, as evidenced by fossils showing that individuals survived with physical conditions that would have proven fatal if they had not had help. They also appeared to believe in an afterlife, as they buried their dead in graves that included useful objects and flowers. Anatomically modern humans appeared in Europe around 30,000 years ago, near the time that Neanderthals seem to have died out. For many years, it was assumed that Neanderthals were direct ancestors of modern humans, but other lines of evidence suggested modern humans had arisen in Africa independently of the Neanderthals, migrated to Europe, and displaced the local Neanderthal population.

The scientists who amplified mitochondrial DNA from the Neanderthal fossil carefully established that the DNA did not come from any of their own cells. They then determined the base sequence of the amplified mitochondrial DNA. The base sequence was very different from the mitochondrial DNA sequences of any known modern human group. So different, in fact, that the researchers concluded that Neanderthals could not be direct ancestors of modern humans.

We stress that the DNA recovered in these processes is extremely fragmentary and is useful only for comparisons with modern sequences to measure the

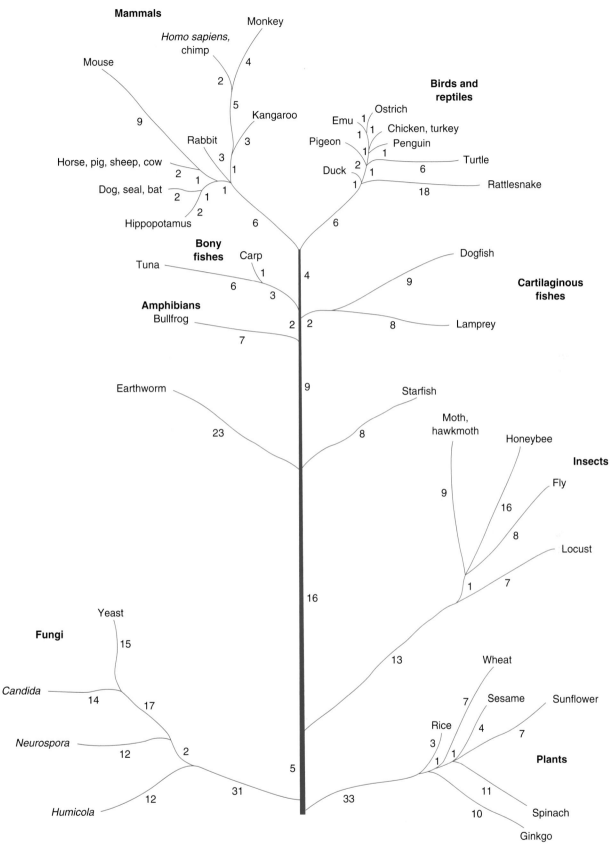

Figure 5.18 An evolutionary tree based on the amino acid sequence of the protein cytochrome *c*. The numbers represent the numbers of amino acid changes between two nodes of the tree. For example, when the common ancestor of cartilaginous fish diverged from the line leading to bony fish, mammals, etc., it evolved and accumulated two amino acid changes in its cytochrome *c* protein before the lines leading to dogfish and lamprey diverged. The dogfish line accumulated nine more amino acid changes in becoming the modern organism; the lamprey accumulated eight changes. Thus, the amino acid sequences of cytochrome *c* in lamprey and dogfish have 17 differences. (From A. Lehninger, D. Nelson, and M. Cox, *Principles of Biochemistry,* 2nd ed., Worth Publishers, Inc., New York, NY, 1993.)

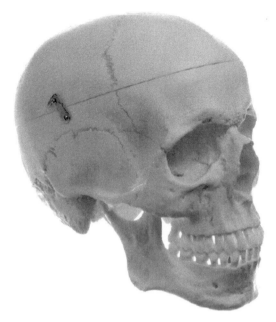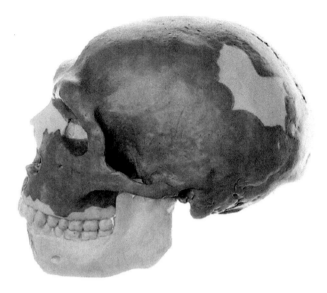

Figure 5.19 Neanderthals (skull cast on right), with their large brains and anatomy similar to ours, were long believed to be direct ancestors of modern humans (skull model on left). Mitochondrial DNA analysis suggests that modern humans instead arose independently from Neanderthals and displaced them. (Photograph courtesy of SOMSO Models, Coburg, Germany.)

amount of change. So far, experience with ancient DNA suggests that using it to resurrect long-extinct species, such as dinosaurs, is impossible.

Overarching Evolutionary Lesson

For scientists who seek to understand the mechanisms behind evolution, molecular biology and biotechnology have provided a gold mine of new information. First, studies of genes and genomes from many organisms have underscored how fundamentally similar we all are. For example, studies using hybridization, restriction fragment length polymorphism (RFLP) analysis (see below), and sequence comparisons suggest only about a 1.6 to 3% difference between the genomes of humans and chimpanzees. From earlier protein and enzyme studies, it was known that all animals produce about the same complement of enzymes and proteins. Now that genes are being mapped to specific locations on chromosomes, scientists are finding that animals' genes are arranged similarly from species to species. Differences in chromosome number and structure seem to have been generated in a process in which chromosomes were cut or broken into large pieces and then put back together in many different ways into different final numbers of chromosomes.

Looking at the DNA sequences of individual genes, we see that nature appears to be quite economical: once a protein function that works has evolved, it is often used over and over. Similar domains are found in many different proteins. (See chapter 4 for more information.) We find proteins that appear to have diverged from two original copies of the same gene, and we find similar proteins filling similar roles in widely disparate species. For example, the proteins that govern body plan in worms, flies, mice, and humans are very similar. The homeotic genes encoding these proteins are even arranged in the same way in the chromosomes of flies, mice, and humans, and they share similar regulatory sequences. Some scientists are now looking at the regulatory sequences associated with homeotic genes as a way to investigate ancestry. The flood of new information about genome structure is giving evolutionary scientists a wealth of fuel for theories about genome evolution.

DNA typing is based on the uniqueness of every individual's DNA sequence

The best-publicized use of genotyping may be its forensic application in criminal cases. Before we had the technologies to look directly at DNA, identifications were made through phenotype: appearance, voice patterns, blood types, dental records, fingerprints, and so on. These methods can be very effective, but they also have significant limitations under certain circumstances. DNA typing, also called DNA fingerprinting, gives both prosecutors and defense attorneys an extremely powerful tool to add to the arsenal of technologies they can use to identify or exonerate an individual suspected of a crime.

It is theoretically possible to identify nearly every individual on earth from his or her DNA sequence, though doing so is not realistic because of the time and money required to sequence an entire genome. It is possible, however, to make very good predictions about individual identity on the basis of a limited and practical examination of highly variable regions of the genome.

One approach to DNA typing uses restriction digestion and hybridization analysis. The hybrization probes match regions of the genome that vary in sequence. The variations give rise to restriction fragments that vary in length from one individual to another, or RFLPs. To perform an RFLP analysis for DNA typing, DNA must first be extracted and purified from a DNA-containing sample. Following restriction digestion, the DNA fragments are separated by gel electrophoresis. The human genome is so large (about 3 billion base pairs) that digestion with most restriction enzymes generates thousands or even millions of fragments, which overlap in the gel lane, forming a smear. Hybridization analysis permits the visualization of only the regions of the genome known to vary when digested with that enzyme. Following electrophoresis, the DNA is transferred to a membrane by blotting and then hybridized to probes for the sequences of interest. The result is an interpretable pattern of bands (Figure 5.20). A similar type of analysis can be performed using PCR instead of restriction digestion. In this approach, primers are chosen that amplify only regions that vary in length from one person to another, giving rise to amplified fragment length polymorphisms.

The regions of interest in DNA typing are usually areas of the genome that vary greatly from individual to individual in that they contain different numbers of back-to-back repeats of the same DNA sequence. When these regions are cut by a restriction enzyme with sites flanking the repeats or amplified by PCR with primers flanking them, the sizes of the resulting products depend on the number of repeats in the individual's DNA. When the products are separated by electrophoresis, the patterns constitute a genetic fingerprint of the individual (Figure 5.20). (See chapter 28 for more information on DNA typing in forensics and paternity cases.)

DNA Typing in Conservation and Ecology
Although DNA typing gets media coverage for its use in forensic cases, the technique is also widely used in conservation biology and ecological research. DNA typing can reveal the degree of kinship of individual animals. This knowledge can be critical to

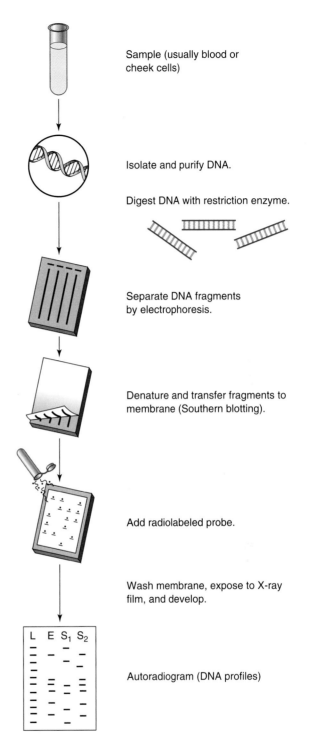

Sample (usually blood or cheek cells)

Isolate and purify DNA.

Digest DNA with restriction enzyme.

Separate DNA fragments by electrophoresis.

Denature and transfer fragments to membrane (Southern blotting).

Add radiolabeled probe.

Wash membrane, expose to X-ray film, and develop.

Autoradiogram (DNA profiles)

Figure 5.20 DNA fingerprinting by RFLP analysis. Radioactive probes are detected by exposing the membrane to X-ray film after hybridization is complete.

the success of captive breeding programs for endangered species, for which there might be only a very few individuals in captivity. For example, DNA typing has been used on whooping cranes so that biologists could select the most genetically different individuals as breeding pairs.

Knowing degrees of kinship between animals in a living group is essential to behavioral ecology studies. In many species, it is impossible to determine paternity simply by watching the living group; DNA typing provides a way to solve this problem. The fundamental social group of African lions is the pride, a group of females and their cubs. Small groups of male lions take up residence with the pride and father cubs, and they can be driven from the pride by an invading group. Sometimes all of the male lions father cubs (typical for groups of two or three males), and sometimes certain males will help with territorial defense of the pride but not father cubs. Scientists have found that male lions that help out other males in territorial defense but do not get to mate with the females are invariably related to the males that do mate. Thus, their behavior helps their own kin to reproduce successfully. By contrast, male lions help unrelated males only if they all get to father cubs.

In another kinship analysis, scientists were able to solve the mystery of the Mexican loggerhead turtles. Pacific loggerheads nest in Japan and Australia, not Mexico, yet young loggerheads could always be found off the coast of Baja California. Many biologists did not believe the young turtles could have come the 10,000-mile distance from Japan (Australia is even farther), and so the origin of the Baja turtles was a mystery. Now, DNA comparisons have established that the Baja population is made up of turtles from both the Japanese and the Australian groups. Apparently the young turtles are carried to Mexico by ocean currents, and they then swim back to Japan or Australia to breed. Quite a swim!

Analysis of genetic variability can provide insight into the status of wild populations. Biologists have assumed that vigorous wild populations contain genetically variable individuals rather than genetically identical ones. DNA typing, together with protein comparisons, provides the means for testing this hypothesis. In a study of lion populations, scientists found that a small population living in an ancient volcanic crater in Tanzania was much less genetically variable than a larger population that roamed the open Serengeti plains. Sperm samples taken from crater lions showed many more abnormalities than comparable samples taken from the plains lions. This result appears to support the greater variability/greater fitness hypothesis.

Gene chips allow simultaneous analysis of many genes

One of the newer technologies in the DNA toolbox allows scientists to test for the presence or expression of many genes at once. The key ingredient of this tool is called a gene chip, or a microarray. A gene chip is a grid of spots of DNA on a tiny glass or silicon slide. Each spot contains copies of one specific DNA molecule, either a synthetic oligonucleotide or a fragment of cloned DNA, that acts as a probe for its complementary sequence. The grid itself may contain thousands and thousands of spots. For example, you can obtain chips with DNA representing every single gene in the yeast genome (about 6,000) on a grid that is less than 1 inch square. Microarray technology has become possible because we have accumulated enough gene sequence information to construct probes for many, many genes from a single organism.

Microarrays allow us to ask questions about the presence of many different forms of genes or the expression of many genes at once. Analysis of the entire genome of an organism, or the expression of many genes from an organism, is often called genomics, as opposed to genetics, which usually refers to the study of one or a few genes at once. (See section E of the activities.)

Two different ways in which microarrays can be used are to ask questions about the forms of genes (alleles) that are present in an individual or about which genes are being expressed in a given tissue under given environmental conditions or in various disease or developmental states. For the genotyping application, a microarray would be constructed with probes that represented all known alleles of a given gene or several genes. A DNA sample from the individual to be tested would be prepared and hybridized to the array. By analyzing which probes hybridized to the sample DNA, researchers could figure out which alleles were present in the individual. (See chapter 30 for more information.)

Genetic Engineering

The term genetic engineering refers broadly to the process of directed manipulation of the genome of an organism, usually a specific gene. The goal of genetic engineering, of course, is not to manipulate an organism's DNA per se but to change something about the proteins produced in that organism: to cause it to produce a new protein, to stop producing an old protein, to produce more or less of a protein, and so on. Manipulating the genome is merely the way to influence protein production. An organism that contains a gene originally from another source is often called a transgenic organism, and the new gene is sometimes called a transgene to distinguish it from the organism's native genes.

Scientists genetically engineer organisms for various reasons. One is to learn about what a given gene does. For example, homologs of the *Drosophila* homeotic genes were identified in mice. To confirm that these genes were important in the development of mice, scientists made genetically engineered mice in which the homeotic genes were nonfunctional and observed the results.

Another reason to make a genetically engineered organism is to obtain large quantities of the protein encoded by the new gene. This kind of approach is used to produce therapeutic proteins, like insulin, from microorganisms. It is also used to produce many of the restriction enzymes and other proteins used as tools in biotechnology. When scientists genetically engineer an organism for protein production, they tinker with the gene to ensure a good output.

No matter what the purpose of genetic engineering is, the gene in question must first be identified and cloned into a plasmid or other vector so that large quantities of a DNA fragment containing the gene can be reproducibly obtained. The cloned gene is manipulated in the laboratory using the tools and techniques described above to get it ready for insertion into the target organism.

Moving a gene into a target organism is a simple phase for a process that can be very complex. Techniques for gene transfer, such as transformation, all move DNA into a single cell. If your goal is to move a gene into a plant or animal, you would like to get the gene into many cells, if not every cell, of that organism. That is a more complicated proposition, and we will look next at some methods for doing so.

Genetic engineering of microorganisms

In the last chapter, we described how to clone DNA using bacteria as host cells. Genetic engineering takes the process one step further, in that it introduces a specific gene or genes into a bacterial host in such a way that the proteins encoded by the genes will be expressed.

The first step in this process is to clone fragments of DNA that contain the gene of interest. If you are cloning a eukaryotic gene into a prokaryote, you will need to tinker a bit with the gene so that it can be expressed in its new host. For example, most eukaryotic genes have introns that are spliced out following transcription in eukaryotic cells (see chapter 4). Bacterial genes do not contain introns, and bacteria lack the means to splice eukaryotic RNA. Therefore, you would use reverse transcriptase to prepare cDNA from mRNA isolated from donor tissue. You would also have to add prokaryotic traffic signals, such as a promoter, so that the gene can be transcribed in its new host. Even if the new host is a single-celled eukaryote, such as yeast, you may need to optimize the traffic signals. Finally, you might want to use a vector that has built-in regulatory signals that permit you to control the expression of the new gene in its new host. One of the simplest of these is based on the *lac* operon (see chapter 4). Remember that the *lac* genes in *Escherichia coli* are expressed only in the presence of lactose, which binds to the *lac* regulatory protein and prevents it from blocking transcription. If you clone a new gene under the control of the *lac* promoter and regulatory regions, that gene will be expressed only when lactose is present in the growth medium (Figure 5.21).

Figure 5.21 Genes are often cloned so that their expression can be controlled. For example, a gene cloned with the *lac* promoter will be expressed only if lactose is present in the growth medium.

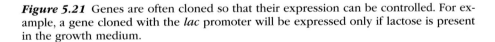
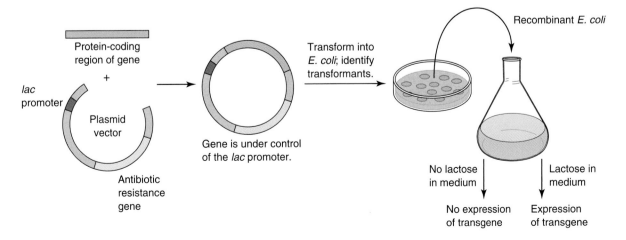

Once you have finished tinkering with the gene's traffic signals to prepare it for its new host, you would use one of the DNA transfer procedures described in the last chapter to introduce the vector and gene. As a final step, you would verify that the transgenic organisms indeed contained the correct new DNA molecule and that they were producing the desired protein.

Microbes have been engineered to make useful proteins, such as human insulin to treat diabetes. They have also been engineered with new enzymes that allow them to make nonprotein products, such as the dye indigo. Microbes used for industrial production of enzymes, such as amylase (see chapter 31), have had additional copies of their own amylase genes added to their genomes, increasing production of the enzyme.

Genetic engineering of plants

The most significant event in the development of human civilization was probably the domestication of plants for agriculture. People have been manipulating the genetic content of crop plants ever since they began selecting seeds from more useful forms for propagation. Plant-breeding techniques, such as forced hybridization, transferred large numbers of unknown genes. Seed mutagenesis introduced random changes into plant genomes in hopes of getting desirable traits, again without knowing what genes were involved. Genetic engineering, in contrast, involves the movement of one or a very small number of known genes into a plant to introduce specific characteristics. A few examples of the ways in which genetic engineering has been used are to:

- Make crop plants resistant to insect pests
- Make crop plants resistant to viral diseases
- Produce edible vaccine proteins in fruits
- Produce other medicinal proteins in plants
- Make plants frost resistant

Making a transgenic plant requires the cloning and engineering of the desired new DNA and the introduction of that DNA into plant cells. One of the most useful vectors for transforming plant cells is the bacterium *A. tumefaciens* and its Ti plasmid. Scientists can clone desired genes into the Ti plasmid and use *A. tumefaciens* to transfer the plasmid into plant cells. (See chapter 22 for more information.) Once scientists have successfully carried out these steps, they have genetically engineered plant cells, not genetically engineered plants. Producing an entire plant from transformed plant cells requires a set of technologies called plant tissue culture.

Plant Tissue Culture

Plant tissue culture allows the regeneration of an entire plant from a piece of a single tissue, or even single cells. If this technology were available for animals, it would be as if you could regenerate a person from a piece of his big toe, liver, or any other tissue. In plant tissue culture, a piece of a plant is surface sterilized and placed in a sterile dish of culture medium containing nutrients and plant growth factors. Under these conditions, the plant piece produces callus cells, an undifferentiated cell type that grows over wounds in plants. Hormone concentrations in the medium can be manipulated to cause the callus to produce shoots, roots, and eventually a tiny, complete plant that is genetically identical to the callus tissue (Figure 5.22). The tiny plant can grow to normal size and reproduce.

Resistance to Viral Diseases

In the 1980s, researchers at the University of Washington in St. Louis paved the way for engineering virus resistance into many important crop plants. Plant growers had known for a long time that infection with a mildly virulent virus often protected plants from infection by more dangerous viruses, just as infection by the mild vaccinia (cowpox) virus was found to protect people from infection by the deadly smallpox virus. The researchers showed that plants transgenic for the coat protein gene of the tobacco mosaic virus were resistant to infection by the virus. Following the publication of their work in 1986, other researchers showed that coat proteins from many other important plant viruses provided protection from infection when cloned into a variety of different plants. This technology has been used to protect numerous crops from diseases that normally take a significant toll.

Fighting Aluminum Toxicity

One of the environmental problems affecting the ability of soil to sustain plant life is aluminum toxicity. More than one-third of the world's soil suffers from aluminum toxicity, and the problem is most severe in the humid tropical climates of many developing countries. Aluminum toxicity is also a consequence of soil acidification as a result of acid deposition, seen in sensitive soils of the United States and other countries. Aluminum ions injure plant root cells, thus interfering with root growth and nutrient uptake. In an effort to make crop plants more resistant to aluminum, Mexican scientists transformed corn, rice, and papaya with a bacterial gene for the enzyme citrate synthase. The transformed plants released citric acid, which binds to soil aluminum and

A
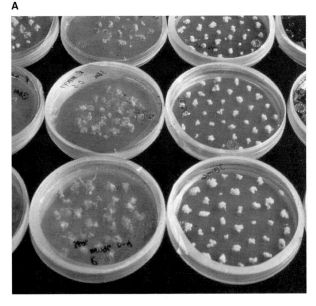

B
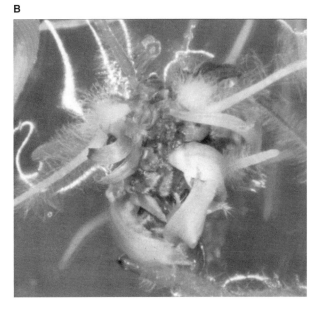

Figure 5.22 Plant tissue culture. (A) Pieces of plant tissue can be induced to form un-differentiated callus cells (plates on right). Manipulation of hormones in the growth medium causes the callus to differentiate into tiny plantlets (plates on left). (B) Close-up of plantlet. Note the fuzzy roots and tiny leaves. (Photographs courtesy of Syngenta.)

prevents it from entering plant roots. The genetically modified plants were able to germinate and develop at aluminum concentrations that were toxic to nontransgenic plants.

Genetic engineering of animals

Making a transgenic animal is similar in concept to making a transgenic plant, but quite different in practice. It is possible to take a piece of a plant leaf and manipulate it in culture to regenerate an entire plant, but it is not possible to take a piece of an animal and regenerate the animal. The only animal cells that can generate entire organisms are fertilized eggs or early embryo cells, and therefore, transgenic animals are made by manipulating these cells.

Microinjection of Fertilized Eggs

To make an animal that expresses a new gene, a recently fertilized egg is microinjected with a DNA fragment containing the desired new gene. The injected egg is implanted into a surrogate mother animal and allowed to develop into a baby animal. After the baby is born, it is tested to determine whether its genome contains the new gene and whether it is expressing the new protein. Making transgenic animals is partly a numbers game. Some percentage of the microinjected eggs do not survive the injection process. Of those that survive,

only a fraction will incorporate and express the new gene. Thus, the procedure has to be repeated with many eggs to increase the chances of success.

Scientists at Virginia Polytechnic Institute and State University (Virginia Tech) have made transgenic pigs that contain the human gene for factor VIII, a protein missing in people suffering from hemophilia A. The blood of hemophilia A patients does not clot properly, and even minor injuries can be life threatening. Hemophilia A patients can be treated with factor VIII protein, but obtaining large amounts of the protein for medical use was problematic. The Virginia Tech scientists engineered the factor VIII gene with a promoter from a milk protein gene that causes the protein to be secreted in the transgenic pigs' milk. The scientists estimate that 300 to 600 milking sows could produce enough factor VIII to meet the world's demand.

Gene Replacement in Embryonic Stem (ES) Cells

In the egg microinjection procedure for making transgenic animals, the new genetic material can be inserted anywhere in the genome. Since the point of the procedure is to obtain an animal that makes a new protein, it often does not matter where the new DNA is located. Sometimes, though, rather than introduce a new gene into an animal, we want to replace an allele of a gene that is already present with a new allele. This application requires precise replacement of an existing gene segment with a new one.

To replace a specific gene, scientists start with ES cells. These cells, taken from early embryos and grown independently in culture, retain the ability to differentiate into every kind of cell in the adult animal. First, the gene of interest is manipulated in the desired way (for example, it may be inactivated by removing a segment or introducing stop codons). The prepared gene is then put into a vector that contains antibiotic resistance genes so that transformed cells can be identified, and the vector with the gene is introduced into ES cells. In this procedure, the vector is linear DNA that cannot be replicated and transmitted independently; it must integrate into the host genome to be stably maintained.

Because the manipulated gene and the host cell's genome contain DNA sequences in common, at some frequency homologous recombination will take place between the newly introduced DNA and the host genome. The recombination event results in the replacement of the intact gene by the inactivated form. Transformed cells are selected by their ability to grow in the presence of the antibiotic and then tested to make sure the new allele is present at the right place in the genome.

Once an ES cell has been engineered with the new allele and allowed to multiply into a clonal population, some of the cells are injected into early mouse embryos. The embryos are implanted into surrogate mothers and allowed to develop into baby mice (Figure 5.23). At some frequency, the ES cells will incorporate into the embryos and develop as part of the babies.

To help identify baby mice that have incorporated the ES cells, scientists usually use stem cells from brown mice and implant them in embryos from white mice. Babies that have incorporated the ES cells will have patches of brown and white fur. Such mice are called **chimeras,** because they are descended from two sets of parents (the brown mice and the white mice). The parts of the mouse with brown fur are descendants of the engineered ES cells and contain the engineered gene. The parts of the mouse with white fur are descendants of the original embryo (Figure 5.23). To get a mouse that has the new allele in all its cells, chimeric mice whose reproductive tissues are descendants of the engineered cells are bred together.

Knockout Mice

One spectacularly useful application of this gene replacement technology has been making so-called **knockout mice.** When scientists identify a gene they believe to be involved in a particular process, the best test of their conclusion is to knock out the gene in a test animal and see if the process is affected. Gene replacement technology is used to inactivate the specific gene in mice by replacing a functional allele with an engineered, nonfunctional one, and the result is called a knockout mouse.

Quite frequently, knockout mice can be used as models for human disease, because the genomes of mice and humans are about 80% similar. For example, by analyzing the DNAs of family members afflicted with hereditary early heart failure, scientists at Harvard University identified a gene for a protein that appeared to be responsible for the condition. The scientists wanted to test whether the mutant gene alone really was the cause of the early heart failure. To do this, they made a knockout mouse in which the gene was disrupted. The knockout mice developed early heart failure, proving that disrupting just that

Figure 5.23 Making a transgenic mouse with ES cells.

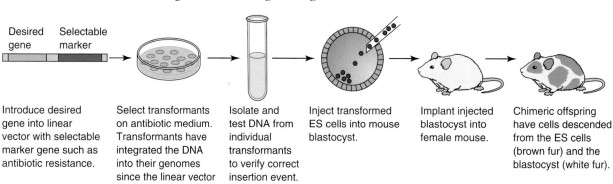

| Introduce desired gene into linear vector with selectable marker gene such as antibiotic resistance. | Select transformants on antibiotic medium. Transformants have integrated the DNA into their genomes since the linear vector cannot be maintained independently. | Isolate and test DNA from individual transformants to verify correct insertion event. | Inject transformed ES cells into mouse blastocyst. | Implant injected blastocyst into female mouse. | Chimeric offspring have cells descended from the ES cells (brown fur) and the blastocyst (white fur). |

one gene caused the disease. Knockout mice are also extremely useful for developing and testing new therapies and drugs before they are tested on humans.

Selected Readings

Anderson, W. F. 1995. Gene therapy. *Scientific American* 273:124.

Bayley, H. 1997. Building doors into cells. *Scientific American* 277:62.

Capecchi, M. 1994. Targeted gene replacement. *Scientific American* 270:52–59.

Ellgren, H. 2005. The dog has its day. *Nature* 438:745–746. A news story/comment about the first publication of a complete dog genome sequence.

Cohen, J. S., and M. E. Hogan. 1994. The new genetic medicines. *Scientific American* 271:76–82.

de la Fuente, J. M., and L. Herrera-Estrella. 1997. Aluminum tolerance in transgenic plants by alteration of citrate synthesis. *Science* 276:1566–1568.

Hall, S. 1995. Protein images update natural history. *Science* 267:620–624.

Jeffreys, A. 2005. DNA fingerprinting. *Nature Medicine* 11:xiv–xviii. A nontechnical retrospective about the development of DNA fingerprinting by the scientist who introduced it to the world.

Kahn, P., and A. Gibbons. 1997. DNA from extinct hominids. *Science* 277:176–178.

Klein, R. 2003. Whither the Neanderthals? *Science* 299:1525–1527. An overview of what is known about Neanderthals, including recent DNA evidence. Written for nonspecialists.

Kreuzer, H., and A. Massey. 2006. *Biology and Biotechnology: Science, Applications, and Issues.* ASM Press, Washington, DC.

Morrell, V. 1997. The origin of dogs: running with the wolves. *Science* 276:1647–1648. News commentary on the evolutionary study of dog DNA.

Mullis, K. 1990. The unusual origin of the polymerase chain reaction. *Scientific American* 262:56–65. How the Nobel Prize-winning surfer-scientist came up with the PCR technique.

Paabo, S. 1993. Ancient DNA. *Scientific American* 269:86–92. Isolating and studying DNA from ancient materials. Anyone with questions about the feasibility of Jurassic Park should read this article.

Pennisi, E. 2004. Genome resources to boost canines' role in gene hunts. *Science* 304:1093–1094. A discussion of how progress in understanding the dog genome and developing molecular tools like microsatellite markers have made purebred dogs an invaluable resource for gene hunters.

Ronald, P. 1997. Making rice disease-resistant. *Scientific American* 277:100.

Roush, R. 1997. Antisense aims for a renaissance. *Science* 276:1192–1193.

Saiki, R. K., S. Scharf, F. Faloona, K. Mullis, G. Horn, H. Erlich, and N. Arnheim. 1985. Enzymatic amplification of beta-globin genomic sequences and restriction site analysis for diagnosis of sickle cell anemia. *Science* 230:1350–1354.

Sarinaga, M. 1997. Making plants aluminum tolerant. *Science* 276:1497.

Scientific American. 1997. Making gene therapy work. *Scientific American* 276(5). This special edition contains several articles about topics in gene therapy, including overcoming obstacles to gene therapy, nonviral strategies, gene therapy for cancer and AIDS, and what cloning means for gene therapy.

Shreeve, J. 1999. Secrets of the gene. *National Geographic* 196:42–75.

Weiner, D., and R. Kennedy. 1999. Genetic vaccines. *Scientific American* 281:50.

White, R., and J. Lalouel. 1988. Chromosome mapping with DNA markers. *Scientific American* 258:40–48.

PART II
Classroom Activities

Classroom activities for wet and dry laboratories are provided. Presented in lesson plan format, these activities are grouped into five sections that stress fundamental concepts. Section A teaches concepts basic to molecular biology: the structure and function of DNA. Sections B and C focus on manipulating DNA and transferring it between organisms. These sections illustrate how a deep understanding of biological systems allows us to use such systems to our advantage. Section D makes the connection between the fields of genetics and molecular biology, illustrating how an understanding of biological processes at the molecular level explains genetic observations that we make at the organism level. Section E, new to this edition, introduces techniques used in the analysis of genomes and how genomic studies are applied to increase our understanding of evolution and individual variation. Section F changes the emphasis to proteins, evolution, and bioinformatics. Students focus on an enzyme common across the kingdoms through wet laboratory activities and an exploration of online bioinformatics resources. See the appendixes and the CD accompanying this book for information on laboratory biosafety, equipment, recipes for solutions, and lists of sources of further information.

(continued)

D. Molecular Biology and Genetics   335

E. Genomics   381

F. Bioinformatics and Evolutionary Analysis of Proteins   439

Classroom Activities

A. DNA Structure and Function

A basic principle of biology is that structure must serve function. We can see examples of this principle when we look at the structure of a bird's wing in relation to flight or at a human eye in relation to sight. The same relationship between structure and function holds true at the molecular level. Nowhere is this relationship more apparent than in the DNA molecule. Its structure is wonderfully suited to its two functions: replication and information translation. It also permits easy regulation through a number of mechanisms.

These lessons focus on the DNA molecule. Models, simulations, and discussions illustrate the structure of DNA, its packaging and storage, DNA replication, the processes of transcription and translation, and aspects of gene regulation, including the exciting field of antisense technology. Simple wet laboratory

procedures for extracting DNA from *Escherichia coli,* plant tissue, animal tissue, and store-bought baker's yeast conclude the section.

In teaching about the structure and functions of DNA, there is no substitute for models, whether you are teaching middle school students or adults. Many excellent models for illustrating DNA structure and replication, as well as transcription and translation, are commercially available. If you already use some of these materials to teach DNA structure and protein synthesis, you may find some ideas in the activities on DNA replication, gene regulation, and antisense technology that could be easily incorporated into your lessons. If you do not have a commercial DNA model set, this section begins with an inexpensive paper model that all of your students can help make.

Classroom Activities

DNA Structure

6

About This Activity

In this activity, students assemble a construction paper model of a DNA molecule. Many students, whether children or adults, often have a difficult time visualizing the structure of DNA. Having students assemble a model of DNA is probably the most useful way of communicating information about DNA structure to the large population of students who learn by doing. If you do not already have an appropriate DNA model kit, this lesson provides an inexpensive model that each student can participate in building. There is also an excellent historical precedent for paper DNA models: Watson and Crick used cut-and-paste models when they were trying to figure out the structure of DNA.

The activity can be used with students of many ages and different abilities. We have used it successfully in classroom settings ranging from junior high school/middle school to college. Even though it may seem too simple for advanced or college students, our experience suggests otherwise. Though advanced students are usually familiar with the base pairs, they often do not understand the antiparallel orientation of the two DNA strands and usually have not thought about the fact that the 5′ and 3′ ends of strands are different. Advanced procedures, such as DNA sequencing and the polymerase chain reaction, depend on these differences, and students need to understand these aspects of DNA structure before they can understand the procedures. Examining the model seems to get the information across better to many students than does looking at pictures in a book.

This simple model can also help correct a common misconception about DNA that drawings that use letters in place of bases (as do many drawings in this book) often promote. That misconception is that DNA can be upside down or backwards, since letters have only one correct orientation. You can use this model to help students see that the only meaningful orientation in a DNA molecule is the 5′- to 3′ orientation of the backbone.

Hanging the DNA model from the ceiling or attaching it to a wall is useful for teaching the rest of the activities in this book, nearly all of which call upon students to visualize a DNA molecule. It gives students a concrete structure on which to base more abstract, difficult concepts. The paper model can be useful in teaching replication, transcription, and restriction of DNA.

Class periods required: *1 or 2 45- or 50-minute classes*

Introduction

The following discussion is intended purely as information for the teacher. It is not necessary to share all of it (or even most of it!) with your students. Use your judgment as to how much you are comfortable with and how much you think your classes can absorb. You also may wish to review parts of chapter 4.

DNA is the blueprint of life. It controls body form, functions, and appearance by coding for the proteins that form the bricks and mortar of tissue, carry out metabolic activity, fight infection, regulate growth, and synthesize fats and pigments. Chemically, DNA is made up of small repeating units. The repeating unit of DNA is quite simple; it is composed of three parts: the sugar deoxyribose, a phosphate group, and one of four organic bases, adenine, cytosine, thymine, or guanine. This unit is called a nucleotide or, more properly, a *deoxynucleotide* (Figure 6.1). (RNA is also composed of nucleotides but uses the sugar ribose; hence, RNA is said to be composed of *ribonucleotides*.) To make a DNA polymer, many deoxynucleotides are joined together by phosphodiester bonds between the phosphate and sugar groups.

The phosphodiester bonds form a bridge between the number 5 carbon of the deoxyribose portion of one nucleotide and the number 3 carbon of the deoxyribose portion of the adjacent nucleotide (Figure 6.1 shows the numbering). Using the carbon numbers, we can talk about direction with respect

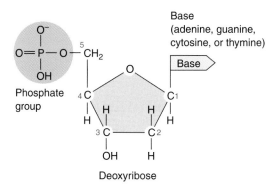

Figure 6.1 The deoxynucleotide.

to DNA polymers. In Figure 6.2, the trinucleotide is shown 5′ to 3′ from top to bottom, because the 5′ carbon of the first nucleotide comes first. Imagine flipping the trinucleotide upside down. Now, the 3′ carbon with the OH group of the "T" nucleotide comes first; this is the 3′-to-5′ orientation.

The four bases of DNA are of two types: purines and pyrimidines. Adenine and guanine are purine bases, and cytosine and thymine are pyrimidine bases. These bases are capable of forming two specific pairs: adenine with thymine and cytosine with guanine. Notice that a purine always pairs with a pyrimidine. Purine bases are larger than pyrimidines. A purine-purine pair would be much larger than a pyrimidine-pyrimidine pair. However, the two purine-pyrimidine base pairs are the same size (Figure 6.3). Since they are the same size, they can fit into the interior of the DNA helix neatly, in any random order. The base pairs are connected by weak

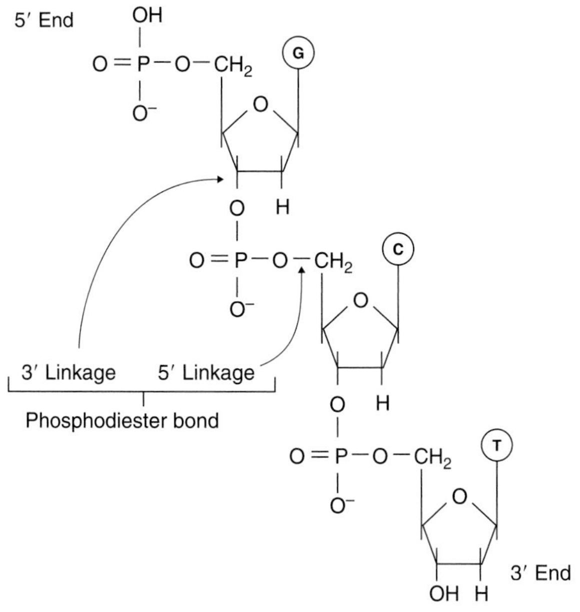

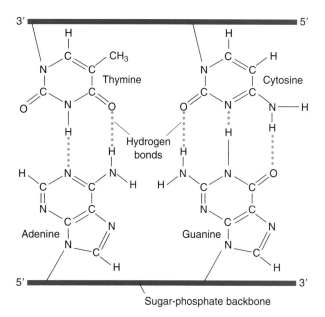

Figure 6.3 Complementary base pairs in DNA.

chemical bonds called hydrogen bonds. (See chapter 4 for more information.) Because of the chemical structures of the bases, adenine and thymine are connected by two hydrogen bonds while cytosine and guanine are connected by three.

In a DNA molecule, two complementary DNA polymers are connected by the hydrogen bonds between these base pairs. The sugar-phosphate backbones of the polymers are oriented in opposite directions: one is 5′ to 3′, and the other is 3′ to 5′. Finally, the two polymers (usually called strands) are twisted around each other to form the famous double helix (Figure 6.4).

The double-helix structure of DNA is ideally suited to the cellular environment of the molecule. The cell is a water-based environment, and molecules that are charged or electrically polar (like the charged phosphate groups and the sugar molecules) interact readily with it. The organic bases of DNA, however, are nonpolar and therefore hydrophobic (water fearing). They are more stable when interacting with other hydrophobic molecules instead of water. In the double helix, the bases are turned toward each other, away from the watery cytosol. Between the sugar-phosphate backbones, the flat hydrophobic base pairs stack together, and these so-called "stacking interactions" further stabilize the double-helical structure.

Even though the base pairs are on the inside of the helix, it is possible for cellular proteins to sense the identities and sequence of those base pairs. The

Figure 6.2 A trinucleotide.

5′ End

3′ Linkage   5′ Linkage

Phosphodiester bond

3′ End

6. DNA Structure • 173

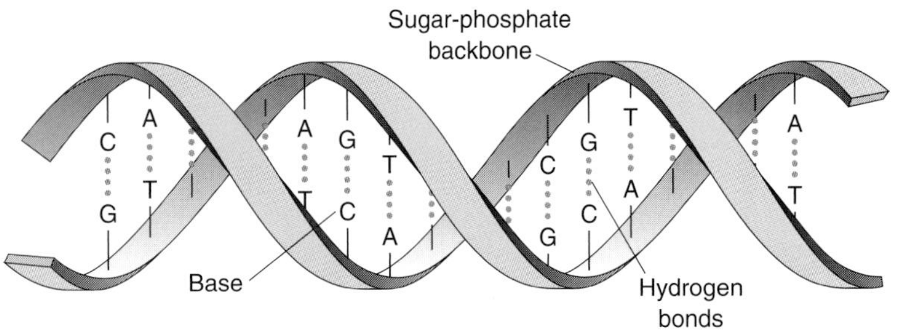

Sugar-phosphate
backbone

Base

Hydrogen
bonds

Figure 6.4 Ribbon model of the DNA helix.

"edges" of the base pairs are "visible" to cellular proteins between the spiraling strands of the sugar-phosphate backbone. The edge of each base presents a different chemical configuration to the outside, so a protein that interacts directly with DNA can "read" the base sequence. In this way, RNA polymerase can recognize a promoter, repressor proteins can find their binding sequences, and DNA replication proteins can identify their "start" sites, the replication origins (see the activity *DNA Replication*).

If you have a three-dimensional model of DNA, you can see the two grooves spiraling up the outside of the helix between the sugar-phosphate backbones. Because of the geometry of the helix, one of these grooves is wider than the other. The wider one is called the *major groove* of DNA; the smaller one is the *minor groove*. (Yes, DNA *is* groovy.) The important thing about the grooves of DNA is that they are the places where proteins interact with specific DNA sequences. Most of the sequence-specific DNA-protein interactions that are currently understood involve the protein binding to DNA in the major groove. (For an example of a specific protein-DNA interaction, see *Lambda and* trp *Repressors: DNA-Binding Proteins* in chapter 4.)

When the structure of DNA was first proposed by Watson and Crick in 1953, it was thought that the helical structure they described was fixed and unchangeable. Since then, we have learned that the DNA helix can assume slightly different forms, can bend and kink, and can even unwind. All of these different forms appear to be important in the functioning of the cell. The "standard" form of DNA is called the B-form helix. It has 10.5 base pairs (bp) per complete turn, and its center of symmetry is down the middle of the base pairs. The B-form helix is the kind that your class will build. However, two additional helix forms called A and Z have also been shown to exist.

A-form DNA has about 11 bp per complete turn, and the center of symmetry is not down the middle of the base pairs but along the outside of them between the "wrappings" of the sugar-phosphate backbone (in

the major groove). Z-form DNA is even more different: A- and B-DNAs are right-handed helices, while Z-DNA is left handed. It has about 12 bp per helical turn, and the geometry of the sugar-base bonds is altered. If you would like details on the structures of A and Z helices, please refer to an advanced text.

The paper DNA model described below allows students to construct a large DNA helix. The model can be used in subsequent lessons on DNA replication and transcription, restriction enzymes, DNA sequencing, etc. Students can make miniature paper DNA models to use in small groups if you reduce the template patterns on a photocopier. (Depending on the size of the models and the stiffness of the paper you use, the smaller models may not twist well, but they would work for DNA replication activities.) The *Tips* section below provides suggestions about adapting the model for younger or more advanced classes.

Objectives

The objective of this lesson is to construct a paper model of the DNA helix. Students will do this by making individual nucleotides. Class members will then join their nucleotides together to form a double helix.

At the end of this activity, students should be able to:

1. Describe the structure of DNA.
2. State which bases form the complementary pairs in DNA.
3. Identify the purine and pyrimidine bases, and state which are larger.
4. Describe what is meant by 5′ and 3′ strand ends (advanced students only).

Materials

- Copies of template patterns from the CD
- Six colors of construction paper or printer paper
- Scissors, ideally for each student
- Glue
- Stapler and staples

Preparation

- Write on the template patterns the color each base, the sugar, and the phosphate should be (for example, cytosine, yellow; guanine, green; adenine, pink; thymine, blue; phosphate, black; and deoxyribose, red).
- Photocopy or print copies of the templates. Printing or copying onto the colored paper will save some classroom time.
- Photocopy the *Student Activity* if you are going to use it and your students do not have manuals (see *Tips* below).

Procedure

To assemble nucleotides (students do this)

1. Cut out the pattern for the assigned nucleotide(s).
2. Place the pattern on the appropriate color construction paper and cut it out.
3. Label the construction paper piece as the pattern is labeled (omit the color name).
4. Glue the nitrogenous base to the sugar molecule by matching up the dots.
5. Glue the phosphate group onto the model by matching up the stars.
6. The teacher will join the nucleotides together to form a helix.

To assemble the helix (teacher does this)

For the helix to come out even, you will need to keep track of how many of what kind of nucleotides are available. An easy way to do this is to use exactly half of all the A's, T's, C's, and G's to make the first strand (put the bases in any order); then, you will be sure to have the right number of complementary bases to assemble the second strand.

To make the first strand, staple the phosphate group to the 3′ carbon position on the deoxyribose molecule (see Figure 6.1 for the carbon numbers; the positions are denoted by squares on the templates). Assemble this one with the sugars "right side up" (the phosphate groups will be pointing upward; this orientation will be 5′ to 3′).

For the second strand, have one of your students hold the first strand for you. Ask your students what nucleotide is needed to base pair with the first nucleotide. When they tell you, select that nucleotide. Turn it "upside down" to simulate the 3′-to-5′ direction (the phosphate group should be pointing down). Remind the class that the two strands of a helix run in opposite directions. Adjust the position of the nucleotide, and staple the phosphate group to the 3′

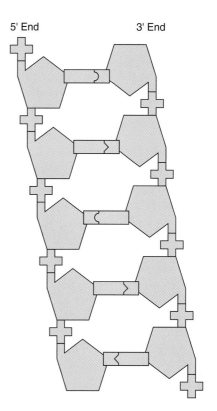

5′ End   3′ End

Figure 6.5 A segment of the assembled model untwisted.

sugar position; allow some overlap for strength. Make the base pair by stapling the two bases together; overlap them to make a strong connection. Ask the students what the next nucleotide should be. Connect it to the previous nucleotide by stapling the phosphate group to the 3′ sugar position. Continue until you have completed the second strand (Figure 6.5).

To make the helix, twist the paper ladder. It will work best to keep the ladder vertical while you twist; have a student hold one end for you, or hang the helix from the ceiling. Some teachers literally festoon their classrooms with the model, letting it run up the walls and across the ceiling. Keep the helix in your room as a twisted model or in the flat form. Although the helix is a better representation of DNA, the flat form illustrates the most important features, the linear sequence of nucleotides and the complementarity of the strands.

Tips

Pass out the *Student Activity* only to younger students, low-achieving classes, or average 9th or 10th grade biology classes. The questions can be given to any class, although some of them are quite easy for advanced students; use your judgment about whether to assign the more advanced questions to your students.

In low-achieving classes, it may be beneficial to cut out models one class period and paste them the second day. Divide the class into groups. Assign each group a nucleotide, the phosphate, or the sugar to cut out. Make sure they label each one.

In an average class, have students cut out all four nucleotides.

In an above-average class, do not tell students how to put the model together (do not give them the activity sheet). Let them discover for themselves. They can be divided into small groups and asked to make miniature DNA molecules using reduced templates. Two 64% reductions of the templates make a nice size, but be aware that a DNA molecule made at this scale out of heavy construction paper does not twist well. You may want to experiment with flimsier paper if you want students to make twisted models to use at their desks. (Flat desk-scale models would be nice for the DNA replication activity.)

Advanced-placement classes can be asked to modify this very simple model to be more accurate. They should label all functional groups on the sugar and number the carbons to determine the 5′ and 3′ termini (see the introductory material for numbering). They can be asked to create better templates for the nitrogenous bases to simulate correctly the single- and double-ringed structures of the purines and pyrimidines. They can also make small models at their desks.

DNA replication can be taught with this model by having students make extra nucleotides, simulate unwinding of the helix, and build new strands (see the next lesson, *DNA Replication*). The large helix model is also useful in introducing restriction enzymes (see the activity *DNA Scissors*); you can build in a palindrome and "cleave" it to create blunt or sticky ends. You can also refer to the large helix during the activity *Sizes of the* Escherichia coli *and Human Genomes* to remind students of the structure of the DNA molecule. Use your imagination to find additional ways to employ this inexpensive model.

Answers to Student Questions

1. Thymine
2. Cytosine
3. The nucleotide (deoxynucleotide)
4. A double helix
5. Adenine and guanine
6. Cytosine and thymine
7. Because of the chemistry of the bases; because of the space requirements of the helix
8. Deoxyribose

Answers to Advanced Questions

9. TCGAGTC
10. C, because the opposite strand is antiparallel to the given strand. The sequence in the answer to question 9 would be in the 3′-to-5′ direction.
11. 55. In 100 bp, there are 200 bases. There are 45 cytosines, so there are also 45 guanines (90 bases accounted for). There are 110 more bases in the molecule, A's and T's. There must be equal numbers of A's and T's, since these two bases pair. Therefore, there are 55 of each.

Constructing a Paper DNA Model   6

Introduction

DNA is the genetic material of every living thing on earth. The information encoded in DNA determines the forms and functions of the cells of which each organism is composed and, ultimately, of the entire organism. DNA fulfills this function because it contains the instructions for making every protein in the organism. Proteins are important because they are the molecules that carry out cellular activities, synthesize nonprotein cellular components, and form many cellular structures. They are what your muscles and tissues are made of; they synthesize the pigments that color your skin, hair, and eyes; they digest your food; they make (and sometimes are) the hormones that regulate your growth; they defend you from infection. In short, proteins determine your body's form and carry out its functions.

The structure of DNA has to allow it to do two things. First, it has to be able to contain instructions for assembling proteins. Second, it has to be easily and accurately duplicated, so that when a cell divides, it can pass on a correct copy of its genetic information to each daughter cell. This activity focuses on the structure of DNA. Subsequent activities will look at how the structure of DNA allows it to fulfill these two important functions.

The DNA molecule is a double helix, which you can imagine as a ladder that has been twisted into a spiral. The outside of the ladder is made up of alternating sugar and phosphate groups. The sugar is called deoxyribose. The rungs of the ladder are made up of four different nitrogen-containing bases: adenine, guanine, cytosine, and thymine. These four bases are of two types: purines and pyrimidines. Purines are large double-ring structures; adenine and guanine are purines. Pyrimidines are smaller single-ring structures; cytosine and thymine are pyrimidines. For more information on DNA and its component parts, see chapter 4.

Inside the DNA ladder, two bases pair up to make a "rung." One base sticks out from each sugar-phosphate chain toward the inside of the ladder. It forms a pair with a base sticking out from the opposite sugar-phosphate chain. Only three rings can fit between the two sugar-phosphate chains, so a pyrimidine (one ring) and a purine (two rings) form a pair. Because of the chemical structures of the bases, adenine always pairs with thymine, and cytosine always pairs with guanine. These two pairs of bases, adenine-thymine and cytosine-guanine, are called complementary base pairs. If you know the identity of the base on one strand of DNA, you also know the identity of the base on the other strand.

Activity

The goal of this lesson is to construct a paper model of a DNA helix. You will do so by making the fundamental unit of DNA. This unit, called a nucleotide, consists of one sugar molecule, one phosphate group, and one nitrogenous base. Each member of the class will make nucleotides and then join them to form the ladderlike helix.

Although making models is fun, building models is also a technique scientists really use to help them figure out how things are put together or how things might work. In fact, Watson and Crick, who discovered the structure of the DNA molecule, used cut-and-paste paper models to help them.

Procedure

1. Cut out the pattern for the nucleotide(s) assigned to you.
2. Place the pattern on the appropriate-color paper.
3. Label your pieces as the pattern is labeled.
4. Glue your nitrogen base to your sugar molecule by matching up the dots.
5. Glue your phosphate group onto your model by matching up the stars.
6. Your teacher will join the nucleotides together to form a helix.

Questions

1. What base does adenine pair with?

2. What base does guanine pair with?

3. What is the smallest unit of DNA called?

4. What is the shape of the DNA molecule?

5. Which bases are purines?

6. Which bases are pyrimidines?

7. Why must a purine pair with a pyrimidine?

8. What is the name of the sugar in the DNA backbone?

Advanced Questions

9. Suppose you know that the sequence of bases on one DNA strand is AGCTCAG. What is the sequence of bases on the opposite strand?

10. Referring to question 9, suppose that the 5′-most base on the given strand is the first A from left to right. What would be the 5′-most base on the opposite strand?

11. Assume that in a 100-base-pair DNA double helix, there are 45 cytosines. How many adenines are there?

Classroom Activities

DNA Replication

7

About These Activities

DNA replication is a topic usually presented in 9th-grade biology. The essential fact of DNA replication is that the base-pairing rules make it very easy to generate two identical new helices from one helix. This basic piece of information is all that is really necessary for young students to know about replication. More advanced students who will be learning about DNA sequencing and the polymerase chain reaction need to know a little more about DNA replication so that they can understand these interesting applications. The approach taken in this lesson is to provide detailed information to the teacher in the introductory material and then give a two-part activity. The first part is appropriate for young students; more advanced students will perform both parts of the lesson.

The first activity described below is a simple (and necessarily inaccurate) paper simulation of DNA replication. It is quite sufficient for most 9th graders, since it makes the point that two strands of the parent double helix are used as templates to synthesize two daughter strands. The paper DNA models described in this section can be used in the exercise. If you are teaching more advanced students, use the basic activity as a starting point and then introduce a more accurate picture of DNA replication through the second activity.

The second activity is a student reading about DNA polymerase, the central DNA replication enzyme. This reading provides the detail students will need to understand subsequent lessons. It contains questions, one of which involves using the information in the reading to identify inaccuracies in the simple model they have just used. You may supply them with additional information as you feel necessary. The background information in the introduction that follows contains far more detail about DNA replication than you need to share with students. It is there for your information and enjoyment. The two aspects of DNA synthesis that your

advanced students need to know (if they will be doing the activities on DNA sequencing and/or the polymerase chain reaction) are that synthesis is unidirectional and that it *absolutely requires a primer.*

Class periods required: 1

Background

All organisms must copy their genetic information, both during cell division and for transmission to their offspring. This critical task is carried out by groups of proteins working together, but the central player is the enzyme DNA polymerase. DNA polymerase selects the correct new nucleotide by checking that the nucleotide pairs correctly with the template base, and then the enzyme forms the new phosphodiester bond linking the new nucleotide to the growing DNA chain. Not surprisingly, the characteristics of DNA polymerase determine the overall features of DNA replication inside the cell and in the test tube.

In the DNA synthesis reaction itself, an incoming deoxynucleoside triphosphate binds to the polymerase. This nucleotide must be in the triphosphate form; the enzyme will not bind mono- or diphosphonucleotides for incorporation into DNA. (A bit of terminology: a nucleotide-like molecule consisting of only the sugar and the base is called a nucleoside; nucleotides are sometimes referred to as nucleoside monophosphates, diphosphates, or triphosphates to specify how many phosphate groups are attached.) Next, the polymerase checks to see whether the incoming nucleotide pairs properly with the template base. If it does, the enzyme forms a bond between the first of the three phosphates on the 5′ carbon of the new nucleotide and the 3′ hydroxyl (OH) group on the last nucleotide of the growing chain (Figure 7.1). The formation of this bond leaves one of the phosphate groups in the growing DNA chain and liberates an inorganic phosphate molecule, a "pyrophosphate," with the other two phosphorus atoms.

Finally, many, but not all, polymerases also possess a proofreading function that checks the new base pair to see whether it is accurate. If the new base does not pair correctly with the template base, then the new nucleotide is removed from the chain, and the enzyme tries again. Polymerases that lack proofreading ability, such as reverse transcriptases, accumulate more mistakes during the synthesis of DNA than polymerases that can proofread.

How frequently do DNA polymerases make mistakes in replicating DNA? According to in vitro studies, the *Escherichia coli* enzyme adds an incorrect base once every 10^5 or 10^6 base pairs (bp), but its ability to proofread and correct its mistakes brings its final error rate down to one error per 10^8 bp. Inside the cell, additional DNA mismatch correction enzymes lower the in vivo error rate to one in 10^{10} bp. *E. coli*'s genome has about 4.6×10^6 bp, so the bacterium makes less than one error in DNA replication per cell division. In contrast, the reverse transcriptase of retroviruses (such as the human immunodeficiency virus that causes AIDS) is a less careful polymerase than *E. coli*'s enzyme and also lacks proofreading ability. These enzymes make an error in replication nearly every 10^4 bp. This error-prone replication of genetic material is thought to be the cause of the rapid mutation rate of the human immunodeficiency virus.

Two important features of the DNA polymerase reaction are the direction of DNA synthesis and the requirement for a primer. Notice (Figure 7.1) that the enzyme always connects the 5′ phosphate of the incoming nucleotide to the 3′ OH group of the growing chain. This feature of DNA replication means that DNA synthesis is unidirectional, from 5′ to 3′. Unidirectionality presents a problem for chromosome replication that is discussed below.

The other feature of the reaction is that DNA polymerase must have a growing new strand to connect with the incoming nucleotide. No known DNA polymerase can begin synthesizing a complementary strand on a naked single-stranded DNA molecule. However, if the single strand has a short, complementary oligonucleotide annealed (base paired) to it somewhere, DNA synthesis can begin at the 3′ end of that oligonucleotide and continue in the 5′-to-3′ direction down to the end of the single strand. In this example, the long single strand is the template strand, and the short complementary oligonucleotide annealed to it is called the primer. In general, a primer is an oligonucleotide (either DNA or RNA; many DNA polymerases use RNA primers) annealed to the template with a 3′ OH group available at the

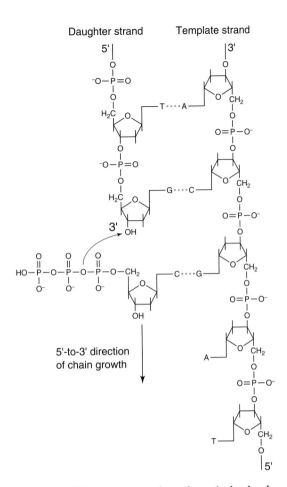

Figure 7.1 DNA polymerase (not shown) checks the pairing of an incoming deoxynucleoside triphosphate with the template base and then forms a new phosphodiester bond between the 5′ phosphate group of the new nucleotide and the 3′ hydroxyl group of the previous nucleotide.

end as a place to begin synthesis. The template molecule usually extends far beyond the end of the primer. Nature has devised many ways of providing primers for DNA synthesis; some are mentioned below.

When working with DNA polymerases outside the cell (as in biotechnology applications), it is necessary to provide a template with an annealed primer, deoxynucleoside triphosphates, and an appropriate buffer to achieve DNA synthesis. The direction of synthesis will always be 5′ to 3′ beginning at the 3′ end of the primer. Scientists use the selection of the primer to determine where or whether DNA synthesis will occur.

Very often, the primer is a synthetic oligonucleotide (see chapter 5) added to the reaction mixture separately. The template DNA and the primer can be annealed by heating and cooling (more about this in chapter 16, *Detection of Specific Sequences: Hybridization Analysis*). Thus, the scientist can decide where she wants DNA synthesis to begin and can synthesize a complementary primer that has its 3′ end at that site (provided she knows the DNA sequence of the region).

Primer specificity can also be used to determine whether a particular DNA molecule is present in a mixture, for example, whether a disease-causing microorganism is present in a tissue sample. In tests based upon the polymerase chain reaction (see chapter 17), primers that hybridize only to the DNA of the organism of interest are synthesized. If the organism is present in the sample, the primers can anneal, and DNA synthesis will occur. If the organism is not present, the primers will not anneal, and no synthesis can occur. The presence or absence of DNA synthesis is used to determine whether the organism of interest is in the sample. The requirement for a hybridized primer is a powerful tool that gives scientists a lot of control over when and where DNA synthesis occurs in vitro.

Cellular DNA does not normally exist as single strands with annealed primers. The chromosomes of organisms are generally completely double-stranded DNA molecules. In vitro, DNA polymerases will not begin DNA synthesis on a perfectly double-stranded DNA molecule. Scientists use manipulations, such as denaturing double-stranded molecules and hybridizing short primers to the single strands, to enable DNA synthesis to occur in the test tube. How do cells get around the problem?

The question of how cells manage to initiate DNA replication (and time and control it) is the subject of ongoing research in many laboratories around the world. For chromosomal DNA replication to begin, three major events must take place: (i) the replication proteins must assemble on the DNA, (ii) the DNA helix must be opened to expose unpaired bases for use as the template, and (iii) a primer must be provided. The following description of DNA replication initiation is a generalization of findings from several systems and does not apply in every single case.

Chromosomal DNA replication is usually initiated at specific sites along the DNA called replication origins. In general, these sites contain specific base sequences at which special replication initiator proteins bind. Origins also often contain a region of mostly A-T base pairs, presumably because it is easier to open a region of DNA helix that is rich in A-T pairs (only two hydrogen bonds pair A and T, but three

bonds pair G and C). The initiator protein or proteins bind to the recognition site in the replication origin, and this binding triggers the assembly of the replication proteins. The double helix is opened at or near the origin, and a primer anneals or is made. The primer is usually RNA. It can be a transcript of the origin region that was synthesized previously, or it can be a special short RNA synthesized on the spot. As you can imagine, the use of RNA for DNA replication primers means that DNA replication initiation is often entangled with transcription, and some of the best-understood initiation mechanisms are fairly complicated. Many scientists are entertaining themselves trying to sort out different methods of DNA replication initiation in different organisms.

After the helix is opened and the primer is available, DNA polymerase can begin to replicate DNA. The polymerase is assisted in its task by helper proteins; typical functions of these proteins are to "unzip" the double-stranded template ahead of the polymerase, to protect any exposed single-stranded regions of DNA (these regions are very vulnerable to degradation by cellular enzymes), and to help solve the problem of replication of the opposite strand.

What is the problem of the opposite strand? The problem is the direction of DNA synthesis and the lack of primers. The primer laid down at the origin enables DNA polymerase to synthesize DNA in the 5′-to-3′ direction away from the primer and the replication origin, using the strand annealed to the primer as the template. What happens to the opposite strand? We know that the opposite strand is replicated along with the first strand as the replication complex moves away from the origin. However, since DNA synthesis can occur only 5′ to 3′, replication on the opposite strand should head back toward the replication origin, in the opposite direction. Nevertheless, electron micrographs of replicating DNA clearly show that DNA replication occurs along both strands at a "fork" (a Y-shaped junction between the unreplicated parental DNA and the two daughter strands) that moves in only one direction along the parental DNA template. How does the cell coordinate DNA synthesis in opposite directions?

We have known since the 1960s that replication of the opposite strand generates short Okazaki fragments. While a complementary strand for the first strand (the *leading* strand) is synthesized in one long piece starting at the origin primer, the partner strand for the second strand (the *lagging* strand) is synthesized in pieces (Figure 7.2). How is this piecewise synthesis of the lagging strand accomplished, and where do the primers come from?

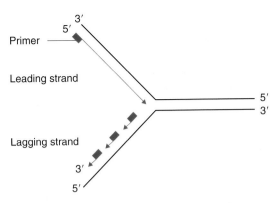

Figure 7.2 The lagging strand is synthesized as Okazaki fragments during DNA replication.

Research on DNA replication in bacteriophage T4 has shown that when the phage's replication proteins assemble, two molecules of DNA polymerase join the complex. One is responsible for synthesis of the leading strand, and the other is responsible for the lagging strand. The way in which these two molecules are believed to work together to replicate both strands of the DNA template is shown in Figure 7.3. This figure is complicated; look at it as you read the explanation in the text. A, B, and C on the DNA strands in the figure are simply position markers to help you follow the movement of the polymerase molecules along the DNA.

The polymerase complex (with its helper proteins) travels down the parental DNA for a few hundred nucleotides, synthesizing the leading strand as it goes. During this time, the other template strand is left single stranded; it is coated with special proteins to protect it. At a certain point, one of the helper proteins in the complex synthesizes a short RNA primer on the second template strand. The second DNA polymerase molecule (the first one is tending to business on the leading strand) begins synthesis of an Okazaki fragment at the 3′ end of this primer.

Look at the first two images in the figure. The top strand of DNA in each image is the template for the leading strand. Imagine the two linked polymerase molecules moving from left to right down the DNA (from A to B along the molecule), synthesizing the leading strand in one smooth piece. Meanwhile, the template for the lagging strand is folded into a hairpin. This orients the template correctly for 5′-to-3′ synthesis. Notice that the lagging-strand DNA is being synthesized from A away from B—in the opposite direction from the leading strand. The polymerase proteins do not move back toward the replication origin during this time; rather, the template for the lagging strand is pulled through the replication complex. Notice that the loop protruding to the

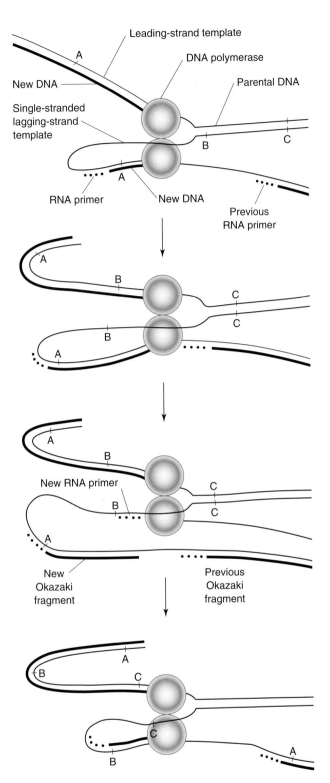

Figure 7.3 This model for the simultaneous replication of both DNA strands (described in the text) has been christened the "trombone model."

left of the polymerase gets larger from the first to the second image. This movement reminds some scientists of the way the slide of a trombone moves out from the instrument.

When the newly synthesized lagging-strand segment reaches the previous primer, the new duplex DNA is released and another primer is synthesized to start the next Okazaki fragment (look at the second and third images in Figure 7.3). This new Okazaki fragment will extend from B back to the primer near A (fourth image). The RNA primers will later be removed and replaced with DNA, and then the Okazaki fragments will be joined together by the enzyme DNA ligase.

This seemingly complicated method of DNA replication is a consequence of the directionality of DNA synthesis and the requirement for primers. Those same constraints govern the use of DNA polymerase enzymes outside the cell. However, these very limitations can give scientists ways of controlling DNA synthesis in vitro (as discussed above).

A final replication problem caused by unidirectionality and the requirement for primers is the "end" problem. Look at Figure 7.2 and imagine what happens when the replication complex reaches the end of the DNA molecule (the right end in the picture). It is clear that the leading strand could be replicated up to the end of a double-stranded helix, but the lagging strand again has a problem. In general, an RNA primer is not synthesized at the extreme end of the chromosome. Instead, several different strategies circumvent the problem created by the specificity of DNA replication enzymes. Some organisms have circular chromosomes. Still other organisms (viruses in particular) use interesting combinations of DNA synthesis and genetic recombination to achieve complete replication of their DNA.

Eukaryotes have structures called telomeres at the ends of their chromosomes. Telomeres contain short repeated base sequences, in vertebrates 5′-TTAGGG-3′. Telomeres are synthesized by telomerase, an intriguing enzyme that carries an RNA molecule that provides a template for the telomere repeated unit. Telomerase is actually a form of reverse transcriptase that reads the RNA template and synthesizes repeats of TTAGGG. In vertebrates, it is expressed only in embryonic cells and germ line cells. Once cells begin to differentiate, every round of cell division results in shorter and shorter telomeres.

The shortening of telomeres is believed to cause cell aging and death. Normally, cells removed from organisms and grown in culture can replicate through a certain number of division cycles, and then they die. In 1998, scientists transformed normal somatic cells with telomerase and showed that they continued to divide in culture. Most cancer cells have been found to express telomerase, perhaps contributing to their ability to divide.

If you have become intrigued by the process of DNA replication and want more information and details, a good discussion of the process can be found in *Molecular Biology of the Gene,* vol. I, by Watson et al. The telomerase experiments are described in a clear commentary by de Lange (*Science* 279:334–335, 1998) and in the original paper by Bodmar et al. (*Science* 279:349–352, 1998).

Objectives

After the simulation activity, all students should be able to:

1. Describe how the complementary base-paired structure of DNA enables two identical new helices to be made from a single template helix.
2. State that DNA replication is carried out by proteins inside the cell.

After the student reading activity, advanced students should be able to:

3. Explain what is meant by the statement "DNA replication is unidirectional."
4. Explain what is meant by the statement "DNA replication requires a primer."
5. Name the central enzyme involved in DNA replication (DNA polymerase).
6. State that purified replication proteins can synthesize DNA in the laboratory under the right conditions.

Materials

1. Simulation of DNA replication
   DNA models showing the sugar, phosphate, and bases separately. These can be colored paper cutouts from the model presented in an earlier lesson, commercially available pop-bead models, or other materials. You will need enough for the students to work with in small groups or as individuals.
2. Student reading activity
   Sufficient copies of the *Student Activity* if students do not have manuals

Preparation

- Assemble kit materials or paper cutouts. If you are using cutouts, photocopy the templates onto different colors of paper and have each student cut out several "sugars," "phosphates," and "bases" (all four) before beginning the activity.

- If you made paper models of DNA and saved them, use them for replication templates with the paper cutouts.
- Print the *Student Activity* if you plan to use that part of the lesson and students do not have manuals.

Procedure

1. Simulation of DNA replication

- Ask your students why a cell might want to make a copy of its DNA. Ask them if they think it would matter whether the copy were an exact copy. (The point is to have them realize that if an organism is to have offspring that are like itself, it must pass on an accurate copy of its genetic material.)
- Ask your students what copies DNA inside the cell. They may not know, but the correct answer is enzymes, and enzymes are proteins. We want them to realize that all body and cellular processes involve proteins and that DNA's central role is to carry the information needed to make all these important proteins.
- Make sure each student (or student group) has its starting DNA molecule (whether it is a paper model they made or one from a kit; the base sequence is totally unimportant). You may enjoy asking younger students how the starting molecule might be used as a pattern to make new DNA.
- Tell students that the DNA building blocks the cell uses are nucleotides (a sugar, a phosphate, and a base). Have them assemble some of each kind of nucleotide from their model components. It is not necessary to bring up the point about nucleoside triphosphates with younger students.
- Explain to your students (if they haven't arrived at the idea already) that the cell uses each strand of the parental DNA molecule as a template (pattern) to make a second strand. If necessary, explain that the parental molecule is "unzipped" (the correct term is denatured or melted) to expose unpaired bases and that new nucleotides are brought in one at a time, in order, and formed into a complementary chain that pairs with the template.
- Have students "unzip" their DNA molecules so that the unpaired bases are exposed. Then have them build a complementary strand on each parental strand, using the nucleotide precursors. Check your students' work to make sure that they have correct base pairs in their new molecules.
- Talk to the students about their new molecules. Are the new ones exactly like the parent molecule? (They should be.) How many "old" strands are in each new molecule? (One.) How many "new" strands? (One.)

This is a sufficient treatment of DNA replication for younger students.

2. Student reading activity

For a more detailed treatment of DNA replication, tell your advanced students that although the exercise they have just completed gets a very basic point across, it is inaccurate in nearly every detail. Have students read the activity *DNA Replication* and answer the questions. Make sure they understand that the stick figure portion of the nucleoside triphosphate in the figure represents the deoxyribose sugar and the base with the 5′ phosphate groups and the 3′ OH group shown (since they are the functional groups that participate in new bond formation). You may want to show them Figure 7.1 from the teacher's introduction.

Ask your students what features of the simple replication exercise are inaccurate (e.g., DNA polymerase binds only nucleoside triphosphates for incorporation, yet the model uses nucleoside monophosphates; the model uses no primers; the model does not take into account the directionality of DNA synthesis). Be sure that they have noted the requirement for a primer and the directionality of DNA synthesis. You may want to use one of their models to show that DNA polymerase would have to synthesize DNA from left to right on one strand but from right to left on the other to maintain the 5′-to-3′ direction. Use your judgment as to how much information to present about mechanisms of chromosomal replication. It is not necessary (and would be very difficult) to try to execute the detailed replication model in Figure 7.3 with paper cutouts, but if you like, you could draw or reproduce the model for students to see.

Answers to Student Questions

1. DNA synthesis will occur on templates B and D. Templates A and C do not have primers. In both B and D, synthesis will begin at the 3′ end of the primer and continue rightward with respect to the picture (only the short strand on the left in template D can serve as a primer, since there is no protruding template beyond the right-hand short strand).
2. Discussed above.

DNA Replication

The accurate copying of DNA is one of the most important jobs an organism must do during its life. Why do you think this statement is true? What would have happened to you if your ancestors' cells had not taken the trouble to make accurate copies of their DNA?

For such an important task, cells employ not one but a whole team of enzymes. However, the star of the team is the enzyme DNA polymerase. This protein builds the new daughter strand from nucleotides in the cell and checks the new base pairs for accuracy. The other protein team members help DNA polymerase do its job.

All bacteria and higher organisms and many viruses have their own DNA polymerase proteins, which are encoded by their DNA. All of these DNA polymerases work in similar ways and even resemble one another. Scientists, therefore, often study the complicated process of DNA replication in simple systems, such as bacterial and animal viruses, and apply what they have learned as they look at higher organisms. From these simple systems, many general facts about DNA polymerases have emerged. Let's look at some of these facts and then consider how they were determined.

1. DNA polymerases use deoxynucleoside triphosphates as precursors for the synthesis of DNA. These molecules are held at a special binding site on the polymerase before they are incorporated into a new DNA molecule. Nucleotides with only one or two phosphate groups will not bind there and are not incorporated into new DNA.

2. DNA polymerases cannot begin synthesizing DNA without a starting point called a primer. A primer is a piece of DNA or RNA base paired to the template strand so that the template strand sticks out past the 3′ end of the primer (see the figure). DNA synthesis begins at the 3′ end of the primer, where DNA polymerase attaches the first new nucleotide.

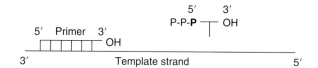

3. DNA polymerases synthesize DNA in one direction only. They start by attaching the 5′ phosphate group of the new nucleotide to the 3′ OH group of the last nucleotide in the primer, and they continue in this manner. They cannot connect the 3′ OH group of a new nucleotide to the 5′ phosphate group of a primer. DNA synthesis is therefore said to occur in the 5′-to-3′ direction (relative to the new strand).

How were these facts deduced? By looking at reactions with purified DNA polymerase proteins in vitro. ("In vitro" means "in glass"; it refers to the use of glass test tubes in the laboratory and basically means "in the test tube," even though almost everyone uses disposable plastic tubes for this kind of work now.) A scientist could take purified DNA polymerase protein, add a DNA template, and then experiment by adding deoxynucleoside monophosphates, diphosphates, or triphosphates and looking for DNA synthesis. By using radioactive nucleotides, a scientist can detect DNA synthesis by the appearance of radioactive DNA strands. In tests like this, it is clear that only nucleoside triphosphates are incorporated into DNA. It is also possible to mix DNA polymerase with radioactive nucleotides and ask whether the radioactivity becomes associated with the protein. In this manner, it is possible to detect binding of nucleotides to DNA polymerase.

Besides looking at the effects of different forms of nucleotides, scientists looked at different forms of

DNA templates. They constructed different types of DNA molecules and asked whether DNA synthesis occurred when polymerase and deoxynucleoside triphosphates were added. Through these experiments, they learned that DNA polymerase must have a primer on the template strand before it can synthesize DNA. Finally, by examining the DNA that was synthesized, they learned that DNA polymerase can add nucleotides only in the 5′-to-3′ direction. The enzyme could never synthesize a complementary strand to a portion of the template that was "upstream" (on the 5′ side) of the primer.

DNA polymerases are now important tools in molecular biology research and in biotechnology applications. They are used in cloning, copying, and sequencing DNA. In later lessons, you may have the opportunity to learn how DNA polymerases can help reveal the base sequence of a piece of DNA or how they are used to make a "DNA Xerox machine" that is extremely useful in detecting organisms that cause disease, among other applications. When you do these activities, notice that in each case, the DNA polymerase is given a primer and that DNA synthesis occurs in the 5′-to-3′ direction.

Questions

1. Given what you have learned about DNA polymerase, on which of the DNA template molecules shown below could the enzyme synthesize a new strand if given nucleoside triphosphates? Show where DNA synthesis would begin on each molecule and in what direction it would proceed.

2. Given what you have learned about DNA polymerase, what is wrong with the simple model of DNA replication that you used earlier?

A

B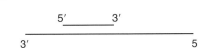

C

D

Expression of Genetic Information

About These Activities

In this lesson, students act out transcription and translation, either at their desks with paper models or by assuming roles and moving around the room. You may simply model the process in prokaryotes or eukaryotes or expand the activity as a means of exploring the differences in genetic traffic signals and gene structures in prokaryotes and eukaryotes. This activity also makes a good lead-in to a discussion of the various means of gene regulation. We have provided additional background information on gene regulation and ideas for gene regulation activities after the description of the transcription/translation activity. This chapter also includes an advanced *Student Activity* about RNA interference (RNAi) and antisense RNA. Background for it follows the material on gene regulation.

Class periods required: *1 to 3*

Background

DNA determines the characteristics of an organism by specifying the amino acid sequences (and therefore the structures and functions) of its proteins (see chapter 4 for a review). In order to direct the synthesis of a protein, the DNA must contain not only codons for each of the amino acids in that protein, but also regulatory sequences that tell the cell's protein-synthesizing machinery where to start and stop. Although prokaryotes and eukaryotes use the same genetic code, they have evolved different regulatory signals.

Transcription and translation are generally covered in adequate to overwhelming detail in high school biology texts. The possible exception to this statement is the frequent lack of discussion of genetic traffic signals and regulation. The question of how RNA polymerase identifies the beginning of a gene is often ignored.

We encourage you to include the traffic signals in your teaching of transcription and translation. It is not necessary to go into details, but we want students to know that transcription and translation do not happen in a vacuum—the signals are there and are part of every organism's genome.

Another singular lack of continuity in many high school texts is the common practice of presenting DNA structure, DNA replication, transcription, and translation with no context for understanding why producing proteins is important. Chapter 4 provides several specific examples of protein structure and function, and the section *Molecular Biology and Genetics* relates several proteins to observable traits.

The first important regulatory region for protein synthesis is the promoter, the sequence of DNA bases that RNA polymerase recognizes and binds to before beginning transcription. Without a promoter, transcription does not occur. Prokaryotic and eukaryotic promoters are different, as are prokaryotic and eukaryotic RNA polymerases. A typical bacterial promoter consists of the sequence 5′-TTGACA separated by 17 bases from the sequence TATATT-3′ (some variation is permitted).

Eukaryotic promoters are more variable. One component of a eukaryotic promoter is the sequence TATA, located about 30 bases upstream of where transcription begins in yeast and 60 bases upstream in mammals. This component is often called the "TATA box." Usually, two more promoter components are found somewhat further upstream of the TATA box. These are the sequence CCAAT and a GC-rich sequence. However, remember from chapter 4 that eukaryotic RNA polymerase alone is not efficient at starting transcription from a promoter. Eukaryotic promoters are associated with a variety of different enhancer sequences, sites on the DNA where various *trans*-activating transcription factors bind and stimulate transcription (see chapter 4).

Transcription begins downstream of the promoter and continues until a transcription terminator is reached. The DNA sequences of terminators are complicated; there are different classes of terminators with different structures, and prokaryotic and eukaryotic terminators also differ. The function of a terminator is to cause RNA polymerase to stop transcribing DNA and to release the DNA template.

The nucleotide sequence of messenger RNA (mRNA) is translated into protein at the ribosome. Translation does not simply begin at the 5′ end of a message and end at the 3′ end. Further regulatory elements are contained within the RNA itself that direct the ribosomes to begin and end translation. The RNA bases at the extreme 5′ end and the extreme 3′ end of the message are not translated into protein.

Foremost among these regulatory elements in bacteria is the signal for ribosomal recognition. In bacteria, this element commonly has the sequence 5′-GAGG-3′ or 5′-AGGA-3′ and is located 8 to 13 nucleotides upstream of the initiation codon. The ribosome recognizes the element and binds to the mRNA there. How does binding occur? The bacterial ribosomal RNA (rRNA) contains a base sequence complementary to the ribosomal recognition element (RRE) on the mRNA, and the rRNA and mRNA hybridize there. Because the RRE is close to the initiation codon, the initiation codon is brought into proper position for translation to begin.

The picture is not as clear in eukaryotes. In the simplest model, eukaryotic ribosomes are thought to bind at the 5′ end of the mRNA and to search down the mRNA for the first initiation codon to begin translation. Experiments in some laboratories suggest that this simple model does not explain all situations.

In both prokaryotes and eukaryotes, translation begins at the initiation codon, AUG. This codon specifies the amino acid methionine. Prokaryotic and eukaryotic ribosomes use a special modified form of methionine (formylated methionine in *Escherichia coli*) to begin protein synthesis. In *E. coli*, initiation of protein synthesis also requires three proteins called initiation factors. In eukaryotes, many more protein factors are required. Once initiated, protein synthesis continues down the mRNA until a stop codon is reached, at which point the ribosome releases the mRNA.

The genetic traffic signals involved in converting DNA sequences into proteins (in prokaryotes) are summarized in Figure 8.1.

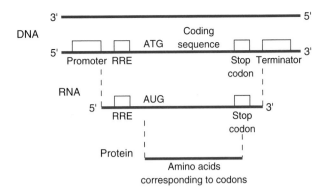

Figure 8.1 Major genetic traffic signals in bacteria. These signals tell RNA polymerase where to begin and end transcription, enable the ribosome to recognize mRNA, and direct the ribosome to start and stop protein synthesis.

In prokaryotes, the product of transcription is mRNA. In eukaryotes, however, RNA polymerase synthesizes a primary transcript that must be processed further before it can function as a message. One of these processing steps is splicing.

Many, but not all, eukaryotic genes contain introns (see chapter 4). An intron is a section of DNA embedded in the protein-coding region of a gene that is not represented in the mRNA or protein sequence. The intron DNA is transcribed into RNA along with the coding regions to form a long precursor RNA. Introns are removed from the precursor RNA by splicing, which occurs in the nucleus.

The mechanisms of splicing are a topic of current research. It appears that there are at least three mechanisms: one for transfer RNA (tRNA), one for mRNA, and one for rRNA. Each mechanism uses a different set of signals to determine precisely what sequences should be removed from the precursor RNA. A generic model of splicing is shown in Figure 8.2. Accurate splicing is very important; several types of thalassemia (a blood disorder) in humans appear to be caused by mutations in the splicing signals of the hemoglobin gene that lead to splicing errors in the hemoglobin RNA and thus incorrect protein synthesis.

For this lesson

The description presented here is superficial; further reading in a molecular biology text will provide more details, should you wish to delve deeper. It is, of course, not necessary to convey all of this information to students. You may teach as many of these regulatory elements as you deem appropriate; the most important is undoubtedly the promoter.

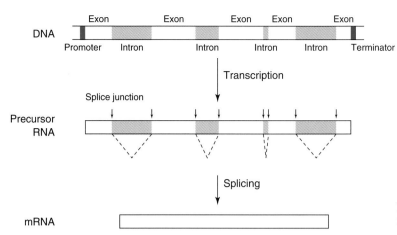

DNA

Promoter   Intron   Intron   Intron   Intron   Terminator

Transcription

Splice junction

Precursor
RNA

Splicing

mRNA

Figure 8.2 Splicing of precursor RNA to produce mRNA. The DNA exons contain the coding sequence for the protein.

The cutouts provided on the CD allow you to simulate protein synthesis in prokaryotes and eukaryotes. The most complete model of bacterial protein synthesis would incorporate the promoter, the terminator, and the ribosome recognition sequence. The most complete eukaryotic model would incorporate the promoter, the terminator, and the splicing start and stop signals (the splice junctions).

Objectives

After this lesson students should be able to:

1. Describe DNA's function as the basic hereditary material controlling cellular activity via control of the cell's enzyme system.
2. Describe transcription and translation.
3. Describe the functions of the promoter and the terminator.
4. (Optional) Contrast prokaryotic and eukaryotic genetic "traffic signals."

Materials

• Scissors
• DNA/RNA genetic-code triplet sheets printed from the CD, along with any desired regulatory-element sheets

Variations of This Activity

Most of the discussion below is related to a form of the activity in which students act out transcription and translation by moving around the classroom. Some teachers use this form of the activity regularly and enjoy it. The activity also works as a demonstration: you can tape the DNA template cards to the blackboard and show how RNA is formed.

Another option is to have students work in pairs at their desks. Reduce the cutout patterns on a photocopier and group them onto a few sheets. Give each team a set of the sheets. Students can cut out the patterns as simple squares of paper to save time and then use the cutouts to simulate transcription and translation at their desks.

Preparation

Modify the preparation according to the variation of the lesson you plan to use.

Before class, cut out the genetic-code cards provided on the CD (or have your students cut them out in class). The DNA cutouts come in complementary pairs; one cutout has filled-in letters; its complement has outlined letters. The outlined-letter strand will represent the coding strand of the DNA; the filled-letter strand is the noncoding strand. Paperclip the complementary DNA cards together in sets of two. Arrange the cutouts into four stacks: DNA, mRNA, tRNA, and amino acids. Put the "start" sequence (TAC) at the second position in the DNA stack and the stop sequence (ATC) on the bottom. The promoter card should then be placed on top of all the DNA cutouts with the terminator on the bottom.

Decide which model (prokaryote or eukaryote) you will use and which traffic signals you want to incorporate. Cut out the traffic signal cards. If you use the RRE card, it should go between the first DNA cutout and the "start" (TAC) card. If you use splicing start and stop cards, they can go anywhere between the "start" and "stop" cards as long as you do not splice out the entire gene! Sample lineups for prokaryotic and eukaryotic models are shown on the next page (Figure 8.3).

Prokaryotic model

Promotor   DSC   RRE   TAC   DSC   DSC . . . DSC   DSC   stop   DSC   terminator

Eukaryotic model

Promotor   DSC   TAC   DSC . . . DSC   splice   DSC . . . DSC   splice   DSC . . . DSC   stop   DSC   terminator

Figure 8.3 Sample DNA sequence lineups for transcription. The splice junctions and splicing step are optional. DSC, DNA sequence card with any three bases; TAC, encoding DNA card with TAC base sequence (the corresponding coding card is ATG); . . ., any number of DNA sequence cards; stop, the stop codon.

Tips

1. Photocopy or print the triplet codes onto colored paper using different colors for DNA, mRNA, tRNA, and amino acids if possible.
2. If possible, laminate the cards so that they can be used over and over.
3. Make sure students understand the importance of proteins in determining the characteristics of an organism before the activity begins (see the introductory material).
4. Some teachers glue the cutouts onto larger squares of paper and then thread yarn through reinforced holes in the squares to make placards. Students can hang them over their shoulders as labels during the simulation. These placards can be reused from year to year.

Procedure for Acting Out Transcription and Translation

Note: These instructions are relevant even if you plan to do the activity as a demonstration or have the students use paper models at their desks.

1. Setting the scene. If you are using the eukaryotic model, it can be helpful to set the scene for students as follows.
   - The classroom's floor, walls, and ceiling are analogous to a cell membrane.
   - Designate one area as the "nucleus," where transcription occurs.
   - Designate another area as the "ribosome," where translation occurs.
   - All other nonnucleus areas in the room are "cytoplasm."

Important: If you want to use the *E. coli* model, remember that *bacteria have no nuclei*. The chromosome and the ribosomes are associated with the cell membrane. Designate a ribosome area accordingly, and let the "DNA" be next to the membrane. Also, remember that *E. coli* uses the RRE but does *not* use splicing. (You do not have to talk about ribosome recognition, but *do not* show splicing in bacteria.)

The eukaryotic model does *not* use the RRE (let the ribosome recognize the 5′ end of the message) and may use splicing. If you are very ambitious and have time, you can have students go through both models to see the similarities and the differences involved.

2. Distribute the DNA sequence card pairs and desired regulatory elements and the matching mRNA codes to the students. The large letters on the cards refer to the first letter of the nucleotide bases (A = adenine, C = cytosine, G = guanine, T = thymine, and U = uracil). Do not distribute the tRNA and amino acids cards until step 10.
3. Review DNA structure as needed. The informational part of DNA is within the paired bases, so transcription does not begin until RNA polymerase unzips the two DNA strands. RNA polymerase catalyzes the pairing of DNA's exposed bases with complementary RNA nucleotides that happen by.

Students may get confused about which strand of DNA is doing what. One strand (in this case represented by cards with the outlined letters) is the *coding strand*. It contains the 3-base codons that will be found in mRNA. To move the coding information to the ribosome for translation, the cell makes an mRNA copy. How can the cell make a copy of the coding strand? By synthesizing mRNA complementary to the noncoding strand.

The process is similar to DNA replication, except that only one strand of the DNA molecule is used as a template. The mRNA generated is a copy of the coding strand. How does the RNA polymerase "know" which strand to use as a template for synthesis? The promoter sequence orients the enzyme properly.

Transcription

4. Students with the DNA card pairs line up in the classroom area designated "nucleus." The student with the promoter card should stand on the left

as the class sees him or her. It does not matter which DNA sequence card is first after the promoter. The student with the filled-letter DNA card labeled "TAC" and the outlined "ATG" (the start sequence) should be the second DNA sequence (the third student, with the first student holding the promoter). The TAC DNA code will produce the mRNA start codon, AUG. The "ATC" (the stop sequence) filled-letter card should be second from the last; the identity of the last card is not important. If you use the RRE for the *E. coli* model, have that student stand between the first DNA codon and the "TAC." All other DNA cards can be arranged in any order. Please refer to the diagram for examples.

Have the DNA students hold their cards in front of their bodies with the letters facing the class. The cards with filled letters should be held above the cards with outlined letters, with the complementary bases facing each other. You should see a row of filled-letter cards across the top and a complementary row of outlined-letter cards across the bottom of two rows of DNA sequence. This represents the base-paired DNA molecule. The student with the promoter should stand to the left.

5. Designate one student to represent RNA polymerase. This student walks to the promoter student and shakes hands, signifying that the RNA polymerase has recognized the beginning of a gene. After the handshake, the DNA students should lower the outlined-letter cards (the lower row) to their sides, exposing a single-stranded template for RNA polymerase. They should hang on to the outlined-letter cards for later comparison with the mRNA.

6. Review the following rules with the students who have the mRNA cards:
   • RNA cytosine always pairs with DNA guanine.
   • RNA uracil always pairs with DNA adenine.
   • RNA adenine always pairs with DNA thymine.
   • RNA guanine always pairs with DNA cytosine.

7. Using these rules, students with mRNA cards must find their DNA matches. Have them pair in order, starting with the first DNA sequence card and progressing down to the terminator, where pairing stops. When the RNA and DNA have correctly paired and the terminator has been reached,

compare the sequence of the RNA to the sequence of the outlined-letter DNA cards that the DNA students are holding (the coding strand of the DNA). The students should see that the base sequences of the outlined-letter cards and the RNA are identical, except for the substitution of U for T in the RNA. The sequence of the RNA is complementary to the filled-letter DNA sequence cards.

8. The DNA students can sit down, leaving a chain of RNA. (You may want to remind the students that in real life the DNA chain extends beyond the terminator to the next gene, so it is important to have a "stop sign" for RNA polymerase.) Tell the class that they have performed transcription. They have made a very short (shorter than in real life) complementary section of RNA that perfectly reflects the sequence of bases in the DNA molecule.

Note: After students match up their cards, they will not necessarily be in the order shown in Figure 8.4, but the DNA/mRNA pairs should match vertically as in the figure.

9. If you are simulating splicing, do it now. To simulate splicing, have the RNA stay in the nucleus. The two splice junctions and everything between them leave and sit down, and the RNA sequences immediately before the first and after the second splice junctions are left adjacent to each other.

Translation

Students will notice that the tRNA cards are arranged in groups of three letters. The 3-base sequence of tRNA represents its anticodon. In addition, each tRNA card has a three-letter abbreviation (in the "arrow" part of the card) for one of the 20 amino acids (Table 8.1). This lesson includes only 7 of the 20 amino acids used for making proteins.

10. Separately distribute the tRNA and amino acid cards, one per student. Students with either type of card should be randomly scattered in the "cytoplasm." If you are going to include ribosome recognition in your simulation, designate one student to be the ribosome and perform the recognition function. That student should stand in the designated ribosome area.

Figure 8.4 DNA/mRNA sequence matches.

DNA (coding)	CCG	ATG	AAT	GTC	GAG	CTA	TCC	TAC	GGC
DNA (noncoding)	GGC	TAC	TTA	CAG	CTC	GAT	AGG	ATG	CCG
mRNA	CCG	AUG	AAU	GUC	GAG	CUA	UCC	UAC	GGC

Table 8.1 Amino acid abbreviations[a]

Abbreviation	Amino acid
Ala	Alanine
Arg	Arginine
Asn	Asparagine
Asp	Aspartic acid
Cys	Cysteine
Gln	Glutamine
Glu	Glutamic acid
Gly	Glycine
His	Histidine
Ile	Isoleucine
Leu	Leucine
Lys	Lysine
Met	Methionine
Phe	Phenylalanine
Pro	Proline
Ser	Serine
Thr	Threonine
Trp	Tryptophan
Tyr	Tyrosine
Val	Valine

[a]These standard three-letter abbreviations are used frequently in scientific literature to denote amino acids.

11. Have the students with the tRNA cards find the students with their specific amino acids and stand together. For example, the tRNA card "GGC" with the letters PRO should find the amino acid proline. In the cell, there is a specific enzyme for each type of tRNA molecule that joins the tRNA to its correct amino acid.

12. Students with mRNA cards should walk to the area designated "ribosome" and stop just outside it. In the eukaryotic cell, the student with the very first (the 5′) mRNA card should shake hands with the ribosome recognizer, to signify that the ribosome has found the 5′ end of the message. In the *E. coli* model, the ribosome shakes hands with the RRE.

13. Review the following rules:
    • tRNA cytosine always pairs with mRNA guanine.

• tRNA uracil always pairs with mRNA adenine.
    • tRNA adenine always pairs with mRNA uracil.
    • tRNA guanine always pairs with mRNA cytosine.

14. The linked tRNA and amino acid students should walk together to the ribosome area. Starting with the AUG codon, have the correct tRNA pair with its codon. When the second tRNA finds its place on the mRNA strand, the first and second amino acids link hands. Then, the first tRNA detaches from its amino acid and returns to the "cytoplasm." The amino acids remain at the "ribosome." Continue down the mRNA molecule, using the tRNAs to link each new amino acid to the growing chain. As each new amino acid is added, the previous tRNA leaves. Translation stops when the "stop" codon is reached. At that point, the amino acid chain is released and the mRNA and the new amino acid chain leave the ribosome.

Amino acids linked in this way are called *peptides*. Long chains of linked amino acids are often called *polypeptides*. Translation is complete when a sequence of mRNA information translates into a polypeptide. A protein can be one polypeptide or an association of more than one polypeptide (see chapter 4).

The students' mRNA-tRNA pairings should match vertically those shown in Figure 8.5.

15. Use the Student Questions as a basis for class discussion or assign them as homework.

Follow-Up (Optional)

• This activity is a good lead-in to a discussion of regulation of transcription. (See chapter 4 for information.) If you plan to talk about transcriptional regulation, be sure to include the promoter in your activity.

• For advanced classes, it is fun to examine all of the steps in the pathway leading from gene to protein and to imagine ways of regulating protein synthesis at every step. It is very likely that natural examples of every imagined regulatory

Figure 8.5 mRNA, tRNA, and amino acid matches. The two codons with no tRNA are unpaired because they come before the start codon and after the stop codon. If these codons were between the start and stop translation signals, they would be paired with GGC/Pro and CCG/Gly, respectively.

mRNA	CCG	AUG	AAU	GUC	GAG	CUA	UCC	UAC	GGC
tRNA	none	UAC	UUA	CAG	CUC	GAU	AGG	AUG	none
Amino acids		Met	Asn	Val	Glu	Leu	Ser	stop	

mechanism actually exist. If you discuss gene regulation here, be sure to bring up the idea of antisense RNA (see the activity *Antisense and RNA Interference*).

- If you and your class like this sort of thing, think about making a videotape of the activity and showing it to a lower grade level class to explain transcription and translation.

Answers to Student Questions

1. The sequences are complementary.
2. The sequences are the same, except that mRNA uses uracil instead of thymine.
3. The noncoding strand is complementary to the mRNA.
4. Certainly the noncoding strand cannot be used to make the same protein as the one encoded by the coding strand; the DNA sequences are different. It is unlikely that the noncoding strand of a gene would contain a promoter and an ATG in proper alignment for transcription and translation. In addition, chances are that even if you could transcribe and then initiate translation, the resulting peptide would be short because of random stop sequences.

(Depending on the sophistication of your students, you may want to discuss the following aspect of transcription with them. The RNA polymerase cannot transcribe the wrong strand after recognizing the promoter. RNA polymerase, like DNA polymerase, synthesizes nucleic acid in the 5′-to-3′ direction. To do so, it must read the template in the 3′-to-5′ direction. Only one strand of the DNA is in the proper orientation to the promoter for use; the other strand essentially points backward.)

5. An extra base inserted near the beginning of a gene would throw off the entire coding sequence. Completely different amino acids would be coded

for, and you would probably encounter a stop codon long before the correct end of the protein. Examples will vary.

6. The obvious answer is that translation would be affected because the deficient amino acid would not be available to bind to its tRNA and could not be incorporated into the peptide chain. A less obvious answer is that, ultimately, transcription would also be affected because RNA polymerase could no longer be synthesized properly on account of the missing amino acid.

7. See Figures 8.6 and 8.7.

Other Ideas for Teaching Transcription and Translation

Here are some activities for teaching transcription and translation that have worked well for us or for our teaching colleagues.

- Extend the paper helix model by using different colors for the sugar ribose and the base uracil, have students put together ribonucleotides as they put together deoxyribonucleotides in chapter 6, and then use the paper DNA helix as a template for synthesizing mRNA. Some teachers have also created paper tRNAs and amino acids and followed the activity through translation.
- A shortcut paper translation
  Cut squares of colored construction paper to represent the amino acids. Use 20 different colors, and assign one color to each amino acid. Give students a DNA sequence to translate into protein. Working in pairs, they should glue or staple the colored squares together in the proper order. Next, assign each pair of students a different mutation. Include insertions, deletions, substitutions, and amber mutations (nucleotide sequence changes that create stop codons within a gene, causing premature termination of translation).

Figure 8.6 Diagram answer to question 7. The noncoding DNA and tRNA sequences are identical, except that tRNA contains uracil instead of thymine. For the amino acid abbreviations, see Table 8.1. C, coding strand; NC, noncoding strand.

DNA	C	ATG	TCC	CCG	GAG	AAT	GTC	GAG	CTA	TCC	GGC	TAG
	NC	TAC	AGG	GGC	CTC	TTA	CAG	CTC	GAT	AGG	CCG	ATC
mRNA		AUG	UCC	CCG	GAG	AAU	GUC	GAG	CUA	UCC	GGC	UAG
tRNA		UAC	AGG	GGC	CUC	UUA	CAG	CUC	GAU	AGG	CCG	none
Amino acid		Met	Ser	Pro	Glu	Asn	Val	Glu	Leu	Ser	Gly	none
		start										stop

DNA	C	ATG	CTA	GTC	CCG	GGC	AAT	TCC	GAG	CCG	GTC	TAG
DNA	NC	TAC	GAT	CAG	GGC	CCG	TTA	AGG	CTC	GGC	CAG	ATC
mRNA		AUG	CUA	GUC	CCG	GGC	AAU	UCC	GAG	CCG	GUC	UAG
tRNA		UAC	GAU	CAG	GGC	CCG	UUA	AGG	CUC	GGC	CAG	none
Amino acid		Met	Leu	Val	Pro	Gly	Asn	Ser	Glu	Pro	Val	stop

Figure 8.7 Answer to question 8. For the amino acid abbreviations, see Table 8.1. C, coding strand; NC, noncoding strand.

Each pair of students now alters the original DNA sequence according to its assigned mutation and then translates the new sequence. Have each pair of students show their new protein chain to the class, identify the mutation, and explain how the mutation caused the alteration (if any) in the protein sequence. The beauty of this system is that changes in the protein sequence are immediately apparent in the appearance of the chain of colored paper squares, and the variety of altered proteins the class will get is striking.

• Using bead models

An alternative to the shortcut model with paper squares is to use beads. Obtain 20 different types of beads (get some highly varying ones), and assign a bead type to an amino acid. Following the model above, have students translate DNA sequences into proteins, stringing the beads in place of amino acids.

The genetic code can also be symbolized in beads. Obtain four colors of pop-beads or other plain beads. These represent the DNA bases. Make a genetic color code using the bead colors. For example, if A is a red bead, T or U is a blue bead, and G is a yellow bead, then red-blue-yellow encodes methionine.

A nice aspect of using plain beads for DNA and highly varied beads for amino acids is that it approximates reality. Nucleotides are very similar, while amino acids vary greatly in structure. DNA is a regular molecule, made of quite similar repeating units. Proteins, on the other hand, are quite different. The regular pop-bead DNA model and the exotic amino acid bead protein models illustrate this fact nicely.

From Genes to Proteins

You Are Your Proteins

You have probably heard many times that "your DNA determines your characteristics." Did you ever wonder how DNA does that? DNA determines all your characteristics (and the characteristics of every plant, animal, fungus, bacterium, etc., in the world) by determining what proteins your cells will synthesize.

Why are proteins so important? Because of the many things they do. There are many types of proteins, and each of these types performs an important kind of job in your body. For example, *structural proteins* form the "bricks and mortar" of your tissues. Two of them, actin and myosin, enable your muscles to contract. Another structural protein, keratin, is the basic component of hair. *Carrier proteins* transport important nutrients, hormones, and other critical substances around your body. One of these proteins is hemoglobin, which carries oxygen through your blood to your tissues.

Another large class of proteins is the *enzymes.* Enzymes are the body's workhorses. They carry out chemical reactions in your body. Enzymes digest your food, synthesize fats so your body can store energy, and carry out the work of making new cells. They make molecules and perform activities necessary for life.

In fact, if you want to sum up the importance of proteins in your body, you could say: *"Nearly every biological molecule in my body either is a protein or is made by proteins."* So, by telling your cells what proteins to synthesize, DNA controls your characteristics.

What Proteins Are

What exactly is a protein? A protein is a biological molecule made of many small units linked together in a chain (rather like beads on a string). The units are *amino acids,* small molecules composed of car-

bon, hydrogen, oxygen, and nitrogen. There are 20 different amino acids that can be used in making up your proteins.

Imagine sitting down to make a string of beads with 20 containers, each filled with a different type of bead. You may use as many or as few of each kind of bead as you wish. You may string them in any order you wish, and you may use any number of beads you wish. This situation is a reasonable representation of the possibilities for protein synthesis within a cell. The beads represent amino acids, and any string of them represents a protein. You can get an idea of the infinite variety of proteins that could be made!

However, in the cell, one thing is different from our example. The cell is handed an instruction sheet that tells it what beads (amino acids) to string to make a particular chain. The instructions come from the cell's DNA.

How DNA Directs Protein Synthesis

DNA contains a *genetic code* for amino acids, with each amino acid being represented by a sequence of three DNA bases. These triplets of bases are called *codons.* The order of the codons in a DNA sequence is reflected in the order of the amino acids assembled in a protein chain. The complete stretch of DNA needed to determine the amino acid sequence of a single protein is called a *gene.*

In your cells, the DNA is located within the chromosomes in the nucleus. DNA contains all the instructions for making every protein in your body. However, your cells make proteins at the ribosomes, located in the cytoplasm. An individual ribosome makes only one protein at a time, so when your cell needs to make a protein, a "working copy" of the instructions for that one protein is copied from the DNA and sent to the ribosome for use. This working copy is "messenger" RNA (mRNA).

After the base sequence of a gene (DNA) is copied into mRNA, the mRNA travels to the ribosome, where its code is "translated" into protein. The translation step is carried out by a second type of RNA called transfer RNA (tRNA). The tRNA matches the correct amino acids to the codons in the mRNA. The amino acids are linked together to make the new protein.

Wait a Minute—Isn't Protein a Kind of Food?

Most of us have heard that "protein is part of a healthy diet." In fact, most people think of food when they hear the word "protein." So what's going on? How does protein in the diet fit into a discussion of genes?

The protein you eat is composed of individual protein molecules made by plant and animal cells.

Animal muscle tissue (meat) is particularly rich in protein (remember actin and myosin, mentioned above?). When you eat protein, regardless of its source, your digestive enzymes break the individual protein molecules down into amino acids (like taking the beads off the string). The individual amino acids are used by your cells to make your proteins—a form of biological recycling—or can be further broken down for energy.

Today's Activity

Today, you and your classmates will act out the steps involved in translating the DNA code into proteins. You will see the roles played by mRNA and tRNA and may be introduced to some of the "traffic signals" that direct their actions, too.

Questions

1. DNA is double stranded. One strand is the coding strand, and the other is the noncoding strand. The noncoding strand is used as the template to make the mRNA. What is the relationship between the base sequences of the coding and the noncoding strands?

2. What is the relationship between the base sequence of the coding strand and the base sequence of mRNA?

3. What is the relationship between the base sequence of the noncoding strand and the base sequence of mRNA?

4. Would there be a problem if the RNA polymerase transcribed the wrong strand of DNA and the cell tried to make a protein?

5. A frameshift mutation is caused by the insertion or deletion of one or two DNA bases. What would be the effect on the amino acid sequence of a protein if one extra base were inserted into the gene near the beginning? Use an example to show what you mean.

6. Suppose an individual has a nutrient deficiency due to poor diet and is missing a particular amino acid. How would transcription and translation be affected?

7. Given below are some tRNA anticodon-amino acid relationships and a stretch of imaginary DNA. Fill in the missing boxes in the chart below by writing the DNA sequence of the coding strand and the correct mRNA codons, tRNA anticodons, and amino acids. Use the following tRNA-amino acid relationships:

GGC UUA CAG CUC GAU AGG CCG

Pro   Asn   Val   Glu   Leu   Ser   Gly

DNA Coding Noncoding											
Noncoding	TAC	AGG	GGC	CTC	TTA	CAG	CTC	GAT	AGG	CCG	ATC
mRNA											
tRNA											
Amino acid	start										stop

What are the similarities between the noncoding DNA sequence and the tRNA sequence?

8. A new and exciting branch of biotechnology is called protein engineering. To engineer proteins, molecular biologists work backward through the protein synthesis process. They first determine the exact sequence of the polypeptide they want and then create a DNA sequence to produce it. Use the rules of transcription and translation to "engineer" the peptide sequence below. Fill in the rows for tRNA anticodons, mRNA codons, and the two DNA strands. Use the tRNA-amino acid relationships given in Question 7.

DNA Coding Noncoding											
mRNA											
tRNA											
Amino acid	Met start	Leu	Val	Pro	Gly	Asn	Ser	Glu	Glu	Pro	Val

Making Antisense: Regulating Gene Expression with RNA

Imagine what you might be able to do if you could prevent or decrease the expression of any gene. Prevent cancer from developing? Prevent viral diseases? Control certain genetic diseases?

Biotechnologists are pondering these and other questions largely because of a new technology for gene regulation: antisense. What is antisense technology, and how does it work? Antisense technology, like the rest of biotechnology, is based on a natural phenomenon. In antisense gene regulation, a cell synthesizes a very short piece of RNA that is exactly complementary to the ribosome recognition region (usually the 5′, or "front," end) of the mRNA of the gene to be regulated. Because of its complementary base sequence, the antisense RNA can base pair with the mRNA. When it does this, the 5′ end of the mRNA becomes double stranded. The ribosome can no longer recognize and translate it. By preventing translation of mRNA, antisense RNA can decrease gene expression.

Antisense technology copies this natural approach, except that the antisense molecule can be RNA, DNA, or even a chemically modified version of either one, just as long as it will base pair with the target molecule. A general word used to refer to these short molecules is antisense oligonucleotides (the prefix oligo means "several"; nucleotide can refer to the building blocks of DNA or RNA). It is theoretically possible to reduce the expression of any gene by introducing an appropriate antisense oligonucleotide into the cell.

Regulation by antisense is very precise because of the specificity of base pairing. For any oligonucleotide, the chance that its complementary sequence will occur randomly is 1 in 4^n, where n is the number of bases in the oligonucleotide. Thus, the chance of a 15-nucleotide antisense molecule accidentally pairing with an unintended target is 1 in 4^{15}, which is less than 1 in 1 billion (multiply it out for yourself). For longer antisense molecules, the odds are even more remote.

Is antisense technology being used for any practical applications? Yes. In 1999, the first antisense drug went on the market in the United States. It is called Vitravene, and it is manufactured by Isis Pharmaceuticals. Vitravene targets the replication protein of cytomegalovirus (CMV). CMV is very common; the odds are excellent that you have been infected by it. CMV infection does not usually present any problems to people with intact immune systems, but in people with impaired immunity, such as AIDS patients, CMV infection can be devastating. One of the most debilitating forms of CMV infection is retinitis, in which the virus gradually destroys the retina. CMV retinitis is the most common cause of blindness in people with AIDS. Vitravene fights CMV by preventing the expression of its replication protein and thereby blocking viral replication.

In the late 1990s, scientists using antisense RNA to study the development of the nematode worm discovered a different but related type of RNA regulation. The scientists were using antisense RNA to block expression of a gene they believed was involved in cellular development, and like good scientists, they included a control experiment so that they could be sure any effects they observed were due to the specific anti-RNA. For their control experiment, they added RNA to the nematodes, but instead of using their anti-RNA, they used the "sense" version, which they reasoned should not base pair with the mRNA and should not block gene expression.

To their surprise, the sense RNA blocked gene expression as well as the antisense RNA did. Rather than give up, the scientists did more experiments to determine how the sense RNA could regulate the gene. They eventually found that both the antisense RNA preparation and the sense RNA preparation, which they had prepared in the laboratory, contained a tiny amount of the double-stranded version of the RNA. It was this double-stranded RNA (dsRNA) that was exerting the powerful gene regulation. They named the phenomenon RNA interference, or RNAi.

Since this initial discovery, the mechanism by which RNAi blocks gene expression was worked out in a series of experiments using cellular extracts and purified RNA molecules. The dsRNA is bound by a cellular enzyme that was named Dicer. To dice a fruit or vegetable means to chop it into small, regular pieces, and the Dicer enzyme does the same thing to dsRNA, clipping it into small (21- to 23-base) dsRNAs called small interfering RNAs (siRNAs). The siRNAs are in turn bound by a complex of proteins that separates the strands and scans cellular mRNA to find molecules with complementary base sequences. When the short single-stranded RNA within the protein complex forms base pairs with an mRNA, the mRNA is degraded, preventing gene expression (see the figure).

It turns out that RNAi probably plays normal and essential roles in the differentiation of cells and tissues. Once scientists realized that short RNA molecules could be regulators of gene expression, they began to look for them within cells. They found hundreds of small RNAs with homology to cellular mRNAs. These so-called micro-RNAs (miRNAs) have base sequences that allow them to fold essentially in half and base pair internally, creating a dsRNA that is processed as shown in the figure. miRNA appears to be very important in regulating the expression of genes that must be switched off at different times or in different tissues. For example, mutations in the Dicer gene in the plant *Arabidopsis* cause developmental abnormalities, as do mutations in specific miRNAs in the nematode worm.

Scientists immediately saw the potential of RNAi for controlling the expression of genes, just as they had recognized the potential of antisense regulation. RNAi has generally proven very fruitful in laboratory experiments, allowing scientists to mimic the effects of gene mutations by shutting off the expression of specific genes. Scientists can introduce the 21- to 23-base-pair dsRNA molecules into cells directly, or they can introduce an artificial DNA gene with a promoter and a self-complementary coding sequence, like the natural miRNAs. Although there were no RNAi drugs on the market at the time of writing, potential uses of RNAi to fight AIDS, certain inherited diseases, and cancer are being explored.

How RNAi works. A dsRNA molecule is clipped into short siRNAs by the Dicer enzyme. The strands of the siRNAs are separated by a protein complex that includes an RNA-cleaving molecule, and the protein-RNA complex scans cellular mRNA. When an mRNA molecule with a complementary sequence is found, the small RNA hybridizes to it, triggering cleavage of the target mRNA.

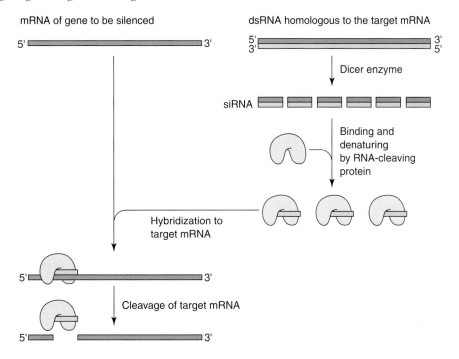

Questions

1. What kinds of things did the Isis scientists need to know to develop Vitravene? What kinds of things would they probably have tested during the development of the drug?

2. Can you think of a disease or a type of disease that would probably not be treatable by antisense or RNAi drugs?

3. *Research question.* Imagine you are part of a research team at a pharmaceutical company. You have been asked to prepare a proposal to develop an antisense or an RNAi drug to control AIDS. What kinds of things would you need to know to begin to design your molecule? Find the information you need and prepare a proposal for your company's management. Explain to them why blocking expression of the target gene should be an effective strategy. Of course, you do not know for certain if it will work—it's a research proposal—but you should be able to justify your idea.

Gene Regulation

All cells regulate the expression of their genes. Essentially every bacterial gene that has been studied is regulated at some step during its expression. A stunning variety of regulation mechanisms are used, including regulation of transcription initiation (as described for the *lac* and *trp* operons in *Regulating gene expression* in chapter 4), regulation of the termination of transcription, degradation of mRNA, control of translation initiation (through antisense oligonucleotides or proteins), and alteration of promoter recognition. It may be safe to assume that any gene regulation mechanism we can imagine is in use somewhere in the natural world.

Specific examples of various gene regulation mechanisms include the following.

Activation and repression of transcription

Regulation of the frequency of transcription is the most important level of gene control exercised in bacterial systems. The most common form of transcriptional control is carried out by specific regulatory proteins that bind within or near the promoter sequence and either prevent or enhance transcription. Two examples of ways that repressors of transcription are used to respond to environmental conditions are found in the *lac* and *trp* operons, as described in chapter 4. Transcriptional activators generally bind near the promoter and interact directly with RNA polymerase to increase the frequency of transcription. This type of interaction appears to be of particular importance in eukaryotes.

Alteration of promoter recognition

Alteration of promoter recognition occurs when a drastic change in gene expression is needed—a new set of genes must be transcribed and/or a currently transcribed set must be turned off. For example, many viruses encode proteins that bind to the host RNA polymerase. The modified RNA polymerase can subsequently recognize only the virus' promoters, which have a different base sequence from the normal host promoters. In this way, the virus switches transcription from the host genes to its genes. Another example of this type of regulation is found in the bacterium *Bacillus subtilis.* This organism forms durable, dormant spores in response to adverse environmental conditions. Spore formation requires the expression of a number of genes that are not active during the normal life cycle of the bacterium. In addition, normal gene activity all but ceases during the spore stage. To achieve this gross change in gene expression, *B. subtilis* synthesizes a special protein that binds to its RNA polymerase and causes it to recognize only the special promoters controlling sporulation genes.

Repression of translation

Regulation of gene expression can be exerted through control of the rate of translation of an mRNA. A good example of translational repression can be found in the synthesis of the ribosomal proteins of *E. coli.* The ribosomes of *E. coli* are made up of large rRNA molecules and several proteins that bind to specific regions of the RNA. The genes for the ribosomal proteins are lined up in operons. A single mRNA is transcribed from each operon and translated into several proteins. As the proteins are translated, they find free rRNA in the cytoplasm and bind to their recognition sites. When all available rRNA is complexed with protein, one of the proteins from that operon instead binds to the translation initiation region of the operon mRNA. The binding of that protein to the mRNA prevents further translation. For each one of the ribosomal protein operons, one of the encoded proteins acts as a repressor of translation. Through this mechanism, a balance between the amount of available rRNA and ribosomal proteins is achieved.

Regulation by antisense RNA

Antisense gene regulation was discovered in the 1980s in the course of studies involving plasmids, transposons, and bacteriophage. Several natural examples of this form of regulation have now been described. Two additional examples of antisense regulation are found in the synthesis of two outer membrane proteins in *E. coli* and the synthesis of the transposase protein of transposon 10. In these instances of antisense gene regulation, a short RNA exactly complementary to the 5' end of an mRNA hybridizes to the translation initiation region of the mRNA and prevents translation. It was originally thought that antisense RNA prevented translation by obstructing the binding of ribosomes to mRNA. In the light of the discovery of the mechanism of RNAi, that conclusion may need to be revisited.

Antisense gene regulation engendered a great deal of excitement because of the perceived potential for artificial gene control. Several biotechnology companies focused on antisense regulation as a means of controlling gene expression in crop plants, controlling viral disease, and treating cancer and other human diseases. They had some success, and an antisense antiviral drug is now on the market. Antisense gene regulation is discussed further in the next activity, along with its cousin, RNAi.

Regulation by RNAi

Attempts to use antisense RNA to regulate gene expression led to the discovery of RNAi. In RNAi, the presence of double-stranded RNA (dsRNA) triggers the enzymatic degradation of mRNA with the same base sequence as the dsRNA. Recent research has shown that cells use the RNAi pathway to regulate gene expression, particularly in a tissue-specific or temporally specific manner. Scientists quickly realized the power of RNAi in experiments to determine gene function, since the introduction of dsRNA into a cell can knock out expression of almost any gene. RNAi has become such an important technique that it is discussed in more detail in the following activity.

After your class has completed the activity *From Genes to Proteins* and understands how the genetic information in DNA is manifested in the structure of proteins, introduce basic concepts of gene regulation. Some of these gene regulation concepts can be illustrated through role playing in an extension of *From Genes to Proteins* (as described below) or simply by classroom discussion.

Ideas for Gene Regulation Activities

- Assign roles and act out the *lac* repressor preventing transcription of the *lac* genes when lactose is not present and allowing transcription when lactose is present (refer to chapter 4). This process could be illustrated simply with the DNA codons, the promoter, RNA polymerase, a repressor protein, and lactose; there is no need to actually make the RNA and go through translation.
- Assign roles and act out the *trp* repressor controlling expression of the tryptophan synthesis genes (refer to chapter 4).
- Have students brainstorm about ways in which a cell might regulate the synthesis of a particular protein. You could use their ideas as a jumping off point for introducing some of the gene regulation mechanisms mentioned above.
- Have students act out translational repression. For this, you could start with the mRNA, the ribosome area (with a "recognizer" in it), and someone assigned to the translational repressor protein.
- Illustrate the regulatory power of changing the promoter recognition by representing a series of genes as lines on the blackboard. Indicate the promoters at the left-hand ends of the lines, with PN for normal promoters and PS for sporulation promoters. Show the students that when RNA polymerase is modified to recognize only PS, all the PN genes are shut off.

Antisense and RNA Interference

Antisense oligonucleotides are segments of DNA or RNA that are exactly complementary in base sequence and antiparallel in orientation to a gene of interest. In antisense gene regulation, an anti-RNA hybridizes to the 5′ end of an mRNA molecule, preventing the ribosome from recognizing the translation initiation site and thus blocking translation of the mRNA. Antisense regulation was discovered in the 1980s during studies of plasmids, bacteriophage, and transposons.

An example of natural antisense regulation can be seen in the replication of bacterial plasmids of the R1 family. The initiation of DNA replication from the R1 origin requires the plasmid-encoded replication protein RepA. Its expression is controlled by an antisense RNA that is transcribed from the strand of DNA opposite to the one that encodes the RepA mRNA. The RepA anti-RNA binds to RepA mRNA and prevents translation, thus preventing plasmid replication. When the concentration of the anti-RNA in the cell is high, little RepA is synthesized. Under these conditions, plasmid replication is limited, and there are few copies of the R1 plasmid in the cell. When the concentration of the anti-RNA is low, more RepA protein is translated. In this case, plasmid replication is more extensive and many copies are present in the cell.

Antisense gene regulation caused great excitement among biotechnologists because of its perceived potential for allowing artificial control of gene expression. Theoretically, the expression of any gene could be suppressed through introduction of a sufficient amount of an appropriate antisense oligonucleotide, allowing the prevention or modulation of any process driven by the expression of specific genes. The first antisense drug on the market, Vitravene, was approved for use in 1999. It targets the replicase protein of cytomegalovirus (CMV), which can cause serious illness in immune-suppressed patients. Antisense cancer drugs targeting growth factors and oncogenes are currently in various stages of clinical trials, as well as antisense drugs against

viral diseases, such as hepatitis C and human papillomavirus, Lou Gehrig's disease (amyolateral sclerosis, or ALS), and Crohn's disease, and all are in various stages of clinical trials.

RNAi was discovered by accident as two researchers were attempting to use antisense RNA to block the expression of genes involved in growth and development in the roundworm *Caenorhabditis elegans*. They were trying to figure out the specific roles played by various gene products in the development of the worm. Their experimental design involved introducing antisense RNA into the worm, essentially by soaking the worm in a solution of the anti-RNA. Their experiment worked; expression of the targeted gene was blocked. However, as a control, they soaked some worms in the "sense" version of the RNA. To their surprise, gene expression was also blocked in these worms. Careful examination of their procedures and reagents revealed that the responsible agent was neither the sense nor the antisense RNA, but rather a double-stranded version of the RNA that was present as a minor contaminant in their preparations. They found that dsRNA triggered silencing of the expression of the matching gene, the phenomenon now known as RNAi. Their initial experiments were published in 1998.

Over the next 2 or 3 years, the mechanism by which RNAi works was figured out. It is illustrated in the figure in the *Student Activity*. The dsRNA is cleaved into short (21- to 23-base-pair) pieces by a nuclease named Dicer. These small RNA molecules, referred to as siRNAs, for *small interfering* RNAs, are bound by a protein complex called RISC that degrades one strand of the dsRNA, leaving a single-stranded RNA bound to the complex. This single-stranded RNA pairs with its complementary sequence within an mRNA molecule, triggering cleavage of the mRNA.

Scientists quickly recognized the potential for RNAi in investigations of gene function, as well as drug development. RNAi allows scientists to silence the expression of essentially any known gene by introducing an appropriate siRNA. This can be done by injecting the siRNA directly or by introducing a cloned DNA molecule containing a promoter and an appropriate base sequence for the production of siRNA.

In 1993, an RNA molecule was discovered in *C. elegans* that was complementary to the 3′ untranslated region (beyond the stop codon) of a specific mRNA. Experiments showed that during the worm's development, this small RNA bound to the complementary mRNA, blocking its translation and triggering the transition of the worm into the next stage of development from larva to adult. The implications of this discovery were not realized for a few years. Since the discovery of RNAi, it has become evident that small RNAs, now called micro-RNAs, or miRNAs, are important in both the normal development of organisms and aberrant development, such as cancer. miRNAs are transcribed as short, self-complementary molecules that fold back upon themselves to form a stem-loop structure reminiscent of a bobby pin (compare this to one arm of the tRNA cloverleaf picture in Figure 4.11). The folded RNA is processed by the Dicer pathway and inhibits gene expression through the same mechanism as siRNA.

Ironically, researchers had known for many years that cellular nucleic acid preparations contained a great many small RNA molecules. They had not known that these small RNAs might play important roles in gene regulation and assumed they were trash. Now, small RNAs are a focus of intense interest and research.

The picture that is gradually emerging is that miRNAs help regulate the expression of genes in a time- or tissue-specific manner and thus appear to play key roles in differentiation and development. Hundreds of miRNAs have been discovered in animal cells, including human cells. A given miRNA is typically expressed only in a specific tissue or group of tissues and/or at a given time in development. The expression of the miRNA decreases the expression of genes whose mRNAs contain complementary nucleotide sequences. Global analyses of mRNA sequences suggest that mRNAs encoding proteins that are needed by all cells at all times (so-called "housekeeping genes") do not contain sequences complementary to any of the known miRNAs and are therefore not regulated by them. The mRNAs of proteins that are expressed in a tissue- or time-specific manner do contain sequences homologous to miRNAs and appear to be the targets of this type of regulation.

Potential applications of antisense or RNAi technology in health care hold great promise for improving treatments for many of our most dreaded illnesses. Cancer and viral infections are among those that result from inappropriate expression of the body's own genes or from expression of foreign genes inside our cells. Current chemotherapy for cancer and viral infections, such as AIDS, interferes with enzymes necessary for cell growth and affects both diseased and healthy cells. Patients undergoing these treatments frequently become seriously ill from "side effects." By contrast, treating these diseases through specific prevention or reduction of the expression of disease-causing genes represents an ideal therapy.

To develop an antisense or RNAi drug or gene regulator, precise understanding of the genetic system to be regulated is essential. For therapeutics against viral diseases, the functions of viral genes in infection must be known. To treat a specific cancer, the genetics of that particular cancer must be worked out. The bulk of molecular biology research over the last decade or more has been directed at answering these genetic and mechanistic questions. Antisense and RNAi technologies now offer the prospect of a direct payoff in a variety of applications.

Answers to Student Questions

Note: There are no "correct" answers to these questions. The questions are designed to get students to think about the way antisense regulation works, the enormous potential of the method, its inherent limitations, and the kinds of things scientists would have to consider in developing a particular antisense gene regulation strategy.

1. Here are some examples of things the Isis scientists would have had to know and test. The scientists had to know what genes CMV has and that knocking out one of these genes would block viral reproduction. They had to know the sequences of the mRNAs of their potential target genes. They had to make antisense oligonucleotides and see if they could block viral replication in a test system, such as a cell culture. They had to make sure the oligonucleotide could block CMV replication in an animal model without making the animal sick. They had to make sure the drug was safe for humans to take and that it was effective against the infection (these steps are required by law for FDA approval).

2. Antisense and RNAi technologies reduce gene expression. They would probably not be effective against diseases that are caused by the lack of a gene product (or lack of a correct gene product),

rather than inappropriate expression of a gene. An example of such a disease is cystic fibrosis, caused by two defective copies of a gene for a cell membrane protein. Turning off expression of these defective genes would leave the patient with no membrane protein at all, which would be worse than a defective protein. Genetic diseases caused by dominant mutations (such as Huntington's disease) may be amenable to antisense drugs. In these diseases, the patient usually has one good copy of the gene and one mutant copy that dominates and causes the illness. If an antisense molecule could be designed that blocked expression of the mutant form of the gene only, it could be an excellent therapy. You may want to discuss this possibility with your class. Ask students what information would have to be available to design such an antisense molecule (examples: the disease gene must be found, it and the normal gene must be sequenced and the differences understood, and it must be determined whether an antisense molecule is effective in blocking only the mutant form).

3. The idea is to get students to think about how much you must know about a system in order to apply antisense or RNAi technology and to investigate a particular system with these technologies in mind. They will have to look up information about human immunodeficiency virus

(HIV), find out about its genes, and learn the roles the proteins play in an HIV infection. They will then have to select a target for antisense therapy and defend it. It is more important that students think through the HIV life cycle and what would be a logical place to block it than that they get any specific answer. Chapter 18, *DNA Sequencing*, contains a reading on some anti-HIV drugs. They are not antisense drugs, but they target the actions of specific HIV proteins. This information is simply for your interest—if students did not select these proteins as targets, it does not mean their answers were wrong.

Selected Readings

Antebi, A. 2005. The tick-tock of aging? *Science* 310:1911–1912.

Gibbs, W. 2003. The unseen genome: gems among the junk. *Scientific American* 289:26–33.

Hammond, S. M., A. A. Caudy, and G. J. Hannon. 2001. Post-transcriptional gene silencing by double-stranded RNA. *Nature Reviews Genetics* 2:110–119.

Lau, N., and D. Bartel. 2003. Censors of the genome. *Scientific American* 289:34–41.

Stix, G. 2004. Hitting the genetic off switch. *Scientific American* 291:91–101.

Zamore, D. Z. 2002. Ancient pathways programmed by small RNAs. *Science* 296:1265–1269.

Modeling Genome Size

9

About This Activity

This activity uses models to show the relative sizes of an *Escherichia coli* cell, the *E. coli* chromosome, a typical plasmid, and a gene. The models demonstrate graphically how much longer the *E. coli* chromosome is than the cell. Calculations included in the student questions use the analogies of letters in a book and miles of railroad track to suggest the size of the human genome. This lesson pairs nicely with the chapter 10 *Student Activity Extraction of Bacterial DNA.* It can be performed as a student activity or as a teacher demonstration.

Class periods required: *1/2 to 1*

Background

DNA is stored in cells in the form of chromosomes and plasmids. The amount of DNA required to store the information necessary for making even a simple organism, such as a bacterial cell, is very large. One of the wonders of biology is that cells are able to store and access the great lengths of DNA needed to encode their hereditary information.

The genome of the bacterium *E. coli* has been completely sequenced. It is 4,639,221 base pairs (bp) in length and contains 4,377 genes. Of these genes, 4,290 encode proteins, and the rest encode RNAs that are not translated (for example, ribosomal RNA and transfer RNA). The "average" protein-encoding bacterial gene is considered to be 1,200 bp in length. A typical plasmid is about 3,000 bp long and carries just a few genes.

What is the physical size of these DNA molecules? The *E. coli* chromosome consists of one large circular DNA molecule that, if stretched out, would be approximately 10^{-3} m (1 mm) in length. By comparison, the *E. coli* cell is only 1×10^{-6} to 2×10^{-6} m in length. The *E. coli* DNA molecule is thus 1,000 times longer than the cell! Even a plasmid of just 3,000 bp would be 10^{-6} m long if it was linear—

approximately the length of the cell. Still, the chromosome of *E. coli* comprises only 2 to 3% of the cell's weight and occupies only 10% of its volume.

DNA occupies such a small fraction of the cell's volume (considering its enormous length) because it is an extremely slender molecule. It is capable of high degrees of folding and coiling, an essential feature for packing it into the cell. Although the degree of folding required to fit the DNA of *E. coli* into the bacterium is impressive, the folding necessary for packaging DNA into a human cell is even more remarkable.

The human cell is approximately 2×10^{-5} m in diameter, and the human genome consists of 3.3 billion bp. If the DNA of a single human cell were stretched out, it would be about 2 m in length—100,000 times longer than the cell! Yet all this DNA not only fits into the cell, it is also accessible to the cell's enzymes for information transfer and replication.

This activity uses models that are 10,000 times life size ($\times 10,000$). It is instructive to note that a $\times 10,000$ magnification is analogous to inflating an ant to the size of a tractor-trailer rig. (It is fun to ask students to guess how big a $\times 10,000$ ant would be.) A convenient model for a $\times 10,000$ *E. coli* cell is a 2-cm (20-mm) gelatin capsule (these can be purchased at pharmacies or health food stores). The capsule is a good model for reasons besides its length; it is rod shaped, like *E. coli,* and it has an outer wall analogous to the outer membrane of the bacterium. If you do this activity in conjunction with DNA extraction, you can use the capsule to illustrate that the cell membrane must be dissolved to release the cell contents, including the DNA.

The DNA models consist of appropriate lengths of thread, yarn, or string. Thread is most accurate, because it is the slenderest, but yarn or string is easier to see if you are doing a demonstration. The $\times 10,000$ *E. coli* chromosome is represented by a

10-m length of thread. Use the thinnest thread you can find, and remind the students that it represents a double helix. The size of a ×10,000 plasmid is represented by 10 mm of thread, yarn, or string. A single ×10,000 gene is only 4 mm in length—a piece of fuzz if you use yarn.

Objectives

At the end of this activity, students should be able to:

1. Describe the relative sizes of the *E. coli* cell, chromosome, genes, and plasmids.
2. Explain that strands of DNA are extremely thin and must be very tightly coiled; an immense length of DNA fits into a relatively tiny cell volume.
3. List some differences between the *E. coli* and human genomes.
4. (Optional) Use scientific notation to make relevant calculations about the dimensions of DNA.

Materials

- Felt tip pen
- Scissors
- Glue
- Meter stick (for measuring thread)

For each laboratory team of four students, you need to gather a letter size envelope and the following:

- A 2-cm gelatin capsule (no. 0; purchase at a pharmacy or health food store)
- 10 m of white thread; 10 m of colored thread
- Two 20-mm strands of white and colored thread, yarn, or twine
- A 4-mm piece of colored thread, yarn, or twine
- Two 3-by-5 index cards with yarn glued to them as described below

Preparation

1. Place one gelatin capsule into each envelope.
2. Use a meter stick and scissors to cut the following for each student team:
   - One 10-m segment of thread
   - One 4-mm segment of colored thread, yarn, or twine
   - One 10-mm segment of colored thread, yarn, or twine
3. Glue the 4-mm piece of colored thread, yarn, or twine to a 3-by-5 card and label it "average length of a single bacterial gene."
4. Glue the 10-mm segment to the second 3-by-5 card and label it "length of typical bacterial plasmid." If you use thread or string, glue it into a

circle to simulate the circular plasmid. You may make a "double-stranded" thread model of a plasmid if you wish.
5. Put a 10-m segment of thread and one each of the two cards into each envelope with the gelatin capsules.

Tips

1. If time allows, have your students prepare the "gene" and "plasmid" cards and measure out the 10-m lengths of thread. This exercise allows them to practice metric measurement.
2. The student questions involving scientific notation should be assigned relative to the mathematical abilities of your students. They can be solved as a class, serving as a review of scientific notation.

Procedure

1. Review or present any background material.
2. For a class activity, split the students into groups of four and have them follow the procedure on the activity sheet. (In step 7, the best way to fit the thread in the capsule is to fold the lengths of thread in half several times until they make a "wad." Insert one end of the wad into the longer section of the capsule, and push the remainder in with a twisting motion.)
3. The students answer the questions in class or at home.

Alternative Procedure

This activity works well as a demonstration coupled with the next activity, *Extraction of Bacterial DNA*. A good time to do it is during the 65°C incubation. It will not hurt the DNA preparation if you incubate the cells longer than 15 minutes.

Answers to Student Questions

1. Each thread could represent one strand of the DNA double helix.
2. If each gene is 4 mm long, there would be 250 genes in a meter, or 2,500 in 10 m. Some *E. coli* DNA is noncoding; the bacterium is estimated to have about 2,000 genes.
3. Real *E. coli* DNA occupies only 10% of the cell volume. The thread occupies a greater portion of the capsule volume; therefore, the thread is too thick for an accurate representation of DNA.
4. 10 m for the *E. coli* genome
   10,000 m for the human genome
   10,000 m divided by 1,609 m/mi = 6.3 mi of thread to represent the human genome

5. $(3.0 \times 10^6) \times (3.4 \times 10^{-10} \text{ m}) = 10.2 \times 10^{-4} \text{ m}$
   $= 1.02 \times 10^{-3} \text{ m}$

6. $(1.2 \times 10^3) \times (3.4 \times 10^{-10} \text{ m}) = 4.08 \times 10^{-7} \text{ m}$

7. $(3.0 \times 10^3) \times (3.4 \times 10^{-10} \text{ m}) = 10.2 \times 10^{-7} \text{ m}$
   $= 1.02 \times 10^{-6} \text{ m}$

8. $(1.02 \times 10^{-3} \text{ m}) \times (1.0 \times 10^4) = 1.02 \times 10^1 \text{ m}$
   $= 10.2 \text{ m}$

9. $(4.08 \times 10^{-7} \text{ m}) \times (1.0 \times 10^4) = 4.08 \times 10^{-3} \text{ m}$
   $= 4 \text{ mm}$

10. $(1.02 \times 10^{-6} \text{ m}) \times (1.0 \times 10^4) = 1.02 \times 10^{-2} \text{ m}$
    $= 10.2 \text{ mm}$

11. $(3 \times 10^9 \text{ ties}) \times (2 \text{ ft per tie}) = 6 \times 10^9 \text{ ft}$
    $6 \times 10^9 \text{ ft} \div 5,280 \text{ ft/mi}) = 1,136,364 \text{ mi}$
    $1,136,364 \text{ mi} \times (1.61 \text{ km/mi}) = 1,829,546 \text{ km}$

12. $1,136,364 \text{ mi} \div (24,000 \text{ mi/circumference}) =$
    47.3 trips around the earth

13. The number will vary, depending on the textbook. In a book with 100 characters per line (a high estimate), 50 lines per page, and 600 pages, there will be 3×10^6 characters. It would take 1,000 of these books to contain enough characters (3×10^9) to represent the human genome.

Note: Discrepancies occasionally appear in discussions of the size of the human genome versus the amount of DNA in the human cell. They arise because human cells contain two copies of each chromosome. Since 3.3×10^9 bp is the amount of DNA in one copy of each chromosome, if you sequence all of these bases, you have sequenced the human genome. Each cell contains a second copy of the genome on the homologous chromosomes. Therefore, the DNA content of a human cell is $2 \times (3.3 \times 10^9 \text{ bp})$, and it would be $2 \times (3.3 \times 10^9) \times (3.34 \times 10^{-10} \text{ m}) = 22.04 \times 10^{-1} \text{ m}$, or 2.2 m in length.

Sizes of the *Escherichia coli* and Human Genomes

9

Introduction

DNA is stored in cells in the form of chromosomes and plasmids. The amount of DNA required to store the information necessary for making even a simple organism, such as a bacterial cell, is very large. One of the wonders of biology is that cells are able to store and access the great lengths of DNA needed to encode their hereditary information.

The genome of the bacterium *Escherichia coli* has been completely sequenced. It is 4,639,221 base pairs (bp) in length and contains 4,377 genes. Of these genes, 4,290 encode proteins and the rest encode RNAs that are not translated (for example, ribosomal RNA and transfer RNA). The "average" protein-encoding bacterial gene is considered to be 1,200 bp in length. A typical plasmid is about 3,000 bp long and carries just a few genes.

What is the physical size of these DNA molecules? The *E. coli* chromosome consists of one large circular DNA molecule that, if stretched out, would be approximately 10^{-3} m (1 mm) in length. By comparison, the *E. coli* cell is only 1×10^{-6} to 2×10^{-6} m in length. The *E. coli* DNA molecule is thus 1,000 times longer than the cell! Even a plasmid of just 3,000 bp would be 10^{-6} m long if linear—approximately the length of the cell. Still, the chromosome of *E. coli* comprises only 2 to 3% of the cell's weight and occupies only 10% of its volume.

DNA occupies such a small fraction of the cell's volume (considering its enormous length) because it is an extremely slender molecule. It is capable of high degrees of folding and coiling, an essential feature for packing it into the cell. Although the degree of folding required to fit the DNA of *E. coli* into the bacterium is impressive, the folding necessary for packaging DNA into a human cell is even more remarkable.

The human cell is approximately 2×10^{-5} m in diameter, and the human genome consists of 3.3 billion bp. If the DNA of a single human cell were stretched out, it would be about 2 m in length—100,000 times longer than the cell! Yet all this DNA not only fits into the cell, it is also accessible to the cell's enzymes for information transfer and replication.

This activity uses models that are 10,000 times life size ($\times 10,000$) to demonstrate the relationship between the sizes of an *E. coli* cell, its chromosome, its plasmid, and a single gene. The $\times 10,000$ *E. coli* is represented by a 2-cm gelatin capsule, the $\times 10,000$ *E. coli* chromosome by a 10-m length of thread, the plasmid by 10 mm of thread, and a gene by 4 mm of thread.

Materials

Obtain from your teacher a letter size envelope containing:

- A 2-cm (20-mm) gelatin capsule
- 10 m of thread
- An index card with 4 mm of thread labeled "average length of bacterial gene"
- An index card with 10 mm of thread labeled "length of a typical bacterial plasmid"

At the direction of your teacher, form groups of four for the next steps.

1. Remove the gelatin capsule from the envelope. It represents a single *E. coli* bacterium that has been enlarged 10,000 times.
2. Remove the two index cards from the envelope. The lengths of thread or string on the cards represent the length (and *not* the diameter) of an *E. coli* gene and plasmid magnified 10,000 times.
3. Remove and stretch out the thread in the envelope. It represents the bacterial chromosome magnified 10,000 times.

4. Two people should make a circle with the threads. The *E. coli* chromosome is circular and is attached to the cell membrane.

5. A third person now holds up the index card with the ×10,000 bacterial gene next to the DNA loop. The average bacterial gene contains about 1,200 bp. Remember that a gene is composed of all the segments of DNA that instruct the cell to make a single protein, whether those segments are continuous or not.

6. With the bacterial-chromosome and bacterial-gene models still in view, the fourth person holds up the index card with the ×10,000 bacterial plas-mid. Plasmids carry one or a few genes necessary for their own replication and stability and often carry genes that add important characteristics, such as antibiotic resistance, to the bacterium. You can see that the plasmid is tiny in comparison to the chromosome.

7. Now that you have compared the sizes of the chromosomes, plasmids, and genes, try to reconstruct the bacterium by inserting the "chromosome" into the capsule. It isn't easy! The real *E. coli* chromosome occupies about 10% of the cell volume.

Questions

1. How could two 10-m lengths of thread represent the *E. coli* chromosome more accurately?

2. How many bacterial genes would fit on your DNA circle (formed in step 4)?

3. Is the thread that you tried to stuff in the capsule too thick to represent the DNA's actual thickness? What is the reason for your answer? (Hint: what percentage of the "bacterial-cell" volume does the thread occupy in your model, and what is the actual volume that DNA occupies in *E. coli*?)

4. In this activity, how many meters of thread did it take to represent the *E. coli* genome? If the human genome is 1,000 times greater than the *E. coli* genome, how many meters would it take to represent the human genome? How many miles of thread would that be?

Mathematical Calculations

- The distance between DNA bp is 3.4×10^{-10} m.
- The *E. coli* chromosome has about 3×10^{6} bp.
- The average *E. coli* gene has 1,200 bp.
- A typical plasmid has about 3,000 bp.

Using the information above, calculate the following:

5. How long (in meters) is the *E. coli* chromosome?

6. How long (in meters) is the average *E. coli* gene?

7. What is the circumference (in meters) of a typical *E. coli* plasmid?

8. If *E. coli* were magnified 10,000 times, how long (in meters) would its chromosome be?

9. If *E. coli* were magnified 10,000 times, how long would its average gene be?

10. If *E. coli* were magnified 10,000 times, how long would a typical plasmid be?

The human genome can be related to a length of railroad track. The railroad ties represent the base pairs, and the rails represent the sugar-phosphate backbone of the DNA molecule. The railroad ties are 2 ft apart.

11. There are 3×10^{9} bp in the human genome. How many miles of track would it take to represent the human genome?

12. The circumference of the earth is 24,000 miles. How many times would the railroad track representing the human genome wrap around the earth at the equator?

13. Another way to represent the size of the human genome is to relate the base pairs to characters on a page in a book. Calculate how many of these books it would take to represent the human genome in the following manner:

- Choose a page in your text that is mostly printed.
- Count the number of characters on five randomly selected lines. Find the average number of characters per line (C). Record C.
- Count the number of lines on the page (L). Record L. Calculate the average number of characters (N) on a page by multiplying C times L. Record N.
- Calculate the number of characters in your text (T) by multiplying N by the number of pages in your text. Record T.
- To determine how many books like your text it would take to represent the human genome if every character represented a base pair, divide the number of base pairs in the human genome (3×10^9) by the number of characters in your text (T). How many books would be required?

DNA Extraction

10

About This Activity

In this activity, students can see a mass of stringy DNA fibers precipitate from bacterial cells or, alternatively, yeast, plant, or animal tissue. The bacterial-DNA extraction is very easy to perform. You can grow your own bacteria and use materials from a drug store, or you can order pregrown cells and solutions from scientific supply houses. The procedure provided here is for home-grown cells; the kits of materials available from scientific supply houses come with their own instructions. Additional procedures for extraction of DNA from yeast, plant, and animal tissue are also provided here.

Class periods required: 1

Background

The preparation of DNA from any cell type involves the same general steps: (i) breaking open the cell (and the nuclear membrane, if applicable), (ii) removing proteins and other cell debris from the nucleic acid, and (iii) final purification. There are several different ways of accomplishing each of these steps, and the method chosen generally depends on how pure the final DNA sample must be and the relative convenience of available options.

If a cell is enclosed by a membrane only (such as *Escherichia coli* or an animal cell), the cell contents can be released by dissolving the membrane with detergent. Cell membranes are made of proteins and fats. Just as detergent dissolves fats in a frying pan, a little detergent dissolves cell membranes. (The process of breaking open a cell is called cell *lysis*.) As the cell membranes dissolve, the cell contents flow out, forming a soup of nucleic acid, dissolved membranes, cell proteins, and other cell contents referred to as a cell *lysate*. Additional treatment is required for cells with walls, such as plant or yeast cells, and those of Gram-positive bacteria. These treatments can include enzymatic digestion of the

cell wall material or physical disruption by means such as blending or grinding.

Detergent treatment provides an additional benefit in DNA preparation: denaturation of proteins. When a protein is denatured, the amino acid chain unfolds and its three-dimensional structure is altered or lost. Denaturation blocks enzyme activity, which is important, because all cells contain DNA-digesting enzymes called deoxyribonucleases (DNases). If the DNases in a cell are not denatured after cell lysis, they will digest the cellular DNA into small pieces. Heat also denatures most proteins.

After cell lysis, the next step in DNA preparation usually involves removing proteins from the nucleic acid. Treatment with protein-digesting enzymes (proteinases) and/or extractions with the organic solvent phenol are two common methods of protein removal. Proteins are soluble in phenol, but DNA is not. Extracting an aqueous DNA-protein mixture (such as a cell lysate) with phenol separates the protein into the phenol and leaves the DNA in the water. Following the removal of the protein, the DNA is usually subjected to additional purification. Final purification methods include precipitation, dialysis, and high-speed centrifugation. The level of purification required depends on what the DNA will be used for.

A different approach to DNA purification involves the use of membranes or matrices to which DNA will bind under specific conditions. Here, cells are lysed in a solution that facilitates selective binding of DNA to the matrix. The lysate is applied to the matrix, and the DNA binds to it. Contaminants are removed from the DNA by rinsing the matrix-bound DNA with buffers. Finally the DNA is eluted from the matrix with a buffer formulated to overcome the attractive force between the DNA and the matrix. These DNA-binding materials are often set up in tiny columns inside centrifuge tubes. The binding, washing, and elution steps are accomplished with the help of centrifugal force by spinning the tubes

In the activity described here, no attempt is made to purify the DNA, since all that is required is to see it. Students will lyse *E. coli* with detergent and layer a small amount of alcohol on top of the cell lysate. Either ethanol or isopropanol (rubbing alcohol) can be used. DNA is insoluble in either alcohol and will form a white, weblike mass (precipitate) at the interface of the alcohol and water layers. By moving a glass rod up and down through the layers, students can collect the precipitated DNA on the rod. This DNA is very impure; the mass contains cellular proteins and other debris, but the stringy fibers are DNA. This easy and fun procedure lets students see DNA with their own eyes and shows its fibrous nature.

This inexpensive laboratory procedure can reinforce discussions in the areas of enzymes, cell structure, lipid membranes, denaturation, solubility, detergent-lipid interaction, density, and the nature of the bacterial genome.

Objectives

At the end of this laboratory procedure, students should:

1. Understand how to extract a visible mass of DNA.
2. Be familiar with certain physical and chemical properties of DNA, such as solubility and high molecular mass.

Materials

Equipment
- Centigrade thermometers
- Hot plate and large pot

Supplies
- *E. coli* cells (You and your students can grow these for yourselves. Alternatively, scientific supply companies sell dehydrated cells for DNA extraction.)
- Uninoculated broth for students' control extraction procedure
- Eyedroppers
- Ethanol (95%) or rubbing alcohol (isopropanol)
- 15-ml culture tubes
- 50% solution of dish detergent (Palmolive works well) or shampoo (without conditioner) in water, or 10% sodium dodecyl sulfate (SDS)
- Glass stirring rods

To grow the cells (see *Preparation* below), you will need:

- Medium, such as tryptic soy broth, Luria broth, or nutrient broth
- *E. coli* culture
- Inoculating loop(s)
- Large flask (if you grow one big culture)
- Incubator (optional)

Note: Medium recipes are on the CD, and procedures for growing *E. coli* can be found in Appendix D.

Preparation

If you grow your own cells, you must start a few days ahead. Streak out a few plates with *E. coli*. Have the students use sterile technique to inoculate 4 ml of tryptic soy broth or Luria broth in a 15-ml tube. Also, have them set up a control uninoculated 4 ml of broth (one tube of each per laboratory team). Allow the cells to grow until they are fairly dense (at least overnight). Proceed with the extraction. The procedure works very well with cells that are suspended in broth.

Alternatively, inoculate one large culture yourself, let it grow up, and dispense 4 ml of cells to the students, along with a second tube containing 4 ml of uninoculated broth when you are ready.

Home-grown cells may be kept in the refrigerator for a day before use.

For the activity
1. Print and copy the *Student Activity* for your class, if the students do not have manuals.
2. If you are using dishwashing liquid or shampoo, make a 50% solution in water (approximately 6 ml per student team). The dishwashing liquid is diluted, because otherwise it is too viscous to pour easily. You can also use 10% SDS (also known as lauryl sulfate), another detergent, in water. To make 10% SDS, dissolve 10 g of SDS per 100 ml of solution.
3. When the students are ready to extract the DNA, they will need a 60 to 70°C hot-water bath. You can use a large kitchen pot, a thermometer, and a hot plate. The only real requirement is to get a volume of water large enough to prevent the temperature from dropping below 60°C during the denaturing process.

Tips

1. Demonstrate the proper procedure for layering alcohol and spooling the DNA while the students' samples incubate in the hot-water bath. Layering is easiest when you slant the tube and let the alcohol run in slowly.

2. The 15-minute incubation is also a good time to remind the students of the length of the *E. coli* DNA (see the chapter 9 *Student Activity, Sizes of the* Escherichia coli *and Human Genomes*) and to make sure they understand that what they will see is the DNA from millions of *E. coli* cells sticking together in a fibrous mass.

3. The precipitation works better if the alcohol is cold. Store it in a freezer if you have access to one. (The procedure works with room temperature alcohol, too.)

4. Question 2 provides a good lead-in to a discussion of how density affects the layering of liquids, if you wish to get into that. Fun demonstrations using different concentrations of sucrose (for example, 0%, 20%, and 50%) colored with different food colorings can further illustrate the point.

Procedure

1. Review or present any necessary background material. This activity makes a nice follow-up to the activity *Sizes of the* Escherichia coli *and Human Genomes*. Explain the basic steps in the DNA extraction (lysing the cell, destroying DNases, and precipitation).

2. Tell the students which procedure they will use. If materials permit, let each student prepare his own DNA, following the instructions on the activity sheet. Demonstrate the layering/spooling technique.

3. Answer questions.

Answers to Student Questions

1. The double-stranded complementary structure makes it easy to replicate the molecule. It also makes it easy to synthesize an RNA copy of one strand by using the other strand as a template. If one strand is damaged, the sequence information is retained on the other strand. The undamaged strand can be used as a template to repair the damaged strand.

2. Alcohol is less dense than water, so it can float on top. If the alcohol were more dense than the broth or the cell lysate, it would sink to the bottom of the tube.

The following procedures were shared with us by experienced high school teachers.

Extraction of DNA from Plant Tissue

Materials

- One medium onion or six strawberries
- Blender
- Liquid detergent, such as Palmolive or Ivory, diluted 50:50 with water
- Two test tubes or other containers, such as a small beaker
- Conical filter (a coffee filter or a folded circle of laboratory filter paper will do)
- Beaker (400 to 600 ml is a good size)
- NaCl
- Glass rod or wooden splint
- Cold 95% ethanol or 70% isopropanol
- Water bath between 60 and 70°C

Procedure

Note: This procedure can be scaled up in size. You can work with more than 7 ml of onion or berry juice, or you can make several aliquots of onion juice for your students to precipitate.

1. Peel and chop a medium-size onion, or cap six medium to large strawberries or the equivalent volume of other berries.

2. Put the chopped onion or berries into a blender with 60 ml of water.

3. Beginning on low speed, blend the onion, increasing the speed to liquefy to thoroughly break up the plant tissue. Blend until there are no chunks left. Pour the juice into a beaker. Since berries are soft, you can grind them in a mortar or even mash them in a sealed plastic bag rather than use a blender, but they need to be thoroughly ground so that no chunks remain.

4. Add a teaspoon of sodium chloride (table salt) and stir to dissolve.

5. Pour 7 ml of the juice into a test tube or other appropriate container containing 7 ml of detergent. Mix gently, without causing foaming.

6. Place in the hot-water bath for 15 minutes. It is important that the temperature be between 60 and 70°C.

7. Filter the blended mixture through the filter into the beaker and discard the solids. Pour the liquid into a test tube.

8. Layer an equal volume of cold ethanol onto the filtrate. Spool the DNA as described for the bacterial-DNA extraction procedure.

Individualized procedure

The berry DNA extraction procedure can be individualized for student groups. For 10 to 12 groups, dissolve 1 tsp of salt in 60 ml of water. Give each group a strawberry or the equivalent volume of other berries and 5 ml of the salt water. The students should thoroughly mash the berry in a mortar, sealed plastic bag, or other container. Once the berry is mashed, add the salt water and mix until the tissue and water are thoroughly blended and no chunks remain. Carefully stir in 5 ml of dilute detergent, avoiding foaming. Incubate in the hot water, and then filter the mixture and spool the DNA as described in steps 6 to 8 above.

Extraction of DNA from Animal Tissue

Materials

- Frozen cat or dog testes, or thymus tissue
- Mortar and pestle
- 10% SDS or 50% detergent solution, as described above
- Cold ethanol or isopropanol
- Glass stirring rod

Contact a local veterinarian and ask him or her to save and freeze cat or dog testes from neutering operations. A butcher might also be able to provide you with gonadal or thymus tissue. After you collect the tissue from the veterinarian or butcher, keep it in the freezer. If you cannot find a source of appropriate animal tissue, insects, with the appendages and exoskeletons removed, make great sources of animal DNA.

To extract DNA, remove one organ or a piece of one organ from the freezer and grind it in the mortar. Add some detergent solution, stir, and transfer the entire contents of the mortar to a test tube. Incubate it in a 65°C water bath for about 15 minutes. Layer cold alcohol, and spool as for the bacterial DNA.

Be sure to ask your students why there is so much DNA in testis tissue.

The same procedure also works with thymus tissue. Calf thymus is often available at butcher shops, referred to as sweetbreads.

Extraction of Yeast DNA

This procedure for extraction of yeast DNA is described fully in an article by Larry Wegmann (*The Science Teacher*, December 1989). It uses household materials that students can bring from home. The procedure is simple and reliable and can be used with 9th graders or younger students.

Materials

- Fleischmann's All Natural Yeast
- Athletic Shoe Cleaner-Deodorizer
- Adolph's 100% Natural Tenderizer, unseasoned
- Ethanol or isopropanol (rubbing alcohol)
- Glass rods
- 250-ml (or similar size) beakers
- Hot plate or other means of heating water

Procedure

1. Heat 100 ml of tap water in a beaker to 50 to 60°C.
2. Mix in one-half package of yeast until it is thoroughly dissolved.
3. Add 5 ml of athletic-shoe cleaner.
4. Maintain solution at 50 to 60°C for 5 minutes, stirring occasionally.
5. Dissolve 3 g of meat tenderizer in the solution. (The meat tenderizer contains papain, a proteinase.)
6. Maintain the temperature at 50 to 60°C for 20 minutes.

At this point the solutions can be stored in the refrigerator overnight.

7. Allow the solution to come to room temperature.
8. Tip the beaker at an angle, and slowly add 100 ml of ethanol or isopropanol so that two layers are formed.
9. Slowly insert the glass rod through the alcohol into the yeast lysate, and stir the layers gently just at the interface. Do not mix the layers. A precipitate of DNA will form at the interface.
10. After stirring it for a few minutes, allow the layered solution to stand for several additional minutes.

The yeast DNA will not spool onto the glass rod. Presumably, there is nuclease activity in the preparation (either from the yeast themselves or from the meat tenderizer) that cuts the DNA into pieces too small to spool. However, the procedure reliably yields a mass of visible DNA.

Extraction of Bacterial DNA

10

Introduction

In this activity, you will extract a visible mass of DNA from bacterial cells.

The preparation of DNA from any cell type involves the same general steps: (i) breaking open the cell (and the nuclear membrane, if applicable), (ii) removing proteins and other cell debris from the nucleic acid, and (iii) final purification. There are several different ways of accomplishing each of these steps, and the method chosen generally depends on how pure the final DNA sample must be and the relative convenience of available options.

If a cell is enclosed by a membrane only, the cell contents can be released by dissolving the membrane with detergent. Cell membranes are made of proteins and fats. Just as detergent dissolves fats in a frying pan, a little detergent dissolves cell membranes. (The process of breaking open a cell is called cell *lysis*.) As the cell membranes dissolve, the cell contents flow out, forming a soup of nucleic acid, dissolved membranes, cell proteins, and other cell contents referred to as a cell *lysate*. Additional treatment is required for cells with walls, such as plant cells and many bacterial cells. These treatments can include enzymatic digestion of the cell wall material or physical disruption by means such as blending or grinding.

After cell lysis, the next step in DNA preparation usually involves purification by removing proteins from the nucleic acid. Treatment with protein-digesting enzymes (proteinases) and/or extractions with the organic solvent phenol are two common methods of protein removal. Proteins dissolve in phenol, but DNA does not. Furthermore, phenol and water, like oil and water, do not mix but instead form separate layers. If you add phenol to an aqueous (water-based) DNA-protein mixture, like a cell lysate, and mix them well, the protein dissolves in the phenol. After you stop mixing, the phenol separates from the aqueous portion, carrying the protein with it. The

DNA remains in the aqueous layer. To remove the protein, simply remove the phenol layer. Following the removal of the protein, the DNA is usually subjected to additional purification.

In this activity, you will not attempt any DNA purification—your goal is simply to see the DNA. You will lyse *Escherichia coli* with detergent and layer a small amount of alcohol on top of the cell lysate. Because DNA is insoluble in alcohol, it will form a white, weblike mass (precipitate) at the interface of the alcohol and water layers. By moving a glass rod up and down through the layers, you can collect the precipitated DNA. This DNA is very impure; the mass contains cellular proteins and other debris, along with the stringy fibers of DNA.

Before you begin the DNA isolation, make sure you know whether to follow procedure 1 or procedure 2. They are essentially the same but differ in the volumes of cells and the volumes and natures of the reagents you will use.

Procedure

1. Obtain from your teacher 4 ml of *E. coli* cells and 4 ml of medium in test tubes. Label the tubes. Shake your *E. coli* culture gently to resuspend the cells. Add to both labeled tubes 3 ml of a 50% solution of dishwashing detergent in water. (Your teacher may substitute other detergent.) Shake each tube to ensure complete mixing.

2. Place each tube in a 60 to 70°C water bath for 15 minutes. *Note:* Maintain the water bath temperature above 60°C but below 70°C. A temperature greater than 60°C is needed to destroy enzymes that degrade DNA.

3. Cool the tubes to room temperature (on ice if you have it).

4. To see the DNA, it must be taken out of solution, or *precipitated*. Watch your teacher demonstrate the following technique. Use a dropper to carefully layer 3 ml of 95% ethanol on top of the suspension in each tube. The alcohol should float on

top and not mix with the cell lysate. (It *will* mix in if you stir or squirt it in forcefully, so be careful.) Water-soluble DNA is insoluble in alcohol and precipitates when it comes in contact with it.

5. A weblike mass (precipitate) of DNA will float at the junction of the two layers (the "interface"). Push a rod through the alcohol into the soup, stir, and turn the rod. The rod carries a little alcohol into the soup and makes DNA come out of solution onto the rod. Keep moving the rod through the alcohol into the cell soup, and each time a little more DNA appears. Do not totally mix the two layers.

Observe and draw both tubes. Indicate the substances in each tube. Answer the questions.

Questions

1. What information storage advantage(s) lies in DNA's double-helix structure?

2. Why does the alcohol stay on top of the cell suspension and the broth in step 3?

Classroom Activities

B. Manipulation and Analysis of DNA

In the last 4 decades, our knowledge of DNA structure and function, and of the biochemical processes cells use to modify the structure and carry out the functions, has grown explosively. This extensive new knowledge has led to the ability to manipulate and analyze DNA outside the cellular environment. For example, we can cut DNA into specific pieces, separate and isolate those pieces, join pieces, copy DNA, and determine its base sequence. We are using our skills at manipulating genes to gather even more knowledge about how basic life processes are carried out and to make useful products for our daily lives.

The activities in this section deal with methods of manipulating and analyzing DNA: restriction digestion, ligation, gel electrophoresis, hybridization analysis, DNA sequencing, and the polymerase chain reaction. Many of these activities are suitable for 9th graders; they can also be used with adult audiences.

The first activity uses paper models to illustrate restriction digestion and gel electrophoresis. Two wet laboratories then let students perform restriction digests and carry out agarose gel electrophoresis for themselves. Next, the students simulate restriction digestion and ligation to construct a recombinant paper plasmid. Worksheets then challenge the students to apply what they have learned about restriction digestion, ligation, and electrophoresis to realistic problems. Finally, the more sophisticated techniques of hybridization analysis, DNA sequencing, and the polymerase chain reaction are illustrated through paper simulations.

Classroom Activities

Restriction Enzymes

11

About This Activity

In the activity in this chapter, students are introduced to restriction enzymes and simulate the activity of restriction enzymes with scissors. They are also introduced to restriction maps and asked to make simple predictions based on a map.

A second paper model activity, *Recombinant Paper Plasmids* (chapter 14), takes students through cloning a gene into a plasmid and making a recombinant plasmid from two different plasmids. *Recombinant Paper Plasmids* also addresses the process of transformation and asks students to think about selecting for various transformed cells. It is very effective to follow these activities with the transformation laboratory and/or a restriction analysis laboratory.

Class periods required: 1/2 to 1

Background

Restriction enzymes

Restriction enzymes, or restriction endonucleases, are proteins that recognize and bind to specific DNA sequences and cut the DNA at or near the recognition site. A nuclease is any enzyme that cuts the phosphodiester bonds of the DNA backbone, and an endonuclease is one that cuts somewhere within a DNA molecule. In contrast, an exonuclease cuts the phosphodiester bonds starting from a free end and working inward.

Restriction enzymes were originally discovered through their ability to break down foreign DNA. Restriction enzymes can distinguish between the DNA normally present in the cell and foreign DNA, such as infecting bacteriophage DNA. They defend the cell from invasion by cutting foreign DNA into pieces and thereby rendering it nonfunctional. Restriction enzymes appear to be made exclusively by prokaryotes.

The restriction enzymes commonly used in laboratories generally recognize specific DNA sequences of 4 or 6 base pairs (bp). These recognition sites are palindromic, in that the 5′-to-3′ base sequences on the two strands are the same. Most of the enzymes make cuts in the phosphodiester backbones of DNA at a specific position within the recognition site, resulting in a break in the DNA. These recognition/cleavage sites are called "restriction sites." Below are some examples of restriction enzymes (with names in a combination of letters and Roman numerals) and their recognition sequences, with arrows indicating cut sites.

		↓				↓	
EcoRI:	5′	GAATTC	3′	**HindIII:**	5′	AAGCTT	3′
	3′	CTTAAG	5′		3′	TTCGAA	5′
		↑				↑	

		↓				↓	
BamHI:	5′	GGATCC	3′	**AluI:**	5′	AGCT	3′
	3′	CCTAGG	5′		3′	TCGA	5′
		↑				↑	

		↓				↓	
SmaI:	5′	CCCGGG	3′	**HhaI:**	5′	GCGC	3′
	3′	GGGCCC	5′		3′	CGCG	5′
		↑				↑	

Notice that the "top" and "bottom" strands read the same from 5′ to 3′; this characteristic defines a DNA palindrome. Also notice that some of the enzymes introduce two staggered cuts in the DNA, while others cut each strand at the same place. Enzymes like SmaI that cut both strands at the same place are said to produce "blunt" ends. Enzymes like EcoRI leave two identical DNA ends with single-stranded protrusions:

5′ G	AATTC 3′
3′ CTTAA	G 5′

Under appropriate conditions (salt concentration, pH, and temperature), a given restriction enzyme will cleave a piece of DNA into a series of fragments. The number and sizes of the fragments depend on the number and locations of restriction sites for that enzyme in the given DNA. A specific combination of 4 bases will occur at random only once every few hundred bases, while a specific sequence of 6 bases will occur randomly only once every few thousand bases. It is possible that a DNA molecule will contain no restriction site for a given enzyme. For example, bacteriophage T7 DNA (approximately 40,000 bp) contains no EcoRI sites. The action of restriction enzymes is introduced and modeled in the activity *DNA Scissors*.

Rejoining restriction fragments

DNA fragments generated by restriction digestion can be put back together with the enzyme DNA ligase, which forms phosphodiester bonds between the 5′ and 3′ ends of nucleotides. As you might expect, any blunt-ended DNA can be ligated to any other blunt-ended DNA without regard to the sequences of the two molecules. Restriction fragments with single-stranded protrusions, like the EcoRI products shown above, are "pickier." For efficient ligation, the single-stranded regions must be able to hybridize to a complementary single-stranded region. The idea of rejoining restriction fragments and the need for complementarity in the single-stranded "tails" are introduced in the activity *DNA Scissors*.

This requirement for complementarity may sound limiting, but an examination of the EcoRI digestion products shown above reveals that two EcoRI "ends" are perfectly complementary. Any two DNA fragments produced by EcoRI digestion can be ligated together, because their single-stranded protrusions are complementary. In fact, fragments with complementary single-stranded protrusions ligate together much more readily than blunt-ended fragments, presumably because hybridization between the single-stranded regions holds the fragments together in the proper position for ligation. Because these single-stranded protrusions actually facilitate the joining of DNA segments with matching protrusions, they are often called "sticky ends."

Restriction enzymes and DNA ligase play starring roles in DNA cloning. To a molecular biologist, "cloning" a piece of DNA means adding that piece to a plasmid or other vector and then putting the plasmid (or the other vector) back into a host cell. One of the simplest methods of cloning is to ligate a restriction fragment into a plasmid that has been cut once with the same restriction enzyme(s). The restriction fragment becomes part of the plasmid when DNA ligase forms phosphodiester bonds between the two formerly separate DNA molecules. This type of cloning is modeled in exercises 1 and 2 of the activity *Recombinant Paper Plasmids* (chapter 14). Exercise 3 models the recombining of two plasmids to make a new plasmid containing both the ampicillin and the kanamycin resistance genes.

Restriction enzymes and genetic engineering

We often read that the discovery of restriction enzymes made genetic engineering possible. Why is that so? It is so because restriction enzymes first made it possible to work with small, defined pieces of DNA. Chromosomes are huge molecules usually containing many genes. Before restriction enzymes were discovered, a scientist might be able to tell that a chromosome contained a gene for an enzyme required to ferment lactose because he knew that the bacterium could ferment lactose and he could purify the protein from bacterial cells. He could use genetic analysis to tell what other genes were close to "his" gene. But he could neither physically locate the gene on the chromosome nor manipulate it.

The scientist could purify the chromosome from the bacterium, but then he had a huge piece of DNA containing thousands of genes. The only way to break the chromosome into smaller segments was to use physical force and break it randomly. Then what would he have? A tube full of random fragments. Could they be cloned? Not by themselves. If you introduce a simple linear fragment of DNA (like those produced by shearing) into most bacteria, it will be rapidly degraded by cellular nucleases. Cloning usually requires a vector to introduce and maintain the new DNA. Could our scientist use a vector, such as a virus or plasmid, to clone his DNA fragments? No. In order to clone DNA into a vector, you have to cut the vector DNA to insert the new piece. Could he simply study the random fragments? No. Every single chromosome from each bacterial cell would give different fragments, preventing systematic analysis. Thus, for many years, physical manipulation of DNA was virtually impossible.

The discovery of restriction enzymes gave scientists a way to cut DNA into defined pieces. Every time a given piece of DNA was cut with a given enzyme, the same fragments were produced. These defined pieces could be put back together in new ways. A new phrase was coined to describe a DNA molecule that had been assembled from different starting molecules: recombinant DNA.

The seemingly simple achievement of cutting DNA molecules reproducibly opened a whole new world of experimental possibilities. Now, scientists could

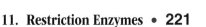

study small, specific regions of chromosomes, clone segments of DNA into plasmids and viruses, and otherwise manipulate specific pieces of DNA. The science of molecular biology literally exploded with the new information that became available, and genetic engineering, which essentially is the directed manipulation of specific pieces of DNA, became possible.

Separating restriction fragments

After restriction digestion, the fragments of DNA are often separated by gel electrophoresis. Background information about electrophoresis, a paper simulation of the process, and two wet laboratories follow the activity *DNA Scissors.*

Objectives

At the end of this activity, students should be able to:

1. Describe a typical restriction site as a 4- or 6-bp palindrome.
2. Describe what a restriction enzyme does (recognize and cut at its restriction site).
3. Use a restriction map to predict how many fragments will be produced in a given restriction digest.

Materials

- Copies of the *Student Activity,* if needed
- Copies of the DNA sequence strips from the CD
- Scissors

Preparation

Copy the *Student Activity* and print and copy the DNA sequence strips for your class, if necessary.

Tips

As students use the paper models, remind them that real DNA is three-dimensional and has no "back" and "front" sides, nor does it matter if the letters representing the bases are upside down. If you made the paper classroom model and still have it, it can be a helpful visual aid. If you have time, you might want to make a short strip of the paper model with a built-in restriction site (such as EcoRI or HindIII) that you can cleave as a demonstration to emphasize that the protruding ends are the same from 5′ to 3′.

Answers to Exercise Questions

The numbers are the item numbers under *Exercises and Questions* in the *Student Activity.* Two of the items, 1 and 5, contain only instructions and no questions and so are not represented below.

2. The DNA should be cut so that 5′ AATT protrudes from each end. The ends are "sticky."
3. The DNA should be cut straight across between the C and G in the middle of the SmaI site. The ends are blunt.
4. The DNA should be cut so that 5′ AGCT protrudes from each end. The ends are "sticky."
6. The two tails are (EcoRI) 5′ AATT 3′ and (HindIII) 5′ AGCT 3′. They are not complementary.
7. Each single-stranded tail has the sequence 5′ AATT 3′. They are complementary. Remember that to look for complementarity, you compare the 5′-to-3′ sequence of one strand with the 3′-to-5′ sequence of the other strand.
8. If the fragments were generated in an EcoRI digest, all of them will have the single-stranded 5′ AATT 3′ extensions on the ends. The ends of all of the fragments will be complementary.
9. Answers may vary. It is easier for DNA ligase to form a phosphodiester bond between two EcoRI fragments because of the complementary single-stranded tails. Hybridization between the bases in the tails brings the backbones into just the right position for resealing. With the noncomplementary tails (HindIII and EcoRI), the noncomplementary base pairs keep the nucleotide backbones from coming into proper position for bond formation. It is very difficult to get two fragments with noncomplementary "sticky ends" to reseal with DNA ligase. Two fragments with blunt ends of any sequence can be connected by DNA ligase. It is more difficult to get blunt-ended fragments to connect than fragments with complementary sticky ends but much easier than for fragments with noncomplementary sticky ends (which hardly ever happens). In a sense, "sticky ends" is a poor name for the ends with single-stranded tails, because these ends are only sticky with respect to complementary sticky ends and very unsticky with respect to noncomplementary "sticky ends."
10. Two linear fragments of 942 and 4,599 (5,541 − 942 = 4,599) bp.
11. Two linear fragments of 2,003 (2,035 − 32) and 3,538 (5,541 − 2,003) bp.
12. Three linear fragments of 2,003, 2,881 (4,916 − 2,035), and 657 [5,541 − (2,003 + 2,881)] bp.
13. The 942-bp fragment contains no PvuII sites and would not be cut. The 4,599-bp fragment would be cleaved into two fragments of 2,305 (3,247 − 942) and 2,294 (4,599 − 2,305) bp, giving three total fragments.

DNA Scissors

Background Reading

Genetic engineering is possible because of special enzymes that cut DNA. These enzymes are called *restriction enzymes,* or *restriction endonucleases.* Restriction enzymes are proteins produced by bacteria to prevent or restrict invasion by foreign DNA. They act as DNA scissors, cutting the foreign DNA into pieces so that it cannot function.

Restriction enzymes recognize and cut at specific places along the DNA molecule called restriction sites. Each different restriction enzyme (and there are hundreds, made by many different bacteria) has its own type of site. In general, a restriction site is a 4- or 6-base pair (bp) sequence that is a palindrome. A DNA palindrome is a sequence in which the "top" strand read from 5′ to 3′ is the same as the "bottom" strand read from 5′ to 3′. For example:

5′ GAATTC 3′
3′ CTTAAG 5′

is a DNA palindrome. To verify this, read the sequences of the top strand and the bottom strand from the 5′ ends to the 3′ ends. This sequence is also a restriction site for the restriction enzyme called EcoRI (pronounced EE ko R [say the name of the letter R] one [say the name of the numeral 1]). Its name comes from the bacterium in which it was discovered, *Escherichia coli* strain RY 13 (EcoR), and "I" because it was the first restriction enzyme found in this organism.

EcoRI makes one cut between the G and A in each of the DNA strands (see below). After the cuts are made, the DNA is held together only by the hydrogen bonds between the 4 bases in the middle. Hydrogen bonds are weak, and the DNA comes apart.

```
                   ↓
Cut sites:   5′  GAATTC  3′
             3′  CTTAAG  5′
                     ↑

Cut DNA:     5′  G          AATTC  3′
             3′  CTTAA          G  5′
```

The EcoRI cut sites are not directly across from each other on the DNA molecule. When EcoRI cuts a DNA molecule, it therefore leaves single-stranded "tails" on the new ends (see the example above). This type of end has been called a "sticky end" because it is easy to rejoin it to complementary sticky ends. Not all restriction enzymes make sticky ends; some cut the two strands of DNA directly across from one another, producing a blunt end.

When scientists study a DNA molecule, one of the first things they do is to figure out where many restriction sites are. They then create a restriction map, showing the location of cleavage sites for many different enzymes. These maps are used like road maps to the DNA molecule. A restriction map of a plasmid is shown in Figure 11.1.

Figure 11.1 Restriction map of pYIP5, a 5,541-bp plasmid. The number after the restriction enzyme name indicates at which base pair the DNA is cut by that enzyme.

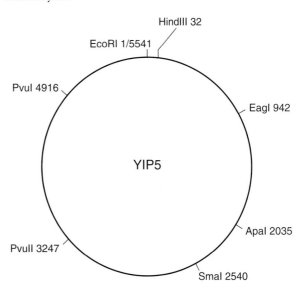

Below are the restriction sites of several different restriction enzymes, with the cut sites shown.

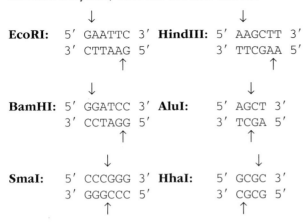

Which of these enzymes would leave blunt ends?

Which would leave "sticky ends"?

Refer to the above list of enzyme cut sites as you do the activity.

Exercises and Questions

Exercise I

Take Worksheet 11.1 with the DNA sequence strips and cut out the strips along the borders. These strips represent double-stranded DNA molecules. Each chain of letters represents the sequence of bases along the phosphodiester backbone, and the vertical lines between the base pairs represent hydrogen bonds between the bases.

1. You will now simulate the activity of EcoRI. Scan along the DNA sequence of strip 1 until you find the EcoRI site (refer to the list above for the sequence). Make cuts through the phosphodiester backbone by cutting just between the G and the first A of the restriction site on both strands. Do not cut all the way through the strip. Remember that EcoRI cuts the backbone of each DNA strand separately.

2. Now, separate the hydrogen bonds between the cut sites by cutting through the vertical lines. Separate the two pieces of DNA. Look at the new DNA ends produced by EcoRI. Are they "sticky" or blunt? Write "EcoRI" on the cut ends. Keep the cut fragments on your desk.

3. Repeat the procedure with strip 2, this time simulating the activity of SmaI. Find the SmaI site and cut through the phosphodiester backbones at the cut sites indicated above. Are there any hydrogen bonds between the cut sites? Are the new ends "sticky" or blunt? Label the new ends "SmaI," and keep the DNA fragments on your desk.

4. Simulate the activity of HindIII with strip 3. Are these ends "sticky" or blunt? Label the new ends "HindIII," and keep the fragments.

5. Repeat the procedure once more with strip 4, again simulating EcoRI.

6. Pick up the "front-end" DNA fragment from strip 4 (an EcoRI fragment) and the "back-end" HindIII fragment from strip 3. Both fragments have single-stranded "tails" of 4 bases. Write down the base sequences of the two tails, and label them "EcoRI" and "HindIII." Label the 5′ and 3′ ends. Are the base sequences of the HindIII and EcoRI "tails" complementary?

7. Put down the HindIII fragment, and pick up the "back-end" DNA fragment from strip 1 (strip 1 cut with EcoRI). Compare the single-stranded "tails" of the EcoRI fragment from strip 1 and the EcoRI fragment from strip 4. Write down the base sequences of the single-stranded "tails," and label the 3′ and 5′ ends. Are they complementary?

8. Imagine that you cut a completely unknown DNA fragment with EcoRI. Do you think that the single-stranded tails of these fragments would be complementary to the single-stranded tails of the fragments from strip 1 and strip 4?

9. There is an enzyme called *DNA ligase* that reforms phosphodiester bonds between nucleotides. For DNA ligase to work, 2 nucleotides must come close together in the proper orientation for a bond (the 5′ side of one must be next to the 3′ side of the other). Do you think it would be easier for DNA ligase to reconnect two fragments cut by EcoRI or one fragment cut by EcoRI with one cut by HindIII? What is your reason?

Exercise II

Figure 11.1 is a restriction map of the circular plasmid YIP5. This plasmid contains 5,541 bp. There is an EcoRI site at bp 1. The locations of other restriction sites are shown on the map. The numbers after the enzyme names tell at what base pair that enzyme cleaves the DNA. If you digest YIP5 with EcoRI, you will get a linear piece of DNA that is 5,541 bp in length.

10. What would be the products of a digestion with the two enzymes EcoRI and EagI?

11. What would be the products of a digestion with the two enzymes HindIII and ApaI?

12. What would be the products of a digestion with the three enzymes HindIII, ApaI, and PvuI?

13. If you took the digestion products from question 10 and digested them with PvuII, what would be the products?

Gel Electrophoresis

12

About This Activity

DNA Goes to the Races is a reading and paper activity that introduces electrophoresis. Students should already be familiar with the activity of restriction enzymes through the activity *DNA Scissors* (chapter 11). This lesson provides enough background for students to continue with activities (hybridization analysis, DNA sequencing, DNA fingerprinting, etc.) that require them to know something about the process. It is also an excellent introduction if you plan to conduct electrophoresis in your classroom. The *Restriction Analysis Challenge* worksheets in chapter 15 illustrate applications of electrophoresis in the laboratory. Students should complete this activity and *Recombinant Paper Plasmids* before using the challenge worksheets.

Class periods required: 1

Background

Please read the introduction to chapter 13 for general background information.

Objectives

At the end of this activity students should be able to:

1. If shown a simple restriction map, predict the number and arrangement of bands in a gel after digestion and electrophoresis.
2. Explain why electrophoresis causes DNA fragments of different sizes to separate.

Materials

- *DNA Goes to the Races Student Activity,* restriction maps, and gel outline

Preparation

- Photocopy the *Student Activity* if students do not have books.
- Print and copy the restriction maps and gel outline (Worksheets 12.1 and 12.2), if necessary.

Procedure

This activity is self-explanatory. Students can read the handout and do the exercises. Show them as much real material (or pictures) as you can: gel boxes, power supplies, an old gel, or a photograph of one. Relate the items to their activity and reading.

Results of the Exercise

If possible, check the students' work as they do the exercise. Make sure they line the fragments up correctly between lanes, as well as within an individual gel lane (for example, that the 2,500-base pair [bp] HindIII fragment is lined up between the 2,000-bp and 3,000-bp BamHI fragments and even with the 2,500-bp EcoRI fragment). Use their own fragment size labels to assist you.

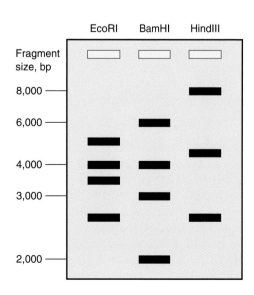

Classroom Activities

DNA Goes to the Races

You have already learned about restriction enzymes and how they cut DNA into fragments. You may even have looked at some DNA restriction maps and figured out how many pieces a particular enzyme would produce from that DNA. But when you actually perform a restriction digest, you put the DNA and the enzyme into a small tube and let the enzyme do its work. Before the reaction starts, the mixture in the tube looks like a clear fluid. Guess what! After the reaction is finished, it still looks like a clear fluid! Just by looking at it, you cannot tell that anything happened.

In order for restriction digestion to be meaningful, you have to be able to see the different DNA fragments that are produced. There are chemical dyes that stain DNA, but obviously it does not do much good to add them to the mixture in the test tube. In the laboratory, scientists separate DNA fragments by a process called *gel electrophoresis* so that they can look at the results of restriction digests (and other procedures).

Gel electrophoresis takes advantage of the chemistry of DNA to separate fragments. Under normal circumstances, the phosphate groups in the backbone of DNA are negatively charged. In electrical society, opposites do attract, so DNA molecules are very much attracted to anything that is positively charged. In gel electrophoresis, DNA molecules are placed in an electric field (which has a positive and a negative pole) so that they will migrate toward the positive pole.

The electric field makes the DNA molecules move, but to cause them to separate and be easy to look at later on, the whole process is carried out in a gel (obviously, the source of the name *gel* electrophoresis). If you have ever eaten Jell-O, you have had experience with a gel. The gel material in Jell-O is gelatin; different gel materials are used to separate DNA. One gel material often used for electrophoresis of DNA is called *agarose,* and it behaves much like Jell-O, but without the sugar and color. To make a gel for DNA (called *pouring* or *casting* a gel), you dissolve agarose powder in boiling water, pour it into the desired dish, and let it cool. As it cools, it hardens (sound familiar?).

Since the plan for agarose gels is usually to add DNA to them, scientists place a device called a *comb* in the liquid agarose after it has been poured into the desired dish and let the agarose harden around the comb. Imagine what would happen if you stuck the teeth of a comb into liquid Jell-O and let it harden. Afterward, when you pulled the comb out, you would have a row of tiny holes in the solid Jell-O where the teeth had been. This is exactly what happens with laboratory combs. When the comb is removed from the hardened agarose gel, a row of holes in the gel remains (look at Figure 12.1). The holes are called *sample wells.* DNA samples are placed into the wells before electrophoresis is begun.

For electrophoresis, the entire gel is placed in a tank of buffer. An electric current is applied across the tank so that it flows through the salt water and the gel. When the current is applied, the DNA molecules begin to migrate through the gel toward the positive pole of the electric field (Figure 12.2). Figure 12.3 shows a scientist loading a DNA sample into an agarose gel sitting in an electrophoresis tank.

At this point, the gel does its most important work. All of the DNA in the gel migrates through the gel toward the positive pole, but the gel material makes it more difficult for larger DNA molecules to move than smaller ones. Thus, in the same amount of time, a small DNA fragment can migrate much further than a large one. You can therefore think of gel electrophoresis as a DNA footrace, where the "runners" (the molecules being separated) separate just like runners in a real race (Figure 12.4). The smaller the molecule, the faster it runs. Two molecules of the same size run exactly together.

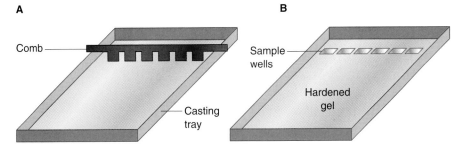

A

Comb

Casting tray

B

Sample wells

Hardened gel

Figure 12.1 Casting an agarose gel. (A) To make a gel, hot liquid agarose solution is poured into a casting tray (any shallow container) and the comb is put in place. (B) After the agarose cools and hardens, the comb is removed, leaving behind pits in the gel called sample wells. Samples are loaded into the wells prior to electrophoresis.

After a time, the electric current is turned off and the entire gel is placed into a DNA staining solution. After being stained, the DNA can be seen. The pattern looks like a series of stripes in the gel ("bands"), where each separate band is composed of one size of DNA molecule. There are millions of actual molecules in the band, but they are all the same size (or very close to it). At any rate, after a restriction digest, there should be one band in the gel for each different-size fragment produced in the digest. The smallest fragment will be the one that has migrated furthest from the sample well, and the largest will be closest to the well, as shown in Figure 12.5.

Activity

In Worksheet 12.1, you have three representations of a DNA molecule and an outline of an electrophoresis gel. The representations show the cut sites of three different restriction enzymes on the same DNA molecule. You will simulate the digestion of this DNA with each of the three enzymes and then simulate agarose gel electrophoresis of the restriction fragments.

1. Cut out the three pictures of the DNA molecule.
2. Simulate the activity of the restriction enzyme EcoRI on the DNA molecule that shows the EcoRI sites by cutting across the strip at the vertical lines representing EcoRI sites. You have now digested the molecule with EcoRI. Put your "restriction fragments" in a pile apart from the other two DNA strips.

Figure 12.3 A scientist is using a micropipette to load a DNA sample into an agarose gel. The gel is in an electrophoresis chamber full of buffer. The power supply for the chamber is on the laboratory bench behind the chamber.

Figure 12.2 In electrophoresis, the gel is placed in a tank of salt solution, and an electric current is applied. The DNA migrates toward the positive pole.

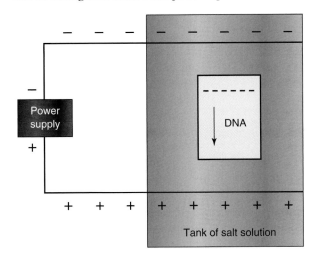

Power supply

DNA

Tank of salt solution

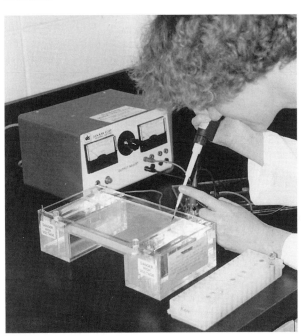

Classroom Activities

12. Gel Electrophoresis • 227

Figure 12.4 In electrophoresis races, the small DNA always wins!

3. "Digest" the second DNA strip with BamHI. Put the BamHI fragments in a separate pile.

4. Now, "digest" the remaining DNA molecule with HindIII. Put these fragments in a third pile.

5. In our imaginary gel electrophoresis, you will separate the EcoRI, BamHI, and HindIII fragments as if you loaded the three sets of fragments into separate but adjacent sample wells. Arrange your fragments as they would be separated by agarose gel electrophoresis. Designate an area on your desk as the end of the gel with the sample wells. Starting with the EcoRI fragments, arrange them from longest to shortest, with the longest one closest to the well.

6. Next, separate the BamHI fragments adjacent to the EcoRI fragments. Be sure to order the fragments correctly by size with respect to the other BamHI fragments and to the EcoRI fragments you have already laid out.

7. Repeat the same procedure for the HindIII fragments. You should now have all three of your sets of fragments arranged in order in front of you.

8. Look at the outline of the electrophoresis gel provided in Worksheet 12.2. Notice that it has a size scale in base pairs on the left-hand side and that sample wells are drawn in. Use the outline and draw the pattern your restriction digest would make in the gel, using the size scale as a guide

Figure 12.5 Gel electrophoresis is used to separate products of restriction digestion. (A) Restriction map, with fragment sizes in base pairs. (B) View of gel after electrophoresis.

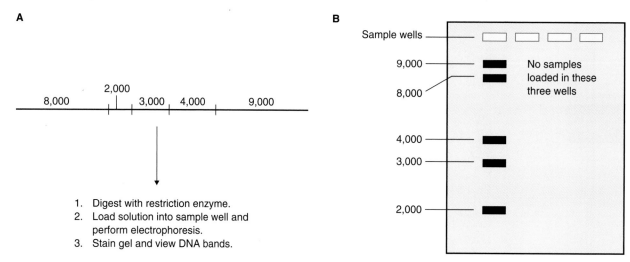

for where to draw your fragments. Use the EcoRI sample well for the EcoRI fragments, and so on.

9. After you draw the bands representing the restriction fragments, use the size information on the paper DNA strips to label the bands on the gel with the sizes of the fragments in base pairs.

10. Use the fragment sizes as a check of your work.

Are all the smaller fragments across all the gel "lanes" in front of all the larger fragments? Did you notice that the size scale does not seem to have regular intervals? The size scale looks the way it does because agarose gels separate fragments that way.

Classroom Activities

Restriction Analysis

13

About This Activity

In this activity, students perform electrophoresis using precut samples of lambda DNA, or they first carry out the restriction digests themselves and then perform electrophoresis with their samples. The instructions are divided into two parts covering restriction digestion of lambda DNA and gel electrophoresis of lambda DNA. It is not necessary to perform the restriction digestion in the classroom, as many supply houses carry predigested DNA. If you choose to use precut DNA samples, simply skip the restriction digestion section of the directions. Prior to this exercise, the students should perform the activities *DNA Scissors* and *DNA Goes to the Races*. *DNA Scissors* includes using restriction maps to predict the sizes of DNA fragments after digestion. Show them the restriction map of bacteriophage lambda in this chapter, and have them predict the sizes of the fragments they will see in their gels.

Class periods required: *1 for the restriction digestion and approximately 2 for the electrophoresis and data analysis. If you have a 3-hour laboratory period or are on block scheduling, you may be able to perform both the restriction digestion and electrophoresis in a single period. If the students can get their gels into the staining solution before the end of the class period, you can transfer the gels to water for destaining and let them sit in water or buffer overnight.*

Background

Please read the introduction to *DNA Scissors* for information about restriction enzymes.

In this introductory material, we provide information about different methods of staining DNA and recording data. The procedure outlined in the *Student Activity* was designed for the safest DNA stains, which are also the least sensitive.

Agarose Gel Electrophoresis

The standard method for separating DNA fragments is electrophoresis through agarose gels. Agarose is a polysaccharide, like agar or pectin, that dissolves in boiling water and then gels as it cools. In electrophoresis, DNA is applied to a slab of gelled agarose, and then an electric current is applied across the gel. Because DNA is negatively charged in the buffer system used for electrophoresis, it migrates through the gel toward the positive electrode.

The rate of migration of DNA through agarose depends on the size of the fragment. The smaller the DNA fragment, the more quickly it can progress through the agarose. The rate of migration of linear fragments through agarose is inversely proportional to the $\log_{10}$ of their molecular weights. Not surprisingly, the rate of migration is also affected by the shape of the DNA molecule; circular molecules (like plasmids) migrate differently from linear fragments with the same molecular weight.

Another important factor in electrophoresis is the concentration of agarose in the gel. The higher the concentration of agarose, the more it retards the movement of all DNA fragments. Therefore, it is better to use relatively high concentrations of agarose to separate and see small DNA fragments and low concentrations if you are interested in large fragments. How large, and what concentrations? Examples of agarose concentrations and the sizes of DNA molecules that they efficiently separate are given in the table.

% Agarose	Range of efficient separation in gel of linear DNA molecules (kilobase pairs)
0.3	60–5
0.6	20–1
0.9	7–0.5
1.5	4–0.2
2.0[a]	3–0.1

[a]A 2.0% agarose gel is a very hard gel.

Agarose gels must be prepared and run in a buffer. The buffer is necessary because ions are generated during electrophoresis that would otherwise cause the anode to become alkaline and the cathode acidic. If a gel is accidentally prepared and run with water, the DNA bands will look bad. Recall that a buffer is a mixture of either a weak acid and its salt or a weak base and its salt. The usual electrophoresis buffer is made from the weak base Tris [tris(hydroxymethyl)aminomethane], and the Tris salt is made by adding boric acid. The metal chelator EDTA (ethylenediaminetetraacetic acid) is also added. The resulting buffer is called TBE, for Tris-borate-EDTA. Other buffers sometimes used for agarose gel electrophoresis are TEA (Tris-acetate-EDTA; made with acetic acid) and TPE (Tris-phosphate-EDTA; made with phosphoric acid).

The voltage applied to the gel affects how quickly the DNA migrates. The higher the voltage, the more quickly the gel runs. There is a trade-off, however. Gels run at high voltages do not separate DNA fragments as efficiently as do slowly run gels. For good separation, gels should be run at no more than 5 volts/cm of gel length. You will have to decide what voltage you want your students to use. Specific recommendations are made in the wet laboratory procedures.

Staining Gels

To make DNA fragments visible after electrophoresis, they must be stained. The favorite DNA stain of researchers is ethidium bromide. Ethidium bromide fluoresces under ultraviolet (UV) light when it is bound to DNA. It is a sensitive stain but has several drawbacks for high school use. The first drawback is that to see it, a UV light must be used. The second is that ethidium bromide is a mutagen and requires very careful handling and disposal.

Methylene blue is an alternative stain for DNA gels recommended for classroom use by the National Association of Biology Teachers. It is not as sensitive as ethidium bromide, so more DNA must be loaded into the gels. It is particularly ineffective for small DNA fragments, such as those of 500 base pairs (bp) or fewer. All laboratory protocols in this manual have been prepared for methylene blue staining.

Staining with methylene blue

Flood the gel with 0.025% methylene blue. Let it stand for 20 to 30 minutes. Use a funnel to return as much stain as possible to a container. The entire gel will probably appear dark blue. Do not be concerned if no DNA bands are visible at this time. Rinse the gel in tap water. Let it soak for several minutes in several changes of fresh water. DNA bands will become increasingly distinct as the gel destains. It will take about 2 hours before the DNA is clearly visible. For best results, continue to destain the gel overnight in a small volume of water. View it over a light box or on an overhead projector, but not too long, since the dye will fade.

Staining with ethidium bromide

If you wish to use ethidium bromide, confine its use to a small area of your classroom. Wear gloves when you prepare the stain, handle stained gels, and dispose of any waste. Do not let students use the stain. The greatest risk is to inhale ethidium bromide powder when dissolving it. The safest route is to purchase a ready-mixed stock solution.

To stain with ethidium bromide, dilute the stock to 1 µg/ml. Soak the gel in this solution for 5 to 10 minutes. Use a funnel to put as much stain as possible back into a storage container (ideally a brown glass bottle). If desired, soak the stained gel in water for 5 minutes or more to clear background ethidium bromide from the gel. View the gel under a UV light source or on a UV transilluminator.

To dispose of ethidium bromide, use the following procedure. If necessary, add sufficient water to reduce the concentration of ethidium bromide to less than 0.5 mg/ml. If you are disabling gels, put them in either water or the staining solution you wish to disable. Add 1 volume of 0.5 M $KMnO_4$ and mix it carefully. Add 1 volume of 2.5 N HCl and mix it carefully. Let the mixture stand at room temperature for several hours (or overnight). Add 1 volume of 2.5 N NaOH and mix it carefully. Discard the disabled solution down the sink drain. Drain the disabled gels and discard them in the trash. *Caution:* $KMnO_4$ is an irritant and a strong oxidizer. It should be handled in a chemical hood.

Commercial DNA stain

Methylene blue staining works, but it is not ideal. It requires high DNA concentrations and a significant amount of time. Ethidium bromide is sensitive and fast but requires UV light for visualization and presents chemical hazards. In general, if the DNA is plentiful (such as commercially supplied bacteriophage lambda DNA or strongly amplified PCR products) and the fragments are over about 800 bp in length, you will be able to visualize them with methylene blue or other visual stains. If you are using miniprep DNA or are visualizing small (500 bp or fewer) PCR products, ethidium bromide will give better results.

Some companies that market biotechnology equipment and supplies to educators offer proprietary visual DNA stains which they state are improvements over conventional methylene blue. Carolina Biological Supply Company's Carolina BLU DNA stain is added to the agarose gel and buffer. Faint DNA bands can be seen after electrophoresis without additional staining. A few minutes of additional staining improves visibility. The company states that this stain can be used with half the amount of DNA required for methylene blue staining, which is an improvement but still not adequate for viewing small PCR products. If you decide to try Carolina BLU (or any additional stains that become available), be sure to follow the manufacturer's instructions carefully.

Recording Data

Photography

One way to record data from electrophoresis gels is to photograph the gels. Biotechnology supply companies sell cameras and camera systems designed for gel photography. They are convenient but can be very expensive.

To photograph a methylene blue-stained gel, use a Polaroid camera with film type 667, an aperture of f/8, and a shutter speed of 1/125 second. For UV light photography of ethidium bromide-stained gels, use Polaroid film type 667 (ASA 3000). Set the camera aperture to f/8 and the shutter speed to B. Depress the shutter for 2 to 3 seconds. Type 667 film can be purchased at photography supply stores.

A related film type, 665, is used in the same way but generates a positive black and white picture plus a negative. The negative must be soaked in a developing solution (inquire at a photography store). Having a negative is important only if you wish to make enlarged prints of the photographs (e.g., for a publication or a display).

Copying gel patterns

The simplest method for copying the band pattern in a stained gel is to lay a piece of clear plastic (such as an overhead transparency sheet) on the gel and carefully trace the bands.

Data from a wet gel can also be preserved in the following manner. Tape a piece of graph paper (centimeter ruled, if possible) on a table. Cover the paper smoothly with plastic wrap, and tape the plastic down. Lay the wet gel on the plastic wrap, and line it up so that the wells are even with a ruled line. Use a needle or pin to carefully pierce through the gel and onto the graph paper at the center of each band's leading edge. When all the bands are marked, remove the gel and plastic wrap. Draw lines to represent the bands in the gel at each pin prick. Mark the ruled line where the wells were. The distance migrated by each fragment can be read directly off the graph paper by counting squares.

If you would like to pursue the mathematical relationship between DNA fragment size and migration, there is an excellent exercise in the book *DNA Science* (see the CD) on pages 269 and 270. If you do not have this book, try plotting the migration distance (the distance from the front of the well to the leading edge of each band) on the x axis and the $\log_{10}$ of the size of the fragment in base pairs (substitute for molecular weight, since they are directly proportional) on the y axis. Talk to the mathematics teacher in your school about ways to use the migration of DNA fragments as an illustration of the use of logarithms or of semilogarithmic paper in plotting data.

Preserving Gels

Wet methylene blue- or Carolina BLU-stained gels can be kept in sealed containers or sealed plastic bags in the refrigerator for a long time. Include a small amount of destaining solution with the gel (a few milliliters)—enough to keep the gel moist but not to allow additional destaining.

Polyacrylamide Gel Electrophoresis

The other gel material used in electrophoresis of DNA is polyacrylamide. Polyacrylamide forms a tighter mesh than does agarose, so polyacrylamide gels can separate smaller molecules. Polyacrylamide also has a higher resolving power, meaning a polyacrylamide gel can separate two molecules whose molecular weights differ by only a small amount. For example, long polyacrylamide gels are used to separate DNA fragments that differ in length by only 1 nucleotide in DNA sequencing. They are used in forensic DNA fingerprinting applications, in which it is critical to determine whether two DNA fragments are exactly the same size. Polyacrylamide gels are also used for protein electrophoresis.

Polyacrylamide is a polymer of the chemical acrylamide. To make a polyacrylamide gel, acrylamide is dissolved in buffer, along with the cross-linking agent bisacrylamide. Catalysts are added to start polymerization. The liquid mixture is then quickly poured into a thin space between two glass or plastic plates. After the gel polymerizes, the plates and gel are clamped to an upright support and run vertically, rather than horizontally. Polyacrylamide gels used to separate DNA are usually made with and run in TBE buffer, just like agarose gels. The chemical urea can be added to the gel to keep the DNA single stranded for applications like DNA sequencing.

Protein Electrophoresis

Proteins are normally separated by polyacrylamide gel electrophoresis, because proteins are much smaller molecules than the DNA fragments commonly separated on agarose. For example, a fairly average-size protein might have a molecular mass of about 60,000 daltons (Da). Electrophoresis should be able to separate it clearly from a protein of 55,000 Da. A single base pair of DNA has a molecular mass of about 660 Da. Therefore, a 60,000-Da protein has the same molecular mass as a 91-bp DNA fragment. A 55,000-Da protein has the same molecular mass as an 83-bp DNA fragment. While gels made with high concentrations of standard agarose can separate large proteins with good resolution, standard agarose does not have the resolving power to separate the mixtures of proteins found in most preparations.

Fine-sieving agaroses can separate proteins much better than standard agarose, but these agaroses do not give bands as sharp as those seen in acrylamide and do not have nearly the same degree of resolution. However, you can use fine-sieving agarose gels with the same gel boxes you use with standard agarose, offering a significant cost saving over implementing polyacrylamide gel electrophoresis. The protein electrophoresis activity in chapter 32 uses fine-sieving agarose gels.

Polyacrylamide Gel Electrophoresis in the Classroom

Protein electrophoresis offers several advantages over DNA electrophoresis. Students typically want to analyze DNA from familiar organisms, often seeking to identify organisms from restriction fragment patterns. Unfortunately, even the genomes of bacteria are too large to give a resolvable number of fragments in a gel. Proteins are a different case. Students

can extract proteins from familiar things, such as different types of fish, meat, or even plants, and see different profiles on polyacrylamide gels. Students can carry out projects, such as identifying sources of meat samples, by comparing protein profiles.

If you wish to perform polyacrylamide gel electrophoresis in your classroom, you will need vertical electrophoresis. Because of the difficulty of casting polyacrylamide gels and the toxicity of monomeric acrylamide, we recommend that you buy precast gels. Precast gels are available for purchase from biological supply houses and are easy to use. Simply clamp them to the electrophoresis chamber, add buffer and the sample, and run it. Since polymerized acrylamide (polyacrylamide) is not toxic, precast gels are not hazardous.

Precast gels are expensive. Be sure that you know which buffer system you need for your gels before you order. The most common buffer used for protein electrophoresis contains Tris base, the amino acid glycine, and the detergent sodium dodecyl sulfate. Sodium dodecyl sulfate is added to the buffer to denature the proteins in the sample and to give all the proteins in the sample a negative charge.

Objectives

At the end of this laboratory, students should be able to:

1. Describe three steps involved in restriction analysis: restriction digestion, electrophoresis, and staining.
2. Analyze stained DNA fragments of lambda DNA in a gel and compare them to a restriction map.

Materials

This procedure is designed for use with nonintercalating DNA stains, and the concentration of DNA needed for visualizing with these stains is much higher than that required for visualizing with ethidium bromide or other intercalators.

Equipment for restriction digestion

Devices to measure microliter volumes, such as:

- 0.5- to 10-μl or 1- to 20-μl micropipettes (the most expensive option)
- One to five 10-μl Wiretrol glass microcapillary tubes with plungers
- The Graduate pipettor with graduated tips
- Plastic 1-cm^3 syringes adapted for use with micropipette tips and graduated tips

- Freezer (*not* frost free, if possible); if you use dehydrated enzymes, a freezer is not necessary.
- Cooler or ice bucket
- 37°C water bath or dry incubator
- Microcentrifuge (optional)

Equipment for gel electrophoresis

- Microwave or hot plate for melting agarose
- Hot-water bath (optional for keeping agarose molten)
- Electrophoresis chamber and gel trays
- Power supplies
- Gel-loading devices, such as:
  - Needle-nose plastic squeeze bulb pipettes (the least expensive option)
  - Wiretrol glass microcapillary tubes with plungers
  - The Graduate pipettor with plain tips
  - Plastic 1-cm³ syringes adapted for use with micropipette tips
  - Digital micropipettes and tips (the most expensive option)
  - Dishes for staining and destaining gels (any small, shallow container will do)
- Light box (optional)
- Polaroid camera with type 667 film (optional)

Supplies for restriction digestion

- Lambda DNA, 0.4 to 0.5 µg/µl; 2 µl per digest
- Restriction enzymes (EcoRI, HindIII, and BamHI)
- Restriction digestion buffer (generally supplied with the enzymes)
- Loading dye
- Sterile water
- Sterile micropipette tips
- Sterile microcentrifuge tubes
- Waterproof marking pens (or other labeling system)
- One holder or rack for microcentrifuge tubes per student team (a Styrofoam cup full of crushed ice will work; the ice is simply to provide support)

Supplies for gel electrophoresis

- Agarose
- TBE buffer
- Flask for melting agarose; each student group will need 30 to 50 ml of melted 0.8% agarose solution (made with 1× TBE buffer)
- Distilled or deionized water
- Masking tape
- Lambda DNA digested with either EcoRI, HindIII, or BamHI; 1.0 to 1.5 µg of total DNA per sample, with loading dye added

- Lambda DNA digested with a second enzyme from the list, with loading dye added
- 50 ml of 0.025% methylene blue solution or other staining solution per group
- Centimeter-ruled graph paper (optional for recording data; the students can bring this)
- Plastic wrap (optional for recording data)

Resource Materials

If you do not have equipment for electrophoresis, Carolina Biological Supply Company's *Exploring Electrophoresis* series offers relatively inexpensive kits that contain gel boxes powered by batteries, along with all the supplies and samples needed for performing an experiment. The gel boxes are reusable. Several different classroom and demonstration kits are available.

Lambda DNA, restriction enzymes, concentrated TBE, and precut lambda DNA are available from many supply houses (see the CD). Some companies offer dehydrated restriction enzymes that can be stored indefinitely at room temperature and are premeasured for individual student experiments to reduce waste and error. Carolina Biological Supply Company markets them as "Instant Enzymes." These greatly simplify the classroom activity, since all that students must do is add the DNA solution to the dehydrated enzymes.

In our experience, Carolina's nontoxic DNA stain "Carolina BLU" works a little better than methylene blue, but it is more expensive. Carolina BLU can be incorporated into the gel itself and begins to stain the DNA during electrophoresis, which speeds up staining. If you choose to purchase it, follow the manufacturer's directions for use.

The following kits are available from Carolina Biological Supply Company.

Restriction Enzyme Cleavage of DNA, no. 21-1149, contains all necessary supplies for gel electrophoresis of precut lambda DNA (refills are available).

Outbreak!, no. 21-1208, contains all necessary supplies for gel electrophoresis of precut DNA, but the exercise is presented as a story in which students must use restriction analysis to distinguish between two types of a fictitious deadly virus (refills are available).

Introductory Gel Electrophoresis Kit, no. 21-1148, supplies materials for students to separate colored dyes on an agarose gel as an introduction to electrophoresis.

DNA Restriction Analysis Kit, no. 21-1103, supplies materials for restriction digestion and for electrophoresis (refills are available).

Preparation for Restriction Digestion

If you ordered dehydrated enzymes, they may be stored at room temperature. If you ordered wet enzymes, place them and the concentrated restriction buffer in the freezer immediately upon arrival. If your freezer is frost free, it is a good idea to keep the enzymes in a Styrofoam box with the frozen ice pack inside the freezer. Frost-free freezers go through cycles of heating to eliminate frost buildup. Keeping the enzymes in a Styrofoam container on an ice pack inside the freezer keeps them cold during these cycles.

Dehydrated DNA may also be stored at room temperature. Keep wet DNA in the refrigerator. Agarose and TBE may be stored at room temperature.

Before restriction digest day

1. (Skip this step if you ordered dehydrated enzymes.) Aliquot the restriction enzyme into several small batches for student use to protect your supply. Since each student digest uses 1 μl of enzyme, it seems like 1 μl of enzyme per student group should be enough. It probably will not be. In fact, you should count on eight groups using about 12 μl. There are probably two reasons for this: pipetting errors by students and (more important) the tendency for the viscous enzyme solution to stick to pipette tips. Keep the enzyme in the freezer or on ice.
2. (Skip this step if you ordered dehydrated enzyme digests.) Thaw the restriction buffers (they are 10×; do not dilute them). Put 5 to 6 μl of 10× restriction buffer in a sterile microcentrifuge tube for each student group. Keep the thawed buffer in the refrigerator or on ice until class.
3. Dilute the lambda DNA to approximately 0.5 μg/μl, if it is not close to this concentration; 0.4 μg/μl is fine. To make the dilution, use the following formula:

(Starting concentration) x = (0.5 μg/μl)
(desired volume),

where x is the amount of DNA you will start with.

Express the starting concentration in micrograms per microliter (for example, if your DNA is 1 mg/ml, use 1 μg/μl as the starting concentration). The desired volume is the amount of the 0.5-μg/μl solution you want, so if you want 100 μl of 0.5-μg/μl lambda DNA and the solution you bought is 2.0 mg/ml:

$$(2 \ \mu g/\mu l) \ x = (0.5 \ \mu g/\mu l) \ (100 \ \mu l)$$
$$x = 25 \ \mu l$$

Take a volume x (in this case, 25 μl) of the starting solution and put it in a sterile microcentrifuge tube. Add (desired volume − x) of sterile TE. In this case, 100 μl is our desired volume, so 100 μl − x (25 μl) is 75 μl of TE. Thus, in our example, 25 μl of starting solution plus 75 μl of TE gives 100 μl of 0.5-μg/μl DNA.

Put 20 μl of uncut 0.5-μg/μl lambda DNA in a sterile microcentrifuge tube for each student group. If you have a precut standard, put 5 μl of it in a sterile microcentrifuge tube for each student group.

If you ordered dehydrated lambda DNA, rehydrate it according to the supplier's instructions.

4. Put 20 μl of sterile water in a sterile microcentrifuge tube, one for each group.
5. Photocopy the *Student Activity* and lambda restriction map for your class, if the students do not have manuals.

Preparation for Gel Electrophoresis

To make your own predigested lambda DNA, simply follow the directions for restriction digestion (including diluting the DNA if necessary), but scale up the reaction mixture size by multiplying everything by, say, 10, *except* the enzyme. You can get great digestion with much less enzyme and avoid waste. For example, use 40 μl of uncut DNA (0.4 to 0.5 μg/μl), 10 μl of 10× buffer (supplied with the enzymes), 45 μl of water, and 5 μl of enzyme. Let the digest incubate for several hours, and then add 15 to 20 μl of loading dye to stop it. Store your precut DNA in the refrigerator or freezer. Alternatively, order predigested DNA from a supply house. Carolina Biological Supply Company offers predigested DNA that can be stored at room temperature.

1. If necessary, dilute concentrated TBE buffer to 1× concentration. For example, dilute 20× TBE to 1× by adding 19 volumes of distilled water to 1 volume of 20× TBE (950 ml of water plus 50 ml of 20× TBE). The 1× buffer solution is used in the gel chamber and for dissolving the powdered agarose.
2. Prepare the agarose solution.
   Determine how much 0.8% agarose solution will be needed to cast your gels. Each gel requires 30 to 50 ml of agarose solution, depending on the size of the gel tray. For example, 0.8 g of agarose powder mixed with 100 ml of 1× TBE buffer (from step 2 above) will yield 100 ml of a 0.8% agarose solution.
   Heat the agarose and buffer in a clean flask using a boiling-water bath (30 to 60 minutes) or a microwave (3 to 6 minutes) to dissolve the agarose

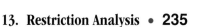

powder. Do not cap the flask during heating. The gel can be cast when the container feels warm but not hot when you touch it to your cheek. If you pour gels when the agarose is too hot, you can warp the casting tray. Melted agarose can be kept in a 55°C (or warmer) water bath until it is time to pour the gels.

Tips

1. Before you begin the activity, go over the outline of the entire activity. Review restriction enzymes and electrophoresis if necessary.
2. Each day, have the students read the procedure for the day very carefully before beginning. Success depends on their carrying out the directions accurately.
3. If the students have not used micropipettes and electrophoresis apparatus or cast gels before, it is a good idea to let them practice pipetting, loading a practice gel, and plugging in the electrophoresis apparatus. For practice loading, make an extra gel ahead of time; you can cut off the loaded wells and melt the rest of the gel for reuse later. Have the students practice loading while their gels harden. Mix loading dye and water in a 1-to-5 or 1-to-10 ratio for a practice loading solution. Alternatively, order Carolina's Practice Pipetting Stations, no. 21-1145, which come with practice loading dye. These can be rinsed with water and reused. They can be stored essentially indefinitely in the refrigerator, as long as you keep them moist.
4. Most gel combs have at least eight wells. This will be enough to let two student groups share one gel, if desired. Make sure the students do a good job of recording which sample went into which well.
5. Agarose gels can be prepared 1 day or more before use and stored covered with $1\times$ TBE buffer at room temperature (right in the gel box, ready to load the next day).
6. Hardened agarose can be remelted and used again. Store the hardened agarose in a closed container at room temperature. Do not reuse gels that have had DNA in them or that have been stained.
7. TBE buffer can be reused a few times.
8. Small gels can be run in less than 1 hour at 130 volts, but this high voltage may cause the bands to be less sharp. If you can adjust the voltage on your power supply, a run at 40 to 70 volts for 2 to 3 hours will give better results with the lambda digests. After you turn the power supplies off, you may simply leave the gels in the boxes covered with buffer until the next day.

You can also run the gels at 12 to 15 volts overnight.

9. If you stain gels with methylene blue, soak them for 20 to 30 minutes. The entire gel will turn dark blue. As you destain the gels, it may seem at first that there is no DNA present. Continue to destain, and the DNA will appear as the background stain diffuses out of the gel. Change the destaining solution frequently to speed up the process. Allowing gels to destain overnight is helpful. If you use Carolina BLU and incorporate it into the electrophoresis gel, faint bands should be visible when electrophoresis is complete. Soak the gels in more Carolina BLU to darken them (as directed), and then destain the gels to remove the background dye.
10. If you are photographing gels, make sure you have supplies on hand. If not, use the graph paper method for recording data (see *Recording Data* above). Have the students bring (or supply for them) centimeter-ruled graph paper. Bring plastic wrap to class, and make sure you have pins or dissecting needles for piercing the gels.
11. The book *DNA Science* by Greg Freyer and David Micklos has a wonderful two-page spread of "bad gels" with the causes of each problem listed.
12. If you like, talk to the advanced-mathematics teacher about using semilogarithmic paper to plot the relationship between fragment migration and fragment size in base pairs (see *Recording Data* above).
13. In a 3-hour laboratory period, students can complete the restriction digestion and electrophoresis, if you run the gels at high voltage. A 90-minute block period is sufficient to complete electrophoresis of precut DNA. Gels can be stored in water or buffer (right in the gel box, if you like) overnight or longer.

Procedure

Review the activity of restriction enzymes, if necessary. Show students the restriction map of lambda DNA. Have them predict the relative positions of the bands from different digests as they might look in a gel.

There are seven HindIII restriction sites within lambda DNA, resulting in eight DNA fragments after restriction digestion with HindIII. One of the fragments is only 125 bp in length and will not be visible on ordinary agarose gels. Another of the fragments is 514 bp in length and may or may not be visible. The remaining six bands should be clearly visible to your students. The five EcoRI restriction sites of lambda DNA yield six fragments of DNA upon digestion. BamHI also cuts lambda DNA at five sites, yielding six DNA fragments. Since there are two pairs of fragments of fairly similar sizes, you may see only four bands in the gel.

Restriction digestion

1. Have the students study the procedure for the day at hand.
2. Students should follow the procedure carefully, keeping the enzymes on ice at all times.
3. Digests should incubate from 30 minutes to several hours at 37°C. After the restriction digestion procedure, store the digests in the freezer if you are not running the gels immediately.

Gel electrophoresis

1. Follow the procedure in the *Student Activity* for casting the gels. If you or the students prepared the DNA, load 10 ml of precut DNA plus loading dye per lane. If you purchased precut DNA, load the amount specified by the supplier. This may take an entire class period.
2. Run the gels at the voltage you choose. When the electrophoresis is finished, the students' gels can simply be left in the gel boxes with the power off until the next day, when the students stain them. Alternatively, you can stain them yourself and put them in a small container, barely covering them with deionized or distilled water overnight for destaining. The gels should be ready for students to analyze the next day, though they may need some additional destaining.
3. Have the students record their results and answer the questions.

Answers to Student Questions

1. See the labeled diagram below. One common difference between actual student gels and the "ideal" pattern is that students' gels may not have run long enough to achieve the same degree of separation. Their bands may be closer together and less well resolved, so the gel may appear to contain fewer, thicker bands.
2. No, because the gels tell you only the sizes of the fragments and give no information on the relative locations of the restriction sites.
3. Please refer to the restriction map of bacteriophage lambda in Figure 13.1 of the *Student Activity*. There are seven HindIII sites and five EcoRI sites in lambda DNA. A HindIII/EcoRI digest of lambda DNA will cleave the lambda DNA at 12 sites, resulting in 13 restriction fragments. These fragments would have the following sizes

in base pairs, according to the restriction map: 21,226, 1,904, 2,027, 947, 1,375, 4,268, 5,148, 564, 125, 584, 4,973, 831, and 3,530. If you wish to test the prediction by performing a double digest, mix 1 µl of each enzyme in a microcentrifuge tube and then add 1 µl of the mixture to a digest. Incubate it for 2 hours if possible. For the clearest interpretation, load the gel with single digests of HindIII, EcoRI, and the double digest. Some of the bands in the double digest will be the same as the single-digest bands. The single-digest bands will also provide a size scale for the double-digest bands. It could be a good exercise for students to predict the migration patterns of the three samples side by side.

4. An experiment to test whether restriction enzymes cut DNA must include an experimental digest with buffer, DNA, and enzyme and a control digest with buffer, DNA, and no enzyme. The two tubes should be incubated under identical conditions and then run side by side in a gel. A good second control sample would be simply to run some of the starting DNA without mixing it with buffer and incubating it, to make sure that the starting DNA is intact. After the gel has run, it should be stained and destained, and the DNA banding patterns should be compared.

Ideal gel (the 125-bp HindIII fragment will not be visible; the 564-bp HindIII fragment probably will not be visible and is not shown).

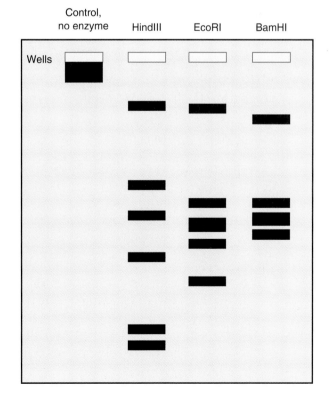

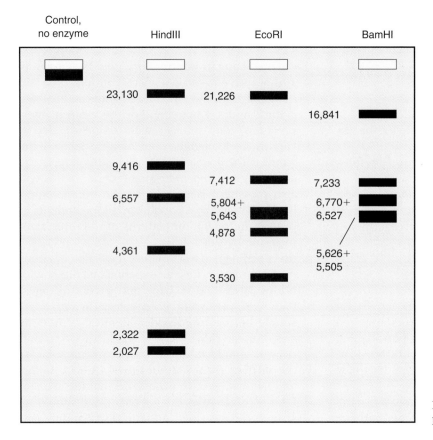

Control, no enzyme	HindIII	EcoRI	BamHI

23,130

21,226

16,841

9,416

7,412   7,233

6,557

5,804+   6,770+
5,643   6,527

4,878

4,361

5,626+
5,505

3,530

2,322

2,027

Ideal gel with fragment sizes in base pairs.

5. The control tube is to show that there are no restriction fragments if the DNA is not treated with an enzyme.
6. The extra microliter of water in the control tube is to make up for not adding 1 μl of enzyme. By adding an extra microliter of water, the final volumes in all of the tubes are made the same.
7. The reason that only five EcoRI fragments are seen (in the ideal student gel) is that the 5,804-bp fragment and the 5,643-bp fragment did not resolve under these electrophoresis conditions. If BamHI was used, the 6,770- and 6,527-bp fragments did not resolve and the 5,626- and 5,505-bp fragments did not resolve.

To resolve these pairs of fragments you could run the gel much longer (which might require that you cast a larger gel) or cast a lower-percentage agarose gel that more efficiently resolves large fragments (see *Agarose Gel Electrophoresis* above).

Classroom Activities

Restriction Analysis of Lambda DNA

Introduction

In this activity, you will be carrying out a series of steps that molecular biologists and biotechnologists use very frequently in their work. You will separate restriction fragments of DNA by electrophoresis through an agarose gel, stain the fragments so that you can see them, and analyze the products of the digest. You will either use predigested DNA samples or set up and carry out the digestions yourself. The process from restriction digestion to analysis of gel patterns is called "restriction analysis."

The procedure has been broken into several parts. The first part is setting up the restriction digests. If you are using predigested samples, you will skip this step. The remaining steps are preparing and loading the agarose gels, staining the gels, and analyzing the data. If you were in a laboratory all day (as scientists are), you could carry out the analysis in a single day. As it is, you may carry out a different part of the procedure each day for several days.

The DNA you will analyze is from bacteriophage lambda. Lambda is a virus that infects *Escherichia coli* and destroys it. Lambda DNA contains about 48,500 base pairs (bp). You will cut lambda DNA with the enzymes EcoRI, HindIII, and possibly BamHI. These enzymes will cut the DNA into a number of different-size pieces called restriction fragments.

You will separate the fragments by agarose gel electrophoresis. The mixture of fragments is loaded into a well in the gel, and then an electric current is applied. Because DNA is negatively charged, the fragments will migrate toward the positive electrode in the gel box. The shorter the fragment, the faster it can progress through the agarose. When you stop the electrophoresis (by turning off the current), the smallest fragment will be closest to the positive electrode and the largest will be furthest away. The intervening fragments will be sorted by size. You will stain the DNA so that you can see the fragments and

analyze your results by comparing them to a restriction map of bacteriophage lambda (Figure 13.1).

Restriction Digests

Each laboratory team will need:

- Device for measuring microliter volumes
- Four sterile microcentrifuge tubes
- Waterproof marking pen or another way to label the tubes
- One holder or rack for the microcentrifuge tubes

Obtain from your teacher microcentrifuge tubes containing:

- 10× restriction buffer
- Uncut lambda DNA
- Sterile water

1. Label your four sterile microcentrifuge tubes as follows: EcoRI, HindIII, BamHI (if you are using the enzyme), and Control. Set up the restriction digests according to the following instructions. Use Table 13.1 to check off each component of the reaction mixture as you add it to each tube.
2. Carefully add 4 µl of sterile water (from the microcentrifuge tube you got from your teacher) to the EcoRI, HindIII, and BamHI tubes. When adding a droplet to the tube, touch the pipette tip to the side of the tube to deposit a small bead on the inside of the tube.
3. Add 5 µl of sterile water to the control tube.
4. Using a fresh tip or capillary tube, carefully add 1 µl of 10× restriction buffer to each of the four tubes, checking them off in the table as you go.
5. Using another fresh tip or capillary tube, carefully add 4 µl of uncut lambda DNA to each of the four tubes, checking them off as you go.
6. Your digests are ready for the enzymes. Your teacher has the enzymes on ice. When you are ready to add the enzymes, prepare your measuring device with a fresh tip or capillary tube, and immediately and carefully add 1 µl of EcoRI to the EcoRI tube. Close the tube.

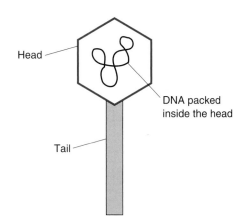

Head

DNA packed
inside the head

Tail

HindIII restriction map, with fragment sizes in base pairs

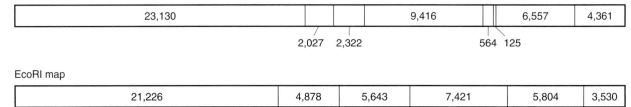

| 23,130 | | | 9,416 | | 6,557 | 4,361 |

2,027 2,322   564 125

EcoRI map

| 21,226 | 4,878 | 5,643 | 7,421 | 5,804 | 3,530 |

BamHI map

| 5,505 | 16,841 | 5,626 | 6,527 | 7,233 | 6,770 |

Figure 13.1 Bacteriophage lambda is the best known of the phages. It has a linear chromosome of approximately 48,500 bp and about 45 known genes. Its restriction map is shown.

7. Change to a fresh tip. Immediately and carefully add 1 µl of HindIII to the HindIII tube. Close the tube.

8. Change to a fresh tip. Immediately and carefully add 1 µl of BamHI to the appropriate tube. Close the tube.

9. Mix the reagents in the closed tubes either by microcentrifuging them for a few seconds or by tapping the bottoms of the tubes gently on your desktop. Tap the tubes gently at the bottom with your finger to ensure good mixing. Make sure the enzyme is mixed into the DNA solution.

10. Place the tightly closed tubes in a 37°C water bath or dry incubator. Incubate them for one to several hours. Your teacher will put them in the freezer for you.

Gel Electrophoresis

Each laboratory team will need:

- 30 to 50 ml of agarose solution
- Gel casting tray
- Gel electrophoresis chamber
- Electrophoresis power supply (one per two groups)
- Gel-loading device
- Your own DNA digests and loading dye or precut DNA samples
- Microliter measuring device if you prepared your own restriction digestions
- Precut lambda DNA standard
- 1× TBE buffer

Table 13.1 Guide for setting up restriction digests

Component	Amount (µl) in tube with:			
	EcoRI	HindIII	BamHI or PstI	Control
Water	4	4	4	5
10× buffer	1	1	1	1
Uncut DNA	4	4	4	4
Enzyme	1	1	1	None

Classroom Activities

1. If appropriate, remove your frozen restriction digests from the freezer and let them thaw while you prepare the gel.
2. Cast the gel.
    - Seal the ends of the tray with masking tape. Place the comb at one end of the tray, making sure it does *not* touch the bottom of the tray but is close to it.
    - Pour enough agarose into the gel tray to cover the lower one-third of the comb. Allow the agarose to cool (it will become whitish and opaque).
    - Remove the tape from the gel tray without damaging the ends of the gel. Do not remove the comb at this point. Place the gel tray into the electrophoresis chamber with the comb nearest the *negative* (black leads) electrode end of the chamber.
3. Fill the electrophoresis chamber with 1× TBE buffer. The buffer must completely cover the gel.
4. Carefully remove the comb from the gel, leaving the wells that you will fill with your restriction digests.
5. If you prepared your own restriction digests, add 2 µl of loading dye to each of your four tubes. Mix them by spinning them briefly and tapping at the bottom of the tube (or by tapping the tube on your desk).
6. Load each sample into a separate well. *Be sure to draw a diagram of the gel and label which well contains which sample.* Also, load one lane with 10 µl of the precut standard (obtained from your teacher). Record its position.
7. Plug the electrophoresis chamber into the power supply by connecting the red (positive) lead to the red electrode and the black (negative) lead to the black electrode. Be sure that the end of the gel with the DNA is connected to the negative pole of the power supply (even if you have to do the colors backwards).
8. Turn on the current to the level indicated by your teacher. Allow the slow marker dye to migrate 2.5 to 3.5 cm into the gel. Depending on the voltage applied to the gel, this could take as little as 40 minutes (170 volts) or as much as 24 hours (10 to 12 volts).

Staining

Each laboratory team will need:

- One gel-staining tray
- Methylene blue solution

1. Carefully remove your gel from the electrophoresis chamber and place it in the staining tray.

2. Add enough stain solution to the tray to just cover your gel. The stain solution will stain your clothes as well as DNA, so be careful. If you get any on your hands, it will wear off in a few hours. There is no hazard from skin contact with blue DNA stains. Soak the gel in the stain for the amount of time indicated by your teacher.
3. Using a funnel, return as much as possible of the stain solution to the stain solution container. Rinse the gel carefully with dechlorinated tap water or deionized or distilled water.
4. Cover the gel with fresh dechlorinated, deionized, or distilled water and soak it for 3 to 5 minutes. Pour off the water.
5. Repeat step 4 until you can clearly see the stained DNA bands or until you run out of time. You will probably need to soak your gel overnight in a volume of dechlorinated, deionized, or distilled water sufficient to cover it.

Data Analysis

Each laboratory team will need:

- Either access to photography equipment or centimeter-ruled graph paper, plastic wrap, and a needle or pin

Make a permanent record of your data in one of the following ways:

1. Photograph the gel with Polaroid film and fasten the picture in your laboratory notebook.
2. Use the graph paper method to make an accurate reproduction of your gel.
3. Lay a transparent film over your gel, and carefully trace the bands.

If you photograph the gel, follow your teacher's instructions exactly.

If you use the graph paper method:

1. Tape a piece of centimeter-ruled graph paper to your desk.
2. Cover the graph paper smoothly with plastic wrap. Tape down the plastic wrap.
3. Lay your gel on the plastic wrap, and line the leading edge of the wells up with a ruled line.
4. Use the needle or pin to pierce through the gel at the center of the leading edge of each band. Also pierce through the leading edge of each well you used. The goal is to generate a pattern of pin pricks on the graph paper that will mark exactly the locations of the bands in the gel.

5. Remove the gel and plastic wrap. Using the pin-pricks as a guide, draw the band patterns on the graph paper. Keep the graph paper in your laboratory notebook as a permanent record of your data.

Look at the restriction map of bacteriophage lambda (Figure 13.1). List in descending order (from largest to smallest) the sizes of the DNA fragments that should be generated in the following digests:

EcoRI   HindIII   BamHI*

*(Do this column only if you used BamHI.)

The predicted gel pattern is shown in Figure 13.2.

Now, sketch the pattern of bands from your gel. Label the gel lanes with the name of the enzyme used in each digest. Compare the pattern of bands in your EcoRI lane to the expected set of fragments. Starting with the largest band in your gel, label the bands with the sizes of the fragments in base pairs. Remember that two fragments very similar in size may not resolve and that very small fragments may have run off the gel or may not be detectable because they do not bind sufficient stain for you to see. Do the same for the HindIII fragments and the BamHI fragments if you used that enzyme. Discuss your band assignments with your laboratory teammates and your teacher.

When you are satisfied that you have made the best possible assignment of sizes to the bands, resketch your diagram with the size labels and put it in your laboratory notebook with the permanent record of your gel.

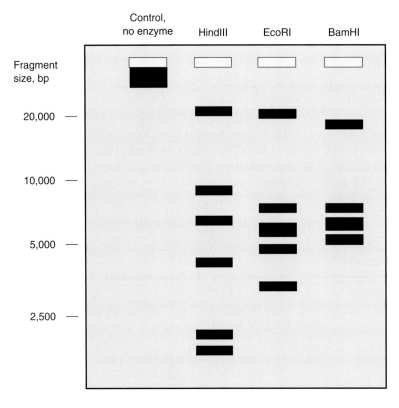

Figure 13.2 Ideal gel pattern. The 125-bp HindIII fragment will never be seen. The 564-bp HindIII fragment is also usually not visible and is not shown in this example.

Questions

1. Label the fragment sizes in the predicted student gel pattern. Compare your fragment patterns to the ideal fragment patterns shown above for a digest of lambda DNA by EcoRI, HindIII, and/or BamHI. Account for differences in separation and band intensities between the experimental gel and the ideal pattern.

2. Can you predict the fragment pattern for a digest using EcoRI and HindIII at the same time with only the information (the location of the bands) from the gels you ran? Why or why not?

3. Use the lambda restriction map in Figure 13.1 to predict the size of fragments that would be generated in a double digest using EcoRI and HindIII.

4. If you used precut DNA, propose an experiment to test whether the restriction enzymes actually cut the DNA. Be sure to describe exactly what you would use for control and experimental treatments. Assume you have access to the restriction enzymes you wish to test.

5. If you did the restriction digest, what was the purpose of the control tube?

6. If you did the restriction digest, why did the control tube receive 5 μl of water instead of 4 μl?

7. The lambda restriction map in Figure 13.1 indicates that six EcoRI fragments should be generated, yet you most likely saw only five in your gel. Why? If you used BamHI, you expected six fragments, yet you likely saw only four. Why? Can you suggest a way to see all of the expected fragments?

Recombinant DNA

14

About This Activity

The *Student Activity Recombinant Paper Plasmids* introduces some basic recombinant DNA techniques to students. Paper models are used to demonstrate how plasmid DNA is cut with restriction enzymes and recombined to create a recombinant DNA molecule. Students should already be familiar with plasmids from *The Sizes of the* Escherichia coli *and Human Genomes* and class discussion. They should also have done *DNA Scissors* as an introduction to restriction enzymes. This activity is very effective when used in combination with a laboratory on bacterial transformation, such as in chapter 19. If students have completed *DNA Goes to the Races* or are comfortable with electrophoresis, the *Restriction Analysis Challenge Worksheets* (chapter 15) make a challenging follow-up assignment.

Class periods required: *1*

Background

"Recombinant" DNA is simply a DNA molecule that has been assembled from pieces taken from more than one source of DNA. Making recombinant DNA became possible with the discovery of restriction enzymes and DNA ligase. Restriction enzymes are used to cut DNA into reproducible fragments, and ligase is used to form new phosphodiester bonds between them. When phosphodiester bonds are formed between DNA pieces from different sources, a recombinant DNA molecule has been created.

The most important application of recombinant DNA technology is gene cloning. To clone a gene, a fragment of DNA containing the gene must be isolated. The gene-containing DNA fragment is then added to a vector DNA molecule that will be replicated inside a cell. The most common type of vector is a plasmid, although other vectors are also used.

Plasmids are (relatively) small circular DNA molecules found in many bacteria and some yeasts.

They are duplicated by the host cell's DNA replication enzymes and transmitted to daughter cells during cell division. Some plasmids maintain a high copy number within their hosts, with 50 or more copies per cell. Others may have a copy number as low as one per cell. The plasmid copy number is a characteristic of the plasmid itself.

In order to be duplicated by the host's replication enzymes, plasmids must contain a replication origin (see the introduction to chapter 7, *DNA Replication*). An origin of replication is a base sequence within a DNA molecule at which the DNA replication process begins. Most plasmids also contain one or a few genes whose products play roles in plasmid DNA replication. In addition, many plasmids contain genes that are useful to the host cell under certain environmental conditions. For example, some plasmids contain genes for toxic products that kill competing bacteria. These plasmids also contain genes that make their hosts immune to the toxin. However, the most familiar plasmid-borne genes are probably antibiotic resistance genes.

Antibiotics are natural substances (or man-made copies and derivatives of natural substances) that kill or suppress the growth of microorganisms. Antibiotics are also made by microorganisms in nature. It is thought that the production of an antibiotic may confer a survival advantage on the producer by allowing it to kill nearby microbes that would normally compete for nutrients.

Since antibiotics arose in nature, it is not surprising that the capacity to resist them also arose naturally. The antibiotic resistance genes found on plasmids are presumed to have evolved in response to the presence of antibiotics in nature. This natural system of antibiotic and resistance gene has provided molecular biologists with a powerful tool for laboratory applications. One example is illustrated below.

After a DNA fragment has been ligated into a plasmid, the recombinant plasmid is introduced into a host cell. The host cell replicates the recombinant plasmid DNA as it divides. Plasmid DNA is often introduced into bacterial host cells by transformation (see chapter 19, *Transformation*). In a typical transformation, billions of bacteria are treated and exposed to the plasmid DNA. Only a fraction (usually less than 1 in 1,000) will acquire the plasmid. Antibiotic resistance genes provide a means to find these transformants among all the nontransformed cells.

After transformation by plasmids containing antibiotic resistance genes, bacteria that acquired the plasmid (transformants) can be detected by plating the bacteria on media containing the antibiotic(s). Only bacteria expressing the new antibiotic resistance genes can form colonies on the antibiotic plates. The only bacteria in the mixture of transformed and nontransformed cells that have the antibiotic resistance genes are those that acquired the plasmid: the transformants. This method of selecting transformants allows scientists to find quite easily the 1 cell in 1,000 that acquired the plasmid.

Because of the ease with which scientists can find cells transformed with plasmids containing antibiotic resistance genes, scientists usually clone their genes of interest into these plasmids. The gene of interest may not change the bacteria in any way that a scientist can readily detect, making it impossible to tell whether a bacterium has been transformed with the gene. However, if the gene of interest is in a plasmid encoding antibiotic resistance, the scientist need only select those bacteria that become antibiotic resistant after transformation. Those bacteria should also contain the gene.

Students usually ask how you can be sure you will get the product you want from a ligation and not alternative products that they see it is possible to form, or why the "right" product forms instead of the "wrong" product. (They will undoubtedly form many alternative products with their paper plasmids.) Be sure that they understand that in any ligation, all possible products will be formed. The ligase enzyme is "blind"; it forms phosphodiester bonds between DNA molecules without regard to the base sequences of those molecules. Thus, after a ligation, you will have a mixture of products in the test tube. How can you separate and identify them? Scientists usually use transformation.

The key to understanding how transformation can help identify ligation products is to realize that transformation is a very inefficient process. Very few DNA molecules actually get into cells, so essentially only

one molecule gets into a given transformed cell. Therefore, that cell now contains an isolated ligation product. If a batch of competent cells is transformed with a ligation reaction mixture, representatives of all the products will likely enter bacterial cells. Those products that contain replication origins will be replicated by their new hosts and transmitted to offspring.

If a scientist performs a ligation with a vector containing an ampicillin resistance gene, transforms the ligated DNA into ampicillin-sensitive cells, and selects for transformants by plating the cells on ampicillin-containing medium, she will get a collection of cells that contain a plasmid with a replication origin, an ampicillin resistance gene, and possibly other components, depending on the ligation reaction. In many cases, therefore, a scientist is faced with a collection of transformants which may or may not contain the plasmid she desires.

To determine whether a given transformant contains the desired recombinant plasmid (as opposed to an alternative ligation product), the scientist often must prepare plasmid DNA from that transformant. The plasmid DNA is then tested (by restriction analysis, DNA sequencing, or other means) to determine if it has the desired structure.

In this activity, students will assemble plasmids carrying genes for ampicillin and kanamycin resistance. Starting with two parent plasmids, one encoding ampicillin resistance and the other kanamycin resistance, the students will simulate digestion of each plasmid with BamHI and HindIII. Next, they ligate pieces together to create a recombinant plasmid containing both resistance genes.

It is very effective to follow this activity with the wet laboratory *Transformation of* Escherichia coli.

Objectives

At the end of this activity, students should be able to:

1. Describe the activity of DNA ligase.
2. Define "recombinant DNA."
3. Describe how to make a recombinant plasmid from two starting plasmids.

Materials

- *Student Activity*
- Plasmid templates (from the CD)
- Scissors, paste, Scotch tape

Preparation

Print and make copies of the *Student Activity* if necessary.

Tips

As students use the paper models, remind them that DNA is three-dimensional and has no "front" and "back," nor does it matter if the letters are upside down. What do matter are complementarity of base pairs and the correct antiparallel orientation of the strands.

Procedure

1. Have the students read the background information and/or discuss the background material.
2. Have them read the instructions, and make sure they understand them. You may want to review the activities of restriction enzymes and DNA ligase.
3. Perform the indicated activities.
4. Answer the questions. These questions provide a good lead-in to the transformation laboratory.

Answers to Student Questions

1. Please refer to the diagram. The restriction fragments (shown underneath the plasmid diagrams)

are labeled. There are six possible combinations. (Remember that DNA is three-dimensional, and it does not matter if the letters in the paper model are upside down.) If the two fragments of pAMP are called 1A and 1B (see the diagram) while the two from pKAN are 2A and 2B (see the following page), the combinations are 1A-1B, 1A-2B, 2A-1B, 2A-2B, 1A-2A, and 1B-2B. You could detect only four of these: 1A-2A, 1A-1B, 1A-2B, and 2A-2B. 1B-2B has no selectable marker (antibiotic resistance), and 2A-1B has no origin of replication. Both of these molecules could form and be transformed into bacteria, but 1B-2B could not be detected on antibiotic media. The 2A-1B molecule would not be copied once it entered the cell and would not be transmitted to the daughter cells that form the colony.

2. The essential elements of the experiment are as follows. Make two samples of bacteria from one culture. Transform one of the samples with the DNA mixture (experimental). Take the other sample through the transformation procedure, but do not add DNA (control). Plate some bacteria from the experimental and control samples on the following media: no antibiotics (are any of the bacteria alive?), ampicillin (none of the control organisms should grow; this is a test of whether the ampicillin is effective; some of the experimental organisms should form colonies), kanamycin

Diagram of the paper plasmids and their restriction products.

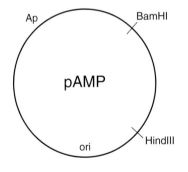

Both plasmids are digested with BamHI and HindIII, producing the following fragments:

pAMP

H   ori   Ap   B (1A)

H   B (1B)

pKAN

(2A)   B   Km   H

(2B)   B   ori   H

Ap: ampicillin resistance
Km: kanamycin resistance
ori: origin of replication
B: BamHI end
H: HindIII end

(similar to the ampicillin plate), and ampicillin plus kanamycin (only cells transformed with the ampicillin-kanamycin recombinant plasmid should grow).

3. The critical thing about these bacteria is that they will survive on one antibiotic plate but not on the other or on the two-antibiotic plate. To find them, plate some of the experimental mixture on an ampicillin plate and some on a kanamycin plate. After colonies grow on the plates, transfer some of the cells from each colony to a plate containing the other antibiotic (or both antibiotics). Those that survive on one antibiotic but fail to produce a colony on plates containing the other antibiotic are the bacteria you seek. Two of the plasmids predicted in question 1 will grow only on ampicillin; one will grow only on kanamycin.

4. The most important reason is that scientists need to be able to identify bacteria that have been transformed with the gene they desire. These bacteria may be very rare among the experimental bacteria. Associating the desired gene with an antibiotic resistance gene gives a tremendous advantage: by plating organisms on an antibiotic medium, a scientist can select only those organisms that have been transformed with the antibiotic resistance gene and, by association, the gene of interest. Another advantage to having the antibiotic resistance gene associated with the desired gene is that it provides a mechanism for ensuring that the desired gene is still there. By propagating the transformed bacteria on antibiotic media, the scientist can continuously select for only those bacteria that contain the antibiotic resistance gene and, by association, the desired gene.

Recombinant Paper Plasmids

Background Reading

Some of the most important techniques used in biotechnology today involve making recombinant DNA molecules. A "recombinant" object has been reassembled from parts taken from more than one source. You could make a recombinant bicycle by disassembling two bicycles and reassembling them in a new way: putting the wheels of one on the frame of the other, for example. Your genome is recombinant in that part of it came from your mother and part from your father. Recombinant DNA molecules are pieces of DNA that have been reassembled from pieces taken from more than one source of DNA. Often, one of these DNA sources is a plasmid.

Plasmids are small circular DNA molecules that can reside in cells. Plasmids are copied by the cell's DNA replication enzymes because they contain a special sequence of DNA bases called an *origin of replication*. The DNA replication enzymes assemble at this special sequence to begin synthesizing a new DNA molecule. As you might expect, bacterial chromosomes and other chromosomes have replication origins, too. Replication origins are essential to heredity; if a DNA molecule does not have a replication origin, it cannot be copied by the cell and will not be transmitted to future generations.

Plasmids often contain genes for resistance to antibiotics. Antibiotics are natural substances produced mostly by soil microorganisms. Antibiotic production allows these microorganisms to kill off competing microbes. Antibiotic resistance is also a natural phenomenon; at the very least, the antibiotic producers must be resistant to the antibiotics they make! We will be working with genes for resistance to the antibiotics ampicillin and kanamycin.

In this activity, you will assemble plasmids carrying genes for ampicillin and kanamycin resistance and then recombine the two plasmids. We call the plasmid with ampicillin resistance pAMP, the plasmid with kanamycin resistance pKAN, and the recombinant plasmid pAMP/KAN. You will use paper plasmid DNA models to go through the process that scientists use when making recombinant DNA. Scissors will substitute for restriction enzymes. The enzyme DNA ligase, which forms phosphodiester bonds between pieces of DNA, is represented by Scotch tape. Your result will be a model of a recombinant DNA molecule. Scientists place real recombinant plasmids back into bacteria, where they are replicated. The multiplying bacteria, carrying the recombinant plasmid, generate millions of copies of the recombinant DNA molecule and the proteins it encodes.

Construction of the pAMP and pKAN Plasmids

Locate the three strips of DNA code on Worksheet 14.1, called "Paper pAMP plasmid model." On each strip, the two rows of letters indicate the nucleotide bases, and the solid horizontal lines indicate the sugar-phosphate backbone of the DNA molecule. The hydrogen bonds between the base pairs are located in the white space between the rows of letters.

1. Use your scissors to cut carefully along the *solid* lines. Cut out each strip, leaving the solid lines intact. Make a vertical cut to connect the open end of the box formed by the solid lines. This cut will remove the 5′ and 3′ from the strip.
2. After all three strips are cut out, glue or tape the "1" end to the "paste 1" area, covering the vertical lines. Connect "2" to "paste 2" and "3" to "paste 3" until you complete the circular model of the pAMP plasmid.
3. Using Worksheet 14.2, called "Paper pKAN plasmid model," cut out and paste together a plasmid containing a kanamycin resistance gene. The procedure is exactly the same as for the pAMP plasmid.

Constructing a Recombinant pAMP/KAN Plasmid

You have now prepared a pAMP plasmid and a pKAN plasmid. In this part of the exercise, you will use them as starting materials to make a recombinant plasmid. You will cut pAMP and pKAN with two specific enzymes: BamHI and HindIII. You will ligate fragments that come from each plasmid, creating a pAMP/KAN plasmid.

1. First, simulate the activity of the restriction enzyme BamHI. Reading from 5′ to 3′ (left to right) along the top row of your pAMP plasmid, find the base sequence GGATCC. This is the BamHI restriction site. Notice that it is a palindrome. Cut through the sugar-phosphate backbone between the two G's, stopping at the center of the white area containing the hydrogen bonds (do not cut all the way through the paper). Do the same on the opposite strand. Cut through the hydrogen bonds between the two cut sites, and open the plasmid into a strip. Each end of the strip should have a single-stranded protrusion with the sequence 3′-CTAG-5′. Mark the ends of the strip "BamHI."

2. Next, simulate the activity of the restriction enzyme HindIII. Reading from 5′ to 3′ (left to right) along the top row of the pAMP strip, find the sequence AAGCTT. This is the HindIII restriction site. It is also a palindrome. Cut the sugar-phosphate backbone between the two A's, stopping at the center of the white space containing the hydrogen bonds. Repeat the cut between the two A's on the opposite strand of the restriction site. Cut the intervening hydrogen bonds. This time you should have two pieces with single-stranded protrusions. One protrusion on each piece is the BamHI end; the other should have the sequence 3′-TCGA-5′. Mark each of these two new ends "HindIII." You now have two pieces of pAMP DNA. Set them aside.

3. Using your pKAN plasmid model, repeat steps 1 and 2. The pKAN plasmid is now in two pieces with labeled ends. Along with the two pieces of pAMP, there are now four pieces of plasmid.

4. Take the piece of plasmid with the ampicillin resistance gene in it, and connect it with the plasmid containing the kanamycin resistance gene using DNA ligase (Scotch tape). Be sure complementary bases are paired where you make the ligations. Notice that the BamHI end will not pair with the HindIII end but will pair with another BamHI end. Likewise, the HindIII end must pair with another HindIII end.

Remember that DNA is three-dimensional. In our model, the letters representing the bases can look upside down, but in real DNA it does not matter, so in this paper simulation, the letters representing the bases do not need to be right side up. All that matters is that the 5′-to-3′ directions match within a strand.

You have now created a recombinant pAMP/KAN plasmid.

Transformation

It is possible to introduce plasmids into bacterial cells through the process of *transformation*. Bacteria that can be "transformed" (can take up DNA) are called "competent." Some bacteria are naturally competent. Others can be made competent by chemical and physical treatment. After the bacteria absorb the plasmid DNA, they express the new antibiotic resistance genes as instructed by the new DNA. Bacteria that express new proteins in this way are said to be *transformed*. New copies of the plasmid are synthesized by the cell's DNA replication enzymes and passed to daughter cells as the bacteria multiply. Because many identical copies of the new genes are generated in this process, you are said to have *cloned* the genes.

After transformation by plasmids containing antibiotic resistance genes, transformed bacteria can be detected by plating them on media containing the antibiotic(s). Only bacteria expressing the new antibiotic resistance genes (the transformed bacteria) can form colonies on the antibiotic plates. This method of "selecting" transformants (because they are the only ones that can grow on the media) is a big advantage, because transformation is usually very inefficient. In a typical experiment, less than 1 cell in 1,000 will become transformed.

Questions

1. A BamHI cut site will ligate only to the matching end of another BamHI site. The same is true for HindIII cut sites and most other restriction enzyme cut sites. Using two fragments at a time, how many possible different combinations could be formed from the four fragments you made in the third part of the exercise? How many of these could you detect in colonies of transformants on antibiotic-containing media?

2. Assume you did the activity with real DNA and attempted to transform bacteria with your new pAMP/KAN plasmid. Describe an experimental procedure for growing the transformed bacteria on plates to find out if the bacteria have actually been transformed by pAMP/KAN. Include controls that would tell you whether any of the experimental bacteria are alive and whether the antibiotics are effective.

3. Referring to question 2, describe an experimental procedure for finding experimental bacteria that were transformed by the other DNA molecules that formed during the ligation step (you described these molecules in question 1).

4. Scientists often combine an antibiotic resistance gene with whatever gene they are trying to clone. The desired gene is then associated with the antibiotic resistance gene. Any bacterium that contains the desired gene is thus resistant to an antibiotic. In most cases, the scientists have no use for the protein that destroys the antibiotic. Why would scientists combine genes in this way?

Classroom Activities

Restriction Analysis Challenge Worksheets

15

About These Worksheets

The restriction analysis challenge worksheets contain three problems that illustrate actual uses of restriction analysis in the laboratory. The problems, particularly the second and third, require multistep analysis. They are most appropriate for advanced classes or for students who enjoy analytical puzzles. Students must be familiar with restriction enzymes, construction of recombinant DNA molecules with DNA ligase, and gel electrophoresis. The required background is found in the activities *DNA Scissors, Recombinant Paper Plasmids,* and *DNA Goes to the Races.*

These worksheets require students to apply their knowledge of the construction of recombinant DNA molecules and restriction analysis to three problems like those encountered every day in research laboratories. The first two problems involve analysis of the molecules produced during a ligation, and the third requires students to generate a restriction map from restriction analysis data.

Students may require assistance getting started with the first two problems, because the problems require that they integrate and use information from several lessons. You may find it helpful to do the first problem together as a class and then let students work on the remaining problems individually or in small groups.

Discussion

When a scientist sets out to make a recombinant molecule, he must plan how he will recognize that molecule when he gets it. He usually looks at the structure of the DNA molecule he wants to make and the structures of other products that will be present in the ligation (such as the starting plasmid) and designs a restriction analysis that will let him identify the molecule he wants. The students will be going through this same process as they do challenges I and II.

In challenge I, the described ligation will produce the regenerated vector (where the ligase has simply resealed the EcoRI site in the plasmid vector), as well as the desired recombinant molecule. The EcoRI fragment can be inserted in either of two orientations, but in this problem, there is no way to distinguish between the two. It is also possible that more than one EcoRI fragment could be inserted into the same vector molecule. In reality, the resealed vector would be the most common product, followed by recombinant plasmids containing one insert. In the laboratory, it is usually uncommon to get plasmids with multiple copies of an insert.

Students may want to know how it is possible to analyze individual products of a ligation when the reaction is done in a single tube. You can explain to them that the process of transformation separates the ligation products (see the introduction to chapter 14). Transformation efficiency is very low—about one in a thousand to a million cells may be transformed in a given experiment. Each transformed cell then contains one plasmid molecule, which can replicate into many copies and is propagated as the cell divides. A transformed cell divides to make a colony on the appropriate antibiotic medium, and every cell in that colony contains a plasmid that is a copy of one of the original product molecules. By analyzing the plasmids present in many different colonies of transformants, it is possible to look individually at different products from the same ligation reaction. Scientists often analyze the plasmids from many transformants when they are looking for a single product from a ligation that could produce many different products.

Challenge II is an amplification of the ideas introduced in challenge I. Here, the vector is ligated in the presence of two different inserts: the one that is to be removed and the one that is desired. It is possible that either insert could be ligated into the vector (in either orientation), and it is also possible that the vector will simply be sealed without an insert. Notice that the reaction products to be drawn are

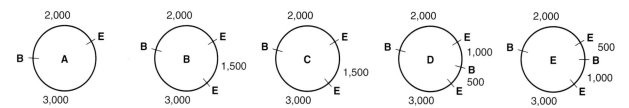

Diagram 15.1 Challenge II. Possible products with one replication origin and a total size of less than 7,000 bp.

restricted to those with a single replication origin (one plasmid molecule) that are less than 7,000 base pairs (bp) in size.

Use this problem to help your students see that constructing a recombinant molecule can be a complicated task and that it requires the use of analytical skills to plan for and identify the various products. These exercises should let your students do some analytical thinking and help them see how restriction analysis is used in real-life problems.

Challenge III should stand on its own. There is no magic method to solve the third problem; students simply have to work with it. If your students enjoy figuring out the restriction map, challenge them to create similar problems for their classmates. They must imagine a DNA molecule with a restriction map and generate the data. Have them draw a gel showing the restriction fragments. Let their classmates construct the maps from the data. If there are problems or ambiguities, the other students will find them. The exercise of imagining a restriction map and drawing the data from it will reinforce all the lessons about electrophoresis.

Restriction Mapping Laboratory Activities

Carolina Biological Supply Company, Restriction Mapping of Plasmid DNA kit (catalog no. 21-1175). Students cast a gel and run predigested DNA samples. Following electrophoresis, they must determine the sizes of the DNA fragments and, using those data, deduce the restriction map of the plasmid. The instructions include practice problems.

Carolina Biological Supply Company, Restriction Enzyme Puzzle kit (catalog no. 21-1190). Students digest lambda DNA with one known and two unknown restriction enzymes. They determine the sizes of the DNA fragments and, using those data, deduce the identities of the enzymes.

Answers to Challenges

I. The plasmid vector can be distinguished from the recombinant molecule either with an EcoRI digest or with a BamHI digest. The vector produces a single 5,000-bp EcoRI fragment, while the recombinant produces that fragment plus the 1,500-bp insert fragment. The vector produces a single 5,000-bp BamHI fragment, while the recombinant produces a single 6,500-bp fragment. There is no way to distinguish the two possible orientations of the insert with the information given. If students have trouble, have them draw the structures of the starting vector and the recombinant plasmid and compare them. Check their gel drawings to make sure they have their fragments labeled properly and drawn in a reasonable manner.

IIA. There are five potential products, given the constraints of the problem (Diagram 15.1). The two plasmids with the original insert in different orientations (B and C) are indistinguishable given the available information.

IIB. The best way to distinguish the products, given the available information, is to perform a digest with BamHI. Product A (regenerated vector with no insert) will give a single 5,000-bp fragment. Products B and C (vector with the original insert in either orientation) will give a single 6,500-bp fragment. Product D will give two fragments of 3,500 and 3,000 bp. Product E will give two fragments of 4,000 and 2,500 bp. An EcoRI digest will not distinguish between the old and new inserts, since both are 1,500 bp in length. An EcoRI-plus-BamHI digest will not distinguish between products D and E. Each will give fragments of 2,000, 3,000, 500, and 1,000 bp.

IIC. If students have good answers for part B, it should be easy for them to draw the gel. Check their gels to make sure they have the fragments correctly labeled and drawn properly with respect to the size scale.

III. See Diagram 15.2. Either orientation is correct. E, EcoRI site; B, BamHI site. The size scale is in kilobases (1,000 bp).

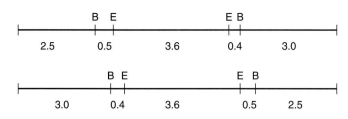

Diagram 15.2 Challenge III. Restriction map of the linear DNA fragment.

Restriction Analysis Challenge

15

The combination of restriction digestion and gel electrophoresis is often called restriction analysis, since the information obtained from these procedures can be used to analyze the structure of a DNA molecule. Restriction analysis is especially important in checking the structures of recombinant DNA molecules and in analyzing an unknown DNA. The following examples show you some typical restriction analysis problems encountered by scientists in the laboratory.

Restriction Analysis Challenge I

You wish to insert a 1,500-base pair (bp) EcoRI restriction fragment into the plasmid vector shown in Figure 15.1. You digest the plasmid DNA with EcoRI, stop the digest, add the 1,500-bp fragment,

and ligate it. You know that this procedure will give you a mixture of the regenerated starting plasmid and the recombinant molecule you desire. Outline a restriction analysis procedure you could use to distinguish between the regenerated vector and the desired product. State the predicted fragment sizes. Use the gel outline provided to show the predicted products from your analysis (one product type per lane). Label each lane with the DNA molecule type (vector or recombinant) and the expected fragment sizes.

Restriction Analysis Challenge II

You wish to remove the 1,500-bp EcoRI fragment from the starting plasmid shown in Figure 15.2 and replace it with a *different* 1,500-bp EcoRI fragment

Figure 15.1 Information and gel outline for restriction analysis challenge I.

Gel outline

Starting plasmid vector

Distance between restriction sites is shown in base pairs.

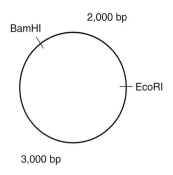

Fragment to be inserted into vector

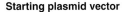

EcoRI   1,500 bp   EcoRI

Starting plasmid   **EcoRI fragment to be inserted**

Distances between restriction sites are shown in base pairs.

Figure 15.2 Information for restriction analysis challenge II.

(also shown). You digest the starting plasmid with EcoRI, stop the digest, add the new fragment to the mixture, and religate it.

A. Draw the products you could generate that would have *one replication origin* and be *less than 7,000 bp in size.* (Hint: what are all the fragments present in the ligation together?) Label the products A, B, etc., and indicate the distances between restriction sites.

B. Design a restriction analysis procedure that will let you identify as many of your proposed products as possible after a single digest. You may use more than one restriction enzyme in the digest. State the predicted fragment sizes for each one of your products from part A.

C. Use the gel outline in Figure 15.3 to draw the predicted gel pattern from your proposed analysis of each of the products shown in part A. Label the gel lanes with the product molecule

Figure 15.3 Gel outline for restriction analysis challenge IIC.

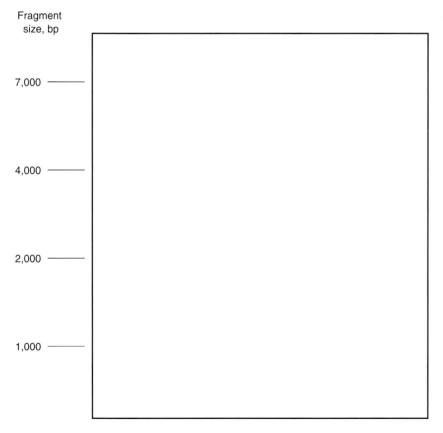

(A, B, etc.). Label the bands with the fragment sizes. (A size scale has been provided.)

Restriction Analysis Challenge III

When scientists study a new piece of DNA, one of the first things they often do is put together a restriction map of that piece. The restriction map is then used as a road map when they study the genes located within the DNA. How is a restriction map put together? Digests with single enzymes and combinations of enzymes are performed. Observing the sizes of fragments produced, scientists determine restriction site locations that would give them the patterns they see. Here is one such puzzle to try for yourself.

A *linear* (not circular) piece of DNA is digested with the restriction endonuclease EcoRI and gives fragments of the following sizes: 3,000, 3,600, and 3,400 bp. When digested with BamHI, the DNA molecule gives fragments of the following sizes: 4,500, 3,000, and 2,500 bp. A double digest with EcoRI and BamHI gives fragments of the following sizes: 2,500, 500, 3,600, 3,000, and 400 bp.

Draw a restriction map of the starting piece of DNA showing the locations of the EcoRI and BamHI restriction sites. Indicate the distances between the restriction sites in base pairs, and label the sites EcoRI or BamHI as appropriate. Also, indicate the distance from each end of the starting piece to the nearest restriction site.

Classroom Activities

Detection of Specific DNA Sequences: Hybridization Analysis

16

About These Activities

The activities in this lesson plan use paper models to illustrate basic concepts of hybridization analysis. Hybridization is a technique that takes advantage of the specificity of DNA base pairing for the detection of specific DNA sequences in a mixed sample. It is one of the fundamental methods for analysis of DNA. The first activity (*Fishing for DNA*) requires only that students be familiar with the structure of DNA and the base-pairing rules. If you are familiar with the second edition of the book, please note that we have added specific information about fluorescent in situ hybridization (FISH) analysis to this section. The second and third activities assume that students are familiar with restriction enzymes and the process of electrophoresis. If you plan to use the subsequent lessons on DNA sequencing, the polymerase chain reaction, or the analysis of human DNA, it is important to introduce your class to hybridization.

Class periods required: *1 to 2*

Introduction

Restriction digestion, electrophoresis, and staining allow scientists to cut DNA molecules into reproducible fragments and to look at the sizes of those fragments. However, it does not give much information about the DNA base sequence within the fragments. We know that the restriction site was present at the end of each fragment, but that is all. Often it is important to know whether a specific DNA base sequence is present in a sample and where it is located with respect to restriction sites. These kinds of questions come up in the screening of DNA libraries (see chapter 5), in testing to determine if a certain gene has been introduced into an organism, and in analyzing the structure of a certain region of DNA (as in DNA fingerprinting). The technique most commonly used to answer these questions is hybridization analysis.

In brief, hybridization analysis involves separating the strands of (denaturing) the DNA molecules to be analyzed and then mixing those separated strands with many copies of a single-stranded DNA or RNA molecule. This single-stranded DNA or RNA molecule has the complement of the base sequence of interest and is labeled for detection (often with radioactive isotopes). It is called a *probe*.

When a probe is mixed with single-stranded sample DNA under the right conditions, hydrogen bonds form between the probe and its complementary sequence in the sample DNA. The formation of hydrogen bonds between two complementary single strands to re-create a double helix is called *hybridization* or *annealing*. If the sample DNA does not contain the base sequence complementary to the probe, no probe molecules will anneal to the sample. After time has been allowed for hybridization, the sample is rinsed to remove unhybridized probe and then tested for the presence of hybridized probe (see below). Hybridized probe indicates the presence of the DNA sequence of interest.

Denaturation of DNA

To separate the two strands of a DNA molecule, the hydrogen bonds between the base pairs must be broken. Breaking these bonds can be accomplished physically with heat or chemically with a base (acid works too, but it destroys parts of the nucleotides in the process). The extremity of the conditions required to separate two strands depends on how stably the two strands are associated. Since G-C pairs are held together by three hydrogen bonds while A-T pairs are bound by only two, it takes higher temperature or pH to separate strands with more G-C pairs. The temperature required to denature a given DNA molecule is called its *melting temperature*, and the melting temperature of a particular molecule can be predicted based on its G-C content. In hybridization applications, denaturing conditions are generally set so that any DNA molecule will melt (for

Classroom Activities

example, 95 to 100°C). The point at which consideration of melting temperature may become important is when the probe is hybridized to the sample (see below).

Hybridization

If a DNA molecule in solution is boiled, it will denature. If the solution is then allowed to cool, the strands will reanneal and the double helix will form again. Annealing will occur between any complementary single strands of DNA under the right conditions. No enzyme is needed. As might be imagined, short complementary molecules tend to hybridize much faster than long ones, because it is easier for them to achieve proper alignment for base pairing through random contact. To facilitate hybridization and subsequent detection of the probe, the DNA to be analyzed is commonly transferred to a membrane, usually nylon or nitrocellulose.

Probes

Any DNA or RNA molecule can theoretically be used as a probe. Short synthetic oligonucleotides of 15 to 30 bases are often employed because they are easy to obtain (in research environments!) and the scientist has complete control over the base sequence. Purified restriction fragments, whole linearized plasmids, or products of in vitro enzymatic DNA or RNA synthesis are also used. Probes can be radioactively labeled by any of several methods, including synthesis in the presence of radioactive nucleotides or attachment of a radioactive terminal phosphate group by the enzyme polynucleotide kinase. Non-radioactive labeling methods are also very widely used. In these methods, the probe DNA is chemically modified with a substituent that results in a colored or luminescent product after it is put through a detection reaction (usually an enzymatic one).

The critical aspect in probe selection is to be certain that the probe will hybridize only to the DNA of interest. The chance of random occurrence of any particular DNA sequence is 4^n, where n is the number of bases in the sequence. Therefore, the chances of random hybridization decrease dramatically as the probe becomes longer. However, it is possible for two single-stranded molecules that are somewhat mismatched to hybridize together. Because of this possibility, the choices of both probe and hybridization conditions are important.

When two slightly mismatched molecules hybridize, the double helix they form is less stable than a perfectly base-paired helix. (Consider that there are no hydrogen bonds between mismatched bases, in addition to the structural disturbance a mismatch can cause.) The imperfectly base-paired helix will denature at a lower temperature than will its perfectly paired counterpart. Thus, at certain temperatures, the perfectly paired helices will form and be stable while mismatched helices will not. By adjusting the hybridization temperature, scientists can either permit or block mismatched annealing of a probe.

Would it ever be desirable to hybridize under conditions that permit annealing of the probe to an imperfectly matched sequence? Yes, often. When scientists "fish" for genes, one technique they often use is to employ a known gene, perhaps a gene from the yeast *Saccharomyces cerevisiae,* to probe for its counterpart in a distantly related organism, such as humans. These scientists do not expect that the yeast gene will be a perfect match to the human gene, but they hope that the two sequences will be similar enough to permit hybridization under mild conditions. This type of approach is often successful. In other applications, hybridization may be carried out under such demanding conditions that only perfect matches will occur.

Two examples of applications of hybridization analysis, with transfer processes, are outlined below.

Screening DNA libraries

A DNA library is a collection of clones that, taken together, represents the entire genome of an organism. A plasmid library can be made by digesting the DNA of the organism with a restriction enzyme and then ligating the resulting fragments en masse into plasmid DNA molecules that were previously cleaved with the same enzyme (see chapter 5). The result is a large number of plasmids containing different restriction fragments of the organism's DNA: a DNA library. There are many ways of making and screening libraries, but the main idea is the same. The example below describes a method for screening plasmids in bacterial colonies.

To find a plasmid containing a specific DNA sequence (for example, from a particular gene), the entire library is transformed into a host, such as *Escherichia coli.* The transformants are plated on agar to produce individual colonies, with each colony containing one particular plasmid (with a unique insert from the organism's DNA). Next, cells from each colony are transferred to a membrane by gently pressing the membrane down onto the agar plate (most of each colony is left on the plate). In this way, the pattern of the colonies is preserved on the membrane (Figure 16.1).

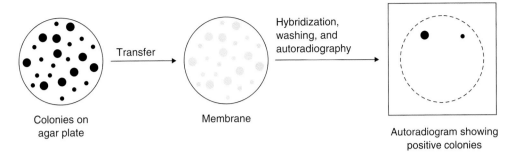

Colonies on
agar plate

Transfer

Membrane

Hybridization,
washing, and
autoradiography

Autoradiogram showing
positive colonies

Figure 16.1 Colony hybridization.

Now the cells must be lysed to expose their DNA, and their DNA must be denatured. This can be accomplished in a single step by soaking the membrane in a solution of a detergent (to lyse the cell membranes) and a base (to denature the DNA). The membrane is then rinsed, and the single-stranded DNA molecules may be fixed to it through heating or exposure to ultraviolet light.

The membrane with the fixed single-stranded DNA is now ready for the hybridization step. It is immersed in a hybridization solution, and labeled single-stranded probe is added. The composition of the solution and the conditions (particularly the temperature) of hybridization are varied to suit the experiment. After the hybridization period has elapsed, the membrane is rinsed repeatedly under conditions that will remove unhybridized probe but will not disrupt hydrogen bonds between any hybridized probe and the sample DNA (the washing conditions can also be adjusted to remove imperfectly hybridized probe).

The membrane is allowed to dry, and the presence of the probe is detected. In the case of a radioactive probe, the membrane is placed against a piece of photographic film. If radioactive probe is hybridized to the DNA from any of the colonies transferred to the membrane, the film next to that region of the membrane will be exposed and will turn dark upon being developed. In this manner, any colony that contained a plasmid with DNA sequences complementary to the probe will show up as a black spot on the film. The process of exposing film by placing it next to a radioactive sample is called autoradiography, and the resulting "picture" is called an autoradiogram, or "autorad" for short (Figure 16.1). For nonradioactive-probe detection, the membrane is usually soaked in a series of solutions. One key solution usually contains an enzyme linked to a molecule that will bind to the labeled probe. The other key solution contains a substrate for the enzyme that yields a colored or luminescent product when cleaved.

In the last step of the process, the autoradiogram is lined up with the original petri plate from which the colonies were transferred. The black spots on the autoradiogram (representing hybridized probe) are matched to specific colonies. The cells in these colonies should contain plasmids with the sequence of interest. To verify that they do, the indicated colonies are transferred to separate cultures. Plasmid DNA is prepared from them and is typically sequenced.

Southern blot hybridization

Another major application of hybridization analysis is the testing of the products of a restriction digest to determine which fragments, if any, contain a certain DNA sequence or to determine the size of a fragment containing a particular sequence. The procedure most commonly used for this purpose involves a process called "Southern blotting," followed by hybridization analysis. Specific applications of Southern blot hybridization can be found in the lessons on genetic diseases and DNA fingerprinting in this book.

To begin the analysis, the sample DNA is digested with restriction enzymes and the resulting fragments are separated by agarose gel electrophoresis in the standard manner. Unfortunately, it is not effective to perform hybridization on DNA molecules embedded in agarose. In 1975, a scientist named Southern published a method for transferring the fragments from an agarose gel to a membrane in a way that preserved the arrangement of the fragments in the gel. This transfer method is called *Southern transfer* or *Southern blotting*.

In Southern transfer, the agarose gel is soaked in a base to denature the DNA fragments. Then, the gel is placed on a long piece of blotting paper whose ends are suspended in a reservoir of salt solution. The membrane is placed directly on the gel, and a stack of dry absorbent paper (such as paper towels) is placed on the membrane (Figure 16.2). The blotting paper acts like a wick, drawing fluid from the reservoir up through the gel into the stack of dry paper. This process is driven by capillary action, just like what is seen when the corner of a paper towel is put into a glass of water. Alternatively, a device that uses an electric current to drive the DNA from the gel can be used. In this case, the procedure is called electroblotting.

As the fluid migrates upward through the gel, it carries the denatured DNA fragments with it. When the fragments reach the membrane on top of the gel, they stick to it and remain there. The assembly of wick, gel, membrane, and paper is put together so that the DNA fragments are transferred straight up, in exactly the pattern that was found in the gel. They stick to the membrane in exactly this pattern, duplicating the pattern in the gel itself.

Once the denatured DNA fragments have been transferred to the membrane, the hybridization procedure is exactly as described above for library screening. Autoradiography of a Southern blot shows dark bands corresponding to the bands in the gel that contained DNA sequences complementary to the probe (Figure 16.3).

Fluorescent (or fluorescence) in situ hybridization

Hybridization with fluorescently labeled probes is used to reveal the location of specific nucleic acid sequences in a procedure called **fluorescent in situ hybridization (FISH)**. In FISH, nucleic acid probes are hybridized to intact chromosomes, cells, or thin sections of tissue. Bound probe is detected by fluorescence microscopy, revealing the physical location of the targeted sequence. The availability of different colors of chromophores (the color-producing portion of the label) makes it possible to hybridize samples to a mixture of probes so that more than one sequence can be visualized in a single experiment.

Some of the more visually spectacular applications of FISH include the detection of chromosome features, such as centromeres, telomeres, and specific repeated DNA sequences, and a procedure called *chromosome painting.* In chromosome painting, probes that hybridize along the length of a given chromosome are all labeled with the same chromophore. Probes to a different chromosome can be labeled with a different color, so that the different chromosomes appear to have been painted different colors. The probes are made by isolating the target chromosome and using it as a substrate for PCR amplification with random primers. An individual chromosome can also be microdissected, and pieces of the dissected chromosome can be used to make probes to individual arms or bands. After application of the fluorescent probe to the sample, the remainder of the chromosome is stained with a nonspecific stain, such as DAPI (4′,6′-diamidino-2-phenylindole) (blue) or propidium iodide (red), so that the entire chromosome is visible under the microscope.

Color plate 1 shows examples of FISH using probes to different chromosome features. Color plate 2 shows an example in which a portion of the wheatgrass (*Thinopyrum intermedium*) chromosome has translocated into the wheat *(Triticum aestivum)*

Figure 16.2 Assembling a Southern blot. (A) The student is placing the membrane on top of the gel. (B) The complete transfer setup. The bottle on top is simply a weight.

A

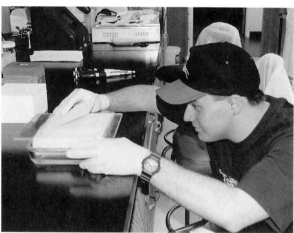

B

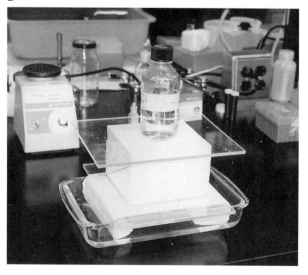

A

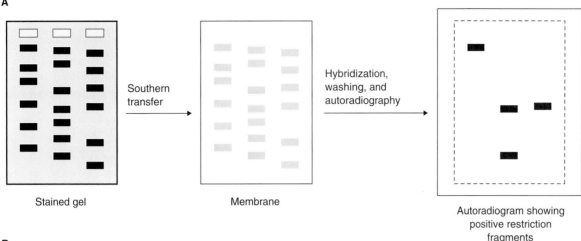

Stained gel

Membrane

Autoradiogram showing positive restriction fragments

Southern transfer

Hybridization, washing, and autoradiography

B

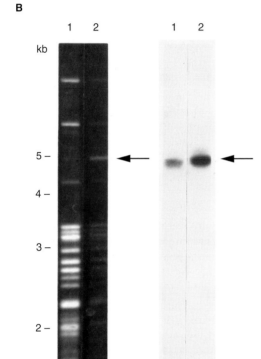

Figure 16.3 Southern hybridization analysis. (A) Diagrammatic representation. (B) Two stained gel lanes and the results of a hybridization to those lanes. kb, DNA fragment size in kilobase pairs.

genome. The researchers who did the pictured experiment were studying resistance to the barley yellow dwarf virus, which attacks wheat, oat, and barley crops. Wheatgrass is resistant to some strains of barley yellow dwarf virus, and scientists at Purdue University have traced the resistance to one of the wheatgrass chromosomes and have introduced that chromosome into the wheat genome. Because the presence of the entire chromosome in wheat is not desirable, they are studying wheat lines in which pieces of the *T. intermedium* chromosomes have translocated onto wheat chromosomes to identify the smallest piece of translocated chromosome that can confer resistance to the virus. In the color plate, FISH was used to paint the translocated piece of the *T. intermedium* chromosome. The *T. aestivum* chromosomes have been counterstained with propidium iodide.

Other uses of hybridization

Hybridization is not always used as an analytical method per se. It plays a critical role in other laboratory procedures, such as DNA sequencing and the polymerase chain reaction (illustrated in subsequent lessons). Both of these procedures involve in vitro DNA synthesis by the enzyme DNA polymerase. DNA polymerase must have a primer at which to begin synthesis (see chapter 7, *DNA Replication*). In vitro, the primer is usually a short synthetic oligonucleotide annealed to the template DNA molecule. Very often, the template molecule starts out as double-stranded DNA. An excess of the short primer is added to it in solution, and then the solution is heated to nearly 100°C to denature the template. As the solution cools, the short primers hybridize to the template strands.

This process of annealing primers is the first step in both DNA sequencing and the polymerase chain reaction. As described in a subsequent lesson, the polymerase chain reaction itself can be used as an analytical technique, and its specificity is actually the specificity of hybridization between the primers and the template DNA (see the chapter 17 activity, *Paper PCR*).

Objectives

After completing part I of the lesson (*Fishing for DNA*), students should be able to:

1. Define hybridization.
2. Describe how single-stranded DNA can be used to detect a DNA sequence of interest.

After completing parts II and III, students should be able to:

3. Describe how to use a DNA probe and restriction analysis to determine what restriction fragment of a given molecule contains a sequence of interest.
4. If shown a probe sequence and the sequence of a target molecule, predict which restriction fragments of the target will hybridize to the probe.

Materials

Part I. Fishing for DNA
- *Student Activity, Fishing for DNA*
- Worksheet 16.I (Print from the CD if students do not have manuals.)
- Scissors

Part II. Combining restriction and hybridization analysis
- *Combining Restriction and Hybridization Analysis* student reading

- Worksheets 16.I and 16.II (Print from the CD if students do not have manuals.)
- Tape

Part III. Southern hybridization
- Worksheet 16.III (Print from the CD if students do not have manuals.)

Preparation

Make any necessary photocopies or printouts from the CD.

Procedure

Part I. Fishing for DNA

1. Remind your students how large the DNA molecule of any organism is. If you did the activity *The Sizes of the* Escherichia coli *and Human Genomes*, get out the scale models of the *E. coli* chromosome and gene. Show the class the models of the chromosome and the tiny gene, and pose the problem to the students that they would like to know if a particular DNA sequence, such as a certain gene, is present in a huge molecule, such as a bacterial chromosome. Sequencing the entire chromosome would give the answer but is not practical; it would take a long time and cost a lot of money.

2. Have the students read the background reading in *Fishing for DNA* and do the activity. It may be helpful to get out your DNA model to show the class single-stranded DNA. Give them as much additional information about hybridization as you feel is appropriate. Talk through the questions.

Fishing for DNA *is a sufficient introduction to hybridization for younger students.*

Part II. Combining restriction and hybridization analysis

After you complete *Fishing for DNA*, have the students read *Combining Restriction and Hybridization Analysis* and do the activity. Give them any additional information about Southern blotting that you deem appropriate. An advanced class should be able to complete parts I and II within a single class period.

Part III. Southern hybridization

The Southern hybridization worksheet can be done in class or for homework. It shows how Southern hybridization can be used to map the location of a gene.

Answers to Student Questions

Part I. Fishing for DNA

1. DNA hybridization is complementary base pairing between two single-stranded DNA molecules to form a double helix.
2. Students should describe the use of probes to find a specific DNA sequence in a sample.
3. Be sure that students have the correct 5′-to-3′ orientation—since the written sequence is 5′ to 3′, the answer sequence should be 3′ to 5′ (omit this requirement for younger students if you think it is too detailed).

Part II. Combining restriction and hybridization analysis

1. The DNA strip is cut into six pieces of 93, 18, 83, 47, 65, and 73 base pairs (bp); the probe hybridizes to the 65-bp fragment.

   Perform Southern hybridization analysis on a sample of virus DNA, using the probe for the desired gene. This analysis will show what restriction fragment contains the gene of interest. Digest a second sample of virus DNA, perform electrophoresis and staining, and then cut the correct fragment out of the gel and purify it for cloning.

Part III. Southern hybridization

Problem A

1. Check the students' pictures for correct labels.
2. The DNA polymerase gene is located in the region where the 10,500-bp BamHI fragment and the 5,500-bp EcoRI fragments overlap.
3. Since the entire DNA polymerase gene was used as a probe and did not hybridize to either the 7,500-bp BamHI fragment or the 4,000-bp EcoRI fragment, the entire virus X DNA polymerase gene is probably inside the indicated area, neither to the right of the EcoRI site nor to the left of the BamHI site. Of course, there is still some uncertainty, since we do not know how similar the two genes actually are, but it is not necessary to discuss this fine point in class unless students bring it up.
4. The point of this question is for the students to realize that proteins that carry out the same function in different organisms often have very similar amino acid and DNA base sequences. The similarity is usually more pronounced the more closely related the organisms are. It is therefore often effective to use a known gene to search for a gene encoding the same kind of protein in a different organism.

Problem B

See the diagram below.

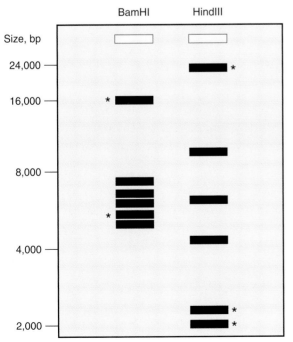

Stained electrophoresis gel

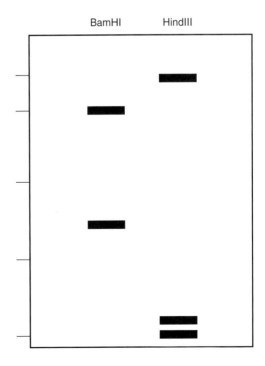

Results of hybridization analysis

Hybridization Wet Laboratories

Hybridization analysis is not hard to do, but it requires quite a few more steps (extra laboratory periods) than restriction analysis, and you have to be careful. The extra steps require extra materials and solutions, too.

If you are familiar with hybridization analysis and know what you are doing, we recommend a labeling/detection system such as the Boehringer Mannheim "Genius" kit for nonradioactive labeling and detection of your probe. If you buy this kit, you will still have to make quite a few solutions to use with it.

If you are inexperienced with hybridization analysis but are determined to try it in your classroom, we strongly recommend that you get a prepared kit for the entire activity. Carolina Biological Supply Company offers a Southern Hybridization of DNA kit, catalog no. 21-1215, that uses a biotin-avidin detection system. It will take your class several days to complete the activity, and students must follow the directions carefully. This kit has been set up so that you have only a few solutions to make by combining and diluting provided materials.

Detection of Specific DNA Sequences
Part I. Fishing for DNA

16

Imagine this: you are a scientist staring at a test tube full of DNA that you just prepared. You want very much to know whether that DNA contains the gene for cystic fibrosis. You know a little about the cystic fibrosis gene, so how can you tell if it is in your sample?

Scientists often need to fish for particular pieces of DNA, such as the cystic fibrosis gene. How do they do it? Once they know a little about the DNA they are looking for, it is not very hard.

When a scientist fishes for a particular piece of DNA, she uses a special DNA "hook" called a *probe*. A probe is usually a form of DNA called single-stranded DNA. Think of pictures of DNA or of models of DNA you have made. The base pairs are in the middle, and the two backbones are on the outside. Single-stranded DNA is only half of this structure: one backbone with bases sticking out from it (rather like an mRNA molecule; look at the picture below). If a piece of single-stranded DNA finds another piece of single-stranded DNA with the right base sequence, it will pair with it to make a regular DNA helix. Scientists call the pairing of two single DNA strands to make a double helix *hybridization* (Diagram 16.I).

To see if a gene is present in a DNA sample, our scientist uses a probe that has the same sequence as part of the gene and a label, such as a radioactive isotope, that allows the probe to be detected. She adds the probe to the DNA sample to see if it will hybridize with anything. How does she know if it hybridizes? If the probe hybridizes to the sample, it will stick to the DNA in the sample. If not, it can be easily rinsed away.

After the scientist rinses her sample, she exposes it to photographic film to determine if the probe has stuck to it. If any radioactive probe has stuck to the sample, it will expose the photographic film at that position. By examining the film, our scientist knows that the sequence of DNA in the probe is also present in her sample. Since the probe has the sequence of part of a gene, that gene is most likely in the sample DNA.

Another way to label and detect probes is to attach a fluorescent dye to the probe molecule. After the probe has been incubated with the sample and unbound probe has been washed away, the scientist can detect the presence of the probe by looking for the fluorescence. Dyes are available that fluoresce in different colors, so that more than one probe can be used at one time.

One type of analysis done with fluorescent probes is called fluorescent (or fluorescence) in situ hybridization, or FISH. A FISH analysis consists of applying fluorescently labeled probes to an intact sample, such as a chromosome, a cell, or a thin slice of tissue, rather than to DNA that has been extracted from the sample. The probes bind to their targets, and when the fluorescence is detected, it shows the location of the binding in the original sample. Color

Diagram 16.I Hybridization of a probe to a sequence within single-stranded sample DNA.

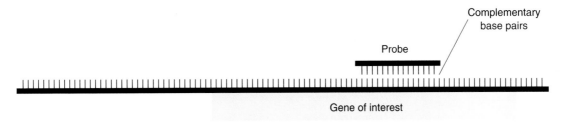

plate 1 shows examples of FISH using probes to different features of chromosomes. Color plate 2 shows an example in which FISH was used to detect a fragment of a wheatgrass chromosome translocated into the genome of wheat.

Activity

On Worksheet 16.I, you have a 379-base pair (bp) DNA sequence representing your sample. Every 10th base is in boldface, and the base at the beginning of each row is numbered. You also have a short sequence that is a probe for the cystic fibrosis gene. You will use the probe to determine if the cystic fibrosis gene is in your sample DNA.

Do you notice anything funny about your long DNA sequence? It is only one strand. To test whether a probe will hybridize to a DNA sample, it is necessary to separate the strands of the sample DNA so that the probe can find complementary base pairs. Your sample DNA has already been prepared.

1. Cut out the probe sequence.
2. Scan the sample DNA sequence to see if the probe will hybridize to it.
3. Is the cystic fibrosis gene present in your DNA sample? If you found a place where the probe could hybridize, mark it on the DNA sequence.

Questions

1. What is DNA hybridization?

2. How can you use hybridization to tell if a certain DNA sequence is present in a DNA sample?

3. Below is a DNA sequence. Write out the sequence of a 10-bp probe that would hybridize to it.

5′ AATGCAGGCCCTATATGCCTTAACGGCATATGCAATGTACAATGCAAGTCCAACCGG 3′

Detection of Specific DNA Sequences Part II. Combining Restriction and Hybridization Analysis

Introduction

The hybridization analysis you learned about in *Fishing for DNA* can indicate whether a given DNA sequence is present in a sample. Sometimes this simple piece of information is all that is needed. However, a positive hybridization test does not give any information about where the sequence of interest is located within the sample DNA molecule.

To get information about the presence and location of a particular DNA sequence, scientists combine restriction analysis with hybridization analysis. Basically, the sample DNA is digested with a restriction enzyme (or enzymes), and the fragments are separated by electrophoresis. After electrophoresis, hybridization analysis is performed on the fragments. Only the fragment or the fragments to which the probe can base pair will be detected (Diagram 16.II).

There are a few important technical details about this whole process. First, it does not work well just to soak an agarose gel in a probe solution. In 1975, a scientist named Southern figured out a way to transfer DNA fragments from a gel directly to a membrane, so that the exact pattern of the fragments in

the gel was preserved. After the fragments were stuck on the membrane, they could be tested for hybridization to a probe. This method of transfer is called "Southern blotting," and the combination of restriction analysis, transfer to a membrane, and hybridization to a probe is called "Southern hybridization analysis."

Activity: Southern Hybridization Analysis

1. Take the sample DNA sheet from the previous exercise and cut out the DNA strips by cutting along the dotted lines. Tape strip 1 to strip 2 to strip 3, etc., to form one long linear molecule. This molecule could be a chromosome, like one of yours.
2. Now, simulate the activity of the restriction enzyme SmaI. This enzyme cuts the DNA sequence 5′ CCCGGG 3′ between the C and G in the middle of the sequence. Digest your chromosome by cleaving it at every SmaI site. (Your fragments are single stranded in anticipation of the hybridization. In reality, the restriction digestion and electrophoresis are performed on double-stranded DNA, and then the two strands of the molecules are separated.)
3. Next, simulate electrophoresis of your fragments. Sort them by size, and lay them on your desktop as if they were in a gel.
4. Using the outlines provided on Worksheet 16.II, draw the arrangement of fragments in your gel.
5. On your drawing of your gel, mark with an asterisk any band(s) that would hybridize to the probe.
6. Keep in mind that the pattern of DNA fragments in your gel will be exactly transferred to a membrane. On the membrane, they will be tested for hybridization with the probe. Afterward, only the fragment(s) that hybridized to the probe will be detected (as in the accompanying diagram of Southern hybridization analysis). Draw what will be detected after hybridization in the box marked "results of hybridization analysis."

Diagram 16.II Southern hybridization analysis shows which restriction fragments hybridize to a probe.

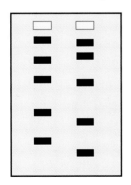
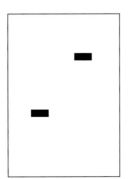

Stained gel   Results of hybridization analysis

Question

1. It is possible to cut a slice from an electrophoresis gel, purify the DNA in that slice, and clone it. Suppose that you have a probe for a viral gene that you would like to clone, but you do not know where it is within the 40,000-bp virus chromosome. How could you use Southern hybridization analysis to help you clone your gene?

Detection of Specific DNA Sequences Part III. Southern Hybridization

Problem A

You are analyzing the chromosome of a newly discovered virus, X. You have already constructed a restriction map of the 26,500-bp linear chromosome, shown in Worksheet 16.IIIA. Since you are interested in DNA polymerase enzymes, you conducted a hybridization analysis of virus X DNA using as a probe a DNA polymerase gene from another virus you believe is closely related to X. To your delight, the probe hybridized to the DNA from virus X.

Now, you would like to determine where the virus X DNA polymerase gene is located, so you perform a Southern hybridization analysis. You digest viral DNA with EcoRI and BamHI separately, separate the fragments in a gel, transfer them to a membrane, and hybridize them with the same DNA polymerase gene probe. Shown on Worksheet 16.IIIA is a picture of the fragments in the electrophoresis gel and the pattern seen after detecting the hybridized probe on the membrane.

Problem B

On Worksheet 16.IIIB is a restriction map of bacteriophage lambda. You digest some lambda DNA with the enzymes BamHI and HindIII separately and then load the fragments into an agarose gel and perform electrophoresis.

Next, you transfer the fragments to a membrane and carry out hybridization analysis using the 4,878-bp EcoRI lambda fragment (refer to the map) as a probe.

1. Draw a picture of the electrophoresis gel, using the outline of the stained electrophoresis gel in Worksheet 16.IIIB (the two smallest HindIII fragments will run off the gel).
2. Indicate which fragments will hybridize to the 4,878-bp probe. (Hint: Even if there is only a partial overlap of the probe with a restriction fragment, the probe will still hybridize to that fragment.)
3. Draw what would be seen after detection of the probe on the membrane in the second outline.

Questions (Problem A)

1. Label the fragments on the gels with their sizes in base pairs.

2. Indicate on the restriction maps the region of the virus X chromosome where the DNA polymerase gene is located.

3. How far to the left, in terms of restriction sites, could the gene lie? How far to the right?

4. Why would a DNA polymerase gene from another virus hybridize to DNA fragments from virus X?

The Polymerase Chain Reaction   **17**

About This Activity

The polymerase chain reaction (PCR) has become a key tool in molecular biology research and in biotechnology applications. Most students—whether school age or adult scientists—will not really grasp the mechanism of PCR until they have gone through all the steps themselves, either by drawing pictures or by manipulating models. In the *Student Activity,* students use paper models to simulate the steps of the PCR. The model exercise demonstrates how DNA polymerase can be used to make multiple copies of a specific DNA fragment and shows how the technique can be used to detect a specific DNA molecule (such as the chromosome of a disease-causing microorganism) in a sample. Students should have performed both activities in chapter 7, *DNA Replication,* and have been introduced to hybridization before performing this activity.

Figures 17.1 and 17.2 are found in the *Student Activity.*

Class periods required: *1 to 2*

Introduction

PCR has become a standard technique in research and forensics and in the clinic. This clever technique allows many copies of a specific DNA segment to be produced from a single copy and is extremely useful as a detection method and as a cloning tool. PCR can be used to detect pathogens in tissue samples, to amplify scarce DNA (such as from small bloodstains) so that it can be analyzed, or even to probe antique DNA from museum specimens. Automated DNA sequencing is also based on PCR technology (see chapter 18).

PCR is a simple technique that combines in vitro DNA synthesis by DNA polymerase and hybridization. It requires a solution containing a starting sample, two primers (many molecules of each), a DNA polymerase enzyme, and free nucleotides. The primers are short single-stranded pieces of DNA, typically 15 to 20 bases long. One primer of the pair hybridizes to each of the two strands of the sample DNA at either end of the segment to be amplified. The primers are mixed in great excess with the sample DNA. To start the reaction, the double-stranded sample DNA in the mixture is denatured into single strands by heating it (usually to 95°C). The mixture is then cooled so that hybridization of complementary strands can occur. Because the primers are in such great excess, one primer molecule will anneal to each single strand before the strands of sample DNA can find one another (Figure 17.1).

In the next step, the DNA polymerase enzyme synthesizes a second DNA strand on each of the two original strands, using the free nucleotides in the solution. The annealed primer serves as a starting point (Figure 17.1). New DNA is synthesized from the 3′ end of each primer, extending in only one direction. The result is two helices where before there was only one. (DNA polymerase enzymes must have a 3′ end from which to start synthesizing, and they can add nucleotides in only one direction; see chapter 7, *DNA Replication.*)

The process of denaturation, hybridization, and synthesis is then repeated, giving four helices (see the diagram). Further rounds yield 8, then 16, then 32, and so on. Typical PCR procedures call for 25 to 40 rounds of amplification, yielding huge numbers of molecules. Nearly all of the new DNA molecules have the primers as their ends and are the same length.

In early PCR methods, new DNA polymerase had to be added after each denaturation step, because the high heat necessary for denaturation destroyed the enzyme. Now however, scientists have purified a DNA polymerase enzyme from a bacterium that inhabits hot springs (*Thermus aquaticus*). This enzyme, the *Taq* polymerase, remains active after being heated to 95°C and does not need to be added after each denaturation step. The *Taq* polymerase has made it simple to automate PCR: all that is needed is a heater with programmable temperature cycles. "PCR machines," called thermal cyclers, have become standard laboratory equipment.

PCR can produce enough product DNA, even from a minute amount of template, to be visible in a gel following electrophoresis. Biotechnologists often use this technique to produce many copies of a DNA segment for cloning. PCR is also used in DNA typing (see chapter 27), and since the PCR process depends on the specific base pairing of primers to the template DNA, PCR can be used as a diagnostic tool.

In PCR-based diagnosis, the primers are chosen so that they hybridize only to a specific portion of the target organism's DNA or so that the distance between the primer hybridization sites is unique to the target. To determine whether the target organism is present, PCR is performed on the sample. If no DNA in the sample can hybridize to the primers, no unique PCR product will be synthesized. On the other hand, if the desired target DNA is present, there will be a unique product representing the segment between the primer hybridization sites. Because PCR produces so many copies of the segment, the product DNA can easily be detected by a variety of methods, including agarose gel electrophoresis and simple staining.

The advantages of PCR as a diagnostic tool are the specificity imparted by the base pairing of primers to the target sequence and the ability of PCR to amplify rare DNA. The amplification allows tiny amounts of specific DNA to be detected. In many clinical specimens, the disease-causing organism is present in very low numbers. PCR can reveal the presence of these rare organisms without taking time for culturing. PCR also permits the use of very small specimens. This advantage has made PCR an invaluable tool in studying ancient DNA samples. Researchers can use minute samples of museum specimens to produce enough DNA for analysis, preserving the original specimen essentially intact. Because the products of PCR are not usually designed to be longer than 2,000 base pairs or so, PCR also works well on samples in which the DNA has undergone extensive fragmentation. (For more information on the use of PCR to study ancient samples, see *Selected Readings* below.)

PCR can also be used to determine how much of a specific nucleotide sequence (DNA or RNA) is present in a mixed sample. One approach to this "quantitative PCR" uses a dye that fluoresces when it binds to double-stranded nucleic acid. As the double-stranded product accumulates, proportionally more dye is bound, and the fluorescent signal increases. The rate of accumulation of product is proportional to the amount of template present in the sample.

Below is a paper PCR activity you can do with your students. Although real PCR primers are at least 15 nucleotides long (to provide greater specificity), the paper primers are only 5 nucleotides in length. In addition, the segments amplified by PCR are usually hundreds of nucleotides long, where this activity amplifies a short segment. However, this paper model very accurately demonstrates the steps of PCR and shows how a specific DNA segment can be amplified from a single copy. The second part of the activity illustrates how PCR is used as a diagnostic tool.

Objectives

After this lesson students should be able to:

1. Describe the steps in the PCR and explain how these steps can generate multiple copies of a specific DNA fragment.
2. Describe how PCR can be used in disease diagnosis.

Materials

- Strips of light- and dark-colored paper (at least eight of each per student group)
- Removable tape (one roll per student group)
- Scissors (one pair per group)
- At least eight light and eight dark primers per student group (Worksheets 17.1 and 17.2)
- One "double-stranded" parental DNA molecule per group (Worksheet 17.1)
- One "sample 1" and "sample 2" page per group (Worksheet 17.3)
- One copy of the student questions per student

Teaching Resource

Carolina Biological Supply Company sells a videotape, catalog no. 21-2734, containing a demonstration of this activity. The demonstration and accompanying explanation of PCR take about 20 minutes. The second half of the videotape demonstrates the manual PCR amplification described below (see *PCR in the Classroom* below).

Preparation

Cut strips of light and dark paper about the width of the primer and template models.

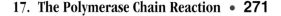

Print as many copies of Worksheets 17.1 and 17.2 with the parental DNA sequence and the primers as needed for your class. For the part of the activity that illustrates the use of PCR in disease diagnosis, make enough photocopies of Worksheet 17.3 for your class (or make transparencies).

Procedure

1. PCR

Refer to Figures 17.3 to 17.8.

Each student group has a parental DNA molecule, primers, and strips of colored paper, which will represent newly synthesized DNA (these strips are somewhat analogous to the free nucleotides in the solution). There is no representative of the DNA polymerase in this model; the DNA "synthesis" will be performed by the students (Figure 17.3).

Step 1. Denaturation

This is the 95°C step. Students should denature their double-stranded parental DNA into two single strands by removing the tape holding the strands together.

Step 2. Hybridization

The temperature is reduced, and hybridization can occur. Because there are so many primers in solution, they will hybridize to the parental DNA before the two parental strands can find each other (ideally, students should have many more than eight primers of each color on their desks). The students should "hybridize" the primers to the long DNA strands by matching the primer sequences to their complementary sequences and taping the primers in place. Be sure that the students hybridize the primers in the correct orientation: if the parental strand is oriented with the 5′ end on the left, the 3′ end of the

primer should be on the left. It does not matter if the letters are upside down! Real DNA strands are twisted into a helix. There is no "upside down." The way the paper primers are designed, a dark primer should always hybridize to a light DNA strand (Figure 17.4).

Step 3. DNA Synthesis

Students will "synthesize" DNA by taping one end of a strip of light-colored paper to the 3′ end of the light primer that is hybridized to the dark single strand. The light-colored strip of paper (the new DNA strand) should be extended to the end of the parental DNA "template," and then any excess paper should be cut off. The new light-colored DNA strand should be taped to its complementary dark strand to represent the new double-stranded DNA molecule. Finally, the students should write the correct DNA sequence on the new strand, beginning at the 3′ end of the primer. Go through the same procedure with the dark primer that is hybridized to the light-colored strand, and synthesize a new dark strand.

After one "round" of synthesis, there will be two double-stranded DNA molecules (one end of each will be uneven [Figure 17.5]).

Figure 17.4 Paper PCR. Primers are hybridized to the denatured template strands.

Figure 17.3 The starting template and primers for paper PCR. The colored paper strips are not shown.

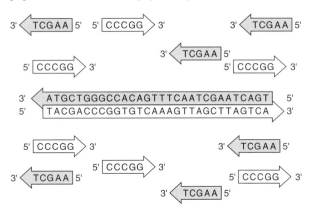

Figure 17.5 Paper PCR, products of the first round of DNA synthesis. Note that the primers are part of the product DNA strands. The base sequence of the product strands has been written on the paper strips.

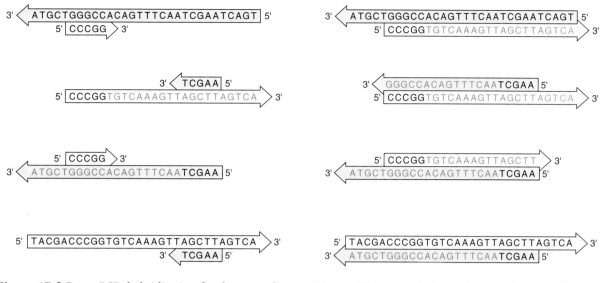

Figure 17.6 Paper PCR, hybridization for the second round of DNA synthesis. It does not matter that the letters on some of the strands are upside down.

Figure 17.7 Paper PCR, products of the second round of DNA synthesis.

Now, repeat steps 1 through 3 with the two DNA molecules. *Note:* The primers that "started" synthesis of strands in the previous round are now part of the new DNA strands. They must be used as part of the new templates, too (Figure 17.6). The products will be four double-stranded molecules (Figure 17.7). Notice how the lengths of the new strands are changing.

Repeat steps 1 through 3 once more. This repetition produces eight double-stranded molecules, two of which stretch only from one primer sequence to the other (Figure 17.8; note the two products with even ends stretching between the primer hybridization sites).

Ask your students to predict the products of another round of synthesis. The predicted products can be drawn on paper or the blackboard or done with paper materials. (If you have students perform another round of paper synthesis, they will need eight additional light and dark paper strips, as well as eight additional light and dark primers.) There will be 16 molecules, and 8 of them will be the short species. Yet another round will give 32 molecules, 22 of which will be the short species. A sixth round of synthesis yields 64 helices, 52 of which are short. The bottom line is that the number of molecules with primer sequences at their ends increases dramatically, forming the vast majority of the products.

Figure 17.8 Paper PCR, products of the third round of DNA synthesis. These products include the first molecules of what will be the major product of many rounds of synthesis: the short products extending exactly from one primer sequence to the other. These molecules are the third one down in the left-hand column and the second one down in the right-hand column.

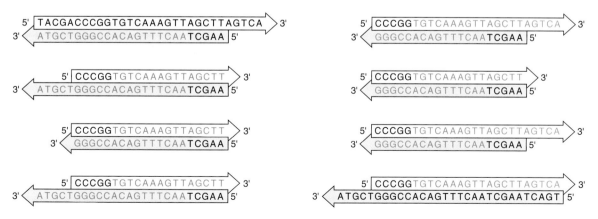

It is important for students to see that the products of PCR are almost all identical and that they are double-stranded DNA segments that begin and end with the two primer sequences.

2. DNA analysis with PCR

Hand out the sample 1 and 2 DNA and primer sequences to your students, if they do not have manuals. Have them cut out the primers. The DNA sequence in each sample represents a single molecule starting with the top row and continuing through the bottom row. Tell the students that the two sample sheets represent DNAs prepared from two different sources. If you think it is a good idea, have the students cut out the strips and tape them together in order from top to bottom. (In this case, they will need tape, and it is critical that the strips go together in the correct order.) The students are laboratory technicians, and they will perform PCR on both samples, using the primers provided. They are to assume that they have large quantities of the primers.

Have the students predict the DNA sequences of the major products from the two separate reactions.

The product from sample 1 is as follows (primer sequences are underlined):

<u>TTCCAGCC</u>AGAGTCTCGGAACTAGCCTTATG
AAGGTCGGTCTCAGAGCCTTGAT<u>CGGAATAC</u>

There is no product from sample 2; the primers do not hybridize.

The students will notice that "something is wrong" with sample 2 and that the primers do not hybridize anywhere. Get them to talk about this. Ask them what they would see if they simply performed the reactions on the two samples and then loaded the results into a gel and electrophoresed and stained them. (They would see a product band in the sample 1 lane but none in the sample 2 lane.) Ask them if they think it would be possible to use PCR to determine if a specific microorganism were present in a sample or if a specific gene were present in an organism.

Explain to the class that PCR can be used to detect the presence of disease-causing microorganisms in medical specimens. DNA is prepared from a sample taken from the patient (blood, tissue, sputum, etc.), and PCR is performed with primers that hybridize *only* to the DNA of the microorganism of interest. Laboratory workers can then tell if the microorganism is present in the sample by noting whether any DNA product is produced. This kind of test can be used in disease diagnosis.

Have the students answer the questions (in class or for homework).

Answers to Student Questions

1. She would have to know enough of the DNA sequence of the virus to design primers that would hybridize to it. She would have to test to make sure that those primers did not hybridize to human DNA or to samples prepared from healthy patients (which might contain small amounts of DNA from other microorganisms). She could use computerized DNA sequence comparisons for some of her primer designing (to make sure the primer sequence was not present in any known organisms other than virus X).
2. Make the primers long. The longer the primer, the less likely it will accidentally hybridize to other DNA molecules, since the odds of a given base sequence occurring randomly are 4^n, where n is the number of bases in the sequence (see chapter 16).
3. 2^n.
4. $2n$.
5. No. With only one primer, you would be limited to synthesizing one strand over and over, and you could never synthesize the complementary strand. After one round, you would have one new single strand. For the second round, you would still have only one template for the primer, so you could generate only one more new single strand, and so on.

PCR in the Classroom

If you do not have a thermal cycler

In theory, any PCR can be carried out without a thermal cycler. In the early days of PCR, there were no such devices, and all PCR was done manually. You will need three constant-temperature water baths set to the temperatures specified in the procedure. In reality, most PCRs would be very tedious for students to carry out. The reactions typically require 30 cycles at perhaps 3 minutes per cycle. If your students are highly disciplined, motivated, and not easily distracted, they can get reasonable results.

Some PCRs are better suited to manual cycling than others. Carolina Biological Supply Company sells materials for an easy manual classroom PCR. It requires two water baths, one boiling and one set at 55°C. Teacher and student directions are included.

You must supply electrophoresis equipment to separate the products. This demonstration kit has been specifically designed to work with the less-than-ideal conditions of the two water baths. The two-water-bath system will not give detectable products with commercial kits that require a thermal cycler.

If you have a thermal cycler

The above-mentioned experiment kit, which was designed for teaching, can be used with a thermal cycler.

Selected Readings

Mullis, K. B. 1990. The unusual origin of the polymerase chain reaction. *Scientific American* 262:56–65.

Paabo, S. 1993. Ancient DNA. *Scientific American* 269: 86–92.

Poinar, G. 1999. Ancient DNA. *American Scientist* 87:446–457.

Classroom Activities

Paper PCR

Introduction

One of the difficulties scientists often face in the course of DNA-based analysis is a shortage of DNA. A forensic scientist may have only a tiny drop of blood or saliva to test. An evolutionary biologist may want to analyze DNA from a museum specimen without destroying the specimen. Even if ample amounts of tissue or numbers of sample cells are available, it takes some work to purify specific DNA fragments in large quantities.

In 1985, a new technique was introduced that changed the whole picture. This technique essentially allows a scientist to generate an unlimited number of copies of a specific DNA fragment. It was invented by a biotechnology industry scientist who had the initial inspiration one night in 1983 as he was driving and thinking about a technical problem he had at work.

The essence of the idea was this. If you set up a reaction in a test tube in which DNA polymerase duplicated a single template DNA molecule into 2 molecules, and then duplicated those into 4, and then duplicated those into 8, then 16, then 32, etc., you would soon have a virtually infinite number of copies of the original molecule. Each round of DNA synthesis would yield twice as many molecules as the previous round—a chain reaction producing specific pieces of DNA. This new technique was called the *polymerase chain reaction,* or PCR.

Of course, the scientist did more than just realize that DNA polymerase can copy one DNA helix into two. You already knew that, too. What he did was to figure out how a chain reaction could be generated in a test tube and how to get the reaction to copy the DNA segment of the scientist's choosing.

PCR relies on the characteristics of DNA polymerase enzymes and on the process of hybridization. Recall that DNA polymerases must have a primer base paired to a template DNA strand so that they can

synthesize the complement to the template strand. Also remember that hybridization is the spontaneous formation of base pairs between two complementary single strands—you can separate the two strands of a helix by heating it, but if you then allow the mixture to cool, the base pairs between the strands will re-form.

Now, how can you use DNA polymerase and hybridization to get a chain reaction of DNA synthesis? Refer to the diagram in Figure 17.1 during this explanation.

First, you decide what DNA segment you wish to amplify (scientists say *amplify* instead of duplicate, because they are making so many copies). Then, you synthesize two short single-stranded DNA molecules that are complementary to the very ends of the segment. These two short molecules must have specific characteristics. Look at the first panel in Figure 17.1 under Round 1. It shows a double-stranded parental DNA molecule with two copies each of the two short single-stranded DNAs. Each of the single-stranded molecules is complementary to only one strand of the parental DNA, and each one is complementary to only one end of the segment. Furthermore, if you imagine these short molecules base paired to the complementary regions in the duplex, their 3′ ends would point toward each other. These short single-stranded molecules are the *primers.*

To begin the chain reaction, a large number of primers are mixed with the template molecule in a test tube containing buffer and many deoxynucleoside triphosphates. (What are they for?) This mixture is heated almost to boiling, so that the two strands of the parental molecule denature (Figure 17.1, Denaturation).

Next, the mixture is allowed to cool. Ordinarily, the two strands of the parental DNA molecule would eventually line up and re-form their base pairs. However, there are so many molecules of primers in the mixture that the short primers will find their

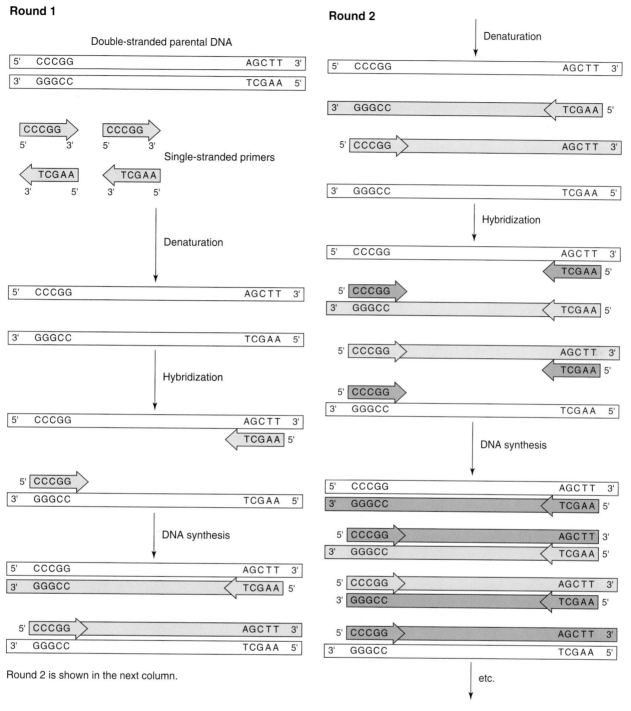

Round 2 is shown in the next column.

Figure 17.1 PCR.

complementary sites on the parental strands before the two parental strands can line up correctly for base pairing. Therefore, a primer molecule hybridizes to each of the parental strands (Figure 17.1, Hybridization).

Now, DNA polymerase enzyme is added. The primers hybridized to the single-stranded parental molecules meet the requirements of the enzyme for

DNA synthesis. DNA polymerase begins adding the correct deoxyribonucleotides to the 3' ends of the primers, forming new complementary strands (Figure 17.1, DNA Synthesis).

After a short time, the mixture is heated up again. Now, the two new double-stranded molecules denature, leaving four single strands (Round 2, Denaturation). The mixture is cooled, and the abundant primer molecules hybridize to the single-stranded molecules (Round 2, Hybridization). DNA polymerase is added again, and new deoxynucleotides are added to the 3′ ends of the hybridized primers, yielding four double-stranded molecules (Round 2, DNA Synthesis). Notice that two of the newly synthesized strands begin and end at the primer hybridization sites.

This process of denaturation, hybridization, and DNA synthesis is repeated over and over, often 25 to 30 times, yielding huge numbers of molecules. The overwhelming majority of the newly synthesized molecules reach exactly from one primer hybridization site to the other, so by choosing the primers, a scientist controls which segment of the parental molecule is amplified. PCR is now used routinely for many different purposes—to amplify a specific fragment of DNA for cloning, to generate a DNA fingerprint from a minute sample, and even to diagnose diseases. Figure 17.2 illustrates the accumulation of the major PCR product during an amplification reaction.

There has been one technical improvement to the process outlined above that has made performing PCR even easier. Did you notice that we added DNA polymerase before each DNA synthesis step? That is because PCR was first carried out with *Escherichia coli* DNA polymerase, which is rendered inactive at the high denaturation temperature, so that more enzyme had to be added for each round of synthesis. However, there are organisms in nature that inhabit the very hot waters of hot springs and thermal ocean vents. The DNA polymerase enzymes from these organisms are not inactivated by the temperatures required for DNA denaturation. Now, PCR is carried out with heat-resistant DNA polymerase, so the enzyme can be added only once at the beginning of the reaction cycles.

When PCR was first developed, scientists preset three water baths to the temperatures required for denaturation, hybridization, and DNA synthesis. They performed PCR by simply moving the reaction tube from one water bath to another. It was not long before enterprising biotechnology companies manufactured incubators that rapidly cycled between the desired temperatures, eliminating the need for manually moving tubes. These *thermal cyclers* have further simplified the performance of PCR. PCR can now be carried out by mixing parental DNA, primers, buffer, deoxynucleotides, and heat-resistant polymerase in a reaction tube; placing the tube in the thermal cycler; programming the thermal cycler to the desired time and

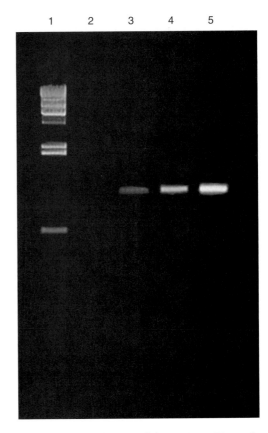

Figure 17.2 Accumulation of the major PCR product with increasing numbers of amplification cycles. The template for this PCR reaction is lambda DNA, and the primers hybridize to it so that the major PCR product is a 1,106-base pair segment. Lane 1 contains DNA fragments from a HindIII digestion of lambda DNA as size markers. Lane 2 shows the PCR mixture before amplification. Lanes 3 to 5 show the PCR mixture after 5, 10, and 15 amplification cycles.

temperature specifications; and waiting for the cycles to finish. It usually takes a few hours for many cycles of amplification.

When we (the authors) and other scientists first learned about PCR, the idea made perfect sense. However, we did not completely get it until we had worked through several reaction cycles ourselves, drawing the parental DNA molecules, the primers, the products, etc. Because we really learned PCR by working through it, we have provided a paper simulation for you to do the same thing. The template and primers are in Appendix A (Worksheets 17.1 to 17.3). Your instructor has directions, though you may be able to figure it out yourself. We have also included materials to simulate a PCR-based diagnostic test.

If you do not do the paper simulation, we highly recommend that you get some clean paper and draw through four rounds of PCR yourself. Figure 17.1 should get you started. We like using different colors of pens to keep track of the parental DNA, the primers, and the newly synthesized molecules (remember, primers will form part of the new molecules).

Here are some questions to think about.

Questions

1. What would a scientist have to know before she could design a PCR-based diagnostic test for virus X?

2. How could a scientist help ensure that her primers will hybridize only to the DNA she wishes to detect?

3. Write an expression that predicts the number of product molecules generated from a single double-stranded DNA molecule after n rounds of synthesis.

4. Predict the number of product molecules that are *not* the short species generated from one double-stranded DNA molecule after n rounds of synthesis.

5. Could you amplify DNA given only one primer? What would the products be after one round? After two rounds? After four rounds?

DNA Sequencing

18

About This Activity

In this lesson, paper-and-paper clip simulations are used to illustrate the most common methods of DNA sequencing. Two approaches are shown, both based on the same biochemistry. The difference is that one employs a heat-resistant DNA polymerase in a modification of the polymerase chain reaction (PCR) ("cycle sequencing"), while the other uses regular DNA polymerase in a single round of reactions ("original approach"). The original approach is the one that was presented in the first and second editions of this book, and it reflects the enzymatic DNA sequencing as it was first published. We have added the cycle-sequencing simulation because it is currently the most widely employed sequencing method. However, either version of the activity correctly illustrates the principle behind DNA polymerase-based sequencing. If you have pop beads for DNA models, they can be substituted for the paper and paper clips.

Class periods required: 1 to 2

Background

The methods most commonly used for sequencing DNA on a small scale employ compounds called *chain terminators,* chemicals that specifically stop the elongation of a new DNA strand by DNA polymerase. For background information on how chain termination methods of DNA sequencing work, please read the introduction to the *Student Activity.*

The dideoxy sequencing method is easily illustrated in the following activity. If you are uncertain as to how the method works, try the activity yourself first. In the activity, your class will "sequence" a piece of DNA. They will synthesize a DNA strand based on a paper template using colored paper clips to represent the new nucleotides. Dideoxynucleotides are represented by colors. The results of the entire class will be pooled and symbolically "run" in a

sequencing gel. This activity will not only illustrate the DNA-sequencing method, but will reinforce base-pairing rules, the mechanism of DNA synthesis by DNA polymerase, and the principle of electrophoretic separation by size.

Several chain terminators are also used as antiviral drugs. The reading at the end of this chapter explains how terminators work to fight human immunodeficiency virus and herpesvirus.

Cycle Sequencing

Cycle sequencing also uses the dideoxy method of sequence determination. Here, the DNA polymerase is the *Taq* polymerase used in PCR, and the sequencing reactions are based on thermal cycles, as is PCR. A major difference between cycle sequencing and PCR is that only one primer is present in cycle sequencing instead of the pair used for DNA amplification. Because only one primer is present, only one strand of the parent DNA molecule is used as a template.

The most common approach to cycle sequencing uses labeled dideoxynucleotides, where each species (A, G, C, and T) is labeled with a fluorescent dye of a different color (for example, A is green, G is purple, T is red, and C is yellow). With this labeling system, any molecule that terminated with a dideoxyadenosine (ddA) would have a green fluorescent dye attached, any that terminated in a C would have a yellow dye, etc. In cycle sequencing, the A, G, C, and T dideoxynucleotides are added to a single reaction tube.

During cycle sequencing, the polymerase synthesizes new DNA complementary to only one strand until it incorporates a terminator. The denaturation step of the PCR separates the strands, and then the second round of synthesis occurs. The same template strand is used, because only one primer was added, so the products of the second round are identical to the products of the first round except that the terminator nucleotide might be incorporated at different positions. Each subsequent round of synthesis is the same, resulting in the accumulation of a population of single-stranded daughter molecules, all identical except that they end at different positions with a fluorescent terminator whose color corresponds to the identity of the base. The fluorescent-dye approach could be used with DNA polymerase enzymes that are not heat resistant, but using multiple cycles of DNA synthesis allows the accumulation of more labeled daughter molecules from the same amount of template. The multiple cycles ensure that detectable amounts of daughter molecules terminated at each nucleotide position extending several hundred base pairs from the $3'$ end of the primer are produced.

The products are loaded into a single gel lane inside a device that includes a laser and a photodetector positioned over the bottom of the gel. As the DNA fragments migrate down the gel, they pass the laser one by one. The laser beam excites the fluorescent-dye molecule, causing it to emit light of its characteristic color (green for A, blue for C, etc.). The photodetector detects the color (wavelength) of the emitted fluorescence, which identifies the terminating base at that position. The output from an automated sequencer is a computer-generated graph showing a series of peaks in the different colors, along with the determined DNA sequence. The peaks correspond to the different wavelengths of light emitted by each DNA fragment as it passed by the photodetector. Software translates the sequence of peaks into base sequences that are stored in data files that can be transferred into bioinformatics software for analysis.

DNA-Sequencing Facilities

In the 1980s and early 1990s, DNA sequencing was usually performed by individual researchers in their own laboratories. Since that time, spurred in large measure by the advances in and increasing emphasis on genomics, DNA-sequencing services have taken over much of the work. These facilities, whether at universities or commercial companies, request that DNA to be sequenced be submitted as clones. At the facility, an array of tubes containing templates cloned into the same vector is set up, and the same reaction mixture of polymerase, nucleotides, labeled dideoxynucleotides, and a primer complementary to the vector is added to all of them (often by means of a robot). The $3'$ end of the primer points toward the cloned inserts, so that the

sequencing reactions read into the desired template. In this way, dozens of different cloned template molecules can be sequenced simultaneously, with very little labor per template. Sequencing has come a long way since graduate students read their own sequencing gels, moving carefully down the X-ray film using a ruler to make sure they read the bands in the right order, recording the sequence by hand, and then carefully typing it into computer files.

Objectives

After this lesson, students should be able to:

1. Explain what a chain terminator is and why it causes DNA synthesis to stop.
2. Explain how chain terminators are used to reveal the base sequence of a DNA molecule.
3. Read a DNA sequence from a sequencing gel.
4. Explain how cycle sequencing is both similar to and different from PCR.

Materials

- Enlarged copies of template and primers (Worksheets 18.1 and 18.2) from the CD (one template and four primers per student for cycle sequencing, four templates and four primers per student for the original approach)
- Colored paper clips
  - For cycle sequencing, 80 silver (regular), 4 red, 4 blue, 4 green, and 4 yellow paper clips for each student
  - For the original approach, five colors, with 20 of each of four colors per student and 4 of the "stop" color per student
- Small paper cups (or other containers), four per student
- Scissors
- Large corkboard or paper sheet on classroom wall; pins; large letters A, G, C, and T (one each); and labels P (for "primer") + 1, P + 2, etc., up to P + 16 (one each)

These materials are for setting up the "sequencing gel" (see below).

Preparation

- Print and enlarge the templates and primers. If possible, laminate them, punch a hole at the end of the primer (to attach the first paper clip), and reinforce the hole. Save these for future use.
- Sort the paper clips by color.
- Set up your sequencing gel on the corkboard or draw it on the blackboard as shown (see below).

For cycle sequencing, just set up the vertical labels (P + 1, etc.). For the original approach, also set up four "gel lanes" labeled A, G, C, and T, as shown below.

	A	G	C	T
P + 16				
P + 15				
•				
•				
•				
P + 4				
P + 3				
P + 2				
P + 1				

Procedure

For either approach

Review the mechanism of DNA synthesis by DNA polymerase. Be sure to emphasize that the $3'$ hydroxyl group is necessary for forming a new bond in the growing DNA chain (Figure 18.2). Show the students the structure of the dideoxynucleotides (Figure 18.1) and explain that they can be incorporated just like regular deoxynucleotides but that then DNA synthesis is stopped.

Cycle sequencing

1. Hand out the DNA templates (both strands), one per student, and four primers per student. Tell them they will sequence the template using the cycle-sequencing method, with paper clips representing the new nucleotides. Silver paper clips represent unlabeled nucleotides, and colored paper clips represent the dideoxynucleotides (ddATP, ddGTP, ddCTP, and ddTTP) labeled with fluorescent dyes.

2. Each student should get four paper cups and label them A, G, C, and T. Put 20 silver paper clips, representing unlabeled nucleotides, into each cup. Add four green paper clips, representing ddATP, to the cup labeled A; four blue or purple paper clips, representing ddGTP, to the G cup; four red paper clips, representing ddTTP, to the T cup; and four yellow paper clips, representing ddCTP, to the cup labeled C. Make sure the paper clips are well mixed.

3. Students should line up their template molecule with complementary base pairs and put the paper cups in easy reach. Direct the students to denature their templates. They should separate the two strands. Next, direct them to hybridize the primer to the correct template strand. The $5'$

end of the primer should be hybridized to the $3'$ end of the template.

4. Now, have the class begin to synthesize new DNA from the $3'$ end of the paper primer. Based on the identity of the first template nucleotide (an A), they should draw out a complementary nucleotide (a T) from the correct paper cup and bond it to the paper primer (by piercing the paper). The drawings must be done at random (eyes closed). If any student pulls out a stop paper clip (in this case, a red one), that student should add the stop paper clip to the primer and then set that template aside. No more nucleotides can be added to it.

5. Repeat step 4 for the next base in the template (the second A), linking the second T paper clip to the first. Again, if a student randomly selects a stop paper clip, that paper clip should be added to the growing DNA strand, but then that template must be set aside and no more nucleotides added to it. Remind the students that the stop paper clips are dideoxynucleotides that lack the $3'$ hydroxyl group for forming a new bond. Continue down the template to the end.

6. Cycle 2. Denature the template and the newly synthesized strand. The primer is part of the newly synthesized strand and stays with its chain of paper clips that represent the newly added nucleotides. Set the newly synthesized daughter strands carefully aside for later.

7. Hybridize a new primer to the template. Make sure the students see that the only molecule on their desks to which the primer can hybridize is the template strand, not its opposite strand (from the original template molecule) or the newly synthesized daughter.

8. Repeat steps 4 to 7 for a total of four cycles of synthesis.

9. Now it is time to simulate loading the gel and performing electrophoresis. DNA-sequencing reaction products are denatured before being loaded, so the students should "denature" their molecules one last time by separating the primer and the last daughter molecule from the paper template, leaving the primer attached to the paper clips.

10. Tell the students that all the molecules the class has synthesized represent products in the reaction tube. All of them would be loaded into a single well of a sequencing gel. Electrophoresis would separate them according to size, and the class will be arranging the bands in the gel accordingly, from smallest to largest.

11. Ask who has a product consisting of the primer plus only one paper clip. Choose a student from among those who respond, and have that

student pin or otherwise attach the product next to "P + 1" on your sequencing-gel outline. Repeat for P + 2, P + 3, etc., up to P + 16. Note that the template molecule and the opposite strand of the template molecule are also size P + 16. They can be placed on the gel, too—only the fluorescent dye will be detected.

12. Now it is time to read and translate the gel. Read off the color of the terminal paper clip, beginning at P + 1 and continuing through P + 16. Record the colors in order, and translate them into bases (green is A, and so forth). The students should see that the order of the terminal nucleotides in the gel corresponds to the complementary sequence of the template.

Original Chain Termination DNA-Sequencing Method

1. Hand out the DNA templates (both strands) and primers, four per student. Tell the students they will sequence the template using the original dideoxynucleotide method, with paper clips representing the radioactive new nucleotides.

2. Have the class count off A, G, C, and T as if they were numbering off. Each student will conduct only one type of reaction on the four template molecules. The reason it is suggested that each student "synthesize" four molecules is to be sure that a stop paper clip is incorporated by someone at every possible position. There should be at least 8 (preferably 10 to 12) new strands synthesized for each type of reaction. You may adjust the number of strands each student synthesizes to fit your class size.

3. Assign one color of paper clip to be A, one to be G, and so on. The fifth color will be the dideoxynucleotide—the "stop" nucleotide.

4. Each student should get four paper cups and label them A, G, C, and T. The "A" students put 20 paper clips of the correct color in the G, C, and T cups. In the A cup, they put 16 of the A color and 4 of the stop color and mix them thoroughly. The stop color represents ddA molecules.

The G students get 20 A, C, and T paper clips and then put 16 G and 4 stop paper clips in their G cups (mixed well). This time, the stop paper clips represent ddG molecules.

The C and T students should set their "reaction mixtures" up appropriately with their paper clips.

5. Students should line up their four template molecules on their desks, with the paper cups in easy reach. The two strands of the template should be lined up with complementary base pairs. Direct the students to denature their templates. They should separate the two strands. Next, direct them to hybridize the primer to the correct template strand. The 5′ end of the primer should be hybridized to the 3′ end of the template.

6. Be sure the "spiked" paper clips are well mixed. Now, have the class begin to synthesize new DNA from the 3′ end of the paper primer. Based on the identity of the first template nucleotide (an A), they should draw out a complementary nucleotide (a T) from the correct paper cup and bond it to the paper primer (by piercing the paper). The students must do this four times, once for each template molecule. The drawings must be done at random (eyes closed). If any student with the "T" reaction pulls out a stop paper clip for a template, that student should add the stop paper clip to the primer and then set that template aside. No more nucleotides can be added to it.

7. Repeat step 6 for the next base in the template (the second A), linking the second T paper clip to the first. Again, if a student randomly selects a stop paper clip, that paper clip should be added to the growing DNA strand, but then that template must be set aside with no more nucleotides added to it. Remind the students that the stop paper clips are dideoxynucleotides that lack the 3′ hydroxyl group for forming a new bond.

8. Continue down the template to the end. Figure 18.5 shows all possible products for the "C" reaction. Notice that there are products that stop at each G in the template plus one full-length product for which the student never drew a stop paper clip from the C paper clips.

9. Now it is time to load the gel and perform electrophoresis. DNA-sequencing reaction products are denatured before being loaded, so the students should "denature" their molecules by cutting the paper primer away from the paper template; leave the paper primer attached to the paper clips.

Point to your gel setup and tell the students that all the A reaction products will be loaded and run in the A lane, the G reaction products in the G lane, and so forth. They should imagine that they have loaded their products into the appropriate lane and that electrophoresis has occurred.

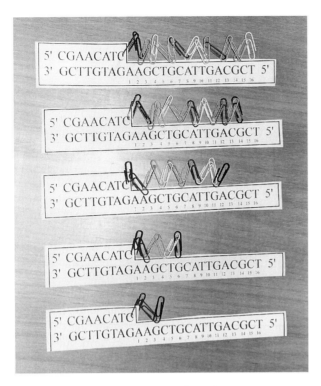

Figure 18.5 Reaction products from the C reaction. The four shorter products were produced when a student drew a stop paper clip at the respective G in the template sequence. The long product resulted when a student happened not to draw a stop paper clip at any of the template G's.

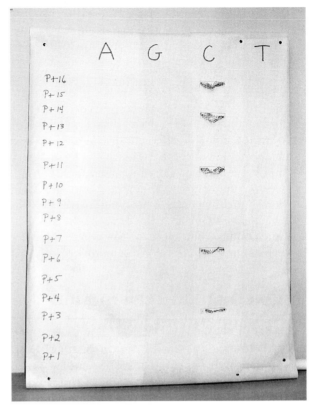

Figure 18.6 The C reaction products have been "loaded" and "separated" on the "gel." Their lengths are P + 3, P + 6, P + 11, P + 14, and P + 16 (refer to Figure 18.5).

10. To see the banding pattern, focus on one lane at a time. For example, ask the A reaction students if anyone has a product molecule with just one new nucleotide (paper clip) added to the primer. (No one should, since the first nucleotide added is a T.) Ask for two, three, four, and five new nucleotides. Some of the "A" group should have a molecule with exactly five new nucleotides, since there is a T at that position in the template. If you are using corkboard, have that student come forward and pin the molecule in the A lane right at the P + 5 level (there may be several such molecules; hang them on pins on top of each other). If you are using a blackboard, simply draw a band in the A lane at the P + 5 level.

Continue asking for molecules up to P + 16. You should get products at P + 5, 9, 10, and 16 from the A reaction students. Repeat for the G reaction students (you should get P + 4, 7, 13, 15, and 16), the C reaction students (P + 3, 6, 11, 14, and 16), and the T reaction students (P + 1, 2, 8, 12, and 16).

Figure 18.6 shows the C lane of the gel loaded with the reaction products pictured in Figure 18.5. This figure also shows how to draw the gel outline. Figure 18.7 shows what your completed "gel" will look like.

11. Now, read the sequencing gel. Starting at P + 1, identify the lane the band is in (T). Write "T" on the blackboard. Continue with P + 2, P + 3, etc., writing the sequence as you go. Compare the base sequence you write on the board to the sequence of the newly synthesized DNA strand (they should match). Compare it to the sequence of the paper template (they should be perfectly complementary).

12. (Optional) Project a transparency of the photograph of the sequencing gel in Figure 18.4 and have students read some of the DNA sequence from it.

13. (Optional) Read and discuss the reading *Chain Terminators as Antiviral Drugs.*

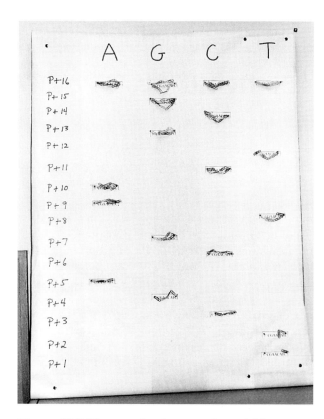

	A	G	C	T
P+16				
P+15				
P+14				
P+13				
P+12				
P+11				
P+10				
P+9				
P+8				
P+7				
P+6				
P+5				
P+4				
P+3				
P+2				
P+1				

Figure 18.7 The completed sequencing gel. The sequence is read from the bottom up: TTCGACGTAACT-GCG. This is a perfect complement to the template sequence.

Answers to Student Questions

1. The template molecules are present in the gel (they are loaded along with the rest of the reaction mixture), but you cannot see them because they are not labeled.
2. DNA replication.
3. Increasing the concentration of dideoxynucleotides in the reaction mixture would increase the chances that the polymerase would incorporate a terminator instead of a normal nucleotide. The average length of the reaction products would decrease.
4. ddC.

Selected Reading

Bartlett, J., and R. Moore. 1998. Improving HIV therapy. *Scientific American* 279:84.

Classroom Activities

DNA Sequencing: The Terminators

Determining the base sequence of a piece of DNA is a critical step in many applications of biotechnology. How is it done? The method most commonly used today employs compounds called chain terminators—a term that describes the lethal effect they have on DNA synthesis—and nature's sequence readers, the DNA polymerase enzymes. These enzymes read the sequence of a single template DNA strand and synthesize a complementary strand. The chain terminators allow us to "look over the shoulder" of the DNA polymerase enzyme: we can see the order in which it adds bases to a new DNA strand, and therefore, we can deduce the sequence of the template strand.

What are chain terminators, and how do they block DNA synthesis? Chain terminators are molecules that closely resemble normal nucleotides but lack the essential 3′ hydroxyl (OH) group (Figure 18.1). In DNA replication, a nucleotide complementary to the template base is brought into position. The DNA poly-

merase then adds it to the growing DNA strand by forming a bond between the 5′ phosphate group of the new nucleotide and the 3′ OH group of the previous nucleotide (Figure 18.2). DNA polymerases cannot synthesize DNA without a preexisting 3′ OH group to use as a starting point (this is referred to as a requirement for a *primer;* see chapter 7). If a DNA polymerase mistakenly adds a chain terminator instead of a normal deoxynucleotide, no further nucleotides can be added, and DNA synthesis is terminated.

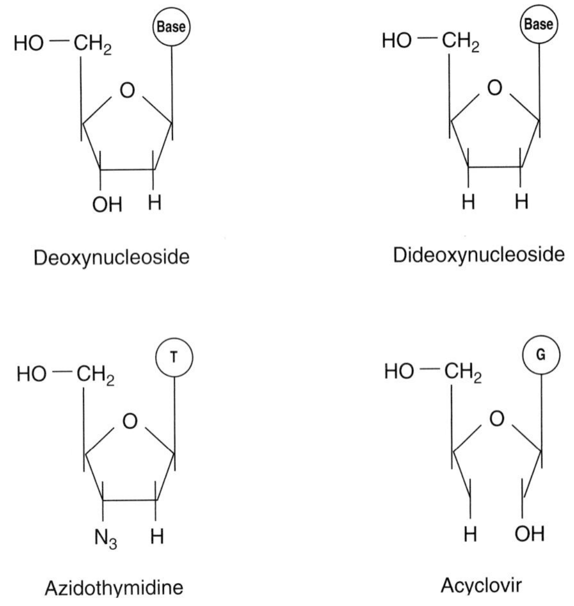

Figure 18.2 DNA replication. Base pairing between an incoming deoxynucleoside triphosphate and the template strand of DNA guides the formation of a new complementary strand.

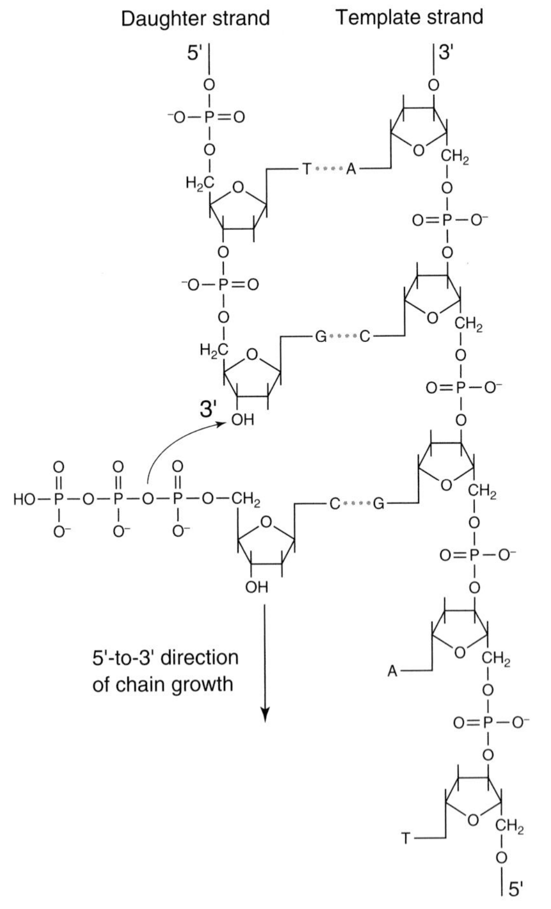

Figure 18.1 Normal deoxynucleoside shown with chain terminators. All are incorporated into DNA from their triphosphate forms.

The chain terminators used in DNA sequencing are the dideoxynucleotides (Figure 18.1). How do dideoxynucleotides let us see the base sequence of a DNA molecule? There is more than one way to go about it. Using the original approach to dideoxynucleotide sequencing, the template DNA molecule to be sequenced is mixed with primers and radioactive normal deoxynucleotides (deoxyadenosine [dA], deoxyguanosine [dG], deoxycytosine [dC], and deoxyribosylthymine [dT]). This master mix is divided into four batches, and each batch is then "spiked" with a different dideoxynucleotide: dideoxyadenosine (ddA), dideoxyguanosine (ddG), dideoxycytosine (ddC), or dideoxyribosylthymine (ddT). DNA polymerase enzyme is added and synthesizes new DNA strands on the template molecules. Occasionally, however, the dideoxynucleotide chain terminators will be inserted in place of the analogous normal deoxynucleotide, and the synthesis of that DNA molecule will be terminated. In the reaction mixture containing ddA (the "A reaction"), some percentage of the new molecules will get a ddA at each place where there is a T in the template. The result is a set of new DNA molecules in which some of the molecules terminate at each T in the template. Similarly, in the G reaction, some of the new molecules will terminate at each C in the template. In the C reaction, some of the new molecules terminate at each G in the template, and in the T reaction, molecules terminate at each template A.

After the synthesis reactions are complete, the A, G, C, and T reaction mixtures are denatured by being heated and are then loaded separately into adjacent lanes of a gel, and a current is applied (Figure 18.3). The newly synthesized molecules are separated by size and then visualized by exposing the gel to photographic film (remember that the new nucleotides are radioactive). The A reaction lane will show bands that correspond in length to the site of each T in the template. The G reaction lane shows bands whose lengths correspond to the site of template C's, and so on. The sequence of the new DNA strand is "read" from the sequencing gel by starting at the bottom (the shortest new molecule) and reading upward (see below). Figure 18.4 shows an autoradiogram of a sequencing gel.

If this written explanation seems confusing, don't worry. The sequencing simulation you will do shows

Figure 18.4 Autoradiogram of a sequencing gel. The scientist who did this sequencing procedure was screening several plasmids from transformants to see if she got a clone she was trying to construct. She did. In fact, every one of the plasmids had the sequence she wanted.

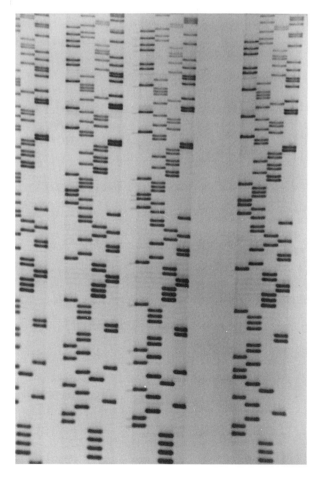

Figure 18.3 A student is loading a sequencing gel. A sequencing gel is what is called a vertical gel—the DNA runs from top to bottom. It is tall and quite thin: less than 1 millimeter thick. Sequencing gels are made with the substance acrylamide instead of agarose. Acrylamide forms a much tighter mesh than agarose, enabling the separation of DNA molecules that are only 1 base pair different in length.

Classroom Activities

how the whole approach works. After you complete the simulation, go back and reread the introductory material and then answer the questions.

Currently, a more common approach to dideoxy sequencing employs a version of PCR. It is called *cycle sequencing* because numerous cycles of DNA synthesis are performed, just as in PCR. In this method, colored fluorescent dyes are used as labels instead of radioactivity. A different colored dye is attached to each of the four dideoxynucleotides (for example, ddA is green, ddG is purple, ddT is red, and ddC is yellow). These labeled dideoxynucleotides are mixed with the four regular deoxynucleotides (unlabeled), and added to the template, *Taq* polymerase, plus a single sequencing primer.

As in PCR, the mixture is put through temperature cycles of denaturation, hybridization, and DNA synthesis. Unlike a PCR, only one primer is present, so DNA synthesis occurs using only one strand of the parental DNA as a template. As the polymerase adds new nucleotides to the growing daughter strand, it might incorporate a labeled dideoxynucleotide. When it does, synthesis stops, leaving a daughter molecule that terminates with a fluorescent label whose color corresponds to the base of the dideoxynucleotide.

At the end of the reaction, the products are loaded into a gel inside an apparatus that contains a laser and a photodetector. As bands migrate past the photodetector, the laser causes them to fluoresce with their characteristic color (wavelength of light), which is detected by the photodetector. A computer records the color of each band as it migrates past the detector.

Again, if this seems confusing, wait until you have performed the simulation and then read the explanation again.

Procedure

Below are two slightly different procedures that simulate dideoxy sequencing using either the original approach or cycle sequencing.

Cycle sequencing

1. You will be performing four cycles of DNA synthesis with your sequencing template (Worksheet 18.1). Obtain a double-stranded template molecule, four primers, and four paper cups. Label the paper cups A, G, C, and T.

2. Put 20 silver (plain) paper clips in each cup. These represent unlabeled deoxynucleotides. Put four green paper clips in the cup labeled A. These represent ddA labeled with a green fluorescent dye. Put four blue or purple paper clips in the cup labeled G. These represent ddG labeled with a blue or purple fluorescent dye. Put four red paper clips in the cup labeled T. These represent ddT labeled with a red fluorescent dye. Put four yellow paper clips in the cup labeled C. These represent ddC labeled with a yellow fluorescent dye. Mix the paper clips in each cup well.

3. Line up your double-stranded DNA template on your desk, forming the complementary base pairs. When directed to do so, denature your template. This is the denaturation step of the first cycle. Next, hybridize a primer molecule to the correct template strand. (How do you know which is correct?) This represents the hybridization step in the cycle.

4. Now, simulate the activity of DNA polymerase by reading the template base next to the 3′ end of the primer and drawing a paper clip out of the correct complementary nucleotide cup (for example, if the template base is an A, draw a paper clip from the T cup). Attach the paper clip to the primer. If you drew out a colored paper clip, you incorporated a dideoxynucleotide into the daughter molecule and you cannot add any more nucleotides, so you must stop for this round. If you drew a silver paper clip, continue with the next base, drawing from the correctly labeled cup.

5. At the direction of your teacher, continue to add nucleotides corresponding to the template base until you incorporate a dideoxynucleotide. At that point, you must stop synthesis for this round.

6. At the end of the first round of synthesis, your teacher will instruct the class to denature their DNA molecules. This represents the denaturation step of the second round of DNA synthesis. Separate the daughter strand (of which the primer is a part) from the template and carefully move it to one side.

7. When instructed, hybridize a second primer to your template and begin another round of synthesis, always drawing the new nucleotide from the correct paper cup. Stop adding nucleotides when you incorporate a labeled dideoxynucleotide.

8. Repeat for a total of four cycles of DNA synthesis.

9. Now it is time to simulate loading the sequencing gel and performing electrophoresis. At the

direction of your teacher, place each daughter molecule in the gel according to its size (primer plus 1 nucleotide, primer plus 2 nucleotides, etc.).

10. Read the gel, remembering that the detector reads the wavelength (color) of the fluorescent dye. Record the colors in order, from the smallest daughter molecule (P + 1) to the largest. Translate the colors into bases. Compare the sequence to that of the template.

Original chain termination sequencing method

1. You will be assigned the base A, G, C, or T. You will conduct the sequencing reactions with a terminator that substitutes for your base. Each base will be assigned a specific color, and the terminator will be assigned a color.

2. Obtain four paper cups, and label them A, G, C, and T. Put 20 paper clips of the correct color in each cup *except* for the one containing your assigned base. In that cup, put 16 of the correct color and 4 of the terminator color and mix them thoroughly. The terminator color represents dideoxy molecules.

3. Line up the four template molecules on your desk, with the paper cups in easy reach (Worksheet 18.2). Be sure the "spiked" paper clips are well mixed. Begin to synthesize new DNA from the 3′ end of the paper primer. Based on the identity of the first template nucleotide (an A), draw out a complementary nucleotide (a T) from the correct paper cup and bond it to the paper primer (by piercing the paper). Do this four times, once for each template molecule. The drawings must be

done at random (eyes closed). If you have the "T" terminators and pull out a terminator paper clip, add the terminator paper clip to the primer and then set that template aside. No more nucleotides can be added to it.

4. Repeat for the next base in the template (the second A), linking the second T paper clip to the first. Again, if you randomly select a terminator paper clip, that paper clip should be added to the growing DNA strand, but then that template must be set aside and no more nucleotides can be added to it. The terminator paper clips represent dideoxynucleotides that lack the 3′ hydroxyl group for forming a new bond.

5. Continue down the template to the end.

6. Now it is time to simulate loading the gel and performing electrophoresis. DNA sequencing reaction products are denatured before being loaded, so "denature" your molecules by cutting the paper primer away from the paper template; leave the paper primer attached to the paper clips.

The A reaction products will be loaded and run in the A lane, the G reaction products in the G lane, and so forth. Imagine that you have loaded your products into the appropriate lane and that electrophoresis has occurred. You will place your reaction products in the correct gel lane according to their sizes.

When all the reaction products are arranged in the gel, read the sequencing gel. Starting at P + 1, identify the lane the band is in (T). Continue with P + 2, P + 3, etc., writing the sequence as you go. Compare the base sequence you write to the sequence of the newly synthesized DNA strand. Compare it to the sequence of the template.

Questions

1. Would the long template molecule (represented by the paper DNA strand) show up in the sequencing gel?

2. If a cell culture were "fed" some dideoxynucleotides in medium, what cellular process, if any, would be most affected?

3. One of the parameters that can be adjusted by the researcher performing sequencing reactions

is the concentration of the dideoxynucleotides in the reaction mixture. What would be the effect on the average length of the products if the concentration of the dideoxynucleotides were increased?

4. (Optional; refer to the reading.) What chain terminator is currently used in DNA sequencing and as an anti-AIDS drug?

Chain Terminators as Antiviral Drugs

Chemicals that were first used as investigative tools in research laboratories often find their way into the pharmacy. Compounds that affect fundamental biological processes are important to basic research and sometimes turn out to be therapeutically useful as well. The chain terminators are a good example of such compounds. As shown in the above activity, chain terminators are essential to the now-favorite method of DNA sequencing. In addition, three of the currently available anti-AIDS drugs and the best available antiherpes drug belong to this class of compounds.

Fighting AIDS

How are chain terminators used in fighting AIDS and herpesvirus infections? They fight viral infections in the same way they assist in DNA sequencing: by terminating DNA synthesis. For a virus to establish an infection, it must replicate its nucleic acid. Blocking this replication is potentially a very effective way to fight the spread of a virus, but blocking DNA synthesis could be as harmful to the patient as to the virus invader. Herpesviruses and the AIDS virus (human immunodeficiency virus, or HIV) are good candidates for anti-replication drugs because they encode their own DNA-synthesizing enzymes with unique properties that can be exploited.

The AIDS virus is an RNA virus. It encodes a special enzyme, reverse transcriptase, that synthesizes DNA using the viral RNA as a template. This step is essential to HIV infection. Chain terminators fight the spread of HIV in the body by interfering at this stage. Reverse transcriptase is a good target for chain terminator drugs, because it is a sloppy enzyme: it is much more likely to incorporate an incorrect nucleotide or a chain terminator than is the human polymerase. Furthermore, reverse transcriptase lacks the ability to proofread its work, so it cannot remove an incorrect nucleotide once it is incorporated.

There are more issues involved in developing a successful chain terminator drug than simply the sensitivity of the viral replication enzymes. The drug must survive in the body and be absorbed by the proper cells. The active triphosphate form of the chain terminators (shown in Figure 18.2 for the nucleotide being added to the growing chain) is not absorbed by cells, so the drugs are given in unphosphorylated form (as shown in Figure 18.1). Once these compounds enter the cell, the human enzymes must add the phosphate groups (phosphorylate the molecules) to activate them. Different compounds are absorbed and phosphorylated with different efficiencies, so the development of a new drug depends not only on its toxicity to the virus and harm to the host, but also on how it is metabolized in the human system.

Three chain terminators employed against the AIDS virus are azidothymidine (AZT), ddC (the same compound used in DNA sequencing), and dideoxyinosine (ddI) (Figure 18.1). AZT, an analog of thymidine, is incorporated into a growing DNA molecule in place of normal thymidine. Likewise, ddC is incorporated in place of cytosine. The compound inosine is a nucleoside analog that is identical to adenosine except for the absence of one amino group. Cells synthesize inosine and convert it directly to adenosine. When dideoxyinosine is given as a drug, it enters cells and is rapidly converted to dideoxyadenosine. The dideoxyadenosine is incorporated by reverse transcriptase in place of normal adenosine.

These drugs have toxic side effects in patients. Although human DNA polymerase is less sensitive to the drugs than is reverse transcriptase, the drugs do affect DNA replication in normal human cells. AZT is particularly toxic to the bone marrow, while ddC and ddI affect the peripheral nerves.

Another class of anti-AIDS drugs that has been developed in recent years targets a second HIV enzyme—protease. HIV infects cells as an RNA genome packaged with the enzymes reverse transcriptase and integrase inside a protein envelope. After infection, reverse transcriptase makes a double-stranded DNA copy of the RNA genome. The integrase enzyme integrates the DNA into the host cell's chromosome. There, the host cell's RNA polymerase transcribes the HIV genes, and the host ribosomes translate the mRNA into protein.

The transcribed protein, however, is not the active form. The HIV protease must cleave the protein into its active pieces for them to function. Several drugs have been developed that block the function of this protease enzyme: these drugs are referred to as protease inhibitors. Current AIDS therapy usually involves taking two nucleoside analogs and at least one protease inhibitor. This therapeutic regimen has extended the life expectancy of AIDS patients and improved its quality.

Herpesviruses

The herpesviruses are DNA viruses with relatively large genomes. They encode many of their own DNA replication enzymes, including a DNA polymerase.

After the initial infection, herpesviruses remain in the body in an inactive (latent) state from which they can be activated and cause subsequent outbreaks of disease. The herpes simplex viruses types 1 and 2 cause fever blisters and genital herpes, respectively. The herpesvirus varicella-zoster virus causes chicken pox when it first infects a person and shingles in subsequent outbreaks.

Herpesvirus diseases can now be treated with the chain terminator acyclovir (Figure 18.1). Acyclovir (marketed under the name Zovirax) is relatively nontoxic to the human host because human cells take up acyclovir but do not phosphorylate it, so the drug remains inactive. Herpesviruses, however, encode an enzyme that does phosphorylate acyclovir. Therefore, acyclovir becomes an active chain terminator only in herpesvirus-infected cells. Acyclovir is an analog of guanidine and is incorporated opposite cytosine residues in the template. Once incorporated, it terminates further DNA synthesis and inhibits activity of the herpesvirus polymerase.

C. Transfer of Genetic Information

The activities in Section B focused on the manipulation and analysis of DNA. Most of these methods depend on the availability of large numbers of identical DNA molecules. We get these identical molecules primarily through DNA cloning. Part of the cloning process was illustrated in the previous activities: the construction of a recombinant DNA molecule and its analysis by restriction and DNA sequencing. The other essential part of the cloning process is gene transfer: the introduction of new genetic information into an organism.

Researchers use gene transfer in cloning as a means of propagating recombinant DNA molecules. The recombinant molecule is replicated inside the host cell as it multiplies, producing an essentially unlimited number of copies. Scientists also use gene transfer to deliberately change the genotype of an organism so that it will have a new trait or make a useful product. For example, researchers have made corn resistant to certain caterpillar pests by introducing a gene for an insecticidal bacterial protein. Scientists have induced *Escherichia coli* to make human insulin by introducing a complementary DNA (cDNA) copy of the human messenger RNA (mRNA).

This section highlights the process of gene transfer. The first three "wet" laboratories illustrate natural methods of gene transfer in *E. coli:* conjugation, transformation, and transduction. Since some people may be concerned about students working with a bacterium that has been associated with severe disease, this section begins with background information about why some strains of *E. coli* cause disease and others (such as the laboratory strains used in these activities) do not. Although it is easy to think of these methods simply as tools of biotechnology (and they are important tools), the natural occurrence of gene transfer by each of these mechanisms is also significant. Introductory material at the beginning of each chapter gives examples of the medical importance of each of these natural gene transfer methods. The *Reading* in chapter 20 focuses on the spread of antibiotic resistance, an example of rapid evolutionary change via gene transfer in which humans have provided the selective pressure.

Following the *E. coli* laboratories is an exercise showing genetic alteration of a living plant by *Agrobacterium tumefaciens.*

Classroom Activities

Gene Transfer, *Escherichia coli,* and Disease

This section focuses on natural mechanisms of gene transfer. Three of the laboratory activities use living cultures of *Escherichia coli.* Use of this organism may raise safety concerns in your mind and may alarm students and their parents because of the many recent reports in the national news media about outbreaks of disease associated with *E. coli.* The most publicized and frightening disease caused by *E. coli* is the potentially fatal hemolytic-uremic syndrome (HUS). *E. coli* also causes nasty but nonfatal travelers' diarrhea and urinary tract infections. No wonder it can be confusing when we claim that *E. coli* is safe for laboratory work. What is going on?

The reason that *E. coli* has so many different faces is that *E. coli* is not just one single type of bacterium. There are many strains of *E. coli,* and they have genetic differences. Some cause disease; most do not. The following information about *E. coli* and disease is provided as background for you. As it happens, the unfolding story of the molecular biology of disease also involves gene transfer; therefore, it is especially appropriate for this section of the book.

Pathogenicity and *E. coli*

The technical term for the ability to cause disease is pathogenicity, and organisms that cause disease are called pathogens. Scientists comparing pathogenic strains of microbes to nonpathogenic strains of the same species find that pathogenic strains contain genes that are completely absent in nonpathogenic strains. These genes confer the ability to cause disease upon the strain carrying them. In general, pathogenicity genes can be divided into two broad categories. One type encodes proteins that help the organism invade a host, for example, by allowing it to attach to specific host cells. The other type of pathogenicity gene encodes proteins that disrupt a normal function of host cells and interfere with normal host body processes, making the host sick. These proteins are often called toxins.

E. coli strains that cause different diseases carry different pathogenicity genes. As an example, let's look at five different general types of diarrheal disease caused by *E. coli.* One of these *E. coli* diarrheal diseases is caused by an organism historically called *Shigella.* DNA analysis has shown that, at the DNA level, *Shigella* is indistinguishable from *E. coli,* and there is overlap among the pathogenicity genes of *E. coli* and *Shigella,* but the historical name persists. The *E. coli* strains and their pathogenicity genes are summarized in Table 1.

Shigella *dysentery*

Dysentery is severe diarrhea in which blood and mucus are present in the stools. *Shigella* strains cause dysentery, and in about 5% of cases, the patient goes on to develop HUS, a potentially fatal complication. HUS can cause kidney failure or permanent kidney damage and is associated with neurological symptoms. *Shigella* invades cells lining the colon and spreads to adjacent cells. The genes responsible for the ability to invade and spread within the intestinal cells are located on a large plasmid. *Shigella* also produces a toxin, the Shiga toxin, which can cross into the bloodstream. Shiga toxin damages blood vessels, is neurotoxic when injected into test animals, and is believed to be the cause of HUS.

ETEC

In one type of *E. coli* diarrhea, the responsible *E. coli* bacteria attach to the cells lining the small intestine and produce two different toxins, a heat-stable toxin (ST) and a heat-labile (heat-sensitive) toxin (LT). The LT toxin is similar to the cholera toxin. These two proteins alter the secretion of salt and water from the intestinal cells, causing diarrhea. The intestinal cells themselves are not altered in appearance. This type of *E. coli* is called enterotoxigenic *E. coli* (ETEC), and its associated diarrheal disease is responsible for a great deal of infant mortality

Table 1 Enteropathogenic *E. coli* strains and their pathogenicity genes

Bacterium	Pathogenicity genes	Pathogenic phenotype
Shigella	Several plasmid-encoded genes	Ability to invade cells lining the colon and spread to adjacent cells
	Shiga toxin gene	Damages blood vessels; is neurotoxic; thought to cause HUS
		Disease: dysentery, sometimes HUS
ETEC	Plasmid-borne genes for ST and LT	Alter salt and water secretion from intestinal cells
	Other plasmid-borne genes	Ability to colonize small intestine; intestinal cells are not disrupted.
		Disease: diarrhea
EPEC	35.6-kb LEE pathogenicity island in chromosome at *selC* locus	Ability to disrupt intestinal epithelium (enterocyte effacement)
	Plasmid-borne pilus gene	Ability of bacterial cells to spread from original attachment site
		Disease: severe diarrhea
EIEC	Several plasmid-borne genes similar to *Shigella* genes	Ability to invade cells lining the intestine and spread to adjacent cells
		Disease: dysentery, no HUS
EHEC	35.6-kb LEE pathogenicity island in chromosome at *selC* locus	Ability to disrupt intestinal epithelium (enterocyte effacement)
	Gene for toxin virtually identical to the Shiga toxin carried within the genome of a lysogenic phage	Damages blood vessels; is neurotoxic; thought to cause HUS
		Disease: dysentery, sometimes HUS
Laboratory *E. coli* (MM294, K12, etc.)	None	None

in poorer countries. In areas where ETEC strains are common, adults usually have immunity to them. Travelers visiting the area, however, do not. In adults, ETEC causes what is commonly known as travelers' diarrhea.

In addition to the genes for the toxins, ETEC carries genes for proteins that enable the bacteria to colonize the small intestine. These genes are usually found on a plasmid, along with the two toxin genes, although they are not always found in the same arrangement. The gene for the ST toxin is within a transposable element, and the different arrangements of the ETEC pathogenicity genes suggest rearrangement mediated by transposable elements.

EPEC

In a second type of diarrheal disease, the responsible *E. coli* strain does not simply attach to the intestinal cells. Instead, the bacteria attach to the epithelial cells and rearrange their entire surfaces,

disrupting the normal microvilli. By disrupting the epithelial cells and microvilli, enteropathogenic *E. coli* (EPEC) causes a more severe form of diarrhea than does ETEC. EPEC is also a cause of travelers' diarrhea and of significant infant mortality in poor countries.

The genes that confer the ability to disrupt the intestinal epithelium are found on a 35.6-kilobase (kb) stretch of DNA inserted into the *E. coli* chromosome, usually at the *selC* gene locus. The insertion is flanked by a short repeated sequence of bases and thus appears to have been inserted by a recombination mechanism analogous to many used in transposition and phage integration. This stretch of DNA and others like it are called "pathogenicity islands."

This particular island is called LEE, which stands for "locus for enterocyte effacement." Enterocyte effacement refers to the disruption of the intestinal epithelial cells. The entire LEE region has been cloned into a plasmid. When that plasmid is transformed into a nonpathogenic laboratory strain of *E. coli,* the transformants gain the ability to disrupt intestinal cells. The LEE element is found in two different types of pathogenic *E. coli*, EPEC and the HUS *E. coli*, O157:H7. In addition to the LEE island, EPEC also contains a plasmid carrying genes encoding a pilus that appears to enable the bacterium to spread out from an original colonization site.

EIEC

Enteroinvasive *E. coli* (EIEC) is able to invade the cells lining the intestine and spread laterally into adjacent cells, as does *Shigella*. The invasion produces an intense inflammatory response and dysentery. The pathogenicity genes of EIEC are found on a large plasmid like those of *Shigella* and are very homologous to the *Shigella* genes. EIEC dysentery resembles *Shigella* dysentery but without the HUS complication. EIEC does not make a toxin like the Shiga toxin.

EHEC

Enterohemorrhagic *E. coli* (EHEC) also causes dysentery. EHEC carries the LEE pathogenicity island and exhibits the same attachment/effacing activity as EPEC. In addition, EHEC carries a gene for a toxin virtually identical to the Shiga toxin. The toxin gene is carried on a lysogenic phage within the bacterial chromosome (for information on lysogeny, see chapter 21, *Transduction*). In about 5% of EHEC infections, the patient goes on to develop HUS. The major EHEC strain is O157:H7, responsible for several highly publicized outbreaks of *E. coli* food poisoning and some associated deaths from HUS.

E. coli O157:H7 appears to be a relatively new strain of *E. coli* present in some cattle. Humans have acquired EHEC infections from eating undercooked contaminated hamburgers, eating apples from trees fertilized with contaminated manure, and drinking lemonade made with water from a well contaminated with animal waste.

Pathogenicity and Gene Transfer

Pathogenicity genes appear to be transferable. The genes described above are found on lysogenic phages, in plasmids, and within transposable elements. In addition, the genes for the cholera toxin of *Vibrio cholerae*, the diphtheria toxin of *Corynebacterium diphtheriae*, and the neurotoxin of *Clostridium botulinum* (the causative agent of botulism) are all carried on phages. Important virulence factors of *Shigella*, *Salmonella*, *Yersinia*, and *Clostridium tetani* (the causative agent of tetanus) are found on plasmids.

Even the chromosomal pathogenicity islands appear to have been introduced by horizontal gene transfer. The evidence behind that statement includes several items. One is the existence of the directly repeated sequences at either end of the islands, characteristic of insertion events. Another is the fact that identical LEE islands can be found at different locations in different *E. coli* strains, as though the different strains were the results of different insertion events. Yet another piece of evidence is the fact that *E. coli* strains causing urinary tract infections (uropathogenic *E. coli*) have pathogenicity islands different from those of enteropathogenic strains, yet one of the islands is inserted at exactly the same site at the *selC* locus as the LEE island in the EPEC strains—down to the base pair. This suggests that the two islands were deposited in their respective strains by the same mechanism.

Finally, a pathogenicity island in *Staphylococcus aureus* has been shown to move in a phage-mediated fashion. The island contains the toxin gene associated with toxic shock syndrome. A specific *S. aureus* phage causes the island to be excised from the chromosome. That particular phage transduces the pathogenicity island at a high frequency into new *S. aureus* hosts.

In short, it seems that horizontal gene transfer may be involved in the creation of new pathogenic strains of bacteria. Obviously, we are not encountering new pathogens on a daily or weekly basis, so these strains are not established with high frequency. The past 2 decades, however, have seen the emergence of what appears to be two new pathogens: *E. coli* O157:H7 and toxic shock syndrome *S. aureus*. In the case of the transfer of antibiotic resistance, the mechanism of selection of the new strains is apparent; selection pressure for new pathogenic strains is less obvious.

E. coli and Laboratory Safety

We hope the above information will enable you to ease fears about working with *E. coli* in the laboratory. The laboratory strains of *E. coli* used in these procedures are well-characterized nonpathogenic strains that lack the pathogenicity genes described above. There is no way that they can spontaneously mutate into any of the pathogenic types. No spontaneous mutation can result in the sudden appearance of many entirely new genes.

Normal safety precautions for working with microbes of any kind should be followed, of course. Even normally nonpathogenic *E. coli* can set up an infection if introduced into an open wound or an eye. Please follow all the safety precautions given in the laboratory procedures, and use aseptic technique as described in the appendixes.

Selected Reading

Hacker, J., G. Blum-Oehler, I. Muhldorfer, and H. Tschape. 1997. Pathogenicity islands of virulent bacteria: structure, function and impact on microbial evolution. *Mol. Microbiol.* 23:1089–1097. This overview of pathogenicity islands may be difficult to read, because it assumes a great deal of background knowledge.

Classroom Activities

Transformation

19

About This Activity

This laboratory is an economical and reliable protocol that teaches students one way to insert new genetic information into bacteria. Students demonstrate that *Escherichia coli* is unable to grow in the presence of the antibiotic ampicillin. They then insert a plasmid with an ampicillin resistance gene into the bacterium. After the transformation process, some of the treated *E. coli* bacteria express the antibiotic resistance gene and grow in the presence of the drug. Students observe phenotypic proof of a genotypic change in *E. coli*.

To observe the transformed phenotype, students spread some of the transformed and untransformed cells on solid media with and without antibiotic. For this procedure, you must make or obtain the appropriate agar plates and either supervise students' use of alcohol and fire or obtain sterile swabs or glass beads.

Class periods required: *1 to 2 50-minute periods or 1 90-minute period, plus observation and data recording on one or two subsequent days. There is an optional break point in the procedure.*

Introduction

Transformation occurs when cells take up free DNA molecules from the environment and express encoded information. This phenomenon is of great importance to experimental molecular biology, because it provides a means of inserting new genes into cells. Transformation was first observed by Frederick Griffith, who found that by mixing a living unencapsulated ("rough") strain of pneumococcus *(Streptococcus pneumoniae,* formerly called *Diplococcus pneumoniae)* with dead cells of an encapsulated ("smooth") strain, the rough cells were "transformed" into the smooth-colony form. In 1944, Oswald Avery, Maclyn McCarty, and Colin McLeod demonstrated that purified DNA from the smooth cells was the substance that caused the transformation.

Bacterial cells in a state that allows them to be transformed are said to be *competent*. Only some bacterial strains, such as pneumococci, are naturally competent. Naturally competent cells contain proteins dedicated to the process of transformation. Proteins on the outside of these organisms bind DNA and transport it into the cell. Internal proteins then compare the base sequence of the new DNA to the genome of the organism. If sufficient similarity (homology) is found, the new DNA is recombined into the genome and expressed. As part of the genome, it is replicated and passed on to daughter cells.

If the new DNA is not similar in sequence to the genome of the organism, it is not incorporated into the genome and is lost. In this way, naturally competent cells have access to genetic variability through transformation but do not waste their time expressing completely irrelevant genes. Natural transformation is believed to be an important mechanism of genetic exchange for a number of species important to humans, notably, *S. pneumoniae,* a causative agent of pneumonia. More information about natural gene transfer and *S. pneumoniae* is given in the *Reading Gene Transfer and the Spread of Antibiotic Resistance* in this section.

Cells that are not naturally competent can often be artificially induced to take up DNA. In 1970, two scientists developed a method that enabled *E. coli* to take up and express DNA. Their method involved treating the cells with cold calcium chloride solution, followed by heat shock. They found that rapidly growing (log-phase) cells took up foreign DNA most efficiently after these treatments. To this day, no one knows exactly why these treatments cause *E. coli* to take up DNA. Obviously, some characteristic of the cell membrane must be altered to allow DNA molecules to enter. The gene transfer method called electroporation, which is similar to transformation, uses high voltage for the same purpose.

Once inside *E. coli,* the DNA must be expressed for transformation to be achieved. This step can be a problem. *E. coli* contains enzymes called exonucleases that degrade linear pieces of DNA, starting at the ends and working inward. In most cases, new linear DNA fragments are destroyed before they even have a chance to recombine into the *E. coli* genome. This is one reason why scientists transform *E. coli* with plasmids. Plasmids are circular and are thus immune to exonuclease attack (because they have no ends).

Another reason scientists use plasmids is that plasmids contain their own origins of replication and so will be replicated and transmitted to daughter cells. It is not necessary for a plasmid to recombine into the *E. coli* genome to be maintained and expressed, so plasmids need not be similar in base sequence to the *E. coli* chromosome. Plasmids therefore make very convenient vehicles (or vectors) for introducing new genetic material into *E. coli*.

Where do plasmids come from? Plasmids have been found in many different species of bacteria and in yeast. Plasmids seem to be extra pieces of DNA; they do not contain any genes that are essential to the life of the organism. However, they often (but not always) contain genetic material that gives the cell survival advantages under certain conditions, such as genes for resistance to antibiotics, for resistance to normally poisonous heavy metals, for degradation of unusual chemicals, or for killing other bacteria.

This laboratory exercise uses the calcium chloride-heat shock method to introduce a plasmid into *E. coli*. The plasmid carries a gene for ampicillin resistance that is used to detect its presence in transformed cells. The antibiotic ampicillin prevents the formation of cell walls in *E. coli* (and many other bacteria), preventing the bacteria from producing new cells. The ampicillin resistance gene encodes an enzyme called beta-lactamase that breaks down the ampicillin molecule, allowing cells to multiply in ampicillin-containing media.

You may notice an interesting phenomenon on the ampicillin-containing plates where transformants are growing. After a period of incubation, tiny colonies begin to grow in a halo around the original ampicillin-resistant colonies of transformed bacteria on the ampicillin plates. These are called "satellite colonies." They form because the beta-lactamase enzyme produced by the transformed cells diffuses into the medium, destroying the ampicillin in the vicinity of the colony. When enough of the antibiotic has been destroyed, nontransformed ampicillin-sensitive cells that were on the plate all the time but could not form colonies because of the antibiotic begin to multiply successfully, forming the satellite colonies.

Note: This activity can also be done with other plasmids, such as pKAN, pUC18, or pBR322. The last two plasmids contain an ampicillin resistance gene and can be substituted in the procedure without any changes. pKAN contains a kanamycin resistance gene instead. If you use pKAN, you must substitute kanamycin for ampicillin in the selective media (recipes are on the CD), but otherwise the procedure is the same.

Kanamycin works differently from ampicillin; instead of interfering with cell wall construction, it blocks translation. Blocking translation is lethal for the cells. There are several different kanamycin resistance genes, but all of them work in basically the same way. They all encode enzymes that modify the kanamycin molecule so that it cannot enter the *E. coli* cell. (The different versions of kanamycin resistance enzymes cause different modifications to the drug.) The kanamycin resistance enzymes stay in the periplasmic space (the area between the inner and outer membranes of *E. coli*) and modify the drug there so that it cannot cross the inner membrane. Because the enzymes stay inside the cell, you will not see satellite colonies on kanamycin plates.

Plasmids that convey a new look to the transformants, such as blue or green color or the ability to glow in the dark, are also available. Although they may require the addition of a chemical indicator to the medium or incubation at a different temperature, the transformation procedure is the same.

Objectives

At the end of this laboratory, students should:

1. Be able to discuss how the transformation procedure enabled some *E. coli* cells to form colonies on antibiotic-containing media.
2. Have reinforced their understanding of genotype and phenotype.
3. Understand the importance of sterile technique when handling bacteria.

Materials

Equipment
- Centigrade thermometers
- 0.5- to 10-μl pipettors or other sterile small-volume measuring devices
- 100- to 1,000-μl pipettors or other sterile measuring devices
- Refrigerator (optional)
- Autoclave or pressure cooker (optional for making plates and sterilizing test tubes)

Supplies

- Overnight culture of *E. coli* streaked on plates (strain MM294 or JM101 or other suitable strain)
- 0.1 M (100 mM) calcium chloride
- Plasmid DNA (pAMP or other)
- Styrofoam cups
- Some type of nutrient agar plates (such as tryptic soy broth plates or Luria broth plates)
- Nutrient plates plus ampicillin at 100 µg/ml
- 15-ml sterile culture tubes
- Marking pencils
- Microcentrifuge tubes
- Sterile tips for 0.5- to 10-µl pipettors
- Sterile tips for 100- to 1,000-µl pipettors
- Ice
- Inoculating loops
- Lysol or bleach sterilizing solution
- Rack(s) for holding microcentrifuge tubes

Recipes for media and solutions can be found on the CD.

Resource materials

Many companies offer transformation activity kits. We have used and are familiar with the products offered by Carolina Biological Supply Company and list them here for your convenience.

- Carolina Biological Supply Colony transformation kits 21-1142, 21-1146, 21-1082, and 21-1088. These kits are all the same transformation activity but use plasmids with different marker genes, including a glow-in-the-dark gene and two different color phenotypes.
- Plasmid DNA, *E. coli,* antibiotics, agar, calcium chloride, and sterile broth can also be purchased separately.
- If you want to avoid pouring plates, try Carolina's kit 21-1078, which uses ready-to-go cards with soluble nutrients bonded to them. When students add their cell suspensions, the medium forms on the spot.

Preparation

Each laboratory team will need the following materials:

- Marking pencil
- 100- to 1,000-µl pipettor and sterile tips or other measuring device
- 0.5- to 10-µl pipettor and sterile tips or other measuring device
- Styrofoam coffee cup for a hot water bath
- Centigrade thermometer
- Cup of ice
- 500 µl of 100 mM calcium chloride solution
- 0.25 µg of plasmid DNA
- Two 1.5-ml sterile microcentrifuge tubes
- Inoculating loop
- Access to flame or other sterilizer for the loop
- 1 ml of sterile broth

For plating the transformation mixtures:

- Two nutrient medium plates (tryptic soy agar [TSA], Luria agar, or nutrient agar)
- Two nutrient medium plates plus antibiotic
- Bacterial spreader (bent glass rod)
- Alcohol in a petri dish (enough to wet the spreader)
- Access to a flame (alcohol or Bunsen burner)

Instead of the spreader, alcohol, and flame, students can also use four sterile cotton swabs and 1 ml of sterile broth or some sterile glass beads.

1. Prepare agar plates with (AMP$^+$) and without (AMP$^-$) ampicillin. You will need four plates per student group plus one plate per group for a starter plate.
2. *E. coli.* Prepare one streak plate for every two laboratory teams the day before doing this laboratory. *This procedure works best with plates incubated for 1 day at 37°C or 2 days at room temperature (20 to 25°C).* If you like, have the student teams prepare one streak plate each the day before the laboratory.
3. Plasmid DNA: The plasmid DNA should be at a concentration of about 0.05 µg/µl (equivalent to mg/ml). Most companies sell DNA in more concentrated solutions, so you may have to dilute the DNA before use. For example, if your DNA is at a concentration of 1 mg/ml (same as 1 µg/µl), you need to dilute it 20-fold. Add 5 µl of plasmid DNA to 95 µl of sterile water or Tris-EDTA buffer. This yields 100 µl of DNA at 0.05 µg/µl, enough for 20 student experiments. Store the diluted DNA in the refrigerator if you need to keep it overnight.
4. Calcium chloride (the recipe is on the CD).
5. Ensure proper treatment of microbial waste. Glass- or plasticware and any other materials that contact cells should be placed in a 10% Lysol solution (or a 20% bleach solution) overnight. Autoclaving the materials at 15 lb for 15 minutes is an alternative method but is not necessary if you use the disinfectant solutions properly. Prepare appropriately designated waste containers for your class.
6. Print a copy of the *Student Activity* for each student if needed.

Tips

1. Students should study the protocol before beginning. Success depends on accurate movement through the steps.
2. In step 4 of the procedure, it is important to use a sufficiently large cell mass. If the plates do not contain enough 3-mm colonies, instruct the students to scrape up several smaller colonies. Plates incubated for 24 hours at 37°C or 48 hours at room temperature should contain sufficiently large colonies.
3. In step 5 of the procedure, it is very important that students achieve good resuspension of their cells. No visible clumps should remain. It is also important that they resuspend the + tube first, because the cells then have time to preincubate in the calcium chloride while the cells in the − tube are being resuspended. The preincubation improves transformation efficiency.
4. After the students have heat shocked their cells and added broth to the tubes, the cells can incubate for up to several hours before being plated or inoculated into liquid media. You will achieve better results with pKAN if you let the cells sit in the broth for at least 30 minutes before plating them; the newly transformed cells need time to express the kanamycin resistance enzyme that will protect them from the lethal effect of the antibiotic. This outgrowth period is not as important with ampicillin. Cells transformed to ampicillin resistance can be plated immediately. If you need to leave the cells until the next day before plating them, let them sit at room temperature or 37°C for a while and then put them in the refrigerator overnight. If your class ends at the end of the school day, put them in the refrigerator then, but get them out in the morning and allow them to incubate at room temperature for a while before class.
5. There should be luxurious (confluent) growth on both TSA plates, some colonies on TSA-AMP⁺ plates (the transformants), and no colonies on TSA-AMP⁻ plates.

Class Discussion and Predicted Results

It is important that students think through what is happening and predict whether cells will grow on each of the four plates on which they spread cells. You can have this discussion with them before or after they do the activity, but do it before they see their results. Some questions to get the discussion started might include the following.

What is necessary for bacterial growth? (Students should be able to think of things like nutrients, suitable temperature, no inhibitors, and proper environment with respect to oxygen.)
What can inhibit bacterial growth? (Lack of necessary nutrient, wrong temperature, presence of inhibitors, etc.)
Do we need controls in this experiment? For what? Why? Which plates are controls (or how could you set up the needed controls)? (See the answer to question 2 below.)

Answers to Student Questions

1. The expected results are given above. If students get other results (such as growth on the TSA-AMP⁻ plate), they should explain.
2. The TSA-AMP⁻ plate is a control to make sure that the ampicillin inhibits the growth of nontransformed cells. There should be no growth. The TSA⁻ plate is a control to check the viability of cells that went through the transformation procedure but received no DNA. The TSA⁺ plate is also a control for cell viability using the cells to which DNA was added. The TSA-AMP⁻ plate demonstrates that ampicillin inhibits the growth of untransformed cells. The TSA-AMP⁺ plate specifically asks whether any cells were transformed by the plasmid DNA. Only cells that express the plasmid-borne ampicillin resistance gene can grow here.
3. No cells grew on the TSA-AMP⁻ plate because the medium contained ampicillin, and the bacteria are not resistant and were not given the opportunity to be transformed by an ampicillin resistance plasmid. Cells grew on the TSA⁺ and TSA⁻ plates because these cultures contained simple nutrient media that supported the growth of transformed or untransformed cells. Cells grew on the TSA-AMP⁺ plate because they had become ampicillin resistant by taking up the pAMP plasmid DNA and expressing the ampicillin resistance gene.
4. The expected outcome for this activity is that colonies will appear on the TSA-AMP plates onto which students plated the "+DNA" cells, and there will be no colonies on the TSA-AMP plates onto which students plated the "−DNA" cells. There should be growth on both plates of TSA without ampicillin. If this is the result obtained, students should state that the colonies on the TSA-AMP plates onto which the "+DNA" cells were plated are transformants. The justification is that the colonies are ampicillin resistant (they are growing in the presence of ampicillin) and they are present only in the cells to which plasmid DNA was added.
5. Answers will vary.

Technical Note

Medium plates and solutions containing antibiotics can be stored in the refrigerator for a month or two without loss of the antibiotic's activity. Antibiotics in media will lose their activity during long-term storage. *Do not keep antibiotic media from one year to another.* If you use antibiotic medium that is too old, the experiment will appear not to have worked; untransformed cells will grow on the "antibiotic" plates (because the antibiotic is no longer active). Stock solutions of antibiotics (see the CD) can be stored in the freezer for years. Make a stock and freeze it in small amounts. When you need antibiotic medium, thaw one of the small tubes and add the drug to some fresh medium.

Classroom Activities

Transformation of *Escherichia coli*

19

Introduction

Transformation is the uptake and expression of foreign DNA by bacterial cells. *Escherichia coli* is one of the many bacterial strains that do not undergo transformation naturally. However, in 1970, a process for increasing the ability of *E. coli* cells to be transformed was developed. Rapidly growing cells were suspended in cold calcium chloride and exposed to high concentrations of plasmid DNA. The cells were then briefly incubated at a relatively high temperature. After this treatment, some of the cells were found to express genes carried on the plasmid—they had been transformed. This laboratory exercise is based on that procedure.

Transformation occurs in nature—some bacteria take up DNA from their environment—and transformation was discovered in an organism that does, *Streptococcus pneumoniae. S. pneumoniae* causes pneumonia, and other members of the genus *Streptococcus* cause other diseases. Transformation is thought to be an important means by which streptococci undergo genetic change, and they are undergoing genetic changes that are important to humans—they are becoming increasingly resistant to antibiotics. More information about the spread of antibiotic resistance among streptococci and other organisms is provided in the *Reading Gene Transfer and the Spread of Antibiotic Resistance,* which follows the conjugation activity.

In this activity, you will cause *E. coli* to take up plasmid DNA. A plasmid is a small circular DNA molecule that acts like a minichromosome in a cell. The cell's DNA replication enzymes duplicate the plasmid DNA just as they duplicate the regular chromosome, so plasmid DNA molecules are inherited by both daughter cells when the bacterium divides. This means that if a single bacterial cell takes up a plasmid molecule, all its descendants will contain the molecule, too. Their genetic makeup will include the genes in the plasmid DNA.

Scientists use plasmids as convenient vehicles for introducing new genes into cells. It is easy to isolate plasmid DNA. New genes can be added to the purified plasmid DNA by using restriction enzymes and DNA ligase. Finally, the new recombinant plasmid can be introduced into host cells by transformation. When plasmids are used in this way, they are called vectors. A vector is any DNA molecule used to deliver new genes to cells.

Here is a problem for you to think about. Even the most carefully conducted transformations of *E. coli* are very inefficient. Only one cell in thousands or millions takes up plasmid DNA. How could you find the few transformed cells among the many untransformed ones?

Scientists usually use plasmids that carry marker genes when they do recombinant DNA work. Marker genes are genes that produce an easily detected phenotype, such as resistance to an antibiotic or a color change when exposed to certain conditions. The plasmid you will be using in this activity carries an antibiotic resistance marker gene. How could you take advantage of this marker to detect cells that take up your plasmid DNA?

Procedure

1. Label one sterile 1.5-ml microcentrifuge tube as "+" (with plasmid) and another "−" (without plasmid). Plasmid DNA (pAMP) will be added to the + tube; none will be added to the − tube, our control.
2. Using a 100- to 1,000-μl pipettor and a sterile tip (or other appropriate device), add 250 μl of 0.1 M calcium chloride solution to each labeled tube (+ and −).
3. Place both tubes in an ice-filled cup.
4. Transfer one or two large (3-mm) colonies from an agar plate to the + tube as follows.

   • Sterilize an inoculating loop in a Bunsen burner flame until it glows red hot. Continue to pass the lower one-third of the shaft through the flame.

- Stab the loop several times into the agar to cool the loop. *Do not touch the bacterial colonies until you have cooled the loop.* If you do, you will kill the cells.
- Scrape into the loop one or two 3-mm bacterial colonies, but be careful not to transfer any agar. Impurities in the agar can inhibit transformation and ruin your experiment.
- Immerse the filled loop in the calcium chloride solution in the + tube, and *vigorously* tap the loop against the tube's wall to dislodge the cells. Hold the tube up to the light to observe if the cell mass fell off the loop.

5. Spin the loop in the tube to suspend the cells. Hold the tube up to the light, and carefully inspect it to see that the suspension is homogeneous. It should be milky white. You cannot hurt the bacterial cells by being too vigorous. *It is important that no visible clumps of cells remain.* Reflame the loop before setting it down.

6. Return the + tube to the ice.

7. Transfer and suspend a second mass of cells to the − tube as described in steps 4 and 5.

8. Return the − tube to the ice. Incubate both tubes on ice for 10 to 15 minutes.

9. Use the 0.5- to 10-μl micropipette and a sterile tip (or other sterile device) to add 5 μl of plasmid DNA to the + tube. Tap the tube with your finger to mix it. Avoid making bubbles in the suspension or splashing the suspension up the sides of the tube.

10. Return the + tube to the ice. Incubate both tubes on ice for an additional 15 minutes.

11. When the incubation period is nearly over, prepare the heat shock bath. Go to the sink and fill your second Styrofoam cup with water that is at 43°C. Use the thermometer to determine the temperature of the water. You will need the water to be at 42 to 43°C at the end of the incubation period.

12. When the incubation on ice is over, heat shock the bacteria. *It is essential that the cells receive a sharp and distinct shock.*

- Make sure your heat shock bath is at 42 to 43°C.
- Remove both tubes from the ice bath; immerse them in the hot-water bath for exactly 90 seconds.
- Immediately return the tubes to the ice bath; let them stand for at least one additional minute.

13. Set the tubes in a rack at room temperature.

14. Use a 100- to 1,000-μl pipettor to add 250 μl of broth to each tube. Gently tap the tubes to mix them.
The cells can be plated immediately if you are looking for ampicillin resistance or left until the next day. If you are looking for kanamycin resistance, incubate the cells for at least 30 minutes before plating them or inoculating them into liquid medium. If the cells are to be left overnight, incubate them for about an hour at 37°C or longer at room temperature and then refrigerate them.

15. Clean up responsibly. Put all waste that has come in contact with bacterial cells in the designated biological-waste containers.

Detection of transformants

Obtain from your teacher the following materials:

- Two plates of tryptic soy agar (TSA) or other nutrient medium, such as Luria broth agar
- Two plates of TSA plus ampicillin (TSA-AMP)
- One spreader (bent glass rod) *or* four sterile cotton swabs and sterile broth *or* some sterile glass beads

1. Label one TSA plate and one TSA-AMP plate "+"; label the other two plates "−."

2. Use the matrix below as a checklist as you spread the + and − cells on each type of plate:

	TSA	TSA-AMP
Control cells (− tube)	100 μl	100 μl
Transformed cells (+ tube)	100 μl	100 μl

3. Use a 100- to 1,000-μl micropipette and sterile tip (or sterile transfer pipette) to add 100 μl of cell suspension from the − tube to the TSA− plate and another 100 μl to the TSA-AMP plate. Do not let the suspension sit on the plate too long before spreading it; if too much liquid is absorbed by the agar, the cells cannot be evenly distributed.
Keep your face away from the tip end while pipetting the suspension culture to avoid inhaling or splashing in your eyes any aerosol that might be created.

4. Spread the cells. The object is to spread the cells out evenly and to isolate them on the agar surface so that each cell can give rise to a distinct colony.

If you are spreading with cotton swabs, use the following procedure. Open the package at the handle end. Withdraw a swab from the package. Have a laboratory partner open the container of sterile broth and hold the lid by the outside. Without touching any other surface, dip the cotton swab into the broth.

Remove the wet swab and use it to spread cells on one plate. Repeat the procedure with a fresh swab for each additional plate.

If you are using glass beads, pour a few sterile beads onto each plate and gently shake the plate back and forth to spread your cells. Let the plates sit for a minute or two and then dump the beads into an appropriate container.

If you are using a spreader, use the following procedure.

- Dip the spreader in ethanol and then pass it through a Bunsen flame only long enough to ignite the alcohol. Remove the spreader from the flame (the spreading rod will become too hot if left in the flame).
- Allow the alcohol to burn off.
- Lift the lid from the TSA⁻ plate, but do not set the lid down on the laboratory bench.
- Cool the spreader by touching it to the agar surface away from the 100-μl cell suspension. *It is essential to cool the spreader before touching the cells.*
- Touch the spreader to the cell suspension and gently drag it back and forth across the surface of the agar.

- Rotate the plate one-quarter turn and repeat the spreading motion. Remember, the object is to spread out the cells as evenly as possible.
  Repeat step 4 with the TSA-AMP⁻ plate and spread the cell suspension.
  Use a new sterile tip to add 100 μl of cell suspension from the + tube to the TSA⁺ plate and another 100 μl to the TSA-AMP⁺ plate.
  Repeat step 4 to spread the cell suspensions on the TSA⁺ and the TSA-AMP⁺ plates.
  Reflame the spreader one last time. Let it cool in the air a minute and then put it down.
5. Place the four plates upside down in a 37°C incubator. Incubate them for 12 to 24 hours. If an incubator is not available, incubate the plates upside down at room temperature for 2 days. After this incubation, move the plates to a refrigerator to preserve them.
6. Clean up responsibly. Put all waste that has come in contact with bacterial cells in the designated biological-waste container. Wipe down your work area and wash your hands before leaving the laboratory.

Last day
Draw the four plates.

Predicted results

Plate	Source of cells	Expected results (indicate many, few, or no colonies)	Observed growth 24 h	Observed growth 48 h
TSA-AMP	+ Tube			
TSA-AMP	− Tube			
TSA	+ Tube			
TSA	− Tube			

Questions

1. Compare the observed growth to what you expected. Account for the similarity or dissimilarity in the results.

2. What was the purpose of the TSA-AMP⁻ plate? Of the TSA⁻ plate? Of the TSA⁺ plate? Of the TSA-AMP⁺ plate?

3. Explain why growth occurred or did not occur on each of the plates or in the tubes.

4. On which plate(s) do you specifically detect transformants? Justify your answer.

5. Using the results of the laboratory, discuss the relationship of genotype to phenotype.

Conjugation

About This Activity

This quick and easy activity allows students to observe conjugative transfer of an antibiotic resistance marker from one strain of *Escherichia coli* to another. It has been used successfully in high school and college classes.

Class periods required: *1 for the experiment; part of another to record results*

Introduction

In 1946, Joshua Lederburg and Edward Tatum discovered that genes could be exchanged between *E. coli* cells in a process that required direct contact between the cells and a special fertility (F) factor in the donor cell. This process was named conjugation, and it is also referred to as bacterial sexuality because of the direct donation of genetic material. Since 1946, other conjugation systems fairly similar to the F system have been discovered.

The basic form of the F factor is the F plasmid, a very large plasmid that contains several genes required for its conjugational transfer. Any cell that contains the F plasmid can synthesize all the proteins needed for conjugation (from the F genes) and so is "fertile." Fertile cells synthesize a special structure called a pilus, a tubelike appendage that protrudes from the outer membrane. The pilus binds to a recipient cell (a cell lacking F) and brings the pair together. In a process that is not completely understood, the F plasmid is then copied from a special replication/transfer origin by the fertile cell, and one copy is transferred to the recipient cell. The recipient thus becomes fertile, and the donor remains so.

Additional genes can be incorporated into the F plasmid by natural recombination processes or through laboratory manipulation. F plasmids that contain other genes are often called F′ (pronounced F prime) plasmids. These bacterial genes are then transferred during conjugation. Finally, the F plasmid occasion-

ally recombines into the *E. coli* chromosome. When these fertile cells begin conjugation, they attempt to replicate and transmit the entire circular chromosome! (They can actually transmit the whole thing if they remain paired with the recipient long enough.)

Many plasmids besides F encode proteins that allow them to be transmitted by conjugation. These conjugative plasmids also occasionally allow the transmission of nonconjugative plasmids, so virtually any plasmid can be transferred at some frequency. Some plasmids can replicate in only a few types of host and only promote conjugation between them. Others, however, have a very broad host range and promote conjugation in hundreds of bacterial species.

The medical implications of conjugation are profound. Virtually every clinically important antibiotic resistance gene is carried on a plasmid. Conjugation allows the spread of plasmids not only between different individuals of the same bacterial species, but also between species and even between genera. Conjugation has been observed to occur in the soil; on plant surfaces; in lakes, rivers, oceans, sediments, and sewage treatment plants; and inside plants, insects, chickens, mice, and humans. It is believed that conjugation is the most important route of transmission of antibiotic resistance in most disease-causing bacteria. A *Reading* on the spread of antibiotic resistance is included at the end of this chapter.

In this activity, students will observe the conjugative transmission of ampicillin resistance to a cell that is already resistant to streptomycin. The antibiotic streptomycin acts by binding to a ribosomal protein and preventing protein synthesis. In the *E. coli* strain used in this experiment, streptomycin resistance is conferred by a chromosomal mutation in the gene for that ribosomal protein. This specific mutation alters the shape of the protein so that streptomycin can no longer bind to it but leaves the protein functional. Since streptomycin resistance in this case is conferred by a chromosomal mutation, it is not normally transmitted by conjugation. Ampicillin resistance, however, is conferred by a plasmid-borne gene and can readily be transmitted on a conjugative plasmid. Conjugation experiments are usually set up so that those recipient cells that receive the desired information can be easily detected. In this experiment, the donor cells are resistant to ampicillin, and the recipient cells are resistant to streptomycin. Only the desired conjugation products will be resistant to both. Thus, by plating bacteria on both antibiotics, we permit only the conjugation products to grow.

In the *Reading* following this activity, students will learn that when scientists first experimented with antibiotic resistance, they found that resistant mutants arose at a rate of about 1 in 10 million (10^7) to 1 in 1 billion (10^9). The scientists therefore concluded that antibiotic resistance would not present a significant clinical problem. The reason the scientists reached this conclusion is that they were usually studying the rate at which organisms grown in pure culture became resistant to streptomycin. In pure culture, with no access to bacterial strains carrying plasmids with resistance genes, the only way an organism could become resistant to streptomycin was to undergo a spontaneous mutation that altered the 30S ribosomal protein in the manner described above. This spontaneous mutation does indeed occur very infrequently. Streptomycin resistance can also be conferred by a plasmid-borne gene that is transferred by conjugation. If scientists had been using strains containing the plasmid, they probably would have reached very different conclusions about the rate at which resistance could arise.

After students have read the accompanying *Reading,* try challenging advanced students to figure out how scientists could have reached the conclusion they did about antibiotic resistance without having made any "mistakes" in their experiments. Students have the information (in Table 20.2 of the *Reading*) that streptomycin resistance can be conferred either by a plasmid-borne gene or by a chromosomal mutation.

Objectives

At the end of this activity, students should be able to:

1. Define the term conjugation.
2. Describe how to set up a bacterial-conjugation experiment and explain how to detect the conjugation products.
3. Give an example of one way in which bacterial conjugation is important to people.

Materials

- Frozen or stab culture of *E. coli* strain cI (streptomycin-resistant recipient cells)
- Frozen or stab culture of *E. coli* strain cII (ampicillin-resistant donor cells)
- Tryptic soy broth, Luria broth, nutrient broth, or other medium
- Ampicillin solution
- Streptomycin solution
- Tryptic soy agar (or other medium, such Luria broth agar) plates, one per student group
- Tryptic soy agar plates plus ampicillin, one per student group
- Tryptic soy agar plates plus streptomycin, one per student group
- Tryptic soy agar plates plus ampicillin and streptomycin, one per student group
- Micropipettes and sterile tips
- Optional: other sterile inoculators (sterile bacterial loops, sterile cotton swabs, etc.)
- Small sterile containers (microcentrifuge tubes, glass test tubes, or other containers)
- Marking pens for writing on petri plates

Note: Recipes for media and antibiotic solutions are given on the CD.

Resources

The bacterial strains cI and cII are available from Carolina Biological Supply Company in the *Bacterial Conjugation Kit Refill* 21-1125A, the complete *Introductory Bacterial Conjugation Kit* 21-1125, and the *Advanced Bacterial Conjugation Kit* 21-1127.

Preparation

Ahead of time

On the day of the experiment, you will need to have freshly grown overnight cultures of cI and cII. Each student group will use 200 to 400 μl of each culture, so plan a convenient volume (10 ml is more than enough for 20 student groups). Broth media can be made and sterilized ahead of time and then stored at room temperature in a closed container until you need it. The day before you plan to do the experiment, add ampicillin to one culture flask and inoculate it with cII. Add streptomycin to the other culture flask and inoculate it with cI. Grow the cells without shaking (or with gentle shaking) at 37°C or at room temperature.

If you prefer, you can have each student group inoculate small overnight cultures of cI and cII for their own use the next day. Depending on your time and resources, you can have them streak out each strain on an appropriate antibiotic medium and then use an isolated colony to inoculate an overnight culture for the experiment.

Make and pour the plates ahead of time. If you have sterile containers, you can make one batch of agar, sterilize it, and then divide it into a total of four sterile containers while it is liquid, using sterile technique. Add antibiotics to three of the containers, and pour your plates. Alternatively, prepare four batches of agar, sterilize them all, and add antibiotics to three of them. Antibiotic-containing agar plates will keep in the refrigerator for a few weeks.

Sterilize anything else you will need.

Print the *Student Activity* from the CD if needed.

Day of experiment

Set up containers of 1:5 dilute bleach or Lysol for disposal of tips and other materials that come in contact with bacterial cells.

Each student group will need:

• One each of the four types of plates
• Three small sterile containers
• Access to a micropipette and sterile tips
• Optional: sterile inoculating tool (such as cotton swabs or inoculating loops)
• Marker
• Copies of the *Student Activity*

Procedure

1. Review the process of conjugation. Have the students look at the diagram in the *Student Activity* and suggest a method for detecting only successful conjugation products. Discuss appropriate controls. We are told that cI is streptomycin resistant and ampicillin sensitive, but what do we need to do to make sure? What about cII?

2. Have the students follow the procedure in the *Student Activity*. Since the cells must be left together for 20 minutes (approximately) to conjugate, you can use the time to continue your discussion. You may want to do question 1 as a class activity at this time. Ask the students what they think happens to their own intestinal bacteria when the doctor prescribes ampicillin to them for an ear infection. They should eventually realize that the ampicillin can kill sensitive intestinal bacteria,

leaving any resistant forms to multiply more freely. These resistant forms could then transmit their resistance genes. The spread of antibiotic resistance among disease-causing microbes is a significant and growing medical problem (see the introduction to this section).

3. Next day: have the students observe and record their results. If the cells have not grown sufficiently, incubate them for another day.

Tips

Your students can use any sterile volumetric device to measure out the cI and cII cells. The exact volume is not important (although the procedure says 100 or 200 µl); just have them mix equal volumes of the two cell cultures.

It is not important that the indicator plates be inoculated with a micropipette, either. Any sterile inoculation device will do. For example, sterile cotton swabs purchased at the pharmacy will work admirably. Have the students use a new swab for each inoculation. The cells may also be spread using bent glass rods flamed in alcohol as in the transformation activity or using sterile glass beads or sterile swabs (please refer to chapter 19 for instructions). If you spread cells on the indicator plates, you will need three plain, three ampicillin, three streptomycin, and three ampicillin-plus-streptomycin plates per student group, since only one sample can be spread on a plate.

Optional Follow-Up

Invite someone from your hospital laboratory (or a knowledgeable physician) to talk to your class about the spread of antibiotic resistance among disease-causing organisms. This topic brings together science and social issues. Much of the spread of antibiotic resistance is blamed on the heavy use of antibiotics in animal agriculture, which ties into public policy and global politics. Antibiotic resistance and its spread is an excellent topic for student research projects.

Answers to Student Questions

1. Make sure that the students fill in the table correctly and understand which cells they are looking at in each case (see Table 20.1 below). Go over the table in class, and refer to it when you discuss the observed results of the experiment.

Table 20.1 Expected answers on student copies of Table 20.1

Medium	Inoculated with:		
	cI	cII	cI + cII
Plain agar Do you expect growth? Reason?	Yes. No antibiotic in medium	Yes. No antibiotic in medium	Yes. No antibiotic in medium
AMP agar Do you expect growth? Reason?	No. cI is ampicillin sensitive.	Yes. cII is ampicillin resistant.	Yes. The mixture contains cII, which is ampicillin resistant, plus conjugation products, which are resistant to both ampicillin and streptomycin.
STREP agar Do you expect growth? Reason?	Yes. cI is streptomycin resistant.	No. cII is streptomycin sensitive.	Yes. The mixture contains cI, which is streptomycin resistant, as well as conjugation products, which are resistant to both streptomycin and ampicillin.
AMP + STREP Do you expect growth? Reason?	No. cI is ampicillin sensitive.	No. cII is streptomycin sensitive.	Yes. Only the conjugation products will grow. They are resistant to both ampicillin and streptomycin.

2. Only the conjugation products (those recipient cells that received the ampicillin resistance gene from the donor) can grow on both antibiotics. Both parental strains will fail to grow.

3. Having the recipient be streptomycin resistant makes it possible to select the conjugation products apart from the parental strains. The conjugation products have a unique combination of resistances that neither parent has. If the recipient strain were streptomycin sensitive (like the donor strain, cII), it would not be possible to tell the conjugation products (ampicillin resistant and streptomycin sensitive) from the ampicillin-resistant, streptomycin-sensitive donor cells.

Selected Readings

Amabile-Cuevas, C., M. Cardenas-Garcia, and M. Ludgar. 1995. Antibiotic resistance. *American Scientist* 83:320–329.

Levy, S. 1998. The challenge of antibiotic resistance. *Scientific American* 278 (3):46.

Miller, R. 1998. Bacterial gene swapping in nature. *Scientific American* 278(1):67.

Price, L., E. Johnson, R. Valles, and E. Silbergeld. 2005. Fluoroquinolone-resistant *Campylobacter* isolates from conventional and antibiotic-free chicken products. *Environmental Health Perspectives* 113:557–560. Content is free on line (http://www.ehponline.org/members/2005/7647/7647.html).

Conjugative Transfer of Antibiotic Resistance in *Escherichia coli*

Introduction

Conjugation refers to the transfer of genetic information from one bacterial cell to another in a process that requires contact between the cells. In conjugation, one cell always donates genetic information to the other; the donor cell is called the "male" cell or the "fertile" cell, while the recipient is the "female."

What makes a bacterial cell "fertile"? Fertile (male) cells contain special plasmids called conjugative plasmids. The first conjugative plasmid discovered is called F (for fertility). Other conjugative plasmids work in a manner similar to that of F.

The F plasmid encodes proteins that enable the host cell to donate DNA. In fact, the DNA that fertile cells donate is the F DNA. F encodes proteins that form a special structure on the outside of the bacterial cell called a pilus. The pilus is a long tubelike ap-

pendage. It binds to a recipient cell (a cell without the F plasmid) and draws the "mating pair" together. The fertile cell then copies the F plasmid and simultaneously transfers it to the recipient cell. The recipient becomes fertile, too (Figure 20.1).

Conjugative plasmids can contain extra genes, such as genes for antibiotic resistance. If you think conjugative transfer of antibiotic resistance cannot possibly have any importance in your life, think again. Public health officials are quite concerned about the rise in antibiotic resistance among disease-causing organisms. For example, *Campylobacter* is a bacterium that causes stomach cramps, diarrhea, and fever. In 1992, 1.3% of cases of *Campylobacter* infection studied in Minnesota were caused by bacteria resistant to a fairly new class of antibiotics called quinolones (ciprofloxacin [Cipro] is an example of a quinolone). By 1998, 10.2% were resistant. In Spain, about 80% of *Campylobacter* specimens isolated from humans in 1998 were resistant to quinolones. A study of

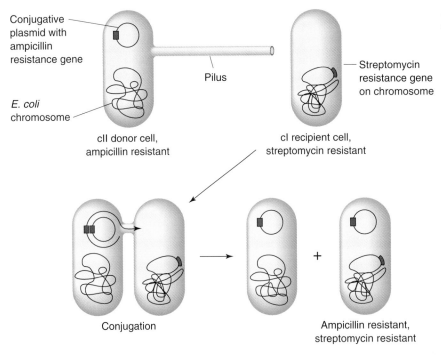

Figure 20.1 Conjugative transfer of ampicillin resistance plasmid.

Campylobacter-infected travelers returning to Finland from trips abroad found that the rate of quinolone-resistant infections increased from 40% to 60% in patients studied from 1998 to 2000 compared to patients studied from 1995 to 1997. Resistance to multiple antibiotics is now common in many organisms. Anywhere you look for information on antibiotic resistance, you will find alarming statistics.

Virtually every clinically important antibiotic resistance gene is carried on a plasmid. Conjugation allows the spread of plasmids, not only between different individuals of the same bacterial species, but also between species and even between genera. Conjugation has been observed to occur in the soil; on plant surfaces; in lakes, rivers, oceans, sediments, and sewage treatment plants; and inside plants, insects, chickens, mice, and humans. It is believed that conjugation is the most important route of transmission of antibiotic resistance in most disease-causing bacteria. A *Reading* on the spread of antibiotic resistance is included at the end of this chapter.

In the experiment you will conduct today, the fertile cell (cII) will transfer a gene for ampicillin resistance on a conjugative plasmid to the recipient cell (cI). The recipient is already resistant to streptomycin but cannot transmit this characteristic to the donor. How might you both detect recipient cells that received the ampicillin resistance gene from cII and distinguish them from their cII "parents"?

Procedure

Obtain from your instructor:

- One agar medium plate (no antibiotics)
- One ampicillin agar plate (label this plate "AMP")
- One streptomycin agar plate (label this plate "STREP")
- One ampicillin-plus-streptomycin agar plate (label this plate "AMP + STREP")
- Three small sterile containers
- Micropipette and sterile tips or other sterile measurement device
- Marker
- Optional: sterile inoculation tools

1. Label the three sterile containers "cI," "cII," and "cI + cII."
2. Following your teacher's instructions, add 100 or 200 μl of cI culture to the tubes marked "cI" and "cI + cII." Use sterile technique. Change micropipette tips between additions. Dispose of the used tips in the containers provided by your teacher.
3. Add the same volume of cII to the tubes marked "cII" and "cI + cII." Change tips between additions.

4. Let the tubes stand at room temperature for approximately 20 minutes. During this time, the cI and cII cells will conjugate.
5. While you wait, mark the back of each of your agar plates with a large Y, so that it is divided into three approximately equal areas. Label one area "cI," another "cII," and the third "cI + cII."
6. When the 20 minutes are up, use a sterile micropipette tip to pipette 5 μl of the cells in tube cI to the area of each agar plate marked "cI." Change tips between plates. Dispose of the used tips properly.
7. Repeat step 6, adding 5 μl of the cells in tube cII to the four plate areas marked "cII."
8. Repeat step 6 once more, using the cells in tube cI + cII and the plate areas marked "cI + cII." *Note:* Your instructor may want you to use an alternative method to inoculate your plates. Listen for directions.
9. Incubate the plates overnight at room temperature or 37°C.

Day 2

Record your results by showing any growth on the plates in Figure 20.2.

Figure 20.2 Diagram for recording results of conjugation experiment.

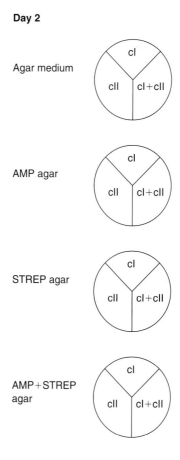

Questions

1. In Table 20.1, indicate where you expect to see growth on your plates. Write why you expect to see growth or no growth. For example, on the AMP plate, you expect to see growth of the cells from tube cII, because strain cII is supposed to be ampicillin resistant. On the other hand, you do not expect to see growth of cII on the STREP plate, because cII is streptomycin sensitive.

2. In this experiment, which cells grow on the AMP + STREP plate?

3. In this experiment, how does having the recipient cell be streptomycin resistant make it possible to detect conjugation products? What would happen in this experiment if the recipient cells were not streptomycin resistant?

Table 20.1 Expected results

Medium	Inoculated with:		
	cI	cII	cI + cII

Plain agar
  Do you expect growth?
  Reason?			

AMP agar
  Do you expect growth?
  Reason?			

STREP agar
  Do you expect growth?
  Reason?			

AMP + STREP
  Do you expect growth?
  Reason?			

Gene Transfer and the Spread of Antibiotic Resistance

The focus in this section of activities is on the transfer of genetic information between organisms other than parent and offspring. Although it is easy to think of transferring genes between organisms as a modern development in biotechnology, microorganisms routinely transfer genes among themselves in nature.

Natural gene transfer has important medical consequences. The spread of antibiotic resistance among disease-causing microbes is an example, but not the only one, of natural gene transfer that has profound effects on human health. The spread of antibiotic resistance is also an example of evolution in action. Because of the importance of this issue, we are taking time here to provide information about it.

The Rise of Antibiotic Resistance

For about 100 years after Charles Darwin published his theory of evolution in 1858, scientists believed that favorable new traits, caused by rare mutations, would spread slowly through the population as the new mutant and its progeny reproduced. Over the past 40 years, the spread of antibiotic resistance among microbes has turned this notion on its head, at least where microbes are concerned.

In the 1940s scientists discovered a microbial enzyme that neutralized the new antibiotic penicillin. They noted that this enzyme might interfere with antibiotic therapy. At the time, it was thought that such resistance would not present much of a problem because of the low rate of mutation to antibiotic resistance observed in the laboratory. The scientists making these predictions could not have been more wrong.

Scientists began to see how wrong they were in the late 1950s and early 1960s. In the 1950s, an outbreak of dysentery occurred in Japan that resisted antibiotic treatment. The culprit, a *Shigella* species, was found to be resistant to four different antibiotics. The Japanese clinicians immediately knew something strange was happening. In the laboratory, mutants resistant to the antibiotic streptomycin arose at a rate of about 1 in 10 million (10^7) to 1 in 1 billion (10^9). The odds against four such mutations arising in the same cell would be at least 1 in 10^{28} [$(10^7)^4$]! Those odds were astronomical, but to make matters even more astounding, *Escherichia coli* isolates from the same dysentery patients were resistant to the same four antibiotics. How could resistance to the same four antibiotics arise simultaneously in two different genera of bacteria? The way antibiotic resistance was arising in nature must be different from what had been observed in the earlier laboratory experiments.

Researchers began to study the phenomenon immediately, and they found that the antibiotic resistance could be transferred among *Enterobacteriaceae* (a group of gut-dwelling organisms that includes *Escherichia, Shigella, Salmonella,* and other genera). They found that organisms isolated by virtue of their resistance to one antibiotic were frequently resistant to multiple antibiotics, like the *Shigella* species described above. They found identical clusters of resistance genes in identical arrangements in different organisms. They also found genes for resistance to veterinary antibiotics in microbes isolated from humans.

To make a long story short, the microbes were sharing genes. We now know that microorganisms acquire new DNA and new traits by several methods other than mother cell-to-daughter cell transmission and spontaneous mutation: transformation, conjugation, and transduction. Transmission of traits through any of these means is often referred to as *horizontal gene transfer* to distinguish it from parent-to-offspring, or vertical, transmission. Horizontal gene transfer has been shown to take place between microorganisms of many different genera.

Examples of the Spread of Antibiotic Resistance

The bacterium *Staphylococcus aureus* can be found in certain places on the bodies of about one-third of all humans and many mammalian pets and in the environment. In healthy individuals, *S. aureus* is usually not a problem. However, it can infect wounds, burned skin areas, and immune-compromised individuals, and untreated, it can kill. Hospitals provide a prime area for transmission of *S. aureus* infection. The organism can be spread by the hands of hospital workers or by equipment, and hospitals are full of wounded, burned, and immune-compromised individuals. Day care centers are also particularly susceptible to outbreaks of "staph" infection, presumably because of the poor hygiene habits of small children.

In 1952, almost 100% of *Staphylococccus* strains were susceptible to penicillin. Because penicillin-resistant infections began to show up with increasing frequency, doctors began routinely treating staph infections with methicillin instead of penicillin in the late 1960s. The switch from penicillin to methicillin increased the cost of a typical drug treatment for a staph infection 10-fold. By 1982, fewer than 10% of all clinical infections could be cured with penicillin. Methicillin-resistant *S. aureus* infections increased in frequency during the 1980s, and by 1993, only one sure-fire treatment for *S. aureus* infections was left: vancomycin. In 1992, around 23 million Americans underwent surgery, and it is estimated that up to 920,000 of them developed postsurgical infections, the majority of which were caused by *Staphylococcus.*

Streptococcus is a family of bacteria that causes infections ranging from "strep" throat to ear infections, pneumonia, and rheumatic and scarlet fevers. In 1941, 10,000 units of penicillin per day for 4 days was enough to cure strep respiratory infections. In the early 1970s, penicillin and erythromycin were still effective against strep infections, and scarlet fever and rheumatic fever were considered diseases of the past. In 1987, a new type of streptococcal infection was described: necrotizing fasciitis. This type of infection is caused by streptococci that contain a toxin gene carried on a lysogenic bacteriophage (see chapter 21), as well as a gene for a proteinase that helps the bacterium break down host tissue. These streptococci have been dubbed "flesh-eating bacteria" by the media.

In the 1980s, a major shift occurred. New, more dangerous strains of *Streptococcus pyogenes* appeared, causing streptococcal toxic shock and necrotizing fasciitis (the "flesh-eating" disease). The streptococci that caused these new, dangerous diseases had some novel weapons in their genomes: specifically, genes for a toxin that were carried on a lysogenic bacteriophage (see chapter 21), in addition to new combinations of proteinases and DNases that help the bacterium break down host tissue. At the same time, penicillin-resistant streptococcal infections, mostly by *Streptococcus pneumoniae,* were on the rise. Streptococcal pneumonia that was resistant to multiple antibiotics broke out in Oklahoma, killing more than 15% of those infected. Streptococcal ear infections in children became increasingly resistant to penicillins, with associated hearing loss becoming an urgent problem. Multiply resistant streptococcal infections were turning up all over the world, and by the 1990s, some strains were resistant to nearly every major class of antibiotics save vancomycin.

Around 1988, vancomycin-resistant strains of *Enterococcus faecalis* and *Enterococcus faecium* emerged. Enterococci are normally harmless inhabitants of the intestinal tract, but in severely ill patients, these organisms can cause severe or even fatal infections, which are usually seen in hospital patients. A 1994 Centers for Disease Control and Prevention survey of major hospitals found that nearly 8% of all *Enterococcus* infections were vancomycin resistant. Rates of resistance among infected intensive-care patients were higher. The first cases of *Staphylococcus* infections that were partially resistant to vancomycin were reported in Japan in May 1996. The first partially resistant cases in the United States were reported in the summer of 1997 in Michigan and New Jersey. In both cases, the patients had been repeatedly treated with vancomycin for infections that were resistant to methicillin. Interestingly, the mechanism of resistance in the partially resistant *Staphylococcus* was different from the mechanism of resistance in *Enterococcus.*

These are only a few examples illustrating the spread of antibiotic resistance among microbial populations. The same patterns have been seen in many other organisms, such as *Neisseria gonorrhoeae,* the causative agent of gonorrhea.

Origins of Antibiotic Resistance Genes

The genes for antibiotic resistance work in many different ways to protect bacteria from drugs. Some encode proteins that act like tiny pumps that actively pump a particular drug out of the bacterial cell. Others encode proteins that chemically modify the antibiotic, turning it into a different chemical that does not harm the bacteria. Still others alter the antibiotic's target within the bacteria so that they are no longer affected by the drug. Table 20.2 lists some examples of specific antibiotics, how they act to kill bacteria, and how resistance genes protect bacteria from them.

Where did all of these different antibiotic resistance genes come from? To understand the answer to this question, you need to know that antibiotics themselves have been around for eons. Most families of antibiotics were not invented by humans but discovered from natural sources, such as soil bacteria. The soil bacteria themselves evolved the ability to produce an antibiotic, presumably to kill other nearby bacteria and thereby gain an advantage in their local environment.

These soil bacteria would be killed by the very antibiotic they produce unless they also evolved some means of protecting themselves. Indeed, this is precisely where at least some antibiotic resistance genes come from—they are the self-protective genes of the organisms that produce the antibiotic. A major group of antibiotic-producing soil organisms, the streptomycetes, are known to exchange genetic material with other types of bacteria via conjugation. In this way, the streptomycetes' resistance genes could be introduced into a wider population of bacteria. Over evolutionary time, the resistance genes must have been picked up by plasmids in the creation of our modern-day multiple-drug resistance plasmids. Our widespread use of antibiotics has simply applied intense selective pressure favoring the increase in frequency of these genes.

Antibiotics are now chemically synthesized, and researchers have created chemically modified versions of the naturally produced antibiotics to increase their effectiveness and overcome bacterial resistance. The beta-lactam antibiotics, penicillins and cephalosporins, are a good example. Resistance to these antibiotics is conferred by beta-lactamase, an enzyme that hydrolyzes (splits by inserting a water molecule) the beta-lactam portion of the antibiotic molecule.

Many different penicillins and cephalosporins have been developed and introduced over the years, and each time, resistance has arisen. When the beta-lactamase enzymes and genes from organisms resistant to the new drugs are compared to the "old" resistance enzymes and genes, it appears that the "new" enzymes are slightly different versions of the "old" resistance enzymes. The base differences in the "new" versions of the gene result in amino acid differences in the beta lactamase protein. The amino acid differences in the protein give the enzyme a different specificity, allowing it to work on the new form of the antibiotic. We put "old" and "new" in quotation marks because the so-called "new" gene may not be any newer than the so-called "old" gene. The use of the new antibiotic (which really is new) may simply have favored the spread of a different ("new") version of the resistance gene that was in the microbial population all along.

Table 20.2 Examples of mechanisms of action of and mechanisms of resistance to antibiotics

Antibiotic	Mechanism of action	Most common mechanism of resistance
Penicillins (such as penicillin G, ampicillin, amoxicillin)	Inhibit enzymes involved in bacterial cell wall synthesis	Plasmid-encoded beta-lactamase enzymes inactivate the drug by hydrolyzing the beta-lactam portion of the molecule.
Erythromycin	Binds to 50S ribosomal subunit and disrupts translation	Plasmid-encoded enzyme that methylates a specific A residue in the rRNA of the 50S subunit, reducing its ability to bind to the drug
Streptomycin	Binds to 30S ribosomal subunit and disrupts translation	1. Plasmid-encoded enzyme that modifies the streptomycin molecule so that it cannot enter the cell. 2. Chromosomal mutation of a protein component of the 30S subunit that alters the streptomycin binding site
Tetracycline	Binds to ribosomes and disrupts translation	Transposon-encoded protein that pumps tetracycline out of cells; usually transmitted on plasmids
Sulfonamides	Block folic acid biosynthesis by competitively inhibiting the enzyme dihydropteroate synthetase	Plasmid-encoded form of dihydropteroate synthetase enzyme is resistant to sulfonamides.
Chloramphenicol	Binds to 50S ribosomal subunit and disrupts translation	Plasmid-encoded chloramphenicol acetyltransferase enzyme modifies the drug and renders it inactive.

Evolution of Resistance Plasmids

Most antibiotic resistance genes are transferred on plasmids. These plasmids, often called R plasmids (for resistance), usually contain several different antibiotic resistance genes close together. Clusters of resistance genes form because most resistance genes are found within transposable elements. Transposable elements, or "jumping genes," are segments of DNA that can move from one DNA molecule to another in a process called transposition (see *Transposable Elements* in chapter 3).

Two features of the transposition process make it an excellent vehicle for creating clusters of resistance genes. First, the ends of the element play a key role in the transposition process, acting as molecular handles for moving whatever DNA is between them. Second, when many transposable elements move, they leave a copy of themselves behind in the original location. These two features allow transposable elements to combine into new, larger elements.

For an example, imagine one transposable element jumping into the middle of a different one. When the target transposable element transposes, it will move all the DNA between its ends, including the transposable element that jumped there. Even if the first transposable element jumps again, it will leave a copy of itself behind in the target element. Thus, the first transposon has effectively become part of the original target transposon, creating a new, larger transposon. In this way, transposable elements recombine with one another to form larger elements carrying multiple resistance genes. The clusters of resistance genes found on R plasmids are complexes of transposable elements that were presumably generated in this way.

Environmental Selection of Antibiotic-Resistant Organisms

The spread of antibiotic-resistant and multidrug-resistant organisms requires more than the formation of clusters of resistance genes on R plasmids. It also requires selection for the drug resistance abilities conferred by the plasmids. Maintenance of a plasmid within a bacterial cell requires energy. In the absence of environmental pressure favoring plasmid-bearing cells, the cell with a plasmid is at a disadvantage compared to plasmid-free cells.

If an antibiotic is added to the medium, however, only resistant cells can continue to divide. If the resistance gene is on the R plasmid, only plasmid-bearing cells will continue to grow. Imagine a culture containing 1 plasmid-bearing cell and 1,000 plasmid-free cells. If you grew this culture in antibiotic-free medium, you might eventually be unable to find any plasmid-bearing cells. However, if you added an antibiotic to which the plasmid-bearing cells were resistant and the plasmid-free cells were sensitive, the situation would be reversed. After several generations of growth, you might be unable to find any plasmid-free cells in the culture. In evolutionary terms, we say the antibiotic is creating selective pressure for drug resistance, which in this case is pressure for maintaining the plasmid.

Our widespread use of antibiotics has created an environment that heavily favors drug-resistant organisms. Imagine the microbial population in the intestinal tract of a human being who begins to take an antibiotic for an infection. The antibiotic will kill not just the intended target, but also all other susceptible microbes. Any resistant microbes will flourish in the face of reduced competition, creating a larger population with antibiotic resistance genes. In patients taking oral tetracycline, for example, the majority of *E. coli* fecal isolates carry tetracycline-resistant R plasmids within 1 week of starting drug treatment.

Unfortunately, modern hospitals are a prime environment for the spread of drug-resistant organisms and genes. In hospitals, many sick people carrying disease-causing organisms are brought together in one place and treated with a variety of antibiotics. Many patients have depressed immunity, whether from a specific disease or general poor health, and are susceptible to infection by organisms that do not cause disease in healthy people. In this environment, a resistant organism from one patient may be transferred to another patient, flourish, and then pass its resistance genes on to that patient's microbial population. Many organisms that are dangerous to severely ill hospital patients (such as *Enterococcus*) are not hazardous to people in good health, but their resistance genes can be transmitted to organisms that could infect healthy individuals.

Inappropriate uses of antibiotics have probably helped the spread of drug resistance. Antibiotics are prescribed frequently for bacterial infections, which is certainly an appropriate use. In many cases, however, antibiotics have been prescribed for viral illnesses, such as the common cold, which cannot be helped by the drugs. Additionally, in many parts of the world, antibiotics can be purchased without a prescription. In these regions, people obtain and take different antibiotics, often without knowing whether the antibiotic is appropriate or in adequate dosage. Taking antibiotics provides selective pressure for increasing the numbers of any resistant organisms that are present—organisms that might disappear from the population in the absence of selection.

Perhaps even more significant, vast amounts of antibiotics are used in farming. It was discovered years ago that feeding antibiotics to animals improved their growth. Now antibiotics are added to animal feed as a routine matter, whether the animals are sick or healthy. Feeding antibiotics to farm animals has been shown to increase the spread of drug resistance plasmids, and these plasmids are transferred from microbes within animals to those within people. Some of the first multidrug-resistant organisms ever isolated contained a gene for resistance to furazolidone, an antibiotic used primarily to treat calves!

Unfortunately, as new antibiotics came on the market, they were introduced into animal food and veterinary medicine, both in the United States and abroad. This practice promoted the spread of resistance to new antibiotics. A case in point is *Campylobacter,* which causes food poisoning. *Campylobacter* infections are usually caused by handling contaminated poultry or eating undercooked contaminated poultry. In 1995, the quinolone antibiotics (Cipro is an example) were the newest class of antibiotics available. The U.S. Food and Drug Administration (FDA) approved the use of quinolones in poultry in 1995. In 1992, a Minnesota study found that only 1.3% of all cases of *Campylobacter* infection in humans were caused by quinolone-resistant bacteria. By 1998, the rate of resistance had increased to 10.2%. Researchers who collected stool samples from *Campylobacter*-infected people before they began taking antibiotics found that the bacterium was usually already resistant to quinolones. A study published in 2003 reported that consumer-ready poultry products in the United States are commonly contaminated with *Campylobacter* and that 35% of the isolates obtained in the study were quinolone resistant.

The problem is, if anything, worse abroad, where there is less regulation of antibiotic use. In Spain, for example, quinolones were introduced into poultry and livestock farming in 1989. At that time, virtually no *Campylobacter* infections in humans were resistant to the drug. By the end of 1991, fully 30% of human *Campylobacter* infections in Spain were resistant, and in 1999, 80% were resistant. In contrast, Australia forbade the use of quinolones in poultry

production, and as of 2003, no poultry-acquired cases of quinolone-resistant *Campylobacter* infection had been reported in that country.

In response to growing concern about quinolone-resistant *Campylobacter,* major poultry producers in the United States pledged to stop adding quinolones to drinking water to treat infections in chickens in 2002. A survey of poultry products from these producers plus two antibiotic-free poultry producers conducted in 2003 found *Campylobacter* in 84% of all the samples tested. Of these isolates, 40% were resistant to quinolones. The rate of quinolone resistance was significantly lower in the chickens produced in an antibiotic-free environment. The researchers concluded that if the producers had indeed abandoned the use of quinolones, then the resistance to the antibiotic was more persistent than expected in the absence of environmental selection.

The increasing ease and speed of global travel have also promoted the spread of drug resistance among microorganisms. Genetic fingerprinting allows epidemiologists to trace the spread of particular microbial strains. For example, a drug-resistant strain of *Streptococcus pneumoniae,* dubbed 23F, emerged in a Spanish hospital in 1978. An infected person carried the microbe to Ohio, where it acquired the ability to resist another antibiotic. Other descendants of the original Spanish isolate turned up in South Africa, Hungary, and the United Kingdom, picking up additional resistances along the way. Eventually, the even more resistant microbial strain made its way back to the United States and to Spain. In 1992, it was possible to trace every known type of *S. pneumoniae* 23F back to the single mutant clone that arose in Spain—a clone that arose from a single cell.

New Antibiotics

As microbes become increasingly resistant to the antibiotics in common use, it is clear that we need to develop new antibiotics. It takes time and lots of money to develop new antibiotics, and the newer antibiotics are more expensive than the older ones. The high cost of the new drugs is a problem in the United States, but it may make the new drugs essentially unavailable to patients in developing countries.

In 1999, the U.S. FDA approved the first member of a new class of antibiotics intended to treat hospital patients infected with organisms that are resistant to all other drugs. The drug, Synercid (quinupristin/dalfopristin), represented an entirely new class of antibiotics—the first new class to appear in the U.S.

Classroom Activities

market in 10 years. Synercid can only be given intravenously, and its administration into a small vein is usually so painful that it must be delivered straight into a large vein of the chest or abdomen. The cost of Synercid treatment is estimated to be about $300 per day—more than 100-fold more expensive than some oral penicillin treatments. Since 1999, two more translation-inhibiting antibiotics have received FDA approval, Zyvox (linezolid) and Tygacil (tigecycline). Like Synercid, Tygacil must be adminstered intravenously. Zyvox can be administered intravenously or by mouth. The cost of Tygacil treatment is estimated to be around $90 per day; the cost of Zyvox is about $140 per day.

Other new antibiotics are currently under development, but we can be certain that microbes will eventually develop resistance to them and that they will share the resistance genes. Clearly, it would be beneficial if we would alter some of our practices that promote the spread of resistance. In response to recent studies on the effects of agricultural use of antibiotics, the FDA is writing regulations that may reserve some drugs exclusively for human use. Unfortunately, these rules will apply only in the United States. It is to be hoped that other countries will follow suit. Many of the problems of drug resistance can be linked to the absence of sanitary infrastructure—waste disposal and water purification systems—and lack of basic medical treatment in poorer countries. Solutions to those problems will require more resolve on the part of richer countries and more cooperation between countries than are currently in evidence.

Summary

Gene transfer is a natural phenomenon. The rapid spread of antibiotic resistance among disease-causing microorganisms is a consequence of their ability to share genes. The spread of resistance is also a clear example of evolution: genetic change followed by selection. In the case of antibiotic resistance, we humans provide the selection force through our widespread use of the drugs.

Selected Readings

To read current reports dealing with drug resistance, visit the Centers for Disease Control and Prevention website at http://www.cdc.gov. Use their browser to search their publication *Morbidity and Mortality Weekly Report (MMWR)* for articles on vancomycin resistance or other topics of interest.

Amabile-Cuevas, C. 2003. New antibiotics, new resistance. *American Scientist* 91:138.

Garrett, L. 1994. *The Coming Plague.* Penguin Books USA, Inc., New York, NY. This is a long, fascinating book about emerging diseases. Chapter 13 deals with antibiotic resistance.

Levy, S. 1998. The challenge of antibiotic resistance. *Scientific American* 278(3):46.

Miller, R. 1998. Bacterial gene swapping in nature. *Scientific American* 278(1):66.

Classroom Activities

Transduction

21

About This Activity

This is an extremely simple wet laboratory activity to show students the transmission of a bacterial antibiotic resistance gene by a bacterial virus. The laboratory is technically easy enough for students in grade 9 to perform, if they can be sufficiently prepared so that they will understand what they are doing and seeing. With the more sophisticated discussion and optional modifications offered in the instructor's introduction, it could be incorporated into college level classes in microbiology, virology, or genetics.

Class periods required: *1, plus observation and data recording on another day*

Introduction

In the process of transduction, a bacterial virus (bacteriophage) carries bacterial genes from one cell to another. Many different bacteriophages are capable of transduction; the details of transduction by any one of them depend on its life cycle.

Viruses recognize their host cells through molecular interactions on the cell and virus surfaces. When the proper interaction has occurred, changes that cause the viral genetic material to be injected into the host cell are triggered. At this point, essentially two different types of infection may take place, depending on the specific virus.

Lytic and latent viral infections

After a lytic virus infects its host cell, it takes over the cellular machinery. Normal cellular metabolism slows or stops. Cellular enzymes are diverted to make many new copies of the viral genetic material and many viral proteins. As the cell is filled with viral components, new virus particles assemble. Finally, the host cell dies, releasing the progeny viruses into the environment. Sometimes this release of progeny is a gradual process; in other cases the infected cell bursts (lyses), releasing all of the new virus particles at once.

The course of a latent infection is very different. After a latent virus injects its genetic material into the host cell, it does not hijack the cellular metabolism. Instead, a few viral proteins are produced that direct the incorporation of viral DNA into the host chromosome. If the host cell is a bacterium, the latently infected host is called a *lysogen*. Bacteriophages that set up latent infections are called lysogenic bacteriophages; lambda is the best known of this group. The viral DNA lies dormant in the host chromosome until a signal directs it to begin an active (often lytic) infection cycle. In most cases, the nature of that signal is unknown. During the active cycle, viral genetic material is reproduced and viral proteins are made. New virus particles are assembled and released.

There are variations on the lytic and latent infection themes. For example, the varicella virus infects its human host and causes the disease chicken pox. During the disease, the virus goes through active infection cycles. However, when the patient recovers, the virus is not gone. Copies of the viral DNA remain in a latent infection, integrated into the chromosomes of certain cells. This viral DNA may remain dormant for the rest of the patient's life, or it may reactivate, causing a second disease known as shingles.

Researchers sometimes use viruses to introduce foreign DNA into cells. The genomes of small viruses are easy to work with, and scientists often use them as cloning vectors. Viruses are especially important in working with mammalian cells. Researchers clone genes of interest into the viral DNA (often after removing viral genes to inactivate the virus) and then package the recombinant DNA into virus particles. The particles inject the DNA into the cells. Some viruses recombine their DNA into the mammalian genome, while others replicate in the cytoplasm in a manner similar to bacterial-plasmid replication.

Transduction

In general, transduction is a result of an error in bacteriophage reproduction. As bacteriophage reproduce, they replicate their genetic material and also produce new virus "coats." The coats (properly called capsids) themselves are assemblies of viral proteins. At some point in the construction of the new virus, the newly replicated viral genetic material must be packaged into the capsids to create a virus particle. Each bacteriophage has a mechanism for packaging its genome into a capsid. Some bacteriophages occasionally make an error and package a piece of the host cell's DNA instead. This is usually a random event, so any bacterial gene could end up inside a virus particle.

Virus particles that contain bacterial DNA instead of viral DNA are completely capable of attaching to a new host cell and injecting DNA (those functions are carried out by the protein capsid and are independent of its contents). Once inside the new cell, the bacterial DNA can recombine with the resident genome and be expressed and transmitted to future generations. When this happens, transduction has occurred.

Sometimes bacterial genes actually become part of a bacteriophage chromosome. These events happen in lysogenic infections in which the phage genetic material integrates into the host genome for a time. Apparently, when the phage DNA pops back out of the host chromosome to reproduce and package itself, it occasionally brings a piece of host DNA with it. This host DNA then acts like part of the viral genome and is replicated and transmitted along with it. Newly infected cells receive that particular fragment of bacterial DNA instead of random fragments as described above.

At one time, it was thought that transduction in nature was fairly rare. However, studies of freshwater and salt water have shown that natural waters teem with bacteriophages. Most infect only one species of bacterium, but a few can infect many different species. Natural transduction has been observed to occur in soil; on plant surfaces; in lakes, oceans, rivers, and sewage treatment plants; and inside organisms, such as shellfish and mice.

Medical importance of transduction

It appears that the transfer of genes by bacteriophages is important in creating new strains of pathogenic bacteria. We know of several specific transduction events that have great medical significance. These involve the second form of transduction described above, in which a bacterial gene has apparently become part of a bacteriophage chromosome and is transferred to other hosts by the phage. Two specific cases involve the transfer of genes for toxins associated with severe diseases, such as the gene for Shiga-like toxin thought to be responsible for hemolytic-uremic syndrome, which is found in the genome of enterohemorrhagic *Escherichia coli,* as described in the *Reading, Gene Transfer,* Escherichia coli, *and Disease,* and genes for the toxin associated with streptococcal toxic shock syndrome.

The usually fatal food poisoning botulism is a phage-borne disease. Botulism is associated with the bacterium *Clostridium botulinum,* but the disease itself is caused by only one particular protein made by that organism, the botulism toxin. (Yes, this is the same toxin that is finding widespread use in cosmetic treatments to paralyze facial muscles and thereby decrease wrinkles.) The gene for the botulism toxin is carried by a bacteriophage that infects *C. botulinum* and is thought to have been transduced from another bacterium. Without the phage, there would be no botulism.

Similarly, *Staphylococcus aureus* food poisoning is associated with a bacterial virus. The most common toxin involved in this disease is encoded by a gene on a lysogenic phage. Another disease associated with *S. aureus* is toxic shock syndrome. This syndrome, too, is caused by a toxin. The gene for the toxic shock syndrome toxin moves from one *S. aureus* strain to another via transduction.

The disease diphtheria is caused by the bacterium *Corynebacterium diphtheriae.* A single toxin produced by the bacterium is responsible for essentially all the symptoms and effects of this severe disease. The gene encoding the diphtheria toxin is carried on a bacteriophage that integrates into the *C. diphtheriae* chromosome. Those strains of *C. diphtheriae* without the lysogenic phage are fairly harmless.

Cholera is also associated with transduction. The most severe symptoms of cholera are caused by a single protein, the cholera toxin. As in diphtheria, the gene encoding this toxin is carried within the chromosome of a lysogenic phage. When the phage infects the bacterium *Vibrio cholerae,* it integrates its DNA into the bacterial chromosome and the bacterium acquires the ability to make the toxin. Interestingly, something about the conditions in the gut of a host organism (such as us) causes the phage to transfer to a new bacterial cell, spreading the toxin gene during an infection.

Demonstration of random transduction

This laboratory activity demonstrates the transmission of a gene for ampicillin resistance to *E. coli* by the bacteriophage called T4. Random transduction actually involves two infection steps: in the first infection, fragments of the host genome are packaged by mistake; in the second infection, new host cells receive the bacterial genes. In this exercise, however, you and your students will perform only the second step. The bacteriophage "lysate" (a suspension of phage particles) you will work with already contains virus particles with packaged bacterial DNA. The suspension of virus particles is called a lysate because the phage is harvested from an infected culture of *E. coli* after it lyses (breaks open) the cells.

The transducing lysate was produced by growing a special strain (see below) of bacteriophage T4 on host cells that contained a plasmid with an ampicillin resistance gene. As the bacteriophage replicated and packaged DNA, it occasionally packaged plasmid DNA by mistake. For every 10,000 virus particles produced in that infection, we can observe one transduction event involving plasmid DNA.

Transfer of the plasmid DNA to a new host will render that host resistant to ampicillin, so transductants (cells that have been successfully transduced) can be selected by plating them on ampicillin-containing media. As stated, the transducing particles represent only 0.01% of the viruses in the lysate. What about the other bacteriophage particles, which normally infect and kill *E. coli?* These normal viruses would usually make it difficult to detect the transductants, because the normal viruses would reproduce and could kill all the host cells present, transduced or not. To avoid this problem, a special strain of T4 was used to make the transducing lysate.

Amber mutations and amber suppressors

The transducing T4 strain carries mutations in two different genes that change an amino acid codon to the stop codon UAG. This particular stop codon has the casual name amber codon, acquired at the time of its discovery, and the mutations are called amber mutations. Amber mutations terminate protein synthesis and usually result in loss of function of the protein (if the amber mutation is very near the end of the protein coding sequence, it may not have much effect). In the transducing T4 strain, the two amber mutations block production of two essential proteins and are therefore lethal.

Amber mutations in bacteriophages are very useful to scientists, because in combination with special host strains, they provide an "on-off" switch for phage growth. Here is how the "on-off" system works: Amber codons (and the other two stop codons) stop translation because there are no transfer RNAs (tRNAs) that have matching anticodons. When an amber codon in messenger RNA (mRNA) reaches the ribosome, no new amino acid is added to the growing peptide chain; instead, protein synthesis is terminated. There is nothing magic about the base sequence of the terminator codons except that no matching tRNAs exist to add amino acids in response to them.

However, tRNA molecules are also encoded in DNA (to make tRNA, the DNA base sequence is simply transcribed into RNA by a special RNA polymerase). The base sequence of any tRNA can be changed by mutations in the gene encoding it. In some bacterial strains, mutations change the anticodons of tRNAs so that they can recognize what used to be one of the stop codons. In these strains, that codon is no longer a stop codon but instead encodes whatever amino acid the mutant tRNA carries. If the mutant tRNA recognizes the amber stop codon and inserts an amino acid at its position, an amber codon no longer terminates protein synthesis. A bacterial strain that makes a tRNA that recognizes the amber codon is called an amber suppressor.

In amber-suppressing strains, amber mutations are essentially erased. The effect of an amber mutation comes from its ability to terminate protein synthesis; in amber-suppressing strains, protein synthesis is not terminated. Since the mutant tRNA does not usually insert the original amino acid at the site of the amber mutation, the protein may still be impaired in function, but often it can still perform adequately. Thus, bacteriophages carrying normally lethal amber mutations may be able to reproduce in bacterial hosts that suppress the mutation through a mutant tRNA.

Our transducing T4 strain fits this description. In bacterial hosts with normal tRNAs, the phage is "dead" (unable to produce new virus particles) as a result of its amber mutations. However, in strains that make a mutant tRNA that suppresses the amber stop codon, the phage can reproduce quite well. Thus, laboratory personnel can control whether the phage reproduces through their choice of the host bacterial strain. The amber mutations in our bacteriophage T4 are an example of *conditional-lethal mutations* —under some conditions, they are lethal; under other conditions, they are not.

For this activity, the transducing lysate was produced by infection of an amber-suppressing strain carrying

the plasmid described above. The suppressing strain allowed the bacteriophage to reproduce. To detect transductants without interference, the lysate is allowed to infect *E. coli* strain B_E, which does not suppress amber codons. The T4 strain cannot grow in B_E, so no bacteriophage plaques will be observed. However, any transducing particles that were made during the first infection can transduce B_E to ampicillin resistance. Thus, when the B_E-plus-T4 mixture is plated on ampicillin media, transductants can be detected without interference from cell killing by the virus.

If you would like your students to see cell lysis of *E. coli* by T4 or to observe a conditional-lethal mutation at work, use the amber suppressor strain CR63 as described below in *Part B*. This strain allows the transducing T4 to grow by suppressing the amber mutations in the two essential genes. If you mix the lysate with CR63 cells and then spread them on a plate, you will be able to see that the phage kills the cells.

Objectives

After completing this activity, students should be able to:

1. Describe how random transduction of bacterial genes occurs.
2. State what happens in the first infection of a transduction and what happens in the second.
3. If you include the material in the lesson, explain what an amber mutation is, why it can be very harmful to the organism, what an amber-suppressing host is, and how it negates the effects of amber mutations.

Materials

Part A: demonstration of transduction

- Copies of the *Student Activity* if needed
- Overnight culture of *E. coli* strain B_E
- Transducing T4 lysate *(Never freeze a T4 lysate.)*
- Sterile tubes; student groups will need four each for mixing cells and phage (0.5-ml total volume); for dilutions, you will need at least three for the entire class (one must be able to hold 10 ml comfortably).
- Micropipettors and sterile tips for measuring 10 µl
- Micropipettors and sterile tips for measuring 100 and 500 µl
- Sterile pipettes for measuring 1 ml and 10 ml
- Tryptic soy agar-plus-ampicillin plates, four per student group (or L agar or nutrient agar; recipes are on the CD)

- Racks, beakers, or ice buckets for holding test tubes upright
- Glass spreader (a glass rod bent into an "L" shape), one per student group, *or* sterile cotton swabs
- Bunsen or alcohol burner (not necessary if you use sterile swabs)
- Beaker with methanol or ethanol large enough for the glass spreader to fit (not necessary if you use sterile swabs)

Part B: optional additional materials for detecting cell lysis

- Overnight culture of *E. coli* strain CR63
- Four sterile test tubes per student group
- Four tryptic soy agar plates without antibiotic

Resources

At this time, the only commercial source of the transducing lysate is Carolina Biological Supply Company. The bacterial strains can also be purchased from them. They also sell a kit of materials for this experiment, *Transduction of an Antibiotic Resistance Gene,* 21-1128.

Preparation

Advance preparation

Prepare the agar plates (directions are on the CD). Sterilize the test tubes and any other necessary equipment. You will also need sterile tryptic soy broth or L broth for the day of the experiment.

Day before class

Start a culture of *E. coli* strain B_E. If you plan to use it, also start CR63.

Day of class

Set up some containers with bleach diluted 1:5 or Lysol for biological waste. The T4 phage lysate is not dangerous and can be treated just like *E. coli* for disposal.

Aliquot small amounts (about 100 to 200 µl) of phage lysate into sterile tubes marked "undilute" for student use. To dilute the transducing lysate, you will need at least one (depending on the final volume of diluted lysate needed) sterile test tube containing 900 µl (0.9 ml) of sterile broth and at least one containing 1 ml of broth. They can be set up the day before and refrigerated overnight.

Make the phage dilutions (referred to in the *Student Activity*). Add 100 μl of phage to the tube containing 900 ml of sterile broth (use a sterile tip). Thump the tube gently to mix it. Mark this tube "1 to 10." With a fresh tip, add 10 μl of undiluted phage lysate to the test tube containing 1 ml of broth. Thump it gently and label it "1 to 100." As you draw the phage suspension into the pipette tip, do it gently. If phage suspension splashes up onto the plunger of the pipettor, the pipettor can become contaminated. For this reason, it is not recommended that students be allowed to prepare phage dilutions.

If you make the dilutions ahead of time, place them in the refrigerator or on ice for storage.

Dilution Mathematics

1-to-10 dilution

100 μl of phage is added to 900 μl of broth.
Initial volume of phage = 100 μl
Final volume of phage = 1,000 μl (broth plus phage)
Initial/final volume = 100 μl/1,000 μl = 1/10

1-to-100 dilution

10 μl of phage is added to 1 ml (1,000 μl) of broth.
Initial volume of phage = 10 μl
Final volume of phage = 1,010 μl
Initial/final volume = 10 μl/1,010 μl = 1/101, or approximately 1/100
If we wanted an exact 1-to-100 dilution, we would add 10 μl of phage to 990 μl of broth. (Do this if you prefer; it will not affect the results significantly.)

Procedure

Hand out the *Student Activity* if needed. Discuss the process of transduction with your class. Make sure they understand what happens. When they add the diluted virus lysate to the host cells, there will be many, many more bacterial cells than virus particles, so if a cell is infected, it will most likely be infected by only one virus. If the virus is a normal virus particle, it will inject its DNA but not be able to reproduce because of the mutations discussed above. If the virus is a transducing particle, it will inject host DNA into the cell, and that cell may be transduced. The class will be looking for host cells that have been transduced with plasmid DNA. Ask the students how transduction of pKK061 DNA might be detected. They should be able to suggest testing for ampicillin resistance.

The figure shows replicated plasmid DNA as a linear molecule. This is correct. When T4 replicates the plasmid DNA, it produces long linear molecules containing many copies of the plasmid DNA—imagine paper towels being pulled off a roll. When this linear DNA enters the new host cell, it recombines with itself and recircularizes. It is possible to isolate circular plasmid DNA from the transductants generated in this experiment. If you wish to do so, simply use the miniprep procedure from *DNA Science* (see *Selected Readings* below).

Follow the procedure in the *Student Activity*. The lysate is diluted so that a reasonable number of transductants will be produced. By testing different dilutions of the lysate, you should get at least one or two plates with a good number of transductants, and you can take the opportunity to talk to your students about dilutions if you wish.

If you wish to look at cell lysis by the bacteriophage, have the students do *Part B* of the *Student Activity*, too. The plates to which the most phage was added may look "trashed" after incubation because of the lysis of cells. If you use soft top agar to plate the cells, the results will be prettier (this procedure is not necessary; do not worry about it if you do not know how). Explain only as much about the amber mutations as you wish. You can simply tell your students that BE does not permit the T4 strain to grow while CR63 does.

Tips

You may want to have students read the *Student Activity* the day before class. Talk through the science, and have the students make up a flow chart for themselves based on the options of the activity you choose. The students could answer several of the questions before doing the experiment.

The activity can be modified for advanced students by having them make their own transducing lysates or by having them recover plasmid DNA from the transductants using the miniprep procedure in *DNA Science*.

Answers to Student Questions

Part A

1. The phage suspension was diluted so that a reasonable number of transductants could be detected on one or two plates. The starting suspension was too concentrated and would have given too many transductants.

2. There should be no colonies on the no-phage plate. This plate has ampicillin in it, and the no-phage tube has only *E. coli* B_E in it. This strain is not resistant to ampicillin and should not grow on the plate. If you see isolated colonies, there has been contamination. Either phage was accidentally added, giving some ampicillin-resistant transductants, or the culture was contaminated with ampicillin-resistant organisms. The purpose of the no-phage plate was to act as a control. This plate should verify that without the addition of transducing phage there are no ampicillin-resistant cells in the *E. coli* culture. If the no-phage plate fails to show colonies, we can conclude that the ampicillin-resistant colonies seen on the phage plates are there because of the phage.

3. The colonies growing on the plates to which phage was added are transductants (unless there were colonies on the no-phage plate; then we cannot be sure what any of the colonies are). They contain plasmid DNA transduced by the phage. There should be approximately 10 times more colonies on the undilute plate than on the 1-to-10 plate, because the undiluted phage suspension should have 10 times more phage in it than the 1-to-10 phage dilution. The 1-to-10 plate should have 10 times more colonies than the 1-to-100 plate, because 10 times more phage suspension (and therefore 10 times more phage) was added to it.

4. Five hundred colonies would be expected on the undilute plate, because 10 times more phage suspension was added to it. Five colonies would be expected on the 1-to-100 plate (see question 3).

Part B

The no-phage plate should have a smooth lawn of cells covering it. The plates to which phage were added may look anywhere from empty (all the cells were lysed) to trashy (some of the cells were lysed) to splotchy (depending on how well the phage was able to spread. They may all look similar, depending on how well the phage was able to spread.

1. The phage plates should be different from the no-phage plate. The difference is that the phage lyses CR63, and new phage may then infect neighboring cells. There may be no cells growing on the phage plates at all because of the spread of reproducing phage. On the other hand, there may be some growth visible, depending on the efficiency of spreading.

2. The purpose of the no-phage plate was to let you see what a normal plate of CR63 would look like so that you could compare it to the plates to which phage was added. The differences between the no-phage and phage plates (which were treated identically except for the phage) are attributable to the phage. The no-phage plate is a control.

Selected Readings

Micklos, D., and G. Freyer. 1990. *DNA Science: a First Course in Recombinant DNA Technology.* Cold Spring Harbor Laboratory Press, Cold Spring Harbor, NY. This very readable (though now aging) text tells the story of the development of molecular biology, explaining the science along the way. It covers all the major techniques of DNA science plus a variety of applications. The book could be used as a text for a molecular biology course for high-achieving high school seniors or for college or community college students. It also makes an easy-to-read reference book for teachers.

Miller, R. 1998. Bacterial gene swapping in nature. *Scientific American* 278(1):67.

Transduction of an Antibiotic Resistance Gene

21

Transduction is a natural method of gene transfer that occurs in bacteria. The key player in transduction is a bacterial virus, or bacteriophage (phage for short). There are many different bacteriophages that infect many different bacteria. You may have already met one of them: lambda. You will meet a different phage today, one that also infects *Escherichia coli*. It is called T4.

How does T4 transfer genetic material between *E. coli* cells? The answer is found in its life cycle. T4 infects *E. coli* by attaching to its outer membrane and injecting its DNA into the bacterial cell. Once inside the cell, the phage DNA takes over. The *E. coli* cell becomes a factory for producing many copies of the T4 genome and for producing large amounts of viral proteins. Some of these proteins help replicate the T4 DNA; others are assembled into new T4 heads and tails. After many copies of the T4 genome have been made and many new heads and tails are floating around in the cytoplasm, still other T4 proteins begin to put together new virus particles. These proteins fill the empty phage heads with T4 DNA and then attach the tails. After many new viruses are assembled, the *E. coli* cell bursts, releasing the virus progeny.

What does this have to do with transferring *E. coli* genes? The critical step is the point at which the new virus particles are assembled. Once in a while, the T4 assembly proteins make a mistake. Instead of filling a phage head with T4 DNA, they fill it with a piece of host DNA. The filled head gets a tail and becomes a virus particle fully capable of injecting the host DNA into a new bacterial cell. However, when it does so, the new host cell receives the bacterial DNA instead of dangerous viral DNA. When the new host expresses the bacterial DNA it received, it is said to have been transduced. (Remember, no virus infection took place, since the virus particle was like a "dummy warhead" filled with harmless bacterial DNA.)

Does transduction occur in nature? Absolutely. It has been observed to occur in soil, lakes, rivers, oceans, and sewage treatment plants and inside organisms, such as shellfish and mice. Studies have shown that natural waters literally teem with bacteriophage particles, and transduction plays a key role in many human diseases. One example is the usually fatal food poisoning called botulism. This disease is associated with the bacterium *Clostridium botulinum*, but the fatal syndrome is actually caused by a single protein produced by that organism. The protein is the botulism toxin. The gene for the botulism toxin is not really a *C. botulinum* gene! Instead, it is carried on a bacteriophage that infects *C. botulinum* and is thought to have been transduced from another type of bacterium. Other examples of human diseases in which transduction plays a role are *Staphylococcus aureus* food poisoning, diphtheria, and cholera.

The Activity

In this activity, you will observe the transmission of an antibiotic resistance gene by phage T4. The T4 viruses you will work with were grown on a plasmid-containing host cell, and some of the virus particles produced from that infection contain plasmid DNA. Your job is to detect some of these plasmid DNA-containing particles by their ability to transduce antibiotic-sensitive *E. coli*. Figure 21.1 summarizes the transduction process.

How do you think you could detect transductants (*E. coli* cells that have received plasmid DNA)?

Procedure

Part A: transduction of E. coli
Obtain the following materials from your instructor:
• Four sterile test tubes
• Two micropipettes, one for small volumes and the other for larger volumes
• Sterile micropipette tips for both
• Four ampicillin medium plates
• Marking pen

1. Infection of plasmid-containing host cell

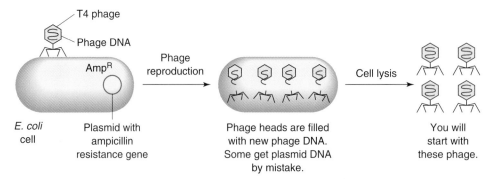

T4 phage

Phage DNA

Amp^R

E. coli cell

Plasmid with ampicillin resistance gene

Phage reproduction

Phage heads are filled with new phage DNA. Some get plasmid DNA by mistake.

Cell lysis

You will start with these phage.

2. Second infection: transfer of plasmid DNA (today's activity)

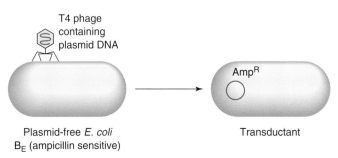

T4 phage containing plasmid DNA

Amp^R

Plasmid-free *E. coli* B_E (ampicillin sensitive)

Transductant

Figure 21.1 Transduction of plasmid DNA by bacteriophage T4.

1. Label all the plates "AMP." In addition, label one plate "no phage," one "undilute," one "1 to 10," and the last one "1 to 100."
2. Label the tubes the same way (it is not necessary to write "AMP" on them).
3. When you have labeled the tubes, use the large micropipette to place 0.5 ml of *E. coli* strain B_E cells in each tube. It is not necessary to change tips for each addition unless you touch the outside of the tube or some other nonsterile surface with the tip.
   Keep your face away from the tip end while pipetting the culture to avoid inhaling or splashing in your eyes any aerosol that might be created.

Your teacher is making 1-to-10 and 1-to-100 dilutions of the phage suspension for you to use. Why do you think the phage suspension is being diluted?

4. Take your tube of B_E cells labeled "undilute" and add 10 μl of the undiluted phage lysate to it, using a sterile tip. Thump the tube gently to mix the phage and cells. To the tube marked "1 to 10" add 10 μl of the 1-to-10 dilution. Mix it gently. Finally, to the tube marked "1 to 100," add 10 μl of the 1-to-100 dilution, using a fresh tip. Mix it gently.

5. Let all of the tubes stand at room temperature for about 15 minutes.

Two methods for spreading the phage-cell mixtures on the ampicillin plates are described below. Your instructor will tell you which option you will use.

Spreading Option 1
6. Obtain four wrapped sterile cotton swabs.
7. Take the tube labeled "no phage" and pour its contents on the appropriately labeled plate. Open one sterile cotton swab. Use the swab to spread the cells over the entire surface of the agar. Do the spreading gently so as not to tear up the agar! Place the used swab and the tube in the biological-waste container provided for you.
8. Repeat for the other tubes and plates. Use a fresh swab each time.

Spreading Option 2
6. You will need a bent glass rod, a large beaker containing alcohol, and access to a flame.
7. Take the tube labeled "no phage" and pour its contents on the appropriately labeled plate. Replace the plate lid.

8. Dip the spreader in the alcohol, and let the excess alcohol drip off. Pass the spreader through the flame only long enough to ignite the alcohol (do not let the spreader heat up in the flame). Remove the spreader from the flame, and allow the alcohol to burn off completely. Repeat.

9. Remove the lid from the plate and cool the spreader by touching it to the agar away from the cells. It is essential that the spreader not be hot, or it can kill the cells.

10. Spread the cells as evenly as possible over the surface of the agar. To do this, drag the spreader back and forth across the plate and then turn the plate one-quarter turn and repeat. Do this until you have turned the plate a full turn.

11. Repeat steps 7, 8, 9, and 10 with the other tubes and plates.

After the plates have sat for a few minutes and the liquid is absorbed, invert them. Incubate the plates overnight at 37°C or at room temperature.

What do you expect to see on the no-phage plate?

Part B: detection of cell lysis by bacteriophage T4

In *Part A,* you set up an experiment to detect transductants. The *E. coli* strain B_E used in this experiment allows the T4 strain to inject its genome but does not permit the virus to multiply inside it. Therefore, you will not see cell lysis (bursting) by the phage. Other *E. coli* strains, such as CR63, will permit the T4 strain to reproduce and will therefore be lysed by the phage. In *Part B* of the activity, you will add T4 phage to CR63 cells to observe the lethal action of the bacteriophage.

To do this optional segment of the activity, you will need:

• Four more sterile test tubes
• Four agar plates *without* antibiotic

1. On each of these tubes, write "CR63." In addition, write "no phage" on one tube, "undilute" on one tube, "1 to 10" on a third tube, and "1 to 100" on the fourth.

2. Label the agar plates in the same manner, writing "CR63" on each plate in addition to the rest of the information.

3. To each of the four test tubes, add 0.5 ml of the culture of *E. coli* strain CR63. It is not necessary to change tips unless you contaminate one by touching some nonsterile surface.

4. Take your tube of CR63 cells labeled "undilute" and add 10 µl of the undiluted phage lysate to it, using a sterile tip. Thump the tube gently to mix the phage and cells. To the tube marked "1 to 10" add 10 µl of the 1-to-10 dilution. Mix it gently. Finally, to the tube marked "1 to 100," add 10 µl of the 1-to-100 dilution, using a fresh tip. Mix it gently.

5. Let all of the tubes stand at room temperature for 5 to 15 minutes.

Spread the cells on the appropriately labeled plates as you did in *Part A.* After the liquid is absorbed, invert the plates and incubate them overnight.

Next Day
Examine your plates. Are there colonies? If the plates were incubated at room temperature, you may need to let them grow for another day. Record your results.

Results

Part A
How many colonies are on the no-phage plate?

How many colonies are on the undilute plate?

How many colonies are on the 1-to-10 plate?

How many colonies are on the 1-to-100 plate?

If there are a reasonable number on the plate, count and record the number. If there are very many colonies on the plate, divide it evenly into four quadrants by making a large "+" on the back of the plate, and count the colonies in one quadrant. Estimate the number on the plate by multiplying by four.

Part B
Record your results (drawing the plates may be helpful).

Describe the no-phage CR63 plate.

Describe the 1-to-10 plate.

Describe the 1-to-100 plate.

Questions

Part A

1. Why was the phage suspension diluted?

2. Did you see colonies on the no-phage plate? Was this what you expected? Why or why not? What was the purpose of this plate?

3. What are the colonies growing on the plates to which you added phage? Are there more of them on the undilute plate than on the 1-to-10 or 1-to-100 plate? Why?

4. Suppose you looked first at the plate to which you added 10 µl of the 1-to-10 phage dilution and counted 50 colonies. How many would you expect to see on the plate to which you added 10 µl of the undiluted lysate? On the plate to which you added 10 µl of the 1-to-100 dilution?

Part B

1. Do the plates to which phage was added look the same as the no-phage plate? If they are different from the no-phage plate, what is causing the difference? If they are the same, why do you think they are?

2. What was the purpose of the no-phage plate?

Gene Transfer in Plants by *Agrobacterium tumefaciens*

About This Activity

In this simple activity, students inoculate plants with the bacterium *Agrobacterium tumefaciens* and observe the subsequent plant tumor formation. Middle school students have performed the exercise successfully, though the science behind it is sophisticated enough for college students. *A. tumefaciens* causes tumor formation because it transfers genes to the plant. This property makes it very useful in plant genetic engineering. The lesson includes information on how *A. tumefaciens* is used in genetic engineering.

Class periods required: *1 to inoculate the plants and a few minutes in several later classes for observation*

Introduction

The common soil bacterium *Agrobacterium tumefaciens* causes crown gall disease in many dicotyledonous plants. Crown gall is a plant tumor, the result of massive proliferation of plant cells. Virulent strains of *A. tumefaciens* contain genes that cause the plant cells to divide. *A. tumefaciens* causes crown gall disease by inserting these and a few other genes into a host plant's genome. Thus, *A. tumefaciens* is a natural genetic engineer. Interestingly enough, *A. tumefaciens,* like many human molecular biologists, uses a plasmid to modify the genomes of plants. The plasmid used by *A. tumefaciens* is the tumor-inducing, or Ti, plasmid.

Chemicals secreted from freshly wounded plant tissue attract *A. tumefaciens* to the wound site. The bacterium binds to the walls of the broken, dead cells and enters them. From there, it injects a segment of Ti plasmid DNA into the adjacent living plant cells.

The injected bacterial DNA diverts the plant cell's machinery to tasks that support the growth and reproduction of *A. tumefaciens.* Living cells surrounding the wound site proliferate into a tumor, or gall, to house and protect the bacteria. These rapidly dividing tumor cells also synthesize novel chemical compounds, opines, that provide nutrients critical to the bacteria but useless to the plant. Opines consist of amino acids bound to common metabolic intermediates, such as pyruvate.

Different strains of *A. tumefaciens* induce the production of different opines. The opine synthesized by a particular tumor can be catabolized (broken down) only by the strain that caused the tumor. The strain-specific opine also promotes the conjugational transfer of tumor-inducing plasmids from that virulent strain to avirulent (plasmid-free) strains of *Agrobacterium.*

The gall induced by *A. tumefaciens* is located at soil level where the roots join the stem (the crown). Strain-specific opines produced by the gall seep into the soil, thus helping that strain of *A. tumefaciens* compete with other strains, as its plasmid is transferred at a higher rate. Are *A. tumefaciens* and the plant tumor it induces simply a plasmid's way of making more plasmids?

The Ti plasmid

The Ti plasmid encodes proteins that enable *A. tumefaciens* to infect plants and induce gall formation. The portion of the Ti plasmid that is inserted into the plant's chromosome is "transferred DNA," or T-DNA. This DNA segment contains genes that code for opine synthesis, as well as those that encourage plant cell proliferation by increasing production of plant hormones, such as cytokinins and auxins. None of the T-DNA genes are involved in the transfer process.

Located to the left of the T-DNA segment are virulence *(vir)* genes encoding proteins that control the steps in the infective process: binding of the bacterium to the plant cell wall and transfer of the T-DNA into the plant cell. Thus, *vir* genes are essential for gene transfer but are not themselves transferred. Wound juices secreted by the plant appear to induce the expression of the *vir* genes.

Genes for opine catabolism and conjugational plasmid transfer are located elsewhere on the Ti plasmid.

The disarmed Ti plasmid

The unique biology of *A. tumefaciens* provides important tools for plant genetic engineers. Before researchers can exploit the natural genetic engineering capabilities of *A. tumefaciens,* however, they much first "disarm" the Ti plasmid by removing its tumor-causing genes. The *vir* genes that control gene transfer ability remain intact. Researchers then replace the T-DNA genes with the foreign gene(s) they wish to transfer to the plant (Figure 22.1). The disarmed plasmid containing the new gene(s) is returned to *A. tumefaciens,* which is grown in culture so that many engineered bacteria are produced. Pieces of plants to be engineered are then placed in solution with the genetically altered *A. tumefaciens.*

Figure 22.1 Transfer of *Agrobacterium* T-DNA to a plant cell.

A

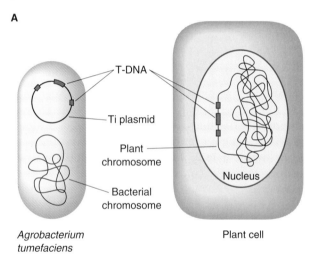

Agrobacterium tumefaciens

Plant cell

B

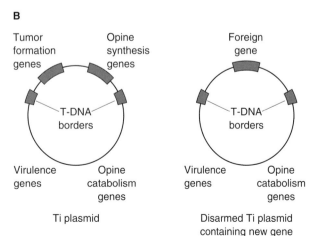

Ti plasmid

Disarmed Ti plasmid containing new gene

To date, plant biologists have used disarmed *A. tumefaciens* plasmids as vectors in the genetic transformation of a wide variety of dicots. More than 20 important agricultural plants have been successfully engineered using *A. tumefaciens.* Engineering a desirable trait into plants usually involves two different biotechnologies. Recombinant DNA technology is used to isolate the desirable gene, to prepare it for moving into its new plant host, and to move it into individual plant cells. However, engineered plant cells are not much use to most farmers and horticulturists; they need whole plants. Once individual plant cells have received new genetic information, the problem becomes how to regenerate whole plants from these cells.

The technology used to regenerate whole plants from individual cells or tissues is called plant tissue culture. Through the use of special media and hormone treatments, it is possible to regenerate a complete plant from a single cell. Plant tissue culture thus makes it possible to genetically engineer single plant cells and then re-create whole engineered plants.

A. tumefaciens is an effective vector for tobacco, petunias, tomatoes, and other dicots—plants with two seed leaves. But *A. tumefaciens* cannot penetrate the cells of most monocots. This is a major limitation, because monocots include most of the important cereal food crops, such as corn, wheat, barley, rice, and rye. This shortcoming forced scientists to look beyond *A. tumefaciens* for ways to deliver foreign DNA into monocots.

Microinjection

Microinjection is a new twist on an old idea. Biologists first used fine glass microtools in the late 1800s to dissect animal tissues. Today, scientists use compound microscopes and micromanipulators fitted with tiny glass pipettes to inject DNA directly into the nuclei of plant cells. Most plant cell walls are so tough, however, that they usually must be stripped with enzymes before DNA can be injected. These wall-less cells, called protoplasts, can then be cultured into whole plants in vitro, i.e., in glass dishes.

Electroporation

Using electroporation, scientists shock protoplasts with electricity until they become receptive to foreign DNA. A high-voltage electrical pulse temporarily opens small holes in the protoplast membranes, allowing the foreign DNA to slip through before the membranes reseal themselves. The protoplasts can then be cultured into whole plants.

Biolistics

When bacterial vectors, microinjection, electroporation, or other techniques are not suitable for a particular plant, scientists increasingly turn to biolistics, a blending of ballistics and biology. With a 0.22-caliber gene gun nicknamed the "bioblaster," plant engineers bombard cells with metallic microprojectiles coated with DNA. Scientists have already used this gun to transform yeast, algae, higher plants, and animal and human cells, and they predict it will become one of the more useful and versatile tools for plant genetic engineers.

Laboratory Activity

In this laboratory, students will wound the stems and leaves of a kalanchoe plant and inoculate them with *A. tumefaciens.* Several different kalanchoe plants are available. One, *Kalanchoe diagremontiana,* is the ornamental succulent known as the pregnant plant and is available from local florists.

Objectives

After completing this activity, students should be able to:

1. Explain the molecular biology underlying crown gall disease and tumor formation.
2. Describe how *A. tumefaciens* can be used to genetically engineer plants.

Materials

Each laboratory team will need:

- Sterile toothpicks
- Test tube rack
- Two squares of Parafilm, 1 cm by 1 cm
- One test tube containing 5 to 10 ml of sterile water
- Two small kalanchoe plants or other dicotyledonous potted plants (but not brassica)
- Access to an agar slant culture of *A. tumefaciens*
- "Lysol water" (30 ml of Lysol in 4 liters of water) or other disinfectant solution
- One inoculating needle
- Alcohol or Bunsen burner
- Rubbing alcohol and cotton swabs or balls for wiping the plant surface

Resources

- *Kalanchoe* spp. and *A. tumefaciens* can be ordered from Carolina Biological Supply Company. *A. tumefaciens* can be ordered from other biological supply houses.
- The *Plant Cancer Kit* from Carolina Biological Supply Company can be used instead of this procedure. In this kit, the experimental organism is the sunflower.

Preparation

- Print the *Student Activity* from the CD if needed.
- Make overhead transparencies of the diagrams if you wish to use them in discussing Ti plasmid biology.
- Assemble materials and prepare disinfectant.

Procedure

Have students read the introductory material in the *Student Activity.* Make sure they understand how the Ti plasmid works and how it can be used as a genetic engineering vector. Have them inoculate their plants, following the procedure given.

Teaching Tips

- Be sure to review safety procedures with your students before proceeding. Remember that rubbing alcohol is flammable; be sure that it has been removed from the work area before burners are lit. The entire procedure should be demonstrated by the instructor before the students begin.
- Show a plant infected with a crown gall to students if one is available. It may be weeks before they see galls form on their plants.
- Different species of plants can be tested for susceptibility. Monocots and plants in the brassica family (e.g., broccoli, cauliflower, and Wisconsin Fast Plants) appear resistant to infection.
- Different portions of different plants can be wounded and tested for susceptibility to infection.

Selected Readings

Chilton, Mary-Dell. 1983. A vector for introducing new genes into plants. *Scientific American* 48(6):50–59.

Ronald, P. 1997. Making rice disease-resistant. *Scientific American* 277(5):100.

Agrobacterium tumefaciens: Nature's Plant Genetic Engineer

22

Would you believe that there exists in nature a genetic engineer that inserts new genes into plant cells using a plasmid as its vector? Would you believe that this natural genetic engineer is a common inhabitant of the soil? It's all true, and the natural genetic engineer is the soil bacterium *Agrobacterium tumefaciens.*

A. tumefaciens infects certain types of plants (most dicots [plants with two seed leaves], but not monocots) at wound sites. Once in the wound, the bacterium injects a segment of its plasmid, called Ti (for tumor inducing), into the adjacent living plant cells. This piece of DNA, called T-DNA (for transferred DNA), is only one region of the plasmid. The T-DNA inserts itself into the plant's genome, where it goes to work hijacking the plant's machinery to support the reproduction of *A. tumefaciens.*

Around the infected wound, living cells proliferate into a tumor, or gall, to house and protect the bacterium. The tumor cells synthesize new chemicals that provide nourishment that is critical to the bacterium but useless to the plant. Both of these effects are driven by genes on the T-DNA.

In the late 1970s, plant scientists realized they might be able to take advantage of *A. tumefaciens'* natural genetic engineering abilities. They developed an important method for plant genetic engineering based on this organism and its Ti plasmid.

In this method, the T-DNA genes that induce tumor formation and nourish the bacterium are removed from the Ti plasmid and replaced with any gene of interest. This is accomplished through the use of restriction enzymes, DNA ligase, and other recombinant DNA techniques. The new plasmid is returned to *A. tumefaciens,* which is grown in culture so that many bacteria carrying the engineered plasmid are produced.

The plants to be engineered are then infected with the bacterium carrying the "designer" Ti plasmid. *A.*

tumefaciens injects the engineered T-DNA into the plant. Instead of receiving genes for tumor formation, the plant gets the genes inserted into the Ti plasmid by the scientist. This method has been used to genetically transform more than 20 important agricultural plants. For example, tobacco has been genetically engineered to produce medically important proteins, such as hemoglobin. In a more lighthearted experiment, scientists have also produced plants that synthesize the protein luciferase—the enzyme that causes the light of fireflies. These plants glow in the dark, though not as brightly as fireflies flash.

Today's Activity

In this laboratory, you will infect a plant with *A. tumefaciens* to start the tumor formation process. Because gall formation is slow, you will need to observe the plant for several weeks to see development of the tumor.

Procedure

1. Observe your teacher's demonstration of the proper inoculation technique.
2. Wipe down the laboratory table with Lysol water or other disinfectant.
3. Use an alcohol swab to wipe the plant surface to be wounded (stem or upper leaf). Put the alcohol away.
4. Pick up a sterile toothpick by touching only one end. Pierce the stem of the plant with the other end; the toothpick should go all the way through the stem.
5. Remove the cap from the test tube of water and flame the tube's mouth. Use a sterile inoculating needle to apply a small amount of sterile water to the wound. Flame the needle and reflame the mouth of the test tube before replacing its cap.
6. Remove the cap from the slant tube of *A. tumefaciens* and flame the tube's mouth. Flame an inoculating needle and then cool the needle. Use the sterile needle to apply a small amount of *A. tumefaciens* from the slant culture to the plant's

wound. Reflame the needle. Reflame the mouth of the test tube before replacing its cap.

7. Remove the paper from the sterile side of the Parafilm (do not touch that side). Apply the sterile side of the Parafilm to the plant's wound. It does not have to adhere tightly to the plant surface, but it should remain in place for a day or so.

8. Repeat these steps with a second plant without *A. tumefaciens.*

9. Wipe down all laboratory surfaces with disinfectant.

10. Water the plants as you normally would.

Observe the plants regularly for gall formation. Evidence of gall formation may appear in 1 to 2 weeks. If this does not occur, however, do not discard the plants. Gall formation sometimes requires several months.

Record your observations on the log provided below. Supplement the log with sketches of the plant. Label each sketch with the date you make it.

Plant Observation Log

Date	Observations
Week 1	
Week 2	
Week 3	
Week 4	
Week 5	
Week 6	
Week 7	
Week 8	

D. Molecular Biology and Genetics

This set of readings and activities ties our growing knowledge of molecular biology to classical (Mendelian) genetics. In preparing this section, we assume your students have had a typical genetics unit in their introductory biology course. Our goal is not to teach classical genetics, but rather to show how classical genetic observations can be explained and understood through molecular biology.

Humans typically are more interested in humans than in any other subject. Unfortunately, humans are a poor system for describing genetics. Compared to experimental systems with purebred strains and a wealth of genetic data, our understanding of human molecular genetics is slight. In fact, most of our knowledge of our own molecular biology comes from comparisons with animal systems. We learn what to look for by studying animals, and then we can find it in ourselves.

We have chosen as our model organism what might be the next-best thing: the dog. This section begins with a discussion of simple cases of dog coat color variations, connecting the genetic observations with the molecular explanation. The explanation draws upon knowledge of the functions of enzymes and receptor proteins and on the effects of mutations on protein function. It makes a good vehicle for review of these fundamental concepts.

After the system has been established with the dog, the following chapter shows examples in human genetics that are based on the same biochemistry. It includes reading about and discussion of questions about human genetic diseases that focus on the molecular bases of these diseases. As one of the activities, students design a DNA test for sickle-cell disease. A short reading on the molecular genetics of cancer is included.

The final chapter in this section is based on an article that first appeared in the *Smithsonian* magazine in February 2006. It tells the story of a medical geneticist's work among the Amish and Mennonite communities in Pennsylvania and brings together many real-life examples of the principles covered in the activities.

Classroom Activities

Mendelian Genetics at the Molecular Level: Dominance and Recessivity

23

About This Activity

This chapter contains a reading and thinking activity that connects the observed genetics of dominant and recessive traits to molecular biology, using coat color in Labrador retrievers as an example. Students read, answer questions, and then work problems. The example in this chapter is simple; chapter 24 builds on it to introduce another layer of complexity.

This activity assumes that students have had a typical classical-genetics unit in their introductory biology course. It does not teach genetics per se; rather, we attempt to explain the observations of classical genetics through the underlying molecular biology. Although we introduce the terms dominant, recessive, genotype, and phenotype, we do so more by way of review. It is essential that students have a grasp of the genetics of gamete formation before beginning this section. They may need to review meiosis first.

If you are familiar with the second edition of this book, you will notice that we have changed how we refer to the genes and proteins involved in pigment production. Since the second edition was written, there has been an explosion of research on the dog genome and dog molecular biology, including the publication of the dog genome sequence. In fact, the purebred dog is now considered a model organism for genetic research. Several articles have appeared in the scientific literature in which scientists have studied coat color genes in dogs. They refer to the tyrosinase-related proteins as tyrosinase-related protein 1 (TYRP1) and TYRP2 rather than TRP-1 and TRP-2. We have changed this edition to reflect that usage.

Class periods required: *1 or 2*

Introduction

This chapter focuses on the molecular biology of black/brown pigmentation in Labrador retrievers,

both because it is simple and because most students will be familiar with the animals. Many students may have or have had Labrador retrievers for pets. The molecular biology of black or brown (called chocolate in the breed) pigmentation in Labrador retrievers is based on the presence or absence of a single enzyme.

The pathway to black and brown pigments (the eumelanins) can be viewed as beginning with the amino acid tyrosine. Tyrosine is synthesized by animal cells and also enters the body through consumption and digestion of protein. In melanocytes (pigment-producing cells), tyrosine is first converted to the chemical dopaquinone, which is then converted by a second enzyme, TYRP2, to a brown form of eumelanin pigment. The brown eumelanin is converted to the black form by a third enzyme, TYRP1.

If the melanocyte cannot produce TYRP1, the brown pigment is not converted into black, and the dog is a chocolate Lab. The genetic basis for brown coat color is a nonfunctional allele for TYRP1. Chocolate is recessive to black because a single functional copy of the gene for TYRP1 leads to the production of plenty of black pigment. To be chocolate, a Lab must have two nonfunctional genes for TYRP1.

Chapter 24 builds on this simple system with an explanation of the molecular biology of the yellow Lab pigmentation. This involves a receptor protein and regulation of the synthesis of TYRP1 and TYRP2. In chapter 25, the molecular biology of pigment biosynthesis is related to two human phenotypes: albinism and the disease phenylketonuria. Both of these conditions are based in the same biochemistry. In chapter 24, we provide a summary of some research on pigmentation in dogs published since the publication of the second edition of this book that correlates perfectly with the examples in these two chapters.

Classroom Activities

Objectives

After completing this activity, students should be able to:

1. Give an explanation of why chocolate Labs are brown based on the biochemistry of pigment production.
2. Explain what the "chocolate Lab" gene is in molecular terms (a nonfunctional allele of enzyme 3, which converts a brown eumelanin to the black form).
3. Explain this statement: to say that a particular allele is dominant does not explain the observation, it only describes it.

Materials

Copies of the *Student Activity* and Worksheets 23.1 and 23.2. These worksheets are in PDF files on the CD.

Procedure

Print and distribute copies of the *Student Activity* to students if needed. They can do this activity independently or in small groups, or you can discuss your way through it as a class.

Answers to Student Questions

1. a. yy, green seeds.
   b. YY, yellow seeds.
   c. Yy, yellow seeds.
   d. Yellow seed color is dominant, because it is the seed color of the offspring.
   e. YY, yy, Yy.
2. a. The blue-colored variety cannot make enzyme Y; therefore, the blue pigment is not converted into purple, and the flowers are blue. In the blue-colored variety, the genes for enzyme Y are defective.
   b. The purple color is dominant, and the blue is recessive.
   c. Plants that contain one functional gene for enzyme Y can produce enzyme Y in their cells. Therefore, the blue pigment is converted into purple pigment, so a plant with one "blue gene" (a nonfunctional allele for enzyme Y) and one "purple gene" (a functional allele of enzyme Y) will be purple. Purple is therefore described as dominant to blue.
3. a. The red variety can produce enzyme Q and can convert the white substance into a red pigment. The white variety does not synthesize enzyme Q and so does not have red pigment.

b. Students will likely predict red to be dominant. This prediction fits with the previous examples.
c. Students will likely predict that they would see red roundbuds from the red × white cross, similar to the purple-flowered geneflower example.
e. The pink roundbuds have one functional allele for enzyme Q and one nonfunctional allele for the enzyme. Therefore, their cells may produce only half as much enzyme Q as would a cell with two functional enzyme Q alleles. It is possible that a lower level of enzyme Q would result in a lower level of red-pigment biosynthesis, and the plant might not make enough red pigment to color its flowers red, only pink.

Answers to Student Worksheets

Worksheet 23.1

Midnight can contribute the black coat color allele to his gametes. Cocoa can contribute the brown coat color allele to hers.

Table 1

Individual	Coat color genes	Coat color
Midnight	BB	Black
Cocoa	bb	Chocolate (or brown)
Puppies	Bb	Black

Table 2 Coat color and TYRP1

Individual	Coat color genes	Do melanocytes make TYRP1?	Color
Midnight	BB	Yes	Black
Cocoa	bb	No	Brown
Puppies	Bb	Yes	Black

Worksheet 23.2

Table 1 Genetics of roundbuds

Variety	Genotype	Enzyme Q produced?	Color
Red	RR	Yes	Red
White	WW (or rr)	No	White

Why might the offspring of a red × white cross be pink instead of red? (The answer is in *Answers to Student Questions,* question 3.e.)

Selected Readings

Barsh, G. 1996. The genetics of pigmentation: from fancy genes to complex traits. *Trends in Genetics* 12:299–305. This is a fairly complicated review of pigmentation genetics.

Ellgren, H. 2005. The dog has its day. *Nature* 438:745–746. A news story/comment about the first publication of a complete dog genome sequence.

O'Brien, S., and W. Murphy. 2003. A dog's breakfast? *Science* 301:1854–1855. An overview of progress in dog genetics and genomics.

Pennisi, E. 2004. Genome resources to boost canines' role in gene hunts. *Science* 304:1093–1094. A discussion of how progress in understanding the dog genome and developing molecular tools, like microsatellite markers, have made purebred dogs an invaluable resource for gene hunters.

Classroom Activities

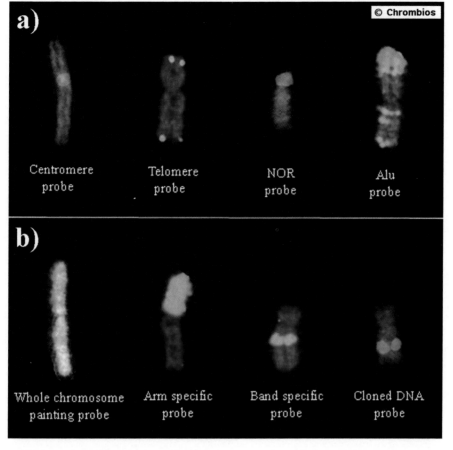

Color plate 1 Fluorescent in situ hybridization of chromosomal DNA. In these examples, chromosomes have been hybridized to various fluorescently labeled probes. NOR, DNA encoding ribosomal RNA, which is present in multiple copies; Alu, a common repeated DNA sequence interspersed throughout the human genome. Following hybridization to the fluorescent probe, the chromosomes were counterstained with either DAPI (blue) or propidium iodide (red). (Courtesy of Chrombios, GmBH.)

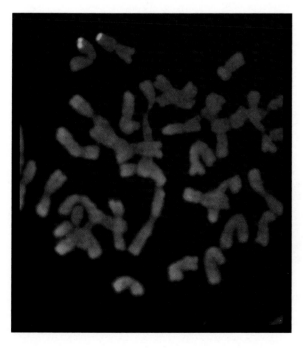

Color plate 2 Fluorescent in situ hybridization showing a DNA translocation. The green fluorescent probe in this image is painting a portion of the wheatgrass (*Thinopyrum intermedium*) that has translocated into the genome of wheat (*Triticum aestivum*). The wheat chromosomes have been counterstained with propidium iodide. (Courtesy of Joseph Anderson, Purdue University; published in M. G. Francki et al., *Genome* 40:716–722, 1997.)

Color plate 3 Cocoa and Midnight. (Cocoa's photo courtesy of Thomas A. Martin and Maria Valencik. Midnight's photo courtesy of the American Kennel Club.)

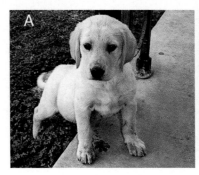

Color plate 4 Two yellow Lab puppies. What color would they have been if their cells could produce functional MC1R protein? (A) Babs lives with her human family, the D'Aubins, and her black Lab buddy, Ike, lives in Scottsdale, Arizona. As you can tell from her black nose, Babs's melanocytes make TYRP1. (B) Angel (the puppy in front) lives with the Morgan family in Cedar Grove, North Carolina. Angel's nose is brown, because her melanocytes do not make TYRP1. (Babs's photo courtesy of Mary Lynn D'Aubin. Angel's photo courtesy of Donna Morgan.)

Color plate 5 Kathleen, who has albinism, lives in Florence, Kentucky. She is active in Girl Scouts and school choir; takes tap, jazz, and ballet lessons; and competes at the state level in the Kentucky High School Speech League in acting and storytelling. Kathleen plans a career in the performing arts. (Photo courtesy of Rick Guidotti.)

Color plate 6 Humans express a wide range of hair and eye color. (Photo A courtesy of Hannah Vaughan. Photos B to I courtesy of Thomas A. Martin.)

Color plate 7 A few examples of dog coat colors and patterns. (A) Labrador retriever. (B) Kerry blue terrier. (C) Basset hound. (D) Maltese. (E) Australian Shepherd. (F) Old English sheepdog. (G) Dachshund. (H) Dalmation. (I) Doberman pinscher. (J) Beagle. (K) English setter. (L) Chow chow. (Photos A, B, D, E, F, H, I, J, K, and L courtesy of the American Kennel Club [photos by Mary Bloom]. Photos C and G courtesy of Helen Kreuzer.)

An Adventure in Dog Hair, Part I   23

Introduction

Cocoa and Midnight are Labrador retrievers. Cocoa is a brown female; Midnight is a black male. Cocoa and Midnight have puppies, and all of them are black, like Midnight. Can we use our understanding of genes and chromosomes to explain this?

On your worksheet, there are pictures of Cocoa and Midnight, with representations of their chromosomes bearing the coat color gene. Actually, dogs have 39 pairs of chromosomes, and the gene that determines black or brown color is on only one of them, chromosome 11. Since we are currently interested in the black/brown color, we will look only at that chromosome pair.

Draw the gametes that Cocoa and Midnight could produce, showing the coat color chromosome. Review the steps of meiosis if you need to. Look at the gametes the parent dogs could produce. Fertilization is a random event, so to model it, we would select one of Cocoa's gametes at random and one of Midnight's gametes at random and combine them.

What chromosome combination could the puppies have? Will all the puppies have the same combination?

As you know, it is not the entire chromosome that affects coat color; it is a specific gene on that chromosome. From now on, we will talk about the gene when we mean the gene and the chromosome when we mean the chromosome.

Fill in Table 1 on Worksheet 23.1 for Midnight, Cocoa, and the puppies, showing their coat color genes. Color the picture of the canine family.

Talk like a geneticist

You know that all of the puppies must have one of their mother's brown coat color genes and one of their father's black coat color genes, yet all the puppies are black, just like their father. Geneticists have

a way to communicate about situations like this one. They say that the black coat color is dominant over the brown coat color. Similarly, they say that the brown coat color is recessive to the black coat color.

Geneticists also have a standard way of abbreviating names of genes when they know which is dominant and which is recessive. They usually use a capital letter to refer to the dominant gene, such as B for black, and the same lowercase letter to refer to the recessive gene: b for nonblack. (It is just a coincidence that the small letter b stands for brown in this case. Geneticists would use the small b no matter what the recessive color was—red, purple, green, etc.) The name of the dominant trait determines what letter will be used.

The dominant black (B) and the recessive brown (b) are two different forms of the same gene. Geneticists call alternative forms of the same gene alleles. Thus, B (black) is an allele of the coat color gene, as is b (brown).

Here is an important point. The terms dominant and recessive do not explain anything, they simply describe an observation, like the one we made with Cocoa and Midnight's puppies. To understand why one version of a gene (black coat color) is expressed while another (brown coat color) is hidden, you have to understand what the proteins encoded by these versions of the gene are doing.

An Adventure in Dog Hair

Let's go on an adventure in dog hair color. We will use the scientific terminology to describe the pigments and the cells that produce them. Don't worry about the terms; instead, think about the logic of what is going on.

Pigmentation in dogs and other mammals (you, too) is caused by the relative amounts and types of two classes of pigment: eumelanin and phaeomelanin.

The eumelanins are the black and brown pigments, while the phaeomelanins are red and yellow. Both eumelanins and phaeomelanins are synthesized in pigment-producing cells called melanocytes.

Since we are interested in Cocoa and Midnight, who are brown and black, let's focus on the eumelanins. A schematic of the synthesis pathways is shown in Figure 23.1. First, the enzyme tyrosinase converts the amino acid tyrosine into a chemical called dopaquinone. If the enzyme called tyrosinase-related protein 2 (TYRP2) is present, it converts the dopaquinone into a version of eumelanin that has a brown color—Cocoa's pigment. If the enzyme called tyrosinase-related protein 1 (TYRP1) is present, it converts the brown version of eumelanin into the final, black-colored pigment.

Remember that enzymes are proteins. How are proteins made? What tells the cell how to put a specific protein together? If you do not remember, go back to chapter 4 and refresh your memory, because you need to know where proteins come from in order to understand the rest of this story.

What would happen to pigment production if the gene encoding the instructions for making TYRP1 was defective, so that TYRP1 could not be made? As you think about this question, assume that tyrosinase and TYRP2 are still present and functioning normally. Would tyrosine still be converted into dopaquinone? Would dopaquinone still be converted into the brown pigment? Would the brown pigment still be converted into the black pigment?

Chocolate Labs like Cocoa cannot make TYRP1. Consequently, their melanocytes produce brown pigment instead of black pigment (Figure 23.2). Look at the photo of Cocoa in Color plate 3. What color is her nose? Her lips? Because chocolate Labs do not make TYRP1, they cannot make black pigment, period. Their noses and lips are brown instead of black, too.

The b allele is actually a nonfunctional allele of the gene for TYRP1. Because Cocoa cannot make TYRP1, she is brown. Midnight's B allele is the functional allele of the gene for TYRP1. Because Midnight's melanocytes make TYRP1, they can convert the brown pigment into the black form, and Midnight is black.

Now, let's think about the situation with Cocoa and Midnight's puppies. They have one recessive allele (b) from Cocoa, the defective gene for TYRP1. They have one dominant allele (B) from Midnight, the functional form of the gene for TYRP1. What is going to happen in the puppies' melanocytes?

Let's think about what we know. We know that all the puppies' cells, including their melanocytes, have one functional allele for TYRP1 (the B allele) and one nonfunctional allele for TYRP1 (the b allele). Will the melanocytes be able to make TYRP1 from the instructions on the chromosome that came from Cocoa? What about the instructions on the chromosome that came from Midnight? What color are the puppies? Do the puppies' melanocytes make TYRP1?

Figure 23.1 Synthesis of black and brown pigments (eumelanins) in melanocytes. First, tyrosinase converts tyrosine to dopaquinone. Next, TYRP2 converts dopaquinone to brown pigment. Finally, TYRP1 converts the brown pigment into black pigment.

Figure 23.2 Pigment synthesis in Cocoa's melanocytes. First, tyrosinase converts tyrosine to dopaquinone. Next, TYRP2 converts dopaquinone to brown pigment. There is no TYRP1 to convert the brown pigment into black pigment. What effect does this lack of TYRP1 have on Cocoa's coloration?

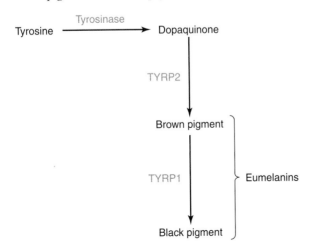

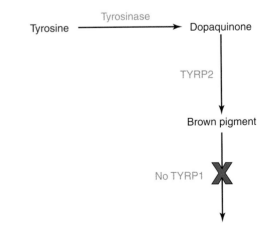

Fill in Table 2, titled "Coat color and TYRP1," on Worksheet 23.1. Use the geneticist's abbreviations B and b when showing the genetic makeup of the dogs.

Talk like a geneticist

Now it's time to introduce two useful terms you probably already know: genotype and phenotype.

The genotype is the nature of an individual's genes. Cocoa's genotype is bb, or brown/brown. The phenotype is the outward expression of the genotype. Cocoa's phenotype is brown. Midnight's genotype is BB, or black/black. What is his phenotype? The puppies' genotype is Bb, or black/brown. What is their phenotype?

Questions

1. A gardener has two strains of peas. One always produces yellow seeds, and the other always produces green seeds. The gardener crosses the two strains of seeds. All of the new plants have yellow seeds.

   a. What is the genotype of the green-seed parent plant? What is its phenotype?

   b. What is the genotype of the yellow-seed parent plant? What is its phenotype?

   c. What is the genotype of their offspring? What is the phenotype?

   d. Is one of the seed color genes dominant over the other? How do you know?

   e. What would be the geneticist's abbreviation for the genotype of the parent yellow plant? The parent green plant? The offspring yellow plants?

2. In the imaginary geneflower, a starting substance is converted by enzyme X into a blue pigment, which is converted by enzyme Y into a purple pigment (Figure 23.3). Most geneflowers

have purple flowers, but there is a variety that always produces blue flowers.

   a. Using what you know about the synthesis pathway for blue and purple pigments in geneflowers, give an explanation for the blue-colored variety.

   b. When gardeners cross the purple-flowered geneflower with the blue-flowered geneflower, they always get purple-flowered geneflowers. Which flower color is dominant? Which is recessive?

   c. Use your knowledge of pigment synthesis in geneflowers to explain why one color is dominant over the other.

3. In a second imaginary flower, the roundbud, red pigment is synthesized from a white precursor by enzyme Q (Figure 23.4). It was believed that all roundbuds were red until a knowledgeable, observant teenager discovered a meadow full of white-flowered roundbuds while hiking with the Genetics Club.

   a. Use your knowledge of the synthesis pathway for red pigment to explain the phenotype of the white-flowered variety.

Figure 23.3 Pigment biosynthesis in the geneflower. Enzyme X converts a colorless precursor into a blue pigment. Next, enzyme Y converts the blue pigment into a purple pigment.

Figure 23.4 Pigment biosynthesis in the roundbud. Enzyme Q converts a white precursor into a red pigment.

b. Would you predict that red color would be dominant or recessive to white?

c. If you crossed the white roundbud with the red roundbud, what would you expect to see, and why?

d. Fill in the table on Worksheet 23.2 for the red and white roundbuds.

The teenaged discoverer of the white roundbud took some of the plants back for the Genetics Club to experiment with. They found that when they crossed white-flowered plants with white-flowered plants, they always got white-flowered plants. When they crossed the white-flowered plants with the red-flowered plants, however, they got a surprise. The offspring plants had neither red nor white flowers but instead had pink flowers.

e. Assume your predictions about the genotype for making enzyme Q in red and white roundbuds and in their offspring (which you probably expected to be red) are all correct and that nothing is going on that you do not know about. Using your knowledge of enzyme Q and its function, can you propose an explanation for why the offspring have pink flowers?

Classroom Activities

Talk like a geneticist

The descriptive term for what the Genetics Club observed with the roundbuds is codominance. When the red- and white-flowered varieties were crossed, neither color was dominant over the other. Instead, the phenotype of the offspring was intermediate between the two parents. In cases of codominance, geneticists use capital letters for both traits. In this case, the red-flowered roundbud is RR, the white-flowered roundbud is WW, and the pink-flowered roundbud is RW.

Remember, codominant is a term that describes an observation. It does not explain why the observed phenotype occurs. In roundbuds, the explanation involves the amount of enzyme Q synthesized in cells with two functional copies of the gene versus in cells with only one functional copy. In other cases, such as human blood types, there is a different explanation for the observation of codominance.

Blood groups A and B are caused by the presence of specific molecules on the surfaces of red blood cells. If an individual has the "A" allele, he has the A molecule on his cells. If an individual has the "B" allele, he has the B molecule on his cells. If an individual has both an A and a B allele, that individual has both molecules on the surfaces of his blood cells and belongs to blood group AB. In this case, the observed codominance is caused by the expression of two different forms of the same gene.

342 • **Classroom Activities: Molecular Biology and Genetics**

Mendelian Genetics at the Molecular Level: An Example of Epistasis

24

About This Activity

This is a second reading and problem-solving activity that connects observational genetics to molecular biology. This installment explains the phenomenon of yellow Labs. It builds on the concept of receptor proteins introduced in chapter 4.

If you are familiar with the second edition of this book, you will note that we have changed the way we refer to the hormone from melanocyte-stimulating hormone to melanocortin 1 (MC1). Since the second edition of the book was prepared, there have been several new publications about the effect of alleles of the melanocortin 1 receptor (MC1R) on animal pigmentation, and they mostly refer to the hormone using that terminology, although the term melanocyte-stimulating hormone still appears. At the end of the introduction, we provide a brief summary of some relevant research into dog coat color published since the second edition of the book was written.

Class periods required: 1

Introduction

The black/brown coloration of Labrador retrievers is determined by whether their melanocytes can synthesize TYRP1 and thus convert brown eumelanin to black eumelanin. The molecular biology of yellow Labs is very different. The yellow coat color is based on the ability of follicle melanocytes to receive a hormone signal that induces production of TYRP2 and TYRP1. These enzymes, which convert dopaquinone to the brown and black pigments, respectively, are not constitutively produced in melanocytes.

Instead, the melanocytes receive a signal from the hormone MC1 (also referred to as melanocyte-stimulating hormone), which causes the cell to produce the enzymes. The hormone signal is transmitted when the hormone binds to a membrane receptor protein, MC1R. If the binding does not occur, no signal will be transmitted, and TYRP2 and TYRP1 will not be produced.

Lack of signaling is what occurs in yellow Labs. The "gene for yellow coat color" is a nonfunctional allele for the MC1R protein. The melanocytes of yellow Labs cannot receive the hormone signal to produce TYRP2 and TYRP1. In their follicle melanocytes, tyrosine is converted to dopaquinone, but the eumelanin synthesis pathway stops there (Figure 24.1). The dopaquinone is converted into another pigment class: phaeomelanin. In yellow Labs, the phaeomelanin is yellow; thus, the hair of the animals is yellow. In chapter 25, students will learn that human phaeomelanin is thought to be responsible for the fire-red hair color seen in some individuals.

Chapters 23 and 24 together provide an excellent opportunity to review the function of genes (directing protein synthesis) and the various roles played by proteins in the cell.

Research Updates

Since the publication of the second edition of this book, there have been some interesting scientific studies that relate to the material in these two genetics chapters. In 2000, a research group in the Netherlands published a study in which they compared the gene sequences of MC1R in black and yellow Labrador retrievers. They found that the yellow Labs were homozygous for a C-to-T mutation that changed an arginine codon to a stop codon and eliminated the 10 carboxy-terminal amino acids from the receptor.

In a study published in 2002, researchers at the University of Saskatchewan genotyped 43 brown dogs (including brown and white), 34 black dogs (including tricolor, black-and-tan, and black and white), and 23 dogs that were described as yellow, red, apricot, gold, or orange. They took skin biopsies and used PCR to determine the protein-coding sequence

Classroom Activities

of the TYRP1 gene. They found two common variants that could account for brown coat color—a premature stop codon in exon 5 and the deletion of a proline, also in exon 5—in addition to one less common variant. The PCR sequencing method they used would not necessarily allow the researchers to know which chromosome (maternal or paternal) carried a given variant, but it did allow them to determine which alleles were present. All 43 of the brown dogs carried two or more of the variant alleles, suggesting that they were homozygous for inactive versions of TYRP1. In contrast, none of the black dogs carried two of the inactivating mutations. Ten of the black dogs carried one variant, so they were heterozygotes with one inactive allele.

The yellow, red, apricot, gold, or orange dogs were all homozygous for the same C-to-T mutation in the MCR1 gene that the Dutch researchers had identified in yellow Labs. This group of 23 dogs contained 13 black-nosed dogs and 10 brown-nosed dogs. The researchers examined their TYRP1 alleles and found that the dogs with brown noses had the same variants in TYRP1 as did the brown dogs.

In natural populations, as opposed to purebred dogs, the association between MC1R and pigmentation is not as absolute, but the association is still strong. Many studies have associated variants in the human MC1R gene (called both MSHR and MC1R in humans) with red hair and also with increased risk of melanoma. This research is summarized in the computer database *Online Mendelian Inheritance in Man* (see chapter 34, *Bioinformatics;* if you are interested, search the database for MC1R).

In a study published in 2006, researchers at the University of California—San Diego and the University of South Carolina genotyped members of natural populations of the mouse *Peromyscus polionotus.* Most populations of this species are grayish brown. However, the population living on Santa Rosa Island off Florida's Gulf coast is much lighter. The ground these animals live on is light-colored sand, and their light pigmentation makes them better able to hide from predators.

Researchers sequenced the MC1R gene in the beach mice versus the mainland mice and found that the beach mice had a single-nucleotide difference in the gene that changed an arginine (a positively charged amino acid) to a cysteine (an uncharged amino acid) in the intracellular domain of the protein. The scientists tested cultured cells engineered to express either the mainland version of the receptor or the beach version of the receptor and found that the beach version had a lower binding affinity for the hormone and also a decreased intracellular response. The MC1R receptor is not the whole story in beach mouse pigmentation, though. Crosses between the beach mice and mainland mice showed that color did not segregate perfectly with the MC1R alleles, and a second light population on Florida's Atlantic coast had normal MC1R alleles.

Objectives

At the end of this activity, students should be able to:

1. Explain why eliminating the MC1R protein causes Labrador retrievers to be yellow.
2. Look at a photo of a yellow Lab, tell what color its fur would have been if the MC1R protein were functioning, and give the dog's genotype at the black/brown locus.
3. Predict genotypes and phenotypes of offspring from crosses with black, chocolate, and/or yellow Labs.

Materials

Student Activity and worksheets

Preparation

Make copies as needed.

Answers to Student Questions

1. The genotype of both parent dogs was Bb. One-fourth of their offspring should be BB (black), one-half Bb (black), and one-fourth bb (brown).
2. These parent dogs had the genotype Rr at the MC1R gene locus. One functional copy of the gene is required to make a functional MC1R protein. When these dogs are bred, one-fourth of their offspring should be RR (black), one-half Rr (black), and one-fourth rr (yellow). Their lips and noses are black because they all have at least one B gene at the black/brown locus, which means they can produce TYRP1 and thus black pigment. The R gene affects only melanocytes in hair follicles.
3. The genotypes and phenotypes of the offspring are predicted as follows: one-fourth BbRr (black Lab), one-fourth bbRr (chocolate Lab), one-fourth Bbrr (yellow Lab with black lips), and one-fourth bbrr (yellow Lab with brown lips).

Selected Readings

Everts, R. E., J. Rothuizen, and B. A. van Oost. 2000. Identification of a premature stop codon in the melanocyte-stimulating hormone receptor gene (MC1R) in Labrador and Golden retrievers with yellow coat colour. *Animal Genetics* 31:194–199.

Hoekstra, H., R. Hirschmann, R. Bundey, P. Insel, and J. Crossland. 2006. A single amino acid mutation contributes to adaptive beach mouse color pattern. *Science* 313:101–104.

Schmutz, S. M., T. G. Berryere, and A. D. Goldfinch. 2002. TYRP1 and MC1R genotypes and their effects on coat color in dogs. *Mammalian Genome* 13:380–387.

Answers to Worksheet 24.1

Male gamete genotype:	BR
Female gamete genotype:	br
Puppies' genotype:	BbRr
Puppies' phenotype:	Black Lab

Answers to Worksheet 24.2

Genotype	No. of occurrences in Punnett square	Does this genotype synthesize:		Phenotype	
		MC1R?	TYRP1?	Color of hair	Color of lips
BBRR	1	Yes	Yes	Black	Black
BBRr	2	Yes	Yes	Black	Black
BbRr	4	Yes	Yes	Black	Black
BbRR	2	Yes	Yes	Black	Black
Bbrr	2	No	Yes	Yellow	Black
BBrr	1	No	Yes	Yellow	Black
bbRR	1	Yes	No	Brown	Brown
bbRr	2	Yes	No	Brown	Brown
bbrr	1	No	No	Yellow	Brown

Summary

Phenotype	No. of occurrences in Punnett square
Black Lab	9
Chocolate Lab	3
Yellow Lab, black lips	3
Yellow Lab, brown lips	1

An Adventure in Dog Hair, Part II: Yellow Labs

Introduction

Many of you have probably already wondered about the third possible coat color of Labrador retrievers, the yellow Lab. The yellow color is determined by an altogether different pair of genes on a different chromosome from the TYRP1 gene. To understand how the yellow color is produced, we must resume our adventure in dog hair.

When we last left our Great Dog Hair Adventure, we had learned that pigments are synthesized in cells called melanocytes. In melanocytes, tyrosine is converted to a chemical called dopaquinone by tyrosinase. Then, TYRP2 converts the dopaquinone to brown pigment, and TYRP1 converts the brown pigment to black. Chocolate Labs are brown because they cannot make TYRP1.

The story with yellow Labs starts from there. In melanocytes, the synthesis of TYRP2 and TYRP1 is regulated by hormones. In particular, a hormone called melanocortin 1 (MC1; sometimes referred to as melanocyte-stimulating hormone) travels through the bloodstream and binds to the surfaces of melanocytes found in hair follicles. MC1 binds to these melanocytes because the melanocytes have a receptor protein (MC1 receptor [MC1R]) on their surfaces that exactly fits the shape of the MC1 molecule.

This receptor protein's job is to receive the hormone's signal and transmit it to the inside of the cell. When MC1 binds to the receptor on the surface of a hair follicle melanocyte, the receptor protein changes shape. The change in shape generates a signal to the inside of the cell, telling it to produce TYRP2 and TYRP1 (Figure 24.1).

What happens if the melanocytes do not get the signal to produce TYRP2 and TYRP1? Tyrosinase is still there, converting tyrosine into dopaquinone. If TYRP2 and TYRP1 are not present in the melanocyte, the dopaquinone is converted into something else: phaeomelanin (Figure 24.2).

The phaeomelanins are the family of red and yellow pigments we mentioned briefly at the beginning of the dog hair adventure and have not mentioned since. The exact color of the phaeomelanin depends on the enzymes available for its synthesis, as is the case with the color of the eumelanin (black or brown). In Labrador retrievers, the phaeomelanin is yellow.

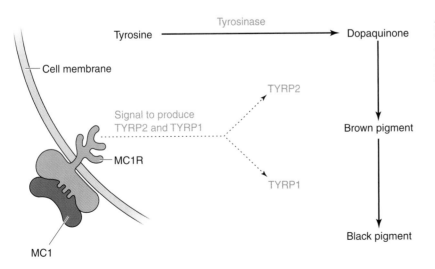

Figure 24.1 In black and chocolate Labs, MC1 binds to its receptor, MC1R, signaling the melanocyte to produce TYRP2 and TYRP1. The melanocyte can then synthesize black or brown pigment.

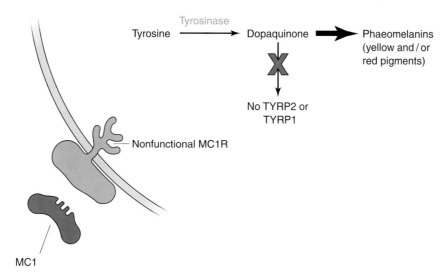

Figure 24.2 In yellow Labs, the MC1R is nonfunctional. MC1 cannot bind to it, and the hair follicle melanocytes never receive the signal to synthesize TYRP2 and TYRP1. The melanocytes therefore do not synthesize black or brown pigment. Instead, dopaquinone is converted into yellow phaeomelanin.

Yellow Labs are yellow because the receptor for MC1 does not work, so even though the dogs' bodies produce MC1 and have functional genes for TYRP2 and TYRP1, the signal to produce the enzymes is never transmitted to the melanocyte. Thus, the melanocyte never makes TYRP2 and TYRP1. All the dopaquinone is converted into yellow phaeomelanin instead of eumelanin.

The gene that causes yellow coat color in Labrador retrievers is actually a nonfunctional allele of the gene for MC1R. We will call the functional gene R for receptor and the nonfunctional allele r. For a Lab to be yellow, it cannot make any functional receptor protein, so its genotype must be rr. Black and brown Labs have at least one functional allele, so their melanocytes receive the hormone signal to make TYRP2 and TYRP1. They have genotype RR or Rr.

If you crossed a Labrador of genotype RR with a Lab of genotype rr, all the puppies would be of genotype Rr. Would these puppies be yellow?

Interestingly, the r allele affects only melanocytes in hair follicles. Melanocytes responsible for pigmentation everywhere except in hair function normally. Now it gets to be fun. You can tell whether a yellow Lab would have been black or brown if its r gene were functional by the color of its nose and lips! Black Labs, who make TYRP1 and produce black

pigment, have black noses and lips, in addition to black fur. Chocolate Labs, who do not make TYRP1 and produce no black pigment anywhere, have brown noses and lips. Look at the photos of Cocoa and Midnight in Color plate 3.

A Lab with genotype RRBB has black fur, black lips, and a black nose. Its melanocytes receive signals normally and can make TYRP1. A Lab with genotype rrBB has yellow fur, but its nose and lips are black. A Lab with genotype RRbb has brown fur, brown lips, and a brown nose. A dog with genotype rrbb has yellow fur, but its nose and lips are brown (see the photos of Babs and Angel in Color plate 4).

Talk like a geneticist

In the case of yellow Labs, the expression of the black or brown genotype is altered by the alleles of a completely separate gene. Geneticists call this phenomenon epistasis (pronounced eppy-STAY-sis). Again, epistasis is an observation, not an explanation for what is observed. Epistasis is defined as the interaction between alleles of different genes that allows an allele of one gene to alter the effects of alleles of a different gene. In our example, the allele r altered the effects of the allele B or b. People had observed the epistatic effect of the r allele on coat color long before any information about the molecular events that produced the coat colors was available.

Questions

1. Is the allele that results in yellow fur in Labrador retrievers dominant or recessive? Explain your answer.

2. What genotype must a Lab have to be yellow?

3. What are all possible genotypes of a yellow Lab with black lips?

4. What are all possible genotypes of a yellow Lab with brown lips?

Genetics with Yellow Labs

To think about the genetics of the B/b gene and the R/r gene at the same time, we can use the same techniques we used above when we were thinking only about B/b. Let's start with a cross between a male dog of genotype BBRR and a female dog of genotype bbrr. What would the phenotypes of these dogs be? Color them in the picture on Worksheet 24.1.

To predict the genotypes and phenotypes of the puppies, we first have to figure out what gametes each parent could produce. Remember that in meiosis, each gamete receives a copy of only one member of each chromosome pair. (Assume the B and R genes are on different chromosomes.) What are the possible gametes the male parent could produce? The female parent? What will be the genotype(s) of the puppies? Fill in the blanks and color the puppies in the picture on Worksheet 24.1.

That was a simple example, because the parents were homozygous. Now, let's have some fun and look at what would happen if two dogs of genotype BbRr were crossed. First, what is the phenotype of the parent dogs? Color them on your worksheet.

To predict offspring genotypes and phenotypes in a cross like this one, Punnett squares are very helpful. Use the following steps to construct one.

1. Assign symbols to the genotype of each parent. We already know what they are in this case: BbRr.
2. Write out the cross, using the genetic symbols: BbRr × BbRr.
3. Determine the gamete possibilities for each parent and fill in the outline of a square. This step is more complicated than it was in the previous example. One way to figure out the possible gametes is to make a table like this one:

Possible gametes for parents with genotype BbRr

Black/brown allele	Receptor allele	Gamete
B	R	BR
B	r	Br
b	R	bR
b	r	br

Since both parents have the same genotype, this table shows the gametes each one could produce. If the parents had different genotypes, it would be necessary to make a table for each one (unless one was very easy, such as the homozygous parents above).

4. Fill in the outline of a square with the parents' gametes. This time, each parent can make four different types of gametes, so the square has to have four slots on each side:

	BR	Br	bR	br
BR				
Br				
bR				
br				

5. Fill in the offspring genotypes produced by combining the different gametes.

	BR	Br	bR	br
BR	BBRR	BBRr	BbRR	BbRr
Br	BBRr	Bbrr	BbRr	Bbrr
bR	BbRR	BbRr	bbRR	bbRr
br	BbRr	Bbrr	bbRr	bbrr

6. Determine the genotype ratios by counting the contents of the boxes. Do this and fill in the table

on your Worksheet 24.2. What are all the different genotypes present in the offspring? How many times does each one appear in the Punnett square?

7. Interpret the genotypes in molecular terms. Fill in the chart to show whether each genotype will produce functional MCR1R protein and functional TYRP1.

8. Determine the phenotype of each genotype. If the dog is a yellow Lab, tell what color its lips will be. How many of each phenotype are predicted from the Punnett square?

The two dominant alleles in this cross are R and B. What are the associated phenotypes? Notice that you get the largest number of offspring (nine) showing these dominant phenotypes. The recessive alleles in the cross are b and r. What are the phenotypes associated with bb and rr? Notice that you get the intermediate number of offspring (three) for both combinations of one dominant and one recessive phenotype. What are these combinations in our cross? The smallest number of offspring (one) shows both recessive phenotypes. What is this phenotype in our cross?

Questions

1. A friend of yours has a pair of black Labs. These two dogs have a litter of puppies. To your friend's astonishment, about one-fourth of the puppies are chocolate Labs. Use genetics to explain this result to your friend, including why approximately one-fourth of the puppies show the chocolate color.

2. A different friend of yours also has a pair of black Labs. These two dogs have a litter of puppies. To this friend's astonishment, about one-fourth of the puppies are yellow (all have black noses and lips). Use genetics to explain this result to your friend, including why approximately one-fourth of the puppies show the yellow color.

3. A black Lab of genotype BbRr is crossed with a yellow Lab that has brown lips. Predict the genotype and phenotype ratios of the offspring. Hint: try using the Punnett square method outlined above.

Human Molecular Genetics

25

About This Activity

This activity extends the discussion of the molecular basis of genetics to human beings. Extensive background information, including a fairly traditional discussion of chromosomal abnormalities and single-gene disorders, is included for teachers. The background information for students focuses on specific single-gene disorders and on the difficulties of studying genetics in humans. We give some information from imaginary cases, but the material requires students to apply to human diseases what they have learned from this book, especially the two preceding chapters.

The background information for students includes some history of how phenylketonuria (PKU) was discovered and how the genetics of sickle-cell anemia was deduced. There are three student worksheets and questions for thought at the end. The questions could be used to extend classroom discussion and to review the relationship between protein structure and function, hybridization analysis, and meiosis. Much of the material in the teachers' chapter is not presented in the student chapter; this information can be shared with students, too. This activity assumes that students know about dominant and recessive traits, are familiar with gamete generation through meiosis, and have completed the DNA-typing activity.

This chapter includes a reading on the molecular genetics of cancer. Chapter 26 tells the story of the work of medical geneticist Holmes Morton's work among the Amish.

Class periods required: *1 or 2*

Introduction

Our genes encode all the proteins present in our bodies: our enzymes, our structural proteins, our hormones, our immune proteins, and others. Some changes in our genes result in changes in these proteins—absence, inappropriate expression, or expression of an altered form—while others have no effect. Any aberration in protein expression can cause problems. Some are fatal before birth, some result in improper development of physical or mental function, and others cause chronic metabolic deficiencies or impaired immune function. Any disorder caused by a change in the genes is a genetic disease. The characteristics of genetic diseases vary widely, depending on the type of change and the gene in which it occurs.

Genetic diseases can be divided into three categories: chromosomal defects, single-gene disorders, and multigenic traits. Chromosomal defects include missing or extra chromosomes and rearrangement or deletion of parts of chromosomes. Single-gene (Mendelian) disorders include dominant and recessive traits that are carried on either the autosomes or the sex chromosomes. Multigenic traits cover a range of conditions presumably influenced by many genes. The prevalence of genetic disorders in the human population is impossible to determine precisely. However, it has been estimated that perhaps 2% of the population is affected by either a single-gene disorder or a chromosomal defect.

Chromosomal defects

Chromosomal disorders include gain or loss of an entire chromosome (aneuploidy), loss of part of one or more chromosomes (deletion), transfer of one segment of a chromosome to another chromosome (translocation), and reversal of a segment of a chromosome (inversion). Considering the large number of genes involved, it is not surprising that chromosomal disorders are usually fatal during early development. It is estimated that up to 50% of early miscarriages involve chromosomal defects. However, there are certain chromosomal disorders that still permit a baby to develop to birth. These people display "syndromes," which are collections of symptoms and phenotypic abnormalities presumably caused by the abnormal number and/or expression of the large number of genes involved in the chromosome abnormality.

The best-known chromosomal disease is probably Down syndrome, caused by an extra copy of chromosome 21. Down syndrome occurs in 1 of 600 to 800 live births and involves mental retardation, the characteristic appearance, and congenital heart defects. Why the presence of an extra copy of the genes of chromosome 21 produces this cluster of effects is not known.

Another relatively common chromosomal disorder is Turner's syndrome. Individuals with this disorder have only a single X chromosome and no Y (or second X). These individuals are phenotypic females but lack functional ovaries and do not develop secondary sexual characteristics, among other abnormalities. One in 1,500 to 5,000 female infants is thought to be affected by this condition.

A second chromosomal disorder involving the sex chromosomes is Klinefelter's syndrome. Klinefelter's syndrome is caused by the presence of two X chromosomes and a Y chromosome. These individuals are phenotypic males but are infertile because their testes fail to form properly. They also suffer other mental, behavioral, and physical abnormalities. This defect is present in 1 of 500 to 1,000 male infants.

Chromosomal deletions, the absence of part of a chromosome, are usually fatal, but some cause chromosomal disease instead. Certain deletions are associated with congenital tumors, such as retinoblastoma and Wilms' tumor. Deletions in the Y chromosome can result in XY individuals who are phenotypic females but infertile. A deletion in chromosome 5 results in the cri-du-chat syndrome, referring to the peculiar catlike crying of babies with this defect. Other symptoms of this syndrome include mental retardation, a characteristic facial appearance, and motor and growth retardation.

Translocations, the relocation of one segment of a chromosome to a different chromosome, may or may not be harmful. In so-called balanced translocations, there is no net gain or loss of genetic material, just a rearrangement. Some of these translocations cause no problems until the altered chromosome is passed to an offspring without its balancing partner (causing the offspring to have extra copies of genes or to be missing copies). Translocations can also be involved in disease, apparently depending on the site of the chromosome breakage and rejoining. Translocations are often observed in cancers that develop later in life. The Philadelphia chromosome associated with chronic myelogenous leukemia is the product of a translocation of a piece of chromosome 22 to chromosome 9. Burkitt's lymphoma (a malignancy of antibody-producing cells) is associated with the translocation of a piece of chromosome 8 to chromosome 14.

Translocations and deletions can give valuable clues to the locations and functions of particular genes. The translocation involved in Burkitt's lymphoma transfers an oncogene (see the *Reading, Molecular Genetics of Cancer*) to a chromosomal region involved in antibody production. The XY "females" mentioned above have helped scientists map the genes on the Y chromosome required for maleness. The retinoblastoma deletion led other scientists to the initial discovery of a tumor suppressor gene (see the *Reading, Molecular Genetics of Cancer*).

Multigene disorders

Multigene disorders are often described as "an inherited tendency to develop. . . ." This sentence could be completed with "high blood pressure," "adult-onset diabetes," "mental illness," or other diseases. Multigene disorders have been difficult to analyze genetically, because they do not show clear inheritance patterns. Environmental factors also contribute to whether disease actually develops.

Cancer is a multigene disease; evidence shows that disruption of the normal functions of several genes is usually required before cancer develops. In general, these alterations happen in individual cells during adulthood and are not inherited. There are families, however, who have congenital defects in specific genes involved in cancer. These families inherit a predisposition to develop cancer, usually at an abnormally early age (see the *Reading, Molecular Genetics of Cancer*).

Single-gene disorders

As implied by the name, single-gene disorders result from alteration of a single gene. Single-gene disorders are usually classified as either dominant or recessive, and the mutation is carried either on the sex chromosomes (X linked) or on an autosome (all the chromosomes except the sex chromosomes are called autosomes; they are the same in males and females). Over 3,000 single-gene disorders have now been characterized.

In general, single-gene mutations result in either the absence of a protein or the production of an altered form of the protein. Therefore, a dominant single-gene disease is one in which the lack of protein function or the production of an altered protein from one copy of a gene is sufficient to cause disease even in the presence of the normal protein made from the second copy of the gene. Conversely, in a recessive genetic disease, the loss of functional protein from one gene (or production of an altered form) can be adequately compensated for by the production of normal protein from the other copy of the gene.

Consider a mutation in a gene encoding an enzyme that renders the enzyme nonfunctional. An individual who is heterozygous for this mutant gene may have half the normal amount of functional enzyme (the other copy of the gene makes functional protein). Whether this individual has a disease will be determined by whether half the normal amount of enzyme is sufficient.

Now, think about a mutation in a gene encoding a structural protein. Suppose that this mutation causes the production of an altered form of the structural protein. Whether this mutation causes disease in the heterozygote depends on whether the single normal gene produces sufficient normal protein to meet the cell's needs and also on whether the presence of the altered mutant protein interferes with assembly of the normal structure. Of the approximately 3,000 genetic diseases described in humans, about 1,000 are dominant. Two dominant disorders whose causes are not understood are polydactyly (extra fingers and toes) and achondroplasia (a type of dwarfism).

Familial hypercholesterolemia

An example of a dominant genetic disorder whose cause is well understood is familial hypercholesterolemia. Affected individuals have very high levels of cholesterol in their blood and usually die of heart attacks. Heterozygous individuals usually die before the age of 60, but homozygotes die much earlier. This disorder is caused by mutations in the gene encoding the low density lipoprotein receptor (LDL-R). Cholesterol is bound to LDL when it travels through the bloodstream. Cells capture cholesterol (which they need for a variety of functions, including building cell membranes and synthesizing hormones) by binding cholesterol-LDL complexes with the LDL-R in the cell membrane. An individual with one nonfunctional LDL-R gene makes only half the normal number of receptors, leading to a high blood cholesterol concentration and increased incidence of heart attack. Thus, the presence of a single mutant copy of the LDL-R gene leads to disease.

Huntington's disease

Huntington's disease (HD) is another autosomal dominant inherited disease. It is characterized by involuntary spasmodic movements and a progressive loss of intellectual function. No symptoms are apparent until middle age, typically in the 40s, so an individual could have children before knowing that he (or even that his parent) was affected and might be transmitting the genetic defect. The gene involved in HD, identified in 1993, lies on chromosome 4.

Studies of affected and unaffected individuals from the same large Venezuelan family showed an interesting and consistent difference in their HD genes. The gene contains a tandem repeat of the sequence CAG, which is a codon for glutamine. The HD protein (huntingtin) contains a tract of glutamine residues as a result of this repeated sequence. In unaffected individuals in this family, both copies of the HD gene rarely had more than 35 copies of the codon. In affected individuals, one copy of the gene had 40 or more copies in tandem. Of the individuals who had one chromosome with 36 to 39 copies of the repeat, some were healthy and some were affected with HD. Huntingtin protein from affected individuals aggregates and forms filaments in vitro. Disrupted protein-protein interactions as a result of the too-large polyglutamine tract are probably responsible for the disease. A single copy of the gene with too many repeats would produce defective protein and must disrupt the function of the normal protein produced from the individual's other copy of the gene, thus accounting for the genetic dominance of HD.

HD is an example of what is called a trinucleotide repeat expansion disease. At least 12 of these diseases have been identified, and all involve an increase in the number of trinucleotide repeats. Two of the better-known trinucleotide repeat expansion diseases are fragile X syndrome and myotonic muscular dystrophy. In all these diseases, expansion of the repeated region is seen from generation to generation of affected families. Increasing length of the repeated region correlates with increasing severity of the disease and also with increasing likelihood of further expansion. The reason for the expansion is not fully understood and is a focus of current research.

Cystic fibrosis

Cystic fibrosis is the most common inherited disease of European Americans. It is a recessive disorder. In cystic fibrosis patients, water and salt secretion is impaired, leading to production of abnormally thick mucus in the lungs, pancreas, and elsewhere. Digestion and breathing are impaired. Lung infections are very common, because the thick mucus prevents normal clearing of bacteria from the lungs. Affected individuals used to die before the age of 30, but improving therapies are allowing many patients to lead longer lives.

The gene involved in cystic fibrosis was identified in 1989. It encodes a protein (called CFTR, for cystic fibrosis transmembrane conductance regulator) that allows chloride ions to cross over the cell membrane. Heterozygous carriers of cystic fibrosis are completely healthy. Therefore, one copy of the CFTR gene produces enough protein to allow normal cellular functioning.

The cystic fibrosis gene is located on chromosome 7, spanning 250,000 base pairs (bp)! The messenger RNA for CFTR, however, is only 6,500 bp long, and the protein has 1,480 amino acids. The most common mutation leading to cystic fibrosis is a 3-bp deletion in the 10th exon that results in a single missing phenylalanine in the CFTR protein (another good example of how the alteration of even 1 amino acid can dramatically affect protein function). To date, approximately 180 different cystic fibrosis-causing mutations in the CFTR gene have been characterized. Some mutations cause a more severe form of the disease than others, presumably because they result in different levels of deficiency in the protein's function.

Sickle-cell anemia

The most common single-gene disease among African Americans is sickle-cell anemia, occurring in 1 of every 600 African Americans. This painful and eventually fatal blood disorder is caused by an altered form of hemoglobin, the protein that carries oxygen in the red blood cells. Hemoglobin is composed of four protein chains. The individual protein chains are called globins and are encoded by the globin genes. People normally have several different globin genes denoted by the Greek letters alpha (α), beta (β), gamma (γ), delta (δ), and epsilon (ϵ). The most common type of hemoglobin in adults is called hemoglobin A and is composed of two α-chains and two β-chains.

Sickle-cell anemia is caused by a mutation leading to a valine at position 6 in the β-globin chain rather than a glutamic acid. This change can be caused by a single base change from A to T, which changes the glutamic acid DNA code GAA to the valine code GTA

(see Table 4.1 in chapter 4). The mutant β-chain causes the hemoglobin to aggregate (see chapter 3), resulting in a change in the shape of the red blood cell from round to sickle shaped. The sickle-shaped cells do not travel well through the capillaries, resulting in impaired circulation.

A heterozygous carrier of sickle-cell anemia does not experience clinical symptoms of sickle-cell disease. However, under special conditions, such as lowered oxygen pressure, sickle-shaped red blood cells can be observed in their blood, since the mutant β-chain is produced in these individuals and is incorporated into hemoglobin, along with normal β-globin chains (produced from the healthy copy of the β-globin gene). These heterozygous carriers are said to have sickle-cell trait.

Interestingly, people with sickle-cell trait are more resistant to malaria than are people with "normal" hemoglobin A. Most African Americans are descended from slaves brought from West Africa, a region where malaria is endemic. For those original Africans, having one copy of the sickle-cell disease gene was probably an advantage.

Gaucher's disease

Gaucher's disease is a recessive autosomal disorder that affects 1 of every 2,500 American Jews. Patients with Gaucher's disease have an enlarged liver and spleen and can have painful bone lesions. Both the severity of the disease and the age of onset vary widely, with the more severe forms appearing earlier in life. Gaucher's disease is caused by a lack of the enzyme glucocerebrosidase, which breaks down the glycolipid (a combination of carbohydrate and lipid) glucocerebroside. The gene encoding this enzyme has been identified and cloned. It is located on chromosome 1, consists of 11 exons, and is about 7,500 bp long. The spliced mRNA is about 2,000 bp long. One good copy of the gene can apparently produce enough enzyme to meet the body's needs, since heterozygotes are healthy.

Studies of patients with Gaucher's disease revealed that they have many different mutations in the glucocerebrosidase gene. The nature of the mutation itself determines whether the form of the disease is severe or mild, although two individuals with the same mutations may still have disease symptoms that differ in severity (possibly because of effects exerted by other genes). A common mutation leading to mild disease is a base change from A to G at position 1226, resulting in an amino acid change from asparagine to serine. This mutation (1226G) accounts for 75% of the disease genes in the Jewish population. It is likely that this amino acid change leaves the protein partially functional, accounting for the mildness of the disease in these cases. Two common mutations that cause severe disease are a single base change from T to C at position 1448, which causes an amino acid change from leucine to proline, and an insertion of a single G at position 84, which causes a reading frame disruption. The reading frame disruption early in the gene would likely result in a complete lack of protein function. Similarly, because the amino acid proline has such different chemical characteristics than leucine does, substitution of a proline for a leucine within the protein probably severely disrupts the protein's structure and therefore its function. These two mutations (serine to proline and the reading frame disruption) would be predicted to impair the function of the enzyme much more than an asparagine-to-serine change. The fact that these two mutations result in a much more severe form of the disease is consistent with this prediction.

Diagnosis of genetic disease and carrier status

In the last 3 decades, the ability of doctors to diagnose genetic diseases has improved dramatically. Two factors that have been critical to these advances are our growing knowledge of the causes of various genetic diseases and the increasing ease of DNA analysis. Advances in analytical chemistry and biochemistry have provided ways to test for metabolic products indicative of enzyme deficiencies and obvious protein abnormalities (such as sickle cells), but not all genetic diseases result in detectable deficiencies. Although we have recently identified the protein whose alteration causes cystic fibrosis, it is not an enzyme that produces a measurable end product. As of now, there is no clinical assay for its function.

Even when we can test for specific defects, there are sometimes problems with prenatal diagnosis. For example, to diagnose sickle-cell anemia by traditional methods, red blood cells must be examined. (Hemoglobin is not an enzyme, but the structural abnormality in sickle-cell disease can be detected through its effects on the shape of red blood cells.) Although it is possible to obtain a sample of fetal blood, the procedure involves significant risk to the fetus. Finally, many of our best procedures cannot detect carrier status. It can be valuable for prospective parents to know if they both carry a recessive disease gene, so that appropriate prenatal tests can be performed.

Now that we can analyze DNA directly, it is possible to bypass tests of protein function and look directly at the affected genes. Using DNA analysis, any disorder for which a gene location is known can theoretically be detected. Prenatal DNA testing can be performed on amniotic cells. Even if the protein in question (such as hemoglobin in the case of sickle-cell anemia) is not produced in the amniotic fluid cells, the genes are there and can reveal whether the defect is present. DNA testing of parents can provide information on carrier status.

Many ethical issues must be considered when contemplating testing for genetic diseases. For example, should populations be screened for common disease genes? Should employers have access to knowledge about employees' genetic predispositions to disease? Should people with certain predispositions (for example, to cancer) be barred from working in occupations that might increase their risk of developing disease (such as one involving exposure to asbestos fibers)? Who should make these decisions?

A good model for confronting ethical dilemmas is presented in chapter 38. This model forces students to analyze their thinking and channels discussion along those lines. We recommend that you refer to this chapter in advance if you would like to raise ethical issues for your class to consider.

Objectives

After completing this activity, students should be able to:

- Explain some of the molecular considerations that determine whether a genetic disease is dominant or recessive.
- Suggest an explanation based on protein function to explain why some mutations in a gene cause mild disease and others cause severe disease.

Materials

- Copies of the *Student Activity* if students do not have books
- Transparency or photocopies of the genetic code (or other access) for reference in answering question 1

Resources

The Howard Hughes Medical Institute publishes a series of reports on biomedical science that began in 1990. The institute provides free copies of their reports to educators. Their address is Howard Hughes Medical Institute, 4000 Jones Bridge Road, Chevy Chase, MD 20815-6789.

The database *Online Mendelian Inheritance in Man* (OMIM) is a catalog of inherited human disorders. For each disorder, an entry of basic information, updated regularly, is provided. Some of the entries are quite short; others are extensive and include such things as descriptions of animal models and gene therapies. To reach OMIM, go to the NCBI home page at http://www.ncbi.nlm.nih.gov and click on OMIM. Once you reach the home page, you can search OMIM by typing terms in the text box near the top of the page or click "Search OMIM" in the list on the left-hand side of the page. The latter option lets you restrict the search to titles, allelic variants, or other options. The search result will usually give a list of matches. Upon selecting one, you will get another menu with choices such as Text, History, and Animal Model. The Text option provides a history of the disorder with links to literature references. The Text entries are technical and sometimes sketchy and can be difficult to understand, but with a little background information, they can be helpful. Browsing this database will be interesting for your students.

Another search option is "Search Gene Map." For these searches to be successful, you must know the correct name of the gene of interest. For example, the gene name for factor VIII, the gene that is defective in hemoglobin A, is F8C. Entering F8C takes you to position q28 on the X chromosome, where the gene is found. However, if you enter "hemoglobin A" in the gene map search, no match will be found. You can enter the name of a chromosome, such as "X," and browse the map of the entire X chromosome.

At http://www.genome.gov you will find the website of the National Human Genome Research Institute, with research news and educational resources. More information about The Cancer Genome Atlas (TCGA) can be found at http://cancergenome.nih.gov.

Procedure

Use the background information and questions as best suits your class. The background information for students continues the focus on the molecular biology that underlies genetic diseases. It is intended to provide a starting point for discussion. The questions can be discussed in class, included in small group projects, or assigned to individuals.

Tips

- It may be necessary to review meiosis and basic genetics before beginning this section.
- Students may wonder why people with PKU have any pigmentation, since they cannot make

tyrosine from phenylalanine. Tyrosine, an amino acid, is a component of dietary proteins, so PKU patients have some tyrosine in their systems from the protein they eat. In the body, tyrosine has other potential fates besides being converted into dopaquinone and pigments. For example, it is used in protein synthesis and can also be metabolized for energy.

- The PKU story can be used as an example of how an observation followed by investigation led to an important discovery. The mother noticed the musty smell in her retarded children's diapers and described it to her relative. He had the curiosity and training to investigate the situation, and PKU was discovered.

- James Neel, the physician who established the genetics of sickle-cell anemia and sickle-cell trait, had a long career in human genetics that included investigating the genetic effects of the atomic explosions on the inhabitants of Hiroshima and Nagasaki and observing the behavior of "natural" humans among native South Americans. His life and work would be an interesting subject for a project. In 2000, journalist Patrick Tierney published a book, *Darkness in El Dorado,* in which he alleged that Neel had committed unethical if not criminal acts during his research in South America. An investigation into the charges by the American Society for Human Genetics, as well as other scholars, found that the charges were unfounded (see the reference in *Selected Readings*). The most inflammatory accusation in the book was that Neel deliberately initiated a measles epidemic, a claim that was repeated worldwide in newspapers and magazines. Scholars who evaluated the claim after its publication concluded that the allegation was "untrue and unjust" and that Tierney had "misrepresented facts to create an inaccurate portrayal of Neel." If students happen upon *Darkness in El Dorado,* be sure that they read this paper, which refutes each of Tierney's allegations.

- The melanocortin 1 receptor (MC1R) gene responsible for the yellow Labrador pigment pattern has a homolog in humans. There is evidence that humans who cannot make MC1R have fire-red hair. These individuals may be the human pigment equivalent of yellow Labs (see the Barsh reference in *Selected Readings*).

Optional Follow-Up

- Research the life and work of James Neel. He has written an autobiography, *Physician to the Gene Pool* (see *Selected Readings*).
- Find out about the PKU diet.

- Invite a speaker from a laboratory that does DNA-based testing for genetic diseases. (The director of your hospital laboratory should be able to tell you where the nearest one is.)
- Invite someone from your hospital laboratory to talk about routine health screening for newborns.
- If you know anyone who has PKU, cystic fibrosis, or another inherited disease, ask him if he would mind sharing information about his disease with your class.
- Invite a genetic counselor to talk to your class. (An obstetrician or hospital laboratory director should be able to tell you where to find one.)

Answers to Student Questions

1. a. Normal sequence: valine, histidine, leucine, threonine, proline, glutamic acid, glutamic acid. Mutant sequence: valine, histidine, leucine, threonine, proline, valine, glutamic acid. The mutation causing sickle-cell disease is the change at position 6 from glutamic acid to valine.

   b. The normal β-globin gene sequence contains an MstII site. The mutation that causes sickle-cell anemia also destroys the restriction site within the gene. Therefore, it is possible to test for the mutation by testing for the presence of the MstII site.

   To carry out the test, digest a DNA sample with MstII, separate the fragments on a gel, and use the Southern blot procedure to transfer the fragments to a membrane. Probe the fragments with a probe that hybridizes to DNA on either side of the MstII site in the β-globin gene. Homozygous normal individuals will have two bands that hybridize to the probe, one approximately 1,150 bp long and the other approximately 200 bp long. Homozygous affected individuals will have a single band of approximately 1,350 bp. Heterozygous carriers will have three bands: 1,350 bp (from the mutant gene), 1,150 bp, and 200 bp.

   Students might design a procedure using MstII and a second enzyme. Such a procedure could work as long as their logic is sound.

   The hybridization step is essential. If students propose simply to cut the DNA with MstII, separate the fragments on a gel, and examine the stained gel, remind them that all they would see is a smear. The human genome contains about 3 billion base pairs and would generate

a vast number of MstII fragments. Hybridization lets them look specifically at the genomic region of interest.

2. a. Whether the offspring is healthy or not depends on which chromosomes he or she inherits. Assume the parent has a balanced translocation in which part of one of his chromosomes 3 is attached to one of his chromosomes 5. That parent also has a normal copy of chromosome 3 and one of chromosome 5, so when the parent's gametes are formed, there are four possibilities: 3, 5 (both normal); 3*, 5* (both abnormal); 3*, 5; and 3, 5*. Two of these combinations would produce healthy offspring: 3, 5 and 3*, 5*. The other two combinations form a gamete with either missing genetic information or extra genetic information and would not produce a healthy offspring.

   b. The most likely reason has to do with the exact point at which the break occurs and where the segment reattaches. If either event disrupts an important gene, the consequences could be bad. Harmless translocations apparently do not interfere with normal gene expression.

3. a. Use this question to make sure students realize that some mutations may have little effect on the function of the protein while others can completely eliminate it. It is likely that mutations causing mild disease leave the enzyme able to function in a diminished capacity, so disease symptoms arise but are not severe. Mutations causing severe disease may disrupt protein function altogether. Supply the class with information about the specific mutations from the introduction, if you wish.

   b. Again, the mild disease mutation presumably results in an enzyme with partial function. That partially functioning protein is present in the heterozygote, who has two different disease genes. Thus, a patient with one mild mutation and one severe one has some enzyme activity, unlike a patient with two severe mutations, and his disease should be milder.

4. a. Recessive. One functional copy of the gene produces enough protein to fulfill the cellular function of the protein, and the nonfunctional copy of the gene does not interfere.

   b. This disease is likely to appear dominant or between dominant and recessive. A person with two defective copies of the gene may not be able to survive, since the description states that a cell that cannot produce the protein will die. If a person with two defective copies of the gene can live, then the disease would appear to be between dominant and recessive. A cell with a single functional copy of the gene can live, but it will be impaired, so a heterozygote will not be as healthy as a person with two functional copies of the gene.

It is possible that this gene could appear to be dominant if fetuses with two defective copies of the gene cannot complete development and die in utero. In this case, the only children born would be heterozygotes, who would be impaired, and homozygous normal children. Since inheritance of a single copy of the disease gene would cause impairment, the disease could appear dominant.

c. This disease will appear dominant. If a person inherits a single nonfunctional copy of the gene, the altered protein product will impair the function of the normal protein product produced by the other copy of the gene. Thus, a heterozygote is just as impaired as a person who is homozygous for the defective allele.

5. Discuss this question with your class if you like.

Answers to Worksheet 25.1

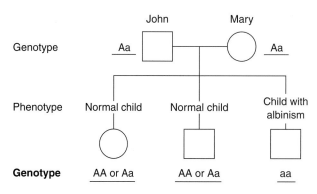

	John	Mary	
Genotype	Aa		Aa

Phenotype	Normal child	Normal child	Child with albinism
Genotype	AA or Aa	AA or Aa	aa

1. The gene itself is the tyrosinase gene. The "albinism gene" is a nonfunctional version (allele) of tyrosinase.

2. Albinism is recessive, so an individual with albinism has two nonfunctional copies of the tyrosinase gene and cannot make tyrosinase. Normally, the tyrosinase enzyme converts tyrosinase to dopaquinone, which is the precursor for pigment biosynthesis. If an individual cannot convert tyrosine to dopaquinone because he or she lacks tyrosinase, then the individual cannot make pigments and has albinism.

3. The albinism gene is recessive, because a single functional copy of the tyrosinase gene can lead to the production of enough tyrosinase enzyme to allow the person to synthesize a normal amount of pigment. Thus, a heterozygote has a normal phenotype, and the condition appears only if the individual has two nonfunctional alleles of the tyrosinase gene.

Answers to Worksheet 25.2

Write the genotypes in the blanks provided.

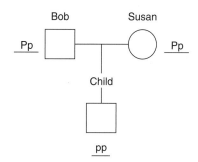

1. A nonfunctional form of the phenylalanine hydroxylase (PH) enzyme.

2. Each of the parents had one nonfunctional PH gene and one normal PH gene. The parent's normal gene promoted synthesis of enough PH enzyme for the parent to be normal. However, the child inherited the nonfunctional allele from each of its parents. The child cannot make PH and so has PKU.

3. It is possible, depending on the genotype of the gametes that form the child. There is a 1-in-4 chance that she will inherit the two normal alleles from Bob and Susan and be PP. There is a 1-in-2 chance that she will inherit one normal allele and one nonfunctional allele and be Pp, with a normal phenotype. There is a 1-in-4 chance that she will inherit two nonfunctional alleles and be pp, with PKU. The fact that Bob and Susan's first child has PKU has no effect on the odds that their second child will have the disorder.

4. See the answer to question 2. The PH enzyme converts phenylalanine to tyrosine. In its absence, phenylalanine is converted to phenylketone and excreted in the urine. Excessive amounts of phenylketone in the system lead to impaired development of the nervous system, so the child is retarded. Lower than normal amounts of tyrosine in the body result in light pigmentation, because tyrosine is the precursor for pigment biosynthesis.

5. The PKU gene is a nonfunctional version of the gene for phenylalanine hydroxylase. It is recessive, because a single functional copy of the PH gene leads to the production of a sufficient amount of PH enzyme for a normal phenotype. Therefore, a heterozygote with one functional and one nonfunctional copy of the PH gene is healthy.

Answers to Worksheet 25.3

1. Sickle-cell anemia is the homozygous condition; sickle-cell trait is the heterozygous condition.

2. A child with sickle-cell disease has two "sickle" versions of the β-globin gene. He or she must inherit one of the alleles from each parent. Thus, each parent must have a sickle allele and would therefore have sickle-cell trait. The siblings would often have sickle-cell trait, because there would be a 1-in-2 chance for each child that the child would inherit one sickle allele and one normal allele.

3. The electrophoresis test was looking at the β-globin protein. Individuals with sickle-cell disease have the sickle form of β-globin. Healthy individuals have the normal form. Individuals with sickle-cell trait are heterozygous for the sickle form and the normal form. Their cells synthesize both forms of β-globin, so both forms are seen in the electrophoresis test.

4. Neel's conclusion was that patients with sickle-cell disease were homozygous for the sickle-cell disease and people with sickle-cell trait were heterozygous. This conclusion predicts that healthy individuals synthesize one form of β-globin, patients with sickle-cell disease synthesize a different form, and people with sickle-cell trait synthesize both (because they have one normal gene and one sickle gene). Thus, the electrophoresis results confirmed his hypothesis.

Selected Readings

American Society of Human Genetics. 2002. Response to allegations against James V. Neel in *Darkness in El Dorado,* by Patrick Tierney. *American Journal of Human Genetics* 70:1–10.

Barsh, G. 1996. The genetics of pigmentation: from fancy genes to complex traits. *Trends in Genetics* 12:299–305. A fairly complex review of pigmentation genetics.

Ellegren, H. 2005. The dog has its day. *Nature* 438:745–746. A perspective on the emerging role of the purebred dog in genetic studies.

Haseltine, W. 1997. Discovering genes for new medicines. *Scientific American* 276:92.

Neel, James. 1994. *Physician to the Gene Pool.* John Wiley & Sons, Inc., New York, NY.

Nemeroff, C. 1998. The neurobiology of depression. *Scientific American* 278:66. This article contains little genetics but much information on the biology of depression.

Pasternak, J. 1999. *An Introduction to Human Molecular Genetics.* Fitzgerald Scientific Press, Bethesda, MD. A new college level text on human molecular genetics, particularly diseases. It is a good reference text.

Plomin, R., and J. DeFries. 1998. The genetics of cognitive abilities and disabilities. *Scientific American* 278:62.

Robbins, L. S., J. H. Nadeau, K. R. Johnson, M. A. Kelly, L. Roselli-Rehfuss, E. Baack, K. G. Mountjoy, and R. D. Cone. 1993. Pigmentation phenotypes of variant extension locus alleles result from point mutations that alter MSH receptor function. *Cell* 72:827–834. This article from the primary scientific literature describes the identification of the gene responsible for yellow Lab pigmentation (and similar pigment patterns in the mouse) as the MSH-R gene. It is a complex scientific paper, and we recommend it only if you are accustomed to reading the primary literature.

Weiner, D., and R. Kennedy. 1999. Genetic vaccines. *Scientific American* 281:50.

Welsh, M., and A. Smith. 1995. Cystic fibrosis. *Scientific American* 273:52.

Human Molecular Genetics

Introduction

Human genetics works the same way as dog genetics in terms of chromosomes, genes, enzymes, and other proteins. There are two basic approaches to learning about human genetics. In the classic approach, geneticists do what can only be described as detective work. First, a condition that seems to be inherited must be recognized. This usually happens through the medical profession: a patient comes in with an unusual condition, and the doctor learns that other family members have it. Geneticists would follow up on this report by visiting the family, interviewing relatives, and gathering as much information as possible. They would construct and analyze a family pedigree diagram to determine the inheritance pattern. This kind of detective work still occurs today, usually through medical centers associated with universities.

The second approach is based on animal studies and molecular biology techniques. The biochemistry of animals is pretty similar from species to species, so scientists do detailed studies of organisms that are small and easy to manipulate, such as mice and fruit flies. When an interesting gene is identified, such as the obesity gene first discovered in mice, scientists look for a homologous gene in humans through hybridization analysis or polymerase chain reaction. They use the gene sequence information from the animal to make probes and/or primers to try on human DNA. In this manner, the human counterpart to the mouse obesity gene was identified. Keep in mind, though, that a homologous gene does not necessarily have an identical function. Scientists must find additional evidence for the role of the homologous gene.

Human Genetic Diseases with Mendelian Inheritance

Here are some human genetics examples. See if you can apply what you learned from the Labrador retrievers to help you understand these stories. In the stories, the people are imaginary, but the diseases and their genetics are real.

John and Mary

John and Mary have brown hair, brown eyes, and medium-toned skin, as do their two children. They are astonished when their third child is born. He has white hair; very light, pinkish skin; and blue eyes. The doctor explains to them that their youngest son has albinism. His body cannot produce pigments (Color plate 5).

The doctor explains to John and Mary that pigments are produced from an amino acid called tyrosine. An enzyme called tyrosinase converts tyrosine to a chemical called dopaquinone, which is converted by other enzymes into all the body's pigments (Figure 25.1). Their young son's cells cannot make tyrosinase. Therefore, his cells cannot convert tyrosine to dopaquinone, and they cannot make pigment. The doctor reassures John and Mary that persons with albinism are otherwise normal and live normal lives, except for an extreme sensitivity to the sun. Why would that be?

This story should remind you of Cocoa and Midnight. Why was Cocoa brown and not black? What was happening in her cells?

Now think about the boy with albinism and his parents. Both his parents have normal pigmentation. How could they have a child with albinism? How could that child have siblings with normal pigmentation? Use what you have learned about genetics and the cellular chemistry of pigment synthesis to

Figure 25.1 Pigment production from tyrosine.

figure this out. Label the pedigree on Worksheet 25.1 (see Appendix A) with the genotypes of the family members (write all possibilities if there are more than one). Answer the questions on the worksheet.

Bob and Susan

Bob and Susan have just had their first child. The baby has undergone all the routine medical tests given to newborns. Afterward, the doctor and a genetic counselor come to talk with them.

The doctor explains that their baby lacks the ability to make an enzyme called phenylalanine hydroxylase (PH). This enzyme normally converts an amino acid called phenylalanine into the amino acid tyrosine. Tyrosine can then be converted to other things (Figure 25.2). The baby's cells cannot convert phenylalanine to tyrosine because they cannot make the PH enzyme. This condition is called phenylketonuria, or PKU. The danger of this condition is that phenylalanine will build up in the baby's cells, since it cannot be converted to tyrosine. High levels of phenylalanine interfere with normal brain development in a child and therefore cause mental retardation.

The doctor quickly reassures Bob and Susan that their baby could develop normally, given proper treatment. Phenylalanine and other amino acids enter the body when protein-rich foods are digested (why would that be?). If Bob and Susan give their baby a strictly controlled diet that is quite low in phenylalanine, the buildup will not occur, and the baby's brain will develop properly.

Bob and Susan are upset about their baby's problem but relieved that the condition is treatable. They have several questions. Susan wants to know how she and Bob, who are both normal, could have a child with PKU. Bob asks the genetic counselor what would happen if he and Susan had another child. Would it also have PKU? Use your knowledge of genetics to answer Bob's and Susan's questions. Label the family pedigree on Worksheet 25.2 (see Appendix A) with the genotypes of Bob, Susan, and their child. Here's a separate question you can answer based on what you know. People with PKU are very lightly

pigmented. Use your knowledge of the biochemistry of pigment synthesis to explain this fact.

People with PKU must avoid phenylalanine-containing foods when they are children and when they are pregnant. Adults are less vulnerable than children because their brains have completed development. The natural source of phenylalanine in the diet is protein, but several years ago a new source was introduced into our food supply. The artificial sweetener aspartame contains phenylalanine. The next time you go to the grocery store, look at cans of diet soda and find the warning message on them.

Discovering PKU

Albinism is very obvious at birth, but PKU is not. The noticeable symptom of PKU is mental retardation, which has many other causes. Although children with PKU are lightly pigmented, they are not so light as to call attention to themselves. PKU was first identified as a specific condition in 1934. An observant mother of two mentally retarded children noticed that her children's diapers had an abnormal, musty smell. She remarked on this observation to a relative named Folling, who was a physician and a biochemist.

Folling was curious and analyzed the urine of the children. He found that it contained a large amount of a substance normally present in only tiny amounts. The substance, phenylketone, was the direct result of the buildup of phenylalanine in the children's bodies. Folling did not know why so much phenylketone was present, but he published his findings about the two children. The condition was called phenylketonuria, which essentially means "lots of phenylketone in the urine."

In the years after Folling's discovery, many institutionalized mentally retarded people were tested for phenylketone in the urine. About 1% of the people tested were found to have PKU.

It was 1954 before the biochemistry of PKU was figured out and 1961 before a reliable way of routinely testing newborns for the condition was developed. This test allowed affected infants to be identified

Figure 25.2 The PH enzyme converts the amino acid phenylalanine into the amino acid tyrosine.

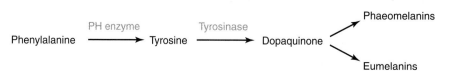

before the damage to their brains began. At that time, researchers knew that PKU was caused by a buildup of phenylalanine, so they reasoned (and hoped) that a diet very low in phenylalanine might protect affected babies' brains from damage. A diet was devised that was very low in phenylalanine but contained enough of the other 19 amino acids to allow the baby to make the proteins it needed to grow. The diet proved to be a miracle cure, but one that was based on scientific reasoning.

A problem with blood cells

Among African Americans whose families originated in West Africa, a specific type of blood disorder occurs more frequently than in other American subgroups. In this disorder, the red blood cells have an unusual shape. Instead of being round, they collapse into flat, curved shapes that resemble an ancient harvesting tool called a sickle. Because of this characteristic shape, the disorder is called sickle-cell anemia.

Sickle-cell anemia is caused by an abnormal form of hemoglobin, the blood protein that carries oxygen. Normally, hemoglobin molecules are distributed throughout the red blood cells. In sickle-cell patients, the abnormal hemoglobin molecules stick together, causing the red blood cell to collapse and assume the sickle shape. Because of their shape, sickled cells do not flow smoothly through capillaries. Instead, they form blockages that prevent adequate circulation and oxygen distribution, causing severe pain and tissue damage.

In the early part of the 20th century, the genetics of sickle-cell anemia was not understood. There appeared to be two forms of the disease: the severe form, which usually led to early death, and a mild form that had little effect on the individual. The severe form was referred to as sickle-cell anemia, and the mild form was called sickle-cell trait. The shape of blood cells in individuals with sickle-cell trait was not as distorted as it was in individuals with sickle-cell anemia, but the blood cells were abnormal (Figure 25.3).

Figure 25.3 (A) Red blood cells (magnification, ×4,600). (Photograph copyright M. Bessir-D. Fawcett/Visuals Unlimited.) (B) Sickle-cell anemia. (Photograph copyright Science VU/Visuals Unlimited.). (C) Normal and sickled red blood cells (magnification, ×943). (Photograph copyright George J. Wilder/Visuals Unlimited.)

A

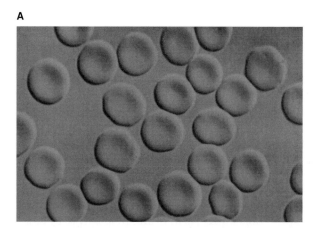

C

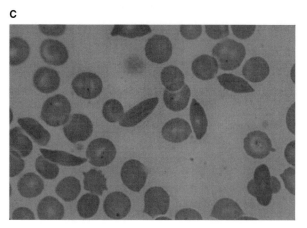

B

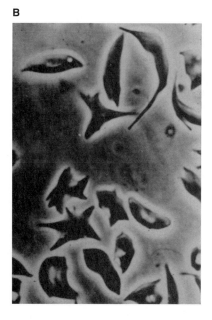

Children with sickle-cell anemia usually appeared in emergency rooms at an early age because of their disease. Their badly impaired blood circulation caused a number of severe problems. Individuals with sickle-cell trait were not affected by their blood cells' unusual shape and so were less commonly identified. The prevailing belief about sickle-cell disease in the 1940s was that it was caused by a gene that did different things in different people, sometimes causing the anemia, sometimes causing the trait.

In 1948, a physician and geneticist named James Neel decided to test an idea he had about sickle-cell anemia and sickle-cell trait. At that time, he was just beginning a job in Detroit, Michigan, which had a large African American population. Neel began to contact the families of sickle-cell patients. With their permission, he tested family members for the presence of sickled red cells.

Neel found, as he suspected, that both parents of children with sickle-cell anemia had sickle-cell trait, though they usually did not know it. Furthermore, many of the siblings of the anemia patients had sickle-cell trait. Neel drew his conclusions and published his data and interpretation in the journal *Science* in 1949. (*Science* is still published today, and important findings are often first described there.)

A few months later, Neel read an article written by scientists who were studying the hemoglobin protein. These scientists had found a way to characterize hemoglobin isolated from blood. They had found that hemoglobin from sickle-cell patients and hemoglobin from healthy individuals migrated differently during electrophoresis under certain conditions. Consequently, they could identify which type of hemoglobin was present in a blood sample (Figure 25.4). These scientists reported that when they examined blood samples from individuals with sickle-cell trait, they found both forms of hemoglobin. When Neel read this report, he concluded that his hypothesis about the inheritance of sickle-cell anemia was correct.

Here is a challenge for you. Put together an explanation that takes all these facts into account: the inheritance of sickle-cell anemia, sickle-cell trait, and the hemoglobin test results. Use Worksheet 25.3 (see Appendix A) as a guide.

It's usually not that simple

You have been learning about the genetics of fairly rare but medically important diseases, but what

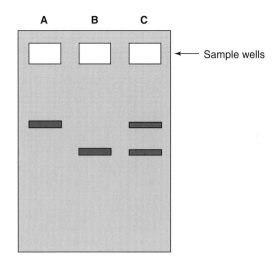

Figure 25.4 Electrophoresis of hemoglobin from a normal individual (lane A), from an individual with sickle-cell anemia (lane B), and from an individual with sickle-cell trait (lane C).

about the inheritance of more normal traits, such as height, general build, and hair color? We wish we knew! We are not being sarcastic when we say that. We say it because the inheritance of most human traits cannot be explained by the action of one or two genes. Inheritance of most traits is complex and poorly understood, if it is understood at all.

Most characteristics and the development of diseases, such as heart disease and high blood pressure, seem to be influenced by many genes and also by the environment in which an individual lives. Here are some simple examples of environmental influence. Suppose an individual had genes that could make him grow very tall but he grew up impoverished and malnourished. His growth might be quite stunted. Despite his genes, he might turn out to be a short or average-size adult. Suppose an individual had genes for normal intelligence but was exposed to high concentrations of alcohol while a developing fetus. Despite his genes, this person might be born with the mental retardation called fetal alcohol syndrome. Environmental conditions—what you eat, what you are exposed to, what diseases you have, and so on—can and do contribute to an individual's traits and fate. By the way, fetal alcohol syndrome is the most common cause of mental retardation in newborns today.

Most traits are influenced by many genes

If you think about most human characteristics, e.g., hair color, height, and eye color, you realize that there is no specific set of categories. We do not come in a specific set of heights, nor do we have a specific set of eye colors. Instead, our heights range from very short to very tall, with everything in between. Our eyes range from colors like light blue, amber, gray, and green to dark brown and everything in between (Color plate 6). Incidentally, blue eyes are not the result of blue pigment. Instead, they are the result of lack of pigment in the iris. Because of the way light is absorbed in the human eye, the iris appears blue if no pigments are present. That is why people with albinism have blue eyes. Hair color, which is the equivalent of coat color in mammals, is the same way. Humans show every shade from fire-red to platinum blond to black. What if you were to try to study hair color inheritance in humans? You could easily observe families like one of ours: a man with black hair marries a woman with medium-brown hair. They have four children: a boy with medium-brown hair, a girl with golden-blonde hair, a girl with white-blond hair that slowly darkens, and a girl with reddish-brown hair. What can you make of that? Probably not much—neither can we.

Why does it look so simple in dogs?

Some of you may be wondering why we can explain coat colors in Labrador retrievers if hair color inheritance in humans is so complicated. The answer to that question involves an important difference between the study of human genetics and the study of animal genetics. While we have genetically pure strains of mice, dogs, fruit flies, etc., there are no "pure breeds" of humans.

It may sound comical, but there is an important point here. We can figure out many of the genes that work in dogs and experimental animals, like mice and fruit flies, because people have bred them for generations and created pure breeds that differ very little in genetic makeup from one individual to another.

Think about the Labrador retrievers. Labs come in three colors: black, brown, and yellow. Does that mean they have only two genes (the TYRP1 gene, which governs black or brown, and the MC1R gene, which governs yellow) that can affect coat color? No, what it means is that Labradors have been bred so purely that all the other coat color genes are the same in every Lab. The only differences are in the TYRP1 gene and the MC1R gene. Thus, in Labrador retrievers, coat color really is determined by only two genes.

If you think about other breeds of dogs, you realize that coat color inheritance in dogs is not so simple, either. Think of examples of different coat colors, such as black-and-white spotted Dalmatians, red

Irish setters, golden retrievers, the typical beagle color pattern, the large splotches of color found in some basset hounds, the tiny spots and blue-gray color of blue tick hounds, the black-and-tan pattern of some Dobermans and dachshunds, and so on. Dogs have many, many genes that affect coat color. However, in pure breeds almost all the genes are the same, so only a very few genes control the coat color in that particular breed (Color plate 7). Coat color inheritance has been extensively studied in mice, and over 100 different genes have been found to affect it! Some genes are more influential than others. The same is probably true of humans.

Man's best friend

In the past few years, researchers have realized that dogs are a fabulous resource for genetic studies. The many different breeds, with their very distinct phenotypes, offer a nearly perfect set of samples for finding genes responsible for particular traits. Because the individual members of one breed are genetically very similar, it is easier to find differences between individuals that can be linked to specific phenotypes. Differences between breeds can also be studied to find genes associated with the differences.

For example, in chapter 27 we describe a study that identified a gene that appears to be responsible for size differences in dogs. Researchers compared small and large members of a single breed, Portuguese water dogs, and found that the small individuals had one version of the gene for insulin-like growth factor 1 *(Igf-1)*. The large individuals had other versions. The researchers then analyzed the sequence of the *Igf-1* gene in a number of tiny and giant dog breeds and found that nearly all of the 18 small breeds they looked at carried the same version of *Igf-1* as did the small Portuguese water dogs and almost none of the giant breeds carried it. The fact that "almost none" of the giant breeds carried the "small" variant of *Igf-1* suggests that other genes are probably also involved in size.

Pedigree and population analyses, studies of animal breeding, and molecular biology point the way to understanding many human genes, and molecular biology now allows us to identify, clone, and study them. The sequencing of the human genome and of the genomes of many other organisms makes it easier and easier to use data gleaned from animal studies to understand human molecular biology. If you were a researcher and identified a gene in an animal model, you would be able to compare the animal gene sequence with more and more of the human genome sequence to look for counterpart genes. For example, humans also produce the insulin-like factor found to be associated with size in dogs. It would be very interesting to compare the sequences of that gene in short humans and very tall humans. Perhaps such a study will be done in the next few years.

Man's genetic best friend?

If we had to identify a specific animal as man's genetic best friend, we would probably choose the mouse. Our choice is based on the fact that mice can be used to test the role of a specific gene. A technique developed mostly in the early 1990s allows researchers to make changes in a gene of their choosing and to substitute the changed copy into the mouse genome (a knockout mouse [see chapter 5]). Scientists can then observe the effect of the specific genetic alteration on the animal. This gives them a way to explore the roles of specific proteins in a whole animal, which is crucial for learning about the roles of the proteins in development and physiology.

Recent determination of the cause of a severe form of hereditary early heart failure illustrates the important role that knockout mice can play in genetic research. Individuals afflicted with this syndrome develop heart failure very early, typically in their 20s,

and it is so severe that they often can be helped only by a heart transplant. The researchers studied several members of a family in which the disease was present and compared the genomes of affected and unaffected family members to identify candidates for responsible genes. They found a gene that was mutated in the affected individuals and normal in the unaffected individuals. Such a result suggests, but does not prove, that the individual gene is responsible. For proof, the researchers turned to mice, which have the same gene within their genomes.

To test whether disrupting the gene caused hereditary heart failure, the researchers made a knockout mouse, replacing the normal mouse gene with the mutant version. The knockout mice developed early heart failure with the same symptoms as the affected humans, proving that the disruption of that single gene caused the disease. With that knowledge in hand, the scientists went back to the laboratory and studied the function of the protein in cultured cells.

With the accumulation of more and more genomic information for comparisons and the availability of knockout mouse technology for testing the roles of proteins, the foreseeable future will be a very exciting time for human genetics.

Questions for Thought and Discussion

1. Sickle-cell anemia is caused by a specific change in the sixth amino acid of the β-globin protein that forms part of hemoglobin. Shown below are a DNA sequence encoding the first 7 amino acids of the normal β-globin protein and a sequence encoding the first 7 amino acids of the sickle-cell protein.

   Normal amino acid sequence:

   5′ GTT CAT CTA ACC CCT GAG GAG . . . 3′

   Sickle-cell sequence:

   5′ GTT CAT CTA ACC CCT GTG GAG . . . 3′

   a. Translate these sequences. What is the amino acid change that causes sickle-cell disease?

   b. The restriction enzyme MstII cleaves the DNA sequence CCTGAGG. In human DNA, there is

   an MstII site 1,150 base pairs 5′ (to the left as shown) of the β-globin-coding region and 200 base pairs 3′ of the region shown. Design a DNA-based test for sickle-cell anemia using the enzyme MstII. Draw a gel outline, and show the expected results of your test on a homozygous normal individual, a sickle-cell anemia patient, and a person with sickle-cell trait (a heterozygous carrier). Provide a scale of base pair sizes to the left of the gel outline based on the expected results from your test.

2. A chromosomal translocation occurs when a segment of one chromosome breaks off and attaches to the end of a different chromosome. These events can occur during formation of the gametes, or they can happen in individual cells during the life of the person. If the gamete carries a translocation, all the cells in the resulting person will have it.

a. Occasionally, healthy adults who carry a translocation in all their cells will be identified. The assumption we make about these people is that they did not gain or lose genetic information during the translocation, and the particular rearrangement (shifting of a segment of one chromosome to another) caused no harm. The translocations in these individuals are called balanced translocations. Would you expect the offspring of a person with a balanced translocation to be healthy?

b. Translocations that happen in individual cells during adulthood can be harmless, or they can lead to cancer. For example, translocation of the end of chromosome 22 to chromosome 9 is associated with leukemia. Why do you think some translocations can be harmless and some can cause cancer or other problems?

3. Gaucher's disease is a recessive genetic disease caused by the lack of the enzyme glucocerebrosidase. By looking at the gene for that enzyme in Gaucher's disease patients, scientists have identified many different mutations that cause the disease. Interestingly, some mutations lead to a mild form of the disease (in homozygotes), and other mutations lead to a very severe form (again in homozygotes). Heterozygotes with one normal copy of the gene are healthy no matter what form of the mutant gene they have on the other chromosome.

a. Why do you think some mutations in the gene can cause mild disease and others can cause severe disease?

b. Some patients with Gaucher's disease have two different mutant forms of the gene. If a person has one copy of the gene with a mutation leading to mild disease and the other copy of the gene with a mutation leading to severe disease, that patient will have mild disease. Propose an explanation for why the disease would be mild.

4. In the following cases, would the associated disease be observed to be dominant, recessive, or in between?

a. The "disease gene" is a nonfunctional form of the gene. No protein product (or a harmless nonfunctional one) is produced from this gene. A cell with one functional copy of the gene can synthesize enough normal protein to take care of its function.

b. The "disease gene" is a nonfunctional form of the gene. No protein product (or a harmless nonfunctional one) is produced from this gene. A cell with one functional copy of the gene cannot synthesize enough normal protein to achieve full function, but it can survive. If none of the protein is present, the cell cannot live.

c. The "disease gene" is a nonfunctional form of the gene. This nonfunctional form encodes the production of an altered form of its protein. The altered form interferes with the functioning of the normal version of the protein, so the cells of a person with one nonfunctional allele and one normal allele cannot carry out the function of the protein any better than the cells of a person with two nonfunctional alleles.

5. Huntington's disease is a dominant genetic disease. Symptoms of this disease do not appear until middle age, typically when the patient is in his or her 40s. However, when symptoms appear, they are devastating. The patient gradually loses the ability to think. At the same time, he or she begins to experience increasingly uncontrolled body movements, usually twitching and shaking. There is no cure. Since the disease is dominant, there is a 50% chance that the child of an affected parent will also develop it. However, since the disease does not show until middle age, an individual can reproduce before he or she knows that the disease gene is present.

Suppose that when you are a teenager, one of your parents develops Huntington's disease. You watch your strong, intelligent parent gradually lose mental and physical function. At first, muscle coordination is slightly impaired, and forgetfulness, confusion, and personality changes develop. As Huntington's disease progresses, these conditions worsen steadily until both voluntary and involuntary movements are uncontrolled, with jerking and writhing. Speech is slurred, thought processes diminish, and psychiatric conditions, such as depression and uncontrolled rage, appear. In the final stages, the Huntington's patient is mute, cognitively nonfunctional, and frozen in a contorted position.

You know that there is a 50-50 chance that you will suffer the same fate and could transmit it to your children. There is now a DNA test that will tell you if you carry the Huntington's disease gene. Would you choose to take the test? What factors would enter into your decision?

Molecular Genetics of Cancer

Few diseases cause as much dread as cancer. In the United States, approximately 500,000 people die from cancer each year. Sometimes, a person's life choices contribute to the development of cancer, such as with cigarette smoking and lung cancer, but often cancer seems to randomly strike apparently healthy individuals. Over 200 types of cancer syndromes have been identified, affecting almost every type of cell, tissue, and organ system.

Our current cancer therapies, though more successful than earlier ones, are actually crude, blunt instruments. Cancer cells are dividing cells, so most anticancer therapies are strategies for killing rapidly dividing cells. Of course, many normal body cells are dividing, too, which is why current cancer treatments have so many difficult side effects. It is hoped that a thorough understanding of the molecular basis of cancer will lead to more precise treatments and even perhaps prevention.

A Genetic Disease in a Single Cell

Most cancers start from a single cell. To become a cancer cell, a normal cell must accumulate mutations in several different genes. These mutations are then transmitted to the cell's cancerous descendants. Cancer can therefore be thought of as a genetic disease at the level of the cell.

A mass of cancer cells descended from a single parent cell is called a *tumor*. If a tumor poses no danger to life or health, it is called benign; if it does pose a danger, it is called malignant. Tumors can sometimes be completely removed by surgery.

Some tumors continue to grow at their original sites, while others stop growing and shrink. In other cases, some of the tumor cells break free of the original tumor, migrate to new sites in the body, and establish many new tumors. The process of spreading throughout the body is called *metastasis*. Metastasis is what makes cancer so lethal. When cancer metastasizes, it may reach many sites at which tumors can cause severe damage, and it lodges in so many places that treatment by surgery alone is impossible.

A cancer is usually classified according to the type of cell from which it originated. For example, 85% of all cancers originate from cells that line organs and form parts of the skin. These cancers are called carcinomas. Cancers originating from cells of connective tissue, bone, or muscle tissue are called sarcomas. Leukemias arise from white blood cells (leukocytes). Cancers of glandular tissue are called adenocarcinomas; cancers of the nonneuronal cells of the brain are called gliomas and astrocytomas.

DNA Damage and Cancer

To become a cancer cell, a normal cell must accumulate mutations in several different genes. These mutations are transmitted to the cell's cancerous descendants. Mutations arise as a result of DNA damage, and cells have many mechanisms for repairing damage. Thus, mutations usually accumulate very slowly, explaining why most cancers develop late in life.

An important and common cause of DNA damage is ultraviolet (UV) light. UV light damages DNA by causing two adjacent T residues in the helix to bond to each other, forming a lesion that distorts the helix and blocks transcription and DNA replication. UV-induced DNA damage can be very harmful: it's possible to kill bacteria and sterilize food by treating them with enough UV radiation.

Normal cells are equipped to repair UV-induced DNA damage, at least most of the time. Proteins of the *excision repair* system detect the distortion in the helix, remove a section of the strand that contains the damage, and use the base sequence of the opposite strand to synthesize a functional replacement piece. Even so, individuals with a long history of exposure to UV light are more likely to develop skin cancer than those without such a history.

Cells that cannot repair damaged DNA accumulate mutations much faster than those that can, and people and animals that cannot repair damaged DNA tend to develop cancer much more frequently than those who can. For example, one of the cell's mechanisms of DNA repair is called excision repair. It removes DNA damage that distorts the DNA helix. Individuals with the inherited disease *xeroderma pigmentosum* lack the excision repair system. Their cells cannot repair thymine dimers; hence, they are extremely sensitive to UV light. Individuals with xeroderma pigmentosum develop multiple skin cancers wherever their skin is exposed to sunlight, illustrating just how well the excision repair system normally works.

Another well-known DNA-damaging chemical is benzopyrene, a substance found in cigarette and other tobacco smoke; smoke from burning leaves, trash, and wood; diesel exhaust; and other sources. Benzopyrene binds to G bases within DNA and can lead to mutations if it is not removed. One place along DNA where benzopyrene apparently binds exceptionally well is within the *p53* gene. From the information presented below about *p53*, you can understand why a chemical that promotes mutation in it would be especially carcinogenic.

A Laboratory Model of Cancer

What kinds of genetic changes are involved in cancer? The first clues came from studies of cultured animal cells. In the 1970s, it was discovered that infection by certain viruses could transform cultured cells. Normal cultured cells behave in a preordained manner. They do not grow on top of one another but instead form a confluent layer in a dish. They usually live for a certain number of life cycles and then die.

When cultured cells are infected by the viruses, however, dramatic changes occur. In some cases, the cells become immortal, losing their predetermined life span. They begin to grow over one another in an uncontrolled manner. Other changes are also observed. The combination of several of these changes is called transformation. (Do not confuse this use of the word with the usage referring to the addition of new DNA to bacteria; they both refer to change, but in different contexts.) In the body, cancer cells lose their normal growth inhibitions and continue to multiply in an uncontrolled and inappropriate manner. Because of the similarities, transformation of cells in culture is considered the equivalent of cancer in a dish.

The ability to produce the cancerlike changes was traced to certain genes the viruses carried. These genes were christened *oncogenes* ("onco" is a prefix meaning tumor). Quite a number of oncogenes were identified; they were given three-letter names, such as *ras, src,* and *myc*. Shortly after the viral oncogenes were discovered, scientists found that copies of these genes, christened *proto-oncogenes,* resided in our chromosomes. The "viral" oncogenes were actually abnormal versions of these cellular proto-oncogenes that the viruses had acquired and transmitted by transduction.

Oncogenes and Proto-Oncogenes

What do oncogenes do? An interesting and logical picture has emerged from over 2 decades of intense study in many laboratories. Proto-oncogenes are normal, essential parts of our genetic material that belong to the group of genes in charge of causing and regulating cell growth and division. It is actually changes in these genes that can lead to the development of cancer. Oncogenes are abnormal forms of proto-oncogenes.

Cell growth and division is a complex system involving many proteins. Growth is usually initiated by a signal, such as a hormone, received by a receptor on the cell membrane. After the receptor receives the "grow" signal, it transmits the message across the cytoplasm to the nucleus. There, the signal initiates the cell division cycle (Figure 25.5). Mutations affecting any step of the signaling pathway could cause a cell to "think" it is constantly being instructed to divide. For example, many receptors respond to outside signals by adding phosphate groups to themselves and/or other proteins. If a receptor were mutated in such a way that it added phosphates whether or not it was being signaled, the cell would "believe" it was constantly being instructed to divide. Such a genetic change would promote uncontrolled cell growth, rather like stepping on the accelerator in a vehicle. Likewise, the cell division cycle itself is heavily regulated. If any of the proteins regulating the cell division cycle were to malfunction, the division cycle could run amok. The oncogenes identified in the transforming viruses had undergone changes with this kind of effect.

The most common oncogene in human cancers is *ras*. Cells can be signaled to divide by a number of different growth factors that bind to receptors in the cell membrane. The Ras protein sits on the inside

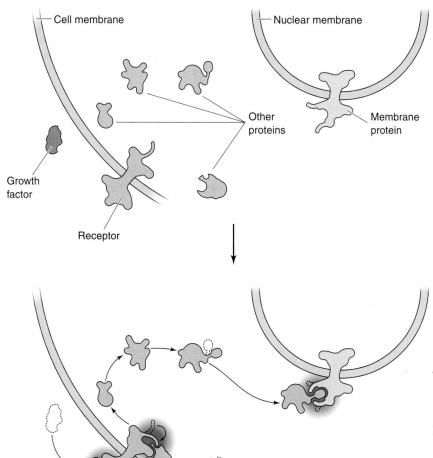

Cell membrane

Nuclear membrane

Other proteins

Membrane protein

Growth factor

Receptor

Figure 25.5 Extracellular growth signals are relayed to the cell nucleus via a series of protein-protein interactions, beginning at the cell membrane. The change in shape triggered by a growth factor binding to its receptor initiates a cascade of changes in associated proteins, resulting in a signal being transmitted to the nucleus.

face of the cell membrane and is activated by a number of growth factor receptors. When the correct growth factor binds to its receptor, the activated receptor in turn activates the Ras protein. The activated Ras protein goes on to activate other proteins, leading eventually to the activation of transcription of proteins involved in DNA replication and cell division.

Cancer-causing mutations in *ras* change the shape of the Ras protein so that it is always activated and therefore always signaling cells to divide. Researchers have identified several different amino acid substitutions that cause Ras to become locked in the active form. Mutant, permanently activated Ras proteins are involved in about 20% of all human cancers.

Tumor Suppressor Genes

Evidence from genetic studies of cancer patients led scientists to believe that another type of gene might also be involved in the development of cancer.

Subsequent research has proved their expectation to be correct. The product of this type of gene is like a brake on cell growth rather than an accelerator. Instead of being part of a signaling chain that tells a cell to divide, this type of protein represses cell growth and division. These repressor proteins apparently must be inactivated before full cancer can develop. The genes encoding these proteins are called tumor suppressor genes.

As might be expected, mutations in oncogenes that lead to cancer are usually different from mutations in tumor suppressor genes that lead to cancer. In oncogenes, the mutations must activate the protein to promote growth inappropriately. This kind of mutational change includes a change in a single amino acid that leads to an altered form of the protein, multiplication of the gene within the chromosome to provide greater activity, or alteration of the control regions of the gene to deregulate its expression or to cause it to be regulated inappropriately. These mutations also tend to be dominant; the presence of one normal copy of the oncogene cannot make up for the mutant, activated form. For example, the chromosome 9,22 translocation that brings about the inappropriate activation of the *abl* gene causes leukemia, even though a normal *abl* gene remains on the other copy of the chromosome. The cancer-causing viruses were able to transform cultured cells even though those cells had normal copies of the proto-oncogenes.

For tumor suppressor genes, however, loss of the protein function is required to promote cancer development. For this reason, mutations in suppressor genes tend to be recessive; one good copy of the gene can provide active protein. You might expect that individuals could inherit one copy of a recessive mutation in a tumor suppressor gene and that these individuals might be more likely to develop cancer during their lives. You would be right.

The first tumor suppressor gene to be identified was the retinoblastoma gene. Retinoblastoma is a tumor of the eye that can be hereditary. Analysis of patients with hereditary retinoblastoma revealed that one copy of a certain gene was inactivated in all of their cells and that both were inactivated in the tumors. A comparison with other patients whose retinoblastoma was not hereditary showed that the same gene, now called the retinoblastoma *(rb)* gene, was also inactivated in their tumor cells. Since loss of function of *rb* was apparently required for the tumor to develop, it was concluded that the *rb* gene product must suppress tumor development. Further study of the *rb* gene has shown that its protein product resides in the cell nucleus. One of its functions is to regulate transcription factors that initiate the expression of proteins required for cells to proceed

from the G_1 stage of the cell cycle to the S phase (Figure 25.6). At present, we do not know why loss of function of the retinoblastoma protein affects retinal cells more than any other type of cell.

Children with hereditary retinoblastoma inherit one inactive copy of the *rb* gene from their parents. Every cell in their bodies has only one good copy of the *rb* gene. In these children, all that is needed is for one retinal cell to suffer a mutation to the other copy of *rb,* and the retinoblastoma tumor can develop. In genetically normal individuals, a single retinal cell would have to suffer two independent mutational events to knock out both copies of *rb,* an unlikely series of events. Retinoblastoma is a rare cancer. Most cases run in families, who apparently transmit the defective gene.

Mutation of another tumor suppressor gene, *p53,* is associated with a different inherited cancer syndrome. The Li-Fraumeni syndrome is a rare inherited syndrome of cancers. Members of Li-Fraumeni families develop cancers (often multiple cancers) of the breast, brain, bone, and other tissues and leukemia, usually at early ages. Patients with Li-Fraumeni syndrome inherit one defective copy of the *p53* gene, so every cell in their bodies has only a single good copy. Apparently, inactivation of this good copy in any number of different cell types can lead to cancer.

The *p53* gene is also important in noninherited cancer. Scientists who analyze genetic alterations in cancerous cells have found that *p53* mutations are present in more different kinds of cancer than any other known cancer-related genetic alteration: fully 50% of all cancers. Needless to say, this finding has sparked enormous interest in the biological role of *p53*.

The *p53* protein is a nuclear protein with several important regulatory roles. It is involved in regulating the cell cycle at both the G_1-to-S phase and the G_2-to-M phase transitions (Figure 25.6). If cells with normal *p53* suffer DNA damage, the *p53* protein prevents the cells from dividing, presumably until the damage is repaired. If the damage is too extensive, *p53* initiates events leading to the death of the cell. Cells that lack normal *p53* continue to divide. The replication of damaged DNA leads to changes (mutations) in the DNA sequences of daughter chromosomes, so a cell that lacks *p53* activity apparently can accumulate mutations at a much higher rate than normal cells do. According to this line of thinking, if one of these mutations leads to activation of an oncogene, the result is cancer.

Figure 25.6 The cell division cycle. The stages were named for what could be seen under the light microscope. G stands for gap, because no visible activity was occurring; S signifies DNA synthesis; and M stands for mitosis. Most nondividing cells are in G_1. Together, G_1, S, and G_2 constitute interphase.

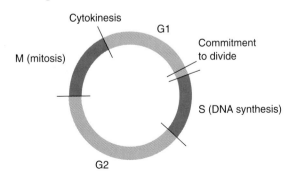

Genetic Complexity of Cancer

Genetic analysis of families with hereditary cancer and individual cancer patients is helping us identify additional genetic loci associated with cancer development. So far, the genetic picture of most common cancers looks very complicated, and no clear interpretation is available. It is clear that cancer is actually many diseases. It is also clear that different sets of mutations can cause similar cancers.

Breast cancer provides a good illustration of the genetic complexity of cancer. Approximately 2% of women who live to age 50 will develop breast cancer, and about 10% of women who live to age 80 will develop it. Most of these cases are sporadic, not associated with the inheritance of predisposing mutations. However, a tendency to develop breast cancer does run in some families. By studying such families, scientists identified two genes, *BRCA1* and *BRCA2,* that are associated with inherited breast cancer. Some of the families had the *BRCA1* mutation, and others had the *BRCA2* mutation. Screening of other families with inherited breast cancer showed that about 50% of them carried the *BRCA1* mutation and 30% of them carried the *BRCA2* mutation, indicating that still other genes associated with inherited breast cancer remain to be discovered. Male breast cancer is associated with *BRCA2* but not *BRCA1*.

Sporadic cases of breast cancer, about 90% of all cases, do not seem to be associated with *BRCA1* or *BRCA2* mutations. Currently, no obvious oncogenes or tumor suppressor genes are consistently associated with sporadic breast cancer. Alterations in a number of genes encoding growth factors, receptors, and cell division proteins have been reported.

Over 100 different oncogenes and tumor suppressor genes have been identified. As the Human Genome Project progresses toward its goal of understanding all human genes, more will likely be found. Table 25.1 lists a few of the known oncogenes and tumor suppressor genes and gives their roles in the control of cell growth.

Does our increasing understanding of cancer genetics hold out any hope for improved therapies in the future? Recent discoveries in cancer genetics have helped to identify several treatments that work by targeting cancer cells with a specific genetic change, such as Gleevec (imatinib mesylate), a drug for chronic myeloid leukemia and gastrointestinal stromal tumors, and Herceptin (trastuzumab), a drug for one form of breast cancer. By targeting specific cancer cells, these treatments avoid the damaging side effects of cancer drugs that simply target all dividing cells. These successful developments suggest that further examination of the molecular origins of cancer will result in the development of additional therapies.

Cancer Genomics

In late 2005, the National Cancer Institute and the National Human Genome Research Institute, both part of the U.S. National Institutes of Health, launched a comprehensive effort to accelerate our understanding of the molecular basis of cancer. The project, called The Cancer Genome Atlas (TCGA), will explore the universe of genomic changes involved in human cancers through the application of genome analysis technologies, especially large-scale genome sequencing. It will begin with a pilot project targeting three cancers: lung cancer, brain cancer (specifically, glioblastoma), and ovarian cancer.

When fully operational, TCGA will consist of four integrated components: a Biospecimen Core Resource (BCR), Cancer Genome Characterization Centers, Genome Sequencing Centers, and a Principal Bioinformatics Resource. The BCR will collect samples of normal and cancerous tissue and distribute them for analysis. The Cancer Genome Characterization Centers will analyze samples from the BCR to identify genomic alterations, such as copy number changes and/or chromosomal rearrangements. Selected genes will be sequenced by the Genome Sequencing Centers to identify small genomic changes, such as single-base mutations and small insertions or deletions. The data will be made publicly available in an integrated database that scientists can use to generate new knowledge through research.

Table 25.1 Some cancer genes and the physiological roles of their products

Gene	Role
Oncogenes	
sis	Growth factor
erbB, fms, neu	Cell surface growth factor receptor
ras, src, abl	Signal transmission within the cell
bcl2	Blocks cell-initiated death
myc, fos, myb	Regulators of transcription
Tumor suppressors	
rb	Regulation of replication and transcription
p53	Regulation of cell division cycle: stops cells from dividing if their DNA is damaged, allowing time for repair; initiates cell death if DNA is too damaged

The TCGA project will apply technologies developed for the Human Genome Project (see the *Reading* in chapter 29), as well as the large-scale, coordinated approach that was so successful in that project, to a molecular analysis of all types of cancers. Data from the TCGA pilot project will provide researchers and clinicians with an early glimpse of what promises to become a comprehensive "atlas" of molecular information describing the genomic changes in all types of cancer. It is envisioned that TCGA will ultimately enable researchers throughout the world to analyze and employ the data to develop a new generation of targeted diagnostics, therapeutics, and preventives for all cancers and will pave the way for more personalized cancer medicine.

Selected Readings

Cancer special issue. 1996. *Scientific American* 275(3). This issue contains articles on how cancer arises and spreads, new therapies, genetic testing for cancer genes, and more.

Cavenee, W. K., and R. L. White. 1995. Genes and cancer. *Scientific American* 272(3):72–79.

Greider, C., and E. Blackburn. 1996. Telomeres, telomerase, and cancer. *Scientific American* 274(2):92.

Pasternak, J. 1999. *An Introduction to Human Molecular Genetics*. Fitzgerald Scientific Press, Bethesda, MD.

Perera, F. 1996. Uncovering new clues to cancer risk. *Scientific American* 274(5):54.

Classroom Activities

Medical Sleuth: A Story of Genetics in Action

About This Activity

In this chapter, we have reprinted an article by Tom Schachtman that was published in the *Smithsonian* magazine in February 2006. It tells the story of D. Holmes Morton's medical genetics work among the Amish and Mennonites in Pennsylvania. It touches on all aspects of genetics, from classical pedigree studies to identification of disease genes to microchips. We provide this article as a compelling humanitarian story that can be used as a basis for integrating genetics and molecular biology information, as well as for putting medical genetics into historical context. We have supplied some questions for research and discussion that will require information not found in the article. Use them, substitute or add your own questions, or ignore them altogether. We provide answers to the scientific questions and pointers for the history questions. We do not intend the answers to the questions to be complete—your students can find much more information on every topic.

Answers to Questions for Research and Discussion

1. The Amish have a high incidence of rare genetic diseases because they are a highly inbred population founded by a small number of individuals. In the entire human population, the incidence of a particular recessive disease gene might be 1 in 1,000, making the odds of two carriers marrying 1 in 1,000,000. However, if a population was founded by 50 individuals and one of them has the gene, then the frequency in that population is 1 in 50. If members of the population marry each other almost exclusively, the odds of two carriers marrying one another would be significantly enhanced.

   You can make a model of the gene frequency in the population. Assume that heterozygotes are as healthy as homozygous normal individuals, that

everyone marries, that all couples have four surviving children, that the genes follow perfect Mendelian distribution, and that generations do not overlap. The first 50 individuals make 25 couples, one of whom contains the individual with the mutation. Half of the four children of that couple will carry the mutation. The next generation will then be 25 couples with 4 children per couple, or 100 children (25 couples × 4 children per couple = 100 children), with two siblings having the mutation. The gene frequency is still 1 in 50. Go through the process again. Everyone marries and has four children. One hundred children make 50 couples, and each has four children. This time, two of the couples contain an individual with the mutation, so half of their collective eight children, or four children, carry the mutation. The gene frequency is still 1 in 50, since we now have 4 carriers out of 200 children. Now we have two sets of sibling carriers that are first cousins to one another. The next generation will produce siblings, first cousins, and second cousins. In an intermarrying population, it becomes possible for some of these individuals to marry one another, bringing about the chance that their children will be homozygous for the mutation. If marriage were a fully random event within the population, the chances of two carriers marrying would be 1/50 × 1/50, or 1/2,500. Those odds are 400 times greater than those in the general population.

2. One reason is the intermarriage—it is much more likely that you will see a recessive phenotype in a small, inbred population than in a large, randomly mating one. Unless you can identify a phenotype, you cannot know that an underlying genetic condition exists. A second reason is the good records. They allow geneticists to reconstruct pedigrees and to deduce the genotypes of individuals. That information can be compared to molecular genetic data, as described in chapter 27, to determine the locations of the responsible genes. Finally, the fact that the individuals all live in a small area makes it easy for the geneticists to

find them, interview them, and ask them for samples for testing.

3. Maple syrup urine disease (MSUD) is an autosomal recessive disease caused by a defect in an enzyme complex required for breaking down the amino acids leucine, isoleucine, and valine. These amino acids and their incomplete breakdown products are secreted in the urine, imparting the sweet odor, and accumulate in the blood, where they cause neurological damage. At birth, infants with this disease appear normal because enzymes in their mother's bodies have broken down excess amino acids. Once the infants can no longer benefit from their mother's enzymes, the breakdown products begin to accumulate. If the disease can be diagnosed quickly, it is managed by carefully controlling the diet to limit the intake of the critical amino acids, rather like phenylketonuria management.

Mr. and Mrs. Hoover must have each been heterozygous for the disease. The affected children were homozygous mutants, and their unaffected children were either homozygous normal or heterozygous carriers.

More information can be found on the National Institutes of Health website at http://www.nlm.nih .gov/medlineplus/ency/article/000373.htm.

4. Morton dreams of making a microchip to determine the genotypes of Amish children. Chapter 27 explains how a microchip can be used to analyze genotypes. In this case, the DNA features on the microchip would represent the fragments of genes containing the mutations that cause the diseases in the Amish population.

5. The four humors theory was established by the ancient Greeks to explain both temperament and disease. The body was believed to be permeated and influenced by four types of fluid, or humors, that corresponded to the four elements of nature: blood (corresponding to air), black bile (corresponding to earth), yellow bile (corresponding to fire), and phlegm (corresponding to water). These humors were believed to influence health, and imbalances in the fluids gave rise to disease. The humors were thought to produce "vapors" that influenced personality, too.

This model of disease gave rise to treatment designed to restore balance among the humors. These treatments included bleeding, emetics (inducing vomiting), and purges (inducing diarrhea).

This approach to medicine was prevalent in U.S. medical practice as recently as the U.S. Civil War.

Louis Pasteur, who lived in France in the 19th century, is credited with the discovery that microbes cause disease. During his lifetime, he fought to convince surgeons that germs existed and carried disease and that dirty hands and instruments could spread infection. Pasteur discovered that weakened forms of bacteria and viruses could be used to protect individuals from future infection, and he invented pasteurization, a process that kills most bacteria in food while preserving the food.

Although we do not know what Terry Sharrer was thinking when he made the comments cited in the article, we presume his comparison of the advent of genetics in medical practice to the advent of germ theory is based on two things. One is the introduction of a new way of thinking about disease: that a person's own genetic makeup would play a pivotal role in how an individual responds to particular environmental stresses, treatments, or other circumstances. The other is that genetic testing is a new technology that is being introduced into medical practice, much as sterilization was introduced as a consequence of Pasteur's germ theory. Sharrer's overarching message is probably that the infusion of molecular genetics into medicine is likely to transform its practice as much as the advent of the germ theory did over 100 years ago. That is indeed a sweeping prediction.

6. Albert Schweitzer was born in 1875 and began his career as a theologian, a minister, and a renowned organist. He decided to go to Africa, not as a minister, but as a doctor, and at the age of 30 (1905) he entered medical school. In 1913, he founded a hospital in French Equatorial Africa. In 1917, he and his wife were made prisoners of war and spent about a year in confinement. After their release in 1918, they returned to Europe for 6 years, where Schweitzer preached, took medical courses, gave lectures and concerts, and wrote books. In 1924, he returned to French Equatorial Africa, where he would spend most of the rest of his life. He spent the royalty money from his book sales, speaking, and concert fees on his mission work and was renowned as a scholar and humanitarian. Schweitzer was awarded the Nobel Peace Prize in 1952 for his humanitarian work. He spent the $33,000 prize money to establish a hospital for leprosy patients. Schweitzer died in 1965.

Genetics in Action

Here, we share with you an article that was originally published in the *Smithsonian* magazine in February 2006. It tells the story of the work of a medical geneticist, D. Holmes Morton, among the Amish and Mennonites in Pennsylvania. Using his knowledge of medicine, some rare conditions he had studied, and basic genetics, Morton has solved several medical mysteries. He has also built a world-class clinic, where he uses genetics to help provide health care to his patients. The article tells his story, the stories of some of the mysteries he solved, and that of the clinic he founded and describes doing genetics among the Anabaptists of Pennsylvania.

You have now had enough exposure to both classical (pedigree-based) genetics and molecular genetics to follow the genetic elements in the story. We have included some questions at the end that will let you test your knowledge. Some can be answered by thinking about information provided in the story; others will require you to look up additional information. Finally, the article refers to two great historical figures in medicine: the scientist Louis Pasteur and the humanitarian Albert Schweitzer. These men have compelling stories of their own, especially if you are interested in medicine or medical science.

Medical Sleuth

To prosecutors, it was child abuse—an Amish baby covered in bruises, but Dr. D. Holmes Morton had other ideas

By Tom Shachtman

It was every parent's nightmare: a few days before Christmas 1999, Elizabeth and Samuel Glick, Old Order Amish dairy farmers in rural Dornsife, Pennsylvania, an hour's drive north of Harrisburg, found their youngest child, 4-month-old Sara Lynn, gravely ill. They rushed her to a local hospital, from where she was soon transferred to the larger Geisinger Medical Center in the next county. There, a doctor noted a hemorrhage in her right eye and extensive bruising on her body, and suspected that her injuries were caused by child abuse.

Alerted to the doctor's suspicion, the police and officials from the Northumberland County Children and Youth Services descended on the Glicks' farm during the evening milking, and took away the couple's seven other children, all boys, ranging in age from 5 to 15. The boys were separated and placed in non-Amish foster homes. Sara died the next day, and when the county coroner found blood in her brain, he declared her death a homicide.

At Sara's funeral, on Christmas Eve, Elizabeth and Samuel were not permitted to speak privately with their sons. By that time Samuel had already contacted the Clinic for Special Children in Lancaster County, and pleaded with its director, pediatrician D. Holmes Morton—the world's leading authority on genetic-based diseases of the Amish and Mennonite peoples—to find the cause of his daughter's death.

The Amish are Anabaptists, Protestants whose forefathers were invited by William Penn himself to settle in Pennsylvania. Today there are almost 200,000 Amish in the United States, of whom 25,000 live in Lancaster County, in southeastern Pennsylvania between Philadelphia and Harrisburg. Some of their customs and religious values have changed little over the past century.

Most people know that the Amish wear conservative clothing, travel mainly by horse and buggy, eschew most modern technologies, and refuse to use electricity from the common grid. The Amish also remove their children from formal schooling after the eighth grade, do not participate in Social Security or Medicare, and in many other ways maintain their sect's separateness from mainstream America.

But most people don't know that the Amish, and their spiritual cousins the Mennonites, experience an inordinately high incidence of certain genetic-based diseases, most of which affect very young children.

Many of these afflictions are fatal or disabling, but some, if diagnosed and properly treated in time, can be managed, enabling the children to survive and lead productive lives.

That possibility—of proper diagnosis and intervention to save children's lives—was what intrigued Morton, then a recently minted M.D. on a postdoctoral fellowship. A colleague at Children's Hospital in Philadelphia asked him one evening in 1988 to analyze a urine sample from a 6-year-old Amish boy, Danny Lapp, who was mentally alert but wheelchairbound because he had no control over his limbs—perhaps from cerebral palsy.

But when Morton analyzed the urine, he saw no evidence of cerebral palsy. Rather, in a diagnosis that must have seemed to others like the amazing deductions of Sherlock Holmes, he recognized the footprint of a genetic-based disease so rare that it had been identified in only eight cases in the world, none of them in Lancaster County. Morton's was an educated guess: he was able to recognize the disease, a metabolic disorder known as glutaric aciduria type 1, or GA-1, because it fit the pattern of diseases he had been studying for almost four years, those that lay dormant in a child's body until triggered into action.

Typically, a child with GA-1 shows no sign of the disorder until he or she comes down with an ordinary childhood respiratory infection. Then, perhaps prompted by the body's immune response, the GA-1 flares up, making the child unable to properly metabolize protein-building amino acids, which in turn causes a buildup in the brain of glutarate, a toxic chemical compound that affects the basal ganglia, the part of the brain that controls the tone and position of the limbs. The result, permanent paralysis of the arms and legs, can resemble cerebral palsy.

Sensing that there might be other GA-1 children in the deeply inbred Amish community—some of them, perhaps, treatable—Morton visited Danny Lapp and his family at their Lancaster County home. Indeed, the Lapps told him of other Amish families with similarly disabled children. "The Amish called them 'God's special children,' and said they had been sent by God to teach us how to love," says Morton. "That idea deeply affected me."

In the following months, Morton and his fellowship supervisor, Dr. Richard I. Kelley of Johns Hopkins University, visited the other families with afflicted children and collected from them enough urine and blood samples to identify a cluster of GA-1 cases among the Amish. "We very quickly were able to add to the world's knowledge base about GA-1," Richard Kelley recalls. "For a geneticist, that's exciting."

Rebecca Smoker, an Amish former schoolteacher who had lost nieces and nephews to GA-1 and now works for Morton's Clinic for Special Children, vividly remembers the sense of relief that began to spread through the close-knit Amish community. Previous doctors, Smoker recalls, had been "unable to tell parents why their children were dying," but Morton was able to identify the disease. That was comforting: "If you can say, 'my baby has this,' or 'my baby has that,' even if it's an awful thing, you can feel better about it," says Smoker.

Later in 1987, Morton began driving out from Philadelphia to Lancaster County to manage the care of children with GA-1. Many of the patients who had been previously diagnosed with cerebral palsy were paralyzed beyond repair, but there were some with less advanced paralysis whom Morton was able to help with a new treatment regimen including a restricted-protein diet and, when needed, hospital care. He also learned, through testing, that some of the affected children's younger siblings—who had not yet suffered paralysis—had the gene mutation and biochemical abnormalities. If he could manage these children through their earliest years, when they were particularly vulnerable to the effects of GA-1, he believed, as he says now, that he could "alter the likely devastating course of the disease."

Several of the children came down with respiratory infections in the months that followed. Morton's strategy—"immediately getting them to a hospital, giving them IV glucose and fluids, anticonvulsants, and reducing their protein intake to get them past the crisis points"—worked, and they escaped without severe injury to their basal ganglia. Morton had gone beyond giving the horror its proper name; he had found ways for Amish parents to help save their other children from the ravages of the disease.

Now, nearly a decade later, Sara Lynn Glick's death presented Morton with a new challenge. He was determined to figure out what had killed her, to exonerate Elizabeth and Samuel Glick, and to help them retrieve their seven sons from non-Amish foster homes.

Morton's first clue to what had actually happened to Sara came in a conversation with her mother. "Liz Glick told me that she had to put socks on Sara's hands, because Sara had been scratching her own face," Morton says. Such scratching, he knew, was a likely sign of an underlying liver disease. Another clue was that Sara had been born at home, where a midwife had not given her a vitamin K shot—standard procedure for hospital-born babies, who are given the shot to ensure that their blood will clot properly.

Morton concluded that Sara's death was due not to child abuse but to a combination of genetic disorders: a vitamin K deficiency, coupled with a bile-salt transporter disorder that he had previously found in 14 other Amish children and some of Sara's cousins.

Convincing the authorities, however, wouldn't be easy. So Morton called a friend, Philadelphia lawyer Charles P. Hehmeyer. "You're always looking for good pro bono cases," Morton remembers telling Hehmeyer. "Well, here's a doozy." Together, they went to see the Glicks in Dornsife, where they sat in a candlelit kitchen, long after dark, as Liz Glick asked through tears if she would be going to jail.

Sure of his diagnosis, Morton went—uninvited—to a meeting between doctors and the district attorney's office at Geisinger Medical Center, hoping to point out that the hospital's own records would conclusively demonstrate that Sara's injuries had not come from child abuse. He was shown the door.

The clinic for Special Children in Strasburg, Pennsylvania, lies only a few hundred miles from Morton's childhood home in Fayetteville, West Virginia. But for him the journey was long and full of unexpected turns. The second youngest of a coal miner's four sons, Holmes flunked all of his science classes in high school, sank to the bottom of his class and withdrew before graduation. "I was never an easy person to teach," he admits. "I was always doubting, questioning, arguing." He took a job in an engine and boiler room of a freighter on the Great Lakes—"my first encounter," he says, "with people who were very intelligent but had little higher education." Focusing on practical shipboard problems and doing plenty of physical labor were a spur to developing his mind: within a few years he passed an examination for a commercial license to operate the boilers, and then completed his high-school equivalency degree.

Drafted in 1970, Morton spent four years "working the Navy's boilers"; off duty he read about, and then took correspondence courses in, neurology, math, physics and psychology. After the Navy, he enrolled at Trinity College in Hartford, Connecticut, volunteered at a children's hospital and set his sights on a medical degree.

At Harvard Medical School Morton developed an interest in what he calls "biochemical disorders that cause episodic illnesses." Like a sudden storm troubling a ship on the Great Lakes, these disorders disrupt in a seemingly static environment and do great damage—maybe irrevocable damage. But afterward everything is calm again. As a resident at Boston Children's Hospital in 1984, Morton met a child who had been diagnosed by the admitting physician as having Reye's syndrome, a buildup of pressure in the brain and an accumulation of fat in the liver and other organs that often occurs during a viral infection such as the flu or chicken pox. Morton thought the diagnosis mistaken, substituted his own—a metabolic disorder—and accordingly changed the child's diet and treatment regimen. The child recovered and now lives a normal life, and the case gave Morton the confidence, three years later, to discount the cerebral palsy diagnosis for Danny Lapp and diagnose him with GA-1 instead.

Another such "episodic" disease, this one not found among the Amish but among the much larger Mennonite community, had piqued his interest in the late 1980s. Like the Amish, the Mennonites are Anabaptists. But they use some modern technologies, such as internal-combustion engines, electricity and telephones in the home.

Enos and Anna Mae Hoover, Mennonite organic dairy farmers in Lancaster County, lost three of their ten children, and had a fourth suffer permanent brain damage, before Morton arrived on the scene. Their ordeal began in 1970 with the birth of their second child. When the child became ill, refusing the bottle and going into spasms, "the doctors had no idea what was wrong," Enos recalls in a low, even voice. When the boy was 6 days old he fell into a coma, and he died a week later at a local hospital. Four years later, when an infant daughter refused to nurse, the Hoovers took her to a larger hospital, where a sweet smell in her diaper finally alerted doctors to what was afflicting her and had killed her brother: Maple Syrup Urine Disease, or MSUD, which prevents the body from properly processing proteins in food. By then, however, the little girl had already suffered irreparable brain injuries. "Even with a later baby, it took three to four days to get a proper diagnosis," Enos says. "We missed the crucial days where better treatment could have made a difference. Then a doctor asked us if we'd like to meet a Doctor Morton. We said yes, and we were amazed when he came to our house. No other doctor had ever come to see us or our babies."

Around the time of Morton's first visits with Enos and Anna Mae Hoover, he was realizing, as he would later write, that the "economic and academic goals of university hospitals" seemed to be "at odds with the care of children with interesting illnesses." He concluded from his work with GA-1 and MSUD children that the best place to study and care for them was not in a laboratory or a teaching hospital but in the field, from a base in the area where they lived. With his wife, Caroline, a fellow West Virginian who holds a master's degree in education and public policy from Harvard and had worked with rural communities and schools, Morton envisioned a free-standing clinic for Amish and Mennonite children who have rare genetic diseases.

Enos Hoover helped raise some money for the Mortons' dream within the Mennonite community, and Jacob Stoltzfoos, grandfather of a child with GA-1 saved by Morton's intervention, did the same among the Amish. Stoltzfoos also donated farmland in the small town of Strasburg for a clinic. Both Hoover and Stoltzfoos eventually accepted invitations to serve on the board of the as-yet unbuilt clinic, where they joined sociologist John A. Hostetler, whose pioneering 1963 book, *Amish Society,* first drew medical researchers' attention to potential clusters of genetic disorders among Pennsylvania's rural Anabaptists.

As Hostetler's book makes clear, says Dr. Victor A. McKusick of Johns Hopkins University, the founding father of medical genetics, the Amish "keep excellent records, live in a restricted area and intermarry. It's a geneticist's dream." In 1978, McKusick published his own compilation, *Medical Genetic Studies of the Amish,* identifying more than 30 genetic-based diseases found among the Amish, ranging from congenital deafness and cataracts to fatal brain swellings and muscular degeneration. Some had never been known before at all, while others had been identified only in isolated, non-Amish cases. "The diseases are hard to identify in the general population because there are too few cases, or the cases don't occur in conjunction with one another, or the records to trace them back are incomplete," McKusick explains. He adds that Morton, by identifying new diseases and by developing treatment profiles for diseases like GA-1 and MSUD, is not only building on the foundation that McKusick and Hostetler laid: he's been able to create treatment protocols that doctors around the world can use to care for patients with the same disorders.

But back in 1989, despite the efforts of Hoover, Stoltzfoos, Hostetler, and Lancaster County's Amish and Mennonite communities, there was still not enough money to build the free-standing clinic the Mortons wanted. Then Frank Allen, a staff reporter for the *Wall Street Journal,* wrote a front-page article about accompanying Morton on house calls to Amish patients, mentioning that Holmes and Caroline were prepared to place a second mortgage on their home to build the clinic and to buy a particularly critical piece of laboratory equipment made by Hewlett-Packard. Company founder David Packard read the article and immediately donated the machine; other *Journal* readers sent in money, and the clinic was on its way.

There was still no building, but the money and machinery were put to use in rented quarters, allowing the screening of newborns for GA-1 and MSUD. And then, on a rainy Saturday in November 1990, dozens of Amish and Mennonite woodworkers, construction experts and farmers erected the barnlike structure of the Clinic for Special Children, stopping only for lunch served by a battalion of Amish and Mennonite women.

Early in the year 2000, pressure from Hehmeyer, Morton and local legislators—and from a public alerted by newspaper stories—pushed the Children and Youth Services to move the seven Glick children from non-Amish foster homes into Amish homes near their farm. In late February the boys were returned to their parents. But Samuel and Elizabeth remained under investigation for child abuse in connection with Sara's death. A week later, the Northumberland District Attorney's office turned over the most important piece of evidence—Sara's brain—to outside investigators. At the Philadelphia Medical Examiner's Office, Dr. Lucy B. Rorke, chief pathologist of Children's Hospital in Philadelphia and an expert on the pathology of child abuse, examined it during a teaching session with other doctors and students, and quickly concluded that Sara had not died of trauma or abuse.

A few weeks later, the Glicks, who had never been formally charged, were entirely cleared of suspicion. The family was relieved, and Morton was inspired: he accelerated his efforts to find the precise genetic locus of the bile-salt transporter disease so the clinic could better identify and treat it. Most newborns in Lancaster County were already being screened for a handful of the diseases that afflict Amish and Mennonite children. Morton wanted to add to the list the disease that took Sara Lynn Glick's life.

"We don't pick problems to research," says the Clinic for Special Children's Dr. Kevin Strauss. "The problems choose us. Families come in with questions—

'Why isn't my child developing properly?' 'Why is this happening?' 'What causes that?'—and we look for the answers." Strauss, a Harvard-trained pediatrician, joined the clinic because he agreed with its operating philosophy. "If you want to understand medicine, you have to study living human beings," he says. "It's the only way to translate advances in molecular research into practical clinical interventions. You can't really comprehend a disease like MSUD, and treat it properly, without involving biology, infections, diet, amino acid transport, brain chemistry, tissues and a lot more."

When Morton began his work among the Amish and Mennonites, fewer than three dozen recessive genetic disorders had been identified in the groups; today, mostly as a result of the clinic's work, some five dozen are known. Cases of GA-1 have come to light in Chile, Ireland and Israel, and of MSUD in India, Iran and Canada.

The clues come from anywhere: working with one Amish family, Morton learned that a 14-year-old girl had kept a diary while caring for a terminally ill sister. Using information from the diary and other patients, the clinic was able to help map the gene mutation for a syndrome responsible for the crib deaths of 20 infants in nine Amish families—with implications, perhaps, for progress in solving SIDS (Sudden Infant Death Syndrome), which kills thousands of children each year in the larger population.

And at a Mennonite wedding two summers ago, family members rolled up their sleeves to have their blood drawn by Morton, Strauss and a clinic nurse. The team was trying to pinpoint a genetic defect that made the males of the family susceptible to a form of meningitis that had killed two of them. The tests revealed that, of the 63 people whose blood was drawn at the wedding, a dozen males were at high risk, and 14 of the women were carriers. The men were put on penicillin, vaccinated and given stashes of antibiotic to take if they became ill. Shortly after the wedding, the combination of antibiotics and immediate hospital care prevented one man from succumbing to a meningitis attack, possibly saving his life. "Genetics in action," Morton comments.

But Morton's approach to identifying and treating a disease is more than mere genetics. On an average morning, the clinic's waiting room looks like any pediatrician's office—albeit with most adults in traditional Amish and Mennonite dress—with children crawling about on the floor, playing with toys or sitting as their mothers read them books. The appearance of normalcy is actually deceiving, says Kevin

Strauss. "Most of the kids here today have genetic diseases that, left untreated, can kill them or lead to permanent neurological disability." Parents have brought their children, some from as far away as India, not only for the clinic's renowned research capabilities but for its treatment. Donald B. Kraybill, one of the foremost scholars of the Amish, and the Senior Fellow of Elizabethtown College's Young Center for Anabaptist and Pietist Studies, praises Morton's "culturally sensitive manner," which he says has won Morton the "admiration, support and unqualified blessing of the Old Order communities."

The communities' support is expressed, in part, through an annual series of auctions to benefit the clinic that are held by the Amish and Mennonites across Pennsylvania. These auctions raise several hundred thousand dollars of the clinic's annual $1 million budget. Another chunk of the budget is covered by outside contributors, and the remainder comes from the clinic's modest fees—"$50 for a lab test that a university hospital has to charge $450 for," notes Enos Hoover.

About two years after Sara Glick's death, Morton, Strauss, clinic lab director Erik Puffenberger, who holds a doctorate in genetics, and researcher Vicky Carlton from the University of California at San Francisco located the precise genetic site of the bile-salt transporter disorder, and devised a test that could tell doctors whether an infant might have it. If the test is done at birth, or at the first sign of a problem, no family will ever have to repeat the Glicks' ordeal.

Or, perhaps, any other ordeal caused by diseases passed on genetically in the Amish and Mennonite communities. Morton and his colleagues believe that they're within a few years of realizing a long-term dream: placing, on a single microchip, fragments of all the known genetic diseases of the Amish and Mennonites, so that when a child is born, it will be possible to learn—from comparing a small blood sample from the child with the DNA information on the microchip—whether he or she may be affected by any of a hundred different conditions, thus allowing doctors to take immediate treatment steps and prevent harm from coming to the child.

The clinic's use of genetic information as the basis of diagnosis and the individualized treatment of patients make it "the best primary care facility of its type that exists anywhere," says G. Terry Sharrer, curator of the Smithsonian's Division of Science, Medicine and Society. And he suggests an analogy: over a hundred years ago, when Louis Pasteur's germ theory of disease replaced the four humors theory, it took decades for a majority of doctors to understand and adopt the new approach. "Most of the switching didn't occur until the next generation came out of medical school. Something similar is happening now with gene-specific diagnoses and treatment, as the aging baby-boom generation demands more effective medicine. The Clinic for Special Children shows that health care can be reasonably priced, highly tailored to patients and conducted in simply managed circumstances."

If Sharrer is right, the clinic may be a model for the future of medicine. Even if it's not, Morton's contribution has not gone unnoticed. Three years after the clinic opened its doors, he received the Albert Schweitzer Prize for Humanitarianism, given by Johns Hopkins University on behalf of the Alexander von Humboldt Foundation. On being notified of the prize, Morton began to read about Schweitzer and found that the great German physician also came to medicine late, after a distinguished career in music and theology—and that he had established his famed hospital in Gabon at age 38, the same age Morton was when he began the clinic in Strasburg. In a speech accepting the award, Morton said that Schweitzer would have understood why the Clinic for Special Children is in the middle of Lancaster County—because that "is where it is needed . . . built and supported by people whose children need the care that the clinic provides." After winning the award, partially in homage to Schweitzer and his love of Bach, Morton took up playing the violin.

Questions for Research and Discussion

1. Why might the Amish have such a high incidence of rare genetic diseases?

2. Victor McKusick said that the Amish are a "geneticist's dream" because they keep excellent records, live in a restricted area, and intermarry. Why would those three characteristics make the Amish such a good population for a geneticist to study?

3. What is the genetic defect in MSUD? What causes the urine of these patients to have a sweet smell? As described in the article, Enos and Annie Mae Hoover lost 3 of their 10 children to MSUD, and a fourth suffered permanent brain damage. What are the likely genotypes of Mr. and Mrs. Hoover? What are the genotypes of their affected and unaffected children?

4. The article states that Holmes Morton has a long-term dream of placing, "on a single microchip, fragments of all the known genetic diseases of the Amish and Mennonites, so that when a child is born, it will be possible to learn . . . whether he or she may be affected by any of a hundred different conditions." To what technology is the article referring? What "fragments of all the known genetic diseases" would be placed on a microchip? How could such a chip be used to reveal whether a child had any of the diseases?

5. Terry Sharrer of the Smithsonian Institution compared the infusion of new understanding about genetics into medical practice to the advent of Louis Pasteur's germ theory and its replacement of the four humors theory. What was the four humors theory? What was the germ theory that replaced it? Why might Sharrer have chosen this for comparison to the advent of medical genetics in medical practice?

6. Who was Albert Schweitzer, and why might he have had a humanitarian prize named in his honor?

E. Genomics

In Section B, we presented techniques for manipulating and analyzing specific segments of DNA. In this section, we expand our focus to whole genomes. This shift of focus parallels the trajectory of molecular biology research: in the 1970s and 1980s, scientists used new tools and techniques to analyze single genes. However, organisms have many genes, and understanding how genotype relates to phenotype, or how gene expression changes during development or disease, requires understanding global gene expression and interplay.

In the 1990s and the beginning of the 21st century, techniques were developed that allowed scientists to begin to address questions at this level, giving rise to a new term and a new field of science: genomics, the analysis of the genome. These lessons focus on approaches for the analysis of genomes: comparing them, typing them with genetic markers, mapping genes and markers, and analyzing global gene expression. All the activities are pencil-and-paper or computer simulations. Genomics is a fast-moving field; our goal for this section is to provide a foundation that will assist you and your students in following and understanding new developments and discoveries.

Comparing Genomes

27

About This Activity

This chapter introduces the concept of using markers, such as repeated sequences, to compare genomes and discusses the various levels at which genomes can be compared: between genera, species, breeds within a species, or individuals. It contains a paper activity that provides the foundation for understanding how restriction analysis and polymerase chain reaction (PCR) can be used for genome comparisons and a *Reading* that discusses the use of mitochondrial DNA in genetic studies. Students will need to have completed the Southern hybridization worksheets and the PCR lessons before doing this activity, since those two lessons introduce the laboratory techniques used in the activity.

The information in the student introduction provides a brief overview of how genomes change and what sorts of questions can be addressed with genome comparisons, explains two approaches to DNA typing, and illustrates their use by focusing on the analysis of dog genomes. We have provided some references about DNA typing and dogs below. The teacher's background reading includes information on the use of DNA comparisons in fields such as conservation biology, evolutionary biology, and behavioral biology.

Chapter 28 narrows the focus to forensic applications of DNA typing. It includes an activity that demonstrates probability calculations used in analyzing forensic DNA data, a reading about an application of DNA typing to an archaeological puzzle, and three short exercises that illustrate specific applications of DNA typing in the form of puzzles: *A Mix-Up at the Hospital, A Paternity Case,* and *The Case of the Bloody Knife.*

Both chapters 27 and 28 discuss the uses of different kinds of DNA markers without addressing how markers are identified. Chapter 29 addresses that question using short tandem repeat (STR) markers as an example and shows how DNA markers are used in the physical mapping of genes.

Class periods required: 1 to 2

Background

There are many reasons to compare genomes, or to analyze the degree to which two or more genomes vary. An evolutionary biologist might want to compare the genomes of different species to assess the degree of relatedness. A mammalogist might want to know just how different the genomes of a Chihuahua and a Great Dane are and whether there are regions where the differences are concentrated. An epidemiologist might want to know whether the bacteria causing disease in two different geographic areas are exactly the same or two different variants of the same strain.

The most complete way to answer these questions is to obtain the sequences of the genomes and compare them. Although scientists are completing genome sequences for more and more organisms, doing so is still a formidable task for creatures with large genomes, such as animals and many plants. The number of completely sequenced genomes is only a miniscule fraction of the genomes that are out there to ask questions about.

In fact, complete genome sequence information is not necessary for many useful comparative genome studies. Scientists typically approach genome comparisons with questions they want to answer, such as how the genomes of individuals within a species vary, how two closely related species are different, or how the genomes of less closely related groups of organisms have diverged during evolution. Answering these different questions requires looking at genomes through lenses of different focal lengths, from the gross to the very fine.

If you are an evolutionary biologist interested in studying the evolution of canids, for example, you would be interested in how the genomes of wolves, jackals, coyotes, foxes, and domestic dogs differ. Regions of the genome that were the same in all canids but differed between canids and, say, cats or bovines would not help you answer your questions. Genome comparisons at that level would have too long a focal length. Similarly, regions of the genome that varied between dachshunds and Great Danes or from one poodle to the next would probably not be helpful to you. The focal length necessary for comparisons of these fine changes would be too short for your questions. Answering your questions would involve finding regions of the genome that were common to dogs but varied between dogs and other canids.

However, if you were interested in identifying genes responsible for the giant size of Great Danes or the short legs of dachshunds, then you would need a shorter focal length. Regions of the genome that varied between dachshunds and Great Danes would be of high interest, while regions of the genome that were common to all dog breeds would not be. And if you were a geneticist who provided a dog paternity testing service (yes, these services exist), you would require an even shorter focal length, focusing on regions that varied from one poodle (for example) to the next. Regions of the genome that all poodles had in common would not be useful for your purposes. Thus, the questions being asked determine the level of variation that is considered.

Techniques for genome comparisons

The introduction to the *Student Activity* describes the use of restriction analysis and PCR to look at polymorphic repeated DNA sequences in the genome. These sequences are an invaluable tool for comparing and mapping genomes. They consist of a repeated sequence of varying length (10 to 15 bases down to just 1 or 2) that occurs at a fixed position (a locus) in the genome, but the number of repeats present on any given chromosome is highly variable. These loci are called microsatellites.

PCR or restriction analysis of microsatellite regions is used both to determine the overall similarity between genomes of different species, strains, or genetic variants (such as different dog breeds) and to compare the genomes of two individuals, as in DNA typing. The focal length of the comparison is determined by the regions selected for study. For comparing genomes of different species or strains, the microsatellites that are most helpful are those that are fairly constant within a given species or strain, so that variation seen between two strains is not just random individual variation but instead represents variation between the two strains. However, for DNA typing of individuals, the microsatellites of greatest utility are those that vary the most from one individual to the next.

How do researchers know what microsatellites are appropriate for making different kinds of comparisons? In the beginning, they don't. When researchers want to study a group of organisms whose genomes have not previously been characterized, they start with brute force, trying out different sets of PCR primers or restriction enzymes and observing the level of variation they see. We describe the process for identifying a microsatellite marker below.

PCR analysis can be used to compare genomes even when nothing is known about the sequences of the genomes in question. To do this, scientists use primers that are 10 bases long with random sequences. Anywhere on the genome that the primers can anneal in the right orientation and close enough together, a product can be generated. Whether a given primer pair yields products and of what lengths is a function of the sequence of the genome. The more similar two genomes are, the more amplification products they will share. Scientists typically test many sets of random primers to identify useful combinations for a given genome. This type of analysis is called **random amplification of polymorphic DNA.**

Another type of variation observed to exist between individuals or strains is single nucleotide differences **(single nucleotide polymorphisms,** or **SNPs);** that is, different individuals have different bases at these particular sites in the genome. Genotyping by SNP analysis involves very short, specific sequencing reactions at just these sites. While microsatellite-based DNA profiling works well for genome comparisons, such as species identification or individual DNA fingerprinting, SNP analysis of specific regions provides much finer-scale information about genome variation, such as which specific alleles of a gene are present.

The development of marker-based genome comparisons

Before the development of methods for studying isolated DNA molecules, scientists had very few options for comparing genomes and little was known about their relative similarities or differences. The technique of whole-genome hybridization and the comparison of the appearance of chromosomes in **karyotypes** were the only methods available for comparing genomes until the 1980s.

Restriction enzymes were first discovered in the 1960s, and in the early 1970s, scientists published descriptions of making the first recombinant plasmids. The method now known as Southern hybridization was published in 1975. Armed with new tools and techniques that let them analyze DNA in a way never before possible, scientists began to look for differences in DNA that could be related to differences in inherited traits. This was not easy research at the time: the chain termination method of DNA sequencing had not yet been developed, no genome maps existed, and very few clones of human DNA had been characterized. One approach scientists could take was to do restriction digests of human DNA and perform a Southern hybridization analysis using available cloned DNA as a probe.

In 1980, Arlene Wyman and Ray White published a paper titled "A highly polymorphic locus in human DNA." This was the first description of a hypervariable locus, and other scientists immediately realized how valuable such loci could be in their studies of inheritance. They began to look for more of them and found them. One of the scientists involved in this early work was Alec Jeffreys, whose laboratory was studying the inheritance and evolution of the globin genes. They realized that some of the newly discovered repeated sequences, which they called minisatellites, shared similar core sequences of 10 to 15 bases.

In the hope of developing a global method for finding more of the polymorphic loci, Jeffreys and his coworkers began to probe human DNA libraries with the minisatellite region from the myoglobin gene. They found additional variable loci that shared the core sequence of the minisatellite. They decided to try a probe that consisted only of the core sequence. They tried it with a Southern blot containing DNA from several family members. On Monday morning, 10 September 1984, they developed the autoradiogram and saw what seemed to be highly variable profiles that were inherited in a simple Mendelian manner. They realized almost immediately how this type of analysis could be applied to criminal cases, paternity disputes, conservation biology, and more. DNA fingerprinting had been born. In fact, Alec Jeffreys himself was the first scientist to conduct a DNA-fingerprinting test, in the course of an immigration dispute over the maternity of an African boy. A nice retrospective article by Jeffreys about these discoveries and the development of DNA fingerprinting is given in the *Selected Readings* below.

Identifying a polymorphic repeat locus

The first STR (or SSR, for simple sequence repeat) in human DNA was described in 1989 and had the sequence $(5'\ CA\ 3')_n$, where n is variable. The sequence was observed to occur at over 50,000 different locations within the human genome. Each one of these locations could potentially be a hypervariable DNA marker. Other microsatellite repeat sequences have also been identified, but we will use $(CA)_n$ as an example to show how polymorphic repeat loci can be identified and mapped.

The first step in identifying a polymorphic STR locus is to find a clone of a specific repeat in a DNA library. To do this, you might screen a human genomic DNA library with a labeled probe consisting of, for example, $(CA)_{20}$. Your screen would pick up many bacterial colonies containing plasmids with inserts having CA repeats. From those positive clones, you could choose one or a few for further characterization.

You would next sequence the cloned plasmid insert to determine the exact number of CA repeats [since your probe could hybridize to many different versions of $(GT)_n$] and also to find the sequence of the flanking regions. For that particular instance of $(CA)_n$ to be a useful marker, the flanking regions must be unique so that unique PCR primers could be designed for them. Once the sequence of the flanking regions had been determined, you would make fluorescently labeled probes using those sequences. You would use the probes to do a fluorescent in situ hybridization (see chapter 16) analysis to a human chromosome spread to determine whether the probes hybridized to only one location and to which chromosome they hybridized. If the flanking regions hybridized to only one site, you would next make PCR primers having the sequence of the flanking regions. A PCR with the primers would result in the amplification of the region with your repeat.

Now you would be equipped to determine whether that particular locus was polymorphic. To do this, you would have to gather DNA samples from many individuals, perform PCR with the flanking-region primers, and analyze the lengths of the products by gel electrophoresis. If the amplified DNA segment was the same length in all the individuals, then that locus would not be polymorphic. Alternatively, if the PCR products were of different lengths, then you would have a polymorphic STR, and the different PCR products would represent different alleles. Studies of large numbers of individuals would reveal the relative frequency of each allele in the population.

Loci of DNA repeats and restriction fragment length polymorphisms (RFLPs) are named according to rules agreed upon by the DNA Committee of the International System for Human Linkage Maps. For example, D1S80 indicates a locus identified by a DNA (D) probe that hybridizes to chromosome 1, to a single-copy (S) sequence. The "80" indicates that the location was the 80th such marker registered on that chromosome.

Physical mapping of chromosome markers

Now that we know we have a polymorphic STR, we would like to know more about where it is located. A fluorescent in situ hybridization analysis can show which chromosome contains the STR, and even which section of a chromosome. However, a section of a chromosome still represents an awful lot of DNA—potentially hundreds of thousands of base pairs. How we go about narrowing the location of our marker depends on the tools we have at hand. If the genome in question has been sequenced, our task is simple. We need only search a computerized database for the unique flanking sequences. What if a genome sequence is not available for comparison? Here again, it depends on what tools exist.

One very valuable mapping tool is a collection of so-called radiation hybrid cell lines. Radiation hybrids are rodent cells that each contain one or more fragments of a human (or other species of interest) chromosome. Such cell lines are made by treating human cells with sufficient high-energy radiation (X rays or gamma rays) to kill the cells and fragment the chromosomes. These dead cells are then fused with rodent cells. A small percentage of the rodent cells take up fragments of the human chromosomes and incorporate them into their genomes. Individual hybrid cells are isolated and propagated in culture, creating a radiation hybrid cell line. The human chromosome fragment carried in a hybrid cell line can be recognized by its staining pattern, just as intact chromosomes can be recognized.

To determine the relative location of an STR marker, the flanking sequences are used as a probe and tested in hybridization analyses to DNA prepared from a panel of radiation hybrid cells. The cell lines to which the probe hybridizes are analyzed to determine the boundaries of their region of overlap.

If clones of smaller chromosome fragments from within that region are available, a similar analysis can be repeated with them. Because of the high level of interest in the human genome, many such tools were available before the sequence was completed, so fairly high-resolution maps could be created. Maps based on analyses of DNA molecules themselves are called physical maps.

Applications of genome comparisons in research

Conservation Biology

Measuring genetic similarity and analyzing kinship can be very important in captive breeding programs for endangered species. Because the captive and wild populations of species in these programs are usually quite small, substantial inbreeding (both in the wild and in captivity) has often taken place, leading to individuals who are very similar genetically. Genetic similarity caused by inbreeding presents problems for populations and individuals. Populations of genetically identical (or nearly identical) individuals are usually less resilient in the face of environmental change. Highly inbred individuals are also frequently less healthy and less reproductively successful than outbred ones. Conservation biologists therefore try to find unrelated individuals to breed together, for the good of the individuals and the species. However, in a substantially inbred population, such as whooping cranes or cheetahs, the fact that two animals live in different zoos does not mean they are genetically different. Conservation biologists have been conducting DNA typing of whooping cranes to determine how genetically similar individuals are. They use the results of the typing to choose the most genetically different birds to form breeding pairs.

Evolutionary Biology

Analysis of genetic variation is used as a clue for deciphering the history of evolution, as well. The reasoning behind this application is as follows. If the accumulation of genetic changes is responsible for phenotypic differences between species, genera, families, orders, and so on, then the amount of evolutionary distance between two groups of organisms should be reflected in the amount of genetic difference. In other words, two groups that diverged long ago will have more genetic differences between them than two groups that diverged recently. Furthermore, an analysis of the nature of the differences might give clues as to what the ancestral genetic pattern was like.

In keeping with this thinking, evolutionary biologists with a molecular bent analyze DNA samples from related organisms and compare the amounts of difference. Using computer programs, they analyze the nature of the differences and produce hypothetical family trees. The trees show how, with a minimum of total genetic changes, the modern patterns could be produced from a single ancestral pattern (similar to the evolutionary tree for the protein

cytochrome *c* discussed in chapter 5). The introduction to the *Student Activity* describes how an international group of researchers used DNA analysis to study the evolution of the dog. Molecular evidence gives evolutionary scientists a wealth of new information to argue about and to incorporate into their ideas of the development of species. At present, there is no way to prove whether trees produced by molecular analysis are correct, but they give evolutionary biologists another line of evidence that supports the theory. We will revisit this notion in chapter 33, in which students will construct an evolutionary tree using protein, rather than DNA, sequence data.

Epidemiology

In some disease outbreaks, it can be important to trace the source of the infectious organism. In 2002, a man and his wife checked into a hospital in New York City complaining of flu-like symptoms, such as fever and swollen lymph nodes. They tested positive for bubonic plague. Were they the first victims of a bioterrorism attack? Or were they themselves terrorists intent on spreading a deadly disease?

Bubonic plague, the Black Death of the Middle Ages, is caused by the bacterium *Yersinia pestis* and is transmitted by bites of fleas that live on rats. *Y. pestis* is endemic among some wild rodent populations in the southwestern United States, including pack rats and prairie dogs. Approximately 10 to 20 cases of human bubonic plague occur each year in the United States. Though it is a very serious disease, plague can be treated with antibiotics.

The man and his wife were tourists in New York City. Their home was near the outskirts of Santa Fe, New Mexico, in an area where wild rodents carried plague. In fact, researchers had tested rodents from the area around Santa Fe where the couple lived and found that many tested positive for *Y. pestis*. Thus, it seemed likely that the couple had contracted plague before they left New Mexico, and the possibility that they were either victims or perpetrators of bioterrorism was deemed remote.

DNA typing of the *Y. pestis* strain that caused the couple's illness further supported the hypothesis that they had contracted the disease at their home. Microsatellite markers can be used to differentiate *Y. pestis* strains. The strain that infected the couple was genotyped and compared to strains from the Santa Fe area; the best matches occurred with bacteria isolated from rodents in the couple's own neighborhood.

Behavioral Biology

In behavioral biology, DNA typing can help biologists determine kinship between members of a group. Determining the effect (if any) of kinship on relationships between animals is a major endeavor in behavioral biology. However, determining kinship is not always easy. Depending on the mating behavior of the species, observation may reveal which individual fathers a particular offspring or brood. However, observation does not always yield the desired information. For example, chimpanzee males and females mate promiscuously, making it impossible to tell by watching which individual is the father of an offspring. This problem long frustrated primatologists who were trying to understand the role of kinship in chimpanzee social interactions. Using blood typing to determine paternity was not a desirable alternative, since it involved tranquilizing the animals, potentially traumatizing them and disrupting their social lives.

In 1994, a group of researchers published a solution to the problem. They developed a PCR-based method of DNA typing using small numbers of hairs. Chimpanzees make impromptu beds in trees at night, abandon them in the morning, and make new beds the next night. The scientists could wait for a chimp to get up in the morning and then climb to its abandoned bed and collect shed hairs for DNA typing without ever disturbing the animal. They are using their new kinship information in analyses of chimpanzee social behavior and hope that wildlife biologists studying other species can also take advantage of their DNA-typing method. In chapter 5, we described an application of DNA typing to understanding the social behavior of lions.

Taxonomy

Who gets excited about taxonomy? Actually, quite a few biologists, and DNA typing is giving them new material to argue about. Traditionally, taxonomists depended on structural, physiological, and behavioral differences to distinguish species. When techniques for protein electrophoresis were developed, taxonomists began including information on enzyme patterns, too. Some molecular scientists now argue that genetic distance should also be grounds for defining species. For example, in the chimpanzee study described above, the researchers analyzed three populations of the chimpanzee *Pan troglodytes*. They compared DNA fingerprints and mitochondrial DNA sequences. Their results showed that the western population of *P. troglodytes* had mitochondrial DNA significantly different from that of

the other two populations—the western sequences were not present at all in the eastern and central groups. The scientists calculated that it would take 1.6 million years for so many differences to accumulate.

Some of the scientists suggested that this amount of genetic difference should make the western chimps a separate species. This is a controversial suggestion; other primatologists point to the similar appearance and many similar behaviors of the two groups and argue that DNA sequences alone do not a species make. There is deep disagreement over what defines a species and what sort of criteria should be used in doing so. Taxonomic argument is going on in many areas of biology as scientists begin to use the new molecular tools to study differences between individuals.

The introduction to the *Student Activity* uses the dog genome as an example of ways in which genome comparisons and genome typing can be used to answer a variety of questions. References to scientific articles and commentary about the dog genome and its analysis are given below. The activity itself illustrates how PCR or Southern hybridization can be used to distinguish microsatellite alleles.

Objectives

After completing this activity, students should be able to:

1. Explain this statement: the level of variation in a genome that a scientist chooses to study depends on the question being asked.
2. Explain that DNA typing is based on an examination of variable regions of the genome.
3. Describe the two technical methods used in DNA typing: PCR and Southern hybridization.
4. Tell what an RFLP is and give one explanation of how an RFLP could arise.
5. Explain why it is necessary to use either hybridization or PCR in DNA typing.

Materials

Copies of the *Student Activity* (reading, exercises, and worksheets), if needed

Preparation

Print the worksheets and other student pages from the CD, if needed.

Procedure

Review Southern hybridization analysis and PCR, if necessary. The *Student Activity* assumes students already understand these procedures.

This activity is self-explanatory. Students read the material and do the exercises. Students often forget that each person has two copies of every chromosome and that the results of any analysis of a person's DNA will be a combination of the results from both chromosomes. Also, since a person's two chromosomes came from different individuals, they are likely to be different in the highly variable regions. Make sure your class is aware of these facts.

Remind students that the exercises show only a very tiny region of a single chromosome (the examples are completely imaginary). In real life, the entire human genome is digested, generating a huge number of fragments. Without the use of hybridization or PCR, every person's DNA would look the same in a gel: a smear.

Hybridization allows laboratory workers to look selectively at certain areas of the genome. PCR specifically amplifies certain regions, producing millions of copies of specific regions of the genome. PCR products can be detected by hybridization, or can sometimes be seen by simple staining, if the original sample DNA is present at a concentration low enough not to interfere (by causing a strong background smear).

This exercise is the foundation for understanding DNA-based paternity tests, forensic DNA typing, and DNA-based diagnosis of genetic disease. Use class discussion and questioning to make sure your students understand the concepts presented in this lesson before going on to chapter 28.

Answer to Student Question

1. The AKC required that Piggy Paws' DNA be submitted along with the suspected fathers' and the puppies' DNAs because the puppies' DNA profiles were a combination of the father's and the mother's. They needed Piggy Paws' DNA profile so that they could subtract her contribution from the puppies' profiles. What was left in the puppies' profiles had to come from their father, and the AKC could match that to either Johnny or Zack.

Note: Johnny has now achieved his own AKC championship.

Answers to Exercise Questions

Exercise 1

1. AGCT or a permutation, such as TAGC.
2. Bob's maternal chromosome gives a fragment of 41 base pairs (bp) (also 3 and 12, but these are the same in every case).

   Bob's paternal chromosome gives a fragment of 53 bp (the difference is the three extra copies of the repeat in the paternal chromosome).

   Mary's maternal chromosome gives a fragment of 61 bp; her paternal chromosome gives a 41-bp fragment.

3. Students must work through several steps to get the solution to this question. They must determine what fragments will be produced in the digest (which they have actually done for the previous question), then determine which of those fragments will hybridize to the probe, and then draw a picture of what the hybridization results would look like. You may need to help them through some of these steps.

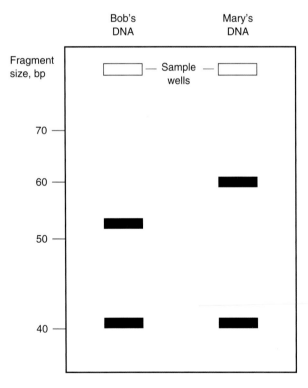

4. The stained pattern of a digest of Bob's or Mary's DNA would be a smear, because so many fragments would be generated from the 3 billion-bp genome. It is necessary to use hybridization to look at a particular region of the genome.

Exercise 2

Note the orientations of the primers. The first primer hybridizes to the complement of the strand that is shown. The second primer hybridizes to the strand whose sequence is shown; remember to take into account the 5′-to-3′ direction.

1. Bob's major maternal chromosome PCR product (53 bp long):

   5′ GGCCTCTAGGACATGTAAAGCTAAAGCTAGCT AGCTAGCTAGCTAAGGCCTAGGTGCG 3′

2. Bob's major paternal chromosome PCR product (65 bp long):

   5′ GGCTCTAGGACATGCTAAAGCTAGCTAGCTAG CTAGCTAGCTAGCTAGCTAAGGCCTAGGTGCG 3′

3. Mary's major maternal chromosome PCR product (73 bp long):

   5′ GGCCTCTAGGACATGCTAAAGCTAGCTAGCTA GCTAGCTAGCTAGCTAGCTAGCTAGCTAAGGCC TAGGTGCG 3′

   Major product from Mary's paternal chromosome (53 bp long):

   5′ GGCCTCTAGGACATGCTAAAGCTAGCTAGCTA GCTAGCTAAGGCCTAGGTGCG 3′

4. PCR products can often be seen by staining them. If a tiny amount of sample DNA is used, there is very little background in the gel lanes to interfere with seeing the products. Note that Bob and Mary share a product.

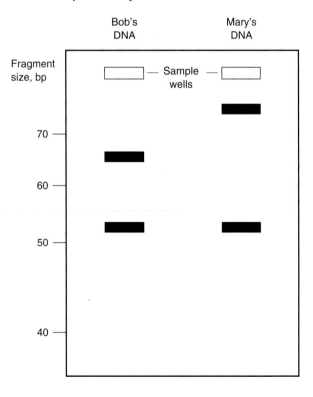

Selected Readings

Ellgren, H. 2005. The dog has its day. *Nature* 438:745–746. A news story/comment about the first publication of a complete dog genome sequence.

Jeffreys, A. 2005. DNA fingerprinting. *Nature Medicine* 11(10): xiv–xviii. A nontechnical retrospective about the development of DNA fingerprinting by the scientist who introduced it to the world.

Morell, V. 1994. Decoding chimp genes and lives. *Science* 265:1172–1173. A nontechnical overview of the chimpanzee studies.

Morrell, V. 1997. The origin of dogs: running with the wolves. *Science* 276:1647–1648. News commentary on the evolutionary study of dog DNA; accompanied the publication of the *Science* article by Vila et al. listed below.

Ostrander, E., and R. Wayne. 2005. The canine genome. *Genome Research* 15:1706–1716. A review by two of the leaders in dog genome research that summarizes studies of dog evolution, breed diversity and origins, disease gene mapping, and behavioral genetics. This article is available free of charge at http://www.genome.org (all articles published in this journal are free 6 months after the date of publication).

Parker, H. G., et al. 2004. Genetic structure of the purebred domestic dog. *Science* 304:1160–1164. An original research paper in which microsatellite markers were used to study relationships among 85 dog breeds.

Pennisi, E. 2004. Genome resources to boost canines' role in gene hunts. *Science* 304:1093–1094. A discussion of how progress in understanding the dog genome and developing molecular tools like microsatellite markers have made purebred dogs an invaluable resource for gene hunters; accompanied the publication of the *Science* article by Parker et al. listed above.

Pollinger, J. P., et al. 2005. Selective sweep mapping of genes with large phenotypic effects. *Genome Research* 2005 15:1809–1819. This article describes the study that identified the region of the dachshund genome near the fibroblast growth factor receptor gene as being invariant (see the introduction to the *Student Activity*). The article is available free of charge at http://www.genome.org.

Vila, C., et al. 1997. Multiple and ancient origins of the domestic dog. *Science* 276:1687–1689. Original research article describing the evolutionary analysis of dog and wolf mitochondrial DNA.

Vila, C., J. Maldonado, and R. Wayne. 1999. Phylogenetic relationships, evolution, and genetic diversity of the domestic dog. *Journal of Heredity* 90:71–77.

Classroom Activities

Comparing Genomes

Introduction

In this set of activities, you will be modeling some of the analyses scientists do when they compare the genomes of two different organisms. In the introduction, we provide information about how genomes change, what kinds of questions can be addressed by comparing genomes, and general approaches to making those comparisons.

How Genomes Change

The raw material for evolution, or genome change, is mutation. In the section *Understanding Evolution* in chapter 3, we described various kinds of mutations that can alter DNA sequences: point mutations, deletions, duplications, inversions, translocations, and transposition. We also described how recombination, the shuffling of genes and parts of genes during gamete formation and sexual reproduction, creates genetic variation in sexually reproducing organisms. Finally, we discussed several means by which asexual organisms can undergo genetic recombination: conjugation, transformation, and transduction. These three phenomena are used extensively in biotechnology and are addressed in detail in section C. For now, we will focus on sexually reproducing organisms.

It is useful to think about genome changes in terms of time scales. For the genomes of two closely related individuals to be different, the changes must have occurred very fast. At this time scale, the process of recombination is extremely important. In diploid organisms, like mammals, each offspring receives one set of chromosomes from its mother and one from its father, and the chromosomal composition of the gamete from each parent is random (see chapter 3 for a review of meiosis, the process by which chromosomes are distributed to gametes). Therefore, even siblings have different chromosomal compositions and can be distinguished by DNA fingerprinting.

But step back a moment. How did the mother's and father's chromosomes come to be different in the first place? The ultimate answer to that question involves the slower processes of genetic change listed above: mutations. Imagine a single ancestor chromosome that is copied and passed down to descendants until it exists in many individuals. Occasionally, a point mutation (a change in a single base pair) will happen as a result of an error during DNA replication or some unrepaired DNA damage. If the altered chromosome is passed on to descendants, they will be genetically different from the rest of the population. If the change makes those individuals better or less able to survive in their environment, natural selection will either favor the change or disfavor it.

However, many changes that do not affect the expression or amino acid sequence of a protein are neutral to natural selection. Such changes, called *neutral mutations,* simply accumulate over time. The longer it has been since two organisms had the same chromosomal ancestor, the more neutral mutations you would expect to have accumulated in the two genomes and the more differences you would see between them. This assumption forms the basis for using DNA sequence comparisons to get clues about evolutionary patterns. You will encounter this logic again in section F when we address the evolutionary analysis of proteins.

Along with point mutations, chromosomes also accumulate duplications, inversions, deletions, and translocations. In comparisons of animal genomes, scientists have found that even between two species that are not particularly closely related, such as mice and humans, you can trace rearrangement events by identifying segments of chromosomes in which the order of genes is the same in the two organisms. Imagine if you took two copies of this page, cut each one into pieces, and stuck the pieces together again. Even though the order of the letters on the two pages would no longer match, careful examination would show that they were formed from the same starting sequence. The mouse and human genomes are about the same size, and careful study of them shows that their large-scale organization has been scrambled by about 180 break-and-rejoin events. Once the break-and-rejoin events have been accounted for, about 90% of the two genomes can still be lined up.

Chromosomal rearrangements accumulate very slowly. It is estimated that the genetic lineages leading to humans and mice diverged about 75 million years ago and that there have been about 180 rearrangements since then. Point mutations occur more frequently. A single gene is estimated to undergo one mutation per 100,000 live births. Repeated sequences undergo change more rapidly. Their repeated nature makes them attractive targets for recombination and also makes it easy for DNA polymerase to "slip" and insert or delete an extra repeat or two. The rate of change of repeat numbers varies among repeat loci, as you will read below.

Putting these kinds of events together, you can imagine a symphony of genetic change. The reshuffling of chromosomes that occurs in every offspring, the occasional gain and loss of repeat units at hundreds of loci, the rarer introduction of a new point mutation, and the very rare event of chromosomal mutations. DNA analyses can be designed to target different levels of change (fast to slow), depending on the question the researcher is asking.

Why Compare Genomes?

Genome comparisons can provide answers, or clues to answers, to many different kinds of questions. To illustrate how genome comparisons can be useful, we have rounded up several examples from studies of dogs. Here are some questions about dogs that researchers have addressed with genome comparisons (we will describe the studies and findings below). Did dogs really descend from wolves, or are they more closely related to coyotes or foxes? Which dog breeds are more closely related? Would it be possible to analyze dog DNA and determine what breed of dog it came from? What DNA differences make breeds so different from one another? What DNA change, for example, causes the legs of dachshunds to be so short? What about identifying individual dogs: could you use DNA fingerprinting to distinguish the DNA of your dog from the DNA of other dogs? Or what if a professional dog breeder had a slip-up at mating time and was not sure which of her pedigreed males had fathered a litter with a pedigreed female?

All of these questions are asking about genetic differences, but at different levels, rather like the way you can look through a microscope using different magnifications. The evolutionary questions address differences between species: how different are the genomes of dogs and wolves, dogs and coyotes, and dogs and foxes? The questions about distinguishing dog breeds and determining how closely related they are ask how different the genomes of boxers, Dobermans, spitzes, and Labrador retrievers are. The question about short legs gets even more specific: what DNA changes are associated with having short legs? Finally, the questions about distinguishing individuals are the most specific of all.

General Methods for Genome Comparison

The most obvious way to compare genomes would be to sequence the genomes in question and use computers to help you understand the differences. However, sequencing the genome of an organism such as the dog is an enormous undertaking requiring years to finish. The sequence of about 75% of the genome of a poodle was published in 2003, and the complete base sequence of a boxer was published in 2005. Both of these achievements required literally years of work, so at this point, comparing the genomes of lots of dogs breeds by determining their complete genome sequences is not a practical approach.

Scientists who seek to compare genomes have developed methods that allow them to sample genetic diversity in the absence of a complete genome sequence. One of the first approaches to comparing genomes was to stain chromosomes and look at them, a specialty sometimes called cytogenetics. Here, scientists collect living cells from an individual, put them into a culture dish, and treat them with chemicals that first induce them to divide and then arrest the division process in metaphase. At this point, the chromosomes are condensed and can be seen with a light microscope. The chromosomes are stained and photographed for examination. If you have seen pictures of chromosomes in which they appeared to be striped X-shaped objects, you were looking at the result of this treatment. The stain reacts differently with DNA that is rich in A-T pairs and DNA that is rich in G-C pairs, and this causes chromosomes to appear banded in specific patterns, according to their base sequences. The chromosomes were X shaped because they were arrested during mitosis—they had already been duplicated, and the

Classroom Activities

duplicated chromosomes had not yet split apart. The number of chromosomes, their sizes, and their specific banding patterns are called a karyotype.

The banding pattern allows cytogeneticists to identify which chromosome is which. It also allows them to compare chromosomes from different organisms. It turns out that the chromosomes of closely related organisms have similar banding patterns, and sometimes you can see where an inversion or duplication has taken place. These comparisons reveal only gross changes in chromosome structure. They are used to diagnose medical conditions that result from chromosomal abnormalities, such as certain cancers and Down's syndrome. Figure 27.1 shows a representation of the banding pattern of chromosomes 1 to 3 in primates. It appears that human chromosome 2 was formed by a fusion of two chromosomes that are separate in the other primates. Now that the genomes of chimpanzees and humans have both been sequenced, researchers have been able to make a detailed count of the number and nature of the differences. The sequence data confirmed what the banding patterns suggested.

Another early method for comparing genomes involved hybridization analysis. Scientists would denature samples of genomic DNA from two different species, hybridize it, and then assess how similar the two genomic sequences were by determining the temperature at which the hybrids separated. The better the match between the two genomes, the higher the temperature that was required. This approach allows the comparison of genomes about which very little is known. If more is known about the genomes in question, scientists can sample genetic diversity by focusing on regions of the genome that differ at the level in which they are interested (individuals, breeds, species, etc.). One of the most useful kinds of difference involved repeated DNA sequences.

The DNA of humans and other eukaryotes contains head-to-tail, or tandem, repeats of different short DNA sequences at many locations within the genome. Scientists use an alphabet soup of acronyms to refer to them: "VNTR" (variable number tandem repeat), **"STR"** (short tandem repeat), and SSR (simple sequence repeat). A general term for any region of short repeated sequences in the genome is **mi-**

Figure 27.1 Structural similarities and differences in chromosomes 1 to 3 of primates. H, human; C, chimpanzee; G, gorilla; O, orangutan.

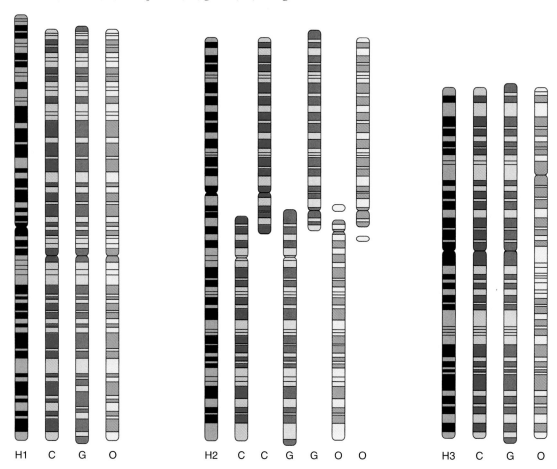

crosatellite. Each place where a microsatellite is found is called a *locus* (from the same root word as location; pronounced low′ cuss); the plural is loci (pronounced low′ sigh).

Different microsatellite DNA loci show different levels of variability. The level that is relevant to a particular type of study depends on the focal length of the question being asked: are you interested in differences between individuals, breeds, species, or something even larger? At some microsatellite loci, the numbers of repeats vary greatly from individual to individual. These loci can be used to generate individual-specific DNA profiles and are useful for forensic DNA typing or paternity studies, where the objective is to distinguish one individual from another.

At other loci, the number of repeats is less variable between individuals but still varies over larger evolutionary differences, such as between dog breeds. If you were interested in identifying regions of the dog genome that made a dachshund a dachshund or a Great Dane a Great Dane, studying regions that varied from one dachshund to the next would not be helpful. You would like to find regions that were common to all dachshunds but different in other dog breeds. Finally, if you were interested in how dogs differ from coyotes, you would not be helped by regions that differ between dachshunds and Great Danes, since both are dogs. Instead, you would want to find regions of the genome that were the same in all dogs but different in dogs and coyotes.

For genome comparisons, several highly variable regions are characterized, and a DNA "profile" is generated. Loci that are highly variable among different individuals are the ones that are useful for DNA typing, while those that are more constant are more useful for comparing breeds or species. A locus used for DNA comparisons is often called a DNA marker.

DNA typing

To generate a DNA profile, you need to focus on variable regions of the genome, such as microsatellites. This requires some specificity in your approach—if you simply digest total genomic DNA from a dog or a person and separate the fragments by electrophoresis, you will see nothing but a smear in your gel because of the large sizes of the two genomes and the huge number of restriction fragments generated. DNA profiling techniques use hybridization to focus on the regions of the genome of interest. The hybridization can be either to a labeled probe, as in Southern hybridization analysis, or to

polymerase chain reaction (PCR) primers. In Southern hybridization, you end up detecting the sizes of restriction fragments to which the probe hybridizes. In PCR analysis, you detect the sizes of the amplified products representing the DNA sequence between the hybridization sites of your primers.

To perform DNA typing by Southern hybridization analysis, genomic DNA is digested with restriction enzymes, separated by gel electrophoresis, and hybridized to specific probes. The probes are chosen so that they hybridize to regions of the genome that vary in sequence, yielding different sizes of restriction fragments, or *restriction fragment length polymorphisms* (RFLPs) (poly, many; morph, form; therefore, polymorphism, many forms).

You can probably imagine how a microsatellite locus could lead to an RFLP. Imagine a microsatellite region with a HaeIII site on either side of it. Obviously, the length of the HaeIII fragment containing the microsatellite depends on how many repeats there are. The more repeats, the longer the fragment (Figure 27.1). To detect microsatellite-based RFLPs, probes that hybridize to a DNA sequence on one side of the microsatellite are designed, revealing the pattern of bands generated (Figure 27.2). Often, the analysis is performed with multiple probes that reveal RFLPs from several different microsatellite loci. The result is a DNA profile. Exercise 1 shows an example of an RFLP created by an STR. Other kinds of chromosome changes, such as rearrangements, insertions, deletions, or single base changes, that create or destroy restriction sites can also generate RFLPs.

Why not just look at the band pattern by staining and skip the hybridization? Remember the size of the human genome: over 3 billion base pairs (bp). If an average restriction fragment were 1,000 bp in length, there would be 3 million of them. There is no way to look at individual bands by staining—you would only see a smear all the way down the gel lane, caused by hundreds, thousands, or millions of overlapping fragments. DNA typing depends on the use of methods, like hybridization and PCR, that allow you to look only at specific regions of the genome.

Microsatellite-based DNA typing can also be performed by PCR (Figure 27.2). PCR primers that hybridize to the DNA sequence on each side of a microsatellite locus are used to amplify it, producing millions of copies. The PCR products are separated in an electrophoresis gel and usually visualized by hybridization. If the amount of background sample

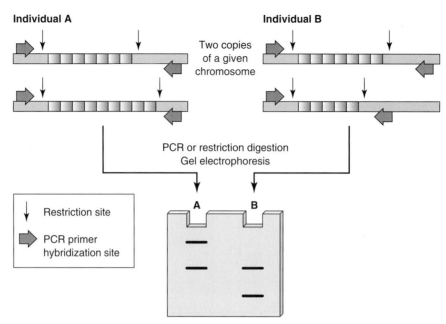

Figure 27.2 Variation in the numbers of tandem (head-to-tail) repeat sequences at many different locations within the genome provides a basis for genetic fingerprinting. To generate a fingerprint, laboratory technicians amplify many different loci using a specific set of primers for each one or they digest the sample DNA with restriction enzymes known to cut it at sites flanking the repeats.

DNA is low, PCR products can sometimes be seen by simple staining, since they are present at such high concentrations. The lengths of the PCR products are determined by the number of repeats present, so the sizes of products vary from person to person. Several sets of primers can be used to look at several STR regions, generating a unique profile for the individual. PCR is particularly useful in situations where the amount of template DNA is very small or the template is fragmented, as it usually is in ancient specimens.

Applications of DNA comparisons

When most people think of DNA typing, they think of solving crimes or paternity cases. This kind of application will be specifically addressed in the next chapter. However, as you saw in the introduction to this chapter, there are many other kinds of applications for DNA typing. In fact, DNA profiling has been applied to address the questions we asked in the first paragraph, and we will take a brief look at some of the information that genome comparisons have revealed about dogs thus far.

Dog Evolution

Charles Darwin himself wondered about the evolutionary origins of dogs. There is so much diversity in their sizes, shapes, and behavior that he wondered

whether different breeds might have been derived from different wild canids (a generic term for members of the Canidae, the group of animals that includes dogs, wolves, foxes, coyotes, and jackals). To address this question, an international team of researchers assembled tissue samples from 162 wolves from North America, Europe, Asia, and Arabia, along with samples from 140 dogs representing 67 pure breeds and 5 mutts. They also collected material from five coyotes and two black-backed, two golden, and eight Simien jackals.

The researchers determined the base sequence of the hypervariable region around the replication origin of the canids' mitochondrial DNA. Mitochondrial DNA is inherited solely from the mother and therefore is very helpful in reconstructing family trees (see the *Reading* for information about mitochondrial DNA). Since there is no shuffling of mitochondrial genes from one generation to the next, the only changes are due to mutations. Thus, mitochondrial DNA sequences are also very useful in reconstructing evolutionary relationships. The more differences there are between two mitochondrial DNA sequences, the more time the two family trees have been separated. Since mitochondrial DNA mutates relatively rapidly, it is useful only for analyzing relationships between relatively closely related groups. The canids were an ideal study set.

Mitochondrial DNA analysis revealed that all the dogs were descended from wolves. The wolf samples altogether contained 26 different mitochondrial sequences. The dog samples had 27. One of the sequences was found in both dogs and wolves. Of the other dog sequences, none differed from the wolf sequences by more than 12 base substitutions, while they differed from coyote and jackal sequences by 20 or more. Interestingly, the dog mitochondrial sequences were not specific to different breeds. For example, in the eight German shepherd samples examined, five distinct mitochondrial DNA sequences were found. In the six golden retriever samples, four mitochondrial sequences were detected. Moreover, the same mitochondrial sequence was found in several breeds. What this means is that the mitochondrial DNA sequences were in the dog gene pool before specific breeds were created and that breeds have more than one female ancestor. (See the *Reading* about mitochondrial DNA for more information.) This finding was not surprising, since most dog breeds have been in existence for only a few hundred years, and breed clubs, such as the American Kennel Club (AKC), with their restrictions on crossbreeding, have existed only since the mid-19th century.

Differentiating Dog Breeds

Scientists interested in the evolution of different dog breeds have used DNA-based comparisons to assess genetic similarity between breeds. Mitochondrial DNA sequences are not useful for these comparisons, as you can tell from the information presented in the paragraph above. Instead, researchers have looked at microsatellite variation. In a study published in 2004, a group of scientists determined the genotypes of 414 dogs from 85 breeds at 96 different microsatellite loci. They hypothesized that dogs from the same breed would have more similar DNA types than dogs from different breeds.

Their hypothesis proved to be correct. They used many computer tools to compare the genotypes in different ways and found that the genotypes of nearly all the dog breeds formed distinct groups that could be differentiated from one another. A few sets of species were very similar: Alaskan malamute and Siberian husky, Belgian sheepdog and Belgian Tervuren, collie and Shetland sheepdog, greyhound and whippet, Bernese mountain dog and greater Swiss mountain dog, and bullmastiff and mastiff. Again, these pairings were not surprising given the known histories of the breeds. Tests showed that they could assign 99% of the dogs to the correct breed based on their genotypes.

Finding Genes for Specific Traits

Other research groups are analyzing genetic variation between breeds through a different lens. They are interested in identifying genes and proteins involved in generating specific traits. One research group put forth the hypothesis that regions of the genome that are involved in producing distinct, breed-specific traits (such as the short legs of dachshunds) will be *less* variable within that breed than the rest of the genome. This group surveyed 302 microsatellite markers in dachshunds and found three markers within a 10 million-base-pair region on chromosome 3 that showed no variation at all. This region contains the fibroblast growth factor receptor 3 gene, the mutation of which is known to cause achondroplastic dwarfism in humans. Although a group of Mexican researchers had previously sequenced that gene in dachshunds and found that it was not mutated, the results of the microsatellite study suggest that the cause of dachshunds' short legs might be associated with the gene after all, perhaps in a regulatory region or a closely linked gene.

Another research group was interested in finding genes involved in the vast size differences between specific dog breeds (think of Chihuahuas and Great Danes, whose body masses can differ by 50-fold). They identified a number of Portuguese water dogs that were either large or small for their breed and looked for differences in their genomes. They reasoned that the genomes of dogs of the same pure breed should be very similar and that consistent differences between large and small individuals might be associated with the size difference. They found that one of the few differences between the large and small dogs' genomes was in a gene for a growth factor called insulin-like growth factor 1, or *Igf-1*. Finding this gene was exciting, because it was already known to influence the size of mice. When *Igf-1* is knocked out in mice, the animals are minimice. The researchers next analyzed the *Igf-1* genes in 75 Portuguese water dogs and 350 other dogs of very large or very small breeds. They also examined the gene in wolves and foxes. Almost all of the 18 small breeds had the same version of the *Igf-1* gene, but none of the giant breeds carried it. This research was published in October 2006—you may be able to find more on the story later on.

A Canine Paternity Case

This is a true story. On Thanksgiving in 2005, a basset hound breeder in Minnesota was killed in a car wreck. Her husband was unable to take care of her dogs because of a health condition, so other basset breeders stepped up and took in the bassets. One of the females, Briarpatch Sweet Peggy Sue, known informally as Piggy Paws, went to stay with the breeder of her sire (father). A few weeks later, it became evident that Piggy Paws was "with pup." Piggy Paws' pregnancy was unplanned, but the only dogs that could possibly have mated with her were all registered basset hounds. When the litter of beautiful, obviously basset hound puppies arrived, Piggy Paws' new caretaker counted back to determine when Piggy Paws was impregnated. The window of time included days before and after her move from Minnesota.

Who was the father of Piggy Paws' pups? It mattered. To register any canine with the AKC, you must list both its sire and dam (mother). In addition, Piggy Paws is a beautifully bred basset, the daughter of a champion (Champion Rebec's Fuzzy Navel, ROM [Register of Merit], CGC [Canine Good Citizen]), and one of the potential fathers was himself a champion basset hound (Champion Citation's Prozac, known informally as Zack). Zack was the prime suspect father, as he had an established history of breeding successes and seemed capable, according to Piggy Paws' new caretaker, of impregnating females just by looking at them from across the room. Another potential father was Rebec's Saturday Night Fever (known informally as Johnny), also a fine show basset on the way to his own championship. Thus, Piggy Paws' puppies would be of interest to serious basset hound breeders and might themselves one day be champion basset hounds. But without the pedigree and registration, they could not even compete in AKC events.

Piggy Paws' caretaker turned to the AKC's parentage determination service for help. You can read about this service at the AKC's website. The AKC sent swabs for collecting loose cheek cells from the mother, the suspected fathers, and the puppies. Piggy Paws' caretaker described a swab as looking like a "teeny tiny bottle brush the size of a Q-tip." She swabbed the inside of each basset's mouth and lips to collect loose cells. She then sealed each swab inside an envelope on which the information regarding the identity of the basset was recorded, and the envelopes were sent to the AKC with the request that the puppies' DNA be compared with Piggy Paws', Zack's, and Johnny's DNAs to determine parentage.

When the results of the paternity testing came back, the caretaker said, "You could have knocked me over with a feather." The DNA testing showed that Johnny, and not Zack, was the father of the litter. The puppies now have pedigrees of their own.

Question

1. Everyone knew that Piggy Paws was the mother of the puppies. Why would the AKC require that her DNA be submitted along with the suspected fathers' and the puppies' DNAs?

Exercise 1. STRs Can Cause an RFLP

Use with Worksheet 27.1

Suppose we are taking a close-up look at a region of chromosomes 8 from two individuals named Bob and Mary. Laboratory workers are focusing on a region of chromosome 8 where they have found an STR microsatellite. Shown on Worksheet 27.1 are the sequences from that region of Bob's and Mary's chromosomes 8. The laboratory will perform Southern hybridization analysis using the restriction enzyme HaeIII and the probe shown on Worksheet 27.1.

Questions

1. What is the tandem repeat sequence in this chromosome region?

2. What size fragments would be generated from the region containing the repeats by digesting Bob's maternal chromosome with HaeIII? By digesting his paternal chromosome? How large would the fragments from Mary's chromosomes be?

3. Suppose you digested samples of Mary's and Bob's DNAs with HaeIII and performed a Southern analysis using the probe shown on Worksheet 27.1. Draw the results on the outline provided.

4. If you were analyzing DNA prepared from Bob's and Mary's white blood cells, why couldn't you simply look at the stained gel pattern and skip the hybridization step?

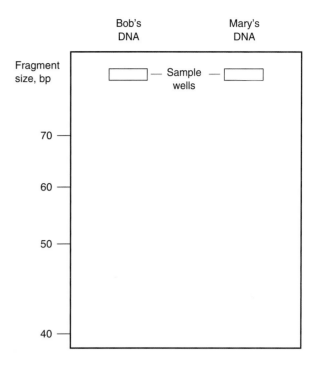

Exercise 2. PCR Can Reveal Differences at Microsatellite Loci

Use with Worksheet 27.2

Another laboratory wishes to look at differences between Bob's and Mary's DNAs. They will also look at the highly variable region of chromosome 8 shown on Worksheet 27.2 (the same region is also shown on Worksheet 27.1), but they will use PCR instead of Southern hybridization analysis.

Laboratory workers prepare DNA from samples of Bob's and Mary's blood and then mix a tiny sample of DNA from each person with DNA polymerase enzyme, deoxynucleotides, and the primers shown on Worksheet 27.2. The PCR was allowed to proceed through 30 replication cycles.

Questions

1. Write the sequence of the major PCR product from Bob's maternal chromosome.

2. Write the sequence of the major PCR product from Bob's paternal chromosome.

3. Write the sequence of the major PCR products from Mary's maternal and paternal chromosomes (label them "maternal" and "paternal").

4. Draw what the products would look like in a stained gel in the outline provided.

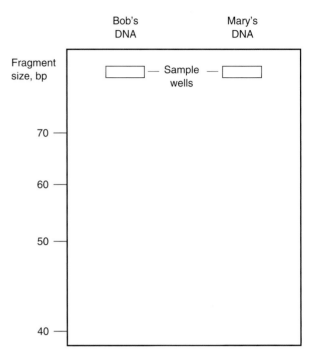

Because PCR yields so many copies of the major product molecule, it is possible to see the major products by simple staining them, even when there is too little original sample DNA to see in a stained gel. PCR products can also be visualized by hybridization to a probe, as in Southern analysis.

Mitochondrial DNA

Human mitochondria contain a circular genome of about 17,000 bp (Figure 27.3). Mitochondrial genes encode some of the proteins that are associated with the electron transport chain and some of the proteins required for mitochondrial protein synthesis. There are hundreds of mitochondria in a typical cell.

Mitochondrial DNA plays a special role in DNA typing because of its inheritance pattern and its high mutation rate. Except in rare instances, mitochondrial DNA is inherited from the mother. The mitochondria present in the ovum become the mitochondria of the zygote after fertilization. In addition, the region of the mitochondrial genome around the replication origin is highly variable and has a high mutation rate. Therefore, two people with the same DNA sequence in that region are highly likely to share a recent female ancestor (Figure 27.4).

Taken together, this information means that an analysis of mitochondrial DNA can show very clearly whether two people are related through the female line. Analysis of mitochondrial DNA can be helpful in reuniting families and in identifying the dead. A particularly poignant application of mitochondrial DNA typing can be found in Argentina. During the Argentine military's brutal rule (1976 to 1983), many families were torn apart. Often, parents were murdered and their children were given away or sold. In other cases, parents were dragged away to prison, unwillingly leaving their babies to uncertain fates. When the dictatorship was overthrown, the relatives of these kidnapped or "disappeared" children tried desperately to find them. Many of the relatives were grandmothers whose children were murdered and who sought their missing grandchildren.

An American scientist, Mary-Claire King, was instrumental in helping Argentine families find their lost relatives. King used mitochondrial DNA in her analyses. Because the parents of the lost children had often been murdered, DNA from more distant suspected relatives was usually the only evidence

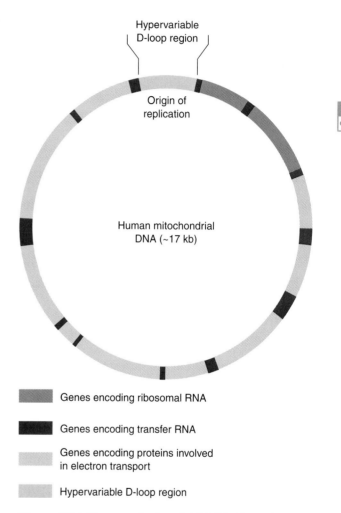

Figure 27.3 Human mitochondrial DNA. The D-loop region around the origin of replication is the hypervariable region.

available for comparison. Since mitochondrial DNA is passed through female family lineages, a child's mitochondrial DNA profile would exactly match the profile from her mother's mother, all of her mother's siblings, and the children of her mother's sisters (Figure 27.4). This approach was successful in identifying many of the missing children.

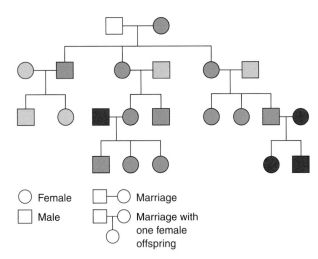

Female ○   Male □   Marriage □—○   Marriage with one female offspring □—○ (with ○ below)

Figure 27.4 Inheritance of mitochondrial DNA follows the maternal line.

Mitochondrial DNA typing was also used to identify the skeletal remains of the royal family of Russia, murdered by Bolshevik soldiers in 1918 and buried in an anonymous grave. Because many of the members of various European royal families are related through Queen Victoria, they share mitochondrial DNA sequences. Czar Nicholas's remains were identified by comparison to mitochondrial DNA sequences of living relatives.

The U.S. Armed Forces has an ongoing program to identify as many previously unidentified remains from military conflicts as possible. Although this mission may seem strange at first, to the families of missing soldiers, it is very important. In 1999, there were approximately 2,200 soldiers unaccounted for from the Vietnam conflict, 8,000 from the Korean conflict, and 78,000 from World War II. Many of these soldiers' bodies were found but could not be identified and were buried as unknowns.

The U.S. Armed Forces DNA identification laboratory uses PCR amplification of mitochondrial DNA to identify human skeletal remains from previous conflicts. Because there are so many copies of mito-chondrial DNA per cell, the chances are greater that it can be amplified even from fragmented remains.

When the laboratory works to identify a set of remains, it first collects all available information about them—where the remains were found, what unit was fighting there, which soldiers were unaccounted for from that engagement, etc. In this way, the laboratory defines a set of possible identities for the remains. Laboratory personnel then contact family members and, with permission, obtain DNA samples for comparison. Under extremely strict conditions designed to minimize contamination, PCR amplifications of mitochondrial DNA are performed, and the products are sequenced. The sequences are compared to sequences from the candidate families. If there is a match, the remains are identified.

In 1998, the remains buried in the Tomb of the Unknown Soldier were brought to the Armed Forces DNA Identification Laboratory under full military honor guard. With great care, the mitochondrial DNA sequence of the remains was determined. The sequence was compared to sequences from seven candidate families, and it matched only one family. The Unknown Soldier was identified as Michael J. Blassie, an Air Force pilot who was shot down in Vietnam in 1972.

The Armed Forces DNA Identification Laboratory now generates a nuclear DNA profile for all soldiers in the hope of eliminating future unknowns.

Selected Readings

Cann, R. L., and A. C. Wilson. 2003. The recent African genesis of humans. *Scientific American* 289:54–61.

King, M.-C. 1990. Genes of war. *Discover* 11(10):46–52. The story of Mary-Claire King's work in Argentina.

National Institutes of Health. http://www.nlm.nih.gov/exhibition/visibleproofs/galleries/cases/index.html. The story of the identification of Michael Blassie, along with several other cases of forensic interest (including one related to the identification of the children of the Argentine victims), can be found on this website.

Wallace, D. C. 1997. Mitochondrial DNA in aging and disease. *Scientific American* 277:40–47.

Forensic DNA Typing

28

About the Activities

This chapter contains three easy activities that illustrate applications of DNA typing plus a reading about applications of DNA analysis to human remains found at an archaeological site. The original reference for the archaeology case is listed below. In Exercise 1, students assign babies to the correct pair of parents based on DNA profiles. In Exercise 2, students analyze DNA typing data to determine if I. M. Megabucks, a recently deceased megabillionaire, is actually the father of any of three children alleged to be his heirs. This activity can be done, with some discussion, by students who have completed *Comparing Genomes*. Students should also have completed *DNA Scissors* (the introduction to restriction enzymes) and *DNA Goes to the Races* (the introduction to electrophoresis) before doing this activity.

Class periods required: *1*

Introduction

DNA-based identification methods focus on highly variable regions of the human genome. Because the genome is so large, it is not possible to look at these regions by restriction digestion, electrophoresis, and staining. A restriction digest of human DNA looks like one giant smear down the gel lane, because so many fragments are generated from 3 billion base pairs (bp).

The techniques of Southern hybridization and the polymerase chain reaction (PCR) allow analysts to look at specific regions and ignore the rest. In Southern analysis, the region you look at is determined by where the probe hybridizes to the sample DNA. In PCR, the specificity is determined by the base sequences of the primers; the DNA sequence between the two sites on the chromosome where the primers hybridize is the region that is amplified in the reaction. If one of these techniques is used to characterize several highly variable regions from one person's genome, the data set generated is referred

to as that person's DNA profile, DNA type, or DNA fingerprint. Since the term "DNA fingerprinting" is now a registered trademark of the biotechnology company Cellmark, we will use "DNA typing" or "DNA profiling" to refer to the process.

To conduct DNA typing, you must have a DNA-containing sample. In paternity cases, blood is drawn from the child, its mother, and the alleged father, and DNA is extracted from the white cells. However, any DNA-containing body fluid or tissue can be used: hair follicles, skin, or semen, for example. DNA is extracted and then analyzed by one of the two approaches outlined above. The results from a DNA-typing analysis are compared to the results from identical analyses of other samples. In criminal cases, the DNA profile of crime scene evidence (such as blood or semen) is compared to the DNA profile of the suspect. In paternity cases, the DNA profile of the child is compared to those of its mother and alleged father.

The first DNA-typing methods used Southern hybridization; now, PCR-based approaches are increasingly popular. PCR has the advantage of greater sensitivity. It can be used with much smaller samples, since it amplifies the DNA to be detected. This advantage is particularly important in criminal cases, where available samples (such as skin fragments under a victim's fingernails) may be exceedingly small. In addition, PCR primers can be selected so that they amplify relatively short segments of DNA—500 bp or fewer—allowing the analysis of DNA that has been partially degraded.

The accuracy of DNA typing used to be a subject of controversy with respect to forensic cases, but the issues have now largely been settled. No one disputed the underlying theory of DNA typing: that everyone's DNA is unique and different and therefore provides a definite means of identification (with the exception of identical siblings). The problems arose over two issues, both of which apply largely to criminal cases.

The first issue was technical. In the past, court cases revealed some instances of sloppiness in laboratory procedure and in data analysis. Accurate laboratory procedure is particularly important in criminal cases. A person's fate may be riding on the outcome of a DNA analysis, and the sample at the crime scene may have been contaminated with other substances, such as fabric dyes; may have been subjected to temperature extremes; or may be otherwise difficult to analyze. Perhaps more important, forensic samples are often vanishingly small. If a result from a DNA analysis is somewhat questionable, it may be impossible to repeat the tests. High-quality laboratory work in these cases is of the utmost importance.

In response to the technical criticisms, the National Academy of Science issued a call in 1992 for laboratory certification, quality controls, and standardization of procedure. An FBI-organized consortium of scientists from academia, crime laboratories, and private industry called the Technical Working Group on DNA Analysis Methods issued a set of guidelines that included laboratory protocols, quality assurance guidelines, and standards for education and training of personnel. Testing laboratories that meet these standards can be certified by professional groups, such as the American Association of Blood Banks and the College of American Pathologists. Reputable testing laboratories adhere to these standards and seek certification. The standards for forensic laboratories are published in the CODIS (Combined DNA Index System) section of the FBI website. It might be an interesting exercise for your students to read them. They refer to training requirements for personnel, critical reagents, facilities, and more.

The second issue concerning the reliability of DNA typing questioned the point at which it could be reasonably concluded that two samples came from the same person. This was a more fundamental question than the technical issues. Here is an example to illustrate the problem. Forensic laboratories have compiled large databases of DNA profiles generated with the probes they use. Suppose a crime scene sample and a suspect's blood are tested with probe A and show the same profile. Suppose that particular profile occurs in 1% of all the profiles in a database. There is then a 1% chance that a sample from any random person would also match the crime scene sample. Suppose that the samples are tested with probes B and C and again reveal identical patterns. Let's say the probe B pattern is present in 5% of the database profiles and the probe C pattern in 10%. What are the chances that the DNA profile of a random person would match all three profiles?

Forensic laboratories use a multiplication rule to calculate this probability. They assume that a person's DNA profile at one locus (for example, the one shown by probe A) is independent of his or her profile at a second locus (such as the one shown by probe B). In this case, the probability of finding all three patterns in a random person can be calculated simply by multiplying the individual probabilities: 0.01 (A) $\times$ 0.05 (B) $\times$ 0.1 (C) = 0.00005. This number predicts that 5 people out of 100,000 would have the same profile with all three probes. If the samples matched using a fourth probe, revealing a pattern that showed up in 2% of the database profiles, the odds of a random match would decrease to 1 in 1,000,000. Obviously, the greater the number of probes used to test two samples, the less likely it becomes that a perfect match for all of the probes could have been a random event. (An activity you can use to demonstrate this kind of probability calculation is described in this chapter immediately following the solutions to the student exercises.)

Here is where the controversy came in. Critics said the chance of a random match might be much higher than the odds calculated by the multiplication rule. For example, they argued that a database containing profiles from Caucasians and African-Americans might not be a reasonable standard for comparison if the relevant ethnic group in a particular case was Polynesian. Although a 1993 report on an analysis of over 70 worldwide databases concluded that there were no meaningful differences in genetic profiles within or between racial groups, this issue still arose in courtrooms, where attorneys argue in an adversarial manner. Also, it is not possible to say that no small ethnic group anywhere has a special, common DNA-typing profile not often seen in any other population—not unless everyone everywhere has been typed. In response to this criticism, public and private DNA-testing laboratories have established DNA profile databases from different racial groups to serve as comparison groups.

Systematic Use of DNA Profiles in Criminal Cases

According to Bruce Budowle of the FBI Academy Forensic Laboratories in Quantico, Virginia, in 1994 alone there were approximately 168,000 rapes in the United States and approximately 149,000 attempted rapes. In about one out of three cases, the perpetrator was unknown to the victim. Most sexual offenders are repeat offenders, and about half of all convicted sex offenders are on probation or parole at a given time.

In response to statistics like these, the FBI established the CODIS, a national database of convicted felons. According to its website, CODIS generates investigative leads in crimes in which biological evidence is recovered from the crime scene using two indexes: the forensic and offender indexes. The forensic index contains DNA profiles from crime scene evidence. The offender index contains DNA profiles of individuals convicted of sex offenses (and other violent crimes), with many states now expanding legislation to include other felonies. States are free to participate in the system, and as of March 2006, all 50 do so. As of March 2006, there were 134,075 profiles in the forensic index and 3,045,761 profiles in the convicted-offender index.

Matches made among profiles in the forensic index can link crime scenes together, possibly identifying serial offenders. Based on a match, police in multiple jurisdictions can coordinate their respective investigations and share the leads they developed independently. Matches made between the forensic and offender indexes provide investigators with the identity of the perpetrator(s). After CODIS identifies a potential match, qualified DNA analysts in the laboratories contact each other to validate or refute the match.

For DNA profiles to be comparable, all the participating states must analyze the same genetic loci when their laboratories conduct DNA typing. Specific genetic markers were therefore selected, and all involved laboratories now agree to look at the same 13 STR (short tandem repeat) loci by PCR. A participating state generates DNA profiles and sends them to their state database. CODIS links the states together. Alternatively, state law enforcement agencies can submit evidence to the FBI laboratory, which will examine it free of charge. The FBI laboratory also provides expert witness testimony regarding the results of their examinations. By March 2006, CODIS had provided 31,400 hits, assisting in 33,100 investigations.

Objectives

After completing the exercises, students should be able to:

1. Explain how to use DNA profiles to determine if a couple are the parents of a particular child. They should be able to give a clear description of what to look for in comparing the profiles.
2. Explain the process of analyzing DNA data to determine if a particular man is the father of a child.

Materials

Copies of the *Student Activities* if students do not have manuals

Preparation

Print and make photocopies of the *Student Activities,* if necessary.

Exercise 1. A Mix-Up at the Hospital

Procedure

For older students, the activity is self-explanatory. A relevant news story would be a good way to relate the activity to the real world. Depending on your class, you may want to suggest the analytical procedure described below. Remind students that half of a person's chromosomes come from each parent and therefore would have counterpart patterns in the parents' DNA profiles. Review as much as you deem necessary for your class before beginning the activity.

To analyze the DNA profiles, students should carefully compare the babies' profiles to the profiles from each couple. For example, students can start with one baby's profile and compare it to the first couple's. First, they should focus on one member of the couple and compare the baby's profile to that person's (for example, the woman's). Each band in the baby's profile should be checked to see if it matches a band in the first woman's profile. Any bands that match should be marked lightly in pencil. If no bands match, the woman cannot be the baby's mother. If some bands match, the student should compare the remaining bands to the man's DNA profile. Every remaining band in the baby's profile should match a band in the man's DNA profile if that man is the father. If some but not all bands match, then the couple are not the parents of that child. The student should go on to the next couple and compare the baby's profile to theirs in the same manner.

If you want to use this activity with younger students who have not completed the Southern hybridization, PCR, and DNA-typing lessons, you will have to decide how much you want to tell them about how the DNA profiles are generated. They will need to be familiar with the activity of restriction enzymes and with gel electrophoresis of restriction fragments to understand the activity. If you think it is necessary, go through one comparison with your class, following the procedure outlined above. As your students work on the puzzle, circulate through the classroom and observe their work to make sure they understand what they are doing.

Answer

Baby 1 is the Stevenson baby. Baby 2 is the Jones baby. Baby 3 is the Smith baby. Check each student's paper to see that they have correctly assigned the maternal and paternal bands.

Exercise 2. A Paternity Case

Procedure

This activity is self-explanatory. You may need to remind the students that half of a person's chromosomes came from the mother and the other half from the father and that not all of the DNA bands seen in the typing data from the mother or the father would be in the child's DNA (fully half of each parent's DNA is not represented in the offspring's DNA).

Answer

Y's child could be Megabucks' child.

Answer to student question

This analysis looks at four VNTR (variable number tandem repeat) regions. Since we have two copies of each chromosome, we have two alleles of each VNTR. For example, if one VNTR region is on chromosome 1, one of our chromosomes 1 might have 30 copies of the repeat while the other of our chromosomes 1 had 20 copies. You would see both of these bands following a Southern hybridization analysis using a probe that hybridizes to that VNTR region.

Since a child inherits one member of each chromosome pair from its mother, the VNTR allele on that chromosome would match one of the mother's alleles. The other member of each chromosome pair would match one of the father's alleles.

Exercise 3. The Case of the Bloody Knife

Procedure

The exercise is self-explanatory.

Answer

The key to this case is the number of bands in the DNA profile from the knife. The chromosome

regions and primers show that each chromosome should generate one band. Since people have two copies of every chromosome, there should be two bands generated per set of primers. Two chromosomal regions are analyzed, and both Milhouse (the victim) and Smink (the suspect) show four bands in their profiles. The profile from the blood on Milhouse's clothes also shows four bands and clearly matches Milhouse's own DNA profile. The DNA profile from the knife, however, shows eight bands. Either something went wrong with the testing (such as contamination of the sample) or the blood on the knife is from more than one person.

A comparison of the banding pattern shows that four of the eight bands in the knife profile are exact matches to the victim's profile, while the remaining four are an exact match to Smink's. Given the circumstances, it seems highly likely that the blood on the knife is a mixture of blood from the two men. The knife was found under the bloody body, so it is possible that the victim's blood could be on it even if the victim was not cut. Smink's hand wound could have been a cut from that knife, possibly as the victim tried to defend himself. On the other hand, poor laboratory practice could have resulted in mixing of the knife sample with Smink's sample during testing. The DNA typing should be repeated with special care taken to make sure no contamination occurs.

Smink should definitely not be released. The current evidence suggests strongly that Smink's blood is on the knife. He should be questioned thoroughly. Additional DNA tests examining other highly variable chromosome regions could further strengthen the conclusion that the "extra" blood on the knife is Smink's.

Selected Readings

Cappellini, E., et al. 2004. Biomolecular study of the human remains from tomb 5859 in the Etruscan necropolis of Monterozzi, Tarquinia (Viterbo, Italy). *Journal of Archaeological Science* 31:603–612.

Jeffreys, A. 2005. Genetic fingerprinting. *Nature Medicine* 11(10):xiv–xviii. The scientist credited with inventing the concept of DNA typing recounts the development of the technology.

Menotti-Raymond, M., et al. 1997. Pet cat hair implicates murder suspect. *Nature* 386:774. The story of the forensic case of Snowball the cat.

Yoon, C. K. 1993. Forensic science. Botanical witness for the prosecution. *Science* 268:894–895. The story of the use of the DNA profile of an individual Palo Verde tree to link a suspect to a murder scene.

Calculating the Odds: A Demonstration Activity

Part I. Shuffling the genetic deck

This exercise provides an excellent opportunity to remind your students that mathematics plays a key role in forensic DNA identification. DNA typing is very much a numbers game. Because absolute certainty in DNA identification is not possible in practice, the next best thing is to claim virtual certainty due to the extremely small probabilities of a coincidental match. This activity is designed to teach students how to calculate allele frequencies and the frequency of a set of alleles simply by dealing playing cards from well-shuffled decks. By virtue of these card games, students should begin to understand and appreciate the strategy forensic scientists adopt to lower the probability of a chance match between evidence and a suspect.

Objectives

After completing this lesson, students should be able to:

1. Determine the predicted frequency of an allele.
2. Apply the multiplication rule to calculate the frequency of a set of alleles occurring together.
3. Explain why probabilities are influenced by the makeup of the database.
4. Discuss whether DNA evidence alone is sufficient to convict a suspect of murder in the absence of supporting evidence.

Materials

* Decks of playing cards. If possible, have everyone in the class bring in their own deck; in this way, everyone is involved and the games go much more quickly.
* Calculator(s)

Procedure

Tell your students they will use decks of cards to demonstrate the multiplication rule as it is applied to calculating probabilities in DNA profiling. Each card dealt represents a locus (a specific site on a chromosome). The color of the card (red or black), *or* the suit (spade, diamond, heart, or club) *or* a specific value (e.g., a king) *or* a combination (e.g., the queen of hearts) represents alleles (specific DNA sequences or profiles that are found at a particular locus; see below).

Here's how the game works. Display a predetermined card or group of cards. Announce to the class that this card (or group of cards) represents the DNA profile of evidence found at a crime scene. Tell your students that they all are suspects in the crime. The question is, how many of them have DNA profiles that match the evidence? Have your students calculate the probability of a match before they deal. Then, have each student deal the same number of cards from their decks as you did. Count how many "suspect" profiles match the evidence profile. If matches occur, compare the experimental results to their calculated probabilities.

Start with high-probability events and work toward lower-probability events. Here is a possible sequence.

1. Turn over a *red* card. The single card is one locus, analyzed by a single probe. The red color represents the allele revealed at that locus by the probe. How many students should also turn over a red card? (Answer: 1 in 2 [half the cards in the deck are red, so the allele frequency for red is 1/2, or 1 in 2].)

   At this point, ask the students how they know the odds are 1 in 2. They will naturally respond that they know half the cards in the deck are red. Explain that criminologists use DNA profile databases so that they can know how frequently a particular pattern turns up at a given locus. If the students were using a collection of cards whose makeup they knew nothing about, they could not calculate the odds of getting a red card. Similarly, criminologists must rely on their databases to make their calculations.

2. Turn over a *club*. Again, the single card represents a single locus; this locus has four alleles, represented by the four suits. How many students should match the *club* allele? (Answer, 1 in 4 [the allele frequency for clubs is 1/4, or 1 in 4].)

   So far, too many "suspects" match the evidence by chance, so let's lower the probability of a coincidental match. You could ask the students how to lower the probability; they are bound to suggest using more cards or using the numbers on them.

3. Turn over four *red* cards in succession. This represents using four different probes to look at four loci, each of which has two alleles (*red* or *black*). How many students should match four red cards? (Answer, 1 in 16 [1/2 × 1/2 × 1/2 × 1/2 = 1/16 = 1 in 16].)

4. Turn over four *red* cards in succession, with one card being an *ace*. How many "suspects" should match that sequence? (Answer, 1 in 208. The probability of getting the three red cards is 1/2 × 1/2 × 1/2. The probability of drawing a *red ace*

is 1/2 for the *red* and 1/13 for the *ace*. The probability of getting the four cards described is thus $1/2 \times 1/2 \times 1/2 \times 1/2 \times 1/13$, or 1/208, or 1 in 208. Note how the inclusion of the relatively low-frequency allele, *red ace*, makes the probability of this match much lower than the one posed in question 3.)

Do a few more examples with more low-frequency alleles. The odds of getting a specific card, like the queen of hearts, is 1/52 (1/4 for the suit $\times$ 1/13 for the queen). The odds of getting four specific cards would be approximately $1/52 \times 1/52 \times 1/52 \times 1/52$.

We say approximately because a clever student may realize that if you draw a specific card from the deck (like the queen of hearts) and then want to know the odds of drawing the ace of spades, there are only 51 cards left in the deck. The odds of drawing the ace of spades from the remaining cards is 1/51. However, in DNA testing, you are not removing anything from the "deck" when you use a probe to look at a particular locus. To be perfectly correct in this simulation, each card should be drawn separately, its identity should be recorded, and then it should be returned to the deck before the second card is drawn. If your students do not bring up this problem, we do not recommend that you do!

5. Remove all of the *hearts* from each deck. Now, answer question 3. (Answer, 1 in 81 [$1/3 \times 1/3 \times 1/3 \times 1/3 = 1$ in 81].) Note how altering the database has significantly changed the probability of this match compared to the allele set posed in question 3.

Tips
- In the case of low-probability sequences, such as the one posed in question 4, students should deal a hand, score it, and then reshuffle those cards back into the deck before dealing another hand. However, the exercise should work fine, especially for high-probability alleles, if you choose to skip the reshuffling step for the sake of time or convenience.
- Sometimes, an "unlikely" combination is dealt in the first two or three hands. When this occurs, have your students continue to deal hands in order to validate the observed frequency. It should become obvious that observed probabilities match calculated probabilities only when sufficient trials are made.

This strategy will help your students understand that lower probabilities of a chance match result from

choosing low-frequency alleles and including more cards (loci) in the game. Point out to your students that including more cards in the game is similar to using more than one DNA probe in a restriction fragment length polymorphism (RFLP) experiment.

Part II. Discussion

The "Prosecutor's Fallacy"
There are two important questions with respect to the interpretation of DNA evidence. First, what is the probability that an innocent individual will match? Second, what is the probability that an individual who does match is innocent? It is the second question that is of direct interest to courts. The two questions are quite different. A common error, the "prosecutor's fallacy," consists of giving the answer to the first question in response to the second, that is, confusing the match probability with the probability of innocence. The probability of innocence depends upon the totality of evidence.

For example, consider a suspect whose DNA profile matches the evidence with a match probability of 1 in 700,000 and yet reliable eyewitnesses swear that he was 1,000 miles from the scene when the crime occurred. Discuss such an example with your students. It is important that you impress upon your students that a small match probability may not, in itself, establish guilt.

1. Explain this statement: there are more differences among people at the DNA level than at the protein level.
2. Explain that DNA typing is based on an examination of variable regions of the genome.
3. Describe the two technical methods used in DNA typing: PCR and Southern hybridization.
4. Tell what an RFLP is and give one explanation for how an RFLP could arise.
5. Explain why it is necessary to use either hybridization or PCR in DNA typing.

DNA Typing in the Classroom

Several companies market "DNA fingerprinting" kits that are simulations in which students digest different samples of DNA (often a "crime scene" and "suspects") with restriction enzymes, run the products on a gel, and then compare the stained banding patterns to "identify" a "criminal." We are not comfortable with these simulations, because they convey a huge misimpression about human DNA typing: that you can do it by simply looking at stained digests of human DNA. This neglects the enormous complexity of the genome and the problem of looking selectively at regions that are informative.

A simple simulated typing wet laboratory that we are comfortable with is the *PCR Forensics Simulation Kit,* no. 21-1210, from Carolina Biological Supply Company. In this kit, samples that represent the products of PCRs from various suspects and a crime scene are provided. The "products" are separated on an agarose gel and examined. This is a realistic simulation, since PCR products can be seen on stained gels.

A slightly different DNA typing scenario is presented in the kit *Outbreak! Fingerprinting Virus DNA,* no. 21-1208. In this scenario, students use restriction digestion to type a virus strain. We are comfortable with this simulation, because virus genomes are small enough that distinct restriction fragments can be resolved.

If you have access to a thermal cycler (a PCR machine), real DNA typing kits can be ordered from Perkin-Elmer Corporation, 761 Main Avenue,

Norwalk, CT 06854 [(800) 762-4002]. The two loci most commonly used are D1S80 and DQ-alpha; kits for both are available. Carolina Biological Supply Company sells a classroom kit *Human VNTR Polymorphism Kit,* no. 21-1233, -1234, or -1235, that uses the pMCT118 (also known as D1S80) locus on chromosome 18. The products from these amplifications are often very small fragments, and they need to be stained with ethidium bromide. Visible stains do not give good results. The kit contains student and teacher instructions. All these kits use cheek cells as a source of DNA.

In cooperation with the DNA Learning Center at Cold Spring Harbor Laboratories, Carolina Biological Supply Company also offers a kit for amplifying human mitochondrial DNA (no. 21-1236, -1237, or -1238). Amplified student DNA samples may be sent to Cold Spring Harbor Laboratories, where for a small fee, they will be sequenced with the results posted online. Student data can then be used to explore online genome databases and for other activities.

Forensic DNA Typing

Introduction

Scientists first began to clone and sequence DNA in the 1970s. One of the early goals of researchers was to relate known genetic differences in humans, such as hemoglobin differences related to various anemias, to base sequence changes in DNA. For studies such as these, researchers had to clone and sequence the same genes from many individuals. An unexpected result of this kind of research was the discovery of tandem repeat sequences in the human genome.

Hypervariable DNA Repeats

In 1980, a research group published a description of a repeated sequence in which the number of repeats varied significantly from one individual to the next. After the publication of this first description of a hypervariable repeated DNA sequence, other researchers quickly reported finding additional variable regions. These repeated sequences were several (10 to 15) bases in length and were called minisatellites.

One research group interested in using minisatellites for genetic studies was headed by Alec Jeffreys, a young English scientist. Jeffreys had been studying the evolution of the globin gene family, including the related myoglobin gene. (The globin genes encode the proteins that comprise hemoglobin.) Jeffreys' group discovered a minisatellite within the myoglobin gene that showed sequence similarities with the other minisatellites that had been described by then. Jeffreys wondered if he might be able to use the core minisatellite sequence as a probe to detect even more minisatellites. He and his group screened a library of human DNA and found more variable loci.

They next decided to try performing a Southern hybridization to human DNA using the core sequence as a probe. They chose DNA from a family group plus some nonhuman species and did the experiment. To their astonishment, they could see patterns in the family group that suggested that the minisatellites were inherited just like genes. DNA typing was born on that day: Monday, 10 September 1984.

The first DNA-typing techniques were based on these minisatellites and used restriction fragment length polymorphism as the typing method. The polymerase chain reaction (PCR) had not been invented at the time. In 1989, a description of microsatellites, or short tandem repeats (STRs), was published. Microsatellites were soon found to be more common in the genome than minisatellites, and because they were much shorter (even when many repeats were present), they could be analyzed in fragmented DNA. Microsatellite markers have become the favored markers for DNA typing, and they have also proven extremely valuable for mapping the locations of genes on chromosomes.

Forensic DNA Typing

In criminal cases, DNA typing is used to exclude the possibility that a given suspect left DNA-containing evidence at a crime scene. For example, if blood is found at a crime scene, a DNA profile can be generated from the blood and compared with DNA profiles from any suspects. If the crime scene profile does not match a suspect's, that suspect cannot have left the sample and is cleared. In fact, about 30% of the FBI's DNA-typing cases have cleared the prime suspect, and DNA typing has cleared people who were in prison serving time for violent crimes.

If the DNA profile of the sample at the crime scene does match a suspect's DNA profile, the relevant question becomes how likely it is that a person other than the suspect could have left it. Since DNA typing looks at only a portion of an individual's genome, two people could have very similar profiles. Scientists look at databases of DNA profiles from many individuals and calculate the frequencies of the patterns in the profile in question. They then use those numbers to calculate the odds that another individual will have the same DNA profile.

For example, if a suspect's DNA profile with probe A matches the crime scene sample but 10% of the population also has that profile, then 1 in 10 people could also have left the sample. As a forensic scientist, what do you do? You use another probe. Suppose the suspect's profile with probe B also matches the crime scene sample profile. Let's say that particular profile is found in 5% of the general population. If the profiles from probe A and probe B are independent, we now have a 0.1×0.05, or 0.005, or 0.5% probability of finding the two matches in a random person. Another way of looking at it is that 1 person in 200 would be expected to have both of these profiles.

The odds of two individuals matching this many probes by chance are very small, though people argue about exactly how small. Originally, many scientists were concerned that individuals belonging to distinct ethnic groups might be shortchanged by this type of calculation. Specifically, they worried that all the members of that ethnic group might have more similar DNA profiles than the general population (just as Scandinavians have blond hair and blue eyes more frequently than the general population). In response to this and other criticisms, the FBI established a Scientific Working Group to advise it on how to address the criticisms. The panel made scientific and procedural recommendations. DNA identification laboratories, whether governmental or private, have now built databases of DNA profiles of individuals from various ethnic groups to serve as comparison data. Studies of DNA profiles of over 70 ethnic groups have not shown great differences between them in terms of profile frequencies. In addition, the panel recommended standards for DNA testing procedures and for the operation of DNA testing laboratories, including training and testing of personnel.

The FBI has now standardized its DNA typing and uses 13 STR loci. To take maximum advantage of the power provided by DNA evidence for solving crime, the FBI established the Combined DNA Index System (CODIS), a national database of convicted felons. According to its website, CODIS generates investigative leads in crimes where biological evidence is recovered from the crime scene using two indexes: the forensic and offender indexes. The forensic index contains DNA profiles from crime scene evidence. The offender index contains DNA profiles of individuals convicted of sex offenses (and other violent crimes), with many states now expanding legislation to include other felonies. States are free to participate in the system, and as of March 2006, all 50 do so. As of March 2006, there were 134,075 profiles in the forensic index and 3,045,761 profiles in the convicted-offender index. You can read about CODIS, including the FBI's standards for DNA testing, on the CODIS section of the FBI's website.

Matches made among profiles in the forensic index can link crime scenes together, possibly identifying serial offenders. Based on a match, police in multiple jurisdictions can coordinate their respective investigations and share the leads they have developed independently. Matches made between the forensic and offender indexes provide investigators with the identity of the perpetrator(s). After CODIS identifies a potential match, qualified DNA analysts in the laboratories contact each other to validate or refute the match.

Forensic Use of Nonhuman DNA

Some crimes have been solved by analysis of DNA that did not come from humans. For example, DNA typing of a tree led to a murder conviction. A woman's body was found buried under a paloverde tree in Arizona. The police had a prime suspect of whose guilt they were certain, but there was no physical evidence linking him to the site. However, the police found paloverde tree pods in the bed of his pickup truck. The prosecuting attorney wondered whether it might be possible to determine whether those pods had come from the tree under which the body was found.

He called a nearby university and asked. The scientists agreed to try the experiment. First they had to develop a profiling method. Using PCR, they tried out many random sets of primers until they were able to find some that yielded unique patterns of products when tested on several different paloverde trees. They were then ready to test the crime scene evidence. They generated profiles from the pods from the suspect's truck, the tree under which the body was buried, and about 15 other paloverde trees. The profile from the pods matched the tree at the crime scene. None of the other trees had the same pattern. The prosecutor had the evidence linking the suspect to the crime scene and obtained a conviction.

Another notable forensic DNA case catapulted a cat into the limelight. In 1994, the body of a woman

Classroom Activities

from Prince Edward Island, Canada, was found in a shallow grave. Her bloody jacket had over 2 dozen white cat hairs on it. The woman's estranged husband owned a white cat named Snowball. The Royal Canadian Mounted Police confiscated Snowball and drew a blood sample for DNA testing, only to find that no one had ever performed forensic DNA typing on a cat before.

They contacted the laboratory of Stephen O'Brien at the National Cancer Institute, who had been studying cat genetic diseases and genetic diversity in felines. Researchers in O'Brien's laboratory had identified hypervariable DNA markers in the cat genome, and the scientists used 10 of them to develop a method for profiling cat DNA. To make sure that the cats of Prince Edward Island did not happen to share DNA profiles, they collected samples from 20 random cats on the island and tested them as well. Their results showed that the hairs on the victim's jacket belonged to Snowball. The estranged husband was convicted of the murder.

Other Uses of DNA Typing

Paternity testing

DNA testing can also be used to establish paternity or maternity. Since an offspring inherits half its chromosomes from its mother and half from its father, its DNA profile should show contributions from both. To establish paternity, profiles of the child, mother, and putative father are generated. The child's DNA profile is compared to the mother's, and the bands that match are subtracted. The remainder of the bands in the child's profile should match bands in the father's.

The very first use of DNA parentage testing was in fact a *maternity* case. A young Ghanaian boy arrived in the United Kingdom in 1985 to join his mother. His passport appeared to be forged, and authorities threatened to deport him. The family lawyer contacted Alec Jeffreys, who had recently made headlines with his newly developed method of DNA typing, and asked if he could help. Jeffreys believed DNA testing could show maternity and paternity, and he agreed to try. Test results confirmed that the boy was indeed the woman's son, and the authorities dropped the deportation case.

Gender determination

DNA testing can also establish whether human remains are male or female. Normally, scientists can make that determination based on measurements of the skeleton (if other evidence is not available), but when the skeleton is badly fragmented (or parts of it are missing), it may not be possible. In humans, gender is determined by the so-called sex chromosome. There are two forms of the human sex chromosome, called X and Y. Females have two X chromosomes, while males have one X and one Y. Analysts can test for the presence of a Y chromosome by doing a PCR assay for base sequences unique to it. In the archaeology case outlined in the *Reading*, researchers tested for the presence of a Y chromosome gene called SRY.

Exercise 1. A Mix-Up at the Hospital

On 6 June at approximately 1:00 p.m., Mrs. Smith, Mrs. Stevenson, and Mrs. Jones each delivered a healthy baby boy at Metropolitan General Hospital. At 1:20 p.m., the hospital's fire alarm sounded. Nurses and orderlies scrambled to evacuate patients. The three new babies were rushed to safety. After the danger had passed, the hospital staff was distressed to find that in the confusion, they had forgotten which baby was which! Since the babies were rescued before receiving their identification bracelets, there was no easy way to identify them. Anne Robinson, head of pediatrics, ordered that DNA typing be performed on the babies and their parents.

The DNA-typing laboratory looked at two different highly variable chromosome regions. The DNA profiles are shown in Figure 28.1. Your job is to decide which baby belongs to which set of parents. To assign a baby to a set of parents, every band in the baby's profile should match a band from either the mother or the father. Not all of the bands in the mother's or father's profiles will have a counterpart in the baby's DNA profile. Hint: use a ruler or straightedge to help you line up the bands.

Which baby belongs to which couple? Show which bands each baby inherited from its mother and from its father by marking them with "M" and "F."

Exercise 2. A Paternity Case

Mr. I. M. Megabucks, the wealthiest man in the world, has recently died. Since his death, three women have come forward. Each woman claims to have a child by Megabucks and demands a substantial share in his estate for her child. Lawyers for the estate have insisted on DNA typing of each of the alleged heirs. Fortunately, Megabucks anticipated

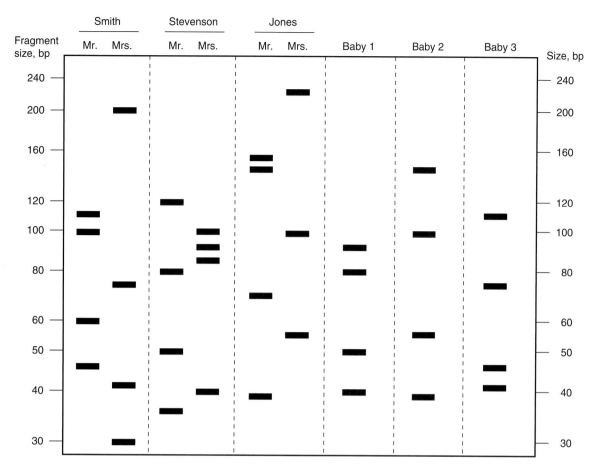

Figure 28.1 DNA profile data from the Smith, Stevenson, and Jones parents and the three infants.

trouble like this before he died and arranged to have a sample of his blood frozen for DNA typing.

Laboratory technicians examined four highly variable STR loci using simultaneous PCR amplifications of the loci. The results of the analyses are shown in Fig. 28.2. Your job is to analyze the data and determine if any of the children could be Megabucks' heir.

Remember that every person has two of each chromosome, one inherited from his mother and one from his father. Half of every person's DNA comes from his mother and half from his father, so some of the DNA bands showing in the children's DNA will come from their mothers and the rest from their fathers. The question is, could that father be Megabucks?

1. For the first child, identify the bands in the DNA profile that came from the mother. (Remember that not all of the mother's DNA is transmitted to the child, just one of each pair of chromosomes.) Mark the bands that came from the mother with an M. Circle the remaining bands.

2. Compare the remaining bands with the DNA profile from Megabucks. If he is the father, then all of the circled bands in the child's profile should have a corresponding band in his profile. Use a straightedge to help you line the bands up accurately. (Remember that only half of the father's chromosomes are transmitted to a child, so not every band from the father would match the child's profile.)

3. Repeat the analysis for the other alleged heirs. Could any of them be Megabucks' child?

Question

Why are there eight bands in every lane, and why do four of each child's bands match corresponding bands in its mother's profile?

Exercise 3. The Case of the Bloody Knife

Late one April night, government agents received an anonymous tip that the National Art Museum was

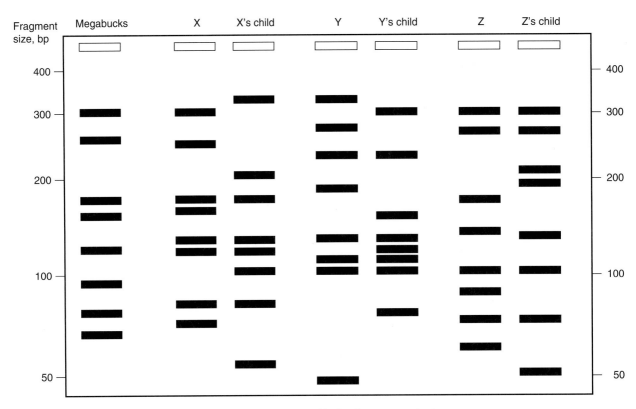

Figure 28.2 Results of hybridization analysis.

about to be robbed of a priceless jewel collection. When they arrived at the museum, they saw they were too late: the jewels were gone. Lying face down on the floor next to the empty jewel case was the body of a man the chief inspector recognized as the international jewel thief Heinrich Milhouse. Milhouse had been shot in the chest at close range; his clothes were saturated with blood. Underneath the body, the inspectors found a bloody knife.

At the airport the next day, police apprehended Englewood Smink, the murdered thief's occasional partner in crime. Smink denied all knowledge of the murder and the theft. When asked about the fresh cut on his hand, Smink said that he had had an accident in the kitchen that morning.

Suspicious, the chief inspector ordered DNA tests on the victim, the blood on the victim's clothes, the blood on the end of the knife found under the victim, and Smink. Police laboratory technicians used PCR to look at two different chromosome regions that contained an STR. They used one set of primers for each region. The chromosome regions and primers are shown below, and the results of the tests are shown in Figure 28.3.

Chromosome region 1:

```
3' AGGCTCGACCTGCACGTC. . .variable number of ATCT repeats. . .CAATGTGCGGACTCAATGCCA 5'
5' TCCGAGCTGGACGTGCAG. . .variable number of TAGA repeats. . .GTTACACGCCTGAGTTACGGT 3'
```

Primer set 1: 5' CCGAGCTGGACGTGCAG + 3' AATGTGCGGACTCAATG

Chromosome region 2:

```
3' GCTGCGAATGCTACAGGTC. . .variable number of CA repeats. . .GCGATCAGCTGCGGTAG 5'
5' CGACGCTTAGCATGTCCAG. . .variable number of GT repeats. . .CGCTAGTCGACGCCATC 3'
```

Primer set 2: 5' GACGCTTACGATGTCC + 3' GCGATCAGCTGCGGTA

What is your interpretation of the data? State your reasons. Should Smink be released? Should other tests be performed?

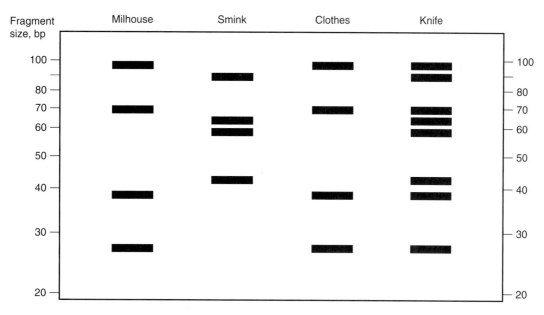

Figure 28.3 Results of PCR analysis.

DNA Forensics: An Archaeological Case

Archaeologists study the remains of ancient humans and their artifacts to learn about early human culture. DNA analysis gives archaeologists an entirely new way to analyze remains from ancient burials and an entirely new kind of data to consider. In one case, published in 2004, a team of scientists studying the Etruscan culture found four skeletons in a tomb in the Monterozzi necropolis near Viterbo, Italy. The Etruscans lived in Italy before the rise of the Roman civilization, and many questions remain about their civilization and culture.

The skeletons were those of a child, a man estimated to be in his 30s, and two people estimated to be 40 to 45 years old. Because of the state of preservation of the skeletons, the gender of the child could not be determined. One of the two 40- to 45-year-olds was identified tentatively as a woman, but the skeleton of the other 40- to 45-year-old was too fragmentary for gender assignment. Near the skull of that individual lay bronze earrings, commonly considered a female ornament. When skeletons are too incomplete for definite gender assignment, archaeologists often rely on artifacts associated with burials to draw conclusions about who is buried there, and the earrings suggested the poorly preserved skeleton belonged to a woman. No information was available as to who the buried individuals might have been or whether they might have been related to each other.

To apply DNA analysis to the remains, the team of Italian scientists extracted DNA from bones and teeth. They used PCR with primers complementary to the SRY gene to determine the genders of the individuals. Males have the SRY gene on the Y chromosome; females lack it. Their results showed that the child, whose gender was previously unknown, was female and confirmed that the 30-something-

year-old was indeed male. Surprisingly, the 40- to 45-year-old individual, tentatively identified as female because the body was buried with earrings, tested positive for SRY, showing that the skeleton was that of a man. The other older individual was indeed a woman.

To determine whether any of the individuals were related through the maternal line, the scientists determined the base sequence of portions of the hypervariable region of mitochondrial DNA from each skeleton (see the *Reading* in chapter 27). The mitochondrial DNA sequences of the woman, the 30-something man, and the child matched, supporting the conclusion that these individuals were related through the maternal line. The scientists hypothesized that the four individuals were a family group, with the older man and the woman being the parents of the younger man and the child. They noted that there was no way to tell whether the individuals had been buried together at the same or different times.

The DNA evidence gave the researchers an entirely different picture of the tomb than they would have had based on physical evidence alone. Before the DNA testing, the skeletons appeared to be a pair of middle-aged women, a young man, and a child, a grouping with no clear family structure. DNA evidence changed the picture to that of a likely set of parents and two children. The DNA-informed interpretation of the identities of the skeletons in the tomb painted a different picture of Etruscan culture than the tentative identifications based on the fragmentary skeletons. The authors concluded their study by stating that adding molecular data to traditional archaeological data could lead to a better understanding of Etruscan culture.

Genome Mapping

29

About This Activity

In chapter 27, we discussed the use of DNA polymorphisms to compare genomes. In this chapter, we illustrate how polymorphic loci are used in genetic mapping, using a human disease gene as an example. The *Student Activity* contains extensive introductory material, including an example of using genetic mapping to determine the relative locations of three chromosomal markers (the ABO blood group locus, the nail-patella syndrome locus, and a hypothetical polymorphic short tandem repeat [STR] locus).

The activity itself is a simulation of part of the process of mapping the physical location of a gene. We provide pedigree data on the inheritance of a genetic marker and a gel for students to analyze to deduce information about the inheritance of an STR marker. They must then combine the genetic and molecular data and decide whether the combined data support a hypothesis of linkage between the STR marker and the gene. Our goal was not to provide a complete discussion of the intricacies of genome mapping. Rather, we sought to provide students with an idea of how scientists figure out where genes are and how physical markers like STRs are used in that process. The activity contains a relatively large amount of data for students to analyze. In comparison to what scientists actually do when they map gene loci, this data set is quite small. In fact, the notion that mapping disease loci is a lot of work is part of the point.

The chapter also includes a reading about the Human Genome Project.

Class periods required: *In contrast to the activities in chapter 28, this activity presents a relatively complex problem that will take time to work and requires integration of a lot of background information. Students may need a review of meiosis and crossing over, and if they are unfamiliar with genetic mapping, you will probably need to work through the example with them or coach them about how to work through it as they read. We hesitate to provide an estimate of the time required. For some classes, it might be sufficient to spend one class period and then assign the problem or its remainder as homework. If you want students to do most of the work in class, it will probably require two or more standard 50-minute sessions.*

Introduction

Please read the background information in the *Student Activity.*

One of the terms we introduce in this chapter is **haplotype,** a contraction of *haploid genotype.* The term haplotype refers to the DNA content of one contiguous molecule, such as a mitochondrial genome, a single chromosome, or part of a chromosome, even down to very small regions. An allele of a gene is a haplotype in that it represents a DNA sequence on a single chromosome. Sometimes the terms allele and haplotype are used interchangeably, but allele refers to a single locus (a gene, part of a gene, or a single repeated region) while haplotype can stretch to the entire chromosome. Occasionally, haplotype is used to refer to half of a genome, in the sense that a gamete contains a haplotype, and it is often used to refer to a specific mitochondrial genotype.

Since microsatellites have become so important in comparing and mapping genomes, many articles you might read refer to haplotypes meaning microsatellite haplotypes. A microsatellite haplotype would be the exact configuration of repeated sequences along a single chromosome. Figure 29.1 shows haplotypes that include both genetic markers and a microsatellite marker.

Objectives

After completing this lesson, students should be able to:

- Explain that genetic maps are made by observing the patterns of inheritance of traits and/or DNA markers.
- Explain that when two features are close together on a chromosome, they are more likely to be inherited together than if they are on different chromosomes or far apart on the same chromosome.
- Explain that when two features are on different chromosomes their chance of coinheritance is 50%.
- Explain how recombination during meiosis separates markers that are on the same chromosome.

Materials

Students will need copies of the *Student Activity*.

Preparation

- Print the *Student Activity* if needed.
- Review Mendelian inheritance and meiosis if necessary.

Answers to Worksheet 29.A

1. Fill in the genotypes of the individuals in Pedigree A based on pedigree analysis and analysis of the molecular data in Gel Diagram A.

Individual	Genotype									
	I1	I2	II1	II2	II3	II4	II5	II6	II7	II8
BFNC	Bb	bb	Bb	bb	bb	bb	Bb	Bb	bb	Bb
STR	6,8	7,9	7,8	6,7	6,9	7,8	8,9	7,8	6,9	7,8

2. What are the father's (individual I1) STR alleles? __6,8__

3. What are the four combinations of BFNC and STR alleles that the father could pass to an offspring? These are the possible paternal haplotypes: __6B__, __6b__, __8B__, __8b__

4. Fill in the paternal haplotype inherited by each member of generation II.

II1	II2	II3	II4	II5	II6	II7	II8
8B	6b	6b	8b	8B	8B	6b	8B

5. Fill in the four paternal haplotypes (from question 3) and record the frequency at which each was seen in the members of generation II.

Haplotype	Frequency in Gen II
6B	0/8
6b	3/8
8B	4/8
8b	1/8

6. If the STR and the BFNC loci are linked, what would be the two nonrecombinant haplotypes? __8B and 6b__. The recombinant haplotypes? __8b and 6B__

7. What is the frequency of these hypothetical nonrecombinant haplotypes in generation II? __4/8 + 3/8 = 7/8__

8. What is the frequency of the recombinant haplotypes? __1/8 + 0/8 = 1/8__

Answers to Worksheet 29.B

1. Fill in the genotypes of the individuals in Pedigree B based on pedigree analysis and analysis of the molecular data in Gel Diagram B.

Individual	Genotype								
	II1	S1	III1	III2	III3	III4	III5	III6	III7
BFNC	**Bb**	**bb**	**bb**	**Bb**	**Bb**	**bb**	**bb**	**Bb**	**bb**
STR	**7,8**	**5,6**	**7,5**	**8,5**	**8,6**	**7,5**	**7,6**	**7,5**	**7,5**

2. If the BFNC locus and the STR locus are linked, what STR allele is linked to the disease allele of BFNC in individual II1 (see results on Worksheet 29.A)? ____**8**____. In Individual II8? ____**8**____

3. What are the nonrecombinant haplotypes in individual II1? ___**8B**___ and ___**7b**___. In II8? ___**8B**___ and ___**7b**___.

4. Fill in the maternal haplotypes of the offspring of II1 and the paternal haplotypes of the offspring of II8.

Offspring of II1					Offspring of II8	
Maternal haplotype					Paternal haplotype	
III1	III2	III3	III4	III5	III6	III7
7b	**8B**	**8B**	**7b**	**7b**	**7B**	**7b**

5. How many haplotypes in generation III are nonrecombinant? ____**6**____

6. How many haplotypes in generation III are recombinant? ____**1**____

7. Adding results from generations II and III together, how many nonrecombinant haplotypes were observed? ____**13**____. What frequency is this? ___**13/15**___

   How many recombinant haplotypes were observed? ____**2**____. What frequency is this? ___**2/15**___

8. Do the data from Worksheets 29.A and 29.B support the hypothesis that the STR locus and the BFNC locus are linked? Explain your answer.
   Yes, the data are consistent with linkage; 13/15 of the haplotypes are nonrecombinant. This result could occur by chance, but it is unlikely.

Answers to Worksheet 29.C

1. Fill in the genotypes of the individuals in Pedigree C based on pedigree analysis and analysis of the molecular data in Gel Diagram C.

Individual	Genotype					
	III6	S3	IV1	IV2	IV3	IV4
BFNC	**Bb**	**bb**	**Bb**	**bb**	**Bb**	**Bb**
STR	**5,7**	**6,6**	**6,7**	**6,5**	**6,7**	**6,7**

None of the individuals in Pedigree C has the same STR alleles that were present in individual I1. Do the data from Pedigree C and Gel Diagram C support the hypothesis that the BFNC and STR loci are linked? Explain your answer.
It supports the linkage hypothesis. None of the individuals in the pedigree has the STR alleles of individual I1 because the afflicted father, individual III6, has a recombinant haplotype, 7B. He passed this haplotype to three of his offspring and passed his other, nonrecombinant haplotype, 5b, to the fourth offspring, so all four of his offspring inherited nonrecombinant haplotypes.

Looking at recombinant and nonrecombinant haplotypes from every generation, we see the following:

Generation	Nonrecombinant haplotypes	Recombinant haplotypes
II	7	1
III	6	1
IV	4	0
Total	17	2

The total frequency of recombination between the STR locus and the BFNC allele is only 2/19, consistent with the linkage hypothesis.

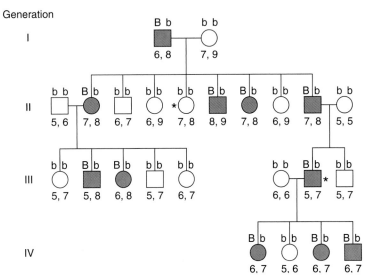

Key to pedigree shown in Figure 29.4. The genotypes of the BFNC gene and the STR locus are shown for each individual. *, recombinant haplotype.

Mapping a Disease Gene

29

Introduction

In chapters 27 and 28, we discussed how polymorphic DNA markers can be used to compare genomes. In this chapter, we will explore how DNA markers can be used to map the locations of genes.

Before it was established that DNA is the material of which genes are made, geneticists observed that certain traits were often inherited together. By analyzing the results of carefully constructed genetic crosses in many individuals, they could deduce the order in which genes were arrayed on a chromosome. Maps showing the positions of genes or other inheritable chromosome features, such as short tandem repeats (STRs), as deduced from genetic experiments, are called genetic maps.

Genetic mapping is still used, but now it can be combined with physical analyses of DNA to yield much more detailed and accurate information about genes. If a scientist is interested in finding the location of a gene for a particular observable trait, she would typically start with genetic mapping to identify the general location of the gene. Once that region was pinpointed, she would examine its base sequence to find the physical location of the gene. Maps showing the positions of genes and other chromosome features that are based on physical analysis of DNA (sequence analysis, restriction analysis, etc.) are called physical maps.

Genetic Mapping

Genetic mapping was invented long before researchers had the ability to analyze DNA (and before they knew that genes were made of DNA). It is based on the observation of inheritance patterns. The idea is that if the chromosomal locations of two different detectable features are close together, the features will be inherited together. Originally, the features were genes, and they were detected by the observation of specific traits in individuals. Scientists are no longer limited to making inferences

about genes and chromosomes by observing the outward expression of traits. They can also analyze physical features of chromosomes, such as specific STR loci. Any feature of a chromosome that has a specific location and whose inheritance can be followed is referred to as a genetic marker. Sometimes markers, like STR and restriction fragment length polymorphism loci, are called physical markers, because they can only be detected by a direct analysis of DNA.

Think about the inheritance of two chromosomal markers (physical or genetic). Chromosomes are inherited in gametes (eggs and sperm) that are created in the two parents' reproductive organs. The process by which chromosomes are distributed into gametes is meiosis (see chapter 3). If the loci of the markers are on separate chromosomes, the two markers will be inherited together 50% of the time as a result of random segregation of chromosomes during meiosis (Figure 29.1A). If the marker loci are on the same chromosome, they may be inherited together or linked (Figure 29.1B).

If that were all there was to it, genetic mapping would be very simple—either two alleles are inherited together 100% of the time (linked) or 50% of the time (unlinked). However, chromosomal inheritance is more complex than that. The process of meiosis includes recombination, in which pairs of chromosomes swap segments (see Figures 3.11 and 3.12). The likelihood of two loci being inherited together depends on how physically close they are on the chromosome. The more physical distance there is between the two loci, the greater the likelihood that a recombination event will occur between them. If the loci are far enough apart, they are no more likely to be inherited together than if they were on separate chromosomes (Figure 29.1C and D).

An example of genetic mapping

A pedigree demonstrating linked inheritance is shown in Figure 29.2. The locus of the gene involved

A. Unlinked: disease gene and STR marker are on different chromosomes.

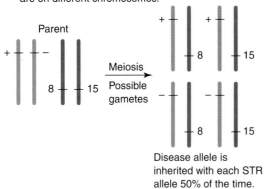

Disease allele is inherited with each STR allele 50% of the time.

B. Linked: disease gene and STR marker are on the same chromosome. No meiotic recombination.

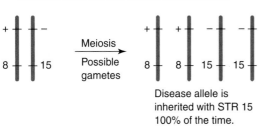

Disease allele is inherited with STR 15 100% of the time.

C. Linkage with recombination. If loci are relatively far apart on the same chromosome, they will frequently be separated by crossing over.

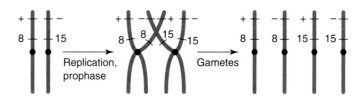

D. Linkage with recombination. If loci are close together, they are far less likely to be separated by crossing over.

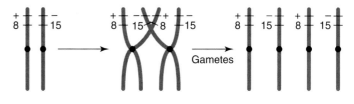

Figure 29.1 Linked and unlinked inheritance of alleles. +, normal allele of gene; −, mutant, disease-causing allele of gene; 8, STR allele with 8 repeats; 15, STR allele with 15 repeats.

in nail-patella syndrome (NPS), a condition involving abnormal growth of fingernails and toenails and a reduced or absent kneecap (patella), is linked to the locus of the gene that specifies the ABO blood type. NPS is dominant over the normal condition, and therefore, the inheritance of a single copy of the altered form of the gene is sufficient to produce the trait.

In Figure 29.2, the shaded symbols represent individuals with NPS. The blood group alleles of each individual are shown below the symbols in the pedigree. Ignore the numbers above the symbols for

now. The mother (generation I, individual 1) has the normal traits (no shading), and therefore she must have two normal NPS alleles (represented by nn). She can donate only a normal allele (n) to her children. The father (generation I, individual 2) has NPS, so he must have at least one disease allele (N). Since some of the children are healthy (nn), the father must be able to donate a normal allele, and therefore he has one n allele.

In this example, we know that the NPS locus and the blood group locus are linked. Given that information, which one of the father's blood group alleles is most

Figure 29.2 Inheritance of NPS and the ABO blood group.

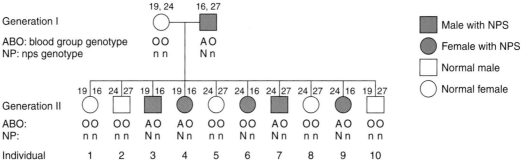

Table 29.1 Paternal haplotypes inherited by members of Generation II in Figure 29.2

Individual	Paternal haplotype		
	NPS allele	ABO allele	STR allele
1	n	O	16
2	n	O	27
3	N	A	16
4	N	A	16
5	n	O	27
6	N	O	16
7	N	A	27
8	n	O	27
9	N	A	16
10	n	O	27

likely on the same chromosome as the disease allele? Of the five affected offspring, four carry the A blood group allele. This pattern supports a hypothesis in which the A allele and the N allele are close together on one of the father's chromosomes. Any combination of alleles on a single chromosome is called a **haplotype,** so we are proposing that the father's haplotypes are NA and nO.

Following this hypothesis, we would propose that a recombination event between the NPS and ABO loci occurred during the generation of the gamete that produced individual II6 (the affected daughter with blood group alleles OO), giving her a paternal haplotype of NO. Since the mother has only one type of allele at both loci, she has only one haplotype, nO, to donate to all her offspring. Even though recombination will occur during production of her gametes, we cannot detect it by looking at these two loci. Thus, we observed one recombination event among the 10 haplotypes inherited from the father.

Genetic mapping hinges on the measurement of recombination frequency. We observed one recombination event between the NPS and ABO loci in 10 paternal haplotypes. Suppose we analyze the same

DNA samples for another known linked marker, this time an STR. We find that the mother has alleles of 19 and 24 repeats, while the father has alleles of 16 and 27 repeats. The results from each individual are given above the pedigree symbols in Figure 29.2. Because the mother has only one allele each at the NPS and ABO loci, we still cannot measure recombination in the maternal haplotypes, so we focus only on the paternal ones.

The alleles the offspring inherited from their father, or their paternal haplotype, can be deduced from the information in Figure 29.2. The offspring's paternal haplotypes are listed in Table 29.1; if you want to test your ability to deduce the haplotypes, do it, and then check yourself by looking at the table.

In Table 29.2, we summarize the frequencies with which alleles of each pair of loci are inherited together. In the first column of the table, we see that altogether the offspring have inherited the 16n combination and the 27N combination once each and the 16N and 27n combinations four times each. These data are consistent with the hypothesis that the father's two haplotypes were 16N and 27n. We have shaded these haplotypes to mark them as the nonrecombinant ones. When we consider the STR

Table 29.2 Frequency of allele combinations in generation II[a]

NPS and STR		NPS and ABO		STR and ABO	
Allele combination	Frequency	Allele combination	Frequency	Allele combination	Frequency
16n	1/10	nA	0/10	16A	3/10
27n	4/10	nO	5/10	16O	2/10
16N	4/10	NA	4/10	27A	1/10
27N	1/10	NO	1/10	27O	4/10
Nonrecombinant haplotypes	8/10		9/10		7/10
Recombinant haplotypes	2/10		1/10		3/10

[a]Nonrecombinant haplotypes are shaded.

and NPS loci, eight of the offspring haplotypes are nonrecombinant and two (16n and 27N) were created by recombination. The recombination frequency between NPS and the STR locus was therefore 2/10, or 0.2. We repeat the process for the combination of the NPS and ABO loci and for the STR and ABO loci.

We know that the three loci are linked, and given that information, there are three possible arrangements of the loci along the chromosome, as shown in Figure 29.3. To distinguish among the possible arrangements, we compare the recombination frequencies between the loci. We observed the greatest amount of recombination between the ABO and STR loci, suggesting that they are the furthest apart. This observation is most consistent with arrangement 3.

Physical maps of chromosomes express distances in base pairs, since they are based on analyses of DNA molecules themselves. Genetic maps are based on recombination frequency and use a unit called the morgan. Genetic-map distances and physical-map distances do not always correspond, because the odds of recombination along the length of a chromosome are not always equal (some chromosome features promote recombination in their vicinity, while others suppress it). However, the relative arrangement of genes revealed by genetic mapping does generally correspond to the physical map.

Figure 29.3 Genetic mapping based on recombination frequency.

Nonrecombinant alleles are n, O, 16 and N, A, 27.

Possible arrangements

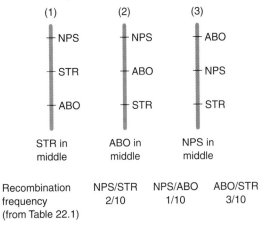

The greater recombination frequency between the ABO and STR loci supports arrangment 3.

A historical note about genetic mapping

The recombination-based technique of genetic mapping was developed by an undergraduate college student, Alfred Sturtevant. In 1911, Sturtevant was doing research in the laboratory of Thomas Hunt Morgan at Columbia University. They were studying inheritance patterns in fruit flies. Morgan's laboratory studied genetics, using the fruit fly as a model organism.

At that time, Morgan had reached the conclusion that genes were on chromosomes, but no one knew that genes were made of DNA. Morgan had observed that while many traits were inherited independently, some were inherited together, or linked. Further, he had observed that the degree of linkage, or the frequency with which different pairs of traits were inherited together, varied. Morgan had the notion that the degree of linkage between traits might be associated with the physical distance between genes on chromosomes, but he had not worked through the idea. He discussed the problem with his undergraduate assistant, Alfred, and challenged him to make sense of some data on recombination between different linked genes.

While talking with his advisor, Alfred suddenly realized that he might be able to use recombination frequencies to figure out the linear arrangement of genes on a chromosome. He went home, worked on the problem nearly all night (neglecting his regular homework), and came up with the method that you have just learned about. Alfred Sturtevant, who went on to become a great geneticist like his undergraduate advisor, named the unit of distance as measured by genetic recombination in honor of Morgan.

Mapping of genes to chromosomal locations

You might think that since the complete sequence of the human genome has been determined, scientists no longer need to hunt for genes. Nothing could be further from the truth. Although researchers have determined the functions of the proteins encoded by many human genes, many more are uncharacterized. If a researcher is interested in finding the gene that produces an observable trait, she usually starts by mapping the gene, narrowing down its location to a small region of a specific chromosome. The way she does this is by testing for coinheritance of the trait and known polymorphic chromosome markers.

The idea behind such a mapping approach is exactly the same as the example of NPS and the ABO blood group. DNA from families expressing the trait of interest is analyzed in the hope of finding a marker that is coinherited with the trait (as was the "A" allele in the above example). When markers that are linked to the trait are identified, the researchers conclude that the gene is close to the markers. The next step is to look for protein-coding sequences near the markers. If the genome in question has been sequenced, the "looking" can be done by searching electronic databases of sequence information.

It is a simple idea, but the process can be a lot of work. In the beginning, a researcher may have no idea on which chromosome the disease gene is located. Because of recombination during meiosis, only markers that are located quite close to the locus of the gene are linked to it with high frequency. This means it is necessary to test many, many markers in the hope of finding one that is inherited with the disease. In addition, there are complexities in data interpretation that do not arise in cases where it has already been established that the markers of interest (such as NPS and ABO) are linked.

What if we had obtained the data in Figure 29.2 but we did not know whether the NPS and ABO loci were linked? This is the situation encountered by researchers trying to map genes for the first time, and it is useful to think about the problem they face. Look at the data again. If the two loci are unlinked, we would predict that 50% of the time both affected and unaffected individuals would inherit the A allele from the father and 50% of the time they would inherit the O allele. Our results show that 100% of the unaffected offspring inherited the O allele and 80% of the affected offspring inherited the A allele. That result certainly does not match the prediction for unlinked loci, but could it have occurred by chance?

Now we come up against the problem that human families are typically small, at least to a statistician. Any time you are counting the occurrence of a random event (like the occurrence of heads in a coin toss or the inheritance of one of two alleles) in a small number of events, you run the risk of being fooled by the statistics of small numbers. You know that there is a 50% chance of getting heads every time you toss a coin, but you also know that it is not at all impossible to get four heads in a row. Now, imagine that you cannot see the coin and you do not know if it has a head and a tail or two heads. If you toss it four times and get four heads, have you ruled out the possibility that the coin has a tail? Of course not. If you could

toss the coin 100 times and get 100 heads, you could be much more certain that it had no tail.

The coin toss analogy is similar to the problem of small human families. If you have only a few offspring to analyze, you may see purely coincidental coinheritance of loci. Thus, the best cases for physical mapping involve large numbers of related individuals, a luxury not always available. (In chapter 26, *Genetics in Action,* you read about a medical geneticist who studies genetic disorders among the Amish and Mennonite populations in Pennsylvania. Note that scientists interviewed in the article state that the Amish are an excellent population for genetic studies, in part because they have large families that tend to stay in the area.)

Geneticists have to deal with the fact that recombination can occur even between closely linked loci plus the reality that human families are often small, magnifying the chances of coincidental results. They cannot solve these problems; they can only deal with them as best they can. The way they deal with them is by making statistical calculations of the likelihood of a given result occurring if two loci are unlinked and the likelihood of the same result occurring if the two loci are linked (allowing for a specified amount of recombination between them). We are not going to go into how the calculations are made, but you can look up the procedure in any human genetics textbook. If you have had advanced algebra, you have had enough math to understand the approach. Getting back to the main point, geneticists calculate the chances of getting a result if two loci are linked and the chances of getting a result if two loci are unlinked. They make a ratio of these two quantities and then take the logarithm of the ratio. This is the logarithm of odds, or LOD, score. It expresses the likelihood that two loci are linked.

In the exercise below, we give you hypothetical data for the inheritance of a single STR marker and a genetic disease and ask you to determine whether the data support linkage between the two loci. We do not ask you to calculate LOD scores, but you will need to look for patterns in the inheritance of alleles, as we did in the NPS example.

Exercise: Mapping a Human Disease Gene

Figure 29.4 shows a hypothetical pedigree of a large family afflicted with an inherited form of epilepsy, benign familial neonatal convulsions (BFNC). The

Generation

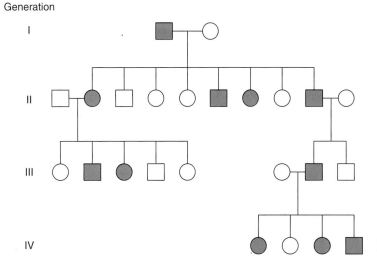

Figure 29.4 Pedigree of a family afflicted with BFNC.

solid symbols indicate individuals who have the disease. Pedigree analysis shows that the disease is dominant, meaning that a single copy of the disease allele is enough to confer the disease on an individual. By examining the pedigree, you can determine which parent of each afflicted individual passed the disease allele to his or her offspring.

DNA was isolated from most of the individuals shown in the pedigree. The DNA samples were typed by polymerase chain reaction (PCR) of 250 microsatellite loci. You will analyze data from a marker on chromosome 20 one generation at a time. This marker has five alleles, consisting of five, six, seven, eight, or nine repeats. The great-grandfather (generation I, individual 1) was deceased, and no sample could be taken from him.

Your assignment is to decide whether the data support linkage between this particular PCR marker and to explain why you think so. Begin with Pedigree A and Gel A, answer the questions on Worksheet 29.A, and then proceed to Worksheets 29.B and 29.C.

Analysis of the pedigree data

Before you tackle the molecular data shown in the gel diagram, figure out what the pedigree tells you. Stop reading right now if you prefer to figure it out without any suggestions from us. Keep reading if you want suggestions about how to start. Every individual has two alleles at the BFNC locus and two alleles at the STR locus. The two alleles can be the same or different. BFNC is dominant, so anyone who does not have the condition has two normal alleles. Write two small b's next to the symbol of every unaffected individual to symbolize their two normal alleles.

Next, consider the affected individuals. They are affected, so you know they have at least one disease allele. Write a single capital B next to the symbol of each affected individual in the pedigree. Now it is time to figure out what each affected individual's other allele is. Do not read any further if you want to figure this out on your own. If you want assistance, keep reading.

Remember that every individual inherited one allele from his or her father and the other from his or her mother. Starting with one affected individual, look at what you know about his or her parents. One of the parents must have a disease (B) allele to pass down to the affected individual. Look at the other parent. If the other parent is unaffected, then his or her alleles are b and b, right? What allele did that parent pass to his or her offspring? Write the second allele next to the B beside the symbols of the affected individuals. You do not have information about the parents of the individuals in generation I. Do you have other information that lets you conclude what the second allele of the affected man in that generation is? If you need help, reread the example with Figure 29.2.

Before you proceed with analysis of the information in the gel, you may want to check with your instructor to make sure you have the BFNC alleles correctly identified.

Analysis of the molecular data

The results of the molecular analyses are shown in the gel diagrams below each pedigree. The molecular size scale provides the number of repeats corresponding to the band size in the gel. Using the number of repeats to symbolize an allele (for example, five), write each individual's two alleles next to the symbols on the pedigree. Remember that an individual may have two copies of the same allele.

The True Story

Although we made up the data in the exercise to simplify it, the trait we describe really was mapped by analysis of inheritance patterns. The researchers tested over 250 markers and found 2 that were linked to the inheritance of the trait. The two linked markers were on chromosome 20. The researchers' data and conclusions were published in the journal *Nature* in 1989, in a paper titled "Benign familial neonatal convulsions linked to genetic markers on chromosome 20."

Pedigree A and Gel Diagram A (generations I and II). Shaded symbols mean afflicted with BNCS. M, molecular size marker.

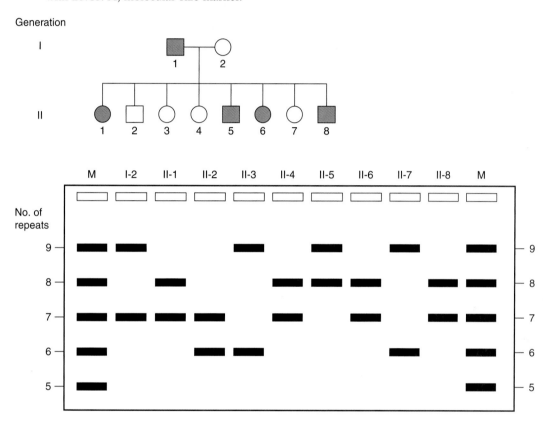

Generation

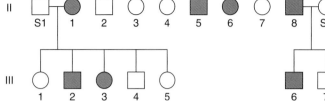

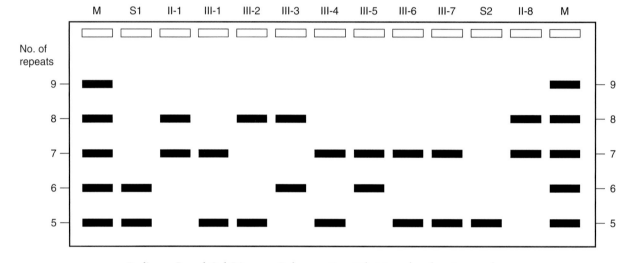

Pedigree B and Gel Diagram B (generation III). M, molecular size marker.

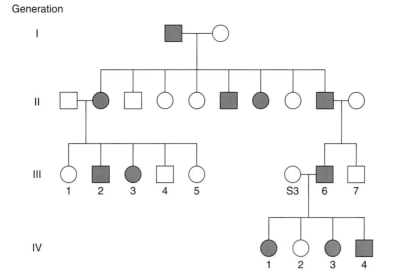

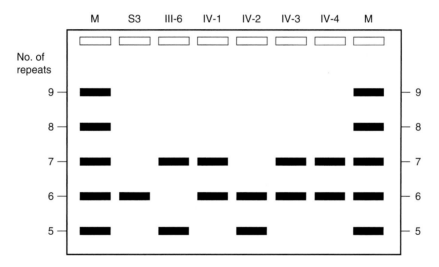

Pedigree C and Gel Diagram C (generation IV). M, molecular size marker.

The Human Genome Project: Its Science, Applications, and Issues

The development of recombinant DNA technology encouraged scientists to dream about the possibility of identifying and locating all of the genes in the human genome. As recently as the mid-1980s, however, such an undertaking seemed like an impossible dream. Using the technologies available at that time, it would have taken 1,000 researchers 200 years to complete the task, even if they devoted all their time to that one project. Excluding salaries, this project would have cost between $15 billion and $20 billion. Such an investment of money and labor in one project seemed foolhardy. However, a few visionary scientists began to consider strategies that would make the job feasible, and the initiative started to take shape.

In 1990, Congress formally initiated the Human Genome Project by providing funding for 5 years at less than $200 million per year. The project was coordinated by the Department of Energy and the National Institutes of Health. The Human Genome Project was the largest project ever tackled by biologists in a systematic, coordinated, interactive way.

The project's initial goals were twofold:

- Develop detailed maps of the human genome.
- Determine the complete nucleotide sequence of human DNA by the year 2005.

Although the Human Genome Project was funded by Congress, research on mapping and sequencing the human genome or genomes of other organisms was not strictly a government initiative, nor was it strictly a U.S. initiative. Scientist-to-scientist communications rapidly built international partnerships, and in the end, the sequence was obtained by researchers at 20 centers in six different countries.

To accomplish these two goals, a number of objectives had to be met. The first involved developing the software and database management systems to support the mapping and sequencing goals. It was clear from the outset that the Human Genome Project would generate massive amounts of data, and researchers needed consistent methods for storing, managing, accessing, and integrating the data before they began to collect significant amounts of it. Correlating, analyzing, and managing genomic data is now a new field in biology—bioinformatics (see Section F).

Organizers of the project also recognized early on that they did not have the technology for such large-scale sequencing, so one of the objectives of the first 5 years was to invent the methods and tools for large-scale sequencing. The chain termination method of DNA sequencing described in chapter 18 remained a fundamental strategy, but only with automation could this become cost-effective. Automated sequencing replaced manual sequencing, and laboratory robots were developed to automate other parts of the process, such as preparation of the DNA templates for sequencing.

Because of technological innovations, the development of new laboratory techniques, and increasing experience, the progress of the Human Genome Project was astonishingly rapid. In fact, the project was officially finished in 2003 for less than the anticipated cost and ahead of the schedule, originally proposed in 1990. Perhaps it was fitting that the Human Genome Project was completed in the year of the 50th anniversary of Watson's and Crick's discovery of the structure of DNA.

Model Organisms

Much of the early sequencing work was done with the genomes of a set of "model" organisms:

- *Escherichia coli:* a common human gut bacterium
- *Saccharomyces cerevisiae:* a yeast, the simplest unicellular eukaryote
- *Drosophila melanogaster:* a fruit fly
- *Caenorhabditis elegans:* a nematode, or roundworm
- *Mus musculus:* the laboratory version of the house mouse
- *Arabidopsis thaliana:* a small plant in the mustard family

Researchers mapped and sequenced genomes of other organisms for a number of reasons. First, most of these genomes were much smaller and therefore easier to analyze than the human genome. In addition, these genomes did not contain as much "junk" DNA as the human genome, so much more useful information could be obtained for each base pair that was sequenced. As a result, researchers could locate specific genes and identify their functions much more quickly. Second, because many genes are conserved as a species evolves, as researchers learned about the nucleotide sequence and function of one of the model organism's genes, they were learning about human genes.

Why were these organisms selected as models and not others? Because researchers already knew a great deal about their genetics, biochemistry, development, and physiology, since most had been intensively studied for decades. The knowledge about these organisms that was already available meant that scientists would have a better context for interpreting and giving meaning to the DNA sequence information obtained from them.

Whose genome is *the* human genome? No one's. There is no "model human." According to information provided by the National Center for Genome Research, DNA for the Human Genome Project was collected from "a number of volunteers whose identity will remain secret to protect their privacy."

Making the Maps

While some scientists were developing better sequencing methods, others focused on the goal of developing detailed maps. To achieve this goal, they needed better mapping techniques and better methods for cloning large amounts of DNA that could be assembled into overlapping fragments.

Geneticists use two types of maps to characterize a genome: physical maps and genetic linkage maps.

A physical map shows the actual physical distance between two points on a chromosome. In most cases, the points are restriction enzyme recognition sites, and distances between points are deduced from the sizes of restriction fragments as measured by how fast they migrate in electrophoretic gels. The actual points may have no biological significance. Distances between points are expressed in numbers of base pairs of DNA. Physical maps can be more or less detailed, depending on how much attention a chromosome has received. Less detailed maps may show only markers on the chromosome, while other

maps may show some of the DNA base sequences between the markers.

Genetic linkage maps, in contrast, are not derived from direct physical measurements. Rather, they are biological maps, obtained by analyzing the frequency with which genes are inherited together. Linkage maps are built by observing the effects of crossing over, or recombination. The closer two genes are on a chromosome, the less likely it is that they will be separated by crossing over during meiosis and the more likely it is that they will be inherited together.

Linkage maps are important because they allow us to locate genes associated with particular diseases or other observable physical characteristics. Such genetic loci are also called markers, but the number of genes whose presence can be deduced by observing the patterns of inheritance of particular characteristics is limited. To make maps that are detailed enough to be useful, scientists need more markers. To be useful for tracing inheritance, markers on a genetic linkage map must vary between individuals.

Mapping a chromosome requires both types of maps. The two different types of maps and their levels of detail serve various needs of researchers. Genetic linkage maps are most useful if researchers are interested in an overall view, while physical maps provide a more detailed view of a segment of the chromosome. Ultimately, the linkage maps and physical maps of a genome will be correlated and used as guides for obtaining the detailed nucleotide-by-nucleotide sequence and for finding out exactly where the genes are located.

Sequencing DNA

The final step in making a physical map is providing the actual DNA sequence between markers. To read a sentence, you begin at the beginning and read one word after another until you reach the end, but DNA sequencing does not work that way. At present, there is no way to hold on to one end of the huge linear DNA molecule that forms a human chromosome and simply start reading.

Because DNA molecules are so large, scientists can only handle them in pieces. Using restriction enzymes, they cut large DNA molecules into shorter pieces, but in doing so, they lose the order of the pieces. By subjecting many copies of the same DNA molecule to different restriction enzymes, however, you have fragments in which certain nucleotide sequences overlap randomly. Eventually, when you have sequenced enough short pieces, you will find that a bit of sequence on one end of one piece is the same as a bit on one end of another piece, as in the following example:

Fragment 1

AATGCCGTAGCTGGGTACCGTATTGCTTG

CCGTATTGCTTGATTGCGCCTTCGAAATTGGGCT

Fragment 2

If you continue sequencing more and more fragments cut with different restriction enzymes and match up the overlaps, you can order the pieces. A set of DNA pieces ordered by overlaps into one contiguous sequence is called a contig.

Once a stretch of DNA has been sequenced and positioned on a physical map, it can then serve as a marker for future work. Such a marker is called a sequence-tagged site, or STS. These markers may or may not be within genes. Most are in noncoding regions, because so much of the human genome is not expressed. A second type of sequencing, based on the use of messenger RNA, provides nucleotide sequences for expressed genes. These sites are called expressed sequence tags (EST).

Societal Issues

While scientists developed better and faster ways of gathering genetic information, nonscientists worried about how the information would be used. From the outset, organizers of the Human Genome Project understood that there was great potential for abuse of genetic information and that many important questions needed to be addressed, such as the following.

Who should have access to your genetic information? How will the information be used by those who have access to it?

Should testing for a genetic condition be performed when no treatment is available?

Should parents have the right to have their minor children tested for adult-onset diseases?

How can we ensure that genetic information is integrated appropriately into medical practice?

Because of these and many other questions, a portion of the congressional funding for the Human Genome Project was earmarked for addressing the ethical, legal, and social issues (ELSI) arising as a result of increasing knowledge of human genetics. The ELSI program became the largest funding source for bioethics research in the United States. Four areas of research were established as priorities: (i) privacy and fairness in the use and interpretation of genetic information, (ii) responsible clinical integration of genetic technologies, (iii) issues surrounding genetics research, and (iv) public and professional education about these issues. Many publications, recommendations, and education programs have been developed as part of the ELSI program. The Human Genome Project was unique among big government science projects in its model of supporting ethics research hand in hand with science.

Mapping and Sequencing Are Not Enough

If the only output of the Human Genome Project consisted of huge stretches of nucleotide sequences and a map of markers on chromosomes, the information would not be very meaningful. Even though those may have been the explicit goals when the project began, the implicit, understood, and widely accepted goals then and now are to identify and fully understand our genes—where they are, what they do, how they are regulated, and what happens when they malfunction. This field of research is called *functional genomics*. With information derived from functional genomics, scientists and physicians will be able to diagnose disorders that have a genetic component even before symptoms appear, develop new pharmaceuticals for treating and preventing diseases, and even correct faulty genes. They will better understand development, gene regulation, and variation in susceptibility to environmental factors.

The best way to discover what a gene does and how it is regulated is to study not genes but proteins. A new discipline within molecular biology, proteomics, focuses on the structures and interactions of proteins. Similar in concept to functional genomics, proteomics is the attempt to identify all proteins in a cell type, determine their functions, and map their interactions. The success of this field, like that of genomics, will depend on the ability of scientists to develop technologies to rapidly identify the types, functions, and amounts of thousands of proteins in a cell.

Researchers have developed a number of new techniques for accomplishing the difficult task of assessing the activity patterns of thousands of genes simultaneously. One technique, serial analysis of gene expression, or SAGE, is based on sequencing gene fragments and builds on the information we have compiled on EST. Another technique relies on an old stand-by, electrophoresis, to separate proteins and fluorescent dyes to determine the amounts of proteins. A mass spectrometer is then used to determine the amino acid sequences. As many as 5,000 proteins can be assayed in a few days.

The U.S. Department of Energy is building on the success and technological innovations of the Human Genome Project with a new program: Genomes to Life (GTL). The GTL project seeks to learn enough about environmental microbiology to enable mankind to use microorganisms to solve many of the environmental problems we now face. Its key goal is to achieve, over the next 10 to 15 years, a basic understanding of thousands of microorganisms in their native environments. It is hoped that microbes can eventually be used to remove excess carbon dioxide from the air, provide clean energy, and clean up environmental pollution. Thousands of researchers are now engaged in various projects designed to contribute to that goal.

Sources for Further Information

Go to http://www.doegenomes.org and click on the Human Genome Project and GTL links.

More information about genetics and public policy can be found at http://www.genome.gov/10002077 and http://www.dnapolicy.org.

Selected Readings

Collins, F., M. Morgan, and A. Patrinos. 2003. The Human Genome Project: lessons from large-scale biology. *Science* 300:286–290.

Frazier, M., G. Johnson, D. Thomassen, C. Oliver, and A. Patrinos. 2003. Realizing the potential of the genome revolution: the Genomes to Life Program. *Science* 300: 290–293.

Classroom Activities

Microarrays and Genome Analysis   30

About This Activity

This activity is an interactive computer simulation of a microarray analysis of gene expression in normal and cancer cells. The simulation is posted on the website of the University of Utah's Genetic Science Learning Center (GSLC), and we refer to it with their permission. The URL for the GSLC is http://learn.genetics.utah.edu/, and that of the activity itself is http://learn.genetics.utah.edu/units/ biotech/microarray/. We have been assured that the activity will remain online, and if its URL changes, you should be able to find it by searching the GSLC website.

The *Reading* looks at an application of genotyping that may one day be part of everyone's medical care: pharmacogenomics.

Class periods required: 1

Introduction

Hybridization analysis allows scientists to ask questions about specific nucleic acid sequences. By hybridizing a probe to DNA, you can determine whether the probe's nucleic acid sequence is present in a sample. By hybridizing a probe to messenger RNA (mRNA), you can ask whether RNA containing that sequence is being synthesized. The key to these analyses is having a probe for a gene of interest.

In the 1990s, the Human Genome Project led to fantastic advances in DNA-sequencing technology and also to an enormous increase in the amount of DNA sequence information available to researchers. Where geneticists used to concentrate on individual genes, it became possible to look at entire genomes.

Microarrays

Microarray technology, developed at about the same time, provides a way to obtain information about thousands of genes at once. Microarrays, also called gene chips, are ordered matrices of single-stranded DNA probes fixed to a glass slide. (GeneChip is a registered trademark of Affymetrix, Inc., a manufacturer of microarrays. A GeneChip is a particular type of array made with the same kind of equipment that is used to manufacture chips for computers.) The principle behind their use is the same as for any other hybridization assay, but they require high-technology tools for manufacturing the arrays and reading the results and they give information about many genes at once.

A microarray resembles a tiny checkerboard on a glass slide about the size of a coverslip. According to one company that manufactures and sells them, each square is about 8 μm square. For comparison, a human hair is about 50 μm in diameter. Within the 8-μm square, thousands of copies of the same DNA probe are attached to the slide. In the vocabulary of microarrays, this square, with its unique DNA probe, is called a feature. The next 8-μm square would contain thousands of copies of a different DNA probe. A gene chip about the size of a thumbnail can contain as many as 400,000 features.

There is not necessarily a one-to-one correspondence between features and genes. Typically, microarray manufacturers will design a number of different probes to different portions of the same gene. This redundant design helps decrease errors that could be caused by cross-hybridization of two different genes to one probe. Manufacturers also go to great pains to design probes to minimize cross-hybridization. To do this, they use a great deal of computing power and the accumulated data from genome-sequencing projects.

Microarray Experiments

As stated above, microarray experiments use the same general principles as any hybridization experiment, but there is one twist. In fluorescence in situ hybridization or Southern hybridization, the probe DNA is labeled. With microarrays, the probes sit on the gene chip and labeled sample is applied to them.

The sample can be either DNA or RNA. When DNA samples are analyzed by microarrays, the goal is to get information about the genotype. The sample is hybridized to probes that represent many different alleles of genes, and the genotype is revealed by the hybridization patterns. When mRNA samples are analyzed, the goal is to get information about what genes are being expressed. In these experiments, mRNA is usually isolated from the cells to be compared. For example, investigators have looked at the suite of genes being expressed when the yeast *Saccharomyces cerevisiae* was being grown on different substrates. They have compared mRNA isolated from tumors and normal tissue. DNA analyses provide information about genotype; mRNA analyses provide information about genome expression.

To conduct a microarray analysis, sample nucleic acid is isolated and labeled, often while it is being amplified. If the sample is mRNA, it is usually converted into complementary DNA (cDNA) and labeled. If the sample is genomic DNA, it is cut into fragments before being labeled or is amplified as fragments. The labeled sample DNA is then applied to the microarray and allowed to hybridize. The chip is washed to remove any unbound sample, and bound sample DNA is detected. The location of the signal from bound sample DNA corresponds to the location of the feature to which it hybridized, providing information about the DNA sequence of the sample. Multiply this by 400,000 features and you have an enormous amount of information to process. Microarray analysis would not be practical without computers. Computerized scanners read the chips after hybridization and interpret the patterns of hybridization into information about specific genes or sequences.

Applications of Microarray Technology

Genotyping by microarray allows an investigator to analyze which allele among many possibilities is present in an individual's genome at many loci simultaneously. One application for which this can be very helpful is genetic mapping. Researchers can prepare DNA from many different family members and hybridize it to arrays that test for the presence of a multitude of single nucleotide polymorphisms (SNPs). In designing a microarray to test for the presence of specific SNPs or other small sequence

changes, scientists typically target many, even dozens, of probes at the same SNP to minimize chances for cross-hybridization and false results. Analysis of the data can show whether any particular SNPs are inherited together with the phenotype (or SNP) of interest. In the previous chapter, students looked at a single gel diagram showing the results of the analysis of a single locus. With a microarray, an individual's DNA can be tested for the presence of many different sequence polymorphisms in one shot. This approach saves an enormous amount of labor compared to analyzing each locus individually.

Another application of microarray genotyping is to determine the allele(s) of a particular gene or genes present in an individual. For example, over 1,000 different mutations in the CFTR gene can cause cystic fibrosis. With a well-designed gene chip, it should be possible to test for the presence of all 1,000 of the mutations at once. One potential application of this kind of genotyping is addressed in the *Reading*.

The other type of experiment that microarrays are used for is to compare gene expression patterns at different times or under different circumstances. In gene expression studies, scientists isolate mRNA, since it represents only those genes being transcribed. The computer simulation activity at the GSLC illustrates this application of microarrays very nicely.

Objectives

After completing this activity, students should be able to:

1. Describe what a microarray is and two different types of experiments (genotyping and gene expression profiling) that can be performed with them.
2. Describe the steps of a microarray experiment.

Materials and Procedure

This is a computer simulation. Students need access to the Internet. The activity is self-explanatory. If students do not have manuals, they will need copies of the *Student Activity* for the background reading.

At the time of writing, the URL for the simulation is http://learn.genetics.utah.edu/units/biotech/microarray/.

We have not put the URL into the *Student Activity*.

Microarray Analysis of Genome Expression

30

Introduction

As students, you have probably been told that science and technology advance hand in hand. Molecular biology illustrates this relationship very well. As scientists learned more about DNA and gene function, they developed tools based on their knowledge that let them ask even more questions. At the same time, the explosion of computer and information management technology made it possible to keep track of, access, and use more and more data. One example of how computing power and molecular biology know-how come together to let scientists do experiments today that they could only have dreamed about in 1990 is the development of microarrays.

The Development of Genomics

The structure of DNA was worked out in the 1950s, and during that decade, enough experimental evidence was accumulated to convince even the skeptics that despite its chemical simplicity, it was indeed the hereditary material. How did it work? That question kept scientists busy into the 1960s, as the relationship between DNA, messenger RNA (mRNA), and protein was discovered and the genetic code was worked out. Restriction enzymes were discovered in the late 1960s, and by the early 1970s, the first pieces of DNA had been cloned into plasmids.

The ability to clone DNA let scientists study individual genes and make probes. Armed with these technologies, they focused on understanding basic processes, such as the control of the expression of individual genes and the replication of DNA. Another focus area was the identification of different versions of specific genes and comparison of specific genes between species. The publication of the chain termination method of DNA sequencing in 1977 gave a big boost to these undertakings. Sequencing genes became routine in the 1980s, and during that decade, the polymerase chain reaction (PCR) was invented.

Table 30.1 Genetics versus genomics

Discipline	DNA	RNA	Protein
Genetics	Gene	mRNA transcript	Single protein
Genomics	Genome	Transcriptome	Proteome

During the 1990s, many scientists began to broaden their focus from individual genes to genomes. It is important to understand the workings of individual genes, but an organism may have thousands of genes. To understand how an organism works, it would be necessary to understand how its genes, cell division, and development are regulated in concert and how the interaction of different genes and cells makes an organism much more than the sum of its parts. The Human Genome Project (see the *Reading* in chapter 29) exemplifies the changing horizons of genetic scientists during the 1990s.

Now, the genomes of dozens of organisms have been completely sequenced. Rather than asking questions about individual genes within an organism, scientists want to understand variation across the entire genomes of individuals. Rather than asking questions about the expression of one or a few genes, they want to know the expression pattern of every single gene. This new focus, along with the accumulation of information from large-scale sequencing efforts, gave rise to a new term: genomics (Table 30.1).

Microarrays

One of the new tools that allow scientists to type many genes at once or to measure the expression of every gene in a cell in a single experiment is the microarray. Microarrays are used in hybridization experiments, and the principles behind microarray experiments are the same ones that you learned when you learned about hybridization: the use of single-stranded probes to find out whether a complementary sequence is present in a sample. The major difference with microarrays is that thousands, even hundreds of thousands, of different probes are used in a single experiment. A minor difference is that in classic hybridization experiments, the probe is labeled and detected, while in microarray experiments, the sample nucleic acid is labeled and hybridized to the thousands of probes.

You can think of a microarray as a tiny grid divided into squares like a checkerboard. Each square of the checkerboard may be as small as 8 μm per side. By comparison, a human hair is about 50 μm thick. Within the tiny square, thousands of copies of a specific single-stranded DNA probe are attached to the surface. The next square contains thousands of copies of a different single-stranded probe. With this configuration, as many as 400,000 different probes can be arranged in a grid about the size of a coverslip for a microscope slide (Figure 30.1). Obviously, it takes specialized equipment to create a grid like that. One company that sells microarrays uses the same kind of equipment that is used to make chips for the computer industry.

It takes lots of computing power to design a good microarray. Say you were interested in changes in gene expression during the development of a mouse embryo. Since your question is about mouse genes, your probes must target them. The genome of the mouse has been sequenced, so you would use enormous amounts of computer power to identify the sequences of all the genes in the mouse genome

Figure 30.1 This handheld device contains a gene chip with thousands of probes. (Photograph courtesy of Affymetrix, Inc.)

and to find unique stretches of base sequence in those genes. Then, you could design PCR primers and amplify mouse DNA to make your probes, or you could synthesize oligonucleotides. Given that mice have about 20,000 genes, this would be a huge task. Once the probes were ready, you would have to test them all to make sure each one hybridized to only its gene, redesign the ones that failed, etc., until you had a good set of probes. Then, you would use an automated instrument to spot tiny amounts of each probe onto a glass slide and use even more computing power to keep track of the location of each probe within the grid. Making a microarray is such a large task that very few individual laboratories do it. Most researchers buy microarrays from companies that design and manufacture them.

Analyzing Gene Expression with a Microarray

Once you had your array, you would obtain mouse embryos at different developmental stages and prepare your samples. Since your question is about gene expression during development, looking at DNA would not do you any good. The DNA content of an organism does not change as it undergoes development; it is the subset of genes that are expressed that differs. To ask about gene expression, you would isolate mRNA. You would need to label it, and typically that is done by conducting global PCR of the mRNA and using a labeled primer.

Finally, you would be ready for the experiment. You would apply the labeled PCR copies of your mRNA from one developmental stage to the array, allow it to hybridize, and then wash away unhybridized sample. Microarrays are too small for visual interpretation of the results, so you would use a computerized scanner to record the intensity of the label at each grid position. The software would also relate the position of each signal to the identity of the probe present there.

As the scientist, you would end up with a massive computer file of data about the intensity of signal at each spot. Typically, you would look for probes that showed distinctly different intensities of signal between the different developmental stages you tested and see what genes they represented. Sometimes the genes identified in experiments have known functions, and sometimes they do not. If your experiment identified genes whose functions had never been identified, your next task as a researcher would probably be to try to find out what they did.

Analyzing Genotype with a Microarray

Another application of microarrays is to analyze genotype. Microarrays let researchers determine which among many specific DNA sequences are present in a genome. For example, over 1,000 different mutations have been associated with the disease cystic fibrosis. A microarray could be designed to test which, among all these different possible mutations, might be present in someone's DNA. It would have to include the normal gene sequences, as well as the different mutated sequences. Microarrays can also be designed to test for base sequence differences that are not necessarily involved in disease but rather are used for mapping the locations of genes, like the short tandem repeat markers you looked at in the last chapter.

The *Reading, Personal Genomics,* looks at an application of microarrays that may find its way into common use in doctors' offices in the next few years.

Exercise

The exercise associated with this reading is an interactive computer simulation of a microarray experiment. You will go through the steps involved in an analysis of gene expression, using normal cells and cancer cells as your subjects. The experiment is posted on the website of the Genetic Science Learning Center of the University of Utah, which gave us permission to refer you to it.

Classroom Activities

Personal Genomics

One of the first broad applications of genotyping will likely concern genes involved in the metabolism of medicines. Physicians have long known that different patients respond differently to the same medicines. Drugs that relieve pain in one patient may not help another patient with the very same painful condition. A medication that is not toxic in one person may cause a life-threatening reaction in another. More often than is comfortable, we hear stories on the news of drugs that appeared safe when tested on a few hundred or thousand individuals being withdrawn from the market because they caused rare but serious reactions when they were widely prescribed.

Research is beginning to reveal the genetic basis for these different responses. Within the body, many medicines are processed by enzymes, and scientists have learned over the past several years that there are different genetic versions of these enzymes in the population. How an individual responds to a given medication depends on what version of the processing enzymes that person has. These differences in responses can determine whether a particular medicine is effective in a given individual and even whether it can cause life-threatening reactions (Table 30.2).

For example, one of the genes involved in processing about a quarter of all medicines, including about 50 of the 100 most commonly prescribed drugs, is called 2D6. About 6 to 10% of Caucasians carry genes for a slow-acting version of this enzyme. When these people take a medication processed by the enzyme, they maintain higher levels of the medicine in their systems much longer than do people with the fast version. Thus, they are more easily overdosed. They are also immune to the pain-killing action of codeine. Codeine is processed by 2D6 into morphine, which exerts most of the pain-killing effect when codeine is taken. People with the slow version of 2D6 do not generate morphine from codeine fast enough to benefit from it.

This marriage of the study of genetics, genomes, and pharmacology has been named **pharmacogenomics.** Pharmacogenomics is an area of intense scientific research and commercial interest. It is estimated that most commonly used medicines now benefit only 30 to 60% of patients who take them and that adverse drug reactions kill about 100,000 hospitalized patients each year in the United States alone. If our medicine-processing genotypes can be associated with responses to specific medications, physicians can more effectively prescribe for individuals, prevent

Table 30.2 Genetic basis of life-threatening adverse drug reactions

Drug	Condition for which prescribed	Potential life-threatening reaction	Gene involved in the reaction
Tofranil	Depression, bed wetting, attention deficit disorder	Heart arrhythmia	CYP2D6
Laniazid	Tuberculosis	Liver toxicity	NAT2
Coumadin	Harmful blood clots	Internal bleeding	CYP2C9
Adrucil	Cancer	Severe immune suppression	DPD
Biaxin (an antibiotic)	Infectious disease	Heart arrhythmia	KCNE2
Imuran	Rheumatoid arthritis	Severe immune suppression	TPMT

30. Microarrays and Genome Analysis • 437

much misery from adverse reactions, and allow proper use of medicines that have been withdrawn because they cause severe reactions in a subset of the population.

Gene chips that contain probes specific to different versions of many genes involved in drug metabolism are being developed. These chips can be used in research to identify genotypes associated with adverse reactions to existing and experimental medicines. In the not-too-distant future, a pharmacogenomic fingerprint may become a routine part of health care, so that prescriptions can be tailored to an individual's genotype. People of the 22nd century (or even the late 21st) may look back on our drug prescription practices, where everyone is prescribed the same medications at the same doses, with the same horror we now view the early days of blood transfusions, when they were given without blood typing, frequently causing fatal reactions.

Classroom Activities

F. Bioinformatics and Evolutionary Analysis of Proteins

In this section of activities, we focus on proteins and protein evolution, using the enzyme amylase as the example. This section provides a vehicle for reviewing protein structure-function relationships, as well as for teaching students how to use online databases. The databases are a living resource, constantly updated through the addition of new data. Once students have learned how to use the resources, they can incorporate database searches into future activities.

In the first activity, students test materials from a variety of sources to determine whether amylase activity is present. This activity establishes that amylase is present in organisms from bacteria to plants to animals. Students then use protein electrophoresis to see the relative sizes of amylases from various sources, identifying the amylase bands in the gel through their enzymatic activity. Next, they compare portions of the amino acid sequences of amylases from a variety of organisms. Using the degrees of difference they find between the amylase protein sequences, they make a hypothesis about the relatedness of the various organisms from which the amylases originated. Finally, the students use online bioinformatics resources (public databases of protein and DNA sequences, as well as analytical software) to identify the organisms, identify relatively constant regions of the proteins, and find other closely related amylase sequences.

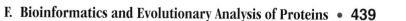

Amylase, an Evolutionarily Conserved Enzyme

31

About This Activity

In this activity, students test a variety of materials for amylase activity, using an iodine test for starch. They start by testing their own saliva and comparing its amylase activity with that of an industrial amylase preparation. They plate samples of soil diluted in water on starch-agar plates to detect amylase activity in the organisms that grow. Finally, they identify several other substances to test, along with a germinating bean, leaves, roots, baker's yeast, and dog saliva.

Class periods required: *2 or 3 over a period of a few days; the days are required for the soil organisms to grow and for students to bring in additional materials for testing.*

Introduction

Plants use energy from sunlight to power the synthesis of glucose, which they can use immediately or store as starch, a polymer of glucose. When plants need the energy stored in starch (for example, when a seed germinates), plant enzymes hydrolyze the starch into sugars. Animals use starch for energy, too, but instead of making it, they get it by eating plants. When an animal eats starch, it breaks down the starch into sugars via the actions of starch hydrolysis enzymes.

Two enzymes that hydrolyze starch are amylase and amyloglucosidase. Amylase attacks a starch molecule like a pair of scissors, clipping it into smaller chains of glucose molecules called dextrins. Some amylases are more specific than others and cut the starch neatly into two-glucose units called maltose, or malt sugar. Starch chains are sometimes branched, and amylase cannot cleave the branches. Amyloglucosidase clips individual glucose molecules from the end of a starch or dextrin molecule, leaving only glucose. For complete conversion of starch to glucose, the action of amyloglucosidase must be added to the action of amylase.

The introduction to the student activity includes a description of the brewing process, which involves amylase action. After the starch source has been treated with amylase (whether from barley malt or industrial sources), there are still dextrins in the solution. Yeast cannot ferment the dextrins. When yeast is added, it ferments the malt sugar released by the amylase and leaves the dextrins in the beer. Although yeast cannot digest dextrins, we can. The dextrins mean additional calories to human consumers. To make lower-calorie "lite" beer, manufacturers use amyloglucosidase to convert the leftover dextrins to sugar. The yeast subsequently ferments the sugars, leaving fewer calories for the drinker.

The student activity also notes that humans have a salivary amylase and a pancreatic amylase, which are encoded by different, though similar, genes. The salivary amylase gene has an element in its promoter that causes it to be expressed in salivary glands. Many starch-eating animals, such as dogs, do not secrete amylase in their saliva. Although dogs clearly eat starch, their saliva tests negative for amylase activity. Researchers testing dog tissue have found amylase messenger RNA in the pancreas, but not in the salivary glands, indicating that dog amylase is not expressed there.

In this activity, students are asked to test dog saliva for amylase. Given a typical dog's diet, students should predict that dogs make amylase. Students may be puzzled by the negative result. If you have them think about human amylase genes, they may be able to come up with the hypothesis that dogs make amylase elsewhere in their bodies (e.g., the pancreas) but do not secrete it in their saliva. Be sure that they record this hypothesis. In the last chapter of this section, students will learn how to use the online bioinformatics resources, which include programs for searching the recent biomedical literature. If students search the literature for "alpha amylase in dog tissue," they will find the article mentioned above. The search results will give them the citation; they should click on the link to reach a summary of the article. (Note that parotid tissue is salivary gland tissue.)

Objectives

After completing these activities, students will be able to:

- Test a variety of substances for amylase.
- Describe the action of amylase.
- List several organisms that express amylase.

Materials

Part 1: Testing human saliva for amylase activity

Every student should conduct the amylase test. Each student will need the following:

- One starch-agar test plate (See the CD for a recipe; you can use starch-nutrient agar plates or potato dextrose-agar plates as well.)
- Two cotton swabs, such as Q-Tips (They do not have to be sterile.)
- Iodine test solution and pipette
- Industrial amylase solution (0.5 g in 500 ml of H_2O)

Part 2: Testing soil organisms for amylase activity

We recommend that students work in pairs or small groups for this portion of the activity. On the day the soil samples are plated, you will need to have enough soil on hand for each group of students to weigh out 1 g.

Each group of students will need the following:

- Five starch-nutrient, starch-Luria broth, or potato dextrose-agar plates
- One clean 200-ml or larger container
- Four clean 15-ml test tubes
- Five clean 1-ml transfer pipettes
- 1 g of soil
- Water, preferably commercially purchased distilled or drinking water
- Iodine test solution and pipette on the day you develop the plates
- Sterile cotton swabs or a fresh pack of Q-Tips (If students handle the Q-Tips carefully, the swabs will be mostly free of microorganisms.)

Part 3: Testing additional materials for amylase activity

We recommend that students work in pairs or small groups for this portion of the activity. The students will test the materials listed in the lesson but will also identify and bring in additional materials to test or take the plates home to test materials available there (such as dog saliva).

Each group of students will need the following:

- A pinto bean or other dry bean soaked overnight in water (Students can set this up.)
- Leaves and roots of a growing plant
- Mortar and pestle (or a blender) for grinding roots and leaves
- Starch-agar test plates
- Baker's yeast
- Small amount of sugar for starting yeast
- Cotton swabs (They do not have to be sterile.)
- Iodine test solution and pipettes

Students will bring in additional materials to test. Here are some of our ideas:

- Yogurt with active cultures
- Plant stems
- *Drosophila melanogaster* (Freeze several fruit flies and grind them up in a little water; test the extract.)
- Saliva from herbivores and carnivores

Resource Materials

Carolina Biological Supply Co. sells a refill kit for an activity called *Caught by a Kiss* that contains materials you need for this activity: melt-and-pour starch agar, petri dishes, and the iodine-potassium iodide solution. The catalog number of the refill kit is 21-2017. You can buy prepoured starch nutrient agar plates as well (catalog no. 82-1996). Carolina also sells industrial amylase (catalog no. 20-2350) and iodine-potassium iodide solution (catalog no. 86-9051).

Preparation

Part 1: Testing human saliva for amylase activity

Prepare starch test plates (see the CD for a recipe). These plates do not need to be sterile, although you should use clean utensils. Disposable petri dishes are convenient containers, but large casserole dishes (for multiple tests), food storage containers, or other containers would also work if students can share large test plates. You can store the hardened plates at room temperature for a day or two; for longer periods, store them in the refrigerator. Make enough for *Part 3* as well.

Prepare or purchase iodine test solution (see the CD for a recipe). If you use Carolina Biological Supply's iodine solution, dilute it 1:40 with water (for example, 5 ml of solution to 195 ml of water). The undiluted solution stains the test plates too deeply. Distribute some of the iodine solution into several containers so that students can have ready access to it.

Set up containers of 10 to 20% bleach for discarding the used swabs.

Part 2: Plating soil organisms for amylase activity

Prepare starch-nutrient agar plates (see the CD for a recipe). These plates must be sterile. The plates can be prepared many days in advance of the laboratory exercise. Store them at room temperature for 2 or 3 days to allow excess moisture to evaporate, and then put them in plastic bags and refrigerate them. The plastic sleeves in which disposable petri dishes are packaged make convenient storage bags and can be sealed with tape or twist ties. As long as the starch-nutrient agar plates do not dry out or develop contamination, they are usable.

Part 3: Testing other materials for amylase activity

Prepare starch test plates and iodine test solution as for *Part 1* of this activity. Plan for each group of students to have access to at least six plates.

Soak dry beans in water overnight before the laboratory. Have students list materials they plan to bring to test for amylase activity, along with any equipment or supplies they will need for processing their samples. For example, they may require a microcentrifuge tube and a microcentrifuge tube pestle for grinding up frozen *Drosophila*.

Tips

The starch-nutrient agar plates used for growing soil organisms must be sterile so contaminants will not grow on them. If you have ready access to sterile water, sterile pipettes, and sterile test tubes, do the entire soil plating with aseptic technique.

In our experience, commercially purchased distilled or drinking water is essentially sterile. We tested this activity with tap water and found that it did not introduce significant contamination, either. The water control plate will show whether the water or equipment (swabs or tubes) is introducing significant contamination into the experiment.

Procedure

Make sure that students read the introduction to the activity. You may wish to make the connection between today's laboratory and any earlier study of the digestive system that students have made.

Part 1: Testing human saliva for amylase activity

Every student should test his or her own saliva and do an industrial-amylase control.

Part 2: Plating soil organisms for amylase activity

Divide students into pairs or small groups. Have them follow the procedure in the *Student Activity*. Incubate the plates right side up for 2 to 4 days at room temperature. The students will observe that the majority of colonies on the plates secrete amylase. Results are clearest on plates with colonies that are separate from one another.

After the experiment is over, soak the plates in 10 to 20% bleach overnight, rinse them, and discard them in the trash.

Part 3: Testing additional materials for amylase activity

Divide students into pairs or small groups. They should first brainstorm about additional materials they could test for amylase activity. Eventually, they should put together a list of the materials, along with any supplies they might need for preparing the materials for testing (such as a mortar and pestle or microcentrifuge tubes and a micropestle). Identify which supplies are available in the laboratory. Some students may need to take home starch plates and iodine solution for testing animal saliva.

Remind the students to use Worksheet 31.1 (and additional pages if needed) to make predictions about the presence of amylase in the materials they are testing and to give reasons for their prediction based on the diets, physiology, or environments of the organisms from which the materials originate.

The predictions are a good topic for small group or class discussions. They can provide a basis for reviewing basic biology or for considering the ecology of an organism (such as the soil organisms).

The next day, the students should bring in the materials. Verify that they have made their predictions and given reasons. Have the additional supplies ready, along with starch test plates, swabs, and iodine solution. The students should be able to carry out the tests independently. Tabulate the results and discuss them in class.

Answers to Student Questions

1. There is no precise answer to this, but a large fraction of the organisms do. Students can look at plates that have a reasonable number of isolated colonies on them and count how many secrete amylase. In our samples, more colonies did than did not.

2. Soil organisms are decomposers. Dead plant tissue often contains starch, so it makes sense that soil organisms would be able to digest it. Starch is a rich energy source for them, since it is broken down into glucose.

3. The bean has starch in it to provide energy for the germinating plant. The germinating plant cannot conduct photosynthesis immediately (it has no leaves; it is usually underground) and so depends on material stored in the seed for nourishment.

Testing for Amylase Activity

Introduction

In nature, plants use energy from sunlight to power the synthesis of the sugar glucose. The plants then break down the glucose to recover the energy and use it for growth and other life processes. Plants store excess glucose by linking the glucose molecules together in a long chain. These long chains of glucose molecules are what we call starch. When a plant needs to use the stored energy, enzymes break the starch molecule back down into its component glucose molecules. Animals also use starch for energy, but instead of making it, they obtain starch by eating plants. When an animal eats starch, enzymes in the animal's body break the starch into glucose molecules, which are used for energy.

All organisms that use starch for energy make enzymes that convert it back into glucose molecules. One of these enzymes is amylase, which breaks the chemical linkages between glucose molecules by inserting a water molecule into the bond, a process known as hydrolysis (Figure 31.1).

Like many other animals, humans use starch for energy: think of bread, potatoes, and rice. Humans secrete amylase in their saliva, so conversion of starch to glucose begins as soon as starch enters the mouth. You can experience this by holding a saltine cracker in your mouth for several seconds and noticing that a slight sweet taste develops. The sweet taste is a result of amylase cleaving starch into sugars.

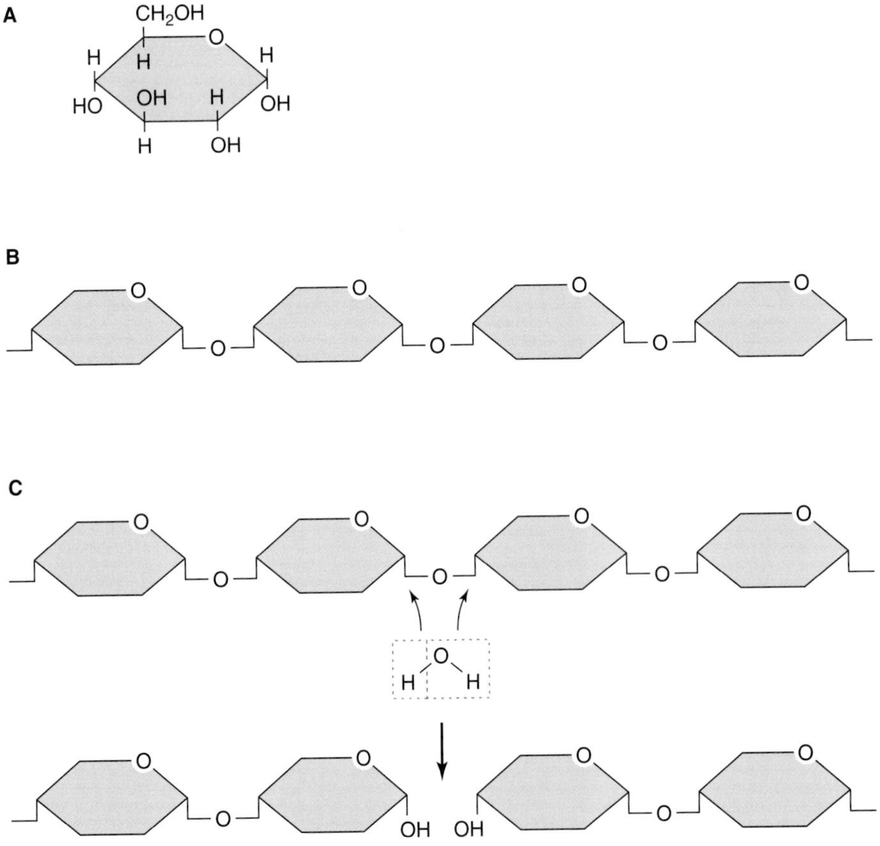

Figure 31.1 (A) The sugar glucose. The corners of the hexagon are carbon atoms. (B) Starch. Starch is a polymer of glucose. In this representation, only the hexagons and the linking oxygen atoms are shown. (C) Hydrolysis of starch. When starch is hydrolyzed, a water molecule splits the bond linking two glucose residues. Complete hydrolysis of starch yields glucose.

Humans also produce amylase in the pancreas, as do many other animals. Amylase is secreted from the pancreas into the small intestine, where digestion of starch continues. Salivary amylase and pancreatic amylase are encoded by separate genes, though the genes are very similar. The gene for salivary amylase has signals in the promoter region that cause it to be expressed in the salivary glands. Production of amylase in the salivary glands seems to be a fairly recent evolutionary development: many animals that use starch for energy produce amylase only in the pancreas, not in saliva.

Industrial uses of amylase

Besides being important for energy production in plants and animals, amylase is an important industrial enzyme. In fact, one of the most common industrial uses of enzymes today involves amylase. The process is the conversion of corn starch into high-fructose corn syrup.

In the 1970s, sugar prices soared. The high cost of cane sugar stimulated the development of an enzyme-based technology for producing a sweet syrup from corn, a crop often produced in surplus in the United States. Starch from corn is converted

into high-fructose corn syrup by three enzymes. First, the enzymes amylase and amyloglucosidase (an enzyme similar to amylase) convert the corn starch to glucose. Glucose syrup does not taste as sweet as cane sugar, so a third enzyme, glucose isomerase, is used to convert some of the glucose to fructose. Fructose is very similar in chemical structure to glucose but tastes about twice as sweet. The result, high-fructose corn syrup, is the sweetener in virtually all soft drinks produced in the United States today, as well as in many other products. Look for it on food ingredient labels and think of amylase.

Amylase and its relatives also play an important role in the brewing industry. In traditional brewing, beer is produced by fermenting barley malt. Barley malt is germinated barley. During germination, the barley grains take in water and swell, activating their amylase enzymes. The barley amylases convert the starch in barley to "malt sugar," a sugar that consists of two glucose molecules linked together. The malt is crushed and steeped in water to extract the sugar. This liquid is boiled with hops for flavor and then cooled. Yeast is then added. Yeast cells cannot use starch for energy, so they use the malt sugar, producing carbon dioxide bubbles and alcohol as products. We call the end result beer.

In modern large-scale brewing, manufacturers want to make large quantities of uniform product at an economical cost. Barley malt is expensive and can vary greatly in quality, so manufacturers use other sources of starch, e.g., corn and rice, and use purified amylase enzyme to convert the starch to sugars for the yeast. These and other industrial uses of amylase are listed in Table 31.1.

Industrial amylase

Where does purified amylase come from? The world's largest producer of industrial enzymes, a company called Novozymes (formerly part of Novo Nordisk), uses the bacterium *Bacillus licheniformis* to produce amylase. *B. licheniformis* naturally produces an amylase enzyme that is very heat stable, a characteristic that is highly desirable in an industrial enzyme. Novozymes scientists genetically engineered *B. licheniformis* to produce more of the enzyme by giving it extra copies of its own gene.

In a Novozymes plant, engineers and plant workers grow *B. licheniformis* in huge tanks, feeding it corn and soy grits. As the bacteria reproduce, they secrete amylase. When growth is complete, the mixture is purified to remove the bacteria and other particles and then concentrated. Five hundred milliliters of

Table 31.1 Industrial uses of amylase

Industry	Use of amylase
Baked goods	Increasing sugar content for yeast fermentation
	Ensuring quality of frozen doughs
	Retarding the staling process
Beverages	Liquefying gelatinized starch
	Increasing sugar content for yeast fermentation
	Clarifying juices by removing starch turbidities
	Producing reduced-calorie beer
Sugars and syrups	Producing high-fructose corn syrup
	Recovering sugar from candy scraps
	Refining sugar from sugar cane
Textiles	Removing the starch adhesive added to cotton and cotton blends to strengthen fibers during weaving (desizing)
Paper	Producing starch-based compounds to strengthen paper base or coat surface of paper
	Producing adhesive binding agents from starch
Environmental	Reducing insoluble carbohydrates in sewage
Household products	Added to laundry detergents to make them capable of removing starch-based stains
Energy	Making ethanol, a fuel additive, from grains

Novozymes' industrial amylase can convert 1 ton of starch into sugars. Although you may think of enzymes, such as restriction enzymes, in little vials, Novozymes ships out its industrial amylase in tank trucks.

Today's Activity

Amylase activity can be easily detected by following the hydrolysis of starch. The presence of starch can be detected with iodine. Because of the bond angles between glucose residues in starch, starch molecules assume a helical form. Iodine molecules nestle inside the coils of the helix, forming a complex that has a dark-blue color (Figure 31.2). It takes at least six turns of a starch helix (36 glucose residues) to form the colored complex with iodine. As amylase cleaves the starch molecule into smaller fragments, the colored complex can no longer form.

The purpose of this activity is for you to test different organisms for amylase activity. You will begin by detecting the activity in your own saliva and comparing it with the activity of an industrial preparation of amylase. You will then test a soil sample to see whether any soil organisms produce amylase, and you will collect and test a variety of other materials

Figure 31.2 The starch-iodine complex. A stylized sketch of the helical starch molecule is shown on the left. On the right, two molecules of iodine (I_2, represented by the dumbbell shapes) nestle inside the helix. Six turns of the helix are required to produce a blue color.

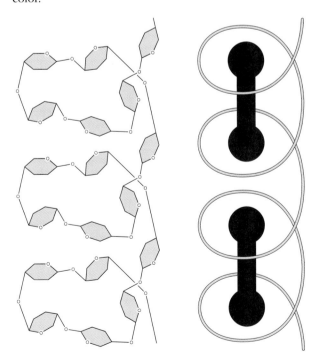

as well. As you prepare the materials for testing, think about the organisms you are testing and how they get energy. Based on your ideas, predict whether you will see amylase activity, and state why you think so.

Part 1: Testing human saliva for amylase activity

Obtain a starch plate, two cotton-tipped applicators, iodine solution, and a sample of industrial amylase enzyme. You should have access to a container of 10 to 20% bleach. Put the tip of the applicator in your mouth to make it moist. Use the applicator to write your name on one-half of the starch plate. Put the used applicator into the bleach. Dip the second applicator into the industrial enzyme preparation. Try to get the tip moist but not drippy. Use the tip to write "amylase" on the other side of the starch plate. Put the used applicator into the bleach. (The industrial enzyme preparation is not dangerous, but the tip may have contacted saliva.)

Let the plate stand at room temperature for about 10 min.

Perform the iodine test for starch by flooding the plate with iodine. You should see your name and the word "amylase" appearing clear against a blue background. The clear area shows that amylase broke down the starch molecules where you touched the plate. The blue background shows the presence of starch in the rest of the plate.

Part 2: Testing soil organisms for amylase activity

Soil contains a very large number of microorganisms. In this procedure, you will dilute soil in water in a series of increasingly large dilutions and plate a sample from each dilution on starch agar. These plates contain starch and other nutrients for supporting the growth of microorganisms. After colonies of microbes grow on the plates, you will test for amylase secretion by flooding the plates with iodine. By plating a series of dilutions, you should have a plate or two on which you can see individual colonies.

This procedure is not written as a sterile procedure. Nevertheless, do not introduce additional contamination by touching the tips of the pipettes, the insides of containers, or the surfaces of the agar plates with your fingers. You will be plating the water as a control to see whether any amylase producers were present in the water (or inside the test tube or pipettes) to begin with.

Obtain five starch-nutrient agar plates, five disposable plastic pipettes, five cotton swabs, one container capable of comfortably holding 100 ml, and four 15-ml test tubes.

1. Label the large container "1:100." Label the four test tubes "water," "1:1,000," "1:10,000," and "1:100,000."
2. Label the five starch-agar plates "water," "1:100," "1:1,000," "1:10,000," and 1:100,000."
3. Put 99 ml of water into the large container.
4. Put 9 ml of water into each of the test tubes.
5. Use a pipette to remove 1 ml of water from the "water" tube and add it to the agar plate labeled "water." Discard the pipette.
6. Dip one cotton swab into the remaining water in the test tube and tap it on the edge of the tube to shake off the excess. Use the cotton swab to spread the water over the surface of the agar plate. Replace the lid on the plate and discard the swab.
7. Weigh out 1 g of soil. Add it to the 99 ml of water.
8. Use a fresh pipette to stir soil in the water. Mix it well. Using the same pipette, add 1 ml of the suspension to the 1:100 plate. Still using the same pipette, add 1 ml of the suspension to the 1:1,000 test tube. Discard the pipette.
9. Dip a fresh cotton swab into the water tube and tap it to shake off the excess. Use the swab to spread the soil-water suspension on the 1:100 plate. Replace the lid on the plate and discard the swab.
10. Use a fresh pipette to stir the suspension in the 1:1,000 tube. Using the same pipette, add 1 ml of the suspension to the 1:1,000 plate. Still using the same pipette, add 1 ml of the 1:1,000 suspension to the 1:10,000 tube. Replace the lid on the plate and discard the pipette.
11. Dip a fresh cotton swab into the water tube. Tap off the excess and use the swab to spread the liquid on the 1:1,000 plate.
12. Use a fresh pipette to stir the mixture in the 1:10,000 tube. Use the same pipette to add 1 ml of suspension to the 1:10,000 plate. Still using the same pipette, add 1 ml of the 1:10,000 mixture to the 1:100,000 tube. Replace the lid on the plate and discard the pipette.
13. Dip a fresh cotton swab into the water tube and use it to spread the sample on the 1:10,000 plate.
14. Use a fresh pipette to stir the mixture in the 1:100,000 tube. Use the same pipette to add 1 ml of the mixture to the 1:100,000 plate. Discard the pipette.
15. Dip a fresh cotton swab into the water tube and use it to spread the sample on the 1:100,000 plate.

16. Incubate the plates at room temperature with the lid side up. It may take 2 or 3 days before many colonies appear.

When colonies have grown up on the plates, flood them with iodine solution and look for areas of starch hydrolysis.

Part 3: Testing other materials for amylase activity

Amylase is a common enzyme, but not all organisms produce it, and multicellular organisms do not necessarily produce it in all their tissues. We are going to suggest a few additional materials for you to test for amylase activity, but the objective is for you and your classmates to come up with a variety of materials to test on your own.

Before you conduct any tests, use Worksheet 31.1 (see Appendix A) to predict whether each material will show activity and why you think so. After you conduct the test, record the results on the worksheet. If the result is different from what you expected, think about the sample organism again and see if you can come up with an explanation for your result. Ideally, the explanation should suggest an additional experiment that could verify it, even if the experiment is something you cannot do at present (such as check various tissues for amylase activity or do a Southern blot to look for an amylase gene).

Germinating Bean
Soak a dry bean, such as a pinto bean or kidney bean, in water overnight. Mash the bean with a little of the water and test the extract. (You will have extracted many substances from the bean into the water; therefore, it can be accurately referred to as an extract.)

Plant Leaves
Remove a leaf or leaves from a plant and grind them up with a little water. Test the leaf extract.

Plant Roots
Grind up or blend a fresh plant root with some water and test the extract.

Yeast
Suspend some baker's yeast in water with a little sugar, and let it sit until the mixture begins to bubble. Test a drop of the culture.

Dog Saliva
If you have a dog, take home two or three cotton swabs, a starch-agar plate, and iodine solution.

Moisten one swab with dog saliva and test it on one-half of the plate. As a control, moisten the other swab with your own saliva and test it on the other half of the plate. Let the plate stand for 10 min, and then flood it with iodine.

Additional Materials

Working with your classmates, identify additional materials that you can test for amylase activity. Record your predictions and the results on Worksheet 31.1.

Questions

1. Approximately what fraction of the soil organisms secreted amylase?

2. Starch is not a component of soil. Where might a soil organism encounter starch? Why do you think so many soil organisms make amylase?

3. Provide a reasonable explanation for the results of amylase testing on the germinating bean, the leaf extract, and the root extract.

Protein Electrophoresis   32

Classroom Activities

About This Activity

In this activity, students will perform agarose gel electrophoresis with amylase samples. Starch will be incorporated into the agarose gels so that students can identify the amylase bands by their activity.

Class periods required: *Gel preparation can be done in advance. Sample preparation requires about 15 minutes (min), followed by 45 to 60 min for electrophoresis. When the electrophoresis is complete, half of the gel is stained for 1 min, and the other half is soaked for 15 min to 2 hours to remove sodium dodecyl sulfate (SDS). Both halves of the gel are then incubated overnight for destaining and detection of amylase activity.*

Introduction

The goal of this activity is to separate the proteins in amylase-containing samples by electrophoresis so that students can see different protein bands and to identify amylase bands through their starch-hydrolyzing activity. Electrophoresis of proteins is different from electrophoresis of DNA in a few important respects. For one thing, DNA has a uniformly charged backbone. Therefore, all DNA molecules will migrate in the same direction in an electrophoresis gel. Proteins, on the other hand, can be uncharged, positively charged, or negatively charged. A mixture of proteins applied to an electrophoretic gel will likely contain some of each. For another thing, proteins assume different shapes, so two proteins of the same molecular weight and charge might migrate differently in a gel.

To make proteins migrate more uniformly, they are often treated with the detergent SDS. SDS is a strong anionic detergent that coats proteins in a blanket of negative charge. In addition, SDS denatures proteins, causing them to assume a more uniform (though not perfectly uniform) shape. SDS loading buffer usually contains a strong reducing agent, such as β-mercaptoethanol, to break disulfide bonds and

further denature proteins. Proteins in SDS buffer migrate toward the positive electrode in an electric field, and the distances they migrate are largely proportional to their molecular weights.

SDS is very helpful in electrophoresis but is detrimental to most protein functions, including that of amylase. Since the goal of this activity is both to separate the proteins in amylase samples and to detect amylase by its activity, the SDS must be removed after electrophoresis. To accomplish this, the gels are soaked in a solution of Triton X-100, a mild, nonionic detergent. The Triton X-100 replaces the SDS, allowing the amylase to resume a more normal configuration. The Triton X-100 is then replaced by water by soaking the gel overnight, which allows the amylase to regain more activity. In addition, the samples in this activity are loaded in a β-mercaptoethanol-free buffer to protect the disulfide bonds.

Traditionally, samples are boiled before SDS gel electrophoresis, which further denatures the proteins. In this case, boiling substantially reduces the amount of amylase activity that can be recovered following electrophoresis. We have eliminated the boiling step of sample preparation in this protocol.

The gels in this activity are made with fine-sieving agarose, which improves the resolution of protein bands. Starch is also added to the gel. Following electrophoresis, the gel is cut in half. One half of the gel is stained to visualize protein bands, and the other half is soaked first in Triton X-100 and then in water to reconstitute amylase activity. As the amylase activity is reconstituted, the enzyme will degrade the starch in its vicinity, creating an area of clearing in the gel. Students will compare the stained half of the gel to the half with amylase activity. The areas of clearing will identify the amylase protein bands.

One of the amylase samples that will be separated is human saliva. Amylase is a major protein component of saliva. Human saliva runs more cleanly through the gel if it is heated first. In fact, saliva gives a slightly different banding pattern if it is heated at 70°C for 2 to 3 min before electrophoresis. Students may be interested in running heated and unheated saliva samples on the same gel. In most cases, the heated saliva will have markedly reduced amylase activity. This result could inspire a discussion about optimal enzyme conditions, irreversible heat denaturation, and electrophoretic mobility. In the case of amylase, the heating presumably causes some irreversible conformation changes that affect electrophoretic mobility.

In chapter 34, students will learn how to use online bioinformatics resources. The example protein in that chapter is pancreatic amylase. After they learn how to use the databases, students can look for salivary amylase. The three-dimensional structure of human salivary amylase has been resolved, and a three-dimensional image is available in the protein structure database (see chapter 34 for more information).

Objectives

At the end of this activity, students will be able to:

- Explain why the detergent SDS is often added to protein samples before electrophoresis.
- Explain how to identify amylase bands in a protein gel.
- Define protein denaturation.

Materials

- Protein gel (fine-sieving) agarose, 1.5 g per student gel
- Low-SDS running buffer
- Starch
- Purified fungal amylase, 0.5 g
- Coomassie protein stain
- Coomassie destain solution
- Triton X-100, 5% solution
- Low-SDS loading dye
- Electrophoresis equipment
- Micropipettes and tips for measuring volumes smaller than 100 ml, or equivalent device
- 1-ml pipettes
- Two weigh boats per student group
- Microcentrifuge tubes
- Dry beans, soaked overnight in water (optional)
- Water bath at 70°C

Each group of students will need the following:

- Gel, running buffer, and gel box
- A few microcentrifuge tubes for sample preparation

- Microcentrifuge tube pestle for grinding beans or a small mortar and pestle
- Microvolume measuring device
- 1-ml pipettes
- Fungal amylase solution
- Two weigh boats
- About 50 ml of 5% Triton X-100 (can be from a shared classroom container)
- About 50 ml of Coomassie protein stain (can be from a shared classroom container)
- About 100 ml of Coomassie destain solution (can be from a shared classroom container)
- Thin plastic ruler or other device for cutting the gel in half

Resources

Carolina Biological Supply Co. sells the following solutions:

- Protein gel agarose (catalog no. 21-7088)
- Low-SDS running buffer (catalog no. 21-1287) (Carolina calls this Enzyme Running Buffer.)
- Low-SDS loading dye (catalog no. 21-1288) (Carolina calls this Non-Reducing Protein Loading Dye.)
- Purified fungal amylase (catalog no. 20-2350)
- Coomassie protein stain (catalog no. 21-9784)
- Coomassie destain (catalog no. 21-9785)
- Triton X-100 (available from Sigma Chemical Co.)

The Carolina *Amylase Electrophoresis Kit* (catalog no. 21-1280) contains all these materials (with melt-and-pour protein gel agarose) in sufficient quantity for six student gels.

Preparation

The starch–fine-sieving agarose gels must be prepared exactly as described. We suggest preparing them the day before the laboratory.

1. Mix 1.5 g of fine-sieving agarose in 40 ml of water in a small flask for each gel to be poured.
2. Mix 0.25 g of starch in 10 ml of 5× Tris–glycine–low-SDS running buffer. Add the starch solution to the flask containing the dissolved agarose solution, and mix it well by swirling it. The starch must be uniformly suspended in liquid before it is added to the agarose solution. Do not simply put starch and agarose powders in a flask and add liquid; clumps that cannot be broken up will form.
3. When both solutions are thoroughly mixed, place the flask into a boiling-water bath to heat and dissolve the agarose and starch—approximately 15 min for a 50-ml sample. The mixture will clear, and numerous small bubbles will form. The

bubbles keep the mixture from appearing perfectly clear. Keep the flask in the boiling-water bath as much as possible to prevent premature gelling. This can be done before class for time considerations.

4. Remove the flask from the water bath, and immediately pour the contents into a casting tray. Add the comb. Using a pipette tip, a pencil, or any other sharp object, try to remove any large air bubbles from the comb area and from the gel itself. Do not worry about the small bubbles. Work quickly so that the bubbles are gone before the gel solidifies.

5. Allow the gel to harden. It should not feel warm to the touch. Remove the comb and place the gel in the electrophoresis chamber. Add 1× Tris–glycine–low-SDS running buffer to the electrophoresis chamber until the gel is covered. Do not store gels overnight in buffer. Instead, cover them tightly with plastic wrap to prevent drying.

If you plan to use the beans, soak them overnight in water.

Prepare 500 ml of 5% Triton X-100 solution by mixing 25 ml of Triton X-100 with 475 ml of distilled or deionized water.

Procedure

Before the students begin the activity, ask them to tell you some differences between DNA and proteins that might affect their behavior during electrophoresis. You may need to review what electrophoresis is and how it works with DNA by asking the students those questions, too. The questions provide an opportunity to review the primary and tertiary structures of proteins. The identities of the amino acids within a protein determine its charge at any given pH. The tertiary structure (shape) affects how quickly the protein can migrate through the gel. At the end of the discussion, students should recognize or remember that proteins can have different charges and different shapes and that these attributes affect how the proteins behave during electrophoresis. This information is given in the student introduction, but students will benefit more from a class discussion.

After the introduction, students should prepare saliva and bean samples. One group can prepare the fungal amylase, or you may decide to prepare it before class. Before they begin, explain to them that they may need to dilute saliva samples more than 1:2 and that they will have to use their judgment when they see how the individual samples behave.

When they have their samples prepared, students should follow the procedure for electrophoresis. When electrophoresis is complete, they should stain one half of each gel for exactly 1 min as described and then pour off the stain and add the destain solution. It is desirable to pour off the destain solution and add fresh solution after the gel has destained for some time. If your class takes place in the morning, a good time to change the solution would be at the end of the day. If the class takes place late in the afternoon, the solution could be changed first thing in the morning.

The other halves of the gels should soak in the 5% Triton X-100 for 15 min to 2 hours. At the end of this soaking period, some slight bands of clearing may be visible. Pour off the Triton X-100, and replace it with water. Soak the gels overnight. Check the gels during soaking if possible. The starch will diffuse from the gel after prolonged soaking, leaving the entire gel clear. Once this happens, the amylase activity can no longer be seen.

When the students examine their gels for zones of clearing, they may need to hold the gels at different angles to the light for best viewing. The clearing around the fungal amylase and the human saliva samples will be very obvious.

Answers to Student Questions

1. The other bond is the disulfide bond formed between two cysteine residues.
2. No. An example is a cooked egg white.
3. The amylase in each type of sample is slightly different. That particular amylase protein might not renature properly after being soaked in Triton X-100.

Electrophoresis of Amylase Samples

Introduction

In the previous activity, you tested a number of substances for amylase activity. In this activity, you will use electrophoresis to separate the proteins in amylase samples, such as saliva, and will identify the amylase bands.

Electrophoresis of DNA is a little different from electrophoresis of proteins. The differences are due to the fact that proteins are quite different from DNA. DNA has a uniform backbone that is negatively charged at the pH of electrophoresis buffers. In addition, restriction fragments of DNA are usually the same shape (linear). Therefore, the major difference between DNA fragments is usually their size. During standard electrophoresis, they all migrate toward the positive pole at rates that depend on their sizes.

Proteins, in contrast, are made of chains of amino acids that can be positively charged, negatively charged, or neutral. Proteins themselves thus have quite different charges, depending on their amino acid compositions (as well as the pH of the environment, which we will not consider here). Proteins also have different shapes, and shape affects how easily a molecule migrates through a gel.

If a protein sample is mixed with the strong ionic detergent sodium dodecyl sulfate (SDS), the detergent molecules will bind to the proteins' backbones and impart a negative charge to all the proteins in the mixture. At the same time, SDS partially denatures the proteins by interfering with hydrophobic interactions and hydrogen bonds, causing the proteins to unfold into more similar shapes. The net effect is that the proteins will now all migrate toward the positive pole during electrophoresis (as do DNA molecules). You will be adding SDS to your amylase samples.

Samples such as saliva and bean extract contain a number of proteins. After you have separated the proteins on a gel, the next task is to determine which band is amylase. You are going to use the same approach to identify the amylase band that you used to detect its presence in different samples: detection of its activity. You (or your teacher) will add starch to the electrophoresis gel. The starch will make the gel cloudy. You will be able to identify the amylase bands because the enzyme will digest the starch in its vicinity, producing an area of clearing.

There is just one problem with this approach. Remember the SDS you will be adding to get the proteins to run in the same direction in the gel? It also partially denatures proteins, which can decrease or abolish their activity. (Do you know why?) To restore the amylase activity, it will be necessary to remove the SDS from the protein.

To do this, you will soak the gel (after electrophoresis is complete) in a solution of a nonionic detergent called Triton X-100. The Triton X-100 will replace the SDS. After letting it soak for 15 minutes (min) to 2 hours, you will move the gel into water and let it incubate for a while and then look for starch hydrolysis.

The entire approach is as follows. You will prepare amylase samples with SDS and then load and run them in duplicate on a starch-agarose gel. After electrophoresis, you will cut the gel in half so that you have two halves, each with a complete sample set. You will stain one half of the gel so that you can see the protein bands. You will soak the other half in Triton X-100 and then in water to restore amylase activity. Then, you will compare the stained half of the gel with the "activity" half of the gel and identify which stained protein bands are amylase.

Procedure

Sample preparation

Purified Fungal Amylase

Dissolve 0.5 g of the powdered amylase in 500 ml of distilled water, and mix them well. Combine 36 μl of this solution with 12 μl of loading dye, and mix them well.

Human Saliva

The preparation of saliva depends on the viscosity of the sample. Usually, a 1:2 dilution in water will dramatically ease the loading and running of the sample. In some cases, you will need to add even more water to the saliva. For a 1:2 dilution, mix 0.5 ml of saliva with 0.5 ml of water. Remove 36 μl of this mixture, and combine it with 12 μl of loading dye.

If the sample is still too viscous, remove 0.5 ml of it to a fresh tube, add 0.5 ml of water, mix, and use 36 μl of this mixture with 12 μl of loading dye. You may need to heat the human sample to 65 to 70°C for 2 to 3 min to make it fluid enough to load. This heating decreases amylase activity but does not eliminate it. If there are several students in your work group, at least one is likely to have saliva that is not too viscous to load. It is normal for people's salivas to vary in viscosity.

Bean

Soak beans overnight in distilled water (5 ml per g of beans). The next morning, crush or grind the beans in the water to make a paste. (More water may need to be added, depending on the absorption by the beans.) Combine 36 μl of bean paste with 12 μl of loading dye, and mix them well.

Electrophoresis

1. Load 20 μl each of the samples into the gel in a repeating pattern. This is best accomplished by visually dividing the gel in half and loading each half in the same order with identical samples (Figure 32.1). After electrophoresis, half the gel will be stained for total protein, and the other half will be assayed for amylase activity.
2. Electrophorese the gel for 45 min to 1 hour at 130 V.
3. Remove the gel from the electrophoresis chamber, and cut it in half vertically with a razor blade, ruler edge, or scissor blade. Make the cut between gel lanes.

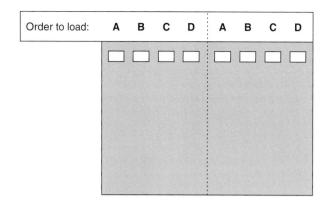

Figure 32.1 Suggested order for loading the samples into the gel.

4. Place one half of the gel in 5% Triton X-100 solution. The gel should incubate in this solution for at least 15 min but can be left there for up to 2 hours.
5. Place the other half of the gel in Coomassie protein stain solution. Incubate it for exactly 1 min. After 1 min has passed, rinse the gel briefly with water and transfer the gel into the destain solution overnight. If possible, change the destain solution once after it has turned blue.
6. After at least 15 min of incubation in Triton X-100, transfer the remaining half of the gel from 5% Triton X-100 to water and incubate it overnight as well, either at 37°C or at room temperature. The amylase activity is higher at 37°C.

The next day, the Coomassie-stained gel should contain fairly discrete bands showing all of the abundant proteins isolated from your samples. Many of the complex samples, such as bean extract or human saliva, will show multiple bands.

To determine which of these bands is amylase, carefully examine the unstained half of the gel for areas of visible clearing. Much of the gel will be opaque because of the starch present in the gel. In areas where amylase activity was present, the enzyme degraded the starch into disaccharides, which then left the gel. The result is a clear spot. Compare the pattern of clearing from this gel with the banding pattern in the other half of the gel to determine which of the bands present in the complex samples corresponds to the enzyme amylase.

Draw the stained half of the gel, labeling your sample lanes. Draw the unstained half, showing areas of clearing. On your drawing of the stained half of the gel, indicate the amylase bands.

Questions

1. The introduction to this chapter states that SDS disrupts hydrophobic interactions and hydrogen bonds. What other type of bond is a major contributor to a protein's three-dimensional structure?

2. Can proteins always be renatured after denaturation? If your answer is no, give an illustrative example.

3. Assume that you saw protein bands in the lane containing one of your samples and knew that the sample initially contained amylase activity, but you could not detect clearing in the unstained half of the gel. Other than an error in procedure, what could be an explanation for this result?

Analyzing Evolutionary Changes   33

About This Activity

In this activity, students count the differences in amylase amino acid sequences from seven different organisms and use the data to construct a simple evolutionary tree. In the next chapter, they will use online bioinformatics resources to identify the organisms from which the sequences came and will conduct further comparisons.

Class periods required: *1 or 2 (constructing the tree could be homework)*

Introduction

Evolution is a unifying theme in biology. In typical biology texts, evolution is treated at an organismal level. Students often learn about the evolution of plants and animals from single-celled progenitors, with discussions of adaptations and niches. This lesson looks at evolution of the protein amylase at the molecular level. Students learned in chapter 4 *(An Overview of Molecular Biology)* that an organism's phenotype is a function of its proteins and their expression timetable and patterns; thus, phenotypic evolution has to be the result of changes at the protein level. This activity allows you to make a connection between various molecular events: mutations, changes in protein structure or expression, and macroevolution.

Macroevolution and Protein Evolution

Proteins from closely related organisms are more similar to one another than are proteins from less closely related organisms. Likewise, the DNAs from closely related organisms are more similar than are the DNAs from less closely related organisms. In fact, given the small percentage of difference in the DNAs of closely related species, such as humans and chimpanzees, it is reasonable to ask how the two species could look and seem as different as they do.

One theory about how small changes at the genetic level can translate into large morphological changes holds that some mutations have a profound effect on the development of the organism. Recall from chapter 4 that the organization of proteins during development is a factor that determines the body plan and that certain proteins direct this organization. A mutation in one of these proteins could result in large morphological changes.

An initial sequence of the chimpanzee genome was published in 2005 (Chimpanzee Sequencing and Analysis Consortium, *Nature* 437:69–87), and the authors of that study made a comparison between the chimpanzee and human genomes. The protein-coding sequences of chimpanzee and human are very similar: 29% of the proteins are identical, and most of the nonidentical proteins differ by only 1 or 2 amino acids. The researchers found 202 genomic elements that were highly conserved in vertebrates but different between chimps and humans. These elements were mostly in noncoding DNA, often near genes associated with transcription and DNA binding, suggesting that they are involved in the regulation of gene expression.

A specific type of change associated with speciation is a change in the timing of body maturity and reproduction. A classic example of this kind of change is the development of the Mexican axolotl, *Ambystoma mexicanum*. The axolotl is an aquatic salamander. Most salamanders have an aquatic larval stage, in which the organism has gills instead of lungs. The larva undergoes metamorphosis into the terrestrial adult form with lungs instead of gills and reproduces at this stage. The axolotl, however, never leaves the aquatic stage and retains its gills. It reproduces in what for other salamanders is a juvenile stage, but it achieves the same body size as its terrestrial counterparts. This change in timing of reproduction and alteration in gene expression is called neoteny.

The axolotl can be induced to develop into a terrestrial form with lungs by injection of thyroid

Classroom Activities

extract. Thus, the axolotl retains all the genes necessary for being a terrestrial organism, including those for thyroid hormone receptors. Some alterations in its ability to produce thyroid hormones may be responsible for its neoteny and thus for many morphological differences between it and terrestrial salamanders.

Since the second edition of this book was published, quite a few molecular studies related to metamorphosis in the axolotl have also been published. Much of the work was done by Randall Voss and his colleagues at the University of Kentucky, who published a genetic linkage map of *Ambystoma* in 2005. In the same year, they published a study identifying a molecular marker associated with neoteny in 99% of the cases examined. The researchers made careful crosses of neotenic and nonneotenic species of *Ambystoma* and compared the presence of the genetic marker with that of the neotenic phenotype. They obtained a strong correlation between the marker and the phenotype when the neotenic parent used for the genetic crosses was a laboratory strain of axolotl. It was less strongly associated when the neotenic parent was a wild-caught axolotl. This study is analogous to the activity in chapter 29, in which students use molecular and phenotypic data to determine whether a hypothetical short tandem repeat marker is associated with a trait.

In 2006, the same authors published the results of microarray studies conducted to identify genes that are involved in the metamorphosis into the terrestrial form. The scientists isolated messenger RNA (mRNA) from *A. mexicanum* epidermis before thyroid hormone treatment and at different time points after thyroid hormone treatment. They used the mRNA to make labeled complementary DNA (cDNA) and probed an *A. mexicanum* microarray. They found that at 2 days posttreatment, only a single gene in the array showed elevated levels of transcription, but after 12 and 28 days, hundreds of genes showed different levels of transcription, both higher and lower. Notably, they observed that several keratin genes showed large differences in transcription (1,000-fold). Keratin gene expression changes coincided with the remodeling of epithelial tissues during the metamorphic process.

The accumulation of genome sequences is making it easier for researchers to trace evolutionary changes. See chapter 27 for more information about comparing genomes and for information about dog evolution gleaned from comparative genomic studies.

Natural selection and proteins

Evolution requires genetic change, such as mutations or chromosomal rearrangements (e.g., duplications, inversions, and breaking and rejoining), and the fixation of the new genotype in a population. According to standard evolutionary theory, genetic changes resulting in an advantage for survival or reproduction will become fixed. Genetic changes that have no effect on survival or reproduction for good or bad can also become fixed by chance. However, we cannot assume that every evolutionary change in a protein that becomes fixed in a population is good or neutral, because the organism carrying the genes is interacting with a complex environment. If a slightly deleterious change in a protein that for a different reason had a survival advantage arose in an organism, that deleterious change could become fixed. The only clear-cut case would be that in which a change in a protein made it difficult or impossible for the organism to survive no matter what. Such a change would be eliminated by natural selection.

Digestion of starch is an important means of obtaining energy in many organisms, so we will assume that there is significant survival advantage in being able to do it. Therefore, we can presume that there is evolutionary pressure to preserve the function of amylase. If we consider changing a protein while preserving its function, it seems logical that there would be regions of the protein, such as the starch binding site, in which changes would be more likely to disrupt function than changes in other regions. Not that these sensitive regions could never be changed, but most changes would likely be deleterious.

Given this kind of logic, scientists compare the sequences of proteins from different organisms and look for regions where the sequence is more highly conserved from organism to organism. These conserved regions are presumed to be important for the structure and function of the protein. Your students will observe conserved regions when they compare amylase sequences. Conserved protein sequences also support the hypothesis that the various proteins are descendants of an ancestral form.

The Activity

Students have sequences from the amylase genes of three mammals (human, pig, and mouse), two insects *(Drosophila melanogaster* and *Drosophila orena),* one bird (chicken), and the paenid shrimp. They do not know the sources of the sequences, and it is important for the activity that they do not. In the next activity (chapter 34), they will perform online searches of protein databases, identify the organisms, and evaluate the evolutionary tree they create.

An evolutionary tree is essentially a hypothesis about how evolution may have occurred. There are numerous methods for constructing these hypotheses from molecular data. Each method requires that certain assumptions about evolution be made. Some methods differ fundamentally in their assumptions and thus differ significantly in how data are analyzed. Most researchers use multiple methods that usually produce the same hypothesis.

The point of this activity is not the method for producing an evolutionary tree. Rather, it is the idea of protein evolution. To keep students focused on the concept that proteins evolve, you will use a very simple analysis method.

The tree, if done properly, will show that the three mammals are most closely related and that the bird is more closely related to them than to the arthropods. The two *Drosophila* species are very closely related, and the shrimp, while not very close to *Drosophila,* is more similar to them than to the bird-mammal branch. After students have performed the online identifications, be sure to ask them if having the shrimp in the same lineage as the insects, rather than with the birds and mammals, makes sense.

Objectives

After completing this activity, students will be able to:

- Use protein sequence data to construct a simple evolutionary tree.
- Explain why conserved regions of protein sequence probably represent functionally important parts of the protein.

Materials

Worksheet 33.1 and the *Student Activity*

Preparation

Students will need to cut the amylase sequences out of the worksheet to facilitate comparison. If it is necessary to preserve the student book, print copies of Worksheet 33.1 from the CD.

If students do not have copies of the student book, print copies of the *Student Activity.*

Procedure

If necessary, review the relationship between DNA sequence and protein sequence, emphasizing how changes in DNA can cause changes in proteins. Remind students that there is more to a gene than its coding sequence: its promoter, any control sequences, and so on. Changes to these regions can also influence protein expression. For example, humans secrete amylase in their saliva because of the insertion of an element upstream of an amylase gene that causes it to be expressed in the salivary gland tissue. Many mammals lack this sequence, and they do not secrete amylase in their saliva.

Students should read the background material.

Work through the two examples in the *Student Activity* in front of the class. Students can refer to the written version of the examples later if necessary.

Before constructing the evolutionary tree, have students look at the seven protein sequences and identify possible conserved regions. You may wish to have them highlight the regions.

Have students construct the evolutionary tree, working alone or in pairs. They can do this assignment as homework.

Results

Here are the data that students should obtain:

Comparison of amino acid sequences for organisms:	No. of differences
A vs A	0
A vs B	9
A vs C	18
A vs D	31
A vs E	33
A vs F	11
A vs G	32

Evolutionary tree

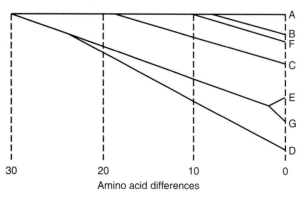

Amino acid differences

Answers to Student Questions

1. To answer this question, students should refer to the diagram and find the point at which the two lineages branch. In the case of B versus C, the lineages branch at the same point where C branches from A, at 18 amino acids. In the other two cases, the lineages branch at the point where D/E/G branches from the others. The differences between the amylase sequences of D, E, or G and organism A are 31, 33, and 32, respectively. The predicted number of differences between organism F and D/E/G would be within this range.

2. See below.

Organism pair	Predicted no. of differences	Actual no. of differences
B vs C	18	17
C vs D	31–33	31
C vs E	31–33	28

Suggested Reading

Culotta, E., and E. Pennisi. 2005. Evolution in action. *Science* 310:1878–1879. A reader-friendly summary of some recent discoveries that shed light on the molecular basis of speciation and evolutionary change.

Classroom Activities

Constructing an Amylase Evolutionary Tree

33

Introduction

Our planet's life forms share the same basic chemistry. For example, we all use DNA as our hereditary material and express the information it contains through transcription and translation. We all make and use DNA polymerase, RNA polymerase, and ribosomes. We all metabolize glucose in fairly similar ways. Our genes encode and our cells produce many similar enzymes.

The theory of evolution holds that existing life forms evolved from earlier life forms through genetic change and natural selection. You are probably familiar with the idea that all organisms with, say, skeletons, are descended from an ancestral creature that introduced the innovation, survived, and prospered. You may be less accustomed to thinking of evolution at the level of proteins, yet proteins are the "stuff" of which we are made and which carries out our functions (see chapter 4, *An Overview of Molecular Biology,* for a review). For a genetic change to affect the form or function of an organism, it has to affect a protein: the shape and function of the protein, the amount of it present, the cells in which it is expressed, the stage in development at which it is expressed, and so on. Thus, proteins are a very reasonable level at which to consider evolution.

Protein evolution

You can think about the evolution of proteins in the same way you think of the evolution of skeletons: at some point, an ancestral form of the protein appeared. Modern versions of a particular protein are descended from an ancestral form of the same protein, just as modern versions of skeletons are descended from an ancestral form. Likewise, the gene for that protein is descended from an ancestral gene. In general, protein evolution tracks organismal evolution. Proteins that share a common ancestor are likely to come from organisms that share a common ancestor.

Protein evolution is not always a simple matter of an enzyme accumulating amino acid differences. For example, proteins can acquire entirely new functions, which is presumably why there are families of similar proteins that do different things. However, for our purposes, we will focus on proteins that retain the same function even as they undergo changes to their amino acid sequences.

Recall what you learned about protein structure in chapter 4. Enzymes have sites that are directly involved in the activity of the protein, and they have other regions that are not directly involved. It seems plausible that a random change in the protein sequence would be more likely to harm the protein's function if the change were in an active site than if it were not. Scientists comparing the sequences of homologous proteins from different organisms often find regions in which the amino acid sequences are particularly similar. They call these regions conserved regions. Conserved regions are assumed to be particularly important to the structure and function of the protein.

You have now observed for yourself that amylase activity is present in a wide variety of organisms. If these amylases are descendants of an earlier, ancestral amylase, then you might expect that there would be some similarity in their amino acid sequences. You might also expect that the more closely related two organisms are, the more similar their amylase proteins will be. These are very reasonable expectations, and they can be used to construct an evolutionary tree.

Constructing evolutionary trees

There is more than one way to approach the construction of an evolutionary tree for a protein. Many evolutionary trees are quite sophisticated and require computers to keep track of all the data. You are going to use a very simple method that can be done by hand. You will count amino acid differences between protein sequences and use the number of differences as a basis for making your tree. For this kind of analysis to be meaningful, you must compare similar proteins.

Here is a simple example using three organisms and an imaginary protein called evolutionase. The evolutionase amino acid sequences from organisms 1 and 2 differ by 30 amino acids. The evolutionase sequences from organisms 2 and 3 differ by 29 amino acids. The evolutionase sequences from organisms 1 and 3 differ by 13 amino acids.

First we will set up a scale of amino acid changes.

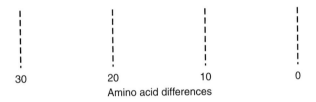

Organisms 1 and 2 differ by the largest number of amino acid changes, 30, so we will begin our evolutionary tree by drawing their lineages diverging at the 30-amino-acid point.

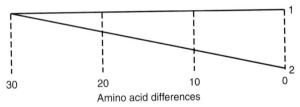

Organism 3's evolutionase differs from organism 2's by 29 amino acids but differs from organism 1's by only 13. Organism 3's evolutionase is clearly more similar to organism 1's, so we make the assumption that organisms 1 and 3 are more closely related than organisms 2 and 3. We draw the organism 3 lineage diverging from the organism 1 lineage by 13 amino acids.

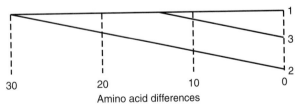

Does this tree take care of the 29-amino-acid difference between organisms 2 and 3? Yes. We will assume, since this is not an exact science, that 29 and 30 are not significantly different. Organism 3's lineage diverged from organism 2's 30 amino acids ago. At that point, organism 3 and organism 1 were still in the same lineage line. This picture is the simplest explanation of the data that we can construct.

Now, let's try an example with a larger data set—seven organisms.

Comparison of amino acid sequences for organisms:	No. of differences
1 vs 1	0
1 vs 2	25
1 vs 3	17
1 vs 4	27
1 vs 5	8
1 vs 6	6
1 vs 7	26

First, draw a scale of amino acid changes. The largest number of amino acid differences in the data set is 27, so we need a scale that covers that range:

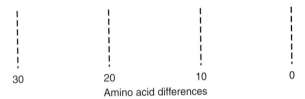

There is a group of three organisms, 2, 4, and 7, that differ most from organism 1. Start by drawing the divergence of the most different one, organism 4.

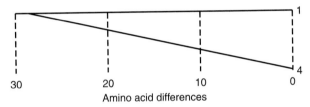

Now, we need to determine whether organisms 2 and 7 are more similar to organism 4 than they are to organism 1. If they are, then they diverged from the organism 1 lineage when organism 4 did. If they are more similar to organism 1 than to organism 4, they diverged from organism 1 after organism 4 did.

Comparison of amino acid sequences for organisms:	No. of differences
2 vs 4	26
2 vs 7	24
4 vs 7	10

The amino acid sequence from organism 7 differs from the sequence of organism 4 by only 10 amino acids, so organism 7 branches from the organism 4 lineage. We will add organism 7 as shown.

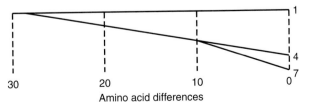

Organism 2's evolutionase, however, differs from the evolutionases of both organisms 4 and 7 by essentially the same number of amino acids as it differs from that of organism 1. Therefore, organism 2's lineage diverges from both organism 1's and organism 4/7's lineages at about the same point on the drawing. The numbers of differences are so close that the diagram could be drawn with the organism 2 lineage diverging from the organism 4/7 lineage at its base or from the organism 1 lineage at its base. We will draw it from the organism 1 lineage, realizing that the case is not compelling.

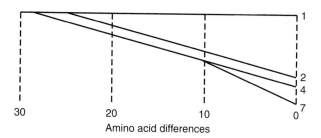

Organism 3's evolutionase differs from organism 1's by 17 amino acids, so we diverge its lineage from the organism 1 lineage at that point.

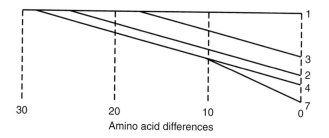

Organisms 5's and 6's evolutionases differ from organism 1's by about the same number of amino acids. We need to compare them to see how many differences there are between them.

Comparison of amino acid sequences for organisms:	No. of differences
5 vs 6	2

The evolutionases from organisms 5 and 6 have fewer differences from each other than from organism 1's, so their lineages were still together when they diverged from that of organism 1. We will add them to the diagram accordingly, and we are finished. Does the diagram account for all of our data? Check and see, comparing it with the data on amino acid differences.

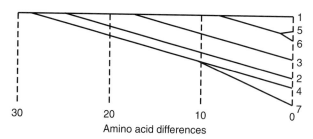

Why did we use organism 1 as a reference and put it on the horizontal line? Would it have been wrong if we had used another organism? No. Using organism 1 was entirely arbitrary. You could do the same analysis using any of the organisms as the horizontal line, and you would get the same pattern of divergence. It would simply be rearranged on the page. Here is an example of the same diagram, using organism 4 as the reference point.

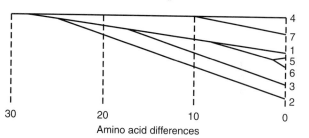

Amino acid differences and the passage of time

Can amino acid sequence changes be equated with the passage of time? In other words, can we use amino acid changes sort of like a clock, in which 10 amino acid changes might equal 5 million years of evolution? In a word, no.

When scientists first started accumulating enough protein and DNA sequence data to do this kind of analysis, they were very excited about this possibility. Further study showed that amino acid sequence changes do not occur at regular time intervals, or even close to them. For example, one study of a single common enzyme found a rate of 6.1 amino acid substitutions per 100 million years in the human-insect comparison, 23.7 changes per 100 million years in the human-mouse comparison, and 69.0 changes per 100 million years in the mouse-rat comparison. Although it is not possible to determine when two organisms diverged on the basis of sequences of a single protein, it is still generally true that the more closely related two organisms are, the more similar their protein sequences will be.

Some scientists still believe that a molecular clock may exist within noncoding regions of DNA. In other words, they believe that nucleotide changes may accumulate at fairly regular rates in these regions. Some of these scientists study differences in the noncoding regions of human mitochondrial DNA in an attempt to determine when various groups of humans diverged from one another. Other scientists are not so enamored of this approach. The data are interesting, but the jury is still out as to what they mean.

Today's Activity

In this activity, you will construct an amylase evolutionary tree. We have provided portions of the amylase amino acid sequences from seven different organisms in Worksheet 33.1 (see Appendix A). The sequences use a standard one-letter abbreviation system for amino acids as follows:

A Alanine	I Isoleucine	S Serine
B Asparagine or aspartate	K Lysine	T Threonine
	L Leucine	V Valine
C Cysteine	M Methionine	W Tryptophan
D Aspartate	N Asparagine	X Unknown or nonstandard
E Glutamate	P Proline	
F Phenylalanine	Q Glutamine	Y Tyrosine
G Glycine	R Arginine	Z Glutamine or glutamate
H Histidine		

The sequences all contain ellipsis points (. . .) to signify where we omitted a portion of sequence. Before you begin constructing the tree, take a moment to look at the sequences. Do you see any conserved regions? Conserved regions do not have to match exactly; they need only be somewhat similar.

As a starting point, use organism A as a reference. Cut out the protein sequences on strips of paper to facilitate comparison. Compare the other six amino acid sequences to sequence A, and record the numbers of differences. Some of the sequences have a hyphen in them. The hyphen was inserted because some of the organisms have an additional amino acid at that point. Count the hyphen and the corresponding amino acid (if the other organism has one) as one difference. If the other organism's sequence also has a hyphen at that point, count it as a match. Make a data table like the one below, and fill it in.

Comparison of amino acid sequences for organisms:	No. of differences
A vs A	
A vs B	
A vs C	
A vs D	
A vs E	
A vs F	
A vs G	

Construct an evolutionary tree, using the examples above if you need a guide. Draw your evolutionary tree neatly, labeling each branch to match the corresponding organism's amino acid sequence. In the next activity, you will perform online searches of protein sequence databases to identify the organisms whose amylase sequences you have just compared.

Questions

1. Based on your tree, how many differences should there be between the amylase sequences of the following pairs of organisms? Fill in the table appropriately.

2. Compare the amylase sequences of the pairs of organisms and fill in the data. How well do the data match the prediction?

Organism pair	Predicted no. of differences	Actual no. of differences
B vs C		
C vs D		
C vs E		

Bioinformatics

34

About This Activity

This activity was designed both to introduce students to the online bioinformatics resources and to reinforce the fact that proteins from many different organisms are quite similar and have evolved from common ancestral proteins. In the first part of this three-part activity, students search online protein databases to identify the organisms whose amylase sequences they used in constructing their evolutionary trees in chapter 33. The second part gives students additional, detailed information about human pancreatic amylase, introduces them to other aspects of the protein databases, and provides them with a good review of protein structure. In the third part of the activity, students search the online biomedical literature to find information about amylase and to learn how to conduct searches.

The number of protein sequences in the database has increased substantially since the publication of the second edition of this book. While this increase is a boon for scientific research, it has made this activity more challenging. For one thing, there are many repetitions of various human amylase sequences in the database now, yielding many repetitive hits when the database is searched with the human sequence. Interestingly, the dog and chimpanzee amylase sequences, which were not available when the second edition was written, are also close matches to human amylase. Because of the similarity of the sequences of the smaller segment of the human amylase gene to those of the dog and chimpanzee, we have switched the instructions so that students search the second, longer portion of the sequence.

Class periods required: *This is an open-ended activity in which students can explore the databases for a long or short while. Identifying the organisms from which the amylase sequences came will require one class period. The other two activities will take at least two more periods.*

Introduction

The U.S. government funds publicly available electronic repositories of scientific data through the National Center for Biotechnology Information (NCBI). NCBI was established in 1988 as a division of the National Library of Medicine (NLM), which along with 18 separate health institutes (such as the National Cancer Institute, the National Institute for Environmental Health Sciences, and the National Institute on Drug Abuse) and the National Center for Complementary and Alternative Medicine, is a division of the National Institutes of Health (NIH). NIH itself is a division of the Department of Health and Human Services. The public health divisions of the Department of Health and Human Services include NIH, the Centers for Disease Control and Prevention, and the Food and Drug Administration, along with a few other agencies.

NIH is a premier medical research organization. Its grant programs support much of the biomedical research carried out in the United States today, thereby paying for many, if not most, of the advances described in this book. Its own laboratories make valuable contributions in many areas of research.

NCBI is a national resource for molecular biology information. NCBI creates public databases and software for analyzing genomic data, conducts research in computational biology, and disseminates biomedical information. The molecular biology databases and search and analysis programs are available through the NCBI website (http://www.ncbi.nlm.nih.gov), along with other features. For example, a feature called "Coffee Break" contains summaries of current research findings, with built-in tutorials showing how to use the various search programs. We strongly encourage you to browse the site. Here, we limit our description to the features involved in this activity, along with a few others.

Information available through NCBI

Nucleotide Sequences

GenBank is the genetic sequence database of NIH. It contains all publicly available gene sequences. GenBank is connected with the DNA DataBank of Japan and the European Molecular Biology Laboratory (EMBL). These three organizations exchange data on a daily basis.

Protein Sequences

The NIH's protein sequence information comes from several databases: translations from GenBank nucleotide sequences; the European database SWISS-PROT, which shares data with NIH through the EMBL; and the three-dimensional structure Brookhaven Protein Data Bank.

Protein Structure

Through the NCBI site, you can access the Brookhaven Protein Data Bank, which contains information on three-dimensional protein structures and three-dimensional images for viewing. Software for viewing and manipulating the images can be downloaded.

Biomedical Literature

The literature database MEDLINE was created and is maintained by the NLM. MEDLINE contains over 10,000,000 citations from more than 3,800 international biomedical journals covering the fields of medical, veterinary, and preclinical science. Journal articles are indexed for MEDLINE by using the NLM's controlled terms, the Medical Subject Headings. A list of Medical Subject Headings terms is available through the NLM website. MEDLINE citations include English-language abstracts of foreign-language articles if one was published with the article.

In addition to MEDLINE, NLM maintains a number of smaller, specialized databases, such as AIDSLINE, which contains bibliographic citations to literature covering research, clinical aspects, and health information on AIDS; BIOETHICSLINE, with bibliographic citations relating to bioethics; and ChemID, a dictionary of chemicals.

Human Genetics

Victor McKusick and his colleagues at Johns Hopkins University maintain a database, *Online Mendelian Inheritance in Man* (OMIM), which can be accessed through the NCBI website. This database is a catalog of inherited diseases. Each entry contains information about the disease and may contain information on gene therapy, animal models, or other topics. The information is not congruent from entry to entry, as different amounts of information are available on different diseases. The information has not necessarily been written to be understandable to the lay public. Nevertheless, it makes interesting browsing. See chapter 25 *(Human Molecular Genetics)* for more information about browsing OMIM.

Taxonomy

If you select the taxonomy database from the pull-down menu and enter the name of an organism in the search term box, you will link your way to complete taxonomic information for the organism you entered. On that page are links to entries for that organism in other databases, such as genes, proteins, structures, and genome projects. There are also links to external sites focused on that organism. For example, if you entered *"Ambystoma mexicanum"* in the search term box and chose Taxonomy as the database, you would see (in 2006) that there are 34,647 nucleotide sequences from *A. mexicanum* in the nucleotide database, 345 protein sequences in the protein database, one genome project, two protein structures, etc. Each of these results is a link that you can follow for more information.

Genome Projects

The genome project database assembles information from genome projects. If you choose it as the database and enter the name of an organism in the search term box on the NCBI website, you will be taken to a summary page of information about the genome project(s) for that organism, including links to external sites. A search of genome projects for *Canis familiaris* (the domestic dog) shows you that in 2006, there are both a *C. familiaris* and a *Canis lupus* (wolf) genome project. Clicking on the link to the dog project takes you to a summary page with diagrams of the dog chromosomes, a photo of a boxer (presumably Sasha, the boxer whose genome was sequenced), links to key publications and to the genome sequence, and links to external sites with more information.

One of our favorite links from this page is to a database called *Online Mendelian Inheritance in Animals*. Click on that link, and you will be taken to a page with a table that summarizes, for one thing, how many traits, "phenes," in a variety of different animals have a known molecular basis. Each of those numbers is a link to a list of the traits. Click on the number for the dog (in 2006, it was 58). The list of traits includes "Coat colour, brown." Clicking on that link takes you to a page that identifies the gene responsible as TYRP1 (Phene ID 2695 Group 001249) and includes a reference to a paper published in 2002 titled "TYRP1 and MC1R genotypes and their effect on coat colour in dogs." MC1R is the melanocortin receptor responsible for the coat color in yellow Labs and other red and yellow dogs.

Search and analysis software

PubMed
PubMed, which is accessed through the NCBI and NLM websites (http://www.ncbi.nlm.nih.gov and http://www.nlm.nih.gov), searches MEDLINE and has links to NCBI's protein sequence, DNA sequence, and protein structure databases. Links specific to each citation appear with the abstract of the article. PubMed now also provides links to background information through an electronic version of *Molecular Biology of the Cell* (B. Alberts, et al., Garland Publishing, Inc., New York, NY, 2002). Terms in article abstracts are linked to relevant portions of the book.

National Library of Medicine Gateway
NLM Gateway, accessed through the NLM website, provides access to MEDLINE and to other NLM databases (see above).

BLAST
BLAST (Basic Local Alignment Search Tool) is a sequence identification and comparison program. BLAST can search protein and nucleotide databases in a number of ways. For example, it can compare an amino acid sequence with the protein databases, a nucleotide sequence with the nucleotide sequence databases, a nucleotide sequence translated in all reading frames with the protein databases, and a protein sequence with the nucleotide sequence databases translated in all reading frames. It finds sequences similar to the query sequence and displays them in decreasing order of matching, calculating statistics related to the quality of the match.

Entrez
Entrez is a search system at the NCBI website that retrieves information from MEDLINE, GenBank, OMIM, the protein sequence and structure databases, and a few other databases. To use Entrez, you select from a menu the database you want to search, and you enter search terms. Entrez is not a sequence comparison program.

Search Tutorials
From the NCBI website (http://www.ncbi.nlm.nih.gov), you can click on Education in the sidebar menu. This takes you to a screen from which you can access PubMed and BLAST tutorials. If you click on Entrez from the NCBI site, you will go to a screen from which you can try an Entrez tutorial.

General searching information

To search MEDLINE with PubMed, go to the NCBI website (http://www.ncbi.nlm.nih.gov) and use the pull-down menu next to the search command on the bar near the top of the page. Select PubMed from the menu and then type your search terms in the window and click on the Go button. For example, let's assume you are interested in the amylase genes of humans. If you enter "human salivary amylase" as search terms, you will be notified that the search retrieved over 1,000 documents (1,272 in October 2006). It is unlikely that you will want to wade through that many citations, so add "gene" to your search. That narrows the results down considerably (to 89 citations in October 2006). If you click on any one of the citations, you will be connected to an abstract of the article.

To search for a protein sequence, protein structure, or DNA sequence, select Proteins, Structures, or GenBank from the pull-down menu and then enter the search terms. For example, if you choose the database Proteins from the pull-down menu and enter the search terms "human salivary amylase," you will get a list of entries in the protein databases that contain your search terms. In October 2006, there were 52 entries. If your search retrieves too many documents, either use the Modify Search command on the same screen to add terms to your search, or go back one screen and redo the search with more restrictive terms.

Each citation has links to several reports: GenPept, FASTA, MEDLINE, and others. Clicking on FASTA takes you to a screen that displays the protein sequence using the one-letter codes for amino acids. Clicking on GenPept takes you to an annotated record. This record tells you the GI number (an important identifier preceded by the letters GI) near the top, lists references and provides links to them, and may give domain information about the protein or specify amino acids in the active site. At the bottom of the page, it gives the amino acid sequence using the one-letter codes.

GenBank search results are similar to Proteins search reports, except that instead of a link to GenPept, the citations have links to GenBank. Clicking on the GenBank link of a citation takes you to a screen similar to the GenPept screen, with a GI number, literature references (if any), more information about the DNA sequence (such as at what point in the nucleotide sequence the start codon for the protein can be found), the amino acid sequence of the protein (if the sequence is a protein sequence), and, finally, the nucleotide sequence.

Searching the Structures database for human salivary amylase in October 2006 retrieved 10 entries. Clicking on the letter and number identifier for an entry takes you to a page with a domain summary and a button that says "View Structure." At the time of writing, the page also has a link for downloading the free viewing software called Cn3D. Once you have downloaded Cn3D, view the structure or save it to the data folder created by Cn3D for later viewing. Cn3D allows you to rotate the structure, zoom in, and select how you want the structure displayed. Structures for both human pancreatic and salivary amylases are available.

Note: This information is far from exhaustive. The best way to get to know the online resources is to spend some time using them. Do some searches, click on links, and look around.

Today's Activity

In this activity, students will use the BLAST sequence comparison tool to identify the organisms from which the amylase sequences in chapter 33 originated. Students will compare their results with the evolutionary tree they constructed. They will use data from their searches to explore other features of the databases and will conduct some literature searches.

Objectives

After completing this activity, students will be able to:

- Use online databases to identify the source of a protein, given its amino acid sequence.
- Search the online biomedical literature database MEDLINE.
- Search online databases to determine whether sequences have been determined for a particular gene or protein.

Materials

- Computers with Internet access
- Worksheet 33.1
- Students' evolutionary trees from chapter 33

Preparation

- Go through the student exercises yourself so that you will be comfortable with them.
- If you have a computer projector, set it up so that you can show students how to do the searches if necessary.

- If you do not have a computer projector, print out relevant screens as you do the activities and make overhead transparencies from them. We have not provided masters, because the appearance of the pages changes from time to time.
- If you want to view three-dimensional protein structures, download Cn3D as described previously. Go to the structure entry for human pancreatic amylase 1HNY (from the NCBI home page, search Structures for human pancreatic amylase), and save the structure in the data file created by Cn3D. Check to make sure the structure displays properly. If possible, download the program to the computers the students will be using.
- Photocopy the *Student Activity* if students do not have manuals.

Procedure

Before you begin, call the students' attention to the cautionary note at the beginning of their instructions. We have provided specific information about how to carry out online searches. However, students must understand that the databases are constantly growing through the addition of new information. They may find information to which we have not referred because the information was added since we wrote this material.

In addition, the people who maintain the NCBI website add features to the site and sometimes change the appearance of Web pages. Your students may get pages that do not look like the figures we have provided. Do not be alarmed if this happens. The databases are there, the links are there, and the data will be there. Students may need to use their good sense to figure out what to do if the appearance has changed. Navigating around the databases is not difficult.

Part 1: Performing a BLAST search to identify amylase source organisms

Introduce the topic as you deem appropriate. The student instructions should enable students to do the searches independently.

They will find that the organism A sequence is human pancreatic amylase.

Answers to Questions from Part 1

1. Human.
2. The second-best match to the second portion of the human pancreatic amylase sequence (organ-

ism A) is chimpanzee amylase (also pancreatic). *Note:* This result may change as the database grows. The E values are a function of the amount of data in the database, and so we cannot provide an answer in a book that will stay valid. The goal of asking students about the E value is to get them to notice the likelihood of a random match to their sequence.

3. The E value represents the chance of finding a random match to the input sequence that is as good as the listed match. The E value of the shorter portion of the sequence is higher because the sequence is shorter. Odds of a random match to a shorter sequence are greater than those of a random match to a longer sequence. However, the odds are vanishingly small in both cases.

4. Organism A, human; organism B, pig; organism C, chicken; organism D, pacific white shrimp; organism E, *Drosophila melanogaster;* organism F, mouse; organism G, *Drosophila orena.* The *Drosophila* results will be very confusing, as there are now amylase sequences from many species of *Drosophila* in the databases. It does not really matter whether students get the species correct—the point is that sequences E and G are from two different species of *Drosophila.*

5. If the students did their evolutionary trees correctly, the results should correlate with patterns of organismal evolution. Shrimp and *Drosophila* are arthropods, and thus it makes sense for them to diverge from the bird-mammal line before they diverge from one another.

Results of Entire Human Pancreatic Amylase Sequence Comparison

1. This search reveals that the closest overall matches to the human pancreatic amylase sequence used for the search are other database entries for human pancreatic amylase. Not very interesting or surprising. The closest nonhuman match is chimpanzee *(Pan troglodytes),* and its match to human pancreatic amylase is better than that between human salivary and pancreatic amylases. The next closest nonhuman match is the monkey *Macaca mulatta,* followed by dog, pig, rat, and mouse (in October 2006; this result will probably change as more sequences are entered into the database). The E value for all these matches is 0, indicating that there is essentially no chance of having a random match from the database.

2. Hypothesis 1. Salivary and pancreatic genes have been separate genes since before the ancestral lineages of humans and rats diverged. Therefore, the pancreatic amylases of rats and humans are less different than the salivary and pancreatic amylases

of humans. This hypothesis predicts that a comparison of the entire human pancreatic amylase sequence with the database would show greater similarity to rat pancreatic amylase than to human salivary amylase.

Hypothesis 2. Rat and human ancestries diverged before the human pancreatic and salivary amylase genes. However, the pancreatic environment puts an evolutionary constraint on some regions of the amylase protein. As a result, pancreatic amylases stay similar in that region (or those regions), and we happened to compare one of those. Provided the constrained region is not too large, comparison of the entire protein sequences should show a better match between human salivary and pancreatic amylases than between rat pancreatic and human pancreatic amylases.

3. There are 419 identical amino acids out of 495 between rat and human pancreatic amylases.

4. There are 481/495 identities between human salivary and pancreatic amylases. Thus, human salivary amylase is more similar to human pancreatic amylase than is rat pancreatic amylase, supporting hypothesis 2.

Part 2: Exploring protein bioinformatics

Exploring the GenPept entry for human pancreatic amylase provides a superb opportunity for reviewing protein structure. Students will be examining detailed information on the structure of human pancreatic amylase. The background information they need is found in chapter 4, *An Overview of Molecular Biology.*

Student instructions for this portion of the activity are also self-explanatory.

Answers to Questions from Part 2

1. D. Burk, Y. Wang, D. Dombrowski, A. Berghuis, S. Evans, Y. Luo, S. G. Withers, and G. Brayer are the authors of the first paper, which was published in 1993 in the *Journal of Molecular Biology.* G. Brayer, Y. Luo, and S. Withers are the authors of the second paper, which was published in 1995 in *Protein Science.*

2. A disulfide bond (—S—S—) is formed between the —SH portions of two cysteine residues. It is a strong stabilizer of protein structure, since it is a covalent bond.

3. Five.

4. Students should connect residues 28 and 86, 70 and 115, 141 and 160, 378 and 384, and 450 and 462.

Viewing the Three-Dimensional Structure of Human Pancreatic Amylase

This portion of the activity requires that you have downloaded the program Cn3D from the NCBI website. Students are essentially asked to play with the program, viewing the structure of pancreatic amylase. If you have time, go through the structure alongside the information from the GenPept file. Identify the helices, beta strands, and disulfide bonds.

Part 3: Exploring online biomedical literature references

Again following self-explanatory instructions, students will be led through a search for references to human salivary amylase in which they will learn how to use the search program and narrow their searches by adding terms. They can also go to the Education page from the NCBI home page and select the tutorial for PubMed.

Answers to Questions from Part 3

1. Meisler and Ting worked in the Department of Human Genetics at the University of Michigan, Ann Arbor. The article was published in 1993.
2. Five.
3. Insertion of a retrovirus upstream of the amylase gene. It contains a salivary gland-specific enhancer.
4. The fact that salivary expression of amylase arose independently in rodents and primates.

Answers to Challenges

1. In October 2006, we found two articles relating to dog amylase gene expression using "dog amylase gene expression" as search terms. They were:

Mocharla, H., R. Mocharla, and M. Hodes. 1990. Alpha-amylase gene transcription in tissues of normal dog. *Nucleic Acids Research* 18:1031–1036.

MacDonald, R., A. Przybala, and W. Rutter. 1977. Isolation and in vitro translation of the messenger RNA coding for pancreatic amylase. *Journal of Biological Chemistry* 252:5522–5528.

Mocharla et al. used Northern blotting and polymerase chain reaction with reverse transcriptase to look for α-amylase RNA in dog tissue. Using human amylase primers, they found α-amylase transcripts in the pancreas, liver, small intestine, large intestine, and fallopian tube, but not in the salivary (parotid) gland. This result explains why dogs can digest starch but have no amylase activity in their saliva.

MacDonald et al. isolated RNA from dog pancreas and translated it in vitro. They found a large protein that reacted with anti-amylase antibodies (gamma globulin) and suggested that it was a precursor protein. They found that almost all amylase RNA was associated with polyribosomes bound to the endoplasmic reticulum (a nice review of cell structure!). They also found that amylase messenger RNA (mRNA) was much larger than it needed to be to encode the protein.

2. Hickey, D. A., B. F. Benkel, P. H. Boer, Y. Genest, S. Abukashawa, and G. Ben-David. 1987. Enzyme-coding genes as molecular clocks: the molecular evolution of animal alpha-amylases. *Journal of Molecular Evolution* 26:252–256.

The search terms "amylase sequences molecular clocks" will find this article. We tried several permutations that did not find it, such as "amylase gene molecular clock."

These authors made a complementary DNA library for the flour beetle *Tribolium castaneum* and screened it with a cloned *Drosophila* amylase gene. They found the beetle amylase gene, sequenced it, and compared it with the *Drosophila* amylase sequence and other published amylase sequences. They found that animal α-amylases are highly conserved over their entire lengths and that parts of the gene are conserved among prokaryotes, plants, and animals.

Although students do not have access to the entire article, we read it and found that the authors concluded that amylase evolution could not be used as a molecular clock! The rates of change were very different between different groups of organisms: 6.1 substitutions per 100 million years for the insect-mammal comparison versus 23.7 for the human-mouse comparison versus 69.0 for the mouse-rat comparison.

Follow-Up

In the activities in the section *Molecular Biology and Genetics,* students learned about the molecular biology of coat color in Labrador retrievers, the molecular biology of several human diseases, and a little about the molecular biology of cancer. They can explore the databases for information related to any of these topics. We suggest they search Taxonomy for *Canis familiaris,* go to the *Online Mendelian Inheritance in Animals* page, and follow the links to dog traits for which the molecular basis has been determined.

We urge you to include online searches in future activities, both to reinforce what students have learned in this chapter and to connect information in their textbook to the outside world. In the second edition of this book, we put this section of activities before *Molecular Biology and Genetics*. In this edition, with the new *Genomics* section, the content did not seem to flow as well. Please provide feedback to ASM Press if you prefer the other order—of course, you can always choose to do these activities before the genetics and genomics sections, since this section does not require any information presented in those chapters.

Classroom Activities

Exploring Bioinformatics

Introduction

One of the consequences of the development of the biotechnology techniques you have been learning about was an explosion in the amount of gene and protein sequence data being collected. For this information to provide the greatest benefit, it needed to be readily accessible and easy to compare with other sequence data.

In 1988, the U.S. government established the National Center for Biotechnology Information (NCBI) to be a national resource for molecular biology information. NCBI creates public databases of information, develops software for data analysis, and conducts research in computational biology. NCBI shares data with other international resources, updating its databases daily.

The information resources include DNA sequence databases, protein sequence databases, and protein structure databases. Powerful search-and-compare programs allow comparison of sequences. The availability of all this information has revolutionized certain aspects of biomedical research.

When a scientist determines a sequence of interest, the first thing she does is to compare it with the sequences in the databases. This comparison could tell her what the sequence is, or what her protein is similar to, or what certain parts of her protein are similar to. Comparison of protein sequences and structures led to the identification of the protein structure motifs described in chapter 4, *An Overview of Molecular Biology*. You can use the sequence comparison programs to search for a particular protein sequence for various motifs, which might provide clues about the function of the protein or about the mechanism of its function.

In fact, the availability of all this molecular information and relatively easy means of analyzing it have

led to the coining of a new word: bioinformatics. The activities in this chapter will provide you with an introduction to it. This chapter is meant to open a door for you. Once you learn how to use the resources, you will be able to go back to them whenever you wish.

Part 1: Performing a BLAST search to identify amylase source organisms

In this activity, you will use the search tool BLAST (Basic Local Alignment Search Tool) to identify the source organisms of the amylase sequences from chapter 33. This is not necessarily a cut-and-dried exercise. The data you are searching for was put into online databases by the scientists who collected it. The databases are constantly growing through the addition of new information. In addition, the people who maintain the NCBI website sometimes change the appearance of Web pages. We have two screen shots of portions of the Web pages as they appeared in October 2006. You may need to use your good sense to figure out what to do if the appearance has changed since then.

Procedure

1. Go to the NCBI home page at http://www.ncbi.nlm.nih.gov (no period after gov). Click on BLAST on the dark bar near the top of the page (Figure 34.1).
2. The next screen asks you to select the type of search you will be doing. You are going to compare the amino acid sequences from chapter 33 to those in the database, so click on protein-protein BLAST (BLASTp).
3. The next screen has a place for you to enter your search sequence. Either you can enter a single-letter amino acid sequence, or if you want to compare a known sequence in the database to other sequences, you can enter the GI number of that protein sequence.

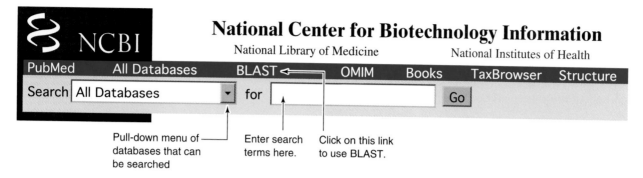

National Center for Biotechnology Information

National Library of Medicine   National Institutes of Health

| PubMed | All Databases | BLAST | OMIM | Books | TaxBrowser | Structure |

Search [All Databases ▼] for [   ] [Go]

Pull-down menu of —— databases that can be searched

Enter search terms here.

Click on this link to use BLAST.

Figure 34.1

4. A box near the place for entering your search sequence should have the letters "nr" in it. If it does not, click on the box and hold down. Select "nr" from the menu. This tells the search engine to search all available nonredundant data.

5. In the data box, enter the second half of the protein sequence from organism A (GRGN-RGFIVFNNDDWSFSLTLQTGLPAGTYCDVISGD-KINGNCTGI) from Worksheet 33.1, using the single-letter codes for amino acids on the worksheet; do not enter the ellipsis points (. . .), and skip over any hyphens in the sequence. Verify that you have entered the sequence correctly and that you have not spanned the ellipsis. The program is not case sensitive. Click on BLAST.

6. The next screen should tell you that your request has been submitted. If "blastp" was not selected, you will get an error message. If this is the case, go back and fix the problem. This screen will also tell you approximately how much time will elapse before your results are ready.

7. Click on Format results.

8. The next screen may be a waiting screen that tells you how much time is still needed, or it may be your results screen. The results screen will come up automatically when the results are ready.

9. The results screen (an edited version is shown in Figure 34.2) presents results first in graphical form with a series of horizontal colored bars. The color of the bar corresponds to the alignment score. The alignment score is influenced by how long the query sequence was: the longer the query sequence, the higher the score can be.

The color groups show the quality of the match in descending order, with the best-match group first.

Below the series of horizontal bars is a list of the match sequences corresponding to the bars. The first entry in the list corresponds to the first bar, and so on. At the right end of each entry is an E score. This is the number of matches of this quality that could be expected by chance in a database the size of the one that was searched. The number in Figure 34.2 is 4e − 18, or 4 × 10^{-18}, over 100,000 times smaller than 1 in 1 trillion (1×10^{-12}). Scroll down below the diagram with the bars to see the amino acid sequence alignments. Part of the header information for each alignment is the number of identical amino acids in the two sequences: for example, Identities = 489/495 98%.

10. What is the best match to the second half of the protein sequence of organism A? What organism gives the second-best match? What are the E values?

11. Go back to the input screen and enter the first half of the protein sequence of organism A (NNNGVIKEVTINPDTTCGND). Verify that the top menu still reads "blastp" and that you have entered the sequence correctly. Click on Search.

12. What is the best match for the first part of the sequence of organism A? What is the second-best match? What are the E values? What is organism A?

13. Repeat the above procedures to identify the source organisms of the other amylase sequences.

Classroom Activities

BLASTP 2.0.11 [Jan-20-2000]

Scroll over the bars with the mouse. As you pass over a bar, its identity will appear in
the box. Identities are listed below the graphic, in order of match quality.

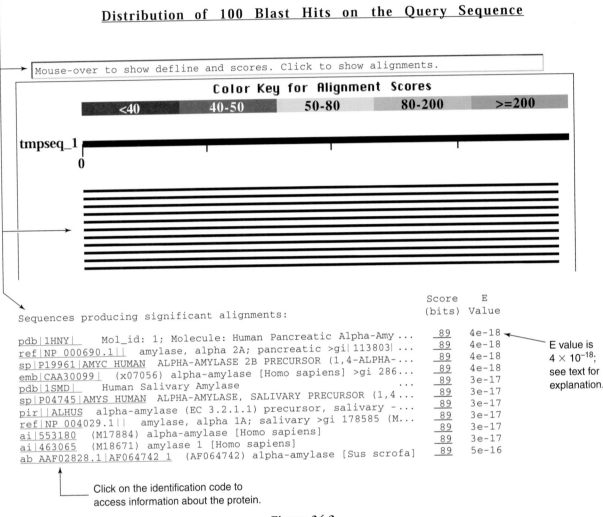

Click on the identification code to
access information about the protein.

Figure 34.2

Questions

1. What is organism A?

2. What were the second-best matches to the two different portions of the sequence from organism A?

3. Why is the E value of the best match to the longer portion of the sequence of organism A lower than the E value of the best match to the shorter portion of the sequence?

4. What are the identities of the other organisms from chapter 32?

5. Discuss your evolutionary tree in the light of the identities of the organisms. How does it compare with what you know about the evolution of the organisms?

Whole-Protein Comparison

The BLAST searches described here used only portions of the amylase sequences. You saw that using the first and second portions of the sequence of pancreatic amylase turned up different second-best matches. What is the closest match to the overall structure of pancreatic amylase?

To avoid typing in the entire amino acid sequence, use the GI number for human pancreatic amylase, 1421331. Enter the GI number in the search box and click on BLAST.

1. Other than different database entries of human pancreatic amylase and its precursors and variants, what is the closest match to human pancreatic amylase?
2. Evidence strongly suggests that the salivary and pancreatic amylase genes are descendants of one original pancreatic amylase gene. A BLAST search using the first portion of the human pancreatic amylase sequence shows that rat amylase is a closer match to the sequence than is human salivary amylase. A BLAST search using the second portion of the sequence finds that human salivary amylase is a better match than rat amylase. Propose two evolutionary explanations for the second-best match result, one in which the genes diverged before the organisms and the other in which the organisms diverged before the genes. Using each hypothesis, predict which entire protein sequence would be a better match to the entire pancreatic sequence from organism A, rat pancreatic amylase to human pancreatic amylase or human pancreatic amylase to human salivary amylase.

To test the hypotheses regarding the evolution of amylase sequences in rats and humans, let's use a slightly modified procedure. Again, type the GI number of human pancreatic amylase (1421331) into the search box. Now locate the menu that lets you choose what organisms to search. The default probably reads "all organisms." In 2006, this menu is on the same screen as the search box, but you may need to click on a link for limiting your search to find it.

From this menu, select *Rattus norvegicus*. Click on BLAST.

3. Your hits should all be rat amylases. How many amino acid identities are there between rat and human?
4. Now, go back and repeat the search, but select *Homo sapiens* from the menu instead of *Rattus norvegicus*. Scroll down among the results until you find salivary amylase. How many amino acid identities are there between human pancreatic amylase and human salivary amylase? Which evolutionary hypothesis does this result support?

Part 2: Exploring the structure of human pancreatic amylase

In this portion of the activity, you will explore protein bioinformatics resources in more depth, using human pancreatic amylase as the example protein.

Procedure

1. Go back to the BLAST search screen and type in a portion of the human pancreatic amylase sequence as you did before (or enter the GI number). Click on BLAST.
2. On the results screen, find the human pancreatic amylase match with the identifiers 1HNY and gi1421331 and click on the link.
3. This takes you to a GenPept display screen with more information about 1HNY. Note the GI number 1421331 near the top of the entry. The GI number is an important general identification number. DBSOURCE indicates that the sequence was retrieved from the Brookhaven Protein Data Base (the abbreviation pdb), a structure database. It indicates that the structure was determined by X-ray diffraction. The source organism *(Homo sapiens)* is given, along with its Linnean classification.
4. Two journal articles are cited, with PubMed links. Click on these links to see abstracts of the published articles that describe the determination of the structure of human pancreatic amylase. Note that the entries are linked to the protein structure and to related articles. Explore the links and then go back to the GenPept screen.

Below the identification information and literature references are FEATURES. This gives structural data for the protein, which is logical, since the sequence comes from a structure database. (For a review of protein structure, see chapter 4, *An Overview of Molecular Biology*.) The phrase to the left indicates the type of structure feature denoted. Source gives the source of the protein and the amino acids that were analyzed (1 to 496).

Region denotes a domain. The first Region entry states that amino acids 1 to 117, together with 163 to 386, constitute domain 1. The second Region entry states that amino acids 118 to 142 constitute domain 2, the catalytic domain. The third and last Region entry identifies amino acids 387 to 496 as domain 3.

SecStr denotes secondary structure. These entries identify regions of alpha helices and beta sheets. The beta sheets and helices are numbered in order. Bond indicates a bond; the bonds indicated are all disulfide bonds. Het denotes bonding to an ion or other cofactor. Human pancreatic amylase binds to a calcium ion.

The GenPept entry ends with the protein sequence.

Questions

1. Who published the structure of human pancreatic amylase? When did they publish the articles, and in what journals?

2. What is a disulfide bond? Is a disulfide bond a strong or a weak stabilizer of protein structure?

3. How many disulfide bonds are in human pancreatic amylase?

4. If possible, print the protein sequence of human pancreatic amylase. Using information from the GenPept entry, find the amino acids involved in the disulfide bonds, and draw a connection between the positions that are bonded together.

Viewing the Three-Dimensional Structure of Human Pancreatic Amylase

1. At the top of the GenPept screen for human pancreatic amylase 1HNY, select Structure from the pull-down menu of databases with the word "Search" beside it. Type 1HNY into the text box and click Go.
2. On the next screen, click on the link to 1HNY. This takes you to a summary page for the protein structure. Click on View 3D Structure.
3. Compare the structure with the structural information from GenPept. Play with the viewing options to get the most useful labels (none, every amino acid, every fifth amino acid, and so on) and the most useful display format (space filling, wire, and so on). Rotate the structure and zoom in on various areas. See if you can locate the calcium ion. See if you can locate the disulfide bridges.

Part 3: Exploring online biomedical literature references

Part of the responsibility of a scientist is publishing your results so that other scientists can use your data and check your work. There are literally thousands of different journals in which scientists can publish their work. Most of these journals focus on specific areas or types of research.

Before the existence of the Internet, it could be quite a challenge to find published biomedical information. You would have to go to a science library, for one thing. Once you got there, it could take hours of work to find some good references, and even then you were not sure you had found all the important ones.

After the Internet was established, the National Library of Medicine revolutionized the lives of researchers by creating an online database of the biomedical literature. Called MEDLINE, it allowed anyone with a computer and Internet access to search thousands of journals with the click of a mouse.

MEDLINE gives you reference information about an article and usually includes a short English language summary of its contents. These summaries are called *abstracts*. Nearly all biomedical journals are now published electronically, and subscribers can access the complete articles directly.

This activity is an introduction to the world of biomedical literature. You will learn how to search the literature, and you can go back and look for information at any time in the future. If you become interested in a particular topic, you can search occasionally to see what new information has been published.

Procedure

1. Go to the NCBI home page (http://www.ncbi.nlm .nih.gov) and use the pull-down menu to select

PubMed from the search menu. Enter "human amylase" in the search window. Note that PubMed appears in the dark bar above the search menu. If you click on this instead, you will get a slightly different screen from the one described below.

2. The next screen shows that over 9,000 documents were identified. We need to make the search more specific to get down to a manageable number of documents. Add the term "salivary" to the search text. Click on Go.

3. Now the number is down to over 1,200. Add the term "gene" to the text box and click on Go.

4. Now over 80 documents are identified. Add the term "evolution" and click on Go.

5. Addition of the last term should reduce the number of hits to nine (this was in October 2006; there may be additional references now).

6. Click on "The remarkable evolutionary history of the human amylase genes" by Meisler et al. and read the abstract.

Questions

1. Where did Meisler and Ting work at the time they published this article? When did they publish it?

2. According to the abstract, how many tandem copies of the human amylase gene are there?

3. What is responsible for the change in tissue specificity from pancreas to salivary gland?

4. What do the authors cite as evidence that there is strong evolutionary selection for amylase to be expressed in saliva?

Challenges

1. Remember that in chapter 31 you tested several substances for the presence of amylase. One substance we asked you to test was dog saliva, which tested negative. By searching the biomedical literature, find out whether dogs have an amy-

lase gene. Can you find why their saliva tested negative for amylase activity?

2. Has anyone ever published an article about using amylase sequences as molecular clocks? If so, give the citation. What did they do, and what did they find?

PART III
Societal Issues

*S*ocietal issues raised by biotechnology, or any technology, are introduced by using background information, readings, and classroom activities designed to provide tools for analyzing difficult issues rationally. The complex interrelationships among scientific understanding, technological development, and societal structure are described and elaborated on in chapters that provide processes for analyzing biotechnology's societal impacts, assessing technology's risks and benefits, and conducting productive, thoughtful analysis and discussion of ethical dilemmas.

Introduction to Societal Issues

Parts I and II provide information on science and technology, or more specifically, on the science of genetics and the technologies that are outgrowths of a scientific understanding of cell and molecular biology. In Part III, we introduce a related body of knowledge that is increasingly relevant and important in today's world: the nature of science and technology—their history, interrelationship, and societal impacts.

In the past, societies often embraced new technologies without making much of an attempt to analyze their possible ramifications. In contrast, modern biotechnology began generating a great deal of discussion about the potential applications—from genetically engineered crops to recombinant microorganisms to gene therapy—at the earliest stages of research, long before any products entered the marketplace and had societal impacts. Debates and discussions about the societal effects of modern biotechnology have continued unabated throughout its development.

This debate is healthy and appropriate, for modern biotechnology will provide societies with novel capabilities. Like all technological applications, even the beneficial applications of biotechnology will be accompanied by some costs. To use the new capabilities provided by biotechnology wisely and to assess costs and benefits rationally, citizens need to be somewhat conversant with the science underlying biotechnology, the applications of the science, and their potential impacts. In addition, they should understand the role of government in encouraging scientific research in some areas but not others and, as a result, driving technological innovation down certain paths and not others. As a teacher, you can play a critical role in shaping the direction of technology development by helping future citizens learn how to grapple with societal issues attendant on science and technology and to appreciate the policy dimensions of technology development.

The background information and critical-thinking tools they learn will enable students to participate in productive discussions.

We begin this part with an overview of the interdependent relationship of science, technology, and society, as well as a reading on the nature of science. Each of these ingredients influences the others and is in turn influenced by them. We then provide tools and information for rational analysis of some of the policy issues and concerns associated with biotechnology that can be addressed through objective analysis of scientific facts. Chapters 36 and 37, on rational analysis of the societal impacts of technology and science-based assessments of technology's risks and benefits, teach skills that will enable students to conduct critical analyses and to participate productively in discussions and debates. Background information on a variety of issues discussed in the context of modern biotechnology—gene flow, cloning, embryonic stem cells, the safety of biotech foods, and harm to nontarget organisms—ensures the discussions are well informed.

Science-based technology assessment methodologies were developed to minimize the contribution of emotion to policy making, when a concern is one that can be addressed with science and data analysis. However, many questions associated with technology development and policy making cannot be answered solely by analyzing technical information. In chapters 38 to 40, we take up highly charged and contentious ethical dilemmas. Even though scientific information alone will not provide answers about moral choices, science can inform these discussions, and rational analysis of these issues is not only possible, but essential. To facilitate productive and informed discussion, we have provided a decision-making model for addressing the bioethical issues that arise in two scenarios relevant to gene-based biotechnology: genetic testing for inherited disorders and gene therapy to treat genetic diseases and to enhance characteristics.

Finally, two additional readings, *Biotechnology Regulation in the United States* and *Science, Law, and Politics,* describe in detail some of the ways in which governments influence the direction of scientific research and encourage or discourage technology development.

Although we have set this section apart from the classroom activities, we suggest you incorporate discussion of the societal impacts into the lessons on a specific technology or scientific discovery, if it is at all feasible.

Societal Issues

Science, Technology, and Society 35

About the Activities in Part III

We cannot overestimate the importance of conducting classroom activities that encourage students to think about the complex interrelationships among science, technology, and society (Figure 35.1). While very few students will pursue biology as a career, all of them will be citizens in a society shaped by science and technology. If they respect the power of these fields to influence society for better or worse, then perhaps they will become concerned citizens, involved in ensuring that science and technology are directed toward maximizing societal benefits and minimizing costs.

However, more than emotion and concern is required for guiding wise deployment of science and technology. Thoughtful technology development and use hinge on each individual's ability to identify the core issues; subject the facts to thoughtful, open-minded analysis; be detached from self-serving perspectives; and compromise with others who may have different priorities. It is not important that your students learn *what* to think about science and technology. It is critical that they learn *how* to think about them. Without these mental tools, students will be vulnerable to interest groups that manipulate through misinformation.

Figure 35.1 Science, technology, and society. In today's world, the relationship among progress in science, technology, and society is circular, not linear. Changes in one lead to changes in the others.

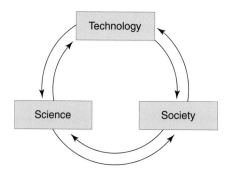

Classroom activities on the interrelationship between science, technology, and society are by their very nature somewhat limited. There are no laboratory techniques to perform and few biology concepts to learn. Instead, a teacher's task is to teach students essential critical-thinking skills: identification of issues, fact-finding, objective analysis, identification of options, and priority setting. Even though societal issues may be the vehicle for teaching thinking skills, they are precisely the same skills that scientists rely on daily and that students must develop for success in any endeavor.

In addition to providing students with skills for analyzing societal issues and participating productively in discussions about technology development and use, we also hope that these activities provide you with a platform for impressing upon students the essential role each citizen must play in a democracy. Whenever possible, try to use examples that demonstrate how interest groups have effectively directed scientific research and technology development down certain paths and not others. It is also very important for students to learn that resources are limited, even in a country as wealthy as the United States. As a result, every choice to pursue a certain area of research or technology development means that others are neglected.

Finally, even though science and technology have been among the most important factors influencing Western civilization, they are rarely given the attention they deserve in social studies classes. Working on the activities in Part III in conjunction with social studies teachers and students would be enriching for all.

Introduction

Science and technology have altered the characteristics of societies throughout history. Some of the forces they have exerted have been revolutionary, and some have been minor. Different aspects of life—social, cultural, ecological, and economic—

have been affected, and different groups have been affected in different ways. Often, one group has benefited at another's expense.

For many years, societies enjoyed the fruits of technology without fully considering the implications. Eventually, people recognized that almost every intended benefit of a technology carried with it an unintended cost: medical advances increase the average life span, as well as the cost of health insurance and the national debt; agricultural advances have provided an abundant food supply and increased leisure time, leading to an increase in obesity and heart disease.

When citizens in industrialized societies began questioning the cost-benefit ratio of technology development, the unintended and unexpected costs that seemed to capture the public's imagination most were harmful impacts of technologies on the environment. Many people began to equate technology with environmental destruction, and for a while, it seemed as if sentiments about technology development had swung from inattentive acceptance of the benefits to cynical disillusionment with the costs. Even today, some people continue to reflexively associate technology solely with adverse impacts on the environment.

Technology does have a profound effect on our species' relationship with the natural world. That, after all, is its intent. Humans use technology as insulation against the realities of nature, and some technologies have had harmful effects on the natural environment. However, other technologies—including some that are often criticized as environmentally destructive—have environmental benefits. For example, chemical herbicides that are used to control weeds also decrease soil erosion because farmers no longer must till the soil for weed control.

Because science and technology have given humans tremendous power over the natural world, it is essential that societies make far-sighted, informed, and judicious choices so they use that power wisely. As the above example of herbicides and soil erosion demonstrates, neither blind acceptance nor unthinking rejection qualifies as acting wisely. Thoughtful development of technology requires rational, dispassionate analysis of a technology's costs and benefits. In addition, because the benefits and costs of technological change are often distributed unequally among different groups, wise and fair technology development also depends upon involving many different people and interest groups. Unfortunately, people increasingly feel powerless to influence the direction of science and technology development; instead, they see themselves solely as recipients of decisions made by others. Consciously or unconsciously, they realize that they have become increasingly dependent on science and technology as their understanding of both has decreased. This paradoxical circumstance generates uneasiness at best, and the tendency is to abdicate influence and control to others, which is both a natural and an unwise response.

Even though it often feels like science and technology are in the driver's seat, each society plays a powerful role in determining the direction of scientific progress and technological change. Societies are not simply on the receiving end of decisions about science and technology made by other, unseen forces. People decide, either actively or passively, the strength and nature of the scientific and technological forces that affect their quality of life. Only certain areas of scientific research are pursued, and certain technologies out of all possible technologies are developed. The best technological solutions to problems may or may not be the ones that are developed and succeed in the marketplace (Figure 35.2).

The agents guiding technology development along certain paths but not others include economics, ethical values, government policies, market opportunities, consumer preferences, and, in democracies, public opinion. Because so many different forces can alter the course of science and technology development, different segments of society with divergent interests have the power to influence the trajectory of scientific research and technological innovation. Therefore, when citizens in democracies feel powerless and relinquish their influence over decision makers, others will fill that void. Those who utilize their power come out as winners, and those who surrender their power typically lose out.

Objectives

After reading this chapter and completing the activities you select, students will:

- Realize the impact technologies have on their daily lives.
- Understand the interrelatedness of science, technology, and society.
- Understand that any technology has many first-order and second-order impacts; some are intended and others are unintended.
- Appreciate the many factors that influence technology development.

Societal Issues

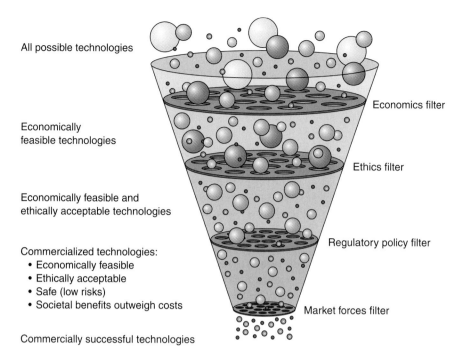

All possible technologies

Economically feasible technologies

Economically feasible and ethically acceptable technologies

Commercialized technologies:
- Economically feasible
- Ethically acceptable
- Safe (low risks)
- Societal benefits outweigh costs

Commercially successful technologies

Economics filter

Ethics filter

Regulatory policy filter

Market forces filter

Figure 35.2 Technology development. Not all technologies that are both scientifically and technically possible make it all the way through the development process and become commercialized. They must pass through a series of filters, created by society, before they become reality. The order of the filters in the figure does not necessarily reflect the actual sequence of barriers every technology confronts during development, except the final filter, market forces. Only certain technologies that are commercialized succeed in the marketplace.

Preparation

Review the *Student Activity,* which contains information not included in the teacher's pages, and the *Reading, The Nature of Science.*

Make copies of the *Student Activity* if the students do not have the student edition of the textbook.

Procedures

Although we have organized the book so that the societal issues raised by science and technology are relegated to a separate part of the text, in a classroom setting, treating the issues as integral to the scientific topic under discussion reinforces the messages that they are interrelated. For example, if you are teaching advanced-placement biology, conduct discussions of the environmental impacts of technology while you are teaching ecological principles. Using the framework of science, technology, and society as mutually reinforcing (Figure 35.1), explain to students how technologies that have a negative impact on the environment can lead to scientific research that assesses impacts and can encourage the development of other technologies that minimize the negative impacts. However, this chain of events is likely only if a segment of society cares enough about the negative environmental impact to apply pressure on policy makers. That pressure can then lead to government funding for new technologies.

Possible formats

A variety of formats create opportunities to develop and practice critical-thinking skills: case studies, research papers, debate, risk-benefit analysis, problem solving, decision making, role playing, brainstorming, team-based research, and classroom discussions. You will know which format is best for the students in your class.

Decisions about whether to have a class discussion or a debate among teams should be dictated by the topic. Usually, the goal of a debate is winning, not necessarily increasing the understanding of those who are observing the debate. In order to win the debate, the participants often omit facts, use inflammatory rhetoric, or skew facts in a certain direction. As a result, debates can confuse untutored audiences more than they enlighten. Therefore, if your goal is greater understanding by all students of all facts and appreciation of all sides of an issue, then a debate format probably will not serve that purpose. Debates, however, can help students analyze facts and develop cogent arguments.

The primary goal of a classroom discussion should not be winning but greater understanding by all parties. Satisfying and productive discussions must be centered on objective facts and not irrational fears or other emotions. The participants should be as clear, specific, and emotionally detached as possible.

Societal Issues

Methodology

Irrespective of the format chosen, it is critical that you help students learn how to:

- Consider one issue at a time.
- Stay focused on the issue.
- Gather accurate information (too often, students believe that if something is in print, it must be true).
- Distinguish science from pseudoscience.
- Assess information critically and unemotionally.
- Draw conclusions.
- Discuss issues fairly and unemotionally, listening to each other so that they acquire the greatest understanding.

Continually remind students that they will learn the most from people whose ideas and opinions differ from their own. Everyone is more inclined to listen open mindedly to arguments and facts that coincide with their own opinions while automatically rejecting others, but doing so is a sure recipe for ignorance.

If the format you choose involves class discussions or debates, have the class establish ground rules. Examples of ground rules for class discussions include the following:

- Everyone must participate.
- Listen to everyone's facts.
- Be specific.
- If necessary, ask others to define and clarify.
- Try to understand each other's points of view.
- Remember, the goal of class discussion is not winning, but greater understanding.

Consider posting the list of ground rules developed by the students somewhere in the classroom so that students can remind themselves and each other of the appropriate behaviors.

Resources

People tend to idealize life before technology primarily because they see it as a simpler time. Some people also seem to assume life was safer then. PBS and the BBC produced a number of series that did an excellent job of showing that life before technology was anything but simple and safe. Titles included *1900 House, Frontier House* (1880s Montana), *Colonial House* (1630), and *1940's House*. Like the activities described below, the resources provided by PBS and the BBC (DVDs, videotapes, websites, and lesson plans) are useful for classes in history and social studies.

Suggested Activities

1. Have students make a list of all of the technologies that influence their activities during a day. Also, have them list all the ways in which they interacted directly with the natural world on that day.

The PBS website on technology at home (http://www.pbs.org/wgbh/aso/tryit/tech/) lets visitors determine how much time they spend on computers, listening to CDs, watching TV, etc. The site also allows people to go back to the beginning of the 20th century and determine when different technologies appeared and how they affected life at home. A related web page (http://www.pbs.org/wgbh/aso/tryit/doctor/) shows the ways that physicians in different periods in the 20th century would diagnose and treat diseases.

2. Have students rank several technologies according to the degree of importance in their daily lives, and then ask them to choose those they could eliminate to reduce consumption of energy or financial resources.

3. Have each student or student team pick a major technology—antibiotics, electricity, or television—and thoroughly analyze its development.

   - What scientific discoveries permitted the development of this technology, and how did they come about?
   - What resources were required to develop the technology?
   - What sort of government bodies oversaw its development?
   - Did that oversight help or hinder development?

Now, ask the students to thoroughly analyze the effects (first order and second order; predicted and unintended; environmental, political, economic, and cultural) of that technology in hindsight.

- What were its intended effects?
- Has it achieved that purpose?
- What were the broader changes in society caused by this technology that we could have predicted would occur?
- What were the unintended consequences we could not have predicted?
- How has the technology changed human activity, our socioeconomic structure, our environment, and our relationships with each other, our government, and other nations?

Societal Issues

- What would life be like without this technology?
- What other technologies did it give rise to?

4. Have students or student teams pick an emerging technology or one application of an all-encompassing technology, like biotechnology. Ask them to:

A. Predict what some of the intended and unintended consequences of its development will be.

- Who will be affected by it, and in what ways?
- What will be the environmental, ethical, and socioeconomic effects of this technological development?
- What sorts of policies, laws, and regulations should be formulated to oversee its development?

B. Explain what scientific knowledge, existing technology, or other resources are required in order for this technology to be fully developed.

C. Prioritize the technologies that have been discussed (first individually and then working as a group).

- Which should be developed first?
- Which should be supported with the most federal funding?

5. Have students discuss the factors that should be used in evaluating technologies. Historically, technologies have been evaluated by three criteria:

- Can it be done?
- Will it sell?
- Is it safe?

Ask your students if they believe these criteria are sufficient. Would they add others? What would their criteria be? How would they implement these new criteria?

6. Assign each student a technology, and ask the student to determine whether that technology has increased society's options.

Technologies are supposed to increase options. You can now choose to be vaccinated against whooping cough or risk getting the disease. You can choose to fly or drive to Atlanta. But do technologies actually increase or decrease available options?

7. Have students visit a senior citizens' center and ask the people there what life was like before antibiotics, the pill, washing machines, or refrigeration.

8. Write a brief essay explaining the following quote by Edward Wenk: "Those who control technology control the future."

9. Ask students to analyze media coverage of a certain scientific breakthrough or technological innovation and contrast the media coverage with primary references or other information sources known to be accurate and thorough.

Selected Readings

Altheide, David L. 2002. *Creating Fear: News and the Construction of Crisis.* Adline Gruyter, New York, NY.

Ausubel, Jesse H., and Hedy E. Sladovich. 1989. *Technology and Environment.* National Academy Press, Washington, DC.

Bennett, W. Lance. 2003. *News: the Politics of Illusion.* Addison-Wesley, New York, NY.

Bronowski, J. 1956. *Science and Human Values.* Harper & Row Publishers, Inc., New York, NY.

Bronkowski, J. 2006. *The Common Sense of Science.* Harvard University, Cambridge, MA.

Burke, James. 2000. *Circles: Fifty Roundtrips through History, Technology, Science, Culture.* Simon and Schuster, New York, NY.

Cardwell, Donald. 1995. *The Norton History of Technology.* W. W. Norton & Co., New York, NY.

Daniels, George H. 1971. *Science in American Society: a Social History.* Alfred A. Knopf, New York, NY.

Jasanoff, Sheila. 2005. *Designs on Nature: Science and Democracy in Europe and the United States.* Princeton University Press, Princeton, NJ.

Postrel, Virginia. 1999. *The Future and Its Enemies: the Growing Conflict over Creativity, Enterprise and Progress.* Free Press, New York, NY.

Sclove, Richard, and Steve Fuller. 1995. *Democracy and Technology.* Guilford Press, Edinburgh, Scotland.

Scott, James C. 1998. *Seeing Like a State: How Certain Schemes to Improve the Human Condition Have Failed.* Yale University Press, New Haven, CT.

Snow, C. P. 1993. *The Two Cultures.* Cambridge University Press, Cambridge, England.

Tenner, Edward. 1997. *Why Things Bite Back: Technology and the Revenge of Unintended Consequences.* Vintage Books, New York, NY.

Societal Issues

Facing the Forces: Science, Technology, and Society

Introduction

Take a minute and try to imagine a world without technology—in the broadest sense of the term, not simply computers and cell phones. You will soon find it is virtually impossible, because technology is woven so tightly into the fabric of your life and your being, you cannot extricate yourself from its influences. Trying to assess the impact of technology on your life resembles a fish stepping back and objectively observing water. Not only does the fish's total dependence on water make this impossible, but the essence of "fishness" is inseparable from the properties of water.

You may resist the idea that technology has penetrated every tiny facet of your life so deeply that you have become technology dependent and are as incapable of living in the natural world as a fish on land. If so, then pay attention to your actions during the rest of the day and notice the many ways technology, both simple and sophisticated, has shaped your world. Some impacts—televisions, cars, and microwave ovens—are immediately obvious. Others are so subtle and pervasive you might not think of them as technology but instead take them for granted as a "natural" part of life. The clothes you are wearing, the chair you are sitting in, the pen you may be holding, the paper this book is printed on, and the food you had for lunch are all products of technology.

As you are paying attention to the ways technology shapes your life, also look at the flip side of the coin. How often do you come face to face with nature? Probably not very often, because humans rejected the idea of living with nature thousands of years ago. Look around the room to see if you can find anything that is natural, such as a spider building its web in a corner or mold on the cheese you wanted for lunch. How do you feel about these calling cards from nature? If you are like most people, you are not pleased they invaded your life. This distaste for the adversities nature indifferently dispenses has driven

society's constant quest for new technologies. People use all technologies for the same purpose: to change the environment so that the natural world suits us better.

Technology and society: a two-way street

Technologies not only alter the environment; they change people. A person's view of the world, the questions they ask, the way they think, and their sense of values, hopes, and fears are all products of life in a world shaped by technology. Through its effects on people, technology changes society as a whole, always in unpredictable ways. Even though technological innovations develop as solutions to specific problems and needs, they also provide doors to unforeseen innovations and unexpected societal changes. As a result, once a technology is introduced to society, predicting the nature and speed of all subsequent technological developments and societal effects is impossible. How could the inventors of the automobile ever guess it would lead to the development of other, extremely sophisticated technologies for extracting and refining oil and, ultimately, to the international petroleum industry with all of its wealth and thousands of products, from plastics to pesticides? Nor could they have possibly known the automobile's social impacts would include fragmented family units; the creation of suburbia, shopping malls, and industrial parks; and, ultimately, the economic decline of some major urban areas.

Beneficial technologies that meet legitimate societal needs can also have negative societal impacts, some of which are also unpredictable. The automobile developers probably expected its widespread use to carry certain risks, such as cars colliding with each other, with horses, and perhaps even with pedestrians. But they never imagined automobiles would cause global air pollution and concerns about climate change from global warming. To complicate matters further, often the risks associated with a technology become apparent only after it has infused society so completely that lives and economies depend on it and rejecting the technology is not feasible. Once society finally realized driving cars causes serious environmental problems, the car was so central to the structure of society and the economy that giving it up was not a viable option. The best society can do when it realizes an essential technology causes problems is devise ways to minimize its negative impacts, which typically means developing another technology, such as the catalytic converter in the case of auto emissions and air pollution.

Even though it may feel like the relationship between technology and society is a one-way street with technology in the driver's seat, each society plays a powerful role in determining its own technological profile. Only certain technologies become realized, because various factors, from technical impediments to politics, encourage the development of certain technologies but not others.

Has it always been this way? Yes, in principle. As soon as humans started fashioning tools from stone, technologies began to influence human societies and our species' relationship to the natural world; on the other hand, societies have always nudged technology development down certain paths but not others. Today, however, technological advances are more pervasive and their effects more potent, and the societal changes they bring occur at an increasing rate. As a result, the impact of technology on our personal lives and societies has greatly intensified. The increased presence, power, and rate of technological change can be traced to the role scientific understanding plays in propelling technology development.

Science and Technology

In today's world, many people equate science with technology, but the two differ in very fundamental ways (Table 35.1). The concept of technology encompasses the practices and products humans

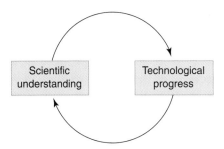

Figure 35.1 The relationship between progress in science and technology development is circular, not linear, so changes in one lead to changes in the other. In addition, their relationship is also reciprocal: technology is as important to scientific advance as scientific understanding is to technological innovation.

develop to modify and control nature for sustenance and comfort. As such, technology predates the birth of science by many centuries, and technologies can exist in the absence of science. Early humans created technologies, such as agriculture and tool making, empirically through trial and error and independent of any understanding of the laws of nature—the domain of science. Manual craft skills, such as lens grinding and metal forging, enabled the creation of tools that allowed scientists in the 1600s and 1700s to observe nature more precisely, so initially, technological tools facilitated scientific progress by opening the door to precise observations and, eventually, scientific experimentation.

As science-based understanding of the natural world broadened and deepened, science and technology began to converge, and progress in one began to drive advances in the other (Figure 35.1). Equipped with a more sophisticated understanding of the world they wanted to control, people began using scientific understanding to drive technology development. In addition, they replaced their empirical and relatively inefficient approach to developing technologies with the methodical, experimental approach of science. Buttressed by both scientific

Societal Issues

Table 35.1 Fundamental differences between science and technology

Science	Technology
The search for knowledge	The practical application of knowledge
A way of understanding ourselves and the physical world	A way of adapting ourselves to the physical world
A process of asking questions and finding answers, then creating broad generalizations	A process of finding solutions to human problems to make our lives easier and better
Looks for order or patterns in the physical world	Looks for ways to control the physical world
Evaluated by how well the facts support the conclusion or theory	Evaluated by how well it works
Limited by our ability to collect relevant facts	Limited by financial costs and safety concerns
Discoveries give rise to technological advances.	Advances give rise to scientific discoveries.

knowledge and methods, the technologies became more effective. With better technologies available to them, scientists could probe deeper into the causes and effects of natural phenomena, which in turn gave rise to new, improved technologies. This science-based technology development led to the increasingly complicated modern technologies that people typically envision today when they use the word "technology": software, DVD players, pesticides, and antibiotics and vaccines.

The positive feedback loop between scientific understanding and technological innovation drives a constantly accelerating rate of technology-related change. This does not simply mean that scientific knowledge increases and leads to more technology, but the rate at which knowledge accrues and technologies develop also increases (Table 35.2). The acceleration of scientific innovation and technological change has important implications for individuals and their governments. Societies struggle to cope with the current breathtaking rate of technological advance, and because of the circular relationship between scientific understanding and technological innovation, the pace will quicken continuously in the future. Will government policies that are intended to minimize the negative social impacts of technologies manage to keep pace with the galloping rate of change?

Science, technology, and society: a three-way street

Science-based technology development accelerates economic growth through its effects on industrial productivity. New technologies create new products, stimulate the creation of new companies and even new industries, improve existing products and processes, and lower manufacturing costs. They also provide industrial researchers with tools and techniques for discovering new products.

During the last century, technological knowledge, or in other words human capital, and physical capital (machines and infrastructure elements, such as roads and widespread electrification) began replacing land, labor, and natural resources as the primary drivers of economic growth. Governments that invested in scientific research, technology development, infrastructure requirements, and education and also created policies and institutions to encourage science and technology development have profited from their investments. The overall economic well-being of their citizens has improved, as have longevity, infant survival rates, and quality of life. The division of the world into industrial nations and developing countries can be explained in large part by the role science and technology have played in stimulating economic growth in the former but not the latter.

However, as described above, advances in technology have unexpected negative effects on society. Some people argue that the economic and social benefits that science-based technology development has provided to some of us are not worth the costs, particularly to the environment, paid by all of us. They link environmental degradation directly to modern technology development. While it is true that some technologies have had negative environmental effects, some of which have been very serious, the relationship between technological advance and environmental quality is more complicated than the simplistic view that technology ruins the environment. New technologies also improve environmental quality and prevent environmental problems. Most people in industrialized nations associate the synthetic chemicals used by farmers solely with environmental pollution, but herbicides also decrease soil erosion, the major contributor to water pollution, while fertilizers and pesticides increase yields, and in doing so, they prevent deforestation. When it comes to technology and the environment, the situation is never as simple as we would like it to be.

Not only do science and technology affect society, but societies also affect science and technology development. Most people would readily agree that decisions to develop certain technologies and not others are often driven by societal factors, such as politics, economics, and cultural values. In other words, the development of certain technologies instead of others is value laden. Fundamental, value-based assumptions drive the quest for more and better technologies spawned by ever-increasing scientific knowledge. One assumption is that science and technology result in progress. A second is that humans should dominate nature.

What most people are not aware of is that science is influenced by these same societal factors of politics, economics, and cultural values. The most obvious influence is the source of research funding. Scientists need money to conduct their research. Most of their

Table 35.2 Time lapse between technology introduction and widespread use[a]

Technology	Infiltration time (yr)
Electric lights	80
Personal computers	16
Internet	4

[a]The speed with which a technology infiltrates a society decreases over time. The measure of infiltration is held constant at 50 million users. (Source: Organization for Economic Cooperation and Development, Paris, France.)

funding comes from government grants, but significant research monies are also provided by companies, nonprofit organizations, special interest groups, and research institutions. All funding sources favor certain research questions over others, and scientists will choose questions that have the greatest chance of being funded. Often, politicians decide that a certain research area deserves more funding, and scientists shift their research accordingly. The politicians may have made their decision based on the needs of all citizens or only a handful. Thus, just as only some technologies are developed from the universe of possible technologies, only certain scientific questions, out of all possible questions, are asked.

Nonscientists view science as an objective search for the truth, and in many ways it is. Honoring the scientific process—systematic observation, generalization construction, and repeated testing—researchers attempt to establish their generalizations objectively and to gather facts through detached, accurate observation and careful, methodical experimentation. The scientific community respects the integrity of the scientific process, and on the whole, scientists do their best to keep that part of the research process value free.

Nevertheless, scientists were raised in society, and that society has shaped the way they view the world. The scientific process cannot shield scientists from the social context in which they conduct research. Society has molded their sense of values and goals since the day they were born. The narrower social context of the scientific world in which they function imposes pressures and erects powerful filters for selectively viewing the world. Facts and findings are interpreted through these filters, and they also play a key role in determining which questions are asked. Consequently, even if scientists could fund their own research, only certain questions from the universe of available questions would be asked. Therefore, while the methodology that scientists use to answer questions may be as objective as possible, the questions asked and the interpretation of the results are not value free.

Science, Technology, and Civic Responsibility

While science and engineering may define the set of potential biotechnology products that are both scientifically possible and economically feasible, a number of factors determine which of those product ideas become realities. One of those factors is you.

Your essential role in technology development may come as a surprise to you. Often, people who are not scientists or engineers feel like technology is in charge and they are powerless to influence the direction of technology development. Although it can often feel like technology is in charge, you have the power to influence technology development. Your power extends well beyond your purchasing power to determine which products on the market become commercial successes. Democratic societies give their citizens the power to influence *every* step in technology development and commercialization (Figure 35.2), including the following:

- Scientific research priorities
- The stringency of the product regulatory approval process
- Postapproval regulatory requirements
- Where and how the product is manufactured
- Product entry into and removal from the marketplace
- Other government-controlled factors that can affect a product's commercial success, such as product labeling

Citizens in democracies exert that power directly, by voting on these issues, or indirectly, by electing representatives and pressuring them to vote in certain ways.

As is clear from this quote by Thucydides, author of *The Peloponnesian Wars,* in Greek democracies the public viewed itself as the best judge of matters that affect the public:

Our ordinary citizens, though occupied with the pursuits of industry, are still fair judges of public matters; for, we alone regard the man who takes no part in public affairs not as one who minds his own business but as good for nothing. We Athenians are able to judge all events, and instead of looking on discussion as a stumbling block in the way of action, we think it an indispensable preliminary to any wise action at all.

Thucydides, 405 B.C.

This philosophy may have worked well in Greece in 400 B.C., but is it appropriate to expect the citizens of pure or representative democracies to be "able to judge all events" in this high-tech world? Can discussion actually *become* "a stumbling block" rather than "an indispensable preliminary to wise action" if the citizens are no longer "fair judges" because they are ill informed? Worse yet, can widespread misinformation lead to unwise action in democracies, and if so, how can this be prevented?

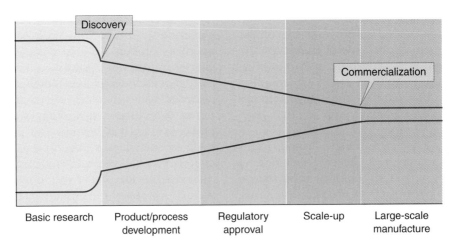

Figure 35.2 Stages of technology development. Basic scientific research leads to discoveries that give rise to ideas about possible technological solutions to problems. For technical and economic reasons, only a few of those ideas become realized as possible products or processes. Of those products and processes that are feasible, only some receive government regulatory approval. Additional attrition of potential technologies occurs during scale-up, because a sufficient amount of product at a saleable price cannot be produced.

A founder of our form of democracy had very clear ideas about how to solve the problem of a misinformed or uninformed public with the power to influence policy makers. He believed the solution could be found in providing information and not in removing citizens' power.

I know of no safe depository of the ultimate powers of the society but the people themselves; and if we think them not enlightened enough to exercise their control with a wholesome discretion, the remedy is not to take it from them, but to inform their discretion by education.

Thomas Jefferson, 1787

The idea of informing citizens so they "can exert their control" over policy makers constructively might have worked well in 1787, but does Jefferson's solution seem a bit antiquated in a world that is so dependent on science and technology? Many of you might feel that his solution is obsolete in today's world. How can someone with very little background in science be expected to make constructive contributions to science and technology policy issues when the meteoric pace of scientific discovery actually precludes scientists in one area of research from understanding discoveries in other areas?

In response to feeling overwhelmed by the speed of scientific progress and technological innovation, the natural tendency for most people is to surrender their power to others rather than to educate themselves. However, opting out on decisions related to the direction of scientific research and uses of technology is a very unwise course of action. While scientific investigation and technological advance may be the province of a few members of society, all citizens in a democracy should be prepared to consider the societal issues spawned by a science and its derivative technologies. In the absence of public understanding and interest, the stage is set for a handful of individuals to direct the course of scientific investigation and technological change and therefore, over time, society's future. The reality is that citizens have the power to influence policy makers in democracies, and that will not change. If you do not exercise your power, someone else will, and there is a high probability they are thinking about what is best for themselves, which may or may not be what is best for you.

Knowledge really is power, so understanding something about two of the primary forces that make today's world go round—science and technology—makes people less vulnerable to manipulation by others. When your car won't start, knowing a few simple things about how internal combustion engines work gives you a bit of protection against unscrupulous auto mechanics, doesn't it? In addition, when a decision involves personal health and well-being, people need to feel they are acting in their own best interest, and increasingly, that means understanding something about science and technology.

This text tries to address the problem by providing you with some basic information about a science, biology, and its derivative technology, biotechnology, both of which will have a great impact on your life in the coming years. Having this information should increase your confidence about making wise decisions in your personal life and "inform your discretion" as you attempt to influence policy makers. But to be perfectly straight with you, we have to admit we can never provide you with all the relevant facts about biology that you will need for making informed decisions, because they change on a weekly basis.

Another one of our goals is to help you understand how science is done (see the *Reading, The Nature of Science*), which may be the most valuable tool we can provide, because that knowledge will *never* become obsolete as science advances. Knowing how science works means learning how scientists think. If you learn how to think like a scientist, you will give yourself the best possible defense against misinformation and manipulation by special interest groups. To think like a scientist, you must be willing to ask questions repeatedly, without being emotionally invested in a certain answer, and then base your conclusions on evidence, not preconceived notions of what ought to be true. You also need to be equally skeptical of all those who are answering your question until the preponderance of evidence makes it clear that one answer is more correct than others. Finally, and probably most surprising to those who think science has all of the answers, you need to be comfortable with ambiguity, uncertainty, and being wrong—a lot.

In democratic societies, the public is a potent force shaping scientific research, technology development, and, ultimately, society's future. Over the course of your life, you will be tempted to let others influence the decisions that will determine which scientific research is conducted and, therefore, which technologies are developed. Dismissing decisions about science and technology as someone else's concern makes it easy for a handful of people to control the most powerful forces shaping the modern world. If only some members of society exert pressure on the political system, then scientific research priorities, the nature of the technologies that are developed, and the uses of those technologies will reflect the priorities of special interest groups and not those of society as a whole.

One of the penalties for refusing to participate in the political process is that you end up being governed by your inferiors.

Plato

Civic responsibility and information sources

Your instructor will ask you to complete one or more activities related to understanding the relationship among science, technology, and society. The primary goal of the activity will be to help you to learn how to analyze issues objectively, or, in other words, to think like a scientist, even if the question under consideration is not a scientific one. We cannot overemphasize the value of critical-thinking skills in all aspects of your life. Smart decision making depends upon your ability to assess information critically and unemotionally, even if the issue you are grappling with triggers strong emotions.

The activity that is assigned to you will be grounded in fact-finding. To analyze issues and make good decisions, citizens need access to accurate information. If you want your decisions to be grounded in sound science, go to primary sources, such as scientific journals or scholarly articles, or to review papers in publications like *Scientific American*. Talk to scientists conducting relevant research. Do *not* rely on news magazines, newspapers, television, or hearsay.

Surveys show that people rely primarily on print and broadcast journalism for information on science and technology. Most reporters believe in the importance of balanced reporting and do their best to meet this criterion for good journalism, but balanced information does not necessarily mean accurate information. In addition, the necessary information for accurately assessing science applications and issues cannot be crammed into a sound bite or a short story in a newspaper. Finally, the media are an industry that needs to sell products to stay in business. The popular media have decided that the attributes that increase their sales are fear and controversy. As a result, their stories often overemphasize risks. In addition, in some stories, they imply there is a controversy within the scientific community when no controversy exists. For these and other reasons, the mainstream media have become a poor source of information on science and technology. The media can alert you to issues, but to formulate thoughtful opinions, you need to seek out other information sources.

Finally, be particularly careful about information you gather from the Internet. Some websites are excellent sources of factual information, but others are full of inaccuracies and opinions. A general guideline in using the Internet might be to use information from primary sources, such as journals, universities, professional scientific societies, and public research institutions, like the National Institutes of Health, the U.S. Department of Agriculture, the National Academy of Sciences, or the National Science Foundation.

Societal Issues

The Nature of Science

Introduction

How often do you read a newspaper article about a scientific discovery that contradicts last week's discovery and think, "Why can't scientists make up their minds?" Contradictory reports can be especially maddening when the debate is related to personal choices you make about your health and well-being. As you read these reports and become increasingly frustrated and distrustful, scientists reading the very same contradictory newspaper articles are thinking, " Hmmm, that's interesting." Why the difference? You might assume the scientists' blasé attitudes stem from knowing a truckload of scientific facts not included in the story, but rarely is that the reason. They are calmly accepting the contradictory reports because they know how science works.

If you are like most nonscientists, you probably have an idealized view of science that is a natural and almost subliminal outgrowth of the way that you were taught science in school. Given the large amount of material teachers must cover, they need to present science as a set of eternal laws, absolute truths, and uncontested facts that represent the final end point of a linear process in which each discovery led, rationally and obviously, to the next.

If only it were that easy! Science is much, much messier than is implied by a retrospective listing of sequential discoveries building logically one upon the other and culminating in a "scientific truth." Your teacher could not share the messiness with you, because describing the mess and mistakes would have consumed every minute of class time, leaving no time to explain the discoveries. Unfortunately, learning about the discoveries and not the mess creates a utopian view of the scientific process and a belief that every scientific discovery represents a concrete, immutable stepping-stone in the path of discovery leading to an infallible truth. Contradictory newspaper reports belie this myth. As is always true with utopian concepts, accidentally glimpsing clay

feet or learning about the mess in the closet makes people recoil from their idealistic belief with a ferocity that equals the strength of their embrace. When it comes to science, neither cynical disillusionment nor blind acceptance is in your best interest, so learning that every scientific finding is always tentative and open to reinterpretation or outright rejection will help you tolerate contradictory results.

Putting scientific facts in perspective

Do you remember when you realized the quickest and most surefire way to figure out how to get the bunny through the maze to the carrot patch was to start at the carrot patch and go backwards? That is how science is taught. Knowing the current thinking about a certain scientific topic, your teacher described the most direct path of discovery that got scientists to that point (the carrot patch).

But that is not how science is conducted. Scientists must enter the maze where the rabbit does, unsure of where they are headed and uncertain how to get there. As a result, the history of science is filled with blind alleys, negations of previous "breakthrough" discoveries, results repudiated in one era that become cornerstones of major scientific theories in another, invalid (even ridiculous) ideas accepted solely because a great scientist proposed them, and seemingly disparate findings made compatible by another finding or a new, improved theory capable of accommodating them all. But don't think of the blind alleys, incorrect theories, and repudiated results as "mistakes" simply because they are not findings that served as stepping-stones on the path to discovery. Even negative results contain useful information; at a minimum, they tell scientists they are on the wrong track.

Doing science is like trying to assemble a jigsaw puzzle without having the picture on the front of the puzzle box. Scientists discover where they were

Societal Issues

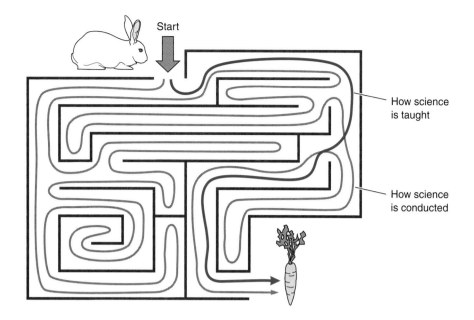

Start

How science is taught

How science is conducted

headed only after they get there. It is only once they arrive that they finally know what the picture on the puzzle box looks like. Once there, they turn around to see how they got there, and the path is so obvious it lights up like a yellow brick road—in retrospect. Interestingly, once scientists have arrived where they were headed, they keep testing whether they are where they think they are, just to be sure.

Scientists report their findings in scientific journals, which are of little interest to most nonscientists. The reported findings include the blind alleys, repudiated results, and invalid ideas described above, though the scientists reporting them do not know their findings are wrong or irrelevant at the time they are reporting them. Recently, however, newspapers have *also* begun announcing scientific findings that might excite their readers. The reporters reading the journal articles and writing the stories are ill equipped to understand where a certain finding fits in the scheme of things. Is it a blind alley in the maze or a key stepping-stone in the path to a breakthrough discovery? A piece that belongs in a different puzzle box? Two puzzle pieces from the correct box that do not seem to fit together? Or is it that magical puzzle piece that makes all of the other pieces lock into place?

The messiness of the scientific process that used to occur outside of the public's view is now splashed across the front pages of major newspapers. It is the scientific equivalent of having neighbors watch a family that is airing its dirty laundry so the family members can understand each other better.

Science is a process

Science is not simply a collection of facts, findings, and theories; it is an ongoing, methodical, iterative process with a specific goal: explaining and understanding the physical world *by way of* facts, findings, and theories. The first step in the scientific process is to gather facts, but only those facts that are relevant to the question or problem being studied are gathered. To a scientist, facts are observations of events or material objects in the physical world. Once the facts are gathered, the scientist, using inductive reasoning (from specific to general), attempts to order them by fitting them together and making a general statement that encompasses all of the facts. For example, a generalization that would provide order to a set of your observations of the natural world is "All birds have feathers and can fly."

Then, the process is turned on its head. The next step in the scientific process is to assume the generalization is valid and, using deductive reasoning, make a prediction—an "if/then" statement—based on the generalization. For example, "If sparrows, eagles, and doves are birds, then they have feathers and can fly." You check and see that indeed they do have feathers and can fly, so your generalization is strengthened by your additional observations. Then, one day you come upon a nest of newly hatched birds. Not only are they featherless, they also cannot fly, so you revise your generalization accordingly. For example, your new generalization might be "All adult birds have feathers and can fly." Later, you take a trip to Australia and New Zealand and see two new bird species, emus and kiwis, neither of which can fly

Societal Issues

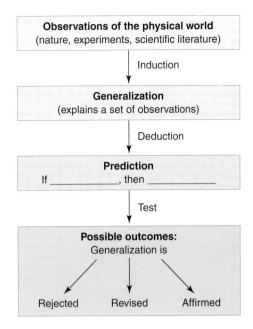

Observations of the physical world
(nature, experiments, scientific literature)

Induction

Generalization
(explains a set of observations)

Deduction

Prediction
If _____, then _____

Test

Possible outcomes:
Generalization is

Rejected   Revised   Affirmed

even though both have feathers. You modify your generalization once again, and now it might read "Most adult birds have feathers and can fly" or "All adult birds in North America have feathers and can fly." However, as a scientist, you would know that this generalization, like the ones before it, is always open to new observations and endless revisions.

After reading this explanation of the scientific process, you probably realize that you follow the very same thought process in your personal life. You order some of your observations into a generalization and work from the generalization until it is proven false. Sometimes, you may even test your generalization to make sure you have got your head on straight. If facts do not fit into your generalization, you may revise your generalization slightly to accommodate the new observations or throw it out completely and start again, but now you have many more facts to incorporate into a generalization, so your new generalization will probably be more accurate than the first one. Thus, the scientific process is not magical, nor is it so demanding that only a few geniuses can master it.

If you think this way every day, does that make you a scientist? If not, then what distinguishes science from your thought processes? One distinguishing

trait is the type of information that can be used to create the generalization. In your personal life, you might let hopes, prejudices, or politics creep into your generalization in addition to facts. Also, you might ignore certain facts that do not fit into your generalization, because you are just sure it is right or you want very badly for it to be true. The only facts a scientist can use in constructing a generalization are observations of the physical world. Scientific observations cannot be subjective, because they must be verifiable by others. Scientists also are not free to ignore facts because they do not fit into their generalizations.

The nature of the generalization may differ, as well. A generalization drawn from your personal life can be subjective and might include a value judgment. In addition, when all of your observations support your generalization, you probably decide it is valid and do not continually revisit the issue. In science, the generalization must be amenable to testing, using the "if/then" process; it also must be open to contradiction by anyone, anywhere, at any time. That is, it must be refutable.

These essential attributes of scientific generalizations are key to understanding what science can and cannot prove. Science can prove a generalization is *not* true, but it can never prove a generalization *is* true. Why? Because scientists know they will never be able to observe all of the relevant facts. In the example given above, if you had never gone to Australia and seen emus, you would continue to think your generalization that all adult birds have feathers and can fly was valid. Therefore, the phrase "scientifically proven to be safe" is inherently false. In addition, if someone says society should never approve the use of a new technology until it is proven to be safe, they are actually saying society should never approve any new technology.

Science accepts a generalization as valid when observation after observation supports it, but that acceptance is tentative and conditional, so nonscientists should not interpret acceptance as proof. This explains why scientists who are quoted in the newspapers are always qualifying their statements. They understand the limitations of science; they also understand that the public does not.

A Framework for Rational Analysis of Issues

About This Activity

This textbook attempts to equip students with information about biology so they can contribute constructively to discussions about biotechnology and make informed decisions in their own personal lives. However, because of the breathtaking pace of scientific progress in the biological sciences, no one can stay abreast of discoveries that will affect human health and well-being. When this reality hits home, letting others make decisions can be tempting, but it is never a smart choice. In this case, it is also not necessary, because making decisions that are in your best interest often depends more on asking the right questions than on having an encyclopedia of facts in your head. Asking the right questions has as much to do with attitude—detached, rational, and open-minded—as with information.

Public debates about biotechnology's applications and impacts provide opportunities for practicing the critical-thinking skills and emotional detachment that are essential in assessing societal issues of technology development. Debates that play out in the media are often driven by issues that have little, if anything, to do with science, such as concerns about misuse of genetic information or the industrialization of agriculture. However, in order to give their opinions credibility, proponents of certain positions will sometimes cherry pick information from the body of scientific evidence to support their stances. This is especially true for agricultural and environmental applications of biotechnology. Those who are opposed to biotechnology pick facts with the highest "fear factor," because they know that controversy plus fear is the perfect prescription for media coverage, which is often their primary goal. On the other hand, proponents of biotechnology sometimes gloss over real issues, and biotechnology companies, like all companies, hype their products. Therefore, analyzing debates about biotechnology issues helps students learn to separate scientific fact from fiction, to appreciate the importance of context to scientific data and analysis, and to develop a certain level of skepticism toward claims made by all interest groups.

In this chapter, we provide an approach and process for analyzing some of the issues associated with biotechnology that are often "debated" through the media. These will assist you and your students by providing the following:

- A few tips about *how* to think about societal issues derived from biotechnology applications
- Generic guidance on identifying and evaluating the real issues to avoid being sidetracked by claims intended to instill fear and anxiety on one hand and unthinking acceptance on the other

We also provide a number of examples of how these critical-thinking skills can be applied to analyzing complicated issues that arise from biotechnology. Through these examples and others, students will review a great deal of biology as they learn how to objectively analyze issues. For example, the issue of gene flow from transgenic crops provides an opportunity to revisit plant reproduction, natural selection and evolution, reproductive isolating mechanisms, and genetic drift.

Two important reference points that you must reiterate continually are the historical and technological continuum that has led to biotechnology and the costs of *no* technology compared to the costs of technology. Because of the way the human mind works, students will continually drift to biotechnology's risks without placing them in context or considering the risks of no new technologies. Your task will be to refocus them repeatedly by asking them to consider (i) what, if anything, is unique about biotechnology and (ii) whether the potential risks of a future technology are greater or less than the known risks of a current technology.

Because the methodology, which is described in more detail in the *Student Activity,* is generic, some steps in the analysis may not apply in all cases; in others, more steps may be necessary. The most important lessons students should take away from these exercises are the essential importance of conducting their own analysis of issues, a commitment to asking questions of all interest groups, and the tone of the analysis: fact based, objective, methodical, specific, detached, and open to all options. In the absence of rational and detached analysis, citizens are susceptible to manipulation through misinformation when they are trying to make decisions about biotechnology development and use.

Finally, by establishing the factual evidence underlying many of the concerns about biotechnology presented in the media that is omitted from news accounts, students will learn that today the popular media, more often than not, create more confusion than clarity about science and technology policy issues. If that understanding is the only thing students take away from these exercises, you will have done them a great service.

Introduction

Biotechnology has remarkable potential to improve our lives and the environment, but that potential can also be abused. While this concept creates anxiety in some, it is not new, nor is it unique to biotechnology. Societies have been faced with the duality of technology use and misuse ever since early humans began to fashion crude tools from stone.

Because the development of biotechnology is still in the first phase of the life cycle pattern that modern technologies exhibit (Figure 36.1), society is at a crucial stage in deciding how best to use biotechnology's potential and prevent its abuse. The early stages of technology development present some of the richest opportunities for thoughtful, productive discussions about how technology could maximize the benefits offered to the greatest number of people. However, the early stages also represent vulnerable times for nascent technologies. Mistakes made early in development are more likely to have undesirable repercussions, pervasive effects, and unintended consequences than those made when a

Figure 36.1 Technology life cycle. In 1981, Arthur D. Little, Inc., developed a simplified model of the economic impacts of a technology during its life span. In the emerging stage, product and process research and development expenditures are high, as is the uncertainty of the technology's economic success. Product commercialization initiates the growth phase. As the technology-based product or process is adopted by various industrial sectors, product sales increase and process improvements reduce costs. In the mature phase, the technology is well accepted, and sales are stable. The cost reductions from process improvements based on the technology have reached a plateau. In the final stage of a technology's life span, growth and acceptance of newer technologies (dotted line) displace the older technology.

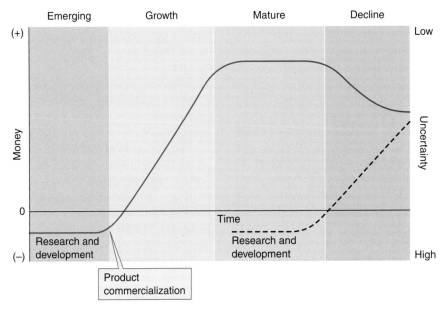

technology is well established. As a result, decisions that governments, scientists, companies, activists, and the public make during these early stages help to set biotechnology on a course that will determine how much of its potential will be realized and who its beneficiaries will be.

Biotechnology's remarkable flexibility generates an enormous number of options for improving existing products and processes and creating new ones. Which products will be developed from the universe of possible products and solutions, and who will be involved in making that decision? Hand in hand with biotechnology's flexibility comes its power and, therefore, a responsibility to make farsighted and informed choices about how these technologies will be used. Where will society draw the line in using biotechnology's power? Will the line be drawn there through active choice or passive acceptance? Who will decide where the line is drawn? Just as important, who will draw the line and ensure that society honors it?

Many perspectives need to inform the discussion about appropriate uses of biotechnology, because like all technologies before and after it, biotechnology will change society in many ways. To ensure that the discussions and decisions are well considered, they must be based on critically evaluating the issues attendant on biotechnology development.

Analyzing Issues

Basic approach

A productive approach to analyzing any issue associated with biotechnology begins with the simple task of placing modern biotechnology within the context of other technologies. Medicine, agriculture, and energy production are age-old attempts to improve human health and control the environment with technology. Biotechnology now gives humans a new set of technologies for shaping the world to their liking. The tools may be different, but the goals remain the same: to improve human health and alter the environment so that our lives and those of future generations are as long and easy as possible. Viewing biotechnology from this perspective provides reference points that help people ground their analysis of the issues. Those having very little, if any, understanding of science can still appreciate that modern biotechnology is one of many technologies humans have developed. Rather than learning about the science underlying biotechnology, they need only compare these technologies to previous technologies.

Unfortunately most public discussions and debates about societal issues raised by biotechnology—especially those carried out through the media—do not begin with a clear articulation of relevant technological or historical context. Observers are left with the impression that these technologies appeared suddenly, with no historical precedents, and that the issues raised by biotechnology are unique to biotechnology. A clear description of the technological ancestry that led to biotechnology cannot substitute, by itself, for responsibly discussing and analyzing the issues raised by biotechnology development. However, *not* incorporating this history into debates, discussions, and decisions while continuing to perpetuate the misconception that the issues are new and unique does a disservice to people who are struggling to make smart choices about biotechnology's implications.

Therefore, throughout these activities, one of your most important tasks will be to repeatedly remind students to think of biotechnology within the context of other technologies. Using other technologies as reference points can be surprisingly difficult, because people are so accustomed to the world in which they live that they tend to think of it as "natural." As a result, only new technologies tend to elicit concern. Asking someone to accept the idea that the unknown risks of future technologies might actually be fewer or less severe than the known risks of existing technologies is easy in theory but difficult in practice.

Knowing that today's biotechnologies are the next step in a continuum of technologies has important, real-world implications for charting its course. Even if people have little knowledge of biology, if they use historical context as a lens for viewing biotechnology issues, it is possible for them to:

- Delineate how the new biotechnologies resemble and differ from earlier technologies.
- Get a better handle on the changes biotechnology might bring and determine which changes are significant and why they are significant.
- Define the set of societal issues specific to the unique capabilities provided by biotechnology.
- Decrease uncertainty about potential impacts on the environment and human health.
- Evaluate the relevance of existing laws and policies, precipitated by earlier technologies, to the products and societal issues of biotechnology.
- Use past mistakes with earlier technologies to develop strategies for doing a better job with these technologies.

Societal Issues

- Compare the safety/risk ratio of these technologies to those of previous technologies for solving the same problem.
- Assess the social costs and benefits of this technology compared with those of past activities directed toward the same goal.

Another important task, related to the one above, is to continually ask students to consider the costs of *no* technology. As we mentioned in the last chapter, in the second half of the 20th century, the general public became aware that using technology can carry hidden costs. As a result, those are the costs associated with technology development that their minds gravitate toward. In doing so, they ignore the costs of no new technology. Just as it is difficult to willingly accept new risks that are unknown and yet to be experienced over those that we live with daily, grasping the concept that the societal costs of no technology are greater than the risks and costs of a technology does not come easily either. While the hidden costs of using a technology are often hidden for many years, more deeply hidden from certain societies are the costs of no new technology. The costs of no technology are difficult to assess, because those costs are usually unknown to people who have lived in a world in which technology has solved certain problems. When comparing the costs of technology and the costs of no technology, they overestimate the cost of having the technology because they have never experienced the costs of not having the technology. This is especially true for people in the industrialized world. The costs of no new technology for people with a high standard of living pale in comparison to those for people in the developing world who are trapped in the poverty cycle. As we discussed in the last chapter, although technology development in the industrialized world has had environmental and social costs, its beneficial legacy includes a comparatively healthy, well-fed, educated, and affluent citizenry.

We provide more information on the costs of no technology in the *Student Activity*.

Analytical process

In addition to the mindset described above, there are also some critical steps that need to be followed in objectively evaluating societal issues associated with an application of biotechnology. In general, proceed very methodically and purposefully in:

- Delineating the issue and defining terms as specifically as possible. Ambiguity is the enemy of rational analysis and greater understanding. Clarity and rigor are essential elements of productive analysis and constructive discussion at all points in the process, but they are especially important as a first step. If they are neglected at the outset, the value of all subsequent steps is diminished.
- Gathering factual information from objective sources. Avoid the mainstream media, which has become a very poor source of information on science and technology. Be particularly careful about information gathered from the Internet. Some websites, Listservs, or discussion groups are excellent sources of factual information, but others are full of inaccuracies and opinions.
- Placing the technology and issue in context. Determine relevant contextual facts, such as the historical development of the technology and similar technologies. Gather information on other technologies used to solve the same problem or meet the same need.
- Assessing what is unique about the technology and what is similar to other technologies currently used. Does it introduce new (science-based) risks to human health and the environment?
- Comparing important attributes of the technology to those of other, similar technologies with respect to its safety, specificity, predictability, widespread availability, economic costs, etc.
- Assessing the benefits of using the technology compared to those of similar technologies for the same purpose, assuming they exist. Another frame for assessing the benefits of the technology is to ask about the costs of not developing the new technology. In other words, are the risks/costs of the current technologies for solving the problem greater than the risks/costs of the new technology?
- Determining any other societal costs that cannot be assessed or measured through scientific or technical analysis, i.e., issues of equity, morality, emotional harm, ethical dilemmas, economic disruption, etc.

This list should not be treated as dogma and followed unthinkingly. Often, many of the steps in the process are not sequential but are intertwined and iterative. Some steps may be appropriate for some technologies and some issues but not others. More important than following every step is the posture and tone that the students should assume when conducting the analysis, which is objective, rational, and, importantly, not emotionally attached to a certain position or answer.

Societal Issues

Objectives

After reading this chapter and completing the activities you select, students will have:

- Learned and utilized a process for objectively analyzing issues related to science, technology, and society
- Practiced critical-thinking skills
- Learned about specific applications of biotechnology in agricultural and medicine and the issues they raise
- Reviewed basic biological concepts in evolution, population genetics, and reproduction and development in plants and animals

Preparation

- Review the *Student Activity,* because it contains information not included here.
- Make copies of the *Student Activity* if students do not have the textbook.

Procedures

If your students are like most of the people whom we have taught, they have misinformation about certain biotechnology issues firmly fixed in their minds. Therefore, because so much of what is required for rational analysis of the societal issues associated with modern biotechnology depends on gathering accurate information on a number of different topics, the most useful formats, initially, are case studies and research papers, including team-based research. Only after the fact-finding part of the analysis is complete are classroom discussions and presentations appropriate. If you open the floor for discussion before students conduct a significant amount of research, little useful discussion will occur and misinformation will be spread. Alternatively, opening the floor for discussion could be the starting point for the fact-finding mission. Creating a master list of the group's conceptions and misconceptions about biotechnology would provide a class-generated set of research topics.

We suggest that you avoid a debate format. There is so much misinformation on biotechnology in the public consciousness that having debates on these sorts of issues only exacerbates the problem. We very much want these activities to correct misinformation, not give it new life. Because the goal of a debate typically is winning, the debaters "spin" the facts to support their position.

Another option that does not entail as much student research and therefore saves time involves you leading a classroom discussion and sharing the factual information provided below as appropriate. You might combine this approach with having students collect information on the topics from newspapers and websites. In this way, they can see the disparity between the correct, factual information and the methodical approach that you provide and information from other sources that is incomplete, inaccurate, or inflammatory. Impressing upon them the need for fact-finding rather than relying solely on mainstream media or the Internet would be an invaluable lesson.

Tips

Depending on the nature of your course, students could conduct the issues analyses when you are teaching the biological principles and concepts that are germane to the issue. For example, the question of ethical issues and the use of human embryonic stem (hES) cells can be taught in the section on human reproduction and development, while the issue of evolution of resistance to pest control measures is relevant to both evolution and community ecology.

You may need to be relentless in asking students for clarity and specificity and challenging why they say certain things. Lazy thinking and sloppy use of terms are so much easier than rigor, and all of us are tempted to let things slide and assume everyone knows what we mean. This natural tendency is exacerbated when we *want* to accept, unthinkingly, one side of a debate because of a built-in prejudice against some interest groups or positions that differ from our own. In our experience, "nagging" may come close to describing the behaviors you might be forced to exhibit to get students to question their assumptions, define terms, and clarify their thoughts.

You might consider posting certain questions to help students maintain their focus, such as the following:

- Is this issue unique to modern biotechnology?
- If modern biotechnology disappeared, would this problem or issue still be there?
- What other technologies have been developed to solve the problem or meet the need? What has our experience with that technology taught us?
- Is there any other well-established technology that raises the same issues or poses the same safety risks? If so, how does biotechnology compare to that technology?

Societal Issues

- What are the benefits of the technology? Are they equally distributed?
- Will society (or the environment, economy, etc.) be better or worse off if society decides not to develop this technology?
- Does society need this technology? Should society spend its limited resources on a different technology?
- Is there a cost to *not* developing the technology?

Not all of these questions may be relevant to the topics you choose to discuss, and many other questions could be added to the list.

Suggested Activities

Virtually any issue that has received media attention as a "biotechnology issue" is not unique to biotechnology and therefore can be analyzed quite productively using the strategy and process described above and in more detail in the *Student Activity*. Below, we provide background information on two of the issues that you might assign to your students: ecological impacts of gene flow from transgenic crops, which are often referred to colloquially as GMOs (for "genetically modified organisms"), to wild plants and ethical issues related to hES cell research. Analysis of a third issue—cloning—is described in the *Student Activity*. Other issues that you might consider include the evolution of resistance to pest control methods in agriculture; the "naturalness" of transgenic crops or any other GMO humans have created; harm to nontarget organisms, such as the monarch butterfly; allergenicity of transgenic food crops; and the use of transgenic plants and animals to produce therapeutic compounds and other useful molecules.

As described above and in the previous chapter, there are a number of possible formats that can be used to teach students *how* to think about societal issues. You will know the best way to help the students in your class develop critical-thinking skills. Irrespective of the approach you select, your emphasis should be on helping them learn how to ask good questions or frame the analysis accurately rather than on providing them with information that they memorize and parrot back to you. For example, in the example below on gene flow from transgenic crops, rather than using the information as the basis of a lecture, during class have students consider all of the things that must happen for gene flow to occur. Then, under cross-pollination, ask the students to create the list of factors that affect the probability that pollen from a crop will fertilize an egg in a wild plant.

Issue: adverse impacts of gene flow

The use of transgenic crops (i.e., "genetically engineered crops," or crops that have been genetically modified with recombinant methods) has triggered questions about adverse impacts caused by the transgene moving from the crop to other, nontransgenic plants, both wild plants and crop plants, via pollination. The task for the students is to take the broad, vague issue "adverse impacts of gene flow" and identify those circumstances under which the issue has relevance.

The take-home messages from the analysis of this issue are best conveyed by assigning different crops and different traits to student groups. Transgenic crops currently on the market in the United States are corn, soybean, canola, cotton, papaya, and squash; the traits incorporated through recombinant DNA technology are herbicide tolerance, insect resistance, and disease resistance. We suggest that you limit detailed analysis of crops to corn, soybeans, and canola, which are widely grown internationally so that abundant information is available. In addition, decades before the development of recombinant DNA technology, farmers worldwide had been growing crops with the same phenotypic characteristics as the transgenic varieties (herbicide tolerance, insect resistance, and disease resistance) and genetically modified by other methods. Transgenic cotton is also grown on millions of acres both in the United States and in a number of developing countries. Because there is a pervasive misconception that large-scale farmers in industrialized countries, especially the United States, are the only farmers growing and benefiting from transgenic crops, the transgenic-cotton story is quite interesting. Small-scale farmers (farming less than 10 acres) in India, South Africa, and China have significantly increased their incomes with transgenic cotton that is resistant to certain insects (Bt [for *Bacillus thuringiensis*] cotton). However, cotton does not lend itself to an easy analysis of adverse impacts of gene flow because the genetics of the wild plants that are related to cotton are quite complicated. You might consider discussing transgenic cotton in class, if you are interested in correcting misconceptions about large-scale farmers and industrialized countries being the sole beneficiaries of transgenic crops. The website of the International Service for the Acquisition of Agri-Biotech Applications (http://www.isaaa.org) has excellent information about the use and impacts of Bt cotton in developing countries.

Define Terms and Delineate the Issue

The first definitional issue students must clarify is "genetically modified crop," genetically modified organism," or "GMO." In popular usage, all three terms have become synonymous with transgenic crops. This ambiguous word use causes confusion in and of itself, because it leads audiences to believe that genetic modification is new and unique to modern biotechnology. As explained in chapter 1, genetic modification of crops began as soon as the first agriculturists began domesticating wild plants. Make sure students understand the many different ways that farmers and plant breeders have genetically changed crops over the centuries. More information on genetic modification techniques is provided in chapter 37. During discussions and in their papers, please make certain students use terminology appropriately. One of the most important lessons you can teach them is the importance of using terms correctly and consistently.

Second, the moniker "the gene flow problem" encompasses two very different problems: gene flow from transgenic crops to wild plants and gene flow from a transgenic crop variety to a comparable non-transgenic, or conventional, crop variety. These two issues differ in all dimensions that are relevant to analyzing potential adverse impacts of gene flow—the probability that gene flow will occur; if it does occur, the nature of the adverse effect (ecological versus economic); and methods to lessen the probability of adverse effects or to mitigate them. Therefore, students must first establish—or you should establish for them—whether the issue they are analyzing is gene flow from crops to wild relatives or gene flow from transgenic to nontransgenic crops. We have provided the analysis of gene flow from crops to wild plants, because it is most consistent with many of the topics covered in biology classes. If, however, you live in a part of the country where students are familiar with agriculture, analyzing gene flow from transgenic to conventional crop varieties may be more interesting for the students, because the impacts are related to agricultural economics, trade, freedom to operate, and other issues that affect the agricultural community. Alternatively, having different groups analyze both issues for a certain crop can be quite instructive, because the contrast makes clear the importance of delineating, quite specifically, the issue being addressed.

For the remainder of the discussion, we focus solely on gene flow from transgenic crops to wild plants.

Gather Information

Students must first assess the possibility, and then the probability, of genes moving from a transgenic crop to an unintended recipient. Then, they must determine whether it matters if crop genes move to wild plants, because gene flow can occur without having adverse effects. If the answer to the first question—can gene flow occur—is no, then there is no need to go further. If the answers to the "adverse impacts" questions are no, then gathering more data is unnecessary, because in that situation, gene flow is not an issue worthy of analysis.

The answers to questions about (i) the possibility and probability of gene flow and (ii) any adverse impacts of gene flow will vary with the crop, the recipient plant, the environmental factors, and the nature of the trait encoded by the transgene. Therefore, to thoroughly assess the issue of gene flow, the students must evaluate each situation systematically, on a case-by-case basis, considering the crop, environment, and trait. First, however, we provide a framework for analyzing adverse impacts of gene flow before describing crop-specific concerns.

Gene flow probability. Gene flow from a transgenic crop to a wild plant depends on the potential for (i) cross-pollination between the transgenic crop (pollen donor) and the wild plant and (ii) successful hybridization.

The students' task is to determine not only whether gene flow is possible, but also if gene flow is probable. They do this by evaluating systematically the factors that affect the potential for each of the three steps required for gene flow—pollen transfer, fertilization, and viable-seed production—for the crop they are analyzing.

Cross-pollination. Cross-pollination includes both the transfer of pollen from one plant to another and subsequent fertilization of the recipient plant's eggs. If transgenic pollen simply lands on the stigma of the flower of the recipient plant, but fertilization does not occur, no gene flow can occur. Students must first determine when/if cross-pollination between the crop and a wild plant is even possible.

Crops and wild plants can potentially cross-pollinate only if the wild plants occurring near the crop field are closely related to the crop. Wild-plant relatives of a crop occur in a circumscribed geographic location known as the crop's center of diversity. Each crop has its own center(s) of diversity, which is usually, but not always, the crop's center of origin (Table 36.1). The center of origin is the geographic area where humans first domesticated the crop from its wild relatives. Therefore, if a transgenic crop is planted in an area containing none of its wild plant relatives, gene flow is not possible. For example, gene flow from corn to wild plants is possible only

Societal Issues

Table 36.1 Probable centers of origin for some important food crops

Center of origin	Crop
Central America	Maize (corn), common bean, papaya
North America	Sunflower, blueberry
Europe	Oats, cabbage, sugar beet
South America	Potato, peanut, sweet potato, peppers, tomato
Central Asia	Soybean, onion, alfalfa, peach, apple
Southern Asia	Asian rice, banana, sugarcane, citrus
Africa	Sorghum, pearl millet, African rice, cowpea, coffee
Middle East	Wheat, barley, rye, pea, lentil

in Mexico and Central America. In analyzing the impact of gene flow from corn to wild relatives in the United States or Europe, the analysis ends at step 1; no gene flow can occur, because no wild relatives of corn are found in the United States and Europe. On the other hand, the mere presence of a wild plant that is a close relative of the crop does not necessarily mean gene flow is possible, nor does cross-pollination inevitably lead to gene flow, because crops can interbreed successfully with only some of the wild plants that are their close genetic relatives. We discuss this in more detail below.

In circumstances where cross-pollination is possible, students must then determine if it is probable. For cross-pollination to occur, viable, mature pollen from the crop must reach the recipient's viable, mature eggs before other pollen fertilizes the eggs. Many factors affect the probability of successful cross-pollination, including the following:

- Physical proximity of the donor and recipient plants to each other
- Tendency of the plants to self-fertilize
- Pollen longevity, which can be minutes for some plants and hours for others
- Degree of synchronicity between pollen shedding of the transgenic crop and the wild plant's receptivity to fertilization
- Relative amounts of pollen produced by the two plants
- Pollination vectors of the two plants—wind, insects, or both

Hybridization. Finally, to assess whether gene flow can occur, students must also determine whether cross-pollination could lead to successful reproduction, or hybridization, which is the production of fertile offspring. Pollen transfer and successful fertilization are necessary, but not sufficient, for gene flow to occur, because there are also many postfertilization barriers to reproduction (Table 36.2). Gene flow

between crops and wild-plant relatives can occur only if postfertilization barriers are overcome and the resulting offspring (seeds) are not only viable, but also fertile. To successfully hybridize with each other (in the absence of human intervention), plants must be very closely related to each other.

To illustrate the necessity of looking at each crop on a case-by-case basis to assess the probability of gene flow, it helps to use specific crop examples.

Corn and the teosintes. The center of origin and center of diversity of corn are the same: Mexico and Central America. Gene flow from corn to wild plants is possible only in Mexico and Central America, because corn's ancestor and close relatives, the teosinte species, are restricted to those areas. Assessing the risk of gene flow from corn is unnecessary outside of Mexico and Central America.

Table 36.2 Reproductive isolating mechanisms in plants

Prefertilization mechanisms
  Populations live in different regions.
  Populations live in same regions but occupy different habitats.
  Populations live in same regions but are sexually mature at different times.
  Cross-pollination is prevented by differences in flower structure.
  Pollen does not germinate or cannot stimulate pollen tube growth.
  Pollen is unable to enter the ovule or sperm cannot fuse with egg.

Postfertilization mechanisms
  Zygotes are inviable and die soon after fertilization.
  Hybrids have early developmental problems: fruits do not form, and seeds do not germinate.
  Hybrid seeds give rise to weak plants that die prior to flower formation.
  Plants (F_1) that develop from hybrid seeds are sterile.
  F_1 hybrids are normal and vigorous, but F_2 hybrids are weak or sterile.

Societal Issues

Gene flow from corn to some, but not all, teosinte species can occur. In fact, botanists have been studying it for at least a century. A corn plant can either self-pollinate or cross-pollinate, and pollen dispersal is driven by wind speed and direction. (Information about pollen dispersal is readily available for all crop plants. Farmers who grow certified seeds need to know how far apart the fields must be to maintain genetic purity. These distances, known as isolation distances, are available on the websites of seed companies, such as Pioneer Hi-Bred Seed, and the commodity groups, such as the National Corn Growers of America or the American Soybean Association.) Because corn pollen does not stay airborne for very long, the probability of cross-pollination is directly related to the distance between the crop and cross-compatible teosintes and the number of teosinte plants near the cornfield (Figure 36.2). Cross-pollination can lead to successful hybridization, but

Figure 36.2 Corn and teosinte. In Mexico, the center of origin for corn or maize, a primitive variety of corn (upper left) readily hybridizes with its wild relatives, the teosintes (lower right). Hybrids are shown between the two parental types. (Photograph copyright Klaus Ammann, University of Bern, Bern, Switzerland.)

the success rate varies according to the direction of the hybridization. A number of studies have shown that genes are much more likely to move from teosinte to corn than in the opposite direction.

Soybean and wild soy relatives. The center of origin for soybeans is northeastern China, and its wild relatives are found only in central/northeast China, Taiwan, and certain regions of Siberia, Korea, and Japan. Therefore, gene flow from soybeans to wild plants is possible only in these areas. How probable is gene flow between soybean and its wild relatives? Not very. Soybean and its relatives are self-fertile and self-pollinated. Pollination typically occurs before the flower even opens, in which case, foreign pollen cannot reach the plant before it fertilizes itself. Cross-pollination, which depends on insects, and hybridization are possible, but not probable. In fact, plant breeders trying to maximize the probability of hybridization between soybean and its relatives by using controlled laboratory conditions have a difficult time creating hybrids.

Canola and its wild relatives. Gene flow from canola is more complicated for a number of reasons. First of all, there is no such thing as a canola plant. Canola is a special type of oil that is produced by any of three *Brassica* species, *Brassica napus*, *Brassica rapa,* and *Brassica juncea,* which are known as the oilseed brassicas. Interestingly, none of these species produced canola oil before plant breeders set to work on them in the 1970s, so there was no such thing as a canola crop until then. Because the question of gene flow now involves three crop species, the number of locations with wild relatives is greater. Various studies describe wild *Brassica* relatives in Canada, Australia, the United States, a number of countries in Africa, Central and South America, and the Mediterranean countries; there may well be more locations where wild relatives of *Brassica* occur.

The oilseed brassicas are both wind and insect pollinated. *B. rapa* is self-incompatible, so it must be cross-pollinated. *B. napus* and *B. juncea* are normally self-fertile, and on average, 80% of the seed arises from self-fertilization. *Brassica* pollen is smaller than corn pollen, but like corn, a large proportion of the pollen falls to the ground within a few feet of the field. Nonetheless, a small percentage (5 to 10%) becomes airborne and, under ideal conditions, may drift 2 to 3 miles away. Insects can transport *Brassica* pollen long distances, as well. All of these factors combine to make cross-pollination between the oilseed brassicas and their wild relatives more probable than it is for corn or soybean.

Cross-pollination does not constitute gene flow, however. Numerous scientists have studied the potential for successful hybridization among the three oilseed brassicas and various wild relatives. Most studies were conducted in the laboratory or under field conditions that maximized cross-pollination. In addition, researchers blocked the self-fertilization that usually occurs in the self-compatible brassicas. As you can see from experiments on only two wild relatives (Table 36.3), no simple generalizations can be made about the possibility of gene flow from "canola" to its many wild relatives.

Few studies representative of natural field conditions have been conducted to assess the *probability,* not the *possibility,* of gene flow from the oilseed brassicas to wild relatives. All show that hybridization can occur with some wild relatives, but it seems to be rare and the direction is asymmetric. For example, French researchers planted a wild relative (wild radish) throughout a *Brassica* field and along its borders, which maximized opportunities for cross-pollination. They found that 1 of 189,084 (5.3^{-6}) wild radish seeds contained a marker gene from the crop, while the genomes of 5 of 73,847 (6.7^{-5}) *Brassica* seeds revealed evidence of gene flow from wild radish to the crop.

Adverse impacts of gene flow. Gene flow (successful pollination plus fertile hybrids) does not necessarily lead to adverse environmental impacts. Your task now is to help students think through the ecological and evolutionary implications of acquiring a certain gene or trait.

Gene flow from certain crops to some wild plants has been occurring for centuries. In agricultural settings, gene flow might make a farmer's life more difficult if a weedy wild plant becomes a worse weed after acquiring a crop gene. However, an adverse impact on agriculture and an adverse ecological impact are not the same thing. We could not find any reports of crop genes transforming a wild relative into a plant that has a competitive edge, leading to an adverse ecological impact. Perhaps this has occurred, but plant scientists may not have studied it or reported it in the literature.

If genes from some crops have been moving to wild relatives in certain locations for centuries, then why hasn't gene flow led to ecological problems that are significant enough to be identified and investigated? After all, for more than a century plant breeders have intentionally incorporated genes into crops to improve crop survival and reproductive success, such as drought tolerance, disease resistance, insect resistance, and higher yields. Why wouldn't these same genes improve the reproductive success of wild plants, which in turn would have an ecological impact, such as displacing another plant species? Here are a few possible explanations for the apparent lack of serious ecological impacts caused by gene flow from crops to wild-plant relatives.

First of all, gene flow from crops to their relatives seems to be a rare event, even in those species and locations where the probability of gene flow is the greatest. The rarity of the event means that crop genes are likely to be lost from the wild plant population due to genetic drift. This is especially true if:

- Crop genes offer no selective advantage to the crop–wild-plant hybrid.

Table 36.3 Cross-pollination of the three oilseed *Brassica* species (canola), *B. napus, B. rapa,* and *B. juncea,* with their wild relatives, *Sinapsis arvensis* and *Brassica nigra,* under laboratory and field conditions

		Results	
Pollen donor	**Pollen recipient**	**Laboratory**[a]	**Field**[b]
B. rapa	*S. arvensis*	No seeds	No hybrids
B. napus	*S. arvensis*	No seeds	No hybrids
B. juncea	*S. arvensis*	No seeds	No hybrids
S. arvensis	*B. rapa*	No seeds	No hybrids
S. arvensis	*B. napus*	No seeds	No hybrids
S. arvensis	*B. juncea*	2.5 seeds/100 pollinations (all sterile)	No hybrids
B. rapa	*B. nigra*	No seeds	No hybrids
B. napus	*B. nigra*[c]	0.1 seed/100 pollinations	No hybrids
B. juncea	*B. nigra*[c]	0.5 seed/100 pollinations	No hybrids
B. nigra	*B. rapa*	No seeds	No hybrids
B. nigra	*B. napus*[c]	0.9 seed/100 pollinations	No hybrids
B. nigra	*B. juncea*[c]	3 seeds/100 pollinations	No hybrids

[a]Hand pollination.

[b]Cocultivation of canola crops and wild relatives for 3 years.

[c]Self-pollination of the two hybrids (*B. juncea* × *B. nigra* and *B. napus* × *B. nigra*) produced no fertile seeds.

- The crop–wild-plant hybrid can no longer interbreed with its wild parent, only with other hybrids.
- A crop gene that does offer a selective advantage, such as drought tolerance, is linked to other crop genes that handicap the hybrid. For example, every time corn hybridizes with a wild teosinte, the hybrid offspring loses its ability to disperse seeds, because all corn plants have the "no seed dispersal" trait. This places the hybrid at a distinct disadvantage, because populations of the wild parent plant disperse their seeds widely; in doing so, they increase their evolutionary advantage over the hybrid.

However, if the crop–wild-plant hybrid interbreeds freely with the wild-plant parent, the crop gene can become established *at a low frequency* in the wild-plant population even if the crop gene offers no selective advantage to the wild plant. Even so, this introgression of crop genes into a wild-plant population does not necessarily constitute an adverse ecological impact. Low levels of introgression have surely occurred between some crops and wild relatives for centuries. The key determinant of adverse effects is not simply the presence of a crop gene in a wild-plant population.

Adverse environmental impacts become much more likely if the hybrid plants not only interbreed freely with one another and the wild parental variety but also have an evolutionary advantage over the parental wild plants, i.e., wild plants with the crop gene survive better and reproduce more than wild plants lacking the crop gene(s). The potential of crop genes to increase individual fitness depends upon many factors, including the nature of the crop genes and the selective pressures acting on the wild-plant population. An increase in individual fitness does not necessarily constitute an adverse environmental impact, however. The potential of gene flow to cause adverse impacts is circumscribed by the ecological factors limiting the wild-plant population size at that time. For example, at first glance, a crop gene for insect resistance might appear to offer a competitive advantage (less herbivory) that would automatically lead to a population explosion of the wild-plant species that has acquired the new gene. But if the factor limiting the size of the wild-plant population is water availability or a plant pathogen and not insect herbivory, the wild-plant population will continue to be held in check by those factors irrespective of this new, advantageous gene for insect resistance. Acquiring a gene for insect resistance might improve the survival and reproductive output of certain individuals within the population; in response,

gene frequencies within the population could change. However, if the wild-plant population size stays the same, then has there been an adverse impact?

In recent experiments, scientists gave the Bt gene (conferring resistance to some insects) to wild sunflowers, and those plants produced significantly more seeds under natural conditions than plants without the Bt gene. If insect herbivory limits wild sunflower populations, then acquiring the Bt gene from cultivated sunflowers could lead to an ecological release of that population (depending on the type of insect that is limiting the sunflower population). If this occurred, would an increase in the population size of wild sunflowers by itself constitute an adverse environmental effect? Maybe the increase in sunflower seeds would increase the number of songbirds whose primary food is seeds. Is that an adverse impact? Why or why not? Maybe the increase in sunflowers would cause the extinction of a rare prairie plant, known to only a few botanists. Is that an adverse effect? Why or why not? We pose these questions not because they have a correct answer, but so that you can help students see the complexity of the issue. Science can help us assess the probability of gene flow, but it offers society no help in assessing the value of songbirds and rare plants.

In analyzing the potential for adverse impacts, we suggest that you focus on herbicide-tolerant crops and insect-resistant crops, because the number of acres of transgenic disease-resistant crops (squash and papaya) is quite small. Therefore, data on impacts are much more limited.

Place in Context
Gene flow can occur with any crop, whether it is a transgenic or a conventional variety. Indeed, evidence shows that gene flow between crops and their wild relatives has been occurring for centuries, long before modern agriculture appeared on the scene (Figure 36.2). Therefore, the answer to the requisite question "is the gene flow issue unique to modern biotechnology" is definitely no. In addition, gene flow occurs among many plants that are closely related but are not crop plants, including plants in different species.

Even though gene flow, i.e., cross-pollination plus successful hybridization, is not a phenomenon that is unique to transgenic crops, what about the question of adverse impacts? Adverse impacts depend, in large part, on the nature of the crop traits that are transmitted to the wild-plant population. Therefore, in determining whether transgenic crops pose unique risks compared to conventional crop varieties, the focus must turn to comparing transgenic crop traits to those of conventional varieties. To date, all of the phenotypic traits provided to crops through recombinant DNA technology are comparable to traits that have been incorporated through other genetic modification techniques, such as selective breeding and mutagenesis. Those traits include insect resistance, disease resistance, and herbicide tolerance. Therefore, if modern biotechnology disappeared tomorrow, the issue of adverse ecological impacts through gene flow from crops to wild plants would still exist.

Compare and Contrast to Other Technologies

Depending upon how far you want to carry the analysis, it can be very instructive to contrast transgenic crops to conventional crop varieties. If nothing else, this exercise erases the misconception that genetic modification of crops is new and unique to biotechnology. Perhaps more important, students are forced to learn about an economic sector on which they are totally dependent and about which they know almost nothing. In learning about the biological problems farmers must continually battle—plant disease, insect pests, and weeds—students also learn more about ecological principles and the structure and function of ecosystems.

When comparing and contrasting transgenic crops to conventional crops, consider assigning some groups the task of comparing genetic modification techniques that have been used over the past few centuries. Although most people commonly think of plant breeding as "natural," many of the techniques used by modern plant breeders bear no resemblance to natural methods of plant reproduction. During the second half of the 20th century, plant breeders developed increasingly sophisticated laboratory techniques for crossbreeding plants in different species and different genera. Table 36.4 provides a few examples of natural physiological barriers to wide crosses and the techniques plant breeders have used to overcome the barriers. Combining research on these unnatural techniques reinforces lessons about reproductive isolating mechanisms discussed in gene flow, because most modern plant-breeding techniques were developed in order to override natural barriers to reproduction.

Other student groups can focus on the nature of the traits and comparing transgenic traits to comparable traits in conventional varieties (i.e., herbicide tolerance) and other methods that farmers use to accomplish the same goal. For example, the gene from a strain of Bt has been engineered into corn plants to control caterpillars that are pests of corn. Prior to the development of Bt corn, how did farmers control those insects? We suggest that you assign the following crop-trait combinations to different student groups in order to maximize the learning experience:

- Herbicide-tolerant soybean
- Bt corn that is resistant to lepidopteran insects
- Bt corn that is resistant to the corn rootworm (*Diabrotica* species), a beetle
- Herbicide-tolerant canola

Table 36.4 Laboratory techniques for crossbreeding distantly related plants.

Barrier	Technique for overcoming the barrier
Prefertilization	
Failure of pollen germination	Remove pistil, then pollinate exposed end. Use recognition mentor pollen.
Slow pollen tube growth	Chemical treatment with organic solvents or growth regulators
Pollen tube growth stops.	In vitro fertilization Use of plant growth hormones and chemicals, like chloramphenicol and acriflavin
Failure to obtain sexual hybrids	Protoplast fusion
Different numbers of chromosomes	Chemically induce chromosome doubling.
Postfertilization	
Embryo abortion (immediate)	In vivo/in vitro embryo rescue/implantation
Embryo abortion (early stages of development)	Culture embryos in petri dishes.
Lethality of F_1 hybrids	Use cell culture to regenerate plants.
Chromosome elimination	Alter genomic ratios of species. Induce chromosomal exchanges with tissue culture or irradiation.
Hybrid sterility	Chemically induce chromosome doubling.

Societal Issues

Issue: ethical issues and embryonic cells

The availability of reliable sources of undifferentiated human cells, or embryonic cells, opens up new avenues for treating, curing, and even preventing human diseases and injuries for which neither effective therapies nor cures exist. Some people, however, object to using embryonic cells for a variety of reasons, most of which are ethical concerns. Using factual information and specific language in discussions of complex technical issues is always important, but it becomes even more essential if the topic elicits strong emotional feelings about moral issues. A structured analytical approach and emotional detachment are especially helpful in this regard. Before you broach questions about the ethical issues surrounding the use of human embryonic cells, you will need to provide students with background information and clear, specific definitions.

Define Terms and Delineate Issues

In November 1998, two different research groups simultaneously reported they had successfully maintained human embryonic cells in culture. One group used the inner cell mass (ICM) of blastocysts to establish hES cell lines (Figure 36.3). The other group derived embryonic cell lines from primordial germ cells that had been taken from aborted fetuses. Embryonic cells derived from primordial germ cells, which are the cells that develop into eggs and sperm in adults, are correctly referred to as embryonic germ (EG) cells.

The distinction between EG cells and ES cells is not trivial, because one of the ethical issues some people find troubling is the original source of human embryonic cells. Certain people object to therapeutic uses of hES cells because they are derived from blastocysts, and blastocysts can develop into a viable fetus if implanted into a human female. Other sources of embryonic cells raise issues that concern some people but not others. Therefore, it is important to delineate the issue of source quite thoroughly when evaluating ethical uses of human embryonic cells. Each source raises unique ethical questions that are more or less problematic to different groups of people.

Gather Information

Blastocysts consist of an outer layer of cells, the trophoblast, and the ICM. The blastocyst implants in the uterine wall 5 or 6 days after fertilization as the trophoblast cells begin to infiltrate the uterine lining. At day 5, the ICM consists of around 30 to 35 cells. After the trophoblast cells become integrated into the uterus, the ICM cells divide and differentiate into the three distinctive germ layers—ectoderm, mesoderm, and endoderm—found in the gastrula (Figure 36.4). Ultimately, the ICM cells develop into the fetus, and the trophoblast cells become the placenta. Even though ES cells or ICM cells can differentiate into any cell type found in the body, neither cell type can develop further if inserted into a female, because they lack the cells required for implantation into the uterine wall. Therefore, neither ES cells nor cells from the ICM could develop into a fetus.

During the earliest stages of human embryonic development, approximately 100 cells that will become the primordial germ cells can be found in one of the extraembryonic membranes known as the yolk sac. When the fetus is approximately 4 to 5 weeks old, the germ cells begin migrating, in an amoeboid fashion, from the yolk sac to the areas where the gonads will eventually develop. By the time the fetus is 6 weeks old, the primordial germ cells have reached their destination but are still undifferentiated. These cells, when extracted from the fetus, are human EG (hEG) cells.

Drawing on decades of stem cell research on other animals, scientists assume that hES and hEG cells maintained in laboratory cultures are pluripotent and

Figure 36.3 ES cell culture. To generate a culture of ES cells, researchers remove the ICM from a blastocyst. If placed into the uterus, neither cells from the ICM nor ES cells will develop into a complete organism because they cannot implant into the uterine wall. Trophoblast cells are required for implantation.

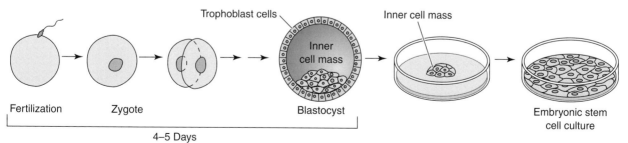

A

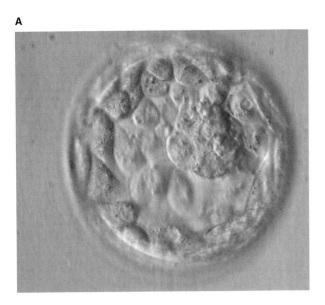

B

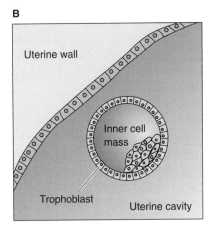

Uterine wall

Inner cell mass

Trophoblast

Uterine cavity

Day 5

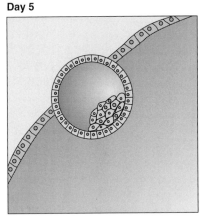

Day 14

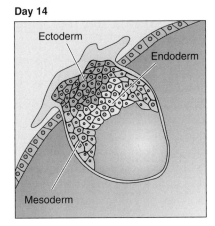

Ectoderm

Endoderm

Mesoderm

Figure 36.4 Early embryonic development. (A) A human blastocyst. (B) Schematic representation depicting human embryonic development from blastocyst to gastrulation. Five days after fertilization, the blastocyst begins to implant into the wall of the uterus. By day 14, gastrulation has occurred, implantation is complete, and the pluripotent cells of the ICM have already differentiated into ectoderm, mesoderm, and endoderm. Therefore, they are no longer pluripotent. The trophoblast cells differentiate into placental tissues. (Photograph courtesy of Michael Vernon, West Virginia Center of Reproductive Medicine.)

can exist almost indefinitely under appropriate laboratory conditions. When implanted in mice, hES cells and hEG cells have differentiated into tissues resembling neural epithelium, bone, cartilage, gut, striated muscle, and kidney. By administering chemical signals known to be released by embryonic cells during development, scientists have created the three tissues that give rise to all cell types, as well as many distinct cell types.

Although we do not discuss safety issues associated with embryonic cells here, it is important to note that if human embryonic cells are implanted into a person for therapeutic purposes, they can become cancerous. When you recall what you have learned about the molecular basis of cancer, you will see that in many ways cancer cells resemble stem cells. To decrease their risk of becoming cancerous, ES cells need to be treated with differentiation factors before they are implanted.

Much more research has been conducted on hES cells than on hEG cells, so hES cells derived from blastocysts will be our primary focus.

Place in Context

To place issues related to research on human embryonic cells in perspective, it is important to gather information on in vitro fertilization (IVF) and other unnatural methods of human reproduction, some of which are discussed in the description of cloning in the *Student Activity.*

Scientists fertilized human eggs under laboratory conditions for the first time in 1968, and the first human baby created through IVF was born in England in 1978.

When couples who are hoping to conceive a child through IVF seek assistance from the clinics specializing in reproductive technologies, physicians begin by administering hormones to the female. The hormones stimulate the development and ovulation of approximately 10 to 20 eggs, which are removed from the female surgically and then fertilized in petri dishes. After fertilization, the eggs are maintained under cell culture conditions for 3 to 5 days so that clinicians can observe the earliest stages of development. Fertilized eggs that develop abnormally are discarded, and the others are saved. At day 5, a blastocyst must be either implanted or frozen, or it will not survive. Typically some, but not all, blastocysts are inserted into the female. Excess blastocysts are frozen for future implantation, if needed, to spare the mother another round of hormone injection and egg removal. After a number of years, IVF clinics request permission from the parents to discard excess frozen embryos. Therefore, a number of embryos are created through IVF and then are discarded, either before or after being frozen, so that a couple can have a single child.

The history of IVF is relevant to discussions concerning human embryonic cell research, because virtually all hES cell lines created to date were derived from excess blastocysts about to be discarded from IVF clinics.

Current sources of embryonic cells. *Blastocysts from fertilized eggs.* Contrary to a pervasive misconception, no hES cell line is derived from aborted fetuses. The source of all ES cell lines created to date has been frozen blastocysts from IVF clinics. At any point, the IVF clinics in the United States collectively have around 400,000 frozen embryos in storage. Researchers obtained frozen embryos that were about to be discarded from IVF clinics after the parents had given written, informed (and unpaid) consent allowing the blastocysts to be used for research. Researchers isolated ICMs from frozen blastocysts and, to encourage growth and maintenance of hES cell lines, they grew the cells on a layer of mouse cells (Figure 36.5).

As we mentioned above, many people object to using the ICM from blastocysts to create ES cell lines. Because creating an hES cell line from a blastocyst necessitates destroying the blastocyst, they believe it is unethical and disrespectful of potential human life. This concern has elicited a variety of political initiatives by governments all over the world. Governmental bodies in different countries have responded in different ways. While the U.S. government has placed significant restrictions on hES cell research, the governments in Japan, South Korea, China, Singapore, France, Switzerland, and the United Kingdom have been supportive, though not without clear, legally mandated restrictions.

Moral arguments and policy decisions rooted in ethical concerns about hES cell research raise other ethical issues. Recall that ES cells have the potential to treat, cure, and even prevent a wide variety of diseases that, when combined, affect over 100 million Americans. Some of these are fatal diseases for which there are no effective treatments. ES cell research would also provide information on developmental

Figure 36.5 Creating hES cell lines. All hES cell lines in the United States were derived from frozen blastocysts from IVF clinics that were being discarded with the parents' permission. The first successful attempts to establish hES cell lines from ICMs (1998 to 2001) relied on a layer of mouse cells to support the human cells. As a result, the early lines are unsuitable for therapeutic uses because of possible contamination with mammalian viruses.

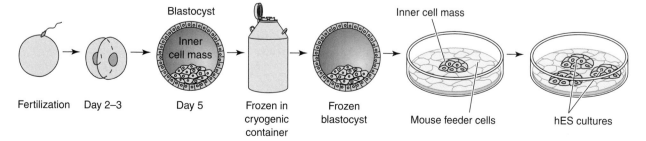

problems that cause birth defects, which might help to prevent them. Although these applications are at least a decade away, the scientific community agrees that the potential is there. What are the ethical implications of denying living human beings these lifesaving cures while focusing on the rights of potential human beings? Is it ethical to value potential human life more than existing human life? Would your answer change if you, your child, or your parent suffered from a disease that might be cured by hES research?

In addition, hES cell lines are derived from frozen blastocysts that are about to be destroyed, with the parents' approval. So, even though you may hear politicians say that scientists are "creating a life to destroy it," the embryos that are the source of virtually all hES cell lines are not being created and destroyed for research purposes. The embryo is about to be destroyed by its parents who no longer need it for the purposes for which it was created. Which of these fates—discarded because it is no longer needed by the parents or salvaged to be used in research—is more respectful of the potential human life? Are both equally disrespectful? If both are unethical, then to be ethically consistent, shouldn't governments that discourage hES cell research also be developing laws or policies that accomplish one or all of the following?

- Close down IVF clinics.
- Ban the practice of discarding frozen embryos from IVF clinics.
- Punish parents who have allowed their frozen embryos to be discarded.
- Require implantation of all frozen embryos currently stored in IVF clinics.
- Restrict IVF of eggs to the number a female is willing to have implanted.

Primordial germ cells. The third source of human embryonic cells does not rely on using blastocysts, and therefore, no embryos are destroyed. Instead, primordial germ cells are extracted from an aborted fetus that is approximately 6 to 8 weeks old. hEG cell lines are then derived from these primordial germ cells. People who do *not* object to using blastocysts, because the embryo is only 5 days old and a hollow ball of cells, may feel very differently about using cells from 6- to 8-week-old fetuses. Those who object to using 5-day-old embryos would surely object to this method, if the fetuses were aborted intentionally. However, people who disapprove of using 5-day-old embryos may approve of using fetal tissue, even though the embryo is much older, if the fetus had been miscarried naturally.

Future sources of embryonic cells. *Blastocysts from somatic cell nuclear transfer (SCNT).* Research on many different mammals, initiated by the birth of Dolly the cloned sheep, has repeatedly demonstrated that nuclei from fully differentiated somatic cells will dedifferentiate if placed in unfertilized, enucleated eggs and subjected to quite specific laboratory conditions. The egg receiving the somatic cell nucleus acts as if it has been fertilized, and it begins the embryonic development process, creating a blastocyst containing an ICM capable of creating an ES cell line.

In February 2004, Korean scientists claimed to have created an hES line from a blastocyst derived from SCNT, or "cloning," as it is known colloquially. After 2 years, stem cell scientists in other countries had not been able to replicate the work of the Koreans. This scandal received international press coverage, and the Korean scientist who perpetrated the deception left in disgrace.

Even so, it is safe to assume that in the near future it will likely be possible to generate ES cells from blastocysts created through SCNT. When this occurs, new ethical questions will be raised. People who may *not* object to using frozen blastocysts that are about to be discarded from IVF clinics as a source of ES cell lines may object to this source of ES cells for at least three reasons.

- Some people fear that allowing the use of SCNT to create ES cells for therapeutic purposes (therapeutic cloning), which they find ethically acceptable, opens the door to reproductive cloning, which they find ethically unacceptable. You will often hear these sorts of ethical concerns characterized as "slippery slope" issues.
- Those who object to females donating their eggs to IVF clinics for reproductive purposes may also object to this method for creating ES cell lines, since it relies on donated eggs.
- In contrast to using IVF blastocysts, in this case researchers would be creating a blastocyst (and therefore a potential life) in order to destroy it for research or therapeutic purposes.

Unfertilized eggs. The right chemical concoction can fool an unfertilized egg into beginning the cell division and differentiation process as if it had been fertilized. The development of an organism from an unfertilized egg is known as parthenogenesis, and the developing organism is not an embryo but a parthenote. In mammals, a parthenote cannot develop into a fetus if placed in a uterus, but researchers have successfully derived pluripotent ES cells from the ICM of animal parthenotes.

As this book goes to press, no researcher has derived hES cells from an unfertilized egg; therefore, we do not know whether it is possible. Using unfertilized eggs as the source of hES, if possible, could circumvent the problem of using embryos, since parthenotes cannot develop into a fetus if implanted, as well as any additional ethical concerns related to SCNT. However, those whose objections hinge on the fear of "opening doors" to future problems and "slippery slopes" would most certainly be upset with the idea of research that devises schemes to trigger parthenogenetic development in a human egg that has not been fertilized by a sperm!

Selected Readings

Bent, Stephen. 2005. Stem cells: under the microscope. *The Scientist* 19(13):22–24.

Chrispells, Martin, and D. E. Sandava. 2003. *Plants, Genes and Crop Biotechnology.* Jones and Bartlett, Boston, MA.

Clarke, Michael, and Michael Becker. 2006. Stem cells: the real culprit in cancer? *Scientific American* 295(1):48–52.

Cookson, Clive, et al. 2005. The future of stem cells. *Scientific American* 292(5):63–96.

Glynn, Kevin. 2000. *Tabloid Culture.* Duke University Press, Durham, NC.

James, Clive. 2006. *Global Review of Commercialized Transgenic Crops: 2006.* The International Service for the Acquisition of Agri-Biotech Applications, Ithaca, NY (http://www.isaaa.org).

Kreuzer, Helen, and Adrianne Massey. 2005. *Biology and Biotechnology: Science, Applications, and Issues.* ASM Press, Washington, DC.

Lewis, Ricki. 2005. Stem cells: an emerging portrait. *The Scientist* 19(13):15–21.

National Research Council. 1996. *Understanding Risk: Informed Decisions in a Democratic Society.* National Academy Press, Washington, DC.

Parson, Anne B. 2004. *The Proteus Effect: Stem Cells and Their Promise for Medicine.* Joseph Henry Press, Washington, DC.

Sankula, Sujatha, Gregory Marmon, and Edward Blumenthal. 2005. *Biotechnology-Derived Crops Planted in 2004: Impacts on US Agriculture.* National Council for Food and Agricultural Policy, Washington, DC. Available at http://www.ncfap.org.

Stewart, Neal, Harold Richards, and Matthew D. Halfhill. 2001. Transgenic plants and biosafety: science, misconceptions, and public perception. *BioTechniques* 29:832–843. Abstract available at http://www.biotechniques.com.

Waterstone, Marvin (ed.). 1996. *Risk and Society: the Interaction of Science, Technology and Public Policy.* Kluwer Publishers, Boston, MA.

West, Darrell M. 2001. *The Rise and Fall of the Media Establishment.* St. Martin's Press, New York, NY.

Wolfenbarger, L. L., and P. R. Phifer. 2000. The ecological risks and benefits of genetically engineered plants. *Science* 290:2088–2093.

Societal Issues

Analyzing the Issues of Biotechnology: Cloning

Introduction

You may have heard that advances in biotechnology will give rise to problems, issues, and concerns humans have never before faced. But what exactly are they? Are the societal issues said to be rooted in biotechnology unique to the new, biologically based technologies, or are they old, unresolved, sometimes unacknowledged problems that are also attendant to earlier technologies?

Many of the societal issues people attribute to the development of biotechnology have actually been around, sometimes in slightly different forms, for many years. While it is totally appropriate for the emergence of biotechnology to trigger questions about its appropriate uses, making biotechnology the sole scapegoat for the same questions raised by earlier technologies makes those concerns somewhat hollow.

Biotechnology began generating public discussion and debate about issues raised by its possible applications at the earliest stages of research, 10 years before any products had been developed. The debate about the potential societal impacts of modern biotechnology has continued unabated throughout its development. Certain questions continually resurface in deciding how best to use biotechnology. Where will society draw the line, and why will it draw the line there? How can biotechnology be used to maximize the gains that accrue to the greatest number of people while minimizing its risks and societal costs? Who will be the beneficiaries of biotechnology's potential? This 30-year discussion is not only healthy, it is essential to responsible technology development, because some potential applications of biotechnology raise difficult questions to which there is no right answer.

Unfortunately, the public discourse about biotechnology has often been uninformed, sensationalistic, and, as a result, counterproductive because it can move society down the path of unwise biotechnology development. Each one of us has a moral obligation to attend to biotechnology's development, but to maximize the good that accrues, that attention must be thoughtful and considered. A primary objective of this book is to equip you with information on biology and biotechnology so that you can do just that. In this chapter, we provide an approach and process for analyzing biotechnology-related issues so that you can contribute constructively to the public discourse on biotechnology and make informed decisions that are in your best interest.

Analytical Approach

When "authorities" espouse disparate views on the risks and benefits of biotechnology, or any technology, deciding which view to accept can pose a problem for ordinary citizens not trained as scientists. Identifying specious arguments, superfluous facts intended to mislead, and intellectual dishonesty requires an understanding of some of the science underlying the issue. How can nonscientists both assess the arguments and engage in productive debates as society struggles to come to terms with the risks and benefits of biotechnology or any other technically complex and potentially volatile issue?

A person does not have to be a scientist to analyze the potential societal costs and benefits of various applications of biotechnology. A user-friendly approach to analyzing any issue associated with biotechnology begins by placing modern biotechnology within the context of centuries of technological change.

As we stressed in chapter 1, modern biotechnology is only one of many technologies humans have developed to improve health and control the environment. Biotechnology now provides a new set of tools, but the goals of using technology remain the same: to make our lives as long and easy as possible. For 2 centuries people in the industrialized world enjoyed the benefits of technologies without

fully considering their potential downside. In the last half of the 20th century, society realized that the solutions and gratification technologies provide often carry some type of cost.

Awareness of the hidden costs of technological change became increasingly pervasive as biotechnology was being born. As a result, many people began to focus more attention on biotechnology at an earlier developmental stage than had been done with any prior technology. Often, however, people view biotechnology in isolation from other technologies and neglect key questions that would enrich their analysis. Most important, how do the potential societal impacts of this technology compare to those of existing technologies currently used to accomplish the same goal? For example, how do the environmental and public health costs of using transgenic microbes to synthesize insulin compare to those of extracting insulin from pancreatic tissue of butchered livestock?

In addition to assessing the costs of biotechnology within the context of other technologies, it is also important to consider the costs of *no* new technology. The costs of no technology are difficult to assess, especially if people have lived in a world in which technology has solved certain problems. When comparing the costs of technology and the costs of no technology, it is only natural to focus on the costs of having the technology, because they have never experienced the costs of not having the technology.

For example, getting immunized against infectious diseases, such as diphtheria and tetanus, carries some risks, including a small chance of dying. During the last decade, the media have focused attention on the risks of immunizations to infants. Some parents have responded by ignoring their pediatrician's advice to immunize their infants against whooping cough, or pertussis. The risks of dying from pertussis are much greater than those of dying from or being harmed by receiving the pertussis vaccine. But parents in today's industrialized societies have lived in a world in which the mortality risks caused by an infant contracting pertussis seem nonexistent. A technology—the widespread immunization of infants against pertussis—has made them blind to the risks of not having that technology.

Analytical Process

To be an active and constructive participant in discussions about biotechnology in society, you actually do not need to understand a lot of the specific scientific details. Being able to grasp scientific concepts and understanding the idea of a technological continuum is essential, but most important, you need to think like a scientist, objective and detached. In assessing the potential impacts of biotechnology, adopt a scientific posture: a willingness to look at all of the data, whether or not they are consistent with your preconceived notions, because you put aside emotional attachment to a certain conclusion.

Wise, judicious, and socially beneficial development of biotechnology depends on informed discussions and critical evaluation of real issues, not hyperbole. All too often, however, discussions about biotechnology issues are merely emotional exchanges of opinion in which participants often seem more interested in winning an argument than in achieving greater understanding. Here are some critical steps in objectively evaluating societal issues associated with applications of biotechnology. In general, proceed very methodically as you:

- Define and delineate the issue as specifically as possible.
- Determine whether the application introduces novel issues.
- Compare the risks/safety of the application to those of similar ways of solving the problem.
- Assess the societal benefits and costs of using this application of technology.

Often many of the steps in the process are not sequential but intertwined and iterative. At each step in the process, you must gather facts, avoid hearsay, and be equally skeptical of *all* voices with a vested interest in the debate. The methodology described below is generic, and therefore, some steps may not apply to certain issues. For analyzing other issues, more steps may be necessary. The most important thing to take away from the description of the analytical process is the tone: fact based, objective, methodical, specific, detached, and open to all options.

Step 1: clarify the issue

Become as clear as possible on the essence of the issue. Because the issues associated with biotechnology development are multifaceted, you need to make sure everyone is talking about the same thing. This problem is exacerbated when discussion participants use broad, ambiguous terms. Making sure everyone is on the same page may seem tedious and unnecessary, but we cannot tell you how many times we have seen discussions devolve into petty shouting matches just because the people were talking about different things and did not know it.

Delineate the Issue Very Clearly and Precisely

Broad, overarching issues need to be dissected. For example, gene flow from transgenic crops to other plants encompasses at least two issues that differ in almost every aspect relevant to assessing environmental risks and economic costs: gene flow from transgenic crops to wild plants and gene flow from transgenic crops to nontransgenic crops. Ambiguous issues need to be clarified. For example, let's assume someone expresses "concern about genetic engineering." Do they mean genetic engineering in plants, animals, microbes, people, or all of the above? If the focus is genetically engineered plants, do they care about the source of the gene (other plants, animals, or microbes) or only about the protein encoded by the gene? Is their concern restricted to genetic engineering, i.e., recombinant DNA technology, or are they also concerned about the many other genetic modification techniques that people have used to change the genetic makeup of the organisms we depend upon? Do they even know that virtually all organisms on which we depend have been changed genetically through purposeful human intervention?

Define the Terms Quite Specifically

In public discussions of issues, if your goal is greater understanding, the value of using discipline in the words you choose cannot be overemphasized. The issues being discussed are almost always technically complex, and frequently they trigger emotional responses. Adding to the confusion by using sloppy language will only make the discussion, or your internal analysis of the issue, less productive.

For example, let's say you attend a debate entitled "The Risks of Biotechnology." Here are a few of the points at which people could be talking about different things and not even know it.

- Biotechnology is an ambiguous term used in a variety of ways by different people. For example, in the 1970s and 1980s, biotechnology meant recombinant DNA technology to American scientists and microbial fermentation to scientists in the United Kingdom. The risks of one technology differ greatly from those of another.
- What type of risk? To some people, the concept of risk relates solely to safety, of either public health or the environment, and is amenable to science-based analysis. Others include social, economic, and ethical concerns when they use the word risk. Science may contribute to discussions of these concerns, but many factors other than science-based facts affect people's concepts of these types of impacts.

- Which facet of biotechnology development is the target of concern? Biotechnology development occurs along a continuum that begins in a research laboratory and extends through product development to widespread commercial use. The risks of biotechnology laboratory research, small-scale field testing, and large-scale commercial use and manufacturing differ from each other both qualitatively and quantitatively.

Perhaps you feel the discussion should include all of these things. That's fine, but discuss each facet sequentially and methodically, making sure everyone knows exactly which facet is under discussion.

Step 2: place the issue in context

Look at the history of using other technologies in similar ways and ask whether the issue or concern is new or unique to biotechnology. Does it derive from biotechnology's novel powers that provide brand new capabilities, or does society currently engage in practices that raise similar or even identical issues? If the practice and issues it raises are unique to biotechnology, do all aspects of the issue raise concerns, or only some? If only some, specifically delineate those areas that trigger concern and focus attention there.

You may determine that the practices that elicit concerns are not at all unique to biotechnology but are accepted as part of life. Does their commonness mean they are not worth analyzing? Absolutely not. Just because people *are* doing something does not mean they *should be*. For example, some people have expressed concern that recombinant DNA technology allows us to genetically modify food by moving genes between different species. However, over a century before the development of recombinant DNA technology, plant breeders genetically modified many food crops by interbreeding different plant species that are incapable of interbreeding naturally. After learning this, some people may still believe scientists should not move genes between species, whether they accomplish it through genetic engineering or crossbreeding. In other words, *what* scientists are doing, not *how* they are doing it, concerns them.

Step 3: consider issues related to safety

Although many people discuss risks and benefits in almost the same breath, that juxtaposition can be confusing. Conceptually these issues need to be separated, because they are fundamentally different considerations. It is clearer to think of risk vis-à-vis safety and benefits as societal benefits compared to societal costs.

Assessing risks is a science-based analysis of the potential to cause harm or injury to the environment or human health. We discuss the formal risk assessment process in detail in chapter 37. In brief, the steps in a risk analysis are to identify the risk, estimate the probability of the risk occurring, and assess the significance of the consequences if the risk occurs. Once you have identified the risks, you must then estimate the probability of the risk occurring. Because much of biotechnology is similar to activities people have been engaged in for many years, there is often abundant data on some products and applications that can be used to estimate general probabilities of specific risks. Others introduce novel risks.

Then, assess the consequences of a particular risk occurring. What is the significance of risk A occurring? If it occurs, will it matter? Under what circumstances does it matter? It may happen on a regular basis anyway without human interference (e.g., insects becoming resistant to chemicals produced by plants and viruses transferring genes across species). Get a sense of the scale of the risk. An adverse effect on an individual organism or a number of organisms may be environmentally, ecologically, or economically insignificant.

Finally, in discussing the risks of any technology, keep in mind that technologies may be used wisely or unwisely, for good or for bad. The risks of a technology per se must be distinguished from the risks of a technology placed in irresponsible hands. The flaws of a technology and the flaws of human nature are two very different considerations.

Step 4: compare the risks

Step 4 formalizes the mental framework described in the previous section. The relevant question to be asked and answered is not "is there a risk" in using this technology or product. Of course there is. There is no such thing as zero risk, because every human activity carries some amount of risk with it. To weigh the value of taking a risk, it needs to be placed in a meaningful context. One way to do this is to compare risks. With biotechnology, that typically means asking how the risks of the new technology compare with those of existing technologies directed toward the same purpose.

Another way to place the risk in a meaningful context is to ask the other essential question: *what are the risks of no new technology?* In industrialized countries people focus almost solely on the *risks of a technology.* They are free to have that as their

singular concern because the risks of no new technology are almost always minor given the quality of life in industrialized countries. As explained above, the same does not hold true for many citizens in developing countries, where the risk (or cost) of no technology may be significantly higher than the risk of the technology.

Step 5: consider societal impacts

Another aspect of placing risks in a meaningful context involves looking at the application's benefits. Only then can society determine if a risk is worth taking. On the other hand, just because a product is safe and has benefits, does that mean it should be developed? Does society need this product? A society's resources are always limited, and as any economist knows, there are always lost opportunities whenever we make a choice about how those resources—time, people, and capital—should be used. In addition, a safe and legal product that benefits a few people may violate one of the fundamental ethical principles of others:

- Do no harm (nonmaleficence).
- Do good (beneficence).
- Do not violate individual freedom (autonomy).
- Be fair (justice).

The benefits and the risks need to be assessed equally objectively, which is difficult, because opinions about what constitutes a benefit vary. In addition, when one group gains, another group often loses, so a benefit for one may be a cost for another.

A Case Study for Issue Analysis: Animal Cloning

In February 1997, scientists and nonscientists alike were shocked by the news that an adult mammal had been cloned. The responses of these two populations were similar on the surface, but the reasons underlying their responses had little in common. The existence of this clone, a sheep named Dolly, unleashed a global media frenzy. This story, which was essentially a story about a scientific breakthrough, received unprecedented coverage in the mainstream media.

Studying the media's coverage of Dolly offers an excellent opportunity for evaluating how reporters and editors handled the story and whether they fulfilled their responsibility to inform the public. It also illuminates a pervasive problem in the way the mainstream media cover science. Even though the key element that made this story "news" was a scientific

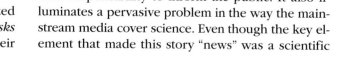

breakthrough, the breakthrough that gave the story its newsworthy status was rarely, if ever, explained in the newspaper and television coverage. Many times it was not even mentioned! Instead, the reporters jumped immediately to *possible* applications in medicine and their ethical implications, which cannot be "news," since they are purely conjectural and many years away.

The issue of animal cloning presents an opportunity to use the strategy described above to analyze an emotionally charged societal issue raised by a scientific breakthrough related to biotechnology development. The discussion below is not meant to be a comprehensive treatment of the issues involved with animal cloning but is included to give you an example of some of the requisite steps in a rational treatment of the issue.

Clarify the issue

When discussing the "ethics of cloning," be sure to separate the ethics of cloning plants, animals, and people. Some people may feel that the issues are the same irrespective of the organism involved; others may not.

What Is a Clone?

A clone is a collection of genetically identical individuals, all derived from a single parent. Cloning is the process of propagating those genetically identical individuals. Biologists use the word "clone" in a number of ways, depending on what is being cloned.

In molecular cloning, the cloned "individual" of interest is a gene or a length of DNA molecule. A clone, in this case, usually refers to both the piece of DNA being copied and, by extension, the collection of organisms (usually bacteria) containing the same piece of DNA. In recombinant DNA work, the clone is a recombinant organism, and the piece of DNA is a recombinant molecule.

Cells can also be the cloned entities. Cellular cloning results in "cell lines" of identical cells. For example, in monoclonal antibody technology, the immune system B cell that is used to manufacture the antibody of interest is separated from a large population of other B cells. The B cells producing the antibody are clones of each other, as are embryonic stem cells derived from a single fertilized egg.

Finally, sometimes the clone is a multicellular organism. This is the type of clone most nonscientists

mean when they use the word. Many plants are propagated through cloning by rooting plant parts taken from adult, fully developed plants. Cloning most animals is more difficult than cloning plants. Except for the simplest animals, such as sponges, scientists cannot simply break off a part of an animal and grow an entirely new organism. In animals, the starting material for cloning a whole organism must be reproductive cells, such as eggs, or cells taken from an embryo at the earliest stage of development. The embryo is split into single cells. Each cell that develops into a complete organism is a clone of the others derived from the original embryo (Figure 36.1).

After reading these different definitions of "clone" and "cloning," it is easy to see why defining terms is essential for an informed discussion. If someone says "clone" and means a mammal that is genetically identical to one of its parents but the listener pictures a petri dish of genetically identical bacteria, misunderstandings will exist from the outset and the discussion will not be very productive.

For the purposes of this discussion, we will use "clone" to mean a collection of genetically identical multicellular organisms. If not explicitly stated otherwise, assume the organisms are mammals.

What Mammalian Cloning Is Not

Cloning mammals is not a matter of creating a copy of an organism instantaneously in the laboratory. If the goal of the work is to produce a whole organism and not simply to maintain genetically identical cells in culture, the clone must go through normal gestation and development. In mammals, the researcher must implant the eggs or embryonic cells into the uterus of a surrogate mother when the embryo is only a few days old. Then, the organism must go through the gestation period typical for that organism. For example, the embryo that became Dolly was implanted into the uterus of a ewe after having been cultured within sheep oviduct tissue for only 6 days. Dolly was born approximately 6 months later, the typical gestation period for sheep.

Nor is a clone a precise copy of the organism that served as the source of the genetic material. Environmental influences, including the uterine environment during gestation in mammals, provide powerful input into the ultimate organism. To assume clones are perfectly identical copies of one another is to ignore the "nurture" component of "nature + nurture = an organism."

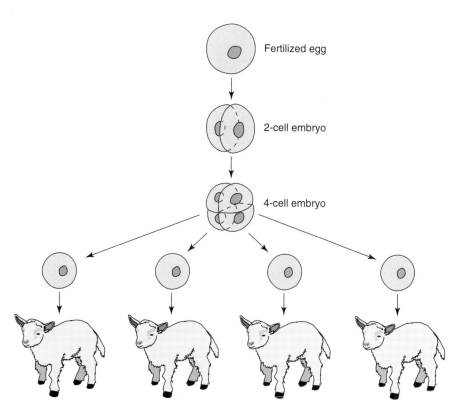

Figure 36.1 Cloning is the creation of genetically identical copies. Livestock breeders have used this form of cloning for approximately 20 to 25 years. A fertilized egg is allowed to develop to the two- or four-cell stage and is divided into single cells. Each cell gives rise to an offspring that is genetically identical to the others.

Fertilized egg

2-cell embryo

4-cell embryo

Gather relevant facts

What Were Some of the Facts about Dolly's Beginnings?

Dolly was the first animal to be cloned from a differentiated somatic cell taken from an adult animal. Scientists removed 277 cells from the udder of an adult sheep, and then they fused those cells with 277 unfertilized egg cells from which the nuclear genetic material had been removed. An egg that has had its nucleus removed is said to be enucleated (Figure 36.2). After culturing the resulting embryos for 6 days, scientists implanted the 29 embryos that appeared to have developed normally into surrogate mothers. Only one produced a live lamb, Dolly, approximately 6 months later.

What Makes Dolly Special?

Since the mid-1980s, researchers have used embryonic cells, such as those shown in Figure 36.1, as sources of genetic material to clone sheep, cows, and other mammals. Until Dolly was born, attempts to use adult cells as sources of genetic material for cloning had ended in failure. Scientists assumed that a cell containing adult genetic material could not develop into a complete organism because the differentiated cell had become specialized into a certain cell type and could not shed its specific role and "remember" how to give rise to a complete organism.

In other words, scientists thought that when certain sections of the cell's DNA "turned off" during cell differentiation, they could not be turned on again. Dolly proved that differentiation was not irreversible and, in being born, challenged a fundamental truth of developmental biology.

That is the scientific achievement that got lost in the hoopla—the scientific breakthrough that made Dolly "news": the despecialization of genetic material that had been committed to a special function and its ultimate reprogramming into embryo-like genetic material capable of directing the development of a complete organism. When you read in news reports that scientists are excited about Dolly, it is this scientific discovery that excites them, not the prospect of making genetically identical copies of a multicellular organism. That is old news. Scientists have been doing that for over 50 years.

View cloning in context

Many people were upset with Dolly's birth because the circumstances surrounding her conception were "not natural." Is this unique to Dolly? No, because unnatural methods of conceiving offspring have been used for over 50 years with animals and 25 years with humans.

Societal Issues

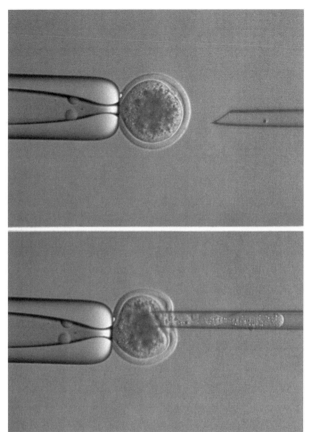

Figure 36.2 Enucleation of eggs. Working at a microscope that is hooked to a computer monitor or television screen, scientists hold the egg steady with a pipette and then insert a much smaller pipette into the egg to withdraw the nucleus. (Photographs courtesy of Roslin Institute, Edinburgh, Scotland.)

Livestock Breeding

Animal breeders began to routinely use techniques that interfered with natural methods of reproduction more than 50 years ago. Some of the techniques listed below are not cloning techniques, but all are "not natural." Livestock breeders developed these techniques to incorporate desired genetic changes into a herd more quickly. As a result, livestock animals are healthier, and the food derived from them is cheaper and of higher quality. In addition, zoos and other conservation organizations use unnatural methods of reproduction to try to save endangered species and to introduce genetic variation into populations.

Artificial insemination. Artificial insemination began to be widely used 45 years ago and is now used routinely in livestock production, thoroughbred horse breeding, and zoo breeding programs. Artificial insemination has allowed U.S. farmers to produce more milk with 10 million cows than they did in 1945 with 25 million. Semen from the best bulls is frozen and used repeatedly to inseminate cows

anywhere in the world. Because of efforts to keep the bull studs healthy, calves produced through artificial insemination are healthier. In addition, its low cost—approximately $10 per breeding—does not place small farmers at a disadvantage compared to large farmers.

Recently, scientists have developed techniques for separating sperm carrying X or Y chromosomes from each other with 80 to 90% accuracy. This allows farmers to predetermine the sex of the offspring, in addition to other economically important genetic traits. Cows on dairy farms are artificially inseminated with semen containing a high percentage of X chromosomes. Beef cattle farmers inseminate cows with a high percentage of Y-chromosome-bearing sperm because male calves grow faster than females.

Multiple-ovulation embryo transfer. Commercial use of embryo transfer to genetically improve livestock herds began in the early 1970s (see Figure 1.28). To produce livestock by embryo transfer, a valuable female with superior genetic attributes is inseminated after being treated with hormones that stimulate ovulation. The resulting four or five embryos, which are not genetically identical to each other or to either parent, are removed from the mother and placed in surrogate mothers who are less valuable than the genetic mother. The genetic mother does not become pregnant, and thus another reproductive cycle can be triggered. Additional technological advances allow breeders to freeze the embryos and implant them later if the first implantation is aborted.

Not only has embryo transfer increased the speed with which superior animals can be produced and bred, it is also an important means for transporting superior genetic material to other countries. Embryos carry fewer diseases than semen or whole organisms, and transportation costs are much lower than for adult animals.

Multiple-ovulation embryo transfer is not a cloning technique.

Embryo splitting. Embryo splitting, or embryo twinning, has been used in cattle breeding for approximately 20 to 25 years and is similar in concept to embryo transfer. At a very early stage in development, an embryo is split in two, and each half is allowed to develop to the blastocyst stage (Figure 36.3). Then, it is implanted into a surrogate female or frozen.

In this case, the resulting embryos are genetically identical, so embryo splitting is a cloning technique.

Nuclear transfer from embryonic cells. Nuclear transfer from embryonic cells is a relatively recent

Figure 36.3 Embryo splitting. At early stages of development, scientists use microsurgical blades to separate livestock embryos into separate cells. (Photograph courtesy of George Seidel, Colorado State University.)

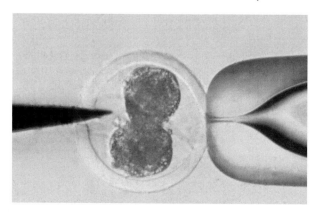

development that bears a resemblance to both embryo splitting and adult-cell cloning. It involves taking the nucleus from an embryonic cell of one individual and placing it in the egg (from which the nucleus has been removed) of another individual. In the mid-1980s a number of research teams successfully produced calves by transferring nuclei from 64-cell-stage embryos into enucleated eggs and implanting these eggs in surrogate females (Figure 36.4). A variation on this theme involves fusing individual cells taken from a 64-cell-stage embryo with 64 eggs whose nuclei have been removed. These embryos are then implanted into surrogate mothers. This technique is called nuclear fusion.

The distinction between nuclear transfer and nuclear fusion, though slight, is relevant to discussions of human reproductive cloning. In nuclear transfer, mitochondrial DNA, which carries approximately 1% of the genetic information, is provided by only one individual, the egg donor. In nuclear fusion, mitochondria from both the embryo cells and the eggs may contribute genetic material to the resulting organisms. Both the nuclear transfer and nuclear fusion techniques are cloning techniques.

Nuclear transfer from adult cells. The trait that distinguishes nuclear transfer from adult cells from the cloning technique just described is the source of the genetic material: adult cells versus embryonic cells. Nuclear fusion is also used to transfer adult genetic material into egg cells. This cloning technology, which was used to produce Dolly, is referred to as somatic cell nuclear transfer (Figure 36.5). The nucleus from a fully differentiated somatic cell is inserted into an enucleated egg with a micropipette or by fusing the two cells.

Laboratory Animals
The application of cloning by nuclear transfer is not unique to mammals. Since the 1950s, developmental biologists have used nuclear transfer from embryonic cells to eggs to study amphibian development.

Humans
In vitro fertilization has been used since 1978 as a method for infertile couples to conceive children. Eggs and sperm are placed into a test tube, where fertilization occurs. The fertilized eggs are kept in culture for 4 or 5 days until the embryo reaches the blastocyst stage (Figure 36.6). A few of the embryos are then implanted into a female who is usually, but not always, the biological mother. Other embryos are frozen for possible implantation at a later date.

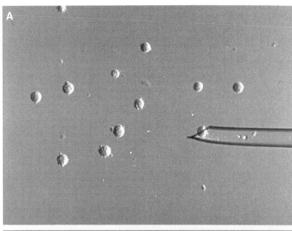

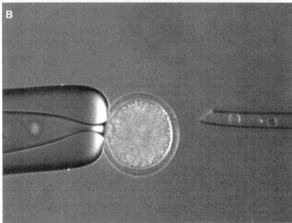

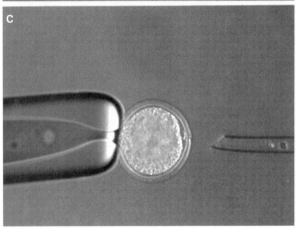

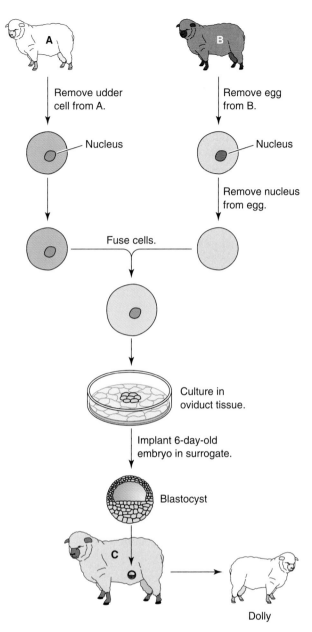

Figure 36.4 Nuclear transfer. (A) Nuclei have been re-moved from either somatic or embryonic cells. (B) Note the nucleus in the small pipette. Each nucleus is in-jected into a separate enucleated egg cell. (C) The newly inserted nucleus is barely visible between the egg cell membrane and the cellular cytoplasm. A small tear in the cell membrane indicates the site where the pipette was inserted. (Photographs courtesy of Roslin Institute, Edinburgh, Scotland.)

Figure 36.5 Somatic cell nuclear transfer. Dolly was produced by a unique type of cloning. The nucleus in a fully differentiated somatic cell from breed A's udder was inserted into the enucleated egg of breed B by fus-ing the two cells. The resulting egg contained breed A's nuclear genetic material and mitochondrial DNA from breeds A and B. The egg developed into a blastula in tissue culture, and the blastula was inserted into the uterus of sheep C, the surrogate mother of Dolly. Note that Dolly's coloring is identical to that of female A, her genetic mother, and does not exhibit any markings of either her surrogate mother or breed B.

Societal Issues

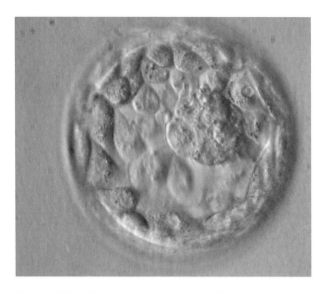

Figure 36.6 Human blastocyst. A fertilized egg develops into a blastocyst in approximately 4 to 5 days, whether it is in cell culture or the female reproductive tract. (Photograph courtesy of Michael Vernon, West Virginia Center of Reproductive Medicine.)

Clearly, in vitro fertilization is not a form of cloning, because genetic material from two sources is combined, creating a genetically unique individual. It is, however, a reproductive technology that is unnatural.

Consider benefits, risks, and other concerns

Benefits of Animal Cloning

Although much media attention has focused on the risks and public concerns about animal cloning, there are a number of benefits.

1. Animal cloning should allow great progress in understanding what turns genes on and off.
2. By using genetically identical animals in experiments, scientists are able to get results more quickly and to use fewer animals, because the variation in experimental results due to genetic variation is eliminated.
3. Improvements in livestock can be incorporated into herds much more rapidly.
4. Animal cloning could help to save endangered species. Researchers have successfully produced a number of cloned embryos of rare or endangered species, but few have survived.
5. When donor cells are in culture, scientists can introduce desirable genetic changes before the cell is fused with an egg to form the embryonic clone. Therefore, animal clones of a consistent quality can be reproduced in large numbers quickly and cheaply. As a result, animal cloning,

used in conjunction with recombinant DNA technology, could provide:

- An alternative and more efficient way of producing transgenic animals that can be used in the production of human therapeutic proteins or for organs to be used in xenotransplantation
- Excellent animal models for studying genetic diseases, aging, and cancer; for discovering new drugs; and for evaluating other forms of therapy, such as gene and cell therapy

Public Concerns about Cloning

The news of Dolly's birth triggered widespread fear and, in many countries, a call for a total ban on cloning humans. A few have suggested that all animal cloning should be banned. Those who want to ban cloning have justified such bans with vague statements that cloning "poses new ethical problems" or "is contrary to nature." To assess the validity of these claims, you need to determine if cloning does pose new ethical problems, and if so, you should define what they are (see below). Above, we provided information for evaluating the naturalness or unnaturalness of cloning and compared cloning with other widely used unnatural means of producing offspring, both human and nonhuman. If governments ban cloning because it is "contrary to nature," then should they consider banning the other unnatural reproductive technologies that couples, farmers, and scientists have used for decades?

Others have sought to ban cloning because of the horrible things they feel could be done if this technology is extended to humans. They have speculated about armies of Hitler clones and brainless cloned duplicates serving as private stocks of replacement parts. Such sensationalist speculation is counterproductive, at best. To assess the costs and benefits of cloning, put concerns in perspective by aligning them with the more immediate and realistic effects of this breakthrough.

Costs or Risks Associated with Animal Cloning

We will limit our discussion to cloning via somatic cell nuclear transfer.

Using the Dolly research as an example, the success rate for animal cloning is very low (1 in 277). Few implantable embryos result from the fusion of an adult cell with an enucleated egg cell. Of those that are implanted, most have developmental abnormalities and do not survive. Of those, most die early in the pregnancy, and a few die in late pregnancy or soon after birth. Since Dolly's birth, researchers have cloned a number of other species by nuclear transfer from adult cells—mice, cows, goats, and cats—and the success rates have been equally low. The inability to produce healthy clones of endangered species can be explained by these low rates of success.

Scientists are not sure about the long-term health implications of this type of cloning. Because "old" DNA is used to produce the clone, the clone may inherit mutations caused by ultraviolet light or other environmental influences, or other genetic problems. This could predispose the clone to premature aging or a higher incidence of environmentally induced cancer.

Other risks related to livestock management may surface. A herd composed of clones would be genetically homogeneous. This lack of genetic variation could result in health problems that affect the entire herd and not just a few individuals. For example, all individuals might be susceptible to the same infectious diseases.

As for ethical issues associated with this research, it is difficult to think of ethical issues that are not associated with animal research in general but are unique to the cloning research.

Risks Involved in Cloning Humans

If you are interested in extending this discussion to the question of cloning humans, it is essential to make a clear distinction between cloning cells taken from an embryo that is only a few days old and cloning to make babies (reproductive cloning). If the type of cloning being discussed involves implanting an embryo into a woman in hopes of producing a cloned offspring rather than simply growing a mass of undifferentiated mammalian cells in culture, for many people the issues will differ. Problems can arise if that distinction is not clearly delineated.

For example, when the birth of Dolly caused governments to move quickly and "ban human cloning," their intent most often was to ban reproductive cloning. However, because they did not consult with scientists before drafting the wording of the law, some of the laws inadvertently made it illegal to give birth, naturally, to identical twins. Others captured cell cloning, such as monoclonal antibody production, or other cell culture techniques used in research.

Our discussion will focus only on the health risks (not other ethical issues) associated with reproductive cloning via somatic cell nuclear transfer from adult cells. It is also important to point out that scientists do not know whether human reproductive cloning with adult genetic material taken from differentiated cells is even possible.

Assuming that the problems associated with cloning in sheep would occur in human cloning, the great majority of the embryos implanted would not survive. Of those that survived, environmentally induced mutations and other genetic problems may pose the same risks as they do for animal clones produced through nuclear transfer from adult cells. Even if experience with cloned animals identifies ways of reducing these health risks in animals, is it ethical to perform similar experiments with women to determine whether the risks in humans are significant?

Potential Benefits of Human Cloning

Under what circumstances, if any, might human cloning be acceptable? Some people believe that there may be situations in the future when nuclear transfer of DNA into an egg cell in hopes of producing a baby could be beneficial.

One example often cited involves people with mitochondrial DNA genetic diseases. Some types of mitochondrial DNA diseases cause blindness and epilepsy. Other evidence suggests that a form of Alzheimer's disease is transmitted via mitochondrial DNA. By removing the nucleus from an embryo produced by in vitro fertilization and inserting it into an egg from a donor, the resulting baby would not have the disorder inherited through mitochondrial DNA, but 99% of its DNA would come from its two parents.

Another example acceptable to some involves cases of infertility that cannot be overcome with in vitro fertilization. The nucleus from an adult cell of one of the parents could be added to an enucleated egg taken from the mother and then implanted. If the donor of the nuclear genetic material is the father, the baby would be the biological offspring of both parents, but 99% of its genes would come from only one parent—the father.

The third example that many people believe is an acceptable use of cloning technology does not deal with reproductive cloning but therapeutic cloning. The value of using nuclear transfer from adult cells is that the therapeutic cells would be a perfect match genetically if the nucleus inserted into the enucleated egg was taken from cells of the person receiving the therapy.

Are Bioengineered Foods Safe?

(Reprinted from *FDA Consumer,* January 2000)

Since 1994, a growing number of foods developed using the tools of the science of biotechnology have come onto both the domestic and international markets. With these products has come controversy, primarily in Europe where some question whether these foods are as safe as foods that have been developed using the more conventional approach of hybridization. Ever since the latter part of the 19th century, when Gregor Mendel discovered that characteristics in pea plants could be inherited, scientists have been improving plants by changing their genetic makeup. Typically, this was done through hybridization in which two related plants were cross-fertilized and the resulting offspring had characteristics of both parent plants. Breeders then selected and reproduced the offspring that had the desired traits.

Today, to change a plant's traits, scientists are able to use the tools of modern biotechnology to insert a single gene—or, often, two or three genes—into the crop to give it new, advantageous characteristics. (See "Methods for Genetically Engineering a Plant" [*FDA Consumer,* January 2000].) Most genetic modifications make it easier to grow the crop. About half of the American soybean crop planted in 1999, for example, carries a gene that makes it resistant to an herbicide used to control weeds. About a quarter of U.S. corn planted in 1999 contains a gene that produces a protein toxic to certain caterpillars, eliminating the need for certain conventional pesticides.

In 1992, the Food and Drug Administration published a policy explaining how existing legal requirements for food safety apply to products developed using the tools of biotechnology. It is the agency's responsibility to ensure the safety of all foods on the market that come from crops, including bioengineered plants, through a science-based decision-making process. This process often includes public comment from consumers, outside experts and industry. FDA established, in 1994, a consultation process that helps ensure that foods developed using biotechnology methods meet the applicable safety standards. Over the last five years, companies have used the consultation process more than 40 times as they moved to introduce genetically altered plants into the U.S. market.

Although the agency has no evidence that the policy and procedure do not adequately protect the public health, there have been concerns voiced regarding FDA's policy on these foods. To understand the agency's role in ensuring the safety of these products, *FDA Consumer* sat down with Commissioner Jane E. Henney, M.D., to discuss the issues raised by bioengineered foods:

FDA Consumer: Dr. Henney, what does it mean to say that a food crop is bioengineered?

Dr. Henney: When most people talk about bioengineered foods, they are referring to crops produced by utilizing the modern techniques of biotechnology. But really, if you think about it, all crops have been genetically modified through traditional plant breeding for more than a hundred years.

Since Mendel, plant breeders have modified the genetic material of crops by selecting plants that arise through natural or, sometimes, induced changes. Gardeners and farmers and, at times, industrial plant breeders have crossbred plants with the intention of creating a prettier flower, a hardier or more productive crop. These conventional techniques are often imprecise because they shuffle thousands of genes in the offspring, causing them to have some of the characteristics of each parent plant. Gardeners or breeders then look for the plants with the most desirable new trait.

With the tools developed from biotechnology, a gene can be inserted into a plant to give it a specific new characteristic instead of mixing all of the genes from

Societal Issues

two plants and seeing what comes out. Once in the plant, the new gene does what all genes do: It directs the production of a specific protein that makes the plant uniquely different.

This technology provides much more control over, and precision to, what characteristic breeders give to a new plant. It also allows the changes to be made much faster than ever before.

No matter how a new crop is created—using traditional methods or biotechnology tools—breeders are required by our colleagues at the U.S. Department of Agriculture to conduct field testing for several seasons to make sure only desirable changes have been made. They must check to make sure the plant looks right, grows right, and produces food that tastes right. They also must perform analytical tests to see whether the levels of nutrients have changed and whether the food is still safe to eat.

As we have evaluated the results of the seeds or crops created using biotechnology techniques, we have seen no evidence that the bioengineered foods now on the market pose any human health concerns or that they are in any way less safe than crops produced through traditional breeding.

FDA Consumer: What kinds of genes do plant breeders try to put in crop plants?

Dr. Henney: Plant researchers look for genes that will benefit the farmer, the food processor, or the consumer. So far, most of the changes have helped the farmer. For example, scientists have inserted into corn a gene from the bacterium *Bacillus thuringiensis,* usually referred to as BT. The gene makes a protein lethal to certain caterpillars that destroy corn plants. This form of insect control has two advantages: It reduces the need for chemical pesticides, and the BT protein, which is present in the plant in very low concentrations, has no effect on humans.

Another common strategy is inserting a gene that makes the plant resistant to a particular herbicide. The herbicide normally poisons an enzyme essential for plant survival. Other forms of this normal plant enzyme have been identified that are unaffected by the herbicide. Putting the gene for this resistant form of the enzyme into the plant protects it from the herbicide. That allows farmers to treat a field with the herbicide to kill the weeds without harming the crop.

The new form of the enzyme poses no food safety issues because it is virtually identical to nontoxic enzymes naturally present in the plant. In addition, the resistant enzyme is present at very low levels and it is as easily digested as the normal plant enzyme.

Modifications have also been made to canola and soybean plants to produce oils with a different fatty acid composition so they can be used in new food processing systems. Researchers are working diligently to develop crops with enhanced nutritional properties.

FDA Consumer: Do the new genes, or the proteins they make, have any effect on the people eating them?

Dr. Henney: No, it doesn't appear so. All of the proteins that have been placed into foods through the tools of biotechnology that are on the market are nontoxic, rapidly digestible, and do not have the characteristics of proteins known to cause allergies.

As for the genes, the chemical that encodes genetic information is called DNA. DNA is present in all foods and its ingestion is not associated with human illness. Some have noted that sticking a new piece of DNA into the plant's chromosome can disrupt the function of other genes, crippling the plant's growth or altering the level of nutrients or toxins. These kinds of effects can happen with any type of plant breeding—traditional or biotech. That's why breeders do extensive field-testing. If the plant looks normal and grows normally, if the food tastes right and has the expected levels of nutrients and toxins, and if the new protein put into food has been shown to be safe, then there are no safety issues.

FDA Consumer: You mentioned allergies. Certain proteins can cause allergies, and the genes being put in these plants may carry the code for new proteins not normally consumed in the diet. Can these foods cause allergic reactions because of the genetic modifications?

Dr. Henney: I understand why people are concerned about food allergies. If one is allergic to a food, it needs to be rigorously avoided. Further, we don't want to create new allergy problems with food developed from either traditional or biotech means. It is important to know that bioengineering does not make a food inherently different from conventionally produced food. And the technology doesn't make the food more likely to cause allergies.

Fortunately, we know a lot about the foods that do trigger allergic reactions. About 90 percent of all food allergies in the United States are caused by cow's milk, eggs, fish and shellfish, tree nuts, wheat, and legumes, especially peanuts and soybeans.

To be cautious, FDA has specifically focused on allergy issues. Under the law and FDA's biotech food policy, companies must tell consumers on the food label when a product includes a gene from one of the common allergy-causing foods unless it can show that the protein produced by the added gene does not make the food cause allergies.

We recommend that companies analyze the proteins they introduce to see if these proteins possess properties indicating that the proteins might be allergens. So far, none of the new proteins in foods evaluated through the FDA consultation process have caused allergies. Because proteins resulting from biotechnology and now on the market are sensitive to heat, acid and enzymatic digestion, are present in very low levels in the food, and do not have structural similarities to known allergens, we have no scientific evidence to indicate that any of the new proteins introduced into food by biotechnology will cause allergies.

FDA Consumer: Let me ask you one more scientific question. I understand that it is common for scientists to use antibiotic resistance marker genes in the process of bioengineering. Are you concerned that their use in food crops will lead to an increase in antibiotic resistance in germs that infect people?

Dr. Henney: Antibiotic resistance is a serious public health issue, but that problem is currently and primarily caused by the overuse or misuse of antibiotics. We have carefully considered whether the use of antibiotic resistance marker genes in crops could pose a public health concern and have found no evidence that it does.

I'm confident of this for several reasons. First, there is little if any transfer of genes from plants to bacteria. Bacteria pick up resistance genes from other bacteria, and they do it easily and often. The potential risk of transfer from plants to bacteria is substantially less than the risk of normal transfer between bacteria. Nevertheless, to be on the safe side, FDA has advised food developers to avoid using marker genes that encode resistance to clinically important antibiotics.

FDA Consumer: You've mentioned FDA's consultative process a couple of times. Could you explain how genetically engineered foods are regulated in the United States?

Dr. Henney: Bioengineered foods actually are regulated by three federal agencies: FDA, the Environmental Protection Agency, and the U.S. Department of Agriculture. FDA is responsible for the safety and labeling of all foods and animal feeds derived from crops, including biotech plants. EPA regulates pesticides, so the BT used to keep caterpillars from eating the corn would fall under its jurisdiction. USDA's Animal and Plant Health Inspection Service oversees the agricultural environmental safety of planting and field testing genetically engineered plants.

Let me talk about FDA's role. Under the federal Food, Drug, and Cosmetic Act, companies have a legal obligation to ensure that any food they sell meets the safety standards of the law. This applies equally to conventional food and bioengineered food. If a food does not meet the safety standard, FDA has the authority to take it off the market.

In the specific case of foods developed utilizing the tools of biotechnology, FDA set up a consultation process to help companies meet the requirements. While consultation is voluntary, the legal requirements that the foods have to meet are not. To the best of our knowledge, all bioengineered foods on the market have gone through FDA's process before they have been marketed.

Here's how it works. Companies send us documents summarizing the information and data they have generated to demonstrate that a bioengineered food is as safe as the conventional food. The documents describe the genes they use: whether they are from a commonly allergenic plant, the characteristics of the proteins made by the genes, their biological function, and how much of them will be found in the food. They tell us whether the new food contains the expected levels of nutrients or toxins and any other information about the safety and use of the product.

FDA scientists review the information and generally raise questions. It takes several months to complete the consultation, which is why companies usually start a dialog with the agency scientists nearly a year or more before they submit the data. At the conclusion of the consultation, if we are satisfied with what we have learned about the food, we provide the company with a letter stating that they have completed the consultation process and we have no further questions at that time.

FDA Consumer: Since genes are being added to the plant, why doesn't FDA review biotech products under the same food additive regulations that it reviews food colors and preservatives?

Dr. Henney: The food additive provision of the law ensures that a substance with an unknown safety profile is not added to food without the manufacturer proving to the government that the additive is safe. This intense review, however, is not required under the law when a substance is generally recognized as safe (GRAS) by qualified experts. A substance's safety can be established by long history of use in food or when the nature of the substance and the information generally available to scientists about it is such that it doesn't raise significant safety issues.

In the case of bioengineered foods, we are talking about adding some DNA to the plant that directs the production of a specific protein. DNA already is present in all foods and is presumed to be GRAS. As I described before, adding an extra bit of DNA does not raise any food safety issues. As for the resulting proteins, they too are generally digested and metabolized and don't raise the kinds of food safety questions as are raised by novel chemicals in the diet. The proteins introduced into plants so far either have been pesticides or enzymes. The pesticide proteins, such as BT, would actually be regulated by EPA and go through its approval process before going on the market. The enzymes have been considered to be GRAS, so they have not gone through the food additive petition process.

FDA's consultation process aids companies in determining whether the protein they want to add to a food is generally recognized as safe. If FDA has concerns about the safety of the food, the product would have to go through the full food additive premarket approval process.

FDA Consumer: Why doesn't FDA require companies to tell consumers on the label that a food is bioengineered?

Dr. Henney: Traditional and bioengineered foods are all subject to the same labeling requirements. All labeling for a food product must be truthful and not misleading. If a bioengineered food is significantly different from its conventional counterpart—if the nutritional value changes or it causes allergies—it must be labeled to indicate that difference.

For example, genetic modifications in varieties of soybeans and canola changed the fatty acid composition in the oils of those plants. Foods using those oils must be labeled, including using a new standard name that indicates the bioengineered oil's difference from conventional soy and canola oils. If a food had a new allergy-causing protein introduced into it, the label would have to state that it contained the allergen.

We are not aware of any information that foods developed through genetic engineering differ as a class in quality, safety, or any other attribute from foods developed through conventional means. That's why there has been no requirement to add a special label saying that they are bioengineered. Companies are free to include in the labeling of a bioengineered product any statement as long as the labeling is truthful and not misleading. Obviously, a label that implies that a food is better than another because it was, or was not, bioengineered, would be misleading.

FDA Consumer: Overall, are you satisfied that FDA's current system for regulating bioengineered foods is protecting the public health?

Dr. Henney: Yes, I am convinced that the health of the American public is well protected by the current laws and procedures. I also recognize that this is a rapidly changing field, so FDA must stay on top of the science as biotechnology evolves and is used to make new kinds of modifications to foods. In addition, the agency is seeking public input about our policies and will continue to reach out to the public to help consumers understand the scientific issues and the agency's policies.

Not only must the food that Americans eat be safe, but consumers must have confidence in its safety, and confidence in the government's role in ensuring that safety. Policies that are grounded in science, that are developed through open and transparent processes, and that are implemented rigorously and communicated effectively are what have assured the consumers' confidence in an agency that has served this nation for nearly 100 years.

Risks and Rationality in Today's Culture of Fear

About This Activity

People often make choices about acceptable levels of risks based on emotion, not facts. As a result, some fears about the risks posed by technology are real, but others are not. In this activity, we describe some pitfalls of making choices based on perceived risks rather than actual risks, compare emotion-based risk perception to science-based risk assessment, and discuss the role that government regulatory agencies play in minimizing the potential risks associated with new products, including those developed using biotechnology. We also provide the data from the 1999 study, conducted by scientists at Cornell University, that triggered a media frenzy about the impact of Bt (for *Bacillus thuringiensis*) corn on monarch butterflies.

After providing students with a detailed description of the science of risk analysis, the activity outlined in the *Student Activity* asks them to determine what questions the regulatory agencies needed to ask in order to determine if Bt corn posed a significant risk to natural populations of monarch butterflies. In addition to describing the steps in a formal risk assessment, the *Student Activity* contains background information sufficient for preparing a list of essential questions without the students having to seek out other information resources. However, if you would like for the students to extend the risk assessment by answering the questions, then additional research work will be required. They will also need to seek outside sources of information if you want them to conduct the second essential component of the risk analysis, the risk management strategy, and to get a complete picture of the relative benefits of Bt corn compared to other methods of pest control.

In the description of the monarch butterfly and Bt corn case study provided below, you will find not only the list of questions that the students are being asked to develop, but also the answers to those questions. At the end of the chapter, we list suggestions for additional activities. Finally, you may also choose

to use the information provided in this chapter and the *Student Activity* that follows to analyze the issue of harm to monarch butterflies, based on the process provided in the previous chapter.

Introduction

Many citizens of industrialized countries seem to suffer great anxiety about the risks involved in simply living. Why is a large segment of society so anxious about risks when factual evidence shows that people in the industrialized world are living longer, healthier lives in the cleanest environments they have experienced for centuries?

A rational explanation is that new risks do arise. A number of emerging infectious diseases, such as human immunodeficiency virus-AIDS and West Nile virus infections, have begun to cause human health problems only recently. Another rational reason is related to scientific advances that can now illuminate risks that have always been there but were previously unknown. For example, viruses that are transmitted in sexual intercourse were a causative agent of cervical cancer centuries before scientists knew viruses existed. Diets high in salt and low in fiber were contributing to health problems long before physicians established the links.

However, much of the current angst gripping many people is irrational. Society's misplaced anxiety over trivial or nonexistent risks is due to a variety of factors.

- Many of the real, high-risk problems have been essentially solved (for now, at least) in the industrialized world. For millennia, microorganisms posed by far the most serious risks to human health through infectious diseases and microbial contamination of food. Refrigeration, better sanitation, and government-mandated food inspection eased the latter problem, and sewage treatment, antibiotics, and vaccination solved the former.

- There is a pervasive misunderstanding about the concept of toxicity. Toxicity is the capacity of a substance to do harm or injury of any kind. In reality, *everything* is toxic if given in high enough amounts. Adults have killed children by forcing them to drink too much water, but no one considers water to be a toxic substance. As Paracelsus said in the 16th century, "All substances are poisons, there is none which is not a poison. The dose differentiates a poison and a remedy." This is a fundamental principle in modern toxicology: the dose makes the poison. However, when the media report that a chemical has been shown to be toxic, they never provide information about dosage that will help the public place the finding in perspective.

- Exacerbating the previous misconception is the fact that the technical ability to measure substances that occur in small amounts has improved by many orders of magnitude. People seem to confuse the mere presence of a substance with its level of risk, but measuring techniques can detect quantities so small they have no real significance for health and safety (see the first sidebar in the *Student Activity*). Were vegetables safer before scientists learned that many of them naturally contain small amounts of toxins?

- Risks are much more publicized. Many people are preoccupied with knowing about anything that might possibly cause them harm, and the media indulge their obsession. At times, some reporters may slant the story by presenting the risks in ways that trigger concerns, often unnecessarily. It is not that they lie; they can tell the truth and mislead simultaneously. For example, a few years ago, major newspapers carried a front-page story about the relationship between hormone replacement therapy and a number of diseases, including breast cancer. All reported that a 10-year study of 20,000 postmenopausal women proved that those who took hormones were 25% more likely to get breast cancer than those who did not. These numbers referred to the *relative risk,* because they compared postmenopausal women who used hormones and those who did not. Few of the media reports provided the actual numbers, or the *absolute risk.* Of the 10,000 women taking hormones, 38 got breast cancer (0.38%); of the 10,000 who did not take hormones, 30 got breast cancer (0.30%). Therefore, approximately one additional woman per year (8 in 10 years) in the hormone replacement group contracted breast cancer. None of the media reports lied, because 38 *is* 25% more than 30, but reporting the actual numbers—0.38% compared to 0.30%, or one more per year—does not sound as scary, does it?

Social scientists have identified a number of psychological factors that skew perceptions of risk, causing people to inflate small risks or trivialize significant risks. They include the following:

- Voluntary versus involuntary risks
- Risks that they can control versus those they cannot
- Familiar versus unfamiliar risks
- Natural versus synthetic substances

We explain these factors in the *Student Activity* and also provide two tables of relative risks posed by familiar substances and activities (see Tables 37.2 and 37.4 in the *Student Activity*). Because of the importance to biotechnology of the misperceptions associated with "natural versus synthetic," we also devote considerable attention to elaborating the fallacy of assuming that natural equals safe and synthetic equals unsafe. We focus specifically on natural plant toxins (see Table 37.1 in the *Student Activity*).

Risk Analysis

The science of risk analysis was developed in the 1970s in response to a new set of congressional acts giving federal regulatory authorities increased oversight and responsibility for protecting consumers and the environment. These laws assign different responsibilities and authorities to the various federal agencies charged with evaluating the science-based risks of products and processes before they are placed on the market. The regulatory agencies can sometimes dictate how certain products can and cannot be used after they are commercialized. They also have the power to withdraw or recall products if circumstances later show the risks were greater than expected. Regulatory agencies are held accountable for the responsibilities that are legally assigned to them. On the other hand, these agencies cannot exceed the authority that the legislation grants to them.

In making decisions about whether to approve research studies, products for commercial sale, manufacturing processes, etc., regulators rely heavily on a formal risk assessment process. According to the National Academy of Sciences, a risk assessment is a "process by which scientific data are analyzed to describe the form, dimension, and characteristics of risk—that is, the likelihood of harm to humans or the environment." In conducting the risk assessment, regulators rely not only on scientific data, but also on empirical knowledge and scientific principles. Even though science and math are involved in assessing risks, there are no absolute answers to questions about risk levels because a risk is a probability statement. This uncertainty often frustrates nonscientists.

The first step in a risk assessment is identifying the hazard, which in risk assessment terminology is anything that could go wrong or might lead to injury or harm. A hazard is the potential to cause harm and is similar, in principle, to the concept of toxicity described above. As such, it is important to mentally distinguish hazard from risk. Just as the probability of harm depends on the dose of a toxin, the risk (probability of loss or injury) increases with the severity of the hazard. Therefore, in addition to identifying a hazard, those conducting a scientific risk assessment must establish the quantitative relationship between the hazard level, or dose, and the adverse effects on human health or the environment. This is known as a dose-response analysis and constitutes the second major part of the risk assessment.

The third step is to determine the likelihood of exposure to the level (dose) of the hazard that will pose a risk. In a risk assessment, the exposure estimate is a statement of the probability that there will be contact between the hazard and the thing that the hazard might harm or injure. Swimming pools are inherently hazardous but are not a risk to someone who has never been near one. The uncertainty of exposure is one of the reasons that risks are discussed in terms of probabilities instead of an absolute assessment of the seriousness of the hazard.

In summary, a hazard, which is the potential for something to cause harm, becomes a risk only if there is exposure; that is:

$$\text{Risk} = \text{hazardous dose} \times \text{exposure to that dose}$$

After identifying the hazard, determining the dose-response relationship, and evaluating the probability of exposure, the regulators put all of that information together in order to determine the overall level of potential risk, which in risk analysis is termed risk characterization, the final product of the risk assessment. Risk characterization is not the final step in the risk analysis, however. Before regulatory officials make a decision to approve a product or process, they evaluate whether and how the risk can be managed. Risk management is the set of activities that can be undertaken to control a hazard, minimize exposure, or both. If effective safeguards can be established, the level of risk can be decreased dramatically, because

$$\text{Risk} = \frac{\text{hazard} \times \text{exposure}}{\text{safeguards}}$$

These two conceptual formulas illustrate the first steps in using rational risk analysis to assess and

manage risks. By no means are these steps perfect or devoid of subjectivity, but they provide a starting point for comparing the relative risks of various products and activities. More information on each of the steps is provided in the *Student Activity*.

Balancing risks with benefits

Regulators can institute safeguards to get the risk level as low as possible, but it will never reach zero, because there is no such thing as risk free. Helping students understand this truth is one of the most important things you can do for them. The essential question to be answered is whether the level of risk is acceptable compared to the benefits. The overall risk level posed by a product or process, when used in ways stipulated by the regulatory agencies, must be very low before the agencies will consider approving it; however, they may be willing to accept a slightly higher level of risk if the benefits are substantial. The most familiar example of this practice is the willingness of the U.S. Food and Drug Administration (FDA) to approve pharmaceutical compounds to treat diseases and vaccines to prevent them even though there is some risk associated with their use. The number of people who benefit greatly exceeds the number harmed.

Assessing risks is a simple task compared with assessing benefits. Benefits are often very difficult to identify in the abstract. People need to experience them. In addition, some eventual benefits of a technology can be very difficult to predict, and even the predictable benefits are typically not distributed equally throughout society. Finally, because many benefits are based on emotions, values, and ethical principles, the concept of a benefit is subjective and varies from one person to another and one circumstance to another. The factors people use to measure benefits include the following:

- Saving lives
- Improving health
- Solving problems
- Improving quality of life
- Increasing emotional well-being
- Saving money

Objectives

After reading this chapter and performing some of the activities listed below, students will:

- Understand how emotions contribute to perceptions of risks.
- Be able to differentiate toxicity and hazard from risk.

- Understand the basic steps of a risk analysis, which include both risk assessment and risk management.
- Be better able to evaluate media coverage of the risks and benefits of technologies.
- Have learned a great deal about Bt, Bt corn, and monarch butterflies.

Preparations

- Read the *Student Activity,* because it contains a substantial amount of important information not included here.
- Make copies of the *Student Activity* if your students do not have the textbook.
- If you plan to go into detail about the government's role in protecting the public and the environment, it will be useful for you to familiarize yourself with various laws that give the regulatory agencies the authority to require safety tests before products can be marketed and to remove unsafe products from the market. The clearest examples are the federal laws governing the approval of pesticides and other chemicals, which are implemented by the U.S. Environmental Protection Agency (EPA), and the FDA's regulatory processes governing drug development (clinical trials) and manufacture. Each state also has companion laws to many federal laws that give the states the authority to act as a partner in the regulatory processes.

Resources

With respect to the monarch butterfly and Bt corn controversy, a number of land grant universities (see http://agribiotech.info for links to all land grant university websites) have created web pages specifically to correct misinformation on this issue. Some of the better ones are those of Cornell University (http://www.nysaes.cornell.edu/comm/gmo), Washington State University (http://www.aenews.wsu.edu), University of Nebraska at Lincoln (http://www.agbiosafety.unl.edu), and Colorado State University (http://cls.casa.colostate.edu/TransgenicCrops). Many of these universities also have excellent descriptions of the federal regulation of products of agricultural biotechnology and other issues associated with food and agricultural biotechnology. The Pew Initiative on Food and Biotechnology (http://www.pewagbiotech.org) also funded a study that summarized the media coverage of the monarch butterfly and Bt corn story.

The federal agencies charged with overseeing the safety and efficacy of products created through biotechnology, primarily the FDA, U.S. Department of Agriculture (USDA), and EPA, also have useful information on risk assessment and regulation on their websites.

A Case Study: Monarch Butterflies and Bt Corn

Companies developed Bt corn so that farmers could successfully control a major pest of corn, the European corn borer (ECB). Background information on the ECB, monarch butterflies, Bt, the Bt protein that kills lepidopteran caterpillars that consume it, and Bt corn can be found in the *Student Activity.*

The regulatory approval process

In the United States, three regulatory agencies required crop developers to provide them with sufficient amounts of test data and information from the scientific literature for their scientists to determine the potential risks of Bt corn to the environment, agriculture, and human health. Two agencies were responsible for assessing environmental impacts on nontarget organisms:

- The USDA has regulatory authority over greenhouses, where the first studies were conducted, and approves or rejects applications for small-scale field tests and interstate transport of transgenic crops. It also must provide final regulatory clearance for large-scale commercial planting.
- The EPA requires approval for field trials larger than 10 acres if the transgenic crop contains a pesticidal compound, such as the Bt protein. Along with the USDA, the EPA also must issue a final permit before Bt crops can be sold or distributed.

For the earliest varieties of Bt corn approved for sale in the United States, the step-by-step regulatory approval process, from submitting a request to the USDA for a permit for a small-scale field test of the Bt gene to inserting the gene into the target crop (corn) and, finally, introducing Bt corn seeds to the commercial market, took approximately 7 to 8 years.

Assessing Impacts on Nontarget Organisms

Before they permit field tests and commercial sale, the USDA and EPA require information on the possible negative effects of transgenic crops on insects and other animals. Companies that sell Bt corn conducted laboratory tests to establish the toxicity of the Bt protein and Bt corn plant material to a number of nontarget organisms, including the honeybee, earthworm, collembola (a soil insect), daphnia (aquatic invertebrate), green lacewing, ladybird

beetle, parasitic wasp, catfish, quail, and mouse. The results confirmed Bt's lack of toxicity to nonlepidopteran insects and other animals.

When the monarch story broke, the federal agencies were chided for not requiring the companies to submit data from research specifically focused on Bt corn and monarch butterflies. The agencies did not require studies on Bt corn and monarchs because they already knew many facts about Bt pesticides, monarch butterflies, and corn growing that were relevant to assessing risks.

Because Bt pesticidal sprays have been tested extensively and used for many decades, they knew the following:

*   Any part of a plant expressing the Bt protein would harm lepidopteran larvae that ate it in sufficient amounts, including monarch larvae.
*   Bt sprays are used widely to control gypsy moth outbreaks in forests and also by organic farmers and home gardeners. In spite of their widespread use in forestry, the impact of Bt sprays on monarchs is minor compared to the other factors threatening monarch populations, such as destruction of their overwintering habitat in Mexico and mowing along roadsides, a common milkweed habitat.

They also knew these critical facts related to potential risks:

*   75% of the North American monarch population does not occur where farmers grow Bt corn (see Figure 37.2A).
*   To harm monarchs, the larvae must consume the Bt protein. Monarch larvae eat only milkweed plants, so the only part of the Bt corn plant that could pose a risk is pollen that landed on milkweed plants.
*   Throughout the breeding range of monarchs in the United States, most milkweeds do not occur

near cornfields. Therefore, across the United States, comparatively few monarch eggs would be laid on milkweeds near Bt cornfields.
*   Monarch females lay eggs before corn releases pollen throughout most of the United States.
*   Lepidopteran larvae are capable of determining the palatability of food and, if given a choice, avoid harmful and distasteful substances.

The regulatory scientists looked at these facts and determined that monarchs were much more likely to be exposed to Bt from large-scale spraying to control outbreaks of gypsy moths and other forest lepidopteran pests than from Bt corn pollen.

The regulators also evaluated the benefits of growing Bt corn to farmers, agricultural ecosystems, the environment, and human health. They projected annual benefits to Midwestern corn growers of anywhere between $38 and $200 million, in spite of the higher seed costs. The range is so wide because predicting when and where ECB infestations will occur is quite difficult, and their impact on yields varies greatly from year to year (Figure 37.1). Because most farmers do not use insecticides to control the ECB, the agencies did not expect to see a large decrease in the total amount of insecticides used per acre. Even so, the scientists at the EPA concluded that because of the large number of corn acres in the United States, even a small percentage decrease in insecticide use could have important positive effects on beneficial insects and farm workers.

In summary, USDA and EPA regulators who reviewed requests for testing and commercializing Bt corn recognized that widespread planting of Bt corn could potentially harm nontarget lepidopteran insects, including monarchs. However, they also felt that the benefits to lepidopterans and all other insects provided by fewer insecticide applications far outweighed the risks.

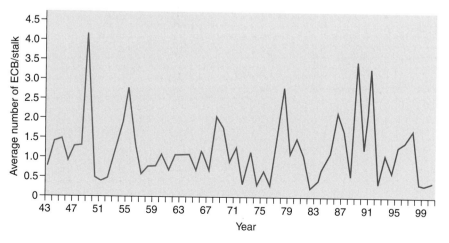

Figure 37.1 European corn borer outbreaks. (Source: Gianessi and Carpenter, National Council for Food and Agricultural Policy, 1999.)

Risk Analysis: Bt Corn and Monarch Butterflies

Before reading this section, please read the *Student Activity,* which provides essential background information. The *Student Activity* revolves around getting students to list the types of questions that must be asked and answered before regulatory officials are able to make definitive comments about the risks of Bt corn to natural populations of monarch butterflies. In deciding whether to grant approvals for field tests and commercial use, the regulatory agencies work carefully and methodically, reviewing all of the data relevant to the question at hand. Their ultimate goal is to use scientific facts to make rational decisions about approving or denying requests. We suspect that as you read this section, you will be struck by the tediousness of a science-based risk assessment, especially compared to those sexy headlines about toxic pollen and killer corn. This is precisely the point we hope to make. Although we are not interested in boring you, the truth is that risk assessments, done well, are *exceptionally* tedious.

Recall that risk depends on both the hazard and probability of exposure to the hazard. To assess the risks of Bt corn pollen to monarchs, regulators must determine:

1. If Bt corn pollen is hazardous to monarch larvae (This step is unnecessary in this case, because it has been widely known for well over 50 years that the Bt protein can kill lepidopteran caterpillars that consume it.)
2. The dose required for harm or injury
3. The likelihood that an individual larva will be exposed to that dose

Finally, in order to have an adverse effect on the monarch population (as opposed to an individual larva), a significant proportion of the larvae must be harmed.

As we said earlier, regulators knew a great deal about monarch butterflies, Bt pesticidal sprays, and Bt corn that allowed them to make the decision to approve Bt corn. However, when the story about monarch butterflies and Bt corn broke, government agencies and the companies that sell Bt corn were under pressure to produce more data to support the sale and use of Bt corn. Many of the data we discuss below were collected through special grants, provided by the USDA and industry, for the sole purpose of clarifying the risks of Bt corn to monarch butterfly populations in the United States.

Determining the hazardous dose

Bt corn contains the gene from *B. thuringiensis* subsp. *kurstaki,* which encodes a protein specifically toxic to lepidopteran insects, including monarch larvae, that ingest any part of the plant expressing the Bt gene. The Cornell study does not provide any information on the number of Bt pollen grains that must be consumed in order to observe the lethal or developmental effects of Bt pollen on 3-day-old larvae, nor does it provide other important information, such as the toxicity of Bt pollen to older, larger larvae, which are known to be less susceptible to Bt sprays. Subsequent research, conducted at Iowa State University (http://www.ent.iastate.edu), provided specific information on the following:

- Bt pollen dose that causes significant harm
- Relationship of the toxic dose to larval age (Table 37.1)

Table 37.1 Toxicities of pollen from different varieties of Bt corn

Corn pollen density (grains/cm)	Variety	Mortality (%) of larvae at age:	
		Less than 12 h	12–36 h
1,300	Bt 176	100	69
	Bt 11	60	56
	Non-Bt	60	12
135	Bt 176	70	37
	Bt 11	60	25
	Non-Bt	0	0
14	Bt 176	—[a]	—
	Bt 11	—	—
	Non-Bt	—	—

[a]No difference among the corn varieties or the ages of the larvae. All survived.

The Iowa State University scientists established the toxicity levels of two varieties of Bt corn (Bt 176 and Bt 11) by forcing monarch larvae to eat milkweed leaves coated with different amounts of corn pollen. A non-Bt corn variety served as the control. The percentages in Table 37.1 refer to percent mortality, measured at a specific time (85 hours after feeding began). Note that newly hatched larvae are much more susceptible to Bt toxins than larvae that are 1 to 3 days old. In addition, the level of hazard posed by Bt pollen varies with the transgenic corn variety, because some are much more toxic than others. Bt 176 is significantly more toxic than Bt 11; interestingly, non-Bt corn pollen can be toxic in large enough amounts. Finally, as expected, the toxicity depends on the dose. At 14 grains of pollen/centimeter, Bt pollen is not toxic to larvae of any age.

In summary, Bt corn pollen can be hazardous to monarch larvae, but the hazard level varies with the dose, the Bt corn variety, and larval age.

Assessing exposure

Not surprisingly, Bt corn pollen can be hazardous to monarch larvae, but what is its risk (probability of harm) to monarch larvae? To be harmed by Bt pollen, a larva must be in the wrong place (on a milkweed plant near a cornfield planted in certain Bt varieties, but not others) at the wrong time (when it is less than 3 days old). The probability that a young larva might be exposed to a sufficient dose of pollen depends on many variables, including the following:

- Corn pollen movement
- Milkweed abundance and distribution
- Female monarch egg-laying preferences between different habitats
- Temporal coincidence of pollen shedding and larval hatching/feeding
- Larval feeding behavior

Therefore, to determine the risk to an individual caterpillar, data must be collected on all of the variables that determine the degree of exposure.

Then, to have a population level effect, a significant proportion of larvae must be in the wrong place at the wrong time. The amount of geographic overlap between Bt corn-growing areas and monarch breeding populations defines the proportion of the population that might be at risk.

Do Monarch Larvae Occur in Areas Where Bt Corn Is Grown?

Yes, monarch larvae do occur in areas where Bt corn is grown. Figure 37.2B shows where most corn (over 90%) is grown in the United States, which also coincides with the regions with the worst infestations of ECB. Farmers in areas with regular outbreaks of ECB are much more likely to plant Bt field corn than farmers in parts of the country where ECB infestations do not occur. Therefore, it appears that only certain monarchs in the eastern population would coexist with the Bt corn varieties currently marketed. Approximately 25% of the U.S. monarch population is concentrated in the Midwest, the primary corn-growing region of the country.

Figure 37.2 Geographic distribution of (A) monarch breeding and (B) corn-growing areas in the United States. (Maps redrawn from the National Biology Information Infrastructure of the U.S. Geological Survey and USDA-NASS.)

A. Monarch breeding

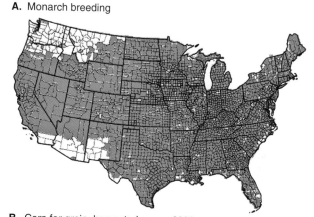

B. Corn for grain, harvested acres: 2002

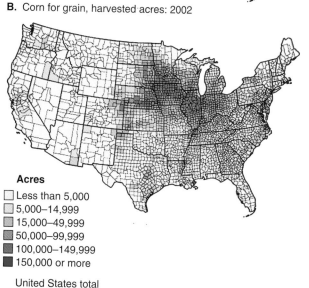

Acres
- ☐ Less than 5,000
- ☐ 5,000–14,999
- ☐ 15,000–49,999
- ☐ 50,000–99,999
- ☐ 100,000–149,999
- ☐ 150,000 or more

United States total
68,230,523

For purposes of our discussion, we will not differentiate between the different varieties of Bt corn, even though that information is essential for accurately estimating the level of risk to monarchs. As you can see in Table 37.1, Bt 176 is significantly more hazardous than all other Bt varieties. When the monarch-Bt corn story broke, Bt 176 was planted on less than 2% of the field corn acres. The Swiss company that marketed Bt 176, Syngenta, has since commercialized other Bt varieties with significantly lower toxicities, and Bt 176 is no longer on the market.

What Is the Likelihood That Milkweed Plants Will Occur near Bt Cornfields?

Monarch larvae eat only milkweed plants, so the degree of exposure of monarch larvae to Bt corn is related to the abundance, geographic distribution, and location of milkweed plants and the percentage of corn acreage planted with a Bt variety. The most common milkweed species are usually found in open areas: along roadsides, in ditches, at the edges of forests, and in open fields, such as pastures. A number of studies have looked at the relative probability that a female monarch butterfly will encounter a milkweed plant near a cornfield compared to other areas.

In Iowa, a female monarch is more likely to encounter roadside milkweeds than milkweeds next to cornfields, because the area around cornfields contains 85% fewer milkweed plants (22 milkweed patches per roadside acre compared to 3 milkweed patches per acre near a cornfield). However, the probability of encountering milkweeds near corn is greater than the 15% implied above, because in Iowa, 19% of the roadside area is next to cornfields. Of those cornfields, approximately 17 to 20% were planted in Bt corn in 1999.

Other studies have compared milkweed densities, instead of the number of patches, in agricultural and nonagricultural areas. In all study sites (central Minnesota, Wisconsin, Iowa, Ontario, and Maryland), milkweed densities are significantly higher in nonagricultural rural areas (prairies, pastures, and old fields) than in agricultural areas, but the magnitudes vary from 4 to 7 times greater in the Midwest to over 100 times greater in Ontario. However, when the relative amounts of agricultural and nonagricultural land are included in the calculations, female monarchs in Minnesota, Wisconsin, and Iowa are more likely to encounter milkweed in agricultural than nonagricultural areas; those in Ontario and Maryland are not.

What Is the Likelihood Bt Corn Pollen Will Occur on Milkweed Plants Growing near Cornfields?

Government regulatory scientists were able to get a general handle on the question of the likelihood that Bt corn pollen would occur on milkweed plants growing near cornfields by looking at many years of data on corn pollen movement. Because corn pollen is relatively large and heavy, it does not move very far outside the borders of the cornfield. Decades ago, agricultural scientists had shown that 70% of the pollen released by corn plants stays within the confines of the cornfield, and only 10% of the pollen travels farther than 3 meters from the edge of the field (Figure 37.3).

Researchers measured the actual density of corn pollen on milkweed plants in and around cornfields and found the average pollen density on milkweeds planted next to cornfields shedding pollen was significantly less than the dose that is hazardous to young monarch larvae.

In addition, rainfall significantly decreases the amount of corn pollen on milkweed plants, especially the upper leaves, which are the preferred feeding sites of young larvae. A single rain removed 54 to 86% of the pollen from milkweed plants, and 55% of the young larvae on milkweed plants are found on the upper leaves.

What Is the Likelihood a Female Will Lay Eggs on a Milkweed Plant with Bt Corn Pollen?

Using only the data on corn pollen movement and milkweed abundance and distribution, insect ecol-

Figure 37.3 Corn pollen dispersal. Many studies to measure corn pollen dispersal from cornfields have consistently demonstrated that pollen levels decrease rapidly as the distance from the cornfield edge increases.

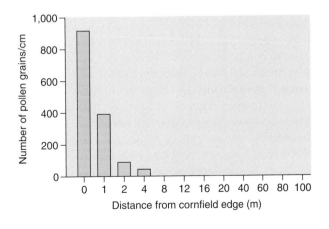

Societal Issues

ogists assumed most female monarchs would lay eggs on milkweed plants too far away from Bt cornfields to have much Bt corn pollen on the leaves. They also assumed females would rarely lay their eggs on milkweeds within cornfields, because milkweeds, which are much shorter than corn plants, are presumably more difficult for females to locate.

Much to their surprise, female egg-laying behavior did not mirror milkweed distribution. Egg densities were higher on milkweeds located in cornfields in Iowa, Wisconsin, and Minnesota but were the same in nonagricultural areas and cornfields in Ontario and Maryland. Even more interesting, the survival rate of larvae in cornfields, including Bt cornfields, was *higher* than those not in cornfields, presumably due to reduced bird predation. Apparently, female monarchs have an easier time seeing milkweeds in cornfields than birds do.

What Is the Likelihood Bt Corn Pollen Will Occur on Milkweeds When Larvae Are Feeding?

The likelihood that Bt corn pollen will occur on milkweeds when larvae are feeding depends on the relative timing of monarch egg laying and corn pollen shedding. A given field of corn sheds pollen for 7 to 10 days sometime between June and August, the corn-growing season in North America. The approximate date of pollen shedding within that 2-month period varies with the latitude at which the corn grows. To be harmed by Bt corn pollen, young larvae must be feeding on milkweeds next to or in the cornfield during the period when pollen is shed.

Because of the migration rates of monarchs that overwinter in Mexico, no overlap occurs between monarch egg laying and corn flowering in the southern parts of the Bt corn-growing region. Some overlap between larval feeding and pollen shedding occurs in central Iowa (15%), and it reaches a maximum in Ontario (60%).

What Is the Probability a Larva Will Consume Enough Bt Corn Pollen To Be Harmed?

Lepidopteran larvae exhibit clear feeding preferences. Data from the Cornell study show that monarch larvae do not readily eat milkweed leaves coated with pollen whether or not that pollen contains the Bt gene. More than likely, if a feeding larva encounters pollen, the larva will avoid it and feed elsewhere. USDA scientists gave monarch larvae a choice between milkweeds with and without Bt corn pollen, and the majority chose the pollen-free leaves.

In addition, a laboratory study gave a different lepidopteran species a choice between insect diets with and without Bt protein, and the larvae consistently avoided the diet with Bt.

What Is the Risk of Bt Corn to Monarch Populations?

Finally, the potential risk of growing Bt corn to the monarch butterfly population has been clarified. The proportion of the monarch population that may be harmed by Bt corn pollen is limited to the Bt corn-growing area that is north of central Iowa and east of the Rocky Mountains (Figure 37.4). In those areas, only very young larvae feeding on milkweed plants within or immediately adjacent to cornfields, planted in certain Bt varieties, are at risk of consuming harmful amounts of Bt corn pollen.

Because of the low probability of exposure of susceptible monarch larvae to hazardous doses of Bt corn pollen, in 2001 the EPA stated: "While there is a small chance that one in 100,000 caterpillars could be affected by Bt corn pollen, research suggests even those larvae will mature into healthy butterflies."

Risk management

The risk level depends on exposure to a hazardous dose and the safeguards that can be put into place to decrease risk. What safeguards might decrease the risk of Bt corn pollen to monarchs even further?

- Because the toxicity levels of pollen from Bt corn varieties differ, in areas where corn pollen shedding and larvae feeding overlap, farmers could plant varieties with the lowest toxicity to monarch larvae.

Figure 37.4 Corn pollen shedding and larval feeding. The timing of corn pollen shedding and monarch egg laying affects the probability that young larvae will be exposed to Bt corn pollen. The blue lines indicate the areas where overlap occurs.

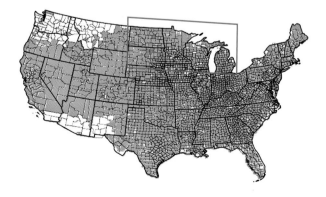

- It is now possible to genetically engineer plants to limit gene expression to specific tissues. Varieties expressing the Bt gene only in the cornstalk would offer protection against the ECB without harming monarchs, because monarchs do not feed on corn plants.
- The EPA requires farmers growing Bt crops, including corn, to plant non-Bt varieties to slow the evolution of resistance to the Bt protein. Planting a border of non-Bt corn around a Bt cornfield would further decrease the probability that Bt pollen would drift to milkweed plants near the field.
- Not mowing roadsides in monarch-rich areas or planting milkweeds at sites distant from cornfields increases the probability that an egg-laying female would encounter milkweeds uncontaminated with Bt corn pollen.
- Milkweeds can be removed from Bt cornfields so that larvae will not be exposed to high pollen levels, but is this wise, since it seems monarch larvae survive better on milkweeds in cornfields, including Bt cornfields?

Benefits of Bt corn

For a fair analysis of Bt corn, students should also assess the potential benefits of Bt corn and contrast these benefits with the risks. Another way to frame the question is to ask them to consider the risks of not growing Bt corn, because Bt corn provides advantages to farmers, farm laborers, consumers, and the environment compared with current corn-growing practices.

- A number of studies have shown that Bt corn has significantly smaller amounts of mycotoxins than non-Bt corn does. These mycotoxins, which are fatal to livestock and harmful to humans, are produced by fungi that invade the corn plant, especially at sites of insect damage (see Figure 1.26). Therefore, decreasing corn borers leads to a secondary benefit of improving food safety.
- Many feeding studies, including those submitted to U.S. regulatory agencies, have shown that the Bt protein, in contrast to chemical pesticides currently used, does not harm beneficial insects, such as ladybird beetles and honeybees, and is safe for freshwater invertebrates, earthworms, other soil organisms, birds, and mammals.
- Various studies have demonstrated that farmers using Bt corn have slightly decreased the amount of insecticides they apply to corn. One example is the work of the National Center for Food and Agricultural Policy (http://www.ncfap.org). This decrease in insecticide use is beneficial to all in-

sects, including monarch butterflies, because spray drifts of pesticides can extend to large areas owing to the small size of droplets. For chemical insecticides, the typical impact on nontarget insects in or adjacent to treated fields approaches 100%.
- Fewer insecticide applications and a reduction in insecticide drift also benefit farm workers.
- Bt corn offers almost 100% protection against the ECB, providing U.S. farmers with an additional $1 billion of income per year.

Additional Activities

1. Select any familiar, pervasive technology on which society depends, and have the students walk through the steps in a risk-benefit analysis, identifying the risks, assessing the probability of each risk occurring, assessing the consequences, and analyzing the benefits.
2. Go back in time and evaluate that same technology before its widespread implementation. What risks and benefits would the students have predicted before that technology's development? What is the relationship between predicted and actual risks and benefits?
3. Select a future application of biotechnology or another emerging technology, and have students conduct a risk-benefit analysis. Pay particular attention to their assessment process, not the results. What information sources would they use to analyze a future development? How applicable is this information?
4. Select a risk that is considered a personal risk, such as skydiving, cigarette smoking, or drug use, and have students analyze the societal ramifications of individual risk taking. What is the appropriate role of the individual, society, or government in those cases?
5. Ask students to analyze the media coverage of a recent scientific discovery or technological advance that was reported in the newspaper, on radio, or on television. Compare and contrast the media coverage with the primary literature on the subject. Many scientific articles on biotechnology that receive popular coverage are published in the journals *Science* and *Nature,* which are available in many public libraries. A less formal, but still informative, assignment is to ask students to monitor media coverage of technology. Is it balanced? Is it accurate? (A balanced story is not necessarily an accurate story.) Is there mention of both the benefits and the risks of the technology in question?
6. Have students develop a case study similar to the one provided for decaffeinated coffee (see the

second sidebar in the *Student Activity*). Aaron Wildavsky, a political science professor at the University of California at Berkeley, asked his students to research the history and current scientific status of some of the major environmental issues of the last half of the 20th century: Love Canal, DDT (dichlorodiphenyltrichloroethane), Alar, etc. Their detailed case studies are published in Wildavsky's book (see *Selected Readings* below), which will serve as an excellent resource for you if you assign this activity to your students.

7. Have students write a short essay explaining either of the following quotes: "To be alive is to be at risk," Daniel Koshland, *Science*, 1987, and "Only the dose makes the poison," Paracelsus, 16th century.

Selected Readings

Altheide, David. 2002. *Creating Fear: News and the Construction of Crisis.* Aldine de Gruyter, New York, NY.

Ames, Bruce, et al. 1987. Ranking possible carcinogenic hazards. *Science* 236:271–279.

Cook, R. J. 2000. Toward science-based risk assessments for the approval and use of plants in agricultural and other environments, p. 123–131. *In* G. J. Persley and M. M. Lantin (ed.), *Agricultural Biotechnology and the Poor.* CGIAR Secretariat, The World Bank, Washington, DC.

Furedi, Frank. 2002. *Culture of Fear: Risk-Taking and the Morality of Low Expectation.* Continuum, New York, NY.

Gilovitch, Thomas, Dale Griffin, and Daniel Kahneman (ed.). 2002. *Heuristics and Biases: the Psychology of Intuitive Judgment.* Cambridge University Press, Cambridge, United Kingdom.

Glassner, Barry. 1999. *The Culture of Fear.* Basic Books, New York, NY.

Glimcher, Paul W. 2003. *Decisions, Uncertainty and the Brain.* MIT Press, Cambridge, MA.

Hahn, Robert (ed.). 1996. *Risks, Costs and Lives Saved: Getting Better Results from Regulation.* Oxford University Press, New York, NY.

Horowitz, Daniel. 2004. *The Anxieties of Affluence.* University of Massachusetts Press. Amherst, MA.

Kahneman, Daniel, and Amos Tversky (ed.). 2000. *Choices, Values and Frames.* Cambridge University Press, Cambridge, United Kingdom.

Losey, John E., Linda S. Raynor, and Maureen E. Carter. 1999. Transgenic pollen harms monarch larvae. *Nature* 399:214.

National Research Council. 1993. *Issues in Risk Assessment.* National Academy Press, Washington, DC.

Slovic, Paul (ed.). 2000. *The Perception of Risk.* Earthscan, London, United Kingdom.

Sustein, Cass R. 2002. *Risk and Reason: Safety, Law and the Environment.* Cambridge University Press, New York, NY.

Wildavsky, Aaron. 1997. *But Is It True: a Citizen's Guide to Environment, Health and Safety Issues.* Harvard University Press, Cambridge, MA.

Societal Issues

The Risks of Technology: Perception, Reality, or Both?

Introduction

Day in and day out, consciously or unconsciously, you take risks. You ride in cars, cross streets, and play sports without thinking twice about the risks involved, because you have "decided," usually without thinking, that the benefits outweigh the risks. Sometimes your choices make good sense and can be supported with factual scientific evidence, but often the level of risk determined through an objective, science-based evaluation and the amount of perceived risk bear little resemblance to each other. People accept huge risks and worry about trivial ones—cigarette smokers protest nuclear power plants; people risk pregnancy and AIDS by having unprotected sex and then worry about pesticide residues on their food, yet smoking and unprotected sex have harmed millions more people than accidents at nuclear power plants or pesticides on food.

Choices about acceptable and unacceptable risks are irrational because they are often based on emotion, not facts. As a result, some fears are real, but others are not.

While some believe in the motto "better safe than sorry," being overly cautious is often not in your best interest. Unnecessary fears encourage people to fret over perceived but nonexistent problems, and stress is harmful to health. Risk misperceptions also lead people to avoid products that are perfectly safe and spend more money on products with no additional benefits. If a significant proportion of the population or a very vocal minority is unnecessarily fearful of trivial risks, new and improved products are not developed, and government agencies spend their limited resources protecting fearful consumers from small or nonexistent risks. In either case, society as a whole suffers.

Risk Perception

What triggers emotional responses and prevents accurate assessments of real risks? Social scientists have identified a number of psychological elements that skew perceptions of risk, causing people to inflate small risks or trivialize significant risks.

- Voluntary versus involuntary. If people consciously decide to take a risk (a voluntary risk, like smoking), they accept a much higher level of risk than if they feel they had no choice in the matter (an involuntary risk, such as the location of a nuclear power plant).
- Control versus no control. People are more fearful of risky situations over which they have no control. Many people choose to ride in their cars rather than fly in airplanes because they fear flying, even though more people are hurt or killed in auto accidents than in plane crashes. Despite the evidence, people perceive the risks of flying as greater than the risks of driving because they are not in control of the plane.
- Familiar versus unfamiliar. Risk perceptions also vary with the degree of familiarity with the risk. People accept the large but well-known risks involved in driving a car, sunbathing, and drinking alcohol. New risks elicit concerns and avoidance, not because the risk is greater, but because they are unfamiliar.
- Natural versus synthetic. Another factor affecting risk perceptions is whether the risk is natural or synthetic, i.e., "man-made." People tend to view nature as benevolent, despite all the evidence to the contrary. Microbes are natural, yet millions of people in developing countries die from diarrhea caused by microbial contamination of water. In developed countries, synthetic products and processes, such as chlorinated water and sewage treatment, prevent this same tragedy from occurring.

Irrational perceptions of the relative risks of natural and synthetic products and processes deserve additional attention, because objectively analyzing the risks of some biotechnology applications requires clearly understanding the pitfalls of equating natural with risk free and synthetic with risk.

Comparing risks: natural versus synthetic

Some of the most instructive examples of problems caused by an unthinking preference for natural products over synthetic products come from issues surrounding food safety. Little has been done to help today's consumer place the risks of synthetic products versus natural products in perspective. As a result, many people have become increasingly anxious about the safety of a food supply that is safer than at any other time in history. Below, we provide a rational discussion of chemicals and food safety to demonstrate how misleading the reflexive assumption "natural is good; synthetic is bad" can be.

Chemical Toxins and Crop Plants

When you read that heading, the first word that popped into your head was probably "pesticide," which is appropriate. However, like most people, you probably think of pesticides solely as synthetic chemicals marketed by companies and sold to farmers, which is only partially accurate. Long before companies began manufacturing pesticidal chemicals synthetically, plants were making them naturally. Most plants naturally contain chemicals that are toxic when tested on laboratory animals; some plant toxins naturally occur at levels high enough to seriously harm grazing livestock.

Plants did not evolve to serve as food for humans, livestock, or any other organism. For millions of years, they have done their very best *not* to be eaten by making hundreds of different chemicals to ward off would-be predators. Using breeding and muta-genesis, plant scientists have decreased the amounts of natural toxins in most food crops. Without their efforts, many of the plants Mother Nature cooked up would not be fit for human consumption. However, despite years of work to strip food crops of these chemicals, many crop plants still contain small amounts of chemical toxins (Table 37.1). According to Bruce Ames, the biochemist who developed the most widely used toxicology test for measuring the carcinogenic (cancer-causing) potential of chemicals, in any given meal, an individual consumes about 100 to 150 natural carcinogens and 10,000 times more natural carcinogens than synthetic carcinogens.

Does this mean you should stop eating plants that contain natural toxins? Of course not. You can see from the crops listed in Table 37.1 that some of the healthiest foods also contain natural toxins. The benefits of eating grains, fruits, and vegetables greatly outweigh the negligible to nonexistent risks posed by consuming the naturally occurring chemicals they contain. Most plant toxins occur in concentrations that are so low they have no effect on human health, unless you eat large amounts of a single fruit or vegetable every day. The issue is not the safety of fruits and vegetables but the public's concept of toxicity, in conjunction with the remarkable capacity scientists now have to measure increasingly trivial amounts of substances (see the first sidebar).

Toxicity is the capacity of a substance to do harm or injury of any kind. In reality, *everything* is toxic in large enough amounts. Adults have killed children by forcing them to drink too much water, but no one

Table 37.1 Natural plant toxins

Chemical	Plant examples	Physiological effects[a]
Cyanogenic glycosides	Almonds, cherries, lima bean	Chewing plant material releases hydrogen cyanide, which causes death by blocking cell respiration.
Glucosinolates	Broccoli, cabbage, peanut, soybean, onion	Inhibit thyroid function, leading to an enlarged thyroid gland, or goiter; low levels of thyroid hormone inhibit growth and reproduction
Glycoalkaloids	Potato, tomato	Interfere with nervous system function, leading to nausea, vomiting, difficulty breathing, and death; also cause birth defects
Lectins	Most cereals, beans, potatoes	Stimulate mitosis, red blood cell agglutination, decreased nutrient uptake/absorption
Oxalic acid	Spinach, rhubarb, tomato	Reduces availability of essential minerals, such as calcium and iron
Phenols	Most fruits and vegetables, cereals, soybean, potato, tea, coffee	Destroy thiamine; raise cholesterol; estrogen mimic
Coumarins	Celery, parsley, parsnips, figs	Light-activated carcinogens; skin irritation

[a]Chemical toxins and carcinogens occur at low levels in most food crops. To assess the physiological effects and determine the toxic and carcinogenic doses, toxicologists feed laboratory animals the chemicals in amounts that are much higher than natural levels in plants. The table contains only a few of the examples of the naturally occurring chemicals known to be toxic or carcinogenic.

How Much Is a Billion?

People living today have become accustomed to hearing the media mention numbers that are so tremendous the human mind is incapable of visualizing them in the abstract: a budget deficit of trillions of dollars; a trade imbalance of billions; 40 million people suffering from AIDS. Putting large numbers in a familiar context can help. A billion seconds ago, World War II ended. A billion minutes ago, St. Paul was writing the Epistles.

Visualizing incredibly small numbers in the abstract presents problems, as well. Part of the difficulty in accurately evaluating statements about the risks of substances found in food, water, and air stems from the extraordinary capacity to measure chemical compounds. Highly sensitive methods for detecting and measuring substances can identify quantities so miniscule they have no real significance for health and safety. The quantities are expressed as parts per million, parts per billion, and, more recently, parts per trillion. What do these values really mean?

What is 1 part per million?
1 cent in $10,000
1 ounce in 62,500 pounds
1 inch in 28,000 yards (280 football fields)
1 pancake in a stack 4 miles high
1 drop in 1,000 quarts of water

How much is 1 part per billion?
1 inch in 16,000 miles
1 minute in 2,000 years
1 cent in 10 million dollars
1 drop in a million quarts of water
1 soybean in a silo 50 feet tall and 30 feet wide

How much is 1 part per trillion?
1 inch in 16,000,000 miles
1 second in 32,000 years
1 needle in a 100,000-ton haystack

---

thinks of water as a toxic substance. Paracelsus, a Swiss physician who reformed the practice of medicine in the 16th century, said it best: "All substances are poisons, there is none which is not a poison. The dose differentiates a poison and a remedy." This is a fundamental principle in modern toxicology: the dose makes the poison. While you probably can accept this way of thinking about certain helpful substances, such as medicines or even vitamins, thinking of water, oxygen, and glucose as poisons or toxins is difficult, isn't it?

Many people are frightened by the use of synthetic chemicals on food crops because they have heard

that these chemicals are "toxic" and "cancer causing," but are all synthetic chemicals more harmful than substances people readily ingest, like coffee and soft drinks, not to mention alcohol? No (Table 37.2). For example, in a study to assess the toxicities of various compounds, half the rats died when given 233 mg of caffeine per kg of body weight, but it took more than 10 times that amount of glyphosate (4,500 mg glyphosate/kg body weight), which is the active ingredient in the herbicide Roundup, to cause the same percentage of deaths as 233 mg of caffeine.

Does that mean that all synthetic chemicals are so safe they can be ignored as health risks? No. Our point is simply to demonstrate that it is in your best interest to keep issues of chemical toxicology and food safety in perspective. Things are never as simple as people would like them to be.

Risk Analysis

The factors driving your concept of risk—emotion or fact—may or may not seem particularly important to you, yet they are. The risks you are willing to assume and the experiences or products you avoid because of faulty assumptions and misinformation affect the quality of your life and the lives of those

Table 37.2 Carcinogenic substances

Substance	Carcinogenic potential[a]
Red wine	5.0
Beer	3.0
Edible mushrooms	0.1
Peanut butter	0.03
Chlorinated water	0.001
Polychlorinated biphenyls (PCBs)	0.0002

[a]The higher the number, the greater the cancer-causing potential. The carcinogenic potential of peanut butter is due to the toxin aflatoxin, produced by a mold that commonly infects peanuts and other crops.

Societal Issues

around you. Thus, even though it may be tempting to let misperceptions and emotions shape your ideas about risky products and activities, there are risks in misperceiving risks.

Fact-based risk analysis provides much needed perspective on questions of real and perceived risks. Although many people understandably equate toxicity with risk, they are quite different and should be thought of as separate. Even though arsenic is known to be highly toxic, it poses no risk to people who do not consume it. Risk is the *probability* of loss or injury, and risk analysis is a science, developed in the 1970s, that uses formal methodologies to both assess risks and develop strategies and tactics to manage risks. Although many different types of risks are amenable to risk analysis, for the purposes of our discussion, risk analysis relates to the safety of biotechnology products and processes.

Assessing and managing risks

In the United States, three regulatory agencies have the primary responsibility for assessing the safety and efficacy of biotechnology products and processes. The responsibilities and authorities of the different agencies were established by acts of Congress that were passed many years before the development of recombinant DNA techniques. Table 37.3 lists the transgenic products over which these regulatory agencies have approval authority, but they have many more responsibilities and types of authority than product approval of transgenic organisms.

In meeting their responsibilities, these agencies conduct formal risk assessments, a component of risk analysis, prior to the development and commercialization of many products, not only biotechnology products. In addition, once the federal government has approved products for commercial sale and consumer use, many of the regulatory agencies also have the power to remove the products from the marketplace. In conducting a risk assessment, government regulators attempt to maximize the contributions of factual information and scientific data to regulatory determinations of product safety. In the case of biotechnology, products and processes are assessed for both environmental and human health impacts and risks.

Regulators rely on empirical knowledge, scientific principles, and scientific data, and they use methodical, objective analysis to characterize the risk associated with a specific activity or product. Risk assessments are often quantitative evaluations, but we will skip the mathematics and limit our discussion to the conceptual aspects. Even though science and math are involved in assessing risks, there are no absolute answers to questions about risk levels because a risk is a probability statement. This uncertainty often frustrates nonscientists.

The first step in a risk assessment is identifying the hazard, which is anything that could go wrong or might lead to injury or harm. A hazard is the potential to cause harm and is similar to the concept of toxicity described above. As such, it is important to mentally distinguish hazard from risk. Identifying a possible hazard is not equivalent to identifying the risk level any more than saying water can be toxic means you should not drink water. Just as the probability of harm depends on the dose of a toxin, risk depends on both the hazard and exposure to the

Table 37.3 U.S. federal regulatory agencies for transgenic products and processes

Regulatory agency	Responsibility	Product approval authority
FDA	To protect consumers by ensuring food/feed safety and quality; ensuring drug safety, quality, and efficacy; and requiring appropriate labels, related to health and safety, on food, feed, and drugs	Food, food additives (including microbes), animal feed, drugs, vaccines, diagnostics, medical devices
USDA	To protect consumers, agriculture, and the environment by assessing impacts of products on agriculture and the environment; ensuring food safety of meat, milk, and eggs; ensuring the safety and efficacy of animal vaccines; and assessing their potential impacts on the environment and human health	Plants, plant parts, animals, some microbes, animal vaccines
EPA	To protect consumers and the environment by assessing potential impacts of products on the environment, human health, and food/feed safety	Pesticides, including insect- and disease-resistant plants, and some microbes

hazard. The risk (probability of loss or injury) increases as the severity of the hazard and/or exposure increases. Therefore, in addition to identifying a hazard, those who conduct a scientific risk assessment must establish the quantitative relationship between the dose (exposure level) and the adverse effects on human health or the environment. This is known as a dose-response analysis.

The task of the risk assessors, then, is to determine the likelihood of exposure to the level (dose) of the hazard that they have determined will cause problems. In a risk assessment, exposure is a statement of the probability that there will be contact between the hazard and the thing the hazard might harm or injure. The uncertainty of exposure is one of the reasons risks are discussed in terms of probabilities instead of an absolute assessment of the seriousness of the hazard. In other words, a hazard, which is the potential for something to cause harm, becomes a risk only if there is exposure. Swimming pools are inherently hazardous but are not a risk to someone who has never been near one.

Therefore, in summary,

$$\text{Risk} = \text{hazardous dose} \times \text{exposure to that dose}$$

After identifying the hazard, determining the dose-response relationship, and evaluating the probability of exposure, the regulators put all of that information together in order to determine the overall level of potential risk, which in risk analysis is termed risk characterization, the final product of the risk assessment. Risk characterization is not the final step in the risk analysis, however. Before regulatory officials make a decision to approve a product or process, they evaluate whether and how the risk can be managed. Risk management is the set of activities that can be undertaken to control a hazard, minimize exposure, or both. Even though the level of risk varies with the severity of the consequences of contacting the hazard and the probability that contact will occur, if effective safeguards can be established, the level of risk can be decreased dramatically, because:

$$\text{Risk} = \frac{\text{hazard} \times \text{exposure}}{\text{safeguards}}$$

In other words, effective safeguards can minimize the hazard level, the degree of exposure, or both. For example, insulation around electrical wires decreases exposure to an electrical current so thoroughly that electrical wires pose almost no risk of electrocution, even though the electricity running through them is clearly a hazard. On the other hand, decreasing the amount of current in the wires (the hazard level) also decreases the level of risk. Both of these risk management tactics allow societies to benefit from a technology—electricity—that clearly poses significant risks.

These two conceptual formulas illustrate the first steps in using rational risk analysis to assess and

The Case of Decaffeinated Coffee

In the mid-1980s, much attention was given to the use of methylene chloride to decaffeinate coffee. Members of the media reported that methylene chloride was a carcinogen in rats. They neglected to mention that it is a carcinogen when inhaled but has a much lower carcinogenic potential when ingested. This distinction is very relevant, because in the case of decaffeinated coffee, the only variable coffee drinkers need be concerned with is the carcinogenic potential when ingested. Nonetheless, in response to consumer fears, some manufacturers began decaffeinating coffee by a different process, and the price increased.

Coffee itself has over 300 chemical compounds, many of which are more toxic and more carcinogenic than methylene chloride. To ingest enough methylene chloride to reach the value shown to cause cancer in rats, a person would have to drink 50,000 cups of coffee a day. In those 50,000 cups, coffee's natural toxins and carcinogens would occur in far greater amounts than methylene chloride. More important, a person would die from drinking 50,000 cups of water a day. Long before the 50,000th cup was consumed, a person's kidneys would shut down.

The message is not that synthetic is safe and natural is unsafe. The situation is not that simple. The point of the story is that everyone works from misconceptions derived from inaccurate or misleading information. Unfortunately, misinformation made worse by emotionally driven evaluations of risk does not yield smart choices very often.

manage risks. By no means are these steps perfect or devoid of subjectivity, but they provide a starting point for comparing the relative risks of various products and activities.

You may ask why regulators worry about all of this. Why don't they overestimate risk to make sure they are fulfilling their responsibilities to protect the public and the environment? As the example of decaffeinated coffee (see the second sidebar) demonstrates, responding to risks that do not exist can be costly, literally and figuratively, for people. Misperceptions about risks can limit your options and increase the money you spend. They also encourage the creation of an unnecessarily prohibitive regulatory process that impedes the development of beneficial products and also raises the costs of goods that obtain regulatory approval. It is possible to be overly cautious. Table 37.4 shows a relative ranking of the risks of familiar activities and products.

Table 37.4 Relative risks

Risk[a]	Source
0.2[b]	PCB[c] in diet
0.3[b]	DDT[d] in diet
1.0[b]	Drinking 1 quart of city water/day
8.0[b]	Swimming 1 hour/day in a chlorinated pool
18	Chance of dying by electrocution in a year
30[b]	Two tablespoons of peanut butter/day
60[b]	12 ounces of diet cola/day
100[b]	3/4 teaspoon of basil/day
367	Accidents in the home
600[b]	Indoor air that contains formaldehyde vapors from furniture
667	Respiratory illness caused by air pollution in the eastern United States
800	Chance of dying in an auto accident in 1 year
2,800[b]	12 ounces of beer/day
12,000[b]	One pack of cigarettes/day
16,000[b]	One tablet of phenobarbital/day

[a]The risks are listed from lowest to highest. Drinking a quart of city water, which contains chloroform, a by-product of chlorination, serves as a reference point. For example, the risk of eating three-quarters of a teaspoon of basil a day is 100 times greater than the risk in drinking a quart of city water a day.

[b]The risks associated with the numbers refer to the lifetime risk of getting cancer. (Source: B. Ames et al., *Science* 236:271–279, 1987.)

[c]PCB, polychlorinated biphenyl.

[d]DDT, dichlorodiphenyltrichloroethane, a pesticide.

Benefit Analysis

Regulators can institute safeguards to get the risk level as low as possible, but it will never reach zero, because there is no such thing as risk free. The essential question to be answered in determining whether a biotechnology product should be developed and commercialized is whether the level of risk is acceptable compared to the benefits.

Assessing benefits

Assessing risks is a simple task compared with assessing benefits. Benefits are often very difficult to identify in the abstract. People need to experience them. Beyond the obvious and intended benefits of electric lights over gaslights, societies would never have foretold the unintended, direct effects of electrification of homes: electric washing machines, water heaters, air conditioners, refrigerators, electric guitars, personal computers, and DVDs. Nor could they have predicted the secondary social impacts of the unintended effects: homes in the desert and frozen north, the Internet, home shopping, more free time, the disappearance of household servants, more women in the workforce, and, as a result, a need for day care centers.

Are these unintended effects of electrification, whether direct or indirect, benefits? It depends on whom you ask. Benefits (and risks) are not distributed evenly throughout society. In addition to being difficult to predict and inequitable, many benefits are very subjective. They are based on emotions, values, and ethical principles and therefore vary from person to person and circumstance to circumstance. Factors people use to measure benefits include the following:

- Saving lives
- Improving health
- Solving problems
- Improving quality of life
- Increasing emotional well-being
- Saving money

Comparing benefits and risks

Defining and assessing the risk level and then comparing it with the perceived benefits are components of a thoughtful, methodical analysis for decision making. If you use this methodology in an informal way, rather than evaluating risks solely based on emotions, you will be in a much better position to make judicious choices on issues involving the risks you are willing to take. However, once you have conducted your own risk-benefit analysis and decided on the acceptability of a risk, the person next to you may have arrived at a different conclusion. You may think that if the decision involves an individual's acceptance or rejection of personal risk, differences of opinion on the acceptability of risks should not present a problem. In other words, if someone wants to accept the risks of riding a motorcycle without a helmet or driving drunk, it is their business. Once again, it is not that simple. When does someone's personal decision about a personal risk they are willing to take infringe upon the rights of others?

For example, people who drive drunk may feel comfortable with their risk-benefit analysis for this behavior, but drivers sharing the road may not. Given that example, most people would probably come to a similar conclusion regarding individual rights versus the rights of other members of society. Driving on public roads is a public matter made possible with public money. An individual's right to accept the risk of driving drunk is secondary compared with the public's right to use roads without fear of drunken drivers.

What about private matters, like safe sex and AIDS? If people risk contracting human immunodeficiency virus infection by having unprotected sex, isn't that their business and only their business? How can that choice infringe on the rights of others? This may not be as simple and straightforward as it first appears. As the cost of providing care for a growing population of AIDS victims increases, everyone's health insurance costs increase. In addition to the increase in health insurance costs that each policyholder would experience directly, society could pay the price in ways that seem completely unrelated to AIDS. For example, many companies pay for their employees' health insurance. As health insurance costs increase and employers pay more money to insure their workers, they must compensate for increased costs by decreasing their other costs or increasing their profits. Cutting jobs is one way to cut costs. To increase income, companies usually raise the prices they charge for goods and services. Thus, an increase in health insurance costs could mean fewer jobs and/or higher prices for consumer goods. Should everyone pay the price for those who have chosen to have unprotected sex and, as a result, have been infected with human immunodeficiency virus?

Issues of individual rights are never as simple as each of us would like. As members of society, all individuals have both rights and responsibilities to themselves and each other. Finding the proper balance between the two can be very, very difficult.

Activity: Assessing Risks and Benefits

In May 1999, a story hit the front pages of major newspapers and popular magazines and got the attention of nature lovers around the world. Researchers at Cornell University had shown that corn pollen from plants genetically engineered to contain a gene from the bacterium *Bacillus thuringiensis* (Bt) killed monarch butterflies. The media coverage left people with the misimpression that the Cornell scientists had made a breakthrough discovery that had been overlooked by the companies that had developed Bt corn and by government regulators who had assessed its ecological risks. However, the companies, agricultural scientists around the world, and regulatory officials were not the least bit surprised by the findings of the Cornell study, which only reconfirmed a discovery that had been made in 1918 and applied commercially for decades. Bt-based pesticidal sprays, which are toxic to only certain insect pests, have been used to control caterpillar pests in crops and forests since 1938. In addition, the Bt gene, the protein it encodes, and Bt corn were thoroughly reviewed and tested before commercialization. The results confirmed that Bt corn and its pesticidal protein are not harmful to most insects or other animals but provide an effective means of controlling caterpillar pests that feed on corn.

Nonetheless, attention-getting headlines, such as "Attack of the Killer Corn," "Genetic Engineering Leads to Toxic Pollen," and "Nature at Risk," triggered strong emotional responses from the public that in turn led to a variety of political reactions. Some activists called for a worldwide ban on agricultural biotechnology. Special interest groups pressured U.S. regulatory agencies to reverse their decision and remove Bt corn from the market. The European Union's regulatory commission, which decides which crops can be grown and imported, halted the regulatory approval process for Bt corn varieties under review. They were also reluctant to allow farmers to plant the Bt corn varieties they had already approved.

In response to media attention and political pressure, U.S. government agencies and the companies that marketed Bt corn provided university scientists with funding for additional research on monarch butterflies and Bt corn. Their results confirmed the regulators' original decision that Bt corn would pose minimal, if any, risks to monarch butterflies. Even though these studies did not receive much media coverage, especially compared to the Cornell study, the uproar eventually died down. If regulators had

Societal Issues

made a decision to ban Bt corn in response to political pressure and not real risks to monarch butterflies, who or what would have benefited: monarch butterflies, other insects, farmers, the environment, the public, environmental organizations, or other special interest groups?

We know of no better case to illustrate the value of rational risk analysis. Your first task is to determine what sort of questions the regulators needed to ask to conduct a risk assessment of Bt corn to monarch butterflies. After answering these questions, you will have characterized the potential risk level of Bt corn to monarchs. You will then need to identify safeguards that could be implemented to further decrease the risks. Finally, your teacher may also ask you to assess what benefits, if any, come from growing Bt corn and who the beneficiaries are.

Background information

Before the regulatory agencies could begin to know what sorts of questions they needed to answer in order to assess the risks of growing Bt corn to monarch butterflies, they first had to gather basic background information on a variety of topics: the natural history of monarch butterflies, use of Bt biopesticides, Bt corn, conventional corn, and the current agricultural practices U.S. corn growers use to control caterpillar pests. Only after they had studied this body of facts would they know the questions they needed to ask and answer to characterize Bt corn's risk to monarch butterflies. So that you can focus on developing the list of questions, we provide some of the essential facts that served as the starting point for the risk analysis.

Facts about Monarch Butterflies
You may already be somewhat familiar with monarch butterflies from the famous monarch-viceroy mimicry relationship. Most introductory biology books contain pictures showing the striking similarity between the unpalatable monarch and the tasty (to birds, at least) viceroy butterfly.

Monarch butterflies are members of the insect order Lepidoptera, which includes butterflies and moths. Like all lepidopterans, monarchs undergo complete metamorphosis, in which there are four distinct stages: egg, larva (caterpillar), pupa (chrysalis), and adult (butterfly) (Figure 37.1). The development from egg to adult takes about 1 month in monarchs. In North America, each year there may be as many as four to six generations of monarchs in the south and one to three generations in the northern states and southern Canada. Monarch butterflies also live

A. Larva  **B.** Pupa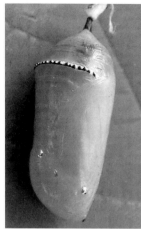

C. Adult monarch

Figure 37.1 Monarch life cycle. (A) Brightly colored black, white, and yellow larvae emerge from eggs and immediately begin to feed on milkweed plant tissues. Like all lepidopteran larvae, monarch larvae molt a number of times as they grow. (Photograph courtesy of the USDA-ARS. Peggy Greb, photographer.) (B) When the larva reaches a certain size, hormonal changes trigger the formation of the pupa. (Photograph courtesy of Herbert A. "Joe" Pase III, Texas Forest Service [http://www.forestryimages.org].) (C) During the pupal stage, the adult butterfly forms. Adult butterflies feed on nectar through highly specialized mouthparts. (Photograph courtesy of the National Biological Information Infrastructure of the U.S. Geological Survey. John Mosesso, photographer.)

throughout Central America and most of South America, in Australia, and in Hawaii and other Pacific islands.

Like some other lepidopterans, monarchs are specialist feeders, because the larvae will eat *only* leaves and flowers from a small set of plants, the milkweeds. It is the chemical toxins in milkweed plants that make monarch larvae and adults unpalatable to many predators. Like virtually all lepidopterans, only

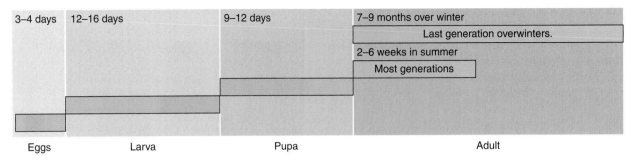

3–4 days	12–16 days		9–12 days		7–9 months over winter
					Last generation overwinters.
					2–6 weeks in summer
					Most generations
Eggs	Larva		Pupa		Adult

Figure 37.2 Monarch life history. Monarchs in most generations in North America live approximately 2 months. The first month consists of immature stages that are not reproductive. The last generation of one summer overwinters and gives rise to the next year's first generation.

the larval stage can feed on plant tissues, because butterflies have highly specialized mouthparts for sucking nectar from flowers. The adults prefer to drink nectar from milkweed flowers, but if they emerge from cocoons before milkweeds bloom, they will take nectar from a variety of flowers. There are approximately 100 species of milkweed plants in North America, and most prefer to grow in open habitats, such as along roadsides.

In North America, the adult (butterfly) stage of most generations lasts only 2 to 6 weeks (Figure 37.2). The butterfly is the only stage capable of reproduction. Females lay eggs, but only on milkweed plants, and then most of them die. However, in North America, the last generation of adults migrates south for the winter, where they aggregate in trees. Millions of butterflies overwinter in huge aggregations for 6 to 9 months before leaving the roost in late February and beginning their return migration north (Figure 37.3). As they migrate, they lay eggs on the newly emerging milkweed plants and then die. The monarchs migrating northward from Mexico usually arrive in the middle portion of Texas and the Gulf states in mid- to late March, where they lay eggs that produce the first new generation of adults by the end of April. Essentially tracking the emerging milkweed populations, these new adults migrate north and east, laying eggs along the way and establishing a large second generation.

Figure 37.3 Monarch migration routes. In North America, at the end of the summer, the last generation of butterflies migrates to overwintering sites in California and Mexico.

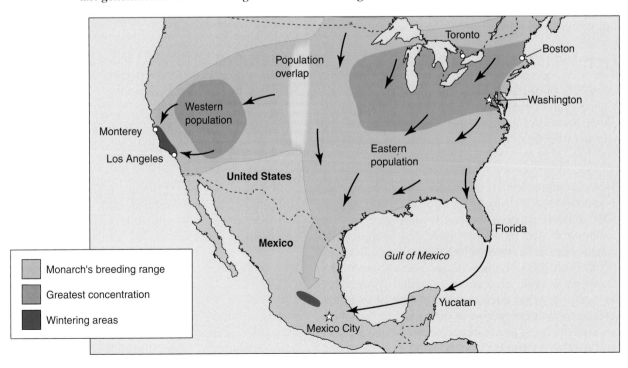

There are two distinct populations of monarchs in North America: an eastern population that overwinters in Mexico and lives and breeds east of the Rocky Mountains and a western population that overwinters in California and lives and breeds west of the Rocky Mountains. The two populations overlap only slightly, so interbreeding between the two populations is much rarer than within a population. Under such conditions, scientists would expect to see genetic differences in the two populations. Surprisingly, mitochondrial DNA studies reveal virtually no genetic variation, either between the eastern and western populations or within either population.

Butterfly lovers have been monitoring the number of monarch butterflies that make the yearly migration south in the winter and back north in the spring. There are wide fluctuations from one year to the next. The greatest threats to monarch populations are destruction of their overwintering habitat, especially in Mexico, and decreasing their sole food source by mowing along roadsides, which is a preferred milkweed habitat.

Facts about Bt

Bt is a naturally occurring bacterium found in soils in forests, savannahs, deserts, and cropland all over the world. It also occurs in enclosed insect-rich environments, such as grain storage facilities, and Bt has been found on tree leaves in some environments.

First commercialized as a microbial pesticide in France in 1938, Bt pesticides were approved by U.S. agencies for sale and use in 1961. Bt produces large amounts of a protein that, when ingested by certain insects, slows development and ultimately kills insects that consume enough of it (Figure 37.4). Unlike most pesticides used in commercial agriculture, the protein must be ingested to be toxic. Bt is harmful to a limited number of insect species but nontoxic to other insects, birds, mammals, and other organisms. Because of its selective toxicity, Bt does not harm beneficial insects, such as pollinators

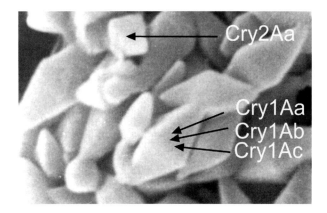

Figure 37.4 Bt proteins. The insecticides in *B. thuringiensis* subspecies *kurstaki* are proteins that form crystals (e.g., Cry1A and Cry2A) under certain conditions. The genes encoding the Bt proteins vary within a single strain of the bacterium, as do the shapes of the crystallized proteins they encode.

and insects that prey on caterpillars. Valued for this targeted pesticidal activity and environmental compatibility, Bt insecticidal sprays have been used by organic growers and home gardeners for decades, and foresters use them to try to control outbreaks of gypsy moths and other lepidopteran pests that feed on tree foliage.

Until the 1970s, scientists thought Bt was toxic only to lepidopteran caterpillars, but they have now identified different strains (or subspecies) of Bt that exhibit equally selective toxicity to beetles, flies, and mosquitoes without harming other insects (Table 37.5). These other strains are also used as biopesticidal sprays to control insect crop pests and insect vectors that carry certain tropical diseases. In the 1980s, Congress required the U.S. Environmental Protection Agency (EPA) to reassess the safety of all pesticides approved before 1984. Companies selling Bt microbial sprays provided the EPA with extensive data on Bt toxicity, environmental fate, and effects on nontarget organisms, which is any organism that is not the target of the pest control

Table 37.5 Specificities of Bt subspecies

Bt subspecies[a]	Insect order	Target pest
kurstaki	Lepidoptera (butterflies and moths)	Gypsy moths
		ECB
tenebrionis	Coleoptera (beetles)	Colorado potato beetle
		Cucumber beetle
israelensis	Diptera (flies and mosquitoes)	Black fly
		Yellow mosquito

[a]Genetic variation also exists *within* each subspecies. In the case of Bt corn for ECB control, scientists isolated various Bt toxin genes from different strains of *B. thuringiensis* subsp. *kurstaki*.

activity or product. After reviewing the data, the EPA issued a finding of no significant adverse effects and approved 180 Bt sprays for use in the United States.

In spite of their many benefits, Bt-based insecticidal sprays have a number of limitations. They break down soon after being applied to the crop, so repeated applications are necessary. Because they do not enter the plant tissue, they cannot control stem borers or root-dwelling insects. Finally, Bt sprays have caused allergic reactions in some people who spray them.

Facts about Bt Corn

Bt corn is a generic name for a number of transgenic corn varieties that contain the gene encoding the insecticidal protein produced by *B. thuringiensis* subspecies *kurstaki*. Agricultural biotechnology companies developed Bt corn to protect field corn from a specific lepidopteran pest, the European corn borer (ECB). In the United States, field corn, which is used primarily for animal feed and secondarily for processed food, such as cornflakes and tortilla chips, is grown almost exclusively in the upper Midwest (Figure 37.5).

The ECB causes an estimated $1 billion worth of damage in lost field corn yields every year. Illinois corn growers alone lose $50 million a year from decreased yields resulting from ECB damage. The

Figure 37.5 Geographic distribution of corn-growing areas in the United States. (Map redrawn from USDA-NASS.)

Corn for grain, harvested acres: 2002

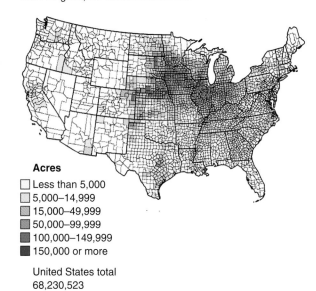

Acres

☐ Less than 5,000
☐ 5,000–14,999
◻ 15,000–49,999
◼ 50,000–99,999
◼ 100,000–149,999
◼ 150,000 or more

United States total
68,230,523

ECB is difficult to control, because it burrows into the corn stem and is inaccessible to insecticides (see Figure 1.9). Little of the insecticide that growers use on corn is for ECB control (approximately 2 to 3%, or 85,000 tons). However, the Bt gene can also provide some protection against other lepidopteran pests that farmers try to control with insecticides, such as the corn earworm, fall armyworm, and southern corn borer. Together, these pests account for approximately 20 to 25% of the insecticides used on corn in the United States (870,000 to 1.2 million tons/year).

Currently, 11 varieties of Bt corn that control pests are available to U.S. farmers. In 2006, farmers planted a Bt variety on approximately 40% of the corn acres in the United States. Globally that year, approximately 41 million acres of corn containing the Bt gene were grown commercially in 12 countries: the United States, Canada, 5 European Union countries, and 5 developing countries.

Before they permit field tests and commercial sale of Bt crops, two federal regulatory agencies, the U.S. Department of Agriculture (USDA) and the EPA, require information on the possible negative effects of transgenic crops on insects and other animals. Companies conducted laboratory tests to establish the safety of the Bt protein and Bt corn plant material for a number of nontarget organisms, including the honeybee, earthworm, collembola (a soil insect), daphnia (an aquatic invertebrate), green lacewing, ladybird beetle, parasitic wasp, catfish, quail, and mouse. The results confirmed Bt's lack of toxicity to nonlepidopteran insects and other animals.

Because Bt pesticidal sprays have been tested extensively and used for many decades, the regulators knew that:

• Any part of a plant expressing the Bt protein would harm lepidopteran larvae that eat it in sufficient amounts, including monarch larvae.
• Bt sprays are used widely to control gypsy moth outbreaks in forests and also by organic farmers and home gardeners. The impact of Bt sprays on monarchs is minor compared to the other factors threatening monarch populations.

Facts Provided by the Cornell Study

In controlled laboratory feeding experiments, researchers at Cornell University clearly demonstrated that 3-day-old monarch larvae that ate milkweed leaves dusted with pollen from a Bt corn variety

(176) ate less, grew more slowly, and suffered a higher mortality rate than larvae that were fed milkweed coated with non-Bt corn pollen or milkweed with no corn pollen. Only 56% of the larvae that were fed leaves with Bt corn pollen survived after 4 days compared with a survival rate of 100% for the other two treatment groups. The larvae feeding on pollen-free leaves ate significantly more leaf material than larvae that were fed leaves with either non-Bt corn pollen (1.61 versus 1.12) or Bt corn pollen (1.12 versus 0.57). After 4 days, the average weight of surviving larvae fed Bt corn pollen was less than half the average weight of the larvae fed leaves with no pollen (0.16 g versus 0.38 g).

Because the harmful effects of Bt on lepidopterans have been well understood for 60 years, these results did not surprise the scientific community or regulatory agencies. The most important question is not whether monarch larvae that are forced to eat Bt pollen will be harmed but whether growing Bt corn will have negative impacts on monarch butterfly populations under natural conditions.

Activity: regulatory decision making

In deciding whether to grant approvals, the regulatory agencies try to consider a new product's potential risks and benefits carefully rather than reacting emotionally, overestimating risk, and reflexively blocking new products because they cannot prove there is zero risk. They work carefully and methodically, reviewing all of the data relevant to the question at hand. Their ultimate goal is to use scientific facts to make rational decisions in approving or denying requests.

If you were being asked to make a decision about approving the sale of a Bt corn variety, what are the:

- Types of questions that must be asked and answered to characterize the risks of Bt corn to natural populations of monarch butterflies
- Potential safeguards (risk management activities) that could be implemented to minimize risks
- Possible or known benefits of using Bt corn to control lepidopteran insects

Biotechnology Regulation in the United States

Modern biotechnology was born under unique social and political circumstances. In 1973, soon after Herbert Boyer and Stanley Cohen described their successful work recombining DNA from very different organisms, a group of scientists who were responsible for seminal breakthroughs in molecular biology sent a letter to the prestigious National Academy of Sciences (NAS) and the widely read journal *Science.* They asked the scientific community to agree to a self-imposed moratorium on certain scientific experiments using recombinant DNA (rDNA) technology. Even though they had a clear view of its extraordinary potential for good and no evidence of any harm, they were uncertain about the risks some types of experiments posed.

In February 1975, 150 scientists from 13 countries, along with attorneys, government officials, and 16 members of the press, met to discuss rDNA research. The conference attendees replaced the self-imposed moratorium with a complicated set of rules for conducting certain kinds of laboratory work with rDNA but disallowed other experiments until more was known. Although it may sound as if the scientists were in general agreement, the debate was quite contentious. At no other time has the international scientific community voluntarily ceased the pursuit of knowledge before any problems occurred, imposed regulations on itself, or been so open with the public. They adopted strict guidelines initially, and as the understanding of rDNA grew and their confidence in its safety became more certain, the guidelines were gradually relaxed. Their approach, supported voluntarily by private and public researchers, ensured the thoughtful, responsible, and very public inspection of this new technology in its earliest stages of development.

A substantial body of knowledge about laboratory research with rDNA, accumulated over a 15-year period, led the NAS to issue a report that set the stage for moving research from basic laboratory work to commercial products. The Academy concluded:

> There is no evidence that unique hazards exist in the use of rDNA organisms or in the transfer of genes between unrelated organisms.
> The risks associated with rDNA organisms are the same in kind as those associated with organisms modified by other genetic techniques.
> Assessments of the risks of using rDNA outside of the laboratory should be based on the nature of the organism, not on the method used to produce it.

During the early 1980s, as biotechnology evolved from basic research into product development, federal regulatory agencies with responsibility for protecting human health and the environment stepped in and based their overall approach to biotechnology regulation on the NAS findings. In June 1986, the federal government issued the *Coordinated Framework for the Regulation of Biotechnology.* The U.S. Congress did not pass any new regulatory laws specifically granting authority over biotechnology development, because the regulatory agencies determined they already had sufficient legal authority to require product developers to seek regulatory approval before testing and commercializing biotechnology products. The three federal regulatory agencies whose mandates captured almost all of the biotechnology products for the foreseeable future at that time were the U.S. Food and Drug Administration (FDA), the USDA, and the EPA. They drew their authority not only from specific laws that gave them the authority and responsibility to protect human health, the environment, and agriculture, but also from the National Environmental Policy Act. This law requires all federal agencies to ensure that any decision they make does not have negative environmental impacts. Therefore, at a minimum, all regulatory agencies must conduct an environmental assessment (EA) for products they approve.

Societal Issues

Follow a Product through Regulatory Review

The regulatory process for approval of biotechnology products is complicated and multifaceted. Specific details vary according to the product in question and how it will be used, so it can be difficult to wrap your mind around the entire regulatory apparatus. Even so, it is important to understand how these products and activities are regulated in the United States, because the regulatory process is a key factor determining which products, from the universe of possible products, make it to the market. Some products may not be approved because their risk-benefit profile makes their value to society questionable, but sometimes the regulatory process impedes development of safe, efficacious, and beneficial products, because the costs of amassing data to meet regulatory requirements are prohibitively expensive for small companies, public institutions, and nonprofit research institutes.

Walking a biotechnology product through the regulatory process allows you to get a handle on the complexity of the process, the roles of the agencies, data requirements, repeated opportunities for public participation, and other details of the extensive and iterative process. We will use the example of a Bt corn variety. While the product developer might be a university, public research institution, or private industry, companies developed the Bt corn varieties on the market. We begin at the point when scientists isolated the Bt gene encoding the pesticidal protein and realized a crop plant with built-in protection against caterpillars would be an excellent product, if they could get the gene to function in a plant. The business development staff did the math and determined that the costs of conducting laboratory and field research, performing tests required for regulatory approval, and marketing the product could be recouped, but only if the company chose to use the Bt gene in a large-volume commodity crop. Two commodity crops, field corn and cotton, suffer significant damage from lepidopteran larvae; our hypothetical company decided to move forward with Bt field corn.

Their first question: who has regulatory jurisdiction over the product and the development process? In the case of Bt corn, three agencies had to sign off on this product before it could enter commerce: the USDA, EPA, and FDA. The combined concerns of these agencies as they relate to Bt corn were as follows:

Its potential impacts on the environment and agriculture

Food safety and nutrition for both humans and animals (74% of the field corn grown in the United States is used in animal feed)

Accurate, meaningful labels

Regulatory review stages

1. Research concept. At the "idea" stage, the company's Institutional Biosafety Committee considered the possible environmental and health impacts of the potential product; determined if the potential risk level was acceptable, especially compared to the benefits to farmers; and notified the USDA, which had to inspect and approve the greenhouse facilities where early research was conducted.

2. Small-scale field trials. Before the company could move its research from greenhouse to small-scale field tests, it needed permission from the USDA, which required detailed information on the host crop (corn), new gene construct (Bt protein, promoter, and selectable markers), new gene products (Bt protein), origin of the new gene (*B. thuringiensis*), purpose of the field trial, and specific precautions the developer would use to prevent escape of pollen, plants, or plant parts. The company's request also contained a large amount of additional technical information on ecological attributes so that USDA scientists could conduct the EA, such as the potential for movement of genetic material to plants of the same or related species (outcrossing) and potential to harm animals near and around the test site.

If the USDA's decision was a "finding of no significant impact" (FONSI), the regulators would issue a field test permit and require that certain risk management procedures be followed, such as maintaining specific isolation distances from other crops, planting borders of nontransgenic corn, and monitoring a number of ecological factors. In addition, federal law requires the USDA to:

- Provide an opportunity for public input during the EA process.
- Make both the EA and FONSI statements available to interested or affected parties.
- Delay action if a member of the public challenges the sufficiency of the EA or FONSI.

3. Large-scale field tests. Before the company could move to large-scale field tests (>10 acres), it needed permission from the EPA. Before granting the permit, the EPA required information that

allowed it to assess impacts on the environment and human health, including:

- Product characteristics, such as the biochemistry and bioactivity of the pesticidal substance, the biology of both the recipient plant and the source of the introduced genetic components, and the expression levels of the pesticidal protein
- Environmental fate of the pesticidal gene product, such as the amount found in soils when plants die, and the length of time the protein persists in soil, water, and air
- Human health effects
- Ecological effects, such as potential for gene flow to wild relatives, increased or decreased disease and herbivore resistance, and effects on nontarget organisms

Federal law requires the EPA to inform the public when it receives a permit request, allow public comment, and publish its decision.

4. Food safety and nutrition tests. The company then moved to the "food safety and nutrition" phase of product development. The large-scale field trials provided enough plant material to address the requisite food safety and nutrition questions required by the FDA and EPA. Around the time the company approached the EPA about a permit for large-scale field tests, it also met with FDA officials to discuss the proposed product so that the FDA could decide the required tests for meeting its statutory mandate of ensuring that food introduced to the market is safe and nutritious. Why isn't there a standard battery of tests for all transgenic crops? Because tests for ensuring food safety and nutrition vary with the host crop, the donor organism, the genetic construct, and the new protein encoded by the gene. However, in all cases, the developer must test the potential toxicity and allergenicity of the new protein that is encoded by the new gene and conduct biochemical analysis of all of the nutritional aspects of the crop to ensure it is as nutritious as conventional varieties.

5. Large-scale production. After the FDA and EPA assessed Bt corn's nutrition and safety for human and animal consumption, the company was ready to start large-scale production and sale of Bt corn seeds. It petitioned the USDA for "nonregulated status" for this Bt corn variety so that it and the farmers did not have to get permits from the USDA every time they planted it. The company had to provide the USDA with data

from its EPA-approved large-scale tests to prove the variety performed as predicted, exhibited the same expected biological and agronomic properties as the same corn variety without the Bt gene, and did not harm agriculture or the environment, especially as it relates to agriculture (e.g., impacts on beneficial insects).

When the USDA receives a request for nonregulated status, it must inform the public, make copies available to those who request them, and give the public 60 days to comment on the request. The USDA must also conduct a final EA, using the more extensive data acquired in large-scale field tests, and inform the public when the EA is available for public review and comment. The USDA scientists grant nonregulated status if they determine the new variety is the same as its nontransgenic parental variety, except for the defined difference, which in this case was the Bt gene and the protein it encodes.

Once the USDA issued a formal determination of nonregulated status, that Bt variety could be produced, marketed, distributed, grown, and used in breeding without getting approval from the USDA. If, however, any problems ever arose, the USDA had the authority to remove it from the market.

6. Final stage. Before the Bt variety could be grown and used commercially, additional reviews by the EPA and FDA were necessary. Before approving a pesticidal product, including Bt crops, for commercial use, the EPA is legally required to demonstrate that it presents "no unreasonable adverse effects on environment or human health when used according to label directions." The EPA looked at data acquired from many large-scale field trials at a number of locations before it made this determination for the Bt corn variety. Once the EPA issued its final approval, the Bt corn variety could be sold, but the EPA also has statutory authority to remove products it has approved from the market if problems arise.

The FDA reviewed the food safety and nutrition data to determine if there were any unresolved issues or unanswered questions. If so, it required additional testing; if not, it issued a letter to the product developer describing its findings and allowing the product to be used in food for humans and animals. In some cases, the FDA requires that the product and foods derived from the product be labeled, but this was not necessary with the Bt corn variety. The FDA has the authority to remove products from the market, as well.

For the earliest varieties of Bt corn approved for sale in the United States, the step-by-step regulatory approval process, from submitting a request to the USDA for a permit for a small-scale field test of the Bt gene to inserting the gene into the target crop (corn) and, finally, introducing Bt corn seeds to the commercial market, took approximately 7 to 8 years.

Bioethical Issues: A Decision-Making Model

About the Activities in Chapters 38 to 40

Experience has demonstrated that while classroom discussions about controversial social issues can be provocative and consciousness-raising, they can also often result in nothing more than emotional exchanges of opinion. As we explained in the previous chapter, having a structured framework for addressing contentious issues helps to keep the discussion on track. This is especially true for highly charged issues related to bioethics. If not handled correctly, these discussions can quickly devolve into rigid and absolute assertions about what should or should not be done in which everyone speaks and no one listens.

Biotechnology triggers strong emotional responses. Its remarkable capabilities raise both hopes and fears. Nowhere is this more apparent than in the area of medical biotechnology. While medical bioethical issues are not derived solely from advances in biotechnology, the new knowledge and powers the biotechnologies provide will force people and governments to wrestle with ethical issues they have never had to address. Our society must begin a discourse on these difficult issues before they become intractable problems.

One effective way to get students to evaluate ethical issues objectively is to use case studies and a highly structured method of analyzing each case. Teachers have found the bioethics decision-making model, which we describe in this chapter, to be helpful in channeling productive thought and debate. Students are presented with specific cases, and they then identify an aspect of the case that poses not just an ethical issue, but also an ethical dilemma. Students then apply ethical principles in a systematic way, following the decision-making model, to formulate a suggested solution. By definition, a dilemma has no single "right" answer, so this frees students from the burden of having to pick the correct solution.

The decision-making model requires students to examine the facts of the case, consider alternatives, and choose the alternative they consider to be the best. They must then justify their choice and defend their decision by citing relevant authorities, such as the Nuremberg Code or the Hippocratic Oath, two widely used documents that set forth standards of conduct for biomedical professionals. Finally, students present their decisions and the justifications for them in classroom presentations or formal reports.

In addition to increasing students' awareness of bioethical issues, the case study approach offers the following benefits.

> Critical-thinking skills. Students must distinguish facts from inferences and personal opinion and emotion from reasoned analysis.
> Experience gathering accurate information. To the extent a teacher requires it, students will read what authorities on ethics have to say about relevant issues. They will also do library research to answer factual questions relevant to the case they are studying.
> Social skills. If rules are enforced by the teacher, students can learn how to discuss a potentially emotional topic, respect different answers to the same question, and agree to disagree.

The nature of the model makes it most readily applicable to situations and decisions involving individuals as opposed to groups, such as corporations or governments. As will be evident, the model can be used to analyze situations in areas other than biotechnology and biomedicine. We introduce the model using ethical issues unrelated to modern biotechnology before turning to bioethical dilemmas posed by the newest gene-based technologies.

Introduction

Chapters 35 to 37 made it clear that no technology exists in a social vacuum. The use of any technology changes the society that uses it, often in unpredicted ways. Not only do technologies shape societies, but societies also determine the trajectory of technological change.

The tremendous and rapid advances in medical technologies since the middle of the 20th century give people reason to rejoice and also to reflect. Initially, the use of these technologies was seen as altogether positive, a must-do choice for the physician. It was not long, however, before difficult questions surfaced. Just because physicians can maintain a person with medical technology, should they? Who should answer such a question: the patient, physician, family, nurse, social worker, hospital administrator, insurance company, ethicist, lawyers, or politicians?

Since the 1960s, questions like these have led to an entirely new discipline, bioethics, which has become an immensely important topic. Few major hospitals are without an ethics committee to assist patients or health care providers when they need help. Colleges offer courses of study in medical humanities or bioethics.

Paralleling the development of mechanical medical technologies, such as kidney dialysis and heart pacemakers, were scientific breakthroughs in cellular and molecular biology. Knowledge of molecular genetics has only recently been applied to specific diseases or patients, but the ethical questions that plague any medical technology also apply to genetic technologies.

How does society decide what is right or wrong, what is better or worse, as genetic technologies offer ever more choices? Should everyone be screened to determine who the carriers of recessive alleles for genetic diseases are? Should genetically engineered products be available to everyone, or should physicians sometimes refuse requests for them (such as requests for human growth hormone)? Who should (and who actually will) have access to a person's DNA fingerprint? Are we on the road to employment and insurance discrimination based on genetic profiles? These are but a few of the bioethical questions that can be posed.

How to identify a dilemma

Not every bioethical case study presents a dilemma; many times, the possible courses of action are clearly right or wrong. A dilemma exists when there is no "right" course of action in a certain situation but, instead, several options, none of which is wholly acceptable. Ethical dilemmas revolve around

trying to find the best solution when no solution is completely good. Students may need help from you in identifying a true dilemma. The following example may clarify what does and does not constitute a dilemma.

Assume that a patient with a certain condition would be an appropriate candidate for a drug research study. The patient's physician places her on this drug without getting her permission. This situation is not an ethical dilemma. It is just plain wrong. Even if the doctor believes the drug will benefit the patient, by all modern medical standards, the physician has the obligation to get the patient's informed consent to include her in the study. The Nuremberg Code and the Declaration of Helsinki, two internationally recognized codes of ethics, specifically address the ethics of research using human beings.

On the other hand, assume that the patient has been given all the information she needs to make a decision. She is told that the drug has the potential to help her but might also have harmful side effects. She sees benefits and costs regardless of which decision she makes. Now we have a dilemma. A dilemma exists when no choice is ideal but all options have benefits and risks that must be carefully assessed.

Once a dilemma has been identified, students should pose the dilemma in the form of a question. Often, more than one question can be formulated. By working through one question at a time, everyone in the group can be sure that everyone else is talking about the same problem. It is also helpful to categorize the issue being discussed. In the example given previously, the category might be humans as research subjects, and the question posed could be, "Should the patient agree to be part of the research study or not?"

Basic ethical principles

To make ethical decisions, members of society must agree on some basic guidelines about what constitutes moral conduct. In the field of biomedical ethics, certain guides are well established. The moral-action guides or principles can be divided into four major principles and several secondary ones. These principles are listed below (the terms in parentheses are those used in the bioethical literature). Before simply listing these for your students, you may want to have the students list guidelines for behavior and decision making that are important to them. Most values can then be categorized into one of the general principles listed here.

Major Ethical Principles

Do no harm (nonmaleficence).

Do good (beneficence).

Do not violate individual freedom (autonomy).

Be fair (justice).

Secondary Ethical Principles

Tell the truth (truth telling).

Keep your promise (fidelity and promise keeping).

Respect confidences (confidentiality).

Use the principle of proportionality: risk-benefit ratio (how much harm can be justifiably risked to effect good).

Attempt to avoid undesirable exceptions (also known as the wedge principle, the slippery slope, or the camel's nose).

Although these rules are simple, they represent fundamental values associated with respect for human dignity that most people agree on. These are the principles to which students should refer when making and justifying their decisions.

Using the Decision-Making Model

When faced with an ethical dilemma, how does someone decide what to do, since there are no right answers? To assist students in learning how to follow a rational decision-making process based on ethical principles, we suggest the decision-making model described below. Following the prescribed, step-by-step procedure laid out in the model greatly reduces the chance of getting off track. The model requires students to stay focused on the issue at hand and helps them develop critical-thinking skills.

Basic steps in the model

1. Identify the ethical dilemma to be addressed. Usually, for any given case, many questions pose dilemmas that could be considered.
2. Identify the category of the issue (e.g., genetic screening, confidentiality, gene therapy, or human research subjects). This step helps in grouping cases and may make it easier for students to find references in the literature.
3. State the facts in the case. Be sure to avoid inferences.
4. Think of as many decisions in the case as possible.
5. Gather additional information as needed.
6. Pick a decision to support.

7. State the ethical principle that supports the decision, which is also known as a claim.
8. Identify an authority(s) that supports the decision/claim. Quote the authority, if possible.
9. Formulate a rebuttal. Under what circumstances might the student abandon the claim?
10. What is the level of confidence in the decision, the qualifier?
11. The student should "box up" the case for reporting the decision (see below for an example).
12. Write a prose argument describing the case and decision.

Two examples using this decision-making model are presented below. The first example is a typical school ethics dilemma, and the second is a case from medical ethics. Different teachers may find one or both effective in demonstrating how to use the model.

Objectives

After reading this chapter and completing the exercises, students will have:

- Learned the basic principles that guide ethical decision making
- Become familiar with internationally accepted codes of conduct
- Learned and practiced a process for making decisions on bioethical issues
- Learned that some issues present dilemmas to which there is no one right answer

Preparation

Review the *Student Activity*.

Make copies of the "box-up" form, Worksheet 38.1, which is found in Appendix A and on the CD.

Resources

A number of universities and associations have lesson plans and other resources for teaching bioethics in high school classrooms. In addition to providing free materials to download, these websites also provide links to other resource-rich websites. Among the best are Iowa State University Office of Biotechnology (http://www.biotech.iastate.edu), University of Pennsylvania (http://www.bioethics.upenn.edu/highschool), Georgetown University (http://www3.Georgetown.edu/research/nrcbl/hsbioethics), and Northwest Association for Biomedical Research (http://www.wabr.org/education/ethicslessons.html).

Procedures

Establish clear rules for the classroom discussion or ask the students to establish them. Since the issues discussed in bioethics are controversial, it is important that rules of etiquette be observed from the very beginning.

The following list has proved effective in the classroom.

1. Only one person at a time speaks after being recognized by the discussion leader.
2. Treat each other with respect; no name calling. Critique the argument, not the author of the argument.
3. Seek clarity by asking questions.
4. Look for gaps in the data.
5. Recognize your own biases.
6. Be true to your own position; do not jump on the bandwagon.
7. Keep emotions in check; use logic.
8. Do not follow authority blindly.
9. You must have a reason, not just an opinion. You can like pepperoni pizza better than anchovy pizza with no reason. You cannot decide bioethical issues based on opinions.
10. Be open minded and willing to be perplexed.

At the beginning, it is best to have the whole class address the same question in the classroom. Working as a large group the first few times is also helpful in modeling the kinds of behaviors you want to encourage. Next, have the students break up into small groups that work on different questions in the same case study. By using this format, they will learn that any one case yields a number of ethical dilemmas. Eventually, work in small groups on different case studies.

The case studies and "right" answers

In the two example cases that follow, as well as in several of the additional case studies, we have supplied you with sample workups and decisions to demonstrate the process. These are included only to show you how to use the model, not because they are the right answers. Remember that a dilemma is a dilemma because there is no perfect solution. The sample solutions may seem to be good resolutions of the situations, but it is critical to remember that your task as teacher is not to guide students to these particular resolutions or even to focus on the same questions. Rather, your task is to guide students to their own resolutions through use of the model.

For example, an alternative decision for case study 1 (Frank and Martin) would be for Martin to report what he saw to the judiciary council. If Frank is not guilty of cheating, he can explain himself to the council. The principles supporting this decision would be justice (being fair in that Martin is adhering to the agreed-upon honor code) and truth telling.

To avoid creating the impression that there are "right" questions to focus on and "right" answers, the sample solutions are not provided in the *Student Activity*.

Example Case Study 1: Frank and Martin

Case study 1 shows an ethical dilemma but does not involve medicine or biotechnology. The case may be useful in introducing students to the method because it involves territory and issues that are very familiar to them. In addition, it serves as a useful example of how the decision-making model can be applied to many problems outside the bioethics arena.

Frank is an 11th-grade student at a small public high school that prides itself on its family atmosphere and strong academic reputation. The school has an honor code. Each student agrees to abide by a published list of rules. One part of the code obligates students to report observations of fellow students who may be cheating. A judiciary council composed primarily of students decides what should happen in each case. Frank works hard at school. He has a B+ average and hopes to go to a good college. He also works part time, runs track, and helps at home with two younger brothers. Both of his parents work. Frank is a close friend of Martin. Martin is very bright and does well in school, although he does not have to put in as much time studying as Frank does. Martin does not have many extracurricular activities or responsibilities. During a history test, Martin notices that Frank appears to be cheating off a note card.

I. Identify the question

Often there are many questions. In this case, some might be as follows.

1. Should Martin report his observation to the judiciary council?
2. Should Martin tell Frank what he saw?
3. Should Frank report himself to the judiciary council?
4. Should other students be asked what they saw?

Other questions could be posed. It is important to encourage students to ask as many questions as possible. List them all, but finally choose one question for the group to focus on first. As students become more adept at using the model, they can be divided into small groups, with each group addressing a different question.

To demonstrate the model using this case, we will use question 1: should Martin report what he saw to the judiciary council?

II. Identify the issue

What general problem does the case demonstrate? In this case, you might say cheating or school rules. This step is an attempt to categorize the case.

III. State facts of case

What are the facts in the case? There are "rights" and "wrongs" at this step of the decision-making model. It is important to help students differentiate between facts and inferences. This skill is one that the students may already have been introduced to in the laboratory, when they tried to draw conclusions based on data collected rather than on inferences. Many students will make unsupported assumptions or jump to conclusions.

In this case, for instance, we are not positive that Frank's note card is actually a cheat sheet. There is a possibility, perhaps remote, that Frank was working on another assignment and had permission to use the card. Students might also propose that everyone cheats and ask why Frank should get punished when others do not. That may or may not be true at this particular school. Nor can we make assumptions about Frank and Martin's friendship. Maybe they had a fight and Martin just wants to get Frank in trouble. It is a possibility but not a fact in this case. Time spent discussing the facts is worthwhile, since it can prevent confusion later in the discussions.

Once the facts have been established, it is helpful to list them in an accurate but concise manner. Facts must be true, relevant, and sufficient.

- Frank was seen using a note card; the honor code requires Martin to report the incident.
- Frank is in 11th grade, works, is a B+ student, and runs track.
- The school is small, with a good reputation.

- Frank plans to go to college.
- Martin does well in school with less effort than Frank.
- Martin does not have many extracurricular activities.

The relevance of the last three facts is worth discussing.

IV. List possible decisions

What are the possible answers to the question of whether Martin should report what he saw? Encourage students to generate as many choices as possible. List them all. Promote creative and lateral thinking by asking leading questions and allowing adequate wait time. Here are some possible choices.

1. No, Martin should not report what he saw. It could harm Frank. Maybe Frank had permission to use the card. How does Martin know whether Frank was really cheating?
2. Yes, Martin should report what he saw. It is his duty under the code. He breaks his own promise to abide by the code if he does not tell. Frank may need help if he did indeed cheat. If he did not cheat, he can explain to the council.
3. Martin should tell Frank's parents and let them decide what to do.
4. Martin should tell his own parents and let them decide what to do.
5. Martin should tell another student and let the other student decide what to do.
6. Martin should tell Frank what he saw and give Frank the opportunity to tell the judiciary council what happened. If Frank refuses to go to the judiciary council, then Martin should tell.

V. Gather additional facts

It is absolutely critical that students learn to seek relevant information and not to make decisions based on uninformed opinion. At the very least, students should recognize when there are gaps in the information they have. This sample case probably will not require gathering additional background information. However, other cases may. For example, to weigh the risks and benefits in a case involving use of an experimental drug to treat a patient, students may want to know just how severe the patient's disease is. It might also be appropriate for them to find out what sort of safety testing a drug must undergo before it is approved for experimental treatment of humans.

VI. Pick decision to support

We have presented six possible decisions, but your class may generate many more. Encourage solutions that represent compromise. Since ethical problems are usually complex, it is important to take time for thoughtful, honest reflection. The students must have a reason for the option they choose, and that reason should be related to one of the principles listed earlier in this chapter. However, students may make different decisions based on different principles.

For our example, we will choose option 6. With this choice, we respect Martin's obligation to tell what he saw, but at the same time, we might be able to avoid harming Frank.

VII. Identify guiding principle

An ethically justifiable decision can be based on alternative principles. In a dilemma, adherence to one principle often results in the breach of another; dilemmas exist because principles often conflict. Application of different basic values can lead to different responses to a situation.

Students should learn what the basic principles are and recognize how they can be applied. One of the goals of using this model in the classroom is for students to develop an appreciation for the fact that two responsible, moral people can make very different decisions in a case because they are guided by two different basic principles.

The logic involved in the choice of principle(s) for these classroom cases should be explicitly discussed and/or written about by the students. They need to be able to justify why they have chosen one principle and why they are willing to breach the alternative principle(s). Although students can choose more than one principle to support their decision, it is helpful to force them to choose only one at the beginning to be certain they understand the differences between principles. In the sample case, the first principle we are adhering to is avoiding harm and the second is truth telling.

VIII. Identify supporting authority

What experts or authorities would back up this position on the case? Normally, people would look to professional codes of ethics. In this case, the honor code itself would be the authority. In bioethical decisions, several codes of ethics are used; these codes include the Hippocratic Oath, the American Medical Association Code of Ethics, the American Hospital Association's Bill of Rights for Patients, and the American Nursing Association Code of Ethics, as well as the Nuremberg and Helsinki statements. Depending on the time available, library research is always desirable. Have your students find these documents in books and bring them to class.

IX. Formulate rebuttal

Under what circumstances would students change their decision about what to do? Here again, it is important to encourage students to think creatively. This section is often difficult, because students are invested in their choice and may not be able to imagine a circumstance that would make them change. In this case, what if Frank were taking a different test, perhaps a more advanced one, and had permission to have the note card? What if the note card was actually an appointment slip that Frank was checking to be sure he was on time for an after-school doctor's appointment? If Martin knew that kind of information, then surely the problem would evaporate.

X. State level of confidence

The student should formulate a one- or two-word statement to describe how strongly he or she believes his or her own argument. Until they have more experience, students tend to believe their own arguments are infallible. One way to assess the strength of the argument is to gauge the likelihood of rebuttal. If a rebuttal is highly unlikely and the rest of the argument makes sense, then the argument is a strong one. Also, if the principle ties the claim and the facts tightly together, then the argument is strong. Forcing students to qualify their arguments is one way to promote self-evaluation. As students work up more and more cases, they become better at constructing an argument and more realistic about its strength. Qualifiers might be "moderately confident," "absolutely confident," or "questionably confident." In this case, we will use "strongly confident."

XI. Box up the case

Figure 38.1 shows a "box" of this example of a case study. Obviously, many other boxes could have been constructed. Students, usually working in groups of three or four, can transfer their boxed-up cases to a transparency and present their case to the class, starting with the facts. Having students critique one another's arguments is a valuable lesson, as well.

Societal Issues

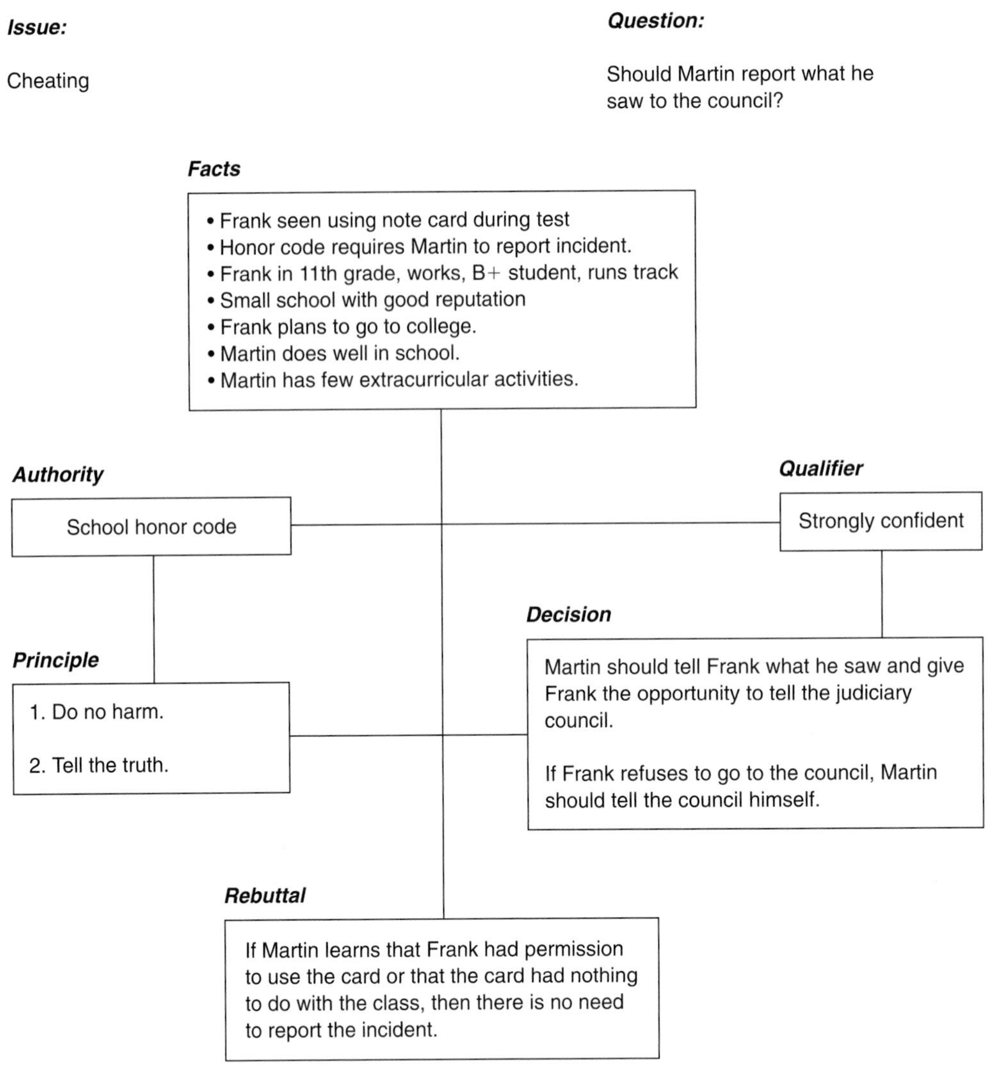

Issue:

Cheating

Question:

Should Martin report what he saw to the council?

Facts

- Frank seen using note card during test
- Honor code requires Martin to report incident.
- Frank in 11th grade, works, B+ student, runs track
- Small school with good reputation
- Frank plans to go to college.
- Martin does well in school.
- Martin has few extracurricular activities.

Authority

School honor code

Qualifier

Strongly confident

Decision

Martin should tell Frank what he saw and give Frank the opportunity to tell the judiciary council.

If Frank refuses to go to the council, Martin should tell the council himself.

Principle

1. Do no harm.

2. Tell the truth.

Rebuttal

If Martin learns that Frank had permission to use the card or that the card had nothing to do with the class, then there is no need to report the incident.

Figure 38.1 "Box" of example case study 1: Frank and Martin.

XII. Prepare argument

Finally, if time permits, students can write up their arguments, using the boxed-up case as an outline to structure their papers. They should produce papers that can be understood by someone unfamiliar with the original case. Students who write well have little difficulty with this task. However, many need help using proper transitions. They should also explain more in the paper than the box can show. For instance, why is avoiding harm more important, at least initially, than telling the truth immediately? Students who are willing to apply this model to other classes often find it works very well in writing papers in English or history class. This reasoning model can be applied to almost any discipline.

Example Case Study 2: Mr. Johnson

Mr. Johnson, age 76, was admitted to a medical unit with pneumonia. He had a history of severe emphysema, for which he had been hospitalized twice in the past year because of secondary pneumonia. During the evening of the third day of intravenous antibiotic treatment, Mr. Johnson's complexion took on an increasingly bluish tint, and he was short of breath. Because the attending physician was out of town, the physician taking calls was notified by the evening nursing supervisor. The physician found the patient to be severely oxygen deficient. He tried unsuccessfully to reach Mrs. Johnson to notify her of his plan to transfer Mr. Johnson to the intensive care

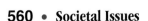

Societal Issues

unit for ventilator support. During the process of ar-ranging for the transfer, a nursing assistant said that she thought that Mr. Johnson and his wife had said that they would refuse life support measures if Mr. Johnson ever needed them. No documentation of these comments could be found.

When Mr. Johnson arrived in intensive care, he was anxious, and his breathing was so labored that he could not talk. He shook his head "no" when the physician explained the procedure for inserting a tube to connect him to the ventilator, and he at-tempted to push the physician away. Mr. Johnson's competence to make decisions about his treatment was questioned because his arterial blood oxygen level was so low. Another unsuccessful attempt was made to reach Mrs. Johnson. Mr. Johnson's condition deteriorated as the physician and staff deliberated about proceeding with the intubation. Finally, the pa-tient was sedated and placed on the ventilator.

Mrs. Johnson arrived the next morning and was shocked that her husband's condition had deterio-rated so rapidly. She was upset that her husband was on the ventilator, explaining that they had agreed that life support measures would not be used for him. When the on-call physician asked, "What do you want me to do?" Mrs. Johnson replied, "I don't know." The physician asked for assistance from the ethics committee.

I. Identify the question

Students will probably raise the most obvious ques-tion about what to do with Mr. Johnson: leave him on the ventilator or take him off. Other questions that they might raise include where Mrs. Johnson was all this time, where the attending physician was, whether Mr. Johnson is competent now, what will happen to Mr. Johnson if he stays on the ventilator, and whether Mr. Johnson has any children who dis-agree with Mrs. Johnson about her husband not wanting to be on a ventilator. These questions are in-teresting, and some of them could certainly add to the data in the case, but the central question about whether to take Mr. Johnson off the ventilator or leave him on would remain, so that is the question we will address here. Should Mr. Johnson remain on the ventilator?

II. Identify the issue

The general category of problem could be called withdrawal of life support, end-of-life decisions, or extraordinary means.

III. State facts of case

What are the facts in this case? Students who have worked through this case are often angry at Mrs. Johnson and the regular doctor for not being there when they were needed, and they attribute all kinds of unworthy motives to both of them. This provides an ideal opportunity to reiterate the differences be-tween facts and inferences and to insist that only facts as they are known may be used.

- Mr. Johnson is 76 years old.
- Diagnosis: pneumonia secondary to severe em-physema
- Third hospitalization in last 12 months
- Treatment: intravenous antibiotics
- Day 3: short of breath, blue, oxygen deficient
- Doctor on call tried to reach Mrs. Johnson twice but could not locate her.
- Nursing assistant reported that she "thought pa-tient and wife would refuse life support mea-sures."
- No documentation of those wishes
- Patient unable to talk on admission to the inten-sive care unit
- Blood oxygen level low; patient anxious
- Condition deteriorating
- Pushed doctor's hands away, shook head "no"
- Sedated and intubated
- Next day, wife shocked at husband's condition
- Wife upset that Mr. Johnson was on ventilator
- Wife said that she and Mr. Johnson had discussed not wanting life support.

IV. List possible decisions

What are the possible courses of action in this case? Students should be encouraged to list as many as possible. For this case, here are some.

1. Mr. Johnson should remain on the ventilator.
2. Assuming Mr. Johnson is competent, discuss the situation with him. If he directs the doctor to re-move the ventilator, then remove it. Keep him clean and comfortable, and treat him with dignity, but do not employ heroic measures. If he wishes to be left on the ventilator, continue the present course.
3. If Mr. Johnson cannot communicate, is deemed to be incompetent, or both, Mrs. Johnson, the at-tending physician, nurses, social workers, etc., should decide what Mr. Johnson would direct were he competent and able to communicate.
4. Assuming that Mrs. Johnson's position is corrob-orated (that Mr. Johnson would not want life sup-port), then Mr. Johnson's ventilator should be re-moved as in option 2.

V. Gather additional facts

Obviously, no further information about the Johnsons is available, but students will probably need to find out more about emphysema to understand Mr. Johnson's situation. They should also be familiar with the laws in your state concerning end-of-life decisions, living wills, and durable powers of attorney. They also may wish to learn more about breathing machines and their uses.

VI. Pick decision to support

For this example, we will choose decision 2: if Mr. Johnson directs the doctor to remove the ventilator, then remove it, but keep him clean and comfortable, and treat him with dignity. Do not employ heroic measures.

VII. Identify guiding principle

Since making the decision involves asking Mr. Johnson what he wants, the decision is based primarily on respect for his autonomy. We are trying to let him decide for himself what he does or does not want. At the same time, we are trying to avoid harming him. Some would argue very convincingly, however, that removing him from the ventilator will result in his death, which is the ultimate harm. Others will argue that the ventilator is merely postponing the moment of death and that in this situation, although removing the ventilator may hasten Mr. Johnson's death, that is not a harm. Herein lies the crux of these decisions. What principles do the individual students think are most important in each case? They need to be able to say why they think one principle takes precedence over another.

VIII. Identify supporting authority

As time permits, students can read about similar cases, and they should examine the codes of medical ethics mentioned above.

IX. Formulate rebuttal

Under what circumstances would the students abandon their decision? For instance, if Mr. Johnson is not competent and no one can corroborate that he did not want life support measures, then they may decide to leave him on the ventilator. Perhaps they discover that Mr. and Mrs. Johnson were having serious marital problems but that Mr. Johnson still has a $5 million insurance policy with Mrs. Johnson as the beneficiary. If he cannot speak for himself, they may be reluctant to discontinue the ventilator.

X. State level of confidence

This case has quite a few unknowns, so the students may not be very confident about their decision. They do not know whether Mr. Johnson is competent; they do not know what his prognosis is without the ventilator; they do not know how he felt about his quality of life before this episode. There are more than enough questions to make students somewhat unsure about their decision. Your task is to make it clear to them that having these questions is, indeed, the "right" response. Often students feel that public uncertainty and confusion is a flaw instead of a virtue.

XI. Box up the case

A box for this case is shown in Figure 38.2.

XII. Prepare argument

As time and interest dictate, students can write up their arguments in paragraph form.

In the next two chapters, the decision-making model will be applied to gene therapy and genetic-screening cases. Blank "box-up" forms for those exercises are provided in Appendix A and on the CD.

Selected Readings

Beauchamp, Tom L., and James Childress. 2001. *Principles of Biomedical Ethics.* Oxford University Press, New York, NY.

Beauchamp, Tom L., and LeRoy Walters (ed.). 1999. *Contemporary Issues in Bioethics.* Wadsworth Publishing Co., Belmont, CA.

Burley, J. (ed.). 1999. *The Genetic Revolution and Human Rights.* Oxford University Press, Oxford, United Kingdom.

Caplan, Arthur L. 1995. *Moral Matters: Ethical Issues in Medicine and the Life Sciences.* John Wiley & Sons, Inc., New York, NY.

Engelhardt, H. Tristram. 1996. *The Foundations of Bioethics.* Oxford University Press, New York, NY.

Harron, Frank, John Burnside, and Tom Beauchamp. 1983. *Health and Human Values: Making Your Own Decisions.* Yale University Press, New Haven, CT.

McGee, Glenn. 1999. *Pragmatic Bioethics.* Vanderbilt University Press, Nashville, TN.

Pence, Gregory. 2003. *Classical Cases in Medical Ethics.* McGraw-Hill, New York, NY.

Peterson, James C. 2001. *Genetic Turning Points: the Ethics of Human Genetic Intervention.* William B. Eerdmans Publishing Co., Grand Rapids, MI.

Veatch, Robert. 2002. *Basics of Bioethics.* Prentice Hall, Upper Saddle River, NJ.

Societal Issues

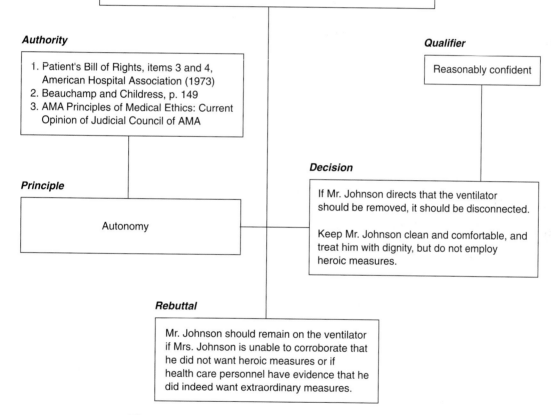

Issue:

Withdrawing life support

Question:

Should the physician disconnect the ventilator?

Facts

- Mr. Johnson is 76 years old.
- Diagnosis: pneumonia, secondary to emphysema
- Third hospitalization in last 12 months
- Intravenous antibiotics prescribed
- Day 3: short of breath, turning blue, oxygen deficient
- Doctor tried unsuccessfully to reach Mrs. Johnson.
- Nurse "thought Mr. J. and wife would refuse life support measures."
- No documentation of those wishes
- Mr. J. unable to talk on admission to the intensive care unit
- Blood O_2 level low, patient anxious
- Condition deteriorating
- Pushed doctor's hands, shook head "no"
- Sedated and intubated
- Next day, wife shocked at husband's condition
- Upset that Mr. J. was on ventilator
- Said she and Mr. J. had discussed not wanting life support

Authority

1. Patient's Bill of Rights, items 3 and 4, American Hospital Association (1973)
2. Beauchamp and Childress, p. 149
3. AMA Principles of Medical Ethics: Current Opinion of Judicial Council of AMA

Qualifier

Reasonably confident

Principle

Autonomy

Decision

If Mr. Johnson directs that the ventilator should be removed, it should be disconnected.

Keep Mr. Johnson clean and comfortable, and treat him with dignity, but do not employ heroic measures.

Rebuttal

Mr. Johnson should remain on the ventilator if Mrs. Johnson is unable to corroborate that he did not want heroic measures or if health care personnel have evidence that he did indeed want extraordinary measures.

Figure 38.2 "Box" of example case study 2: Mr. Johnson.

Making Decisions When There Is No Right Answer

Introduction to Bioethics

Since the middle of the 20th century, rapid improvements in technology have changed the practice of medicine profoundly. Technologies often allow physicians to restore or supplant basic bodily functions. Mechanical ventilators, heart pacemakers, kidney dialysis machines, exotic drugs, organ transplantation, and artificial nutrition and hydration are some of the life-extending tools available to the modern practitioner.

Initially, the use of these techniques was seen as altogether positive, a must-do choice for the physician. It was not long, however, before difficult questions surfaced. Just because physicians can maintain a person with artificial ventilation and nutrition, should they? Who should answer such a question: the patient, physician, family, nurse, social worker, hospital administrator, insurance company, ethicist, lawyers, or politicians?

Since the 1960s, questions like this have provided an entirely new discipline or, more accurately, interdiscipline: bioethics. Bioethics, or biomedical ethics, has become an immensely important topic. Few major hospitals are without an ethics committee to assist patients or health care providers when they need help. Colleges offer courses of study in medical humanities or bioethics. High school students, too, must be aware of the questions bioethicists study.

Paralleling the medical developments described above were striking advances in the scientific understanding of cellular and molecular biology. Knowledge of molecular genetics has only recently been applied to specific diseases or patients, but the ethical questions that plague society about any medical technology apply to genetic technologies and other applications of medical biotechnology.

How does society decide what is right or wrong, what is better or worse, as medical biotechnology offers ever more choices? Should everyone be screened to determine who the carriers of recessive alleles for genetic diseases are? Should genetically engineered products be available to everyone, or should physicians sometimes screen requests for them? Who should have access to a person's DNA fingerprint? What should be the avenues of recourse if someone who should not have access to genetic information gains access legally? Is society on the road to employment and insurance discrimination based on genetic predisposition to certain diseases? These are but a few of the ethical questions that can be posed.

Basic ethical principles

To make ethical decisions, society must agree on some basic guidelines about what constitutes moral conduct. In the field of biomedical ethics, certain guides are well established. The moral-action guides or principles can be divided into four major principles and several secondary ones. These principles are listed below. The terms in parentheses are those used in the bioethical literature. You may encounter them in your library work.

Major Ethical Principles
1. Do no harm (nonmaleficence).
2. Do good (beneficence).
3. Do not violate individual freedom (autonomy).
4. Be fair (justice).

Secondary Ethical Principles
1. Tell the truth (truth telling).
2. Keep your promise (fidelity and promise keeping).
3. Respect confidences (confidentiality).
4. Use the principle of proportionality: risk-benefit ratio (how much harm can be justifiably risked to effect good).
5. Attempt to avoid undesirable exceptions (also known as the wedge principle, the slippery slope, or the camel's nose).

Although these rules are simple, they represent fundamental values associated with respect for human dignity that most people agree on. These are the principles to which you should refer when making and justifying decisions in the exercises below.

The Bioethics Decision-Making Model

In attempting to answer difficult ethical questions, it is helpful to focus on specific cases and follow a step-by-step procedure. Below, you will be presented with specific case studies to analyze and a decision-making model to be used in these analyses. Part of the decision-making process will be to gather any additional background facts you need to evaluate the situation and to find out what sort of ethical standards that apply to your case have been established. Finally, you will make a decision as to the best course of action and justify it in terms of basic ethical principles.

Identifying a dilemma

The case studies we have provided raise bioethical dilemmas. A dilemma exists when there is no "right" course of action in a certain situation but, instead, several options, none of which is wholly acceptable. Ethical dilemmas revolve around trying to find the best solution when no solution is completely good. Not every situation presents a dilemma; many times, the possible courses of action are clearly right or wrong. For example, assume a patient would be an appropriate candidate for a drug research study, and her physician places her on this drug without getting her permission. This situation is not a dilemma. It is just plain wrong. Even if the doctor believes the drug will benefit the patient, by all modern medical standards, the physician has the obligation to get the patient's informed consent to include her in the study.

Now, assume that the patient has been given all the information she needs to make a decision. She is told that the drug has the potential to help her but might also have harmful side effects. She sees benefits and costs regardless of which decision she makes. Now we have a dilemma. A dilemma exists when no choice is ideal but all options have benefits and risks that must be carefully assessed.

Basic steps of the decision-making model

Once a dilemma has been identified, the next step is to pose the dilemma in the form of a question about a specific case. Often, more than one question can be formulated. You will need to choose one

question at a time for analysis, although you may consider several questions, one after the other, about a single case. It is also helpful to categorize the kind of issue being discussed. In the example given previously, the issue might be humans as research subjects and the question posed could be, "Should the patient agree to be part of the study or not?"

Here is a summary of the steps in the decision-making model. Each step will be explained more fully later on.

1. Identify the question you want to address. Usually, for any given case, many questions could be considered. Choose the one you want to explore.
2. Assign the issue you are exploring to a category (e.g., genetic screening, confidentiality, gene therapy, or human research subjects). Categorizing the issue will help you in the search for relevant literature.
3. State the facts in the case clearly and concisely.
4. Think of as many possible decisions in the case as you can.
5. Gather additional information as needed.
6. Pick the decision you want to support.
7. State the ethical principle that supports your decision. Your decision is also known as your claim.
8. Identify an authority that supports your decision. Quote the authority, if possible.
9. Formulate a rebuttal. Under what circumstances would you abandon your claim?
10. How strongly do you believe your claim? What is your level of confidence, the qualifier?
11. "Box up" the case for reporting your decision. The "box-up" form is provided in Appendix A (Worksheet 38.1).
12. Write a prose argument describing the case and your decision.

Example Case Study 1: Frank and Martin

This case shows an ethical dilemma but does not involve medicine or biotechnology.

Frank is an 11th-grade student at a small public high school that prides itself on its family atmosphere and strong academic reputation. The school has an honor code. Each student agrees to abide by a published list of rules. One part of the code obligates students to report observations of fellow students who may be cheating. A judiciary council composed primarily of students decides what should happen in each case. Frank works hard at school. He has a

B+ average and hopes to go to a good college. He also works part time, runs track, and helps at home with two younger brothers. Both of his parents work. Frank is a close friend of Martin. Martin is very bright and does well in school, although he does not have to put in as much time studying as Frank does. Martin does not have many extracurricular activities or responsibilities. During a history test, Martin notices that Frank appears to be cheating off a note card.

I. Identify the question

Think of as many questions as you can. It may help you to think of questions that start with "Should. . . ." For example, one question you might ask could be, "Should Martin ask other students if they saw anything?" List as many questions as you can. After you have done so, choose one question for analysis. Remember, you can consider other questions later.

II. Identify the issue

What general problem does the case demonstrate? In this case, it might be cheating or school rules. Although the principal figures in the case are students, students per se are not the issue. This step is an attempt to categorize the case in a way that would help you find additional information when you search the bioethical literature.

III. State facts of case

What are the facts in the case? Most of us have a tendency to draw conclusions based on some information and then believe that our conclusions are facts, too. Be sure to distinguish between the facts and your inferences. Once the facts have been established, list them in an accurate but concise manner.

IV. List possible decisions

What are the possible answers to the question you are addressing? Now is the time to be creative. Think of as many possible answers as you can. At this point, do not worry about which answer is best; that comes later.

V. Gather additional information

Obviously, you do not have access to any additional information about Frank and Martin, but often, background information on the important issues involved in a case study is available. Frank and

Martin's case will probably not require gathering additional background information. However, other cases will. The bottom line is that you should be as well informed as possible before you make an ethical decision.

VI. Pick decision to support

Consider all the possible decisions listed in step IV. Since ethical problems are usually complex, it is important to take time for thoughtful, honest reflection. You must have a reason for the option you choose, not just an opinion. Your reason should be related to one of the principles listed at the beginning of this chapter.

VII. Identify guiding principle

An ethically justifiable decision can be based on alternative principles. In a dilemma, adherence to one principle often results in the breach of another; dilemmas exist because principles often conflict. Application of different basic values can lead to different responses to a situation. Part of the decision-making process is realizing to which principle you are giving preference.

For example, someone who believes it is more important in this case to do no harm (principle 1) might make a different decision from someone who believes it is more important to uphold justice (principle 4). Neither person would be wrong. Both are adhering to high moral principles, but their principles are different.

Contrast these decisions to those of a person who decides to do nothing because someone might get angry or because he does not want to get involved. Keeping people from getting mad at you and keeping from getting involved are not high moral principles.

VIII. Identify supporting authority

What experts or authorities would back up our position on this case? Normally, it is appropriate to look to professional codes of ethics. In this case, the honor code itself would be the authority. In bioethical decisions, several codes of ethics are used; they include the Hippocratic Oath, the American Medical Association Code of Ethics, the American Hospital Association's Bill of Rights for Patients, and the American Nursing Association Code of Ethics, as well as the Nuremberg and Helsinki statements.

IX. Formulate a rebuttal

Under what circumstances would you change your decision about what to do? Try to imagine a circumstance or new information that could make you change your mind.

X. State level of confidence

Use a one- or two-word statement to describe how strongly you believe in the argument you have made for your decision (your claim). One way to assess the strength of the argument is to gauge the likelihood of rebuttal. If a rebuttal is highly unlikely and the rest of the argument makes sense, then the argument is a strong one. Also, if the principle ties the claim and the facts tightly together, then the argument is strong.

Indicate the degree of confidence you have by using one of the following qualifiers or a similar one: "moderately confident," "absolutely confident," or "questionably confident."

XI. Box up the case

Your teacher will show you how to summarize your case in the box outline provided.

XII. Prepare argument

Write up your argument, using the box as an outline. You should produce a paper that can be understood by someone who is completely unfamiliar with the original case. You should also explain more in the paper than the box can show. For instance, you should elaborate on why you selected the decision you chose and why the ethical principle justifying that decision was most important in this incident.

Example Case Study 2: Mr. Johnson

Mr. Johnson, age 76, was admitted to a medical unit with pneumonia. He had a history of severe emphysema, for which he had been hospitalized twice in the past year because of secondary pneumonia. During the evening of the third day of intravenous antibiotic treatment, Mr. Johnson's complexion took on an increasingly bluish tint, and he was short of breath. Because the attending physician was out of town, the physician taking calls was notified by the evening nursing supervisor. The physician found the patient to be severely oxygen deficient. He tried unsuccessfully to reach Mrs. Johnson to notify her of his plan to transfer Mr. Johnson to the intensive care unit for ventilator support. During the process of arranging for the transfer, a nursing assistant said that she thought that Mr. Johnson and his wife had said that they would refuse life support measures if Mr. Johnson ever needed them. No documentation of these comments could be found.

When Mr. Johnson arrived in intensive care, he was anxious, and his breathing was so labored that he could not talk. He shook his head "no" when the physician explained the procedure for inserting a tube to connect him to the ventilator, and he attempted to push the physician away. Mr. Johnson's competence to make decisions about his treatment was questioned because his arterial blood oxygen level was so low. Another unsuccessful attempt was made to reach Mrs. Johnson. Mr. Johnson's condition deteriorated as the physician and staff deliberated about proceeding with the intubation. Finally, the patient was sedated and placed on the ventilator.

Mrs. Johnson arrived the next morning and was shocked that her husband's condition had deteriorated so rapidly. She was upset that her husband was on the ventilator, explaining that they had agreed that life support measures would not be used for him. When the on-call physician asked, "What do you want me to do?" Mrs. Johnson replied, "I don't know." The physician asked for assistance from the ethics committee.

I. Identify the question

What question will you address about this case? List possible questions ("Should . . . ?") and then select one.

II. Identify the issue

The general category of problem could be called withdrawal of life support or end-of-life decisions.

III. State facts of case

Only the facts as they are known may be used. List the facts in a concise manner, avoiding inferences or conclusions you may have drawn.

IV. List possible decisions

What are the possible courses of action in this case? List as many as you can.

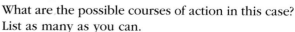

V. Gather background facts

No further information is available about the Johnsons, but you may have other questions that are important to your decision. What is emphysema? What is secondary pneumonia? What is arterial blood oxygen, and why would low oxygen levels make a doctor think Mr. Johnson was not competent to make decisions? What is a ventilator? Are there laws in your state that make certain decisions impossible? Be sure you have a good understanding of the case before you attempt to make a decision.

VI. Pick decision to support

VII. Identify guiding principle

VIII. Identify supporting authority

As time permits, read about similar cases. What do recognized codes of ethics say about these situations?

IX. Formulate a rebuttal

Under what circumstances would you abandon your decision?

X. State level of confidence

XI. Box up the case

XII. Prepare argument

Write up your argument in paragraph form so that someone who has never seen the case before can understand it and the reasons for your decision.

Societal Issues

Bioethics Case Study: Genetic Testing

39

About This Activity

Now that the students have familiarity with the decision-making model and have practiced using it on ethical issues with which society has many years of experience, it is time to introduce some of the newer bioethical issues raised by modern biotechnology, especially genetic technologies. In this chapter, students are asked to use the decision-making model to analyze ethical dilemmas related to newer capacities to test for genetic defects. In the next chapter, we address ethical issues related to gene therapy.

As more and more genetic tests become available, it is incumbent on biology teachers to help students examine the pros and cons of testing. The students in your classrooms right now will have many choices to make with respect to genetic testing. Like most people, they will probably believe that if a diagnostic test is available they should ask their physician to perform it. They are not aware that there is definitely a downside to large-scale testing just because the test is available, especially the risk of receiving a false positive, which is inherent in every medical test. If you enable them to make a truly informed choice as they contemplate issues such as genetic testing, you will have served a worthy goal.

Before your students tackle ethical dilemmas posed by genetic testing, you should provide information that allows them to place ethical issues associated with genetic testing in context. As is true of all issues associated with modern biotechnology, many of the concerns that people express about genetic testing are not unique to the new genetic-testing technologies. Some, however, are novel issues that are derived from the increased breadth and power of the new technologies. As is always the case, it will be essential for you to help your students define what is and is not unique to recent developments in genetic testing. By viewing these issues in context, students will be able to see not only the benefits of genetic testing but also the pitfalls. One can only

hope that familiarity with past mistakes will help society avoid future ones.

Previous sections in Part II provide essential background in the techniques of genetic testing and the science underlying them. Below, we offer additional information that will allow you to assist your students in analyzing the issues of genetic testing, using the process described in chapter 36. This information also provides ideas for additional bioethics case studies you could develop and will also help students in applying the decision-making model to some of the ethical dilemmas associated with new testing technologies.

Discussion of new and old genetic testing technologies that are used on fetuses (fetal, or prenatal, genetic testing) and newly fertilized eggs (preimplantation testing) invariably brings up issues related to abortion and "right to life." Depending on the particular school system, classroom, and teacher, discussion of these issues can be handled well by using the same critical-thinking tools introduced earlier, as well as the bioethical decision-making model. If you are willing to confront these difficult issues, we suggest you include a brief presentation (or review) of human development, using filmstrips or the *Nova* tape "Miracle of Life," so that students connect trimesters to a meaningful picture of the embryo or fetus. No bioethical dilemma has an absolute answer; bringing logic and reason to this highly politically and emotionally charged debate is beneficial. If a teacher has a good relationship with his or her students, the classroom is a good place to promote reasoned discussion.

Introduction to Genetic Testing

Although the relationship between genes and health is receiving more public attention now than ever before, using genetics as a component of health care is not new. For centuries, physicians have used a person's hereditary history to shed light on current and future health, because many medical conditions and diseases are familial, which means they "tend to" run in families. By the middle of the 20th century, genetics researchers had identified more than 500 inherited genetic diseases through careful analysis of medical pedigrees.

Nor is diagnostic testing for genetic disorders a new development, made possible only recently with modern biotechnology. Physicians have used cell morphology to identify carriers of the gene mutations for sickle-cell anemia and Tay-Sachs disease for over 30 years, and fetal testing for genetic disorders, such as trisomy 21 (Down's syndrome), became common for older women who became pregnant after a British medical researcher developed amniocentesis in 1952.

Even so, physicians, ethicists, policy makers, and the general public have begun to voice a number of interrelated and complex concerns associated with the new genetic testing technologies. Some of these concerns are not unique to new developments in biotechnology but have existed for a number of decades. Others are brand new. Whether new or old, some of the issues are not easy to resolve in ways that are equally acceptable to all viewpoints. Because the concerns are often inextricably interrelated, not to mention technically complex and often emotionally charged, analyzing them rationally can be difficult. In chapter 36, we discussed the necessity of defining terms quite specifically, using them appropriately, clearly delineating the issue being discussed, and staying focused on that issue. All of these tactics for ensuring disciplined and productive discussions are especially helpful when it comes to genetic testing.

A first step to informed discussion of genetic testing is establishing a semantic distinction between genetic testing, gene (or genetic marker) identification, and genetic screening. These terms are often used interchangeably. Because this topic is so emotionally charged, we suggest that you try to use the terms consistently and ask your students to do the same.

Genetic testing

For purposes of this chapter's discussion, we use genetic testing to refer to a diagnostic test, usually ordered by a physician, to test for a specific genetic disorder because the physician thinks there is a good chance the person might have the defective gene. Perhaps there is a history of the genetic disorder in the family or the patient's ethnic group, or the patient has begun to display symptoms associated with a mutation in that gene.

Until very recently, diagnosing the great majority of inherited disorders depended on observing certain phenotypic traits. For certain disorders, the phenotypic correlate of the genetic defect might be a biochemical or cellular abnormality that could be identified prior to any clinical manifestations. In a few of those cases, such as phenylketonuria, early detection led to early intervention that prevented the development of clinical symptoms. These same tests also allowed physicians to identify carriers of some genetic disorders that are prevalent in certain populations (Tay-Sachs disease and sickle-cell anemia). However, in most cases, the phenotypic trait that allows physicians to diagnose genetic disorders or identify disease carriers is a clinical symptom of the disease. As you learned in Part II, modern genetic testing technologies now allow physicians to identify patients with genetic disorders and carriers long before any clinical symptoms appear.

Gene identification

Before a physician can test for a certain genetic disorder, medical researchers must identify the defective gene or locate a genetic marker that is associated with having that disease. For example, scientists located genetic markers for Huntington's chorea (Huntington's disease [HD]) in 1983, for Duchenne muscular dystrophy in 1987, and for cystic fibrosis (CF) in 1989. The causative genetic defects for all of those diseases have now been identified and mapped. Genetic markers and genes that contribute to multifactorial traits, like colon cancer, Alzheimer's disease, and multiple sclerosis, have also been identified.

Genetic screening

Finally, genetic screening is large-scale genetic analysis of populations, whether or not the individuals display any symptoms of the genetic disorder or have a family history of genetic disorders. Large-scale genetic analysis of many individuals is typically a first step in gene or genetic marker identification and therefore is a type of genetic screening. However, for the purposes of our discussion, we will use genetic screening to refer to routine tests, conducted after genes and genetic markers have already been identified, for a number of genetic defects on many asymptomatic individuals.

Screening for genetic disorders is also not a brand new development of modern biotechnology. We have already mentioned one form of genetic screening: prenatal testing of fetuses using amniocentesis or a newer and safer technique, chorionic villus sampling, to collect fetal cells. Because physicians recommend these tests for all pregnant women over a certain age, this is a form of genetic screening. Prior to development of the newest genetic testing technologies, biochemical and microscopic tests on fetal cells collected from amnionic fluid or chorionic villi could detect more than 200 diseases, congenital disorders, and chromosomal abnormalities.

Two other forms of screening for genetic disorders have also been widely used for a number of decades, prior to the development of the new genetic testing technologies. In neonatal screening, all newborns are tested for certain genetic disorders by using biochemical analysis. Currently in the United States, all states require testing for three genetic disorders, and approximately 35 to 40 states require neonatal testing for five additional genetic disorders (Table 39.1). Eight states require that hospitals test newborns for as many as 30 genetic disorders. A final form of screening for genetic diseases, preimplantation genetic testing, has also been conducted at in vitro fertilization (IVF) clinics for a number of decades. All embryos produced by mixing sperm and eggs in a petri dish are screened for a variety of genetic defects prior to implantation. This screening improves the probability the fetus will be carried to term.

Context and issues

As we mentioned earlier, for many decades, physicians have used genetic information to assess a patient's current and future health, but new gene technologies have novel attributes, some aspects of which introduce novel issues.

Predictive Powers

With the older testing technologies, in order to diagnose a genetic disease, physicians had to have a measurable or observable phenotype that was consistently associated with the genetic defect. Now, however, with only partial DNA sequences of disease genes or nearby marker sequences, physicians can diagnose genetic diseases long before symptoms appear. At first, using sequence information for predictive diagnosis of genetic defects will be limited to disorders caused by one or a very few genes, such as CF or early-onset Alzheimer's disease. Over time, as researchers use genomics to identify genes involved in multigenic disorders, genetic tests will help people learn their disease tendencies.

Traditionally in medicine, early disease diagnosis is considered desirable. For diseases influenced by both the environment and genes, such as emphysema and high blood pressure, early detection could save lives. By identifying those people who have genetic propensities to diseases with strong environmental components, genetic testing could provide them with an opportunity to make appropriate lifestyle changes, such as avoiding tobacco or changing their diet, to lessen the probability they will contract the disease. Therefore, not only would genetic

Table 39.1 Legally mandated neonatal screening for genetic disorders

Disorder	U.S. incidence	Description	Potential effect	Treatment
Phenylketonuria[a]	1/13,947	Deficiency of phenylalanine breakdown enzyme	Mental retardation; seizures	Low-phenylalanine diet
Hypothyroidism[a]	1/3,004	Deficiency of thyroid hormone	Mental retardation, stunted growth	Thyroid hormone
Galactosemia[a]	1/53,261	Deficiency of galactose breakdown enzyme	Brain and liver damage, cataracts, death	Galactose-free diet
Sickle-cell disease[b]	1/3,721	Hemoglobin abnormalities	Organ damage, stroke, delayed growth	Penicillin, vaccinations
Adrenal hyperplasia[b]	1/18,987	Deficiency of cortisol and aldosterone	Death due to salt loss, growth difficulties	Hormone replacement, salt replacement
Biotinidase deficiency[b]	1/61,319	Biotin-recycling deficiency	Mental retardation, seizures, hearing loss	Biotin supplements
Maple syrup urine disease[b]	1/230,028	Deficiency of leucine, valine, isoleucine breakdown enzyme	Mental retardation, coma, seizures, death	Dietary management and supplements
Homocystinuria[b]	1/343,650	Deficiency of homocysteine breakdown enzyme	Mental retardation, skeletal abnormalities, stroke	Dietary management and vitamin supplements

[a]Required in all states in the United States.

[b]Required in 35 to 40 states.

Societal Issues

testing push disease diagnosis to the earliest possible stage (predisposition), it also has the potential to shift the focus of clinical medicine from disease treatment to prevention (Figure 39.1).

But what if there is no way to treat or prevent the disease? The wealth of genomics information made available by the Human Genome Project is providing diagnostic tests for early detection of many hereditary diseases for which there is no treatment or cure. Some people fear that providing people with information about personal genetic disorders could harm them psychologically if no cure for the disease is available at the time the diagnosis is given. In addition, they feel that telling parents they had conceived a child who suffered from an incurable genetic disease would force parents into a psychologically and emotionally difficult decision about terminating the pregnancy.

What about this issue is new or unique to the new genetic testing capability? First, let's consider what is not new, because people and their physicians have grappled with many of the same issues for many decades. If a routine physical examination reveals you have an incurable cancer, do you expect your physician to share that information with you? Probably so. But just a few decades ago, physicians did not tell their patients they had cancer because they thought not knowing was best for the patient's emotional well-being. The tension between withholding and disclosing information is not new to medicine and varies with the disease, the physician, the patient, and the prevailing practice. It is not specific to new genetic testing capabilities.

The same can be said about prenatal (fetal) testing for inherited disorders. Physicians can treat or prevent a few, such as phenylketonuria and the other seven disorders in Table 39.1, once the baby is born. However, most of the 200 disorders detected by these tests are incurable. Their progress cannot be subverted by any known treatments. For many decades prior to the development of the new genetic testing methodologies, hundreds of thousands of parents have voluntarily had these tests and been faced with difficult decisions about pregnancy termination.

Parents who know they are carriers of serious genetic diseases can avoid having to make decisions about aborting fetuses through preimplantation testing for genetic diseases. However, this "solution" to the problem of fetal genetic testing that ends in abortion provides no solace for people who believe life begins when egg and sperm fuse.

You may be asking why these issues are potential problems that society as a whole must contemplate. Can't people avoid the potential psychological trauma posed by these difficult decisions by simply *not* subjecting themselves or their unborn offspring to genetic testing? After all, no one requires them to take the tests, do they? Unfortunately, history shows that technologies that are supposed to increase a person's options sometimes increase their obligations. A choice at one time may become a requirement later on. As described above, neonatal screening for at least three genetic disorders is required *by law* in all states, and some states screen all newborns for as many as 30 different genetic disorders. Even

Figure 39.1 Progress in disease diagnosis. A disease process begins with molecular and cellular changes, moves through a series of stages, and, in the later and most severe stages, becomes manifested as visible clinical symptoms. Technological advances in the 20th century made it possible to diagnose diseases earlier, before the patient showed clinical signs of having the disease. Technological advances in the coming century will make it possible to identify diseases at increasingly earlier stages. In certain cases, the diagnosis may occur before the disease process has begun, which increases the prospect of disease prevention.

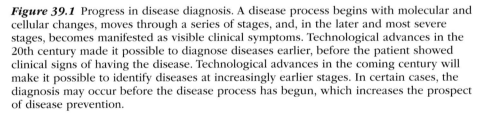

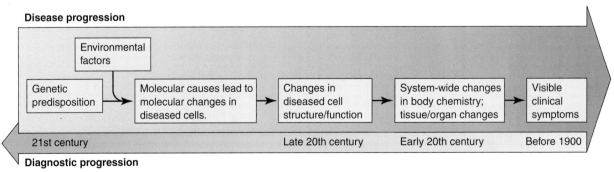

though it seems inconceivable that prenatal genetic testing would ever be required in this country, it is important to go through the mental exercise of analyzing the potential uses and abuses of a technology to ensure the technology develops in the most responsible way.

Ambiguity of Genetic Information

Some people object to the large-scale use of genetic testing because the nature of genetic information means that the test results are often ambiguous. The presence of certain genetic defects does not guarantee that an individual will actually develop the disease. In many cases, the presence of a defective gene means only that the person is a carrier. In other, more ambiguous cases, the presence of a defective gene means only that they have a genetic predisposition to that disease. In other words, it tells them that they are at risk.

But how does this differ from current practices? Every time someone has their blood pressure taken or cholesterol measured, they are assessing risk factors. When someone describes their family's medical history to a physician, they are providing data on genetic predisposition. Is the information gained through genetic testing any more ambiguous than the information physicians already gather?

Both bodies of information *are* ambiguous, but while most people are prepared to accept the ambiguity of blood pressure readings and cholesterol measurements, they do not realize that genetic testing data can be equally ambiguous. The nature and power of genes are thoroughly misunderstood by the public. An aura of predestination surrounds a gene. People believe that if they have a gene for a given trait, then all is determined, and nothing is left to chance.

Scientists and physicians know that this is not true. The path from genotype to phenotype is an indirect one filled with opportunities for myriad factors other than a single gene to exert an influence (see Figure 3.6). The final expression of a gene is affected by many variables: the allelic forms of the gene in question, other genes, and environmental influences. In addition, having a genetic defect that has been consistently linked to a disease does not necessarily guarantee the person will actually develop the disease. These facts are often unknown to people who are assessing the meaning of genetic test results.

The link between having disease genes and displaying disease symptoms is strongest when the

disorder results from a single dominant allele that is relatively impervious to environmental influences and has 100% penetrance. However, even with this seemingly absolute degree of genetic predetermination, the severity of a genetic disorder can vary considerably among people having precisely the same genetic mutation (variable expressivity). The link between genes and a certain disorder becomes increasingly vague and ambiguous when (i) many genes contribute to the disorder (multigenic disorder) or (ii) genes and environmental factors interact and lead to the disorder (multifactorial disorder). As ambiguity in the link increases, the predictive power of genetic testing in medical diagnoses lessens. Unfortunately, the overwhelming majority of diseases and disorders that are responsible for mortality in the industrialized world are both multigenic and multifactorial.

Researchers have made remarkable progress linking certain mutant genes with certain diseases, but the DNA sequence, encoded protein, or phenotypic trait associated with thousands of other genes will remain unknown for decades. Any of those unidentified genes might interact with identified genes in ways that significantly alter the expected phenotypic trait expressed by the identified gene. Without knowing how the identified disease genes interact with other, unidentified genes or the environment, the results of a diagnostic test for the disease gene could lead to unnecessary heartache or unwarranted relief. On the other hand, if the research community waits until all genes have been identified and their interactions elucidated before it develops a diagnostic test for a genetic disease, thousands of opportunities to save lives will be missed.

For example, let's say researchers have linked a nucleotide substitution mutation in gene A with disease A, so they develop a test capable of identifying that gene A mutation. If you received a test result indicating you had that mutation, you might well think you would eventually develop that disease. But think back on the genetics of Labrador retriever coat color, in which one gene prevents the expression of genes at another locus. What if a certain form of gene B interacts with gene A and *prevents* disease A, but researchers have not identified gene B, much less determined its function? If you have the form of gene B that blocks phenotypic expression of A, you would not get the disease. On the other hand, is it unethical not to develop a diagnostic test for the A mutation until the entire human genome is decoded?

Misuse of Genetic Information

In the 1970s, screening for sickle-cell anemia resulted in confusion and fear. People who merely carried the sickle-cell gene were threatened with loss of jobs and insurance. The airline industry and the armed forces argued that low air pressures would cause heterozygotes to experience symptoms. Some airlines eliminated these people as prospective employees. Although the hiring decisions and refusal of insurance coverage had no scientific basis, it took many years to end the discrimination against carriers of the sickle-cell mutation. The initial benefit of helping people understand their likelihood of bearing children with sickle-cell disease was lost in a mire of misunderstanding.

As more and more genetic tests become available, the opportunity for similar misuse of genetic information increases. This is exacerbated by the public's lack of understanding about the meaning of genetic information, as well as the inherent ambiguity of genetic information described above. Some people fear that all of these factors, when combined, will lead to the use of genetic testing results to justify discrimination, deny health insurance coverage, and withhold jobs.

What is unique about these concerns? Again, let's first consider what is not unique. Many companies screen potential employees for health-related factors. For example, at least 50% of the employers in the United States require that job applicants have medical examinations. In those examinations, many of the tests conducted are for indirect measures, i.e., phenotypic manifestations, of genetically based traits. As early as 1982, 23 of 366 companies surveyed by the U.S. Congress Office of Technology Assessment were using genetic tests in screening applicants. Historically, insurers have used medical technologies to identify preexisting conditions as a basis for denying coverage or raising premiums. They have also excluded people from coverage because their family medical history put them in a high-risk category.

When insurance companies require medical examinations to determine the health of their applicants, the physicians measure physiological and morphological (phenotypic) correlates of health and disease. That is, if the person has begun to develop cardiovascular disease, the physician will be able to discern the problem through blood tests, electrocardiograms, and blood pressure measurements. The same is true of a lung disease, such as emphysema. Measurable factors, such as lung capacity and force of exhalation, indicate that the applicant/patient is

developing or has developed the disease. Genetic information is unlike this type of data. As we explained above, some genes foretell the likely future. Some tests allow physicians to predict the *possible* appearance of cardiovascular or lung disease in the absence of symptoms.

Will insurance companies classify these telltale markers as determinants of preexisting conditions, thus allowing insurance coverage to be denied? This, of course, is exacerbated by the public's tendency to view genes as the final arbiter of traits rather than one of many variables that contribute to the presence or absence of a disease.

How about employment issues? Should someone not be hired because genetic tests reveal they are predisposed to certain diseases? On the other hand, is it unethical for a company to hire someone if the genetic tests reveal the job candidate is especially susceptible to hazards that they would encounter in that working environment?

Genetic Privacy

As you know, the process of developing genetic markers for diseases and identifying and mapping disease genes depends on acquiring blood samples from many members of a family and analyzing their DNA. As a part of this gene discovery process, a person who is being screened discovers information about their genetic makeup. What if someone discovers they have a disease gene, but they do not want anyone to know? This poses an ethical dilemma between a person's right to genetic privacy and a responsibility to family members. Where would you draw the line in balancing your need for privacy with their need to know? Would your answer vary depending upon the nature of the disease, its genetic basis, or whether it can be treated? If you feel you have a responsibility to tell family members, how broadly would you define family? Siblings, parents, cousins, second cousins?

Once a disease gene is identified and screening of asymptomatic populations becomes available, some people will want to know if they have that disease gene, and others will not. If people share their test results with certain family members, people who do not want to know their probability of having disease genes may well end up knowing. How does someone balance their right to know with the right of others not to know? At first, the answer may seem obvious. People being tested should keep the information to themselves. But what if the genetic screening reveals that a person has a disease gene but getting the disease depends on environmental factors related to individual choices, such as diet or smoking? Is it unethical *not* to tell a family member they or their child may have a gene for a disease that could be avoided through behavioral changes?

Objectives

After reading the materials and completing the activities, students will have:

- Obtained more experience in analytical treatment of ethical issues
- Learned how to recognize ethical dilemmas
- Learned more about genetic testing methodologies and applications and the issues they raise

Preparation

- Review the *Student Activity.*
- Make copies of the blank box-up form, which can be found in Appendix A and on the CD.
- Make copies of the *Student Activity* if your students do not have the textbook.

Case Study 1: James and Carol H.

Refer to the step-by-step method for using the model. To follow the methodology outlined in the model, students must have a basic knowledge of CF, including its genetic basis, symptoms, and treatments and the prognosis for patients who suffer from the disease. The Cystic Fibrosis Foundation is an excellent source of information.

Alice is the 6-year-old daughter of James and Carol H. Alice was diagnosed with CF at age 2. Alice has a moderately severe case of CF and receives postural drainage from her mother three times a day. In addition, she is on a regimen of vitamins and enzymes that requires her to take about 25 pills each day. Carol spends much of her day caring for Alice. At least twice each year, Alice requires hospitalization to fight respiratory tract infections.

Alice is cared for at a public clinic associated with a medical school that has a research project on CF. Scientists there have learned that there are many different mutations in the gene encoding the protein that leads to CF and that the different mutations lead to varying degrees of severity of CF.

The clinic physicians ask if James and Carol would undergo genetic testing to determine the nature of the mutations in the CF gene that each of them

carries. James and Carol agree to be tested. They have been talking about having another child. They hope that understanding the nature of their mutations will help researchers develop better ways to treat the disease.

When the results come back to the clinic, Alice's physician is quite surprised to learn that Carol, but not James, carries the CF gene. Since James is not a carrier, it is virtually impossible for him to be Alice's biological father.

I. Identify the question

Is there a dilemma, and if there is, what is it? Questions that might be posed include the following.

1. Was the physician morally justified in requesting the testing in the first place?
2. Should the physician divulge the test information to James and/or to Carol?
3. If the physician chooses not to divulge the information but James and/or Carol requests the results of the test, should the physician tell the truth or withhold it?

For purposes of this discussion, we select question 2.

II. Identify the issue

What is the general topic in this case? Genetic testing.

III. State facts of case

What are the facts in the case? It is particularly important here to remind students about the difference between facts and inferences. Many students will make unsupported assumptions or jump to conclusions. For instance, in this case, we cannot be sure that Carol and James are not already aware of Alice's biological father. We simply want to list the facts in an accurate but concise manner.

- James does not carry the CF gene.
- Carol does carry the CF gene.
- Alice (6 years old) has moderately severe CF.
- James and Carol are considering having another baby.
- Alice requires postural drainage three times each day.
- She is hospitalized twice a year for respiratory infections.
- She requires 25 pills a day.
- The mother is the main caretaker.

The facts must be true, relevant, and sufficient. If a different question had been selected, then the facts chosen from the case might be different.

IV. List possible decisions

What are some possible solutions to the problem? List as many as possible.

1. The physician should not tell James or Carol.
2. The physician should tell James only.
3. The physician should tell Carol only.
4. The physician should wait until Alice is 18 and tell Alice only.
5. The physician should tell only if asked.
6. The physician should tell Carol and direct her to tell James.
7. The physician should tell the minister of James and Carol.
8. If Carol will not tell James, then the physician should tell him.

These are eight of the possible decisions. Encourage solutions that represent compromise. Again, we need to choose a decision that is supported by one of the major ethical principles listed earlier.

V. Gather additional facts

Students may or may not need to gather additional information on CF, the accuracy of genetic tests, and so on.

VI. Pick decision to support

For our example, we choose number 8: the physician should tell Carol and then tell James if she will not. We need to know what principles we are abiding by and what principles we are breaching. First of all, we are telling Carol the results of James's test, and that is a breach of his right to confidentiality. We justify it by the principle of trying to avoid harm to Alice, Carol, and James. We fear the information may endanger their family structure and might even result in a divorce. Because Alice requires a lot of support and this family appeared to be happy before this event, threatening the stability of the family is seen as wrong in this argument.

On the other hand, we respect James's right to know. The information is about James, so he does have the right to know according to the principle of respect for individual freedom. Harm could be done to James by withholding this information. For example, if James were widowed and remarried, he might want more children but not be willing to take a risk

that does not, in fact, exist. By giving Carol the chance to tell James, we try to respect the rights of the parties involved. It is not a perfect solution, but there would be no dilemma if a perfect solution existed.

A student's (or student group's) decision should be stated in the form of an "I" or a "we" statement. For instance, "I believe the physician would be morally justified in telling Carol the results of the test first. But if she does not inform James, then the physician should reveal the results of the test to James."

VII. Identify guiding principle

The principle we have chosen is to avoid harm.

VIII. Identify supporting authority

What authorities would back us up on this point? One of the first places to look is in professional codes of ethics. For instance, the Hippocratic Oath states that the physician should do no harm. Similar statements are found in the American Medical Association Code of Ethics. Students can also do library research to find similar cases to support their position.

IX. Formulate rebuttal

Under what circumstances would we change our minds about what to do? Here again, it is important to encourage students to think freely. What if the physician found out that the test results were false? What if James came to the physician and said he already knew that he was not a carrier? In these two circumstances, the physician might change her mind.

X. State level of confidence

In this case, the qualifier might be "moderately confident."

XI. Box up the case

Use Worksheet 39.1 in Appendix A.

XII. Prepare argument

Case Study 2: Angela

Angela is a healthy 32-year-old certified public accountant who is married with no children. She has been working for 8 years in a small accounting firm.

She has decided she would like to move to a nearby large city and work for a major accounting firm.

Five years ago, Angela's mother was diagnosed with HD, an autosomal dominant disease. (Depending on student background, discussion of the disease may be as brief or as extensive as the teacher chooses.) Angela's mother is in a nursing home.

Angela's job interviews go well, and she is offered an excellent position that she readily accepts. However, as Angela is filling out the required paperwork in the personnel office, she notices that she must sign a consent form for a physical that includes a genetic screening test of her DNA.

When Angela questions whether she must have the tests, she is told that genetic screening is required by the company for all new employees.

The issue here could be called genetic screening.

A number of questions could be raised about this case. Here are a few examples.

1. Should the accounting firm require genetic screening?
2. Who should have access to Angela's results if she agrees to be tested?
3. Should Angela's employment be contingent on the results of her DNA test?
4. Should Angela tell her sisters and brothers the results of her tests if genetic predispositions to other diseases are found in the screening?

For the purposes of illustration, we will choose question 1. Should the accounting firm require the testing?

Question 1 provides a good opportunity to illustrate a major source of contention in all beginning ethics discussions: the difference between ethical behavior and legal behavior. Students must understand that their discussions are to focus on what is ethical, that is, what is right or wrong. An ethical decision may not necessarily be one that is supported by the law. Learning that what is legal may not necessarily be ethical is a lesson in itself for most students.

Even if the accounting firm is not legally prevented from DNA testing, that does not mean that such testing is right or ethical. In posing the questions listed above, the use of "should" rather than "can" implies that we are interested in the ethical rather than the

legal questions. Although the legal perspective cannot be ignored, it must be clearly separate from an ethical decision.

Helping students differentiate between what is an actual fact and what is an inference, a judgment from the facts that may or may not be true, is essential. To make this model work, we must use the available facts. Appropriate research should be encouraged. Students can be assigned library research, or they can interview local physicians, lawyers, insurance agents, employers, and so on. The March of Dimes is also an excellent source of information on genetic disorders.

What are the facts in this case? Try to list them as briefly as possible without inferences.

• Angela is a 32-year-old healthy female accountant.
• Mother has HD.
• Angela was offered position with new firm.
• Firm requires physical, including DNA screen.

What are some possible solutions or choices in this case? Again, here is a partial list.

1. The accounting firm is not morally justified in requiring the test.
2. The company should require the test only if it is trying to prevent possible harm to its employees (e.g., to prevent exposure to a chemical to which a person may have a genetic susceptibility).
3. The company should request the screen, but only for insurance purposes, not for employment decisions.
4. Angela should go somewhere else for a job.
5. Angela should go to an independent laboratory for the test and be sure that only she receives the results.

At this point, a class can be divided into small groups to try to construct arguments for their choices, or they could be assigned a position.

After choosing the option, the next step is to decide why that option is morally best. Each participant must decide which principle best supports the choice he or she has made. If our claim is that the accounting firm is not morally justified in requiring the DNA test, then we must be prepared to explain which moral-action guide is being violated by such a request.

In this case, more than one principle can be called on to support the choice. Next, locate backing for the position. Formulate a rebuttal, and state the level of confidence in the decision.

Analyzing Ethical Dilemmas in Genetic Testing

Background Information

Genetic testing is examining a person's genetic makeup to determine whether he or she carries a gene(s) of medical interest. It is usually thought of in the context of determining whether someone has a genetic disorder or is a carrier of a recessive disease gene, such as a gene mutation that leads to cystic fibrosis (CF).

You have already learned about various ways that physicians tested for genetic disorders before scientists developed the capacity to look directly at a person's DNA. Some involve analysis of genetic material at a microscopic, but not a molecular, level by examining a person's karyotype. For example, Down's syndrome is caused by the presence of an extra chromosome 21, and other genetic diseases, such as Burkitt's lymphoma and a form of leukemia, can be identified through changes in chromosomal banding patterns and sizes. Other longstanding genetic tests for disorders and carriers rely on phenotypic manifestations of the genetic defect. For example heterozygous carriers of diseases such as Tay-Sachs and sickle-cell anemia have been identifiable for many decades because their cells display characteristics that distinguish them from normal cells. Physicians diagnose other genetic disorders using biochemical phenotypes, such as high serum levels of the amino acid phenylalanine in phenylketonuria and the absence of an enzyme that breaks down galactose in the genetic disease galactosemia.

As society enters the new age of molecular genetics, it will be possible to test for more and more of the 3,000 to 4,000 genetic disorders physicians have identified in the past century. The broadened capacity results from the new ability to analyze DNA directly, from the abundance of sequence data provided by the Human Genome Project, and from research that is identifying more and more of the defective genes that result in genetic disorders. In the past decade, genes responsible for CF, muscular

dystrophy, hereditary breast cancer, Huntington's disease (HD), and many other disorders have been identified. Once a disease gene has been identified, it is usually possible to devise a test that will reveal whether a person carries a disease-causing mutation.

Many other diseases, such as colon cancer, late-onset Alzheimer's disease, multiple sclerosis, emphysema, and diabetes, are not thought of as strictly genetic disorders, but all have a genetic component. They result from an interaction among a number of genes associated with the disease, a person's genotypic background, and the environment and, as such, are known as multifactorial diseases. Research is currently focused on identifying genes that predispose individuals to these conditions. Physicians already have the capability to test for genetic predisposition to some multifactorial disorders, and it is reasonable to assume that in the future they will be able to determine (through genetic testing) a person's genetic risk of developing many more of these diseases, as well.

Societal Impacts

Society has considerable experience with genetic testing; some of that experience has made it clear that we must proceed carefully, conduct widespread education about genetic diseases, and develop clear policies to prevent misuse of genetic information. In the 1970s, testing for sickle-cell anemia resulted in confusion and fear. Many people did not understand the difference between being a carrier and having the disease, and because of this widespread misunderstanding, people who merely carried the sickle-cell gene were threatened with loss of jobs and insurance. The initial beneficial purpose of helping people understand their risk of having children with sickle-cell disease was lost in a mire of misconceptions.

Another problem with genetic testing is that physicians can identify many more diseases than they can treat. HD is a good example. People who carry this dominant genetic disease gene live their early lives normally and do not develop symptoms until middle age. At that point, however, the individual begins to lose the ability to think and develops uncontrollable body movements, such as twitching and shaking. These conditions become worse and worse over a period of years until the victim dies. There is no treatment or cure. Unfortunately, since the disease is dominant, the children of an HD victim know that they have a 50% chance of inheriting the disease gene, and since the disease does not reveal itself until middle age, most victims have already had children before they know whether they themselves have the gene.

Therefore, should a 20-year-old be able to find out that he carries a dominant gene for a disease that will first debilitate and finally kill him in middle age? If people should be given this information, at what age is it best revealed? Does respect for human dignity in this situation mean that physicians should try to prevent harm by not telling the patient or try to promote autonomy by telling the person their genetic status? Because of the experiences with sickle-cell testing 25 years ago, screening for HD is being done under very strict guidelines at 14 centers in the United States. The centers require clients to undergo extensive psychological testing before screening and counseling after screening. If a cure or even a treatment were available, then these questions would not be nearly as difficult, but as mentioned earlier, neither treatment nor cure is available. In the case of sickle-cell anemia, the screening test has been available for almost 40 years, but there is still no cure.

Because the ethical questions surrounding genetic testing are apparent, 3% of the $3 billion budget for the Human Genome Project was earmarked for studying ethical considerations. The Working Group of Ethical Legal and Social Implications of the Human Genome Project proposed nine areas that merit attention:

1. Fairness in the use of genetic information
2. Impact of knowledge of genetic variation on the individual
3. Privacy and confidentiality
4. Impact on genetic counseling
5. Impact on reproductive decisions
6. Issues raised by the introduction of genetics into mainstream medical issues
7. Uses and misuses of genetics in the past
8. Questions raised by commercialization
9. Conceptual and philosophical implications

It is easy to see that different genetic conditions present different problems. With HD, there is currently no medical benefit to the individual in knowing that he or she has the disease gene (though such knowledge might affect the choices that person makes about how to live his or her early life and whether to have children). On the other hand, for conditions that have both environmental and genetic causes, such as high blood pressure and emphysema, early detection of susceptibility may save many lives and significant dollars. If people do make appropriate lifestyle changes, good has been achieved.

But who will be tested for these kinds of traits? Will the tests be voluntary or required? Who will pay for the test? (Remember that when anyone says, "The government pays," they really mean you pay, since the government gets its money from taxing people who work for a living. Middle class people pay about one-third of their incomes in various taxes.) Will companies deny insurance or employment to people who refuse testing or to people who refuse to make lifestyle changes after tests indicate they have increased risks for certain conditions?

These are not easy questions. The medical profession and insurance industry are considering them now. You will, no doubt, be affected by their decisions during your lifetime.

Case Study 1: James and Carol H.

Alice is the 6-year-old daughter of James and Carol H. Alice was diagnosed with CF at age 2. Alice has a moderately severe case of CF and receives postural drainage, a carefully specified, time-consuming program of thumping to loosen mucus in the lungs, from her mother three times a day. In addition, she is on a regimen of vitamins and enzymes that requires her to take about 25 pills each day. Carol spends much of her day caring for Alice. At least twice each year, Alice requires hospitalization to fight respiratory tract infections.

Alice is cared for at a public clinic associated with a medical school that has a research project on CF. Scientists there have learned that there are many different mutations in the gene encoding the protein that leads to CF and that the different mutations lead to varying degrees of severity of CF.

The clinic physicians ask if James and Carol would undergo genetic testing to determine the nature of the mutations in the CF gene that each of them carries. James and Carol agree to be tested. They have been talking about having another child. They hope that understanding the nature of their mutations will help researchers develop better ways to treat the disease.

When the results come back to the clinic, Alice's physician is quite surprised to learn that Carol, but not James, carries the CF gene. Since James is not a carrier, it is virtually impossible for him to be Alice's biological father.

I. Identify the question

First of all, is there a dilemma, and if there is, what is it? List some possible questions (Should . . . ?)

II. Identify the issue

What is the general issue in this case?

III. State facts of case

What are the facts in the case? It is easy to jump to some conclusions in this case, so try hard to stick to what you actually know (as opposed to what you have concluded or what you suspect). Depending on the question you are considering, different facts may be more relevant.

IV. List possible decisions

What are some possible solutions to the problem? Brainstorm as many as possible.

V. Gather additional facts

Do you need additional background information to inform your decision?

VI. Pick decision to support

Formulate your decision (or your group's decision) in the form of an "I" or a "we" statement. For instance, "I believe the physician would be morally justified to. . . ."

VII. Identify guiding principle

Which of the major ethical principles justifies your decision?

VIII. Identify supporting authority

What authorities would back you up on this point? One of the first places to look is in professional

codes of ethics. For instance, the Hippocratic Oath states that the physician should do no harm. Similar statements are found in the American Medical Association Code of Ethics.

IX. Formulate rebuttal

Under what circumstances would you change your mind about what to do? Think freely.

X. State level of confidence

Formulate a one- or two-word statement to describe how strongly you believe your argument.

XI. Box up the case

Draw up a box as described previously (chapter 38). Use Worksheet 39.1.

XII. Prepare argument

Write out an argument as described previously (chapter 38).

Case Study 2: Angela

Angela is a healthy 32-year-old certified public accountant who is married with no children. She has been working for 8 years in a small accounting firm. She has decided that she would like to move to a nearby large city and work for a major accounting firm.

Five years ago, Angela's mother was diagnosed with HD, an autosomal dominant disease. Angela's mother is in a nursing home. There is a 50% probability that Angela has inherited the genetic defect that causes HD, but she has chosen not to be tested for the HD gene.

Angela's job interviews go well, and she is offered an excellent position that she readily accepts. However, as Angela is filling out the required paperwork in the personnel office, she notices that she must sign a consent form for a physical that includes a genetic screening test of her DNA for a variety of genetic defects. When Angela questions whether she must have such a test done, she is told the company requires genetic screening of all its employees.

A number of questions could be raised about this case. Here are a few examples.

1. Should the accounting firm require genetic screening of all potential employees?

2. Who should have access to Angela's results if she agrees to be tested?
3. Should Angela's employment be contingent on the results of the genetic screening?
4. If the tests reveal a number of genetic defects, such as mutations that predispose Angela to emphysema or colon cancer, should Angela tell her sister?

Think of more questions, and choose one for analysis. Apply the model as before.

Societal Issues

Science, Law, and Politics

In addition to ensuring that medical products on the market are safe and efficacious through their regulatory laws and policies, governments are responsible for determining how the products of medical biotechnology can and cannot be used. With respect to genetic testing, in 1996, the Joint Working Group on Ethical, Legal and Social Implications of Human Genome Research and the National Action Plan on Breast Cancer published a set of recommendations for policies that would protect against genetic discrimination in insurance coverage and employment practices. In general the recommendations prohibit:

- Insurance providers from requiring genetic tests, disclosing genetic information, or using genetic information to affect insurance premiums or deny or limit coverage
- Employers from requiring genetic tests as a condition of employment, disclosing genetic information, or using genetic information in hiring or promotion decisions

To date, the U.S. Congress has not enacted federal legislation related specifically to genetic discrimination in the workplace or insurance coverage. Over the past decade, several bills were introduced. Some were new bills based on the Working Group on Ethical, Legal and Social Implications of Human Genome Research-National Action Plan on Breast Cancer recommendations. Others attempted to amend existing civil rights statutes and labor laws, because some attorneys believed existing antidiscrimination laws, such as the Americans with Disabilities Act (1990), could be interpreted to cover genetic discrimination.

The only federal policy explicitly prohibiting genetic discrimination is an executive order signed by President Clinton in 2000. However, the order protects only current or potential employees of the federal government by prohibiting federal employers from requiring or requesting genetic tests as a condition of being hired or receiving benefits or using protected genetic information to deprive employees of advancement opportunities. Various state legislatures have enacted laws to prevent genetic discrimination, but none are comprehensive with regard to coverage, protections afforded, or enforcement authorities.

Government Policies and Scientific Research

Most people agree that the federal government has an important role to play in ensuring the safety, efficacy, and fair use of products and processes derived from science and technology. What should be the role, if any, of politicians in determining which paths of scientific inquiry are pursued and therefore which medical breakthroughs are most likely to occur? This question is not unique to biotechnology research. Below, we focus primarily on embryonic stem (ES) cell research to describe some of the ways government bodies influence which scientific questions are pursued and, therefore, which potential technological applications become commercial realities.

Providing funding for research

Governments use a variety of tools—both carrots and sticks—to encourage some areas of research while discouraging others. The most obvious is the carrot of money. The U.S. government or, more accurately, the U.S. taxpayer is far and away the primary funding source for basic research and a major contributor to applied research. This is one of the reasons policy makers have a right and a responsibility to determine research priorities. Both the executive and legislative branches have mechanisms for altering, relatively quickly, allocations of research monies to respond to political pressure, unforeseen problems (bioterrorism or emerging infectious diseases), and shifts in philosophies and priorities that often accompany changes in the makeup of Congress or new administrations. Funding agencies, such as the National Institutes of Health (NIH), are overseen by cabinet-level agencies, headed by presidential appointees who must carry out the administration's policies. In addition, every year, when Congress passes its appropriations bills that divvy up grant monies among the various federal funding agencies, funds are earmarked for research areas favored by certain members and their constituents.

Appropriations bills can also use money as a stick and explicitly prohibit funding for some research areas that are perfectly legal but out of favor. Most research scientists rely solely on public funds to finance their research, so they often alter their research programs to varying degrees in response to government allocation of funds and any policy shifts being executed by the various funding agencies.

Influencing research with regulations

Governments can also influence the course of scientific discovery by establishing regulatory requirements that govern certain types of research but not others. In the reading on biotechnology regulation in the United States, we mentioned that the financial costs of obeying regulatory requirements can deter some areas of academic research, often unintentionally. However, sometimes bans on certain areas of research or strict regulatory guidelines can advance research by creating clarity about what is and is not permissible. A well-established and fair regulatory regime, no matter how strict, is preferred by both academic and industrial researchers, because it creates a known, predictable environment within which they can function. For example, the United Kingdom has strict government guidelines regarding ES cell research: the ES cells must be derived from frozen blastocysts about to be discarded, no research can be conducted on human embryos older than 14 days, and so forth. ES cell research is flourishing in the United Kingdom, which is viewed internationally as a leader in the field.

Controlling research through funding bans

Soon after the 1973 *Roe v. Wade* Supreme Court decision legalizing abortion, the executive and legislative branches of the U.S. government began issuing a series of policies, moratoria, and funding bans that discouraged or prohibited research on embryos, IVF, and fetal tissue.

In 1993, President Clinton issued an executive order to remove the funding ban on fetal-tissue research. Later that year, the U.S. Congress enacted the NIH Revitalization Act, which permitted funding for research on both fetal tissue transplants and IVF. However, in 1995, Congress attached a ban to the NIH appropriation bill (and each successive NIH appropriation bill) that prohibits use of federal funds for "the creation of a human embryo or embryos for research purposes; or research in which a human embryo or embryos are destroyed, discarded, or knowingly subjected to risk of injury or death greater than that allowed for research on fetuses *in utero*." As a result, since 1974, private companies and research foundations have financed almost all research on IVF and fetal tissue in the United States, as well the human ES (hES) cell and human embryonic germ cell work described in chapter 36.

In response to the 1998 scientific breakthroughs that allowed researchers to create and maintain hES cell lines in culture, the NIH director, recognizing the potential medical benefits of research on hES cells, asked for a legal opinion regarding research on hES cell lines. He wanted to know if the congressional ban on research that destroyed human embryos applied to hES cell lines, because no embryos were destroyed once the cell lines were established. In 1999, the Office of the General Counsel of the Department of Health and Human Services, which oversees the NIH, ruled that hES cell lines should be excluded from the funding ban because they lack the natural "capacity to develop into human beings" (i.e., lacking trophoblast cells, they would not be able to implant) and therefore "are not embryos."

Following the determination that the NIH could legally fund some types of hES cell research, the NIH began drafting guidelines to specify requirements under which it would fund such research. In August 2000, the NIH released guidelines that placed the following stipulations on hES research that it could fund.

1. The *derivation* of hES cell lines was *not* fundable, because derivation of ES cells from blastocysts destroys embryos.
2. Research on hES cells derived and established by privately funded entities was acceptable, but only if:

   - The private entity derived the hES cell lines from frozen blastocysts that were about to be discarded.
   - The frozen blastocysts were donated with the informed, written consent of the parents.
   - The parents were not compensated for donating the blastocysts.
   - The private entity did not profit from the sale of blastocysts used for stem cell derivation.

Soon after the NIH established these guidelines, George W. Bush was elected president. A few months after he took office in 2001, his administration published a *Notice of Withdrawal of the NIH Guidelines for Research Using Pluripotent Stem Cells* in the *Federal Register.* On 9 August 2001, President Bush announced a more restrictive funding policy than the one established by the previous administration. Federal funding is now allowed only on the hES cell lines that had been derived prior to 9 August 2001. No federal research funds can be spent on any hES cell line derived after his announcement. In his announcement, President Bush also promised to increase federal funding for adult stem cell research.

Controlling research with federal laws

The past four sessions of Congress have witnessed a flurry of legislative activity related to advances in hES cell research and the closely related topic of somatic cell nuclear transfer (i.e., cloning). Various members of Congress in both houses have introduced dueling bills. Some attempt to expand the ban established in the Bush administration's 9 August 2001 policy statement, while others try to counter the policy with a bill expressly supporting hES cell research.

The thorny scientific, technical, and ethical issues surrounding hES cell research become even more intractable when they are linked to those associated with somatic cell nuclear transfer (SCNT). Even though everyone in Congress is opposed to SCNT for reproductive purposes, they have been unable to pass a law prohibiting it. The bills that attempt to ban "reproductive cloning" also include either a ban on therapeutic cloning or support for both therapeutic cloning and research on hES cells derived from IVF blastocysts destined for destruction.

Therefore, because members of Congress are divided on the issues of therapeutic cloning and use of hES cells, they have not banned reproductive cloning, as have other governments throughout the world, even though every member of Congress is opposed to it. The standoff in Washington leaves the United States open to a practice that virtually everyone on Earth opposes, because it has no law that bans cloning or specifically gives a federal agency authority over activities in private research laboratories, IVF clinics, or any commercial operation dealing with human embryos.

At least 10 bills related to hES research and cloning were introduced and referred to various congressional committees between 1999 and 2006, but as of August 2006, only one had passed. In 2001, and again in 2003, the House of Representatives passed bills criminalizing SCNT for research or therapeutic purposes. Any scientist engaged in such research would be imprisoned for up to 10 years and fined a minimum of $1 million. A companion bill has been introduced in the Senate every year since 2001, but to date, none has passed.

Many members of Congress who are antiabortion do not support funding restrictions or bans on hES cell research and therapeutic cloning. In May 2004, over 200 members of the House, including members of President Bush's party and more than 30 abortion opponents, sent a letter urging him to reverse his August 2001 decision and reinstate federal funding for research that adheres to the guidelines the NIH issued in August 2000.

Bioethics Case Study: Gene Therapy

About This Activity

These case studies introduce a few ethical dilemmas associated with gene therapy. Many of these dilemmas result from society's increased capacity to test for genetic mutations. As a result, it is best if your students are familiar with the potential applications and issues of genetic testing before considering gene therapy.

As was true of issues in genetic testing, gene therapy raises issues that are both old and new. Helping students place issues in context will be essential for informed discussions. Gene therapy also provides a gradation of issues and, as such, excellent case studies for addressing the inescapable question surrounding technology development: where should we draw the line? Does providing a hemophiliac with injections of the functional gene they lack differ from injecting them with the protein (clotting factor VIII) encoded by the missing or mutated gene? Instead of regular injections of the gene, what about a one-time, very costly, painful bone marrow transplant in which the patient's blood stem cells are given a correct copy of the missing or mutated gene? What if the gene is provided to an embryo prior to implantation? Instead of providing the correct gene to an existing embryo, what about the parents using in vitro, preimplantation testing to select the embryo that does not have the hemophilia genetic disorder and discarding the rest of the embryos? Since hemophilia can be managed with regular injections of the missing protein, should limited public research funds for gene therapy be provided only for diseases that cannot be managed with other therapies?

None of these questions has an easy answer. Helping your students accept that fact and teaching them how to think about these issues rationally should be your primary objectives.

Introduction

At first glance, gene therapy, like any medical treatment that proposes to benefit the patient, seems free of any ethical implications. Closer examination raises a number of important questions regarding the acceptability of different kinds of gene therapy. Leroy Walters, of Georgetown University's Kennedy Institute of Bioethics, divided gene therapy into four possible categories.

1. Somatic-cell gene therapy for the cure or prevention of disease. *Example:* insertion of a DNA sequence into a person's cells to allow production of an enzyme like adenosine deaminase.
2. Germ line gene therapy for cure or prevention of disease. *Example:* insertion of an adenosine deaminase sequence into early embryo or reproductive cells, which would affect not only the individual but all of his or her offspring.
3. Somatic-cell enhancement. *Example:* insertion of a DNA sequence to improve memory, increase height, or increase intelligence, which would affect only that individual.
4. Germ line enhancement. *Example:* insertion of a DNA sequence for enhancement into a blastocyst, sperm, or egg, which would affect future generations.

One of the ongoing issues concerning gene therapy is whether any or all of the four types of manipulation are ethically acceptable. At present, only somatic cell gene therapy for the cure or prevention of serious disease is considered ethically appropriate. Germ line therapy (altering disease genes so that the individual not only will be healthy but will pass on the healthy genes to his or her offspring) is considered desirable by some, but the techniques used to alter animal embryos have far too high a failure rate to consider their application to humans at this time. Nonetheless, human germ line therapy will probably be feasible in the future.

Enhancement therapy (whether somatic cell or germ line) is viewed as much less acceptable. French Anderson, a pioneer in this field, has stated, "I will argue that a line can and should be drawn to use gene transfer only for the treatment of serious disease, and not for any other purpose. Genetic transfer should never be undertaken in an attempt to enhance or 'improve' human beings."

While virtually everyone agrees with the spirit of Anderson's position, how will society determine what is a serious disease and what is enhancement? Some situations, such as inherited diseases that are incurable or medical problems with no other options for medical intervention, will be obvious. In the future, however, physicians will be able to detect many more defective genes than those that cause illnesses that are incurable and untreatable. Some genetic defects that are life threatening will be controllable with therapeutics (e.g., hemophilia). Other genetic defects will simply indicate a propensity to develop a disease. Still others will cause defects that make life difficult and cannot be cured but are not life threatening (e.g., color blindness and albinism). In other words, when is a genetic defect truly a defect? At what point will eliminating genetic defects inch toward selecting desirable characteristics, such as tall stature or more attractive physical appearance?

Guidelines for Gene Therapy

In considering any application of gene therapy, basic respect for human dignity is, as always, the underlying moral principle. Like any other experimental medical treatment, gene therapy should be used to benefit the patient. Harm should be avoided. Certain factors must be considered. A National Institutes of Health (NIH) committee suggested the following considerations in 1985.

1. What is the disease to be treated?
2. What alternative treatments for the disease exist?
3. What are the potential benefits of gene therapy for human patients?
4. How will patients be selected in a fair and equitable manner?
5. How will a patient's voluntary and informed consent be solicited?
6. How will the privacy of patients and the confidentiality of their medical information be protected?

These six questions address some ethical concerns that have been considered in evaluations of experimental proposals. Encourage students to consider extensions of these questions or completely new ones.

Sample Issues Related to Gene Therapy

Considering ethical issues related to gene therapy is like opening the proverbial can of worms. Using specific case studies will focus your students' discussion on one set of circumstances at a time. Students can consider broad issues in the context of one individual decision. The following paragraphs contain some of the issues and questions you may want to address with your students.

What kind of diseases should be treated with gene therapy? How does one define a serious disease? What is the difference between a serious genetic disease, a moderate genetic defect, and "normal" variation? Will there be subtle or even outright pressure on people with genetic diseases to be treated so they can be more nearly "perfect"? Are the procedures really safe? Could gene therapy have long-term consequences for the patient that cannot be predicted? Who will have access to this kind of treatment—only the rich or the insured?

When germ line therapy becomes feasible, wouldn't it be much less costly to use such therapy to eliminate the bad genes from a whole family than to treat just one person at a time? Are scientists knowledgeable enough for society to consider changing the gene pool by doing germ line therapy? Are all alleles that physicians consider harmful really harmful in the long run and under all conditions? We only need to remember the sickle-cell allele's effect on malaria infection to answer that question.

As knowledge about genetic risk factors and the ability to manipulate the human genome increase, even more issues will be raised. For example, if physicians could lower a person's susceptibility to an environmental toxin, like a common water pollutant, should they? Who will decide? If such therapy were available, would insurance companies and employers begin to discriminate against people who have not had that therapy?

What about genetic enhancement? Presumably, many people would want to be genetically altered to increase longevity or intelligence. In fact, students are usually quick to condemn the idea of genetic enhancement in general, but they frequently change their tune when asked if they would like to receive treatment to become smarter or stronger, to have perfect teeth or a perfect complexion, to excel at athletics, etc. Similarly, adults view the issue differently when asked if they would like to ensure that their children will be intelligent, strong, and resistant to disease. Will standards be set for which traits are desirable and which are sufficiently undesirable to warrant therapy? When we consider enhancement gene therapy for height, for instance, how will we decide how short is too short? Many traits we might consider enhancing are multifactorial, a situation that increases the risks, since many genes may be involved.

In all of these discussions, the specter of eugenics cannot be ignored. Eugenics, the study of hereditary improvement through genetic control, was widely embraced by scientists, politicians, and other well-respected people in the United States in the 1920s. However, the horror of the way eugenics was practiced in Nazi Germany brought the U.S. efforts to a halt. By 1939, to prevent "racial deterioration," Germany had sterilized nearly 400,000 people who had genetic "defects," including alcoholism and feeble-mindedness. It was not long before people who were costing the state too much money because of their genetic defects were killed. Seventy thousand such deaths are documented between 1939 and 1941. A thorough and clear account of the history of the eugenics movement in both the United States and Germany can be found in Edwin Black's 2003 book, *War against the Weak: Eugenics and America's Campaign to Create a Master Race* (Four Walls Eight Windows, New York, NY). (Teaching about the Holocaust in this unit is an opportunity to bring in the history department to implement interdisciplinary study.)

Even if society is naive enough to believe that the eugenics movement could never happen again, we still have numerous and difficult questions to answer about gene therapy. Obviously, you cannot provide your students with answers, but you can provide them with essential skills. First, teach them to ask as many questions as possible. Then teach them to find the data they need to try to answer these questions rationally by using the critical-thinking skills you are helping them develop.

Objectives

After reading the chapter, reviewing background materials, and completing the activities, students will have:

- Learned about the costs and benefits of gene therapy
- Gained additional experience in identifying ethical dilemmas, using the decision-making model, and analyzing issues associated with modern biotechnology critically

Preparation

- Review or present any necessary background material.
- Read the *Student Activity,* which contains material not included here.
- Make copies of Worksheet 40.1 in Appendix A to box up the case studies.
- Make copies of the *Student Activity* if your class does not have the textbook.

Procedure for Using the Model

1. Have students read the case study.
2. Have students pose questions about the case that represent potential dilemmas ("Should . . . ?")
3. Select a specific question for analysis.
4. List the facts of the case. Seek additional background information.
5. Identify the relevant primary and secondary (if any) ethical principles.
6. List possible courses of action that could be taken.
7. Discuss how the possible solutions are related to the general ethical principles.
8. Have students select their recommended courses of action.
9. Discuss sources of authoritative opinion; assign students to look for authorities that support their chosen actions. (This can be homework.)
10. Have students brainstorm about rebuttal conditions.
11. Have students decide their levels of confidence in their recommendations.
12. Prepare a summary. (If students have been assigned outside research, the summary should be written after that research is completed.)
13. Individual students or groups may present their summaries to the class. Be sure to examine how several different ethically defensible positions can result from one case.

Case Study 1: Anne B.

Students should have a basic understanding of enzyme activity and single-gene inheritance patterns. If they have that background and have used the decision-making model, they should be able to work through this case in two 50-min classes. All of these steps can be done as a whole class or in groups. The class may start out as a whole class and then break up into groups to decide courses of action. Alternatively, the decisions may be made by individual students.

This case study deals with Gaucher's disease, which is described in detail in the *Student Activity* and in earlier chapters on genetic diseases. Briefly, the disease results from the production of a structurally altered form(s) of the enzyme β-glucocerebrosidase, which breaks down a lipid-like substance, glucocerebroside. Failure to properly metabolize this substrate results in its accumulation in the spleen, liver, and bone. The disease varies widely in its severity: some afflicted individuals die in early childhood, while others are diagnosed on autopsy after death from other causes in old age. Because of the variation in severity, Gaucher's disease is classified into forms I, II, and III. Studies suggest that these forms of the disease are caused by different mutations within the same gene. Those patients with the most common form (form I) of Gaucher's disease may first produce symptoms in their late teens, with an enlarged spleen often being an early sign (Figure 40.1).

In 1992, Joan B., age 20, died from the secondary effects of Gaucher's disease. Joan's 25-year-old sister, Anne B., also has the condition. However, she has been receiving enzyme replacement therapy each week since April 1994.

The biotechnology firm Genzyme began selling the β-glucocerebrosidase enzyme commercially in the early 1990s. Anne receives the enzyme every week by intravenous infusion at the office of her physician. Her physician is considered one of the most knowledgeable practitioners with regard to this particular disorder. The infusion process takes about 4 h each week, including 2 h of travel time. The efficiency of the enzyme replacement therapy is not completely known, but early studies have been very encouraging. There have been no harmful side effects.

Anne recently read about a new experimental treatment that involves removing some of her own cells, treating them with a retrovirus that carries the correct sequence for glucocerebrosidase, and replacing the cells. The cells should reproduce, creating a colony of cells that make the proper enzyme. The hope is that this treatment could be done several times, resulting in enough cells to permanently produce adequate amounts of normal enzyme. The technology is experimental, and there may be risks, some known and some unknown. For instance, since the sequence is added by using a retrovirus, some of Anne's normal DNA might be altered. This treatment is also more expensive than Anne's current treatment. Anne requests the experimental treatment from her physician.

Implementing the decision-making model

First of all, does an ethical dilemma exist? Anne has requested an experimental treatment that may harm her or may help her and improve her quality of life. Should her physician help her get the treatment she wants? Is it medically indicated? Is her disease as serious as French Anderson says it should be for her to qualify for gene therapy? There are certainly enough questions to indicate that this case does present bioethical issues.

I. Identify the Question
What question do you want to focus on to illustrate this case? Any question can be selected, depending on your interests and those of the students. In our example, we will focus on the question of whether Anne's physician should help her get gene therapy.

II. Identify the Issue
We might choose to call this an issue of experimental treatment or humans as research subjects. (This is a good opportunity to discuss the ethics surrounding the experimental treatments for human immunodeficiency virus infection.)

III. State Facts of Case
For the box, we want to restrict ourselves to the true, relevant, sufficient facts from the case itself. All the background material should not be included, since it will make the box very unwieldy.

- Anne, 25 years old, has Gaucher's disease.
- Anne's sister died from the disease.

Figure 40.1 Enzyme replacement therapy for Gaucher's disease. In 1991, Roscoe Brady of NIH isolated enough glucocerebrosidase from human placentas to give injections of the missing enzyme to 12 patients. Even though the injections had dramatic effects, isolating the enzyme from placental tissue was not cost-effective. Using recombinant DNA techniques, the gene encoding the enzyme was engineered into yeast cells, making glucocerebrosidase replacement therapy a viable option. (Photographs courtesy of Roscoe Brady, NIH, NINDS.)

Before

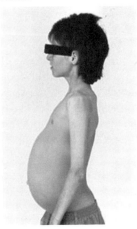

After

- Anne has been on enzyme replacement therapy since April 1994.
- Anne is requesting experimental gene therapy.
- Anne thinks the therapy will improve the quality of her life.
- Anne has read extensively about the new treatment.
- Anne's physician is considered an expert.

IV. List Possible Decisions

Here are some possible options. There could be others that are more appropriate. None is intended to be the "right" answer, since there are no right answers in these issues, only more or less justifiable choices.

1. Anne's physician should arrange for her to have the treatment.
2. Anne's physician should refuse to help her, since she is already being treated with an effective, safe method.
3. Anne's physician should try to discourage her from having experimental treatment, but if she insists, he should help Anne.
4. Anne should not have any treatment, since both are experimental.
5. Anne should not have any treatment, since all treatments are very expensive.
6. Anne should have both kinds of treatment.

Although the last three options may seem inappropriate, it is very important to let students suggest all alternatives and discuss them. Otherwise, students are misled into thinking that there are "right" answers to these questions. Furthermore, they will think that the teacher has the right answer! Be sure to have students consider the relevant ethical principles as they discuss the alternatives.

V. Gather Additional Facts

Do students need additional information to make a good decision? For example, what is a retrovirus, and what is known about the risk of cancer from retrovirus DNA insertion? Has the retrovirus approach been successful with other patients? Depending on the question that the students address and the issues they focus on, these questions may be very relevant.

VI. Pick Decision To Support

In this example, the claim will be that the physician should try to discourage Anne because gene therapy is even more experimental than her current therapy. Gene therapy may pose unknown, unnecessary risks for her.

VII. Identify Guiding Principle

Using option 3, the physician would be trying first to prevent harm from coming to Anne. However, if he cannot convince her to change her mind, then the physician, by helping Anne get what she wants, would be respecting her autonomy. Also, for Anne's autonomy to be respected, she would have to give informed consent for her treatment. Before she was treated, she would need to be able to understand her choices and the consequences of her choices.

VIII. Identify Supporting Authority

What experts would support this option? You can turn to the various codes of medical ethics or to the extensive literature in bioethics to answer that question. For this case, the Hippocratic Oath, the Nuremberg statements, and the Patient Bill of Rights all provide sources of ethical support.

IX. Formulate Rebuttal

Under what circumstances would someone abandon this choice? What if Anne's physician had information unavailable to the public that the gene therapy was ineffective but not harmful or that it was harmful and ineffective? What if Anne's diagnosis of Gaucher's disease was incorrect to begin with? What if the cost of the gene therapy was beyond Anne's ability to pay? What if Anne was very depressed about her condition and her physician questioned her competence to make an informed choice? Again, encourage as much creative thinking as possible.

For this example, we chose the possibility that the physician had information that the treatment was harmful. If that were true, then the physician could try to talk Anne out of the treatment and should refuse to help her get it. If she were still insistent, Anne could fire her doctor and find another.

X. State Level of Confidence

How strongly do you believe in your position? In the example we provide, we believe the rebuttal is so unlikely that we have strong confidence in the argument.

XI. Box Up the Case

Use the blank box-up form (Worksheet 40.1) in Appendix A.

XII. Prepare Argument

Putting the argument in paragraph form gives students an opportunity to explain on paper why they chose one option rather than the other or why the principle supports the option. The written argument should reflect the thinking necessary to construct the box.

Case Study 2: Bobby K.

In contrast to the previous example, case study 2 deals with a situation that raises questions about genetic enhancement rather than using gene therapy to treat a serious disease. In this case, parents are considering gene therapy for their son, who will achieve an adult height of 5 feet 3 inches. The April 2004 issue of *Popular Science* contained an article by a young woman, Jenny Everett, whose brother is being treated with human growth hormone. This article can be downloaded for free from the magazine's website (http://www.popsci.com).

Bobby K. is a healthy 10-year-old boy. He is very agile and quick and loves sports. The coach of his city league basketball team has told Bobby's parents that their son's skills on the court are astounding for a child of his age.

Bobby's father, Mr. K., is a healthy man who is 5 feet 3 inches tall. Bobby's mother is also short, only 5 feet. Bobby's pediatrician has predicted that Bobby will attain an adult height of about 5 feet 3 inches but has emphasized to his parents that he is a normal, healthy boy. Mr. K. remembers being teased constantly about his size and recalls that his lack of height kept him off all the varsity sports teams at his high school. He has often wondered if his shortness is a disadvantage to him in business dealings, too. Mr. and Mrs. K. both anticipate that their son will be the recipient of more and more pointed teasing as he reaches his teenage years. They also fear that he will not be selected for the basketball team when he reaches junior high or high school.

Bobby's coach has heard of a program at the local university in which gene therapy is being conducted on children who have a disease that results in the inadequate production of growth hormone. The children are being given working copies of the gene for human growth hormone, and the levels of growth hormone in their bodies have increased. These children's growth rates have also increased.

Bobby's coach tells the K. family about what he has heard. Bobby is thrilled, because he might be able to keep playing basketball, maybe even on a professional level. Bobby's parents are more cautious but would like their son to be spared the pain of being much smaller than his classmates.

The K.'s go to Bobby's pediatrician and request that Bobby be given the growth hormone gene therapy.

Using the model

As you use the model to discuss this case, you could head the discussion in any of several directions. For example, you could have the students do library work to try and learn whether any studies have determined whether it is safe to give additional growth hormone to healthy children. Sample questions include the following.

- Should Bobby's doctor allow him to have the gene therapy?
- Should Bobby's parents let him have the therapy?
- Should Bobby's coach have told the K.'s about the program?

Sources of Additional Case Studies

In the *Student Activity,* we present a case study that is based on an NIH study of growth hormone in children of short stature who were not growth hormone deficient. Beginning in 1988, injections of the hormone, not the gene encoding the growth hormone protein, were given to children. Children were eligible for enrollment in the study if they were two or more standard deviations below the average height for children of their age. The results of the study were published by the principal investigator, Ellen Leschek, and her colleagues in 2004 in the *Journal of Clinical Endocrinology and Metabolism.* Their paper can be downloaded for free from the journal's website at http://jcem.endojournals.org/cgi/content/full/89/7/3140. The bottom line: injections of human growth hormone to children of short stature who were not growth hormone deficient increased their height by 3 cm to 5 cm (1.1 to 1.9 inches). Allowing growth hormone to be administered to children of short stature who are not growth hormone deficient would have far-reaching economic and social ramifications. According to NIH's 1996 calculations, if growth hormone were to be provided to the approximately 11,000 children who are growth hormone deficient, the annual cost would be $155 million. If, however, the Food and Drug Administration approved growth hormone for healthy children who are two or more standard deviations below the mean height, 1.3 million children would be eligible, and the cost would jump to $20 billion per year.

Many bioethics texts contain case studies. News stories and magazine articles offer another rich source. Have your students watch out for potential material, but remind them that factual information must come from primary sources, not newspapers and magazines. No matter which case you are considering, make sure your students have access to real information about the issues. Library work is very important!

Analyzing Ethical Dilemmas in Gene Therapy

Background Information

Advances in genetic engineering technology offer hope to people afflicted with a number of genetic diseases. In the first approved human gene therapy experiment, W. French Anderson attempted to use genes to treat the genetic disease called severe combined immunodeficiency disease (see Figure 1.16). You may be familiar with the movie (starring John Travolta) about the "bubble boy," a young man with this disorder. The disease is caused by the lack of the enzyme adenosine deaminase (ADA).

In 1990, Anderson inserted the correct sequence for ADA into the white blood cells of a young child who was not producing it. Before this experimental treatment could begin, Anderson's proposal went through extensive review by the Recombinant DNA Advisory Committee at the National Institutes of Health (NIH), the government agency that funds most of the country's biomedical research. In fact, Anderson's proposal was reviewed 15 times. The need for a standard review procedure for proposed human gene therapy experiments was made clear by an incident in 1979–1980, when a University of California—Los Angeles researcher tried a human gene therapy experiment on two patients without getting approval from the appropriate review committee at his institution. He was eventually demoted, and his case demonstrated the need for a national policy on human gene therapy experiments.

A standard procedure is now in place, and researchers are using gene therapy to treat not only genetic diseases, but also other diseases, such as various cancers and AIDS (see chapter 1 for more details).

What is gene therapy, and what ethical issues, if any, does it raise? At first glance, gene therapy, like any medical treatment that proposes to benefit the patient, seems free of any ethical implications. Closer examination raises a number of important questions regarding the acceptability of different kinds of gene therapy. Leroy Walters, of Georgetown University's Kennedy Institute of Bioethics, divided gene therapy into four possible categories.

1. Somatic-cell gene therapy for the cure or prevention of disease. *Example:* insertion of a DNA sequence into a person's cells to allow production of an enzyme, like ADA (Figure 40.1).
2. Germ line gene therapy for the cure or prevention of disease. *Example:* insertion of an ADA sequence into early embryo or reproductive cells, which would affect not only the individual but all of his or her offspring.
3. Somatic-cell enhancement. *Example:* insertion of a DNA sequence to improve memory or increase height, which would affect only that individual.
4. Germ line enhancement. *Example:* insertion of a DNA sequence for enhancement into a blastocyst, sperm, or egg, which would affect future generations.

One of the ongoing issues concerning gene therapy is whether any or all of the four types of manipulation are ethically acceptable. At present, only somatic-cell gene therapy for the cure or prevention of serious disease is considered ethically appropriate, even by researchers like Anderson. Germ line therapy (altering disease genes so that the individual not only will be healthy but will pass on the healthy genes to his or her offspring) is considered desirable and ethical by some, but the techniques used to alter animal embryos have far too high a failure rate to consider their application to humans at this time. Human germ line therapy will probably be feasible in the future.

Enhancement therapy (whether somatic or germ line) is generally viewed as much less acceptable. Anderson has stated, "I will argue that a line can and should be drawn to use gene transfer only for the treatment of serious disease, and not for any other purpose. Genetic transfer should never be undertaken in an attempt to enhance or 'improve' human beings." Anderson, a pioneer in this field, has maintained his position since 1980.

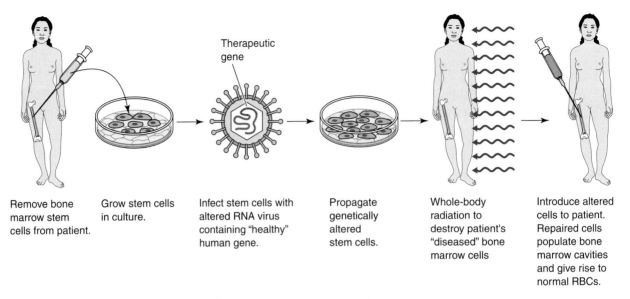

Remove bone marrow stem cells from patient.

Grow stem cells in culture.

Infect stem cells with altered RNA virus containing "healthy" human gene.

Therapeutic gene

Propagate genetically altered stem cells.

Whole-body radiation to destroy patient's "diseased" bone marrow cells

Introduce altered cells to patient. Repaired cells populate bone marrow cavities and give rise to normal RBCs.

Figure 40.1 Gene replacement therapy. For genetic defects that affect cells derived from bone marrow stem cells, such as severe combined immunodeficiency disease, bone marrow is removed from the patient, and the stem cells are multiplied in cell culture. A correct copy of the gene is inserted into a viral vector, using recombinant DNA techniques. The virus is cultured with the bone marrow cells. It infects the cells and inserts the replacement gene into some of them. Radiation destroys the patient's defective bone marrow cells, and the physician injects cultured cells, now containing the correct gene, into empty bone marrow cavities. To date, only a few of the gene replacement trials have been therapeutic.

Guidelines for gene therapy

In considering any application of gene therapy, basic respect for human dignity is, as always, the underlying moral principle. Like any other experimental medical treatment, gene therapy should be used to benefit the patient. Harm should be avoided. Certain factors must be considered. An NIH committee suggested the following considerations in 1985.

1. What is the disease to be treated?
2. What alternative treatments for the disease exist?
3. What are the potential benefits of gene therapy for human patients?
4. How will patients be selected in a fair and equitable manner?
5. How will a patient's voluntary and informed consent be solicited?
6. How will the privacy of patients and the confidentiality of their medical information be protected?

These six questions address some ethical concerns that have been considered in evaluations of experimental proposals submitted by medical researchers to NIH. You may think of additional ones.

Case Study 1: Anne B.

Gaucher's disease, first identified in the late 19th century, is an autosomal recessive disorder. The disease results from the production of a structurally altered form(s) of the enzyme β-glucocerebrosidase, which breaks down a lipid-like substance, glucocerebroside. Failure to properly metabolize this substrate results in its accumulation in the spleen, liver, and bone. The disease varies widely in severity: some afflicted individuals die in early childhood, while others are diagnosed on autopsy following death from other causes in old age. The severest form (the infantile form) follows a course somewhat similar to that seen in a related lipid disorder, Tay-Sachs disease.

Because of the variation in severity, Gaucher's disease is classified into forms I, II, and III: adult noncerebral, acute neuropathic, and subacute neuropathic, respectively. Studies suggest that these forms of the disease are allelic; that is, they are caused by different mutations within the same gene. The resulting structural defect in the protein is related to the severity of the disease.

The most common of the three types is form I (the least severe adult form). Form I Gaucher's disease is known to occur frequently in Ashkenazic Jews; approximately 1 in 13 members of this population is a carrier (heterozygote). Patients with type I disorder have about 15% of the enzyme activity of healthy individuals. There are estimated to be 20,000 cases in the United States. Although form I Gaucher's disease is the least severe of the three forms of the disease, it, too, varies in severity: some afflicted individuals die in early adulthood, while others live essentially symptom free to old age. Form I Gaucher's disease is therefore subdivided into acute, moderate, and mild courses.

Form I Gaucher's disease may first produce symptoms in a patient's late teens, with an enlarged spleen often being an early sign. Blood studies show reduced platelet, white blood cell, and red blood cell counts. Anemia and enlargement of the spleen may require removal of the spleen. Although this surgery decreases the anemia and relieves abdominal stress, the unmetabolized material is deposited in the bone more rapidly after splenectomy. Bone pain and pathologic fracture may occur.

In 1992, Joan B., age 20, died from the secondary effects of Gaucher's disease. Joan's 25-year-old sister, Anne B., also has the condition. However, she has been receiving enzyme replacement therapy each week since April 1994.

The biotechnology firm Genzyme began selling the β-glucocerebrosidase enzyme commercially in the early 1990s. Anne receives the enzyme every week by intravenous infusion at the office of her physician. Her physician is considered one of the most knowledgeable practitioners with regard to this particular disorder. The infusion process takes about 4 h each week, including 2 h of travel time. The efficiency of the enzyme replacement therapy is not completely known, but early studies have been very encouraging. There have been no harmful side effects.

Anne recently read about a new experimental treatment that involves removing some of her own cells, treating them with a retrovirus that carries the correct sequence for glucocerebrosidase, and replacing the cells. The cells should reproduce, creating a colony of cells that make the proper enzyme. The hope is that this treatment could be done several times, resulting in enough cells to permanently produce adequate amounts of normal enzyme. The technology is experimental, and there may be risks, some known and some unknown. For instance, since the sequence is added by using a retrovirus, some of Anne's normal DNA might be altered. This treat-

ment is also more expensive than Anne's current treatment. Anne requests the experimental treatment from her physician.

Implementing the decision-making model

First of all, does an ethical dilemma exist? Anne has requested an experimental treatment that may harm her or may help her and improve her quality of life. Should her physician help her get the treatment she wants? Is it medically indicated? Is her disease as serious as French Anderson says it should be for her to qualify for gene therapy? There are certainly enough questions to indicate that this case does present bioethical issues.

I. Identify the Question
Several possible questions are listed above. List as many others as you can think of, and then choose one for consideration.

II. Identify the Issue
What is the issue you chose? Experimental treatment? Medical necessity? Something else?

III. State Facts of Case
There is a lot of background material in this case. For the box summary, try to select only the information that is necessary for presenting the dilemma. You could reasonably omit background information about Gaucher's disease from the box, for example.

IV. List Possible Decisions
List as many alternatives as you can without trying to decide at this time which is best.

V. Gather Additional Facts
Do you need additional information to make a good decision? For example, if you are trying to decide whether the doctor should give Anne the treatment, do you need to know more about retroviruses, how they work, and how they might alter Anne's DNA? Do you need more information about Gaucher's disease to decide whether it is serious enough to warrant gene therapy?

VI. Pick Decision To Support
Choose what you believe to be the best solution to your question.

VII. Identify Guiding Principle
Which of the major ethical principles justifies your decision?

Societal Issues

VIII. Identify Supporting Authority

What experts would back up your option? Find relevant passages in documents such as the Hippocratic Oath, the Nuremberg statements, and the Patient Bill of Rights.

IX. Formulate Rebuttal

Under what circumstances would you abandon your choice?

X. State Level of Confidence

How strongly do you believe in your position?

XI. Box Up the Case

Use Worksheet 40.1.

XII. Prepare Argument

Explain on paper why you chose one option rather than the other and why the principle supports the option. Your written argument should reflect the thinking necessary to construct the box and be complete enough that someone who has never read the case study can understand the situation and your analysis.

Case Study 2: Bobby K.

Bobby K. is a healthy 10-year-old boy. He is very agile and quick and loves sports. The coach of his city league basketball team has told Bobby's parents that their son's skills on the court are astounding for a child of his age.

Bobby's father, Mr. K., is a healthy man who is 5 feet 3 inches tall. Bobby's mother is also short, only 5 feet. Bobby's pediatrician has predicted that Bobby will attain an adult height of about 5 feet 3 inches but has emphasized to his parents that he is a normal, healthy boy. Mr. K. remembers being teased constantly about his size and recalls that his lack of height kept him off all the varsity sports teams at his high school. He has often wondered if his shortness is a disadvantage to him in business dealings, too. Mr. and Mrs. K. both anticipate that their son will be the recipient of more and more pointed teasing as he reaches his teenage years. They also fear that he will not be selected for the basketball team when he reaches junior high or high school.

Bobby's coach has heard of a program at the local university in which gene therapy is being conducted on children who have a disease that results in the inadequate production of growth hormone. The children are being given working copies of the gene for human growth hormone, and the levels of growth hormone in their bodies have increased. These children's growth rates have also increased.

Bobby's coach tells the K. family about what he has heard. Bobby is thrilled, because he might be able to keep playing basketball, maybe even on a professional level. Bobby's parents are more cautious but would like their son to be spared the pain of being much smaller than his classmates.

The K.'s go to Bobby's pediatrician and request that Bobby be given the growth hormone gene therapy.

Using the decision-making model

This case obviously presents many possible questions for consideration. For example:

- Should Bobby's doctor allow him to have the gene therapy?
- Should Bobby's parents let him have the therapy?
- Should Bobby's coach have told the K.'s about the program?

There are probably many background questions you will need to investigate. For example:

- Have there been any studies to determine whether it is safe to give additional growth hormone to healthy children?
- What height range is considered normal? How do physicians determine whether a short person is a normal height for someone that age?

Choose a dilemma question for analysis, and use the model to generate a solution.

Case Study from Real Life

This case study also revolves around human growth hormone but concerns administration of human growth hormone as a drug, not gene therapy.

Doctors normally prescribe growth hormone to children who fail to produce enough of it naturally and who are expected to reach an adult height of 4 feet or less. These children are referred to as growth hormone deficient. The hormone has been on the market for over 15 years, and to date, no safety problems associated with its use in growth hormone-deficient children have been observed.

In 1988, researchers at the NIH initiated a study of children, none of whom suffered from growth hor-

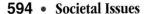

mone deficiency but who were expected to be in the bottom third percentile when they reached adult height. Without treatment, the boys in this study were expected to reach 5 feet 6 inches or less, and the girls were expected to reach 5 feet or less as adults. Half of the children were given human growth hormone (the protein, not the gene), and the other half received a placebo. Neither the parents nor the children knew whether they were in the experimental (human growth hormone) or control (placebo) group. The physician who conducted the study wanted to learn how much additional growth the hormone injections would produce in healthy short children.

The study received a great deal of attention in the media, because there were questions about the appropriateness of conducting studies on healthy children who happened to be short.

Scientists defended the study on several points. One was that many parents would obtain human growth hormone for their children illegally. Physicians needed to know whether giving healthy children injections of human growth hormone would have harmful effects and, if so, what the effects would be. Another was that there needed to be studies to determine the safety and efficacy of this treatment for short stature that was not caused by growth hormone deficiency. Another was that our society is preoccupied with height, and short stature can cause anxiety to the point that mental health is compromised. One of the scientists who defended the study was Arthur Levine, scientific director of the National Institute of Child Health and Human Development. At 5 feet 4 inches, Levine said he probably would have enrolled himself in the study if it had been available to him as a child.

Critics of the study argued that shortness is not a disease and should not be defined as requiring treatment. They questioned the use of a relatively experimental treatment in healthy children for apparently cosmetic reasons. They were concerned that virtually any cosmetic defect could eventually be redefined as illness.

Using the model

Apply the decision-making model as outlined previously. Questions that could be asked include the following.

- Should this study be conducted?
- Should a parent enroll a short child in this study?
- Should doctors prescribe growth hormone to short but otherwise healthy children?
- Should people be able to make their own decisions about taking growth hormone?
- Should any treatment be available for shortness?
- Should the NIH prevent children who want to enroll from participating?

You may be able to think of more questions. Choose one for analysis, using the decision-making model.

Societal Issues

PART IV
Appendixes

The appendixes contain information to assist you in incorporating biotechnology into your curriculum. Appendix A contains worksheets associated with the student activities found in Parts II and III. The other appendixes provide information on laboratory biosafety and basic techniques used in microbiology laboratories.

Appendixes

Appendix A: Worksheets

This appendix contains worksheets that can be cut out from this section or photocopied as needed for use with the *Student Activity* pages. Electronic versions of these worksheets are also on the CD provided with this book.

Worksheet 11.1: DNA sequence strips for *DNA Scissors*

```
┌─────────────────────────────────────────────────┐
│ 1                                             1   │
│                                                   │
│   5'-TAGACTGAATTCAAGTCA-3'                        │
│      | | | | | | | | | | | | | | | | | |         │
│   3'-ATCTGACTTAAGTTCAGT-5'                        │
│                                                   │
└─────────────────────────────────────────────────┘


┌─────────────────────────────────────────────────┐
│ 2                                             2   │
│                                                   │
│   5'-ATACGCCCGGGTTCTAAA-3'                        │
│      | | | | | | | | | | | | | | | | | |         │
│   3'-TATGCGGGCCCAAGATTT-5'                        │
│                                                   │
└─────────────────────────────────────────────────┘


┌─────────────────────────────────────────────────┐
│ 3                                             3   │
│                                                   │
│   5'-CAGGATCGAAGCTTATGC-3'                        │
│      | | | | | | | | | | | | | | | | | |         │
│   3'-GTCCTAGCTTCGAATACG-5'                        │
│                                                   │
└─────────────────────────────────────────────────┘


┌─────────────────────────────────────────────────┐
│ 4                                             4   │
│                                                   │
│   5'-AATAGAATTCCGATCCGA-3'                        │
│      | | | | | | | | | | | | | | | | | |         │
│   3'-TTATCTTAAGGCTAGGCT-5'                        │
│                                                   │
└─────────────────────────────────────────────────┘
```

Worksheet 12.1: restriction maps for *DNA Goes to the Races*

Below are three representations of a 15,000-base-pair DNA molecule. Each representation shows the locations of different types of restriction site, with vertical lines representing the cut site. The numbers between the cut sites show the sizes (in base pairs) of the fragments that would be generated by digesting the DNA with that enzyme.

EcoRI sites

4,000	3,500	2,500	5,000

BamHI sites

6,000	4,000	3,000	2,000

HindIII sites

8,000	4,500	2,500

Worksheet 12.2: gel outline for *DNA Goes to the Races*

EcoRI   HindIII   BamHI

Sample — [ ]   [ ]   [ ]
wells

Size scale
in base pairs

8,000 ——

6,000 ——

4,000 ——

3,000 ——

2,000 ——

Worksheet 14.1: paper pAMP plasmid model for *Recombinant Paper Plasmids*

1

Ampicillin resistance gene

5' AATTCGATGAATTCXXXXXXXXXXXXXXXXXXXXXXXXXXGAATTCTGAAGGTTCGAAGCGCTAT

3' TTAAGCTACTTAAGXXXXXXXXXXXXXXXXXXXXXXXXXXXCTTAAGACTTCCAAGCTTCGCGATA

paste 2

2

5' GTCGGATCCAGATCCGAAGTCTCTCTAGGACCTTGCGAAGCCACGTAGTTCAGATTAATGCCTGAT

3' CAGCCTAGGTCTAGGCTTCAGAGAGATCCTGGAACGCTTCGGTGCATCAAGTCTAATTACGGACTA

paste 3

3

Origin of replication

5' CGCTACAAGCTTATAGCGGCCXXXXXXXXXXXXXXXXXXXXXXXXXAATATTGCGCAGTCTTAGCACTCC

3' GCGATGTTCGAATATCGCCGGXXXXXXXXXXXXXXXXXXXXXXXXXTTATAACGCGTCAGAATCGTGAGG

paste 1

Worksheet 14.2: paper pKAN plasmid model for *Recombinant Paper Plasmids*

1

Origin of replication

5' TACTCGATGAAATCXXXXXXXXXXXXXXXXXXXXXXXXXXXXXAGCTATGTTCTGAAGGATCCATATAGCGC

3' ATGAGCTACTTTAGXXXXXXXXXXXXXXXXXXXXXXXXXXXTCGATACAAGACTTCCTAGGTATATCGCG

paste 2

2

Kanamycin resistance gene

5' ATGACCGTCAGATCCGATGCTTCXXXXXXXXXXXXXXXXXXXXXXXXXXXXXXXXXTCGAACGTACGGGTCCGA

3' TACTGGCAGTCTAGGCTACGAAGXXXXXXXXXXXXXXXXXXXXXXXXXXXXAGCTTGCATGCCAGGCT

paste 3

3

5' GATCACATGCTTATAAATATTGCGAAGCTTCAGTCAGCGGTAGCACTCCTTAACGGCGATGCATTAA

3' CTAGTGTACGAATATTTATAACGCTTCGAAGTCAGTCGCCATCGTGAGGAATTGCGCTACGTAATT

paste 1

Worksheet 16.I: single-stranded DNA sample sequence and probe for *Detection of Specific DNA Sequences: Part I. Fishing for DNA*

Probe:

3´ GGATGCTACCATAGC 5´

3´ GGATGCTACCATAGC 5´

Sample DNA sequence, written 5´ to 3´

```
1    GATCAGACTTCTAGCAGGCTCTTGACCAATGATCACAGCTTCCGATCTCTAGAGCTCGATCTCTTGATCTCGTGTGCGGAATCTAG

91   CCGGGGGTGAATTCTAGCCCGGGTCAGCTATGCTAAGATAGACCGGAATCGAGAATTCCGGATATCGATTGTGCGACCGCATTAT

181  CGATCGTTTGCCCGGGATCCTAGCTTTCCGATCTAGCTGTGTGGGCGATCTGGGATCGATTCCCGGGATCTAGGCCTACGATGGTATCGTTAG

271  TAGCTCTCTAGCTTAGCTCTTCAAGTGATCTCACCCGGGTAGATCTAGTATATTGTATCGATATTTGGGCCCCCTAGCTCGAGCTAGCT

361  TCTCTAGCTAATAGATAG
```

Worksheet 16.II: outlines for gel and results of hybridization analysis for *Detection of Specific DNA Sequences: Part II. Combining Restriction and Hybridization Analysis*

Stained electophoresis gel

Results of hybridization analysis

Fragment size, bp

Sample well

100 —
90 —
80 —
70 —
60 —
50 —
40 —
30 —
20 —

100 —
90 —
80 —
70 —
60 —
50 —
40 —
30 —
20 —

Worksheet 16.IIIA: restriction maps and hybridization analysis of virus X for *Detection of Specific DNA Sequences: Part III. Southern Hybridization*

The map at the top shows the EcoRI restriction sites. The bottom line shows the BamHI restriction sites. Fragment sizes are in base pairs. The stained electrophoresis gel shows EcoRI and BamHI fragments of the virus. The hybridization analysis shows which of the bands in the stained gel hybridized to the probe.

EcoRI map

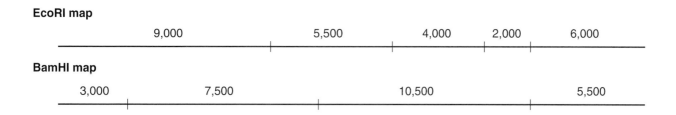

BamHI map

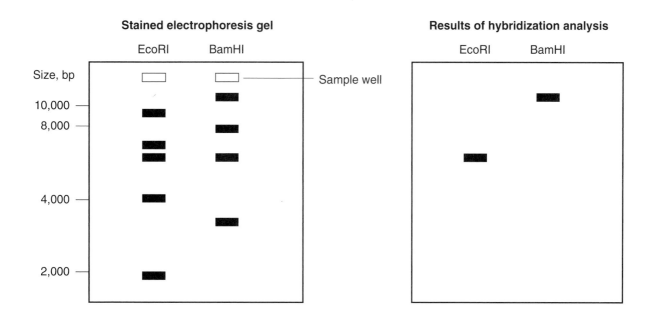

Worksheet 16.IIIB for *Detection of Specific DNA Sequences: Part III. Southern Hybridization*

Shown below are restriction maps for bacterio-phage lambda, showing the BamHI, HindIII, and EcoRI sites. Fragment sizes are in base pairs. The gel and hybridization analysis outlines are provided for your use.

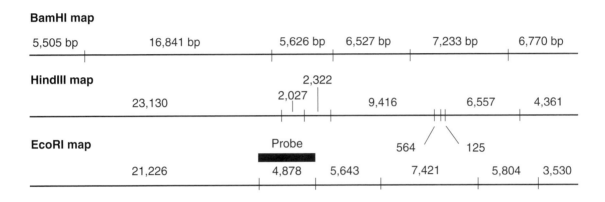

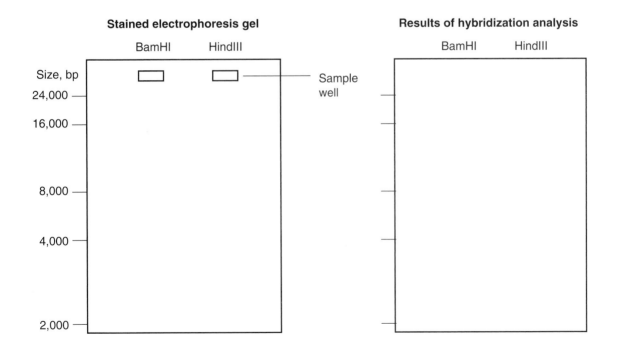

Worksheet 17.1: parental DNA molecule and primers for *Paper PCR*

5' **TACGACCCGGTGTCAAAGTTAGCTTAGTCA** 3'

5' **TACGACCCGGTGTCAAAGTTAGCTTAGTCA** 3'

5' **TACGACCCGGTGTCAAAGTTAGCTTAGTCA** 3'

5' **TACGACCCGGTGTCAAAGTTAGCTTAGTCA** 3'

5' **CCCGG** 3'   5' **CCCGG** 3'

5' **CCCGG** 3'   5' **CCCGG** 3'

Worksheet 17.2: sample DNAs and primers for PCR-based diagnosis for *Paper PCR*

3' ATGCTGGGCCACAGTTTCAATCGAATCAGT 5'

3' ATGCTGGGCCACAGTTTCAATCGAATCAGT 5'

3' ATGCTGGGCCACAGTTTCAATCGAATCAGT 5'

3' ATGCTGGGCCACAGTTTCAATCGAATCAGT 5'

3' TCGAA 5'   3' TCGAA 5'

3' TCGAA 5'   3' TCGAA 5'

Worksheet 17.3: simulation of PCR diagnostic test for *Paper PCR*

Primers: 5´TTCCAGCC 3´   3´CGGAATAC 5´

Sample 1

5´GCGTAATCGGATGCCGTAATAGGTATGCGCGAATTTGTG
3´CGCATTAGCCTACGGCATTATCCATACGCGCTTAAACAC

5´ATGCATCCGATAGCGCGGGCCTATATTGTAACTGGCATC
3´TACGTAGGCTATCGCGCCCGGATATAACATTGACCGTAG

5´ATGCGTTAACGCGTAATGGCCTAATCTATGTAGCGCGAA
3´TACGCAATTGCGCATTACCGGATTAGATACATCGCGCTT

5´AATGCGCTTCCAGCCAGAGTCTCGGAACTAGCCTTATG
3´TTACGCGAAGGTCGGTCTCAGAGCCTTGATCGGAATAC

Sample 2

5´GAATTCCTCATGATCCAGGTCACTAATGCACGGTTACAC
3´CTTAAGGAGTACTAGGTCCAGTGATTACGTGCCAATGTG

5´GGGCCCTATAGCTACTCTAGAATCTAGCGAATATTGCGC
3´CCCGGGATATCGATGAGATCTTAGATCGCTTATAACGCG

5´GAATTACGCTAGCGATCGGCTATTCGAATTCGGTTATCG
3´CTTAATGCGATCGCTAGCCGATAAGCTTAAGCCAATAGC

5´AGGCCTCGCATGAATCTCGATTTAAATGCGCATCGATAT
3´TCCGGAGCGTACTTAGAGCTAAATTTACGCGTAGCTATA

Worksheet 18.1: template and primers for cycle sequencing for *DNA Sequencing: The Terminators*

Template

5' CGAACATCTTCGACGTAACTGCGA 3'

3' GCTTGTAGAAGCTGCATTGACGCT 5'

Primers

5' CGAACATC 3'

5' CGAACATC 3'

5' CGAACATC 3'

5' CGAACATC 3'

Worksheet 18.2: template-primers for original approach to DNA sequencing for *DNA Sequencing: The Terminators*

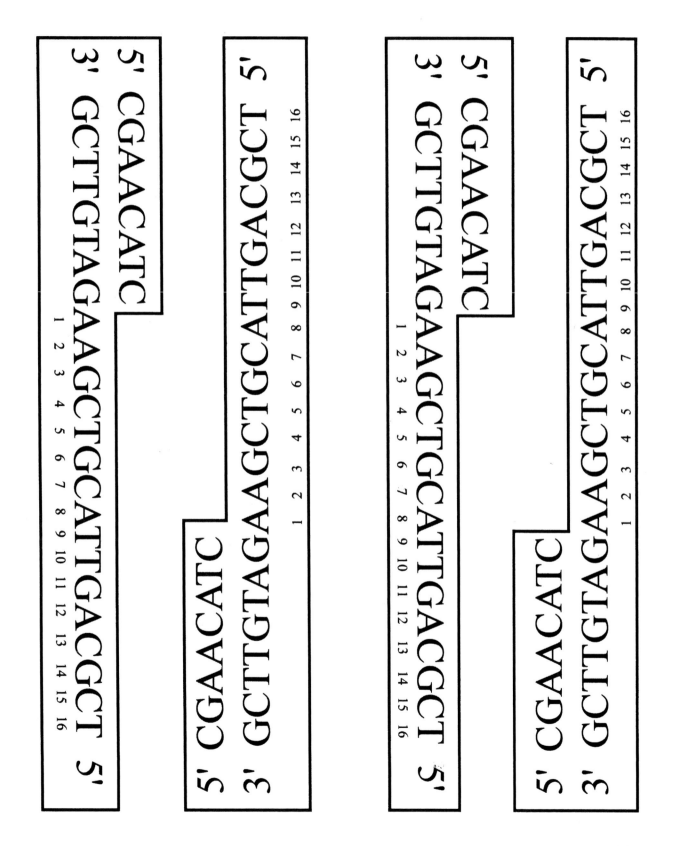

Worksheet 23.1 for *An Adventure in Dog Hair, Part I*

Label the coat color alleles that Cocoa and Midnight could contribute to their gametes on the diagram. Draw the puppies' chromosomes with coat color genes, and label the genes. Color Cocoa, Midnight, and the puppies.

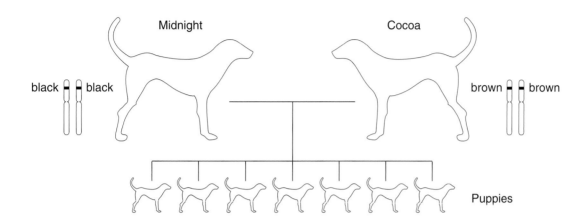

Table 1

Individual	Coat color genes	Coat color
Midnight		
Cocoa		
Puppies		

Table 2. Coat color and TYRP1

Individual	Coat color genes	Do melanocytes make TYRP1?	Coat color
Midnight			
Cocoa			
Puppies			

Worksheet 23.2 for *An Adventure in Dog Hair, Part I*

Genetics of roundbuds

Variety	Genotype	Enzyme Q produced?	Color
Red			
White			

Why might the offspring of a red × white cross be pink instead of red?

Worksheet 24.1 for *An Adventure in Dog Hair, Part II: Yellow Labs*

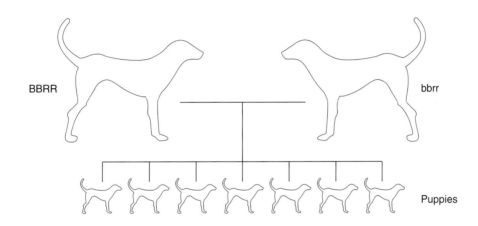

BBRR

bbrr

Puppies

Fill in the blanks, and then color the parents and puppies the correct colors for their genotypes.

Male gametes' genotype(s) _____

Female gametes' genotype(s) _____

Puppies' genotype(s) _____

Puppies' phenotype(s) _____

Worksheet 24.2: Punnett square for *An Adventure in Dog Hair, Part II: Yellow Labs*

Parents: BbRr × BbRr

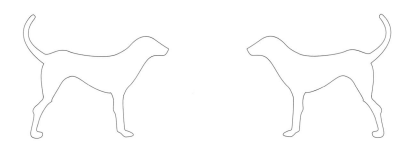

Genotype	No. of occurrences in Punnett square	Does this genotype synthesize:		Phenotype	
		MC1R?	TYRP1?	Color of hair	Color of lips

Summary

Phenotype	No. of occurrences in Punnett square
Black Lab	_____
Chocolate Lab	_____
Yellow Lab, black lips	_____
Yellow Lab, brown lips	_____

Worksheet 25.1 for *Human Molecular Genetics*

Fill in the individuals' genotypes in the blanks. If the
individual could have more than one genotype, write
each genotype that applies.

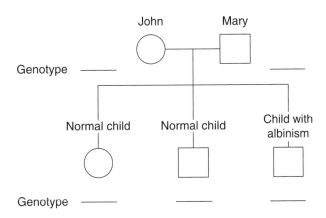

1. What protein does the "albinism gene" encode?

2. Explain how the "albinism gene" causes albinism.

3. Is the "albinism gene" dominant or recessive? What is the molecular explanation of your answer?

Worksheet 25.2 for *Human Molecular Genetics*

Write the genotype of the individuals in the blanks
provided. If an individual could have more than one
genotype, write all that apply.

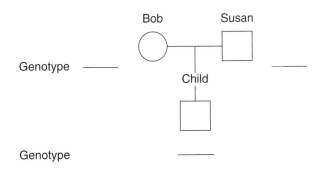

1. What protein does the "PKU gene" encode?

2. How can two normal parents have a child affected with PKU?

3. If Bob and Susan have another child, will it also have PKU?

4. Explain how the "PKU gene" causes PKU and why Bob and Susan don't have the disease but their child
   does. Use your knowledge of cellular biochemistry.

5. Is PKU a dominant or recessive genetic disease? What is the molecular explanation of your answer?

Worksheet 25.3: genotypes in sickle-cell anemia for *Human Molecular Genetics*

1. What is the relationship between sickle-cell anemia and the sickle-cell trait?

2. Why do the parents of children with sickle-cell disease have the sickle-cell trait? Why do many of the siblings of these children have the trait?

3. Why did individuals with sickle-cell disease have one form of hemoglobin in the electrophoresis test, normal individuals have a second form, and people with the sickle-cell trait have both?

4. Based on the story in the text, what do you think Dr. Neel's conclusion about sickle-cell disease was, and why did the protein electrophoresis results confirm it?

Worksheet 27.1

Use with Exercise 1: STRs can cause an RFLP

Shown below are the probe sequence and the DNA sequence from the highly variable region of Bob's and Mary's chromosome 8 homologs. Use the sequence information to answer the questions in Exercise 1.

The restriction enzyme HaeIII recognizes the sequence 5′ GGCC 3′ and cuts between the G and C.

Probe sequence: 3′ GGAGATCCTGTACGATTT 5′.

Highly variable region of Bob's chromosome 8s

Maternal chromosome:

5′ AGGCCTCTAGGACATGCTAAAGCTAGCTAGCTAGCTAGCTAAGGCCTAGGTGCGAT 3′

3′ TCCGGAGATCCTGTACGATTTCGATCGATCGATCGATCGATTCCGGATCCACGCTA 5′

Paternal chromosome:

5′ AGGCCTCTAGGACATGCTAAAGCTAGCTAGCTAGCTAGCTAGCTAGCTAAGGCCTAGGTGCGAT 3′

3′ TCCGGAGATCCTGTACGATTTCGATCGATCGATCGATCGATCGATCGATTCCGGATCCACGCTA 5′

Highly variable region of Mary's chromosome 8's

Maternal chromosome:

5′ AGGCCTCTAGGACATGCTAAAGCTAGCTAGCTAGCTAGCTAGCTAGCTAGCTAGCTAAGGCCTAGGTGCGAT 3′

3′ TCCGGAGATCCTGTACGATTTCGATCGATCGATCGATCGATCGATCGATCGATCGATTCCGGATCCACGCTA 5′

Paternal chromosome:

5′ AGGCCTCTAGGACATGCTAAAGCTAGCTAGCTAGCTAAGGCCTAGGTGCGAT 3′

3′ TCCGGAGATCCTGTACGATTTCGATCGATCGATCGATTCCGGATCCACGCTA 5′

Worksheet 27.2

Use with Exercise 2: PCR can reveal differences at microsatellite loci

Shown below are the sequences of the PCR primers and the highly variable regions from Bob's and Mary's chromosome 8 homologs. Use the information to answer the questions in Exercise 2.

PCR primers:

5′ GGCCTCTAGGACATGCTAAAGC 3′ and 3′ TCGATTCCGGATCCACGC 5′

Remember to hybridize the 5′ to 3′ primer to the 3′ to 5′ DNA strand, and vice versa.

Highly variable region of Bob's chromosome 8s

Maternal chromosome:

5′ AGGCCTCTAGGACATGCTAAAGCTAGCTAGCTAGCTAGCTAAGGCCTAGGTGCGAT 3′

3′ TCCGGAGATCCTCTACGATTTCGATCGATCGATCGATCGATTCCGGATCCACGCTA 5′

Paternal chromosome:

5′ AGGCCTCTAGGACATGCTAAAGCTAGCTAGCTAGCTAGCTAGCTAGCTAGCTAAGGCCTAGGTGCGAT 3′

3′ TCCGGAGATCCTGTACGATTTCGATCGATCGATCGATCGATCGATCGATCGATTCCGGATCCACGCTA 5′

Highly variable region of Mary's chromosome 8s

Maternal chromosome:

5′ AGGCCTCTAGGACATGCTAAAGCTAGCTAGCTAGCTAGCTAGCTAGCTAGCTAGCTAGCTAAGGCCTAGGTGCGAT 3′

3′ TCCGGAGATCCTGTACGATTTCGATCGATCGATCGATCGATCGATCGATCGATCGATCGATTCCGGATCCACGCTA 5′

Paternal chromosome:

5′ AGGCCTCTAGGACATGCTAAAGCTAGCTAGCTAGCTAGCTAAGGCCTAGGTGCGAT 3′

3′ TCCGGAGATCCTGTACGATTTCGATCGATCGATCGATCGATTCCGGATCCACGCTA 5′

Worksheet 29.A

Use with Pedigree A and Gel Diagram A

1. Fill in the genotypes of the individuals in Pedigree A based on pedigree analysis and analysis of the molecular data in Gel Diagram A.

Individual	Genotype									
	I1	I2	II1	II2	II3	II4	II5	II6	II7	II8
BFNC										
STR										

2. What are the father's (individual I1) STR alleles? —————

3. What are the four combinations of BFNC and STR alleles that the father could pass to an offspring? These are the possible paternal haplotypes: —————, —————, —————, —————

4. Fill in the paternal haplotype inherited by each member of generation II.

II1	II2	II3	II4	II5	II6	II7	II8

5. Fill in the four paternal haplotypes (from question 3) and record the frequency at which each was seen in the members of generation II.

Haplotype	Frequency in Gen II

6. If the STR and the BFNC loci are linked, what would be the two nonrecombinant haplotypes? ——————— and ———————. The recombinant haplotypes? ———————

7. What is the frequency of these hypothetical nonrecombinant haplotypes in generation II? ———————

8. What is the frequency of the recombinant haplotypes? ———————

Worksheet 29.B

Use with Pedigree B and Gel Diagram B

1. Fill in the genotypes of the individuals in Pedigree B based on pedigree analysis and analysis of the molecular data in Gel Diagram B.

Individual	Genotype								
	II1	S1	III1	III2	III3	III4	III5	III6	III7
BFNC									
STR									

2. If the BFNC locus and the STR locus are linked, what STR allele is linked to the disease allele of BFNC in individual II1 (see results on Worksheet 29.A)? _____. In individual II8? _____

3. What are the nonrecombinant haplotypes in individual II1? _____ and _____. In II8? _____ and _____.

4. Fill in the maternal haplotypes of the offspring of II1 and the paternal haplotypes of the offspring of II8.

Offspring of II1					Offspring of II8	
Maternal haplotype					Paternal haplotype	
III1	III2	III3	III4	III5	III6	III7

5. How many haplotypes in generation III are nonrecombinant? _____

6. How many haplotypes in generation III are recombinant? _____

7. Adding results from generations II and III together, how many nonrecombinant haplotypes were observed? _____. What frequency is this? _____

    How many recombinant haplotypes were observed? _____. What frequency is this? _____

8. Do the data from Worksheets 29.A and 29.B support the hypothesis that the STR locus and the BFNC locus are linked? Explain your answer.

Worksheet 29.C

Use with Pedigree C and Gel Diagram C

1. Fill in the genotypes of the individuals in Pedigree C based on pedigree analysis and analysis of the molecular data in Gel Diagram C.

Individual	Genotype					
	III6	S3	IV1	IV2	IV3	IV4
BFNC						
STR						

None of the individuals in Pedigree C has the same STR alleles that were present in individual I1. Do the data from Pedigree C and Gel Diagram C support the hypothesis that the BFNC and STR loci are linked? Explain your answer.

Worksheet 31.1 for *Testing for Amylase Activity*

Use additional sheets if needed.

Material	**Prediction:** Will there be amylase activity and why do you think so?	**Result:** Was there activity?	Possible explanations for results different from expectation
Germinating bean			
Leaf extract			
Root extract			
Dog saliva			

Worksheet 33.1: amylase sequences for *Constructing an Amylase Evolutionary Tree*

A NNNGV–IKEVTINPDTTCGND ... GRGNRGFIVFNNDDWSFSLTLQTGLPAGTYCDVISGDKINGNCTGI

B NNNGV–IKEVTINADTTCGND ... GTGNRGFIVFNNDDWQLSSTLQTGLPGGTYCDVISGDKVGNSCTGI

C NSDGS–TKSVTINADTTCGND ... GRGDRGFIVFNNDDWYMNVDLQTGLPAGTYCDVISGQKEGSACTGK

D HDGSFNIISPSFNADGSCGNG ... CRGNKGFLAINNDGWDLKETLQTCLPAGTYCDVISGSKNGGSCTGK

E TTDGHNIASPIFNSDNSCSGG ... SRGSRGFVAFNNDNYDLNSSLQTGLPAGTYCDVISGSKSGSSCTGK

F NNNGK–TKEVSINPDSTCGND ... GRGNKGLIVFNNDDWALSETLQTGLPAGTYCDVISGDKVDGNCTGI

G TTDGQNIASPVFNSDSSCSGG ... SRGSRGFVAFNNDNYDLNSSLQTGLPAGTYCDVISGSKSGSSCTGK

653

Worksheet 38.1: "box-up" form for *Making Decisions When There Is No Right Answer*

Issue:

Question:

Facts

Authority

Qualifier

Principle

Decision

Rebuttal

Worksheet 39.1: "box-up" form for *Analyzing Ethical Dilemmas in Genetic Testing*

Issue: *Question:*

Facts

Authority

Qualifier

Principle

Decision

Rebuttal

Worksheet 40.1: "box-up" form for *Analyzing Ethical Dilemmas in Gene Therapy*

Issue: *Question:*

Facts

Authority   *Qualifier*

Principle

                        Decision

Rebuttal

Appendix B: Laboratory Biosafety

Handling Microorganisms in the Laboratory

Escherichia coli is a normal inhabitant of the digestive tract. Of the many strains of *E. coli,* some inhabit the human gut and others reside in animals. A few strains of *E. coli* cause significant disease in humans and have received attention from the media in recent years. Please refer to our lengthy discussion of *E. coli* and pathogenicity in *Gene Transfer,* Escherichia coli, *and Disease,* if you or others have any concerns about the activities involving *E. coli.*

The laboratory strains of *E. coli* discussed in this book, MM294, cI, cII, CR63, and B$_E$, have been used in the laboratory for years and do not normally cause disease. MM294 is reported to be ineffective at colonizing the human digestive tract and so is especially harmless in that respect. However, all of these strains could cause infection if introduced into an open wound or into the eye. It is therefore very important to use aseptic technique (see Appendix C) when handling the organisms. Students should never eat, drink, smoke, or apply cosmetics in the laboratory. Many instructors require that students wear protective goggles while working in the laboratory.

When transferring cultures of *E. coli,* keep pipette tips away from the face to avoid inhaling any aerosol that might be created. If anyone contaminates his or her hands with a culture, wash them immediately. Avoid contaminating any cuts with bacterial culture, and keep all bacteria away from the eyes. If someone believes he or she may have contaminated an area of broken skin, wash that area immediately. If someone gets bacteria in his or her eyes, use the eyewash fountain to rinse the eyes, and call a physician's office for further advice.

Agrobacterium tumefaciens does not cause disease in humans. However, the same precautions should be taken with it as with *E. coli.*

Bacteriophage T4 is harmless to humans. It cannot infect human cells. It should be handled in the same manner as *E. coli,* mostly to avoid inadvertent contamination of laboratory *E. coli* cultures.

Disinfect all plates and cultures before disposing of them.

Disinfecting

Keep disinfectant solutions available in the laboratory in squeeze bottles. These solutions can be 2% Lysol, 70% ethanol, rubbing alcohol, or other special disinfectants.

Before carrying out any experiments involving bacteria or phage, wipe down the laboratory bench with disinfectant solution. At the end of the laboratory period, wipe down the bench top with disinfectant again. Clean up any spills involving organisms immediately, and disinfect the area thoroughly.

After the laboratory period, disinfect all materials that have come in contact with bacteria or phage (micropipette tips, pipettes, agar plates, culture tubes, flasks, etc.) either with pressurized steam or by soaking them in concentrated disinfectant. To use steam, place all biological waste in an autoclave bag, and sterilize it in an autoclave or pressure cooker. To use disinfectant, soak all contaminated materials for at least 15 min in either 10% Lysol or 15 to 20% chlorine bleach. Drain the liquid. Place items to be disposed of in a plastic bag, and put the bag in the trash. Thoroughly rinse any containers to be reused. Clean the containers as usual, and then sterilize them as needed.

It is not necessary to disinfect materials that have come in contact only with DNA and restriction enzymes (for example, from the gel electrophoresis laboratories).

Regulations for Recombinant DNA Work

The National Institutes of Health (NIH) oversee and regulate all research involving transfer of DNA between species. Rules for conducting recombinant DNA research are published as the *NIH Guidelines for Research Involving Recombinant DNA Molecules.* Certain kinds of recombinant-DNA research are designated as exempt from these guidelines. Provision III-D-3 states that "the following molecules are exempt from these guidelines . . . those that consist entirely of DNA from a prokaryotic host, including its indigenous plasmids or viruses when propagated only in the host (or a closely related strain of the same species) or when transferred to another host by well-established physiological means." *Under this guideline, all the experiments and DNA molecules used in this book are exempt and may be conducted in a high school setting.*

If students wish to pursue further research involving recombinant DNA, instructors should make sure that the work involves only exempt molecules and procedures or should make arrangements for the students to work in an NIH-approved laboratory.

Appendix C: Aseptic Technique

General Information

In experiments with microorganisms, it is essential to avoid contamination of the experiment by other microbes. Contamination could cause the procedure to fail. More important, contaminating organisms cannot be guaranteed to be harmless. Avoiding contamination is most important when cultures are being inoculated. If a contaminating microbe should find its way into the growth medium at the beginning of the growth period, the contaminant could grow along with the intended experimental organism.

To prevent microbiological contamination, a system of laboratory practice called "sterile technique" or "aseptic technique" is used. Proper aseptic technique in the laboratory minimizes the risk of contamination.

There are a few principles to remember when learning aseptic technique.

- An object or solution is sterile only if it contains no living thing.
- In general, objects and solutions are sterile only if they have been treated (autoclaved, irradiated, etc.) to kill contaminating microorganisms.
- Any sterile surface or object that comes in contact with a nonsterile thing is no longer considered sterile.
- Air is not sterile (unless it was sterilized inside a closed container).

Acting on these principles, scientists first sterilize all containers and solutions to be used when culturing microorganisms. Thereafter, material is transferred between sterile containers with sterile tools (sterile pipettes, micropipettes with sterile tips, sterile inoculating loops, etc.) in such a way as to minimize exposure to outside air and avoid contact with nonsterile surfaces and items.

The keys to good aseptic technique are as follows.

- Keep lids off sterile containers for the shortest time possible.
- Pass the mouths of open containers through a flame. Flaming warms the air at the opening, creating positive pressure and preventing contaminants from falling into the tube. Even plastic items can be flamed briefly.
- Hold open containers at an angle whenever possible to prevent contaminants from falling in.

Specific Techniques

Use of glass or plastic pipettes

Glass pipettes are put into containers or wrapped and then autoclaved. Plastic pipettes are purchased presterilized in individual wrappers. To use a pipette, remove it from its wrapper or container by the end opposite the tip. Do not touch the lower two-thirds of the pipette. Do not allow the pipette to touch any laboratory surface. Draw the lower length of the pipette through a Bunsen burner flame. Insert only the untouched lower portion of the pipette into a sterile container.

Using test tubes or culture tubes

Sterilize test tubes with lids or caps on. When you open a sterile tube, touch only the outside of the cap, and do not set the cap on any laboratory surface. Instead, hold the cap with one or two fingers while you complete the operation, and then replace it on the tube. This technique usually requires some practice, especially if you are simultaneously opening tubes and operating a sterile pipette. If you are working with a laboratory partner, one person can operate the test tube, and the other can operate the pipette.

After you remove the cap from the test tube, pass the mouth of the tube through a flame. If possible, hold the open tube at an angle. Put only sterile objects into the tube. Complete the operation as quickly as you reasonably can, and then flame the mouth of the tube again. Replace the lid.

Inoculating loops and needles

Inoculating loops and needles are the primary tools for transferring microbial cultures. Loops and needles are sterilized by flaming them. Put the business end of the tool directly into a Bunsen or alcohol burner flame, and hold it there until the end glows bright red. Withdraw the tool from the flame. The tool is now hot and sterile. Count to 5 or 10 to let it cool, and then transfer the organisms.

If you are moving organisms from an agar plate, touch an isolated colony with the transfer loop. Be sure your inoculating loop is cool before you do this. Replace the plate lid. Open and flame the culture tube, and inoculate the medium in it by stirring the end of the transfer tool in the medium. If you are removing cells from a liquid culture, insert the loop into the culture. The loop may hiss. If so, wait until the hissing stops, move the loop a little in the cul-

ture, and then withdraw it. Even if you cannot see any liquid in the loop, there will be enough cells there to inoculate a plate or a new liquid culture.

Transferring large volumes

If you do not have to be careful about the volume you transfer, a pure culture or sterile solution can be transferred to a sterile container or new sterile medium by pouring. Remove the cap or lid from the solution to be transferred. Thoroughly flame the mouth of the container, holding it at an angle as you do so. Remove the lid from the target container. Hold the container at an angle and flame it, if possible (you may not be able to if you are holding many items in your hands). Quickly and neatly pour the contents from the first container into the second. Flame the mouth of the second container. Replace the lid.

If you must transfer an exact volume of liquid, use a sterile pipette or a sterile graduated cylinder. When using a sterile graduated cylinder, complete the transfer as quickly as you reasonably can to minimize the time the sterile liquid is exposed to the air.

Appendix D: Basic Microbiological Methods

The Care and Feeding of *Escherichia coli*

Escherichia coli is not a fussy microbe and will grow on a variety of media at a variety of temperatures. Although *E. coli* cultures grow faster if aerated, they will also grow without aeration. All of this means that you have many options for growing the organism and can choose methods that suit your circumstances.

Three medium recipes are given in the *Laboratory Resources* folder on the CD: Luria broth (and agar), nutrient broth (and agar), and tryptic soy broth (and agar). Any of these will do, and they can be freely substituted for each other in the procedures in this book.

Liquid cultures

Use an inoculating needle or loop to touch an isolated colony on an agar plate. Introduce the cells into sterile broth. The cells will multiply fastest at 37°C with constant shaking for aeration. If you have an incubator but no shaker, grow them at 37°C and shake the flask yourself whenever it is convenient. If you cannot shake the flask at all, the cells will still grow. If you do not have an incubator, they will grow at room temperature, but do not expect heavy growth overnight. It is a good idea to test how long it will take *E. coli* to grow under your conditions if you need to be certain of having the cells ready at a particular time.

When microbiologists need to grow large volumes of cells, they usually grow a small overnight culture first and then use that culture to inoculate the larger volume of medium. You do not have to follow this procedure, but if you do, you will be able to tell much sooner that your cells are actually growing.

If you have a magnetic stirrer, you can use it to aerate your cultures. Sterilize your medium in a loosely capped or cotton-plugged flask with a stir bar in it. Introduce the cells, replace the cap or plug, put the flask on the stirrer, and turn the stirrer on. The stirring action will keep the culture aerated.

Dispose of liquid cultures as soon as you reasonably can after using them. Contaminants can grow in cultures that have been opened and used.

Growing E. coli *on agar plates*

The factors that apply for growth in liquid also apply for growth on agar but without the question of stirring. The warmer you keep the plates (up to 37°C), the faster the cells will grow. Once the colonies have grown, keep the plates in the refrigerator. Chilling the plates prevents the growth of bacterial and fungal contaminants. Plates can be stored in the refrigerator for a month or so, and colonies taken from them will grow in fresh media. When you start a series of experiments using *E. coli,* it is a good idea to streak a "master plate" to use as a source of isolated colonies for inoculating liquid cultures. Keep this plate in the cold.

Long-Term Storage of *E. coli*

The best long-term storage method is to add glycerol to 20% (vol/vol) to a fresh liquid culture and keep the culture in a −80°C freezer. Given the frequency with which we find −80°C freezers in schools, it is good that other methods will work.

Frozen cultures

Frozen cultures can be stored in a regular freezer, preferably not a frost-free one. If your freezer is frost free, put the cells in a place away from the heating coils. Even better, place a small Styrofoam box inside the frost-free freezer, and keep the frozen cultures inside the box to protect them from defrosting cycles.

To make a frozen culture, grow up the desired cells in any medium. With a sterile pipette, remove a specific volume of cells. Add sterile glycerol so that it will be 20% of the final volume. For example, to 2 ml of culture, add 0.5 ml of sterile glycerol. Label (all frozen cultures look alike), shake well to mix in the glycerol, and freeze.

Glycerol is very viscous and therefore extremely difficult to deliver from pipettes with accuracy, because so much clings to the pipette walls. It is more accurate to measure glycerol by weight. The density of glycerol is 1.25 g/ml. The method we use is to add 0.625 g (0.5 ml) of glycerol to several autoclavable screw-cap plastic tubes, autoclave them, and then add 2 ml of culture to one tube whenever we need to make a frozen stock. The extra tubes with sterile glycerol can be kept tightly closed at room temperature.

Stab cultures

E. coli cultures can be kept for a year or longer in stab cultures at room temperature. A stab culture is a screw-cap test tube containing agar medium. To make these cultures, mix up one of the solid-medium recipes listed in the *Recipes* file in the *Laboratory Resources* folder on the CD. Heat it to boiling to melt the agar. Let the medium cool a little, and then fill clean glass screw-cap test tubes or vials about one-third full. The size of the test tubes is not important. Autoclave the test tubes, and let the medium harden. Use an inoculating needle to stab some *E. coli* cells into the agar. Incubate the stab tubes with the caps loose until you see growth. Then, screw the caps on tight, label the tubes, and put them in a drawer or other convenient place.

Working cultures

The laboratory strains of *E. coli* recommended in this book are safe for student use with reasonable precautions. Because no one can say the same thing for random contaminants of cultures, it is important for safety, as well as for the success of your experiments, that you do not contaminate your cultures. It is especially important to protect your long-term storage cultures. The best way to do this is to use them as little as possible and as carefully as possible.

As directed above, when you begin a series of *E. coli* experiments, use your long-term storage culture once to streak an agar plate for isolated colonies. Store the plate in the refrigerator after it grows up, and use this plate as a source of cells for 2 to 4 weeks. This approach minimizes risks of contamination to your long-term storage cultures.

Streaking Agar Plates To Obtain Isolated Colonies

Using isolated colonies is another method to protect yourself from contaminated cultures. Microbiological procedures recommend starting with an isolated colony because an isolated colony is considered to be a pure culture. When an isolated single cell on an agar plate multiplies repeatedly, it forms a nice, round, isolated colony. This colony is a pure culture of descendants of that founder cell. If you touch it to inoculate another culture, you can be assured that you are introducing only one type of organism into that culture. On the other hand, if you scrape up cells

Figure D.1 Schematic diagram showing how to streak an agar plate to obtain isolated bacterial colonies (refer to the text).

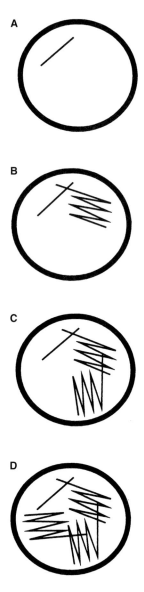

from a region in which cells are growing in large streaks, you are introducing descendants of many cells and may even be introducing contaminants that you could not detect in the thick growth.

Thus, isolated colonies are best for starting cultures. If you do not have any such colonies, you can start your cultures from streaks of growth or directly from long-term storage cultures and hope for the best.

Procedure

1. Flame your inoculating loop. Allow it to cool (you may touch it to the surface of the sterile agar to speed this process), and then touch the source of cells (liquid or solid culture). Do not try to get a large volume of culture into the loop; it will only make the rest of the job harder. The goal is to isolate single *E. coli* cells on the agar plate.
2. On the fresh plate, make a single streak with the loop (Figure D.1A).
3. Reflame the loop, and allow it to cool.
4. With the loop lightly touching the surface of the agar plate, drag the loop once across the streak of cells you just made. This line should be at somewhat more than a 90° angle to the streak. You are dragging some of those cells farther out across the plate.
5. Without lifting the loop from the agar surface, zigzag the loop back and forth in a tight pattern across about a quarter of the plate. The idea is to spread those cells out (Figure D.1B).

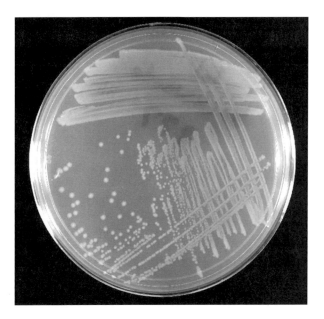

Figure D.2 A streaked agar plate.

6. Reflame the loop.
7. Drag the cooled loop through the area that you last streaked (one drag). Repeat the zigzag pattern in a new area of the plate (Figure D.1C).
8. Reflame the loop.
9. Repeat the "drag and zigzag" (Figure D.1D).
10. Incubate the plates until visible colonies form. Figure D.2 is a photograph of a streaked plate. You can see the zigzag pattern and some isolated colonies.

You may want to practice this technique a few times.

Glossary

Abzyme An antibody designed to catalyze a specific chemical reaction. Also called a catalytic antibody.

Active site The portion of an enzyme that binds to its substrate.

Adenine A nitrogen-containing base found in DNA and RNA.

Adenosine monophosphate (AMP) A nucleotide composed of adenine, ribose, and one phosphate group, formed by removing two phosphate groups from adenosine triphosphate (ATP). In its cyclic form, AMP plays an important role in cellular responses to external chemical cues.

Adenosine triphosphate (ATP) The nucleotide that serves as the energy currency for cellular metabolism.

Adult stem (AS) cell A stem cell is a cell that divides to produce daughter cells that can either differentiate or remain as stem cells, i.e., undifferentiated. In fully differentiated mammals, certain tissues maintain a population of stem cells to replenish the supply of a number of cell types. For example, human bone marrow cells are a type of adult stem cell that can differentiate into red blood cells, white blood cells, or platelets.

Agrobacterium tumefaciens A common soil bacterium that causes crown gall disease by transferring some of its DNA to the plant host. Scientists alter *A. tumefaciens* so that it no longer causes the disease but is still able to transfer DNA. They then use this altered organism to ferry desirable genes into plants.

Allele One of several alternative forms of a specific gene that occupies a certain locus on a chromosome.

Alpha helix One of a small number of stable arrangements of the peptide backbone within proteins.

Amino acid The fundamental building blocks of a protein molecule. A protein is a chain of hundreds or thousands of amino acids. Our bodies can synthesize most of the amino acids from their component parts (carbon, nitrogen, oxygen, hydrogen, and sometimes sulfur). However, eight amino acids, known as essential amino acids, must be obtained from food.

Anabolism The combination of metabolic reactions involved in synthesizing molecules.

Antiangiogenesis Angiogenesis is the growth of new blood vessels.

Antibiotic resistance marker A gene encoding a protein that renders a cell resistant to an antibiotic. The organisms, usually bacteria, that have this gene are not susceptible to that antibiotic. It is often used to identify organisms that have been successfully transformed.

Antibody A protein produced in response to the presence of a specific antigen.

Anticodon A triplet of nucleotides in a transfer RNA molecule that is complementary to a codon in a messenger RNA molecule.

Antigen A foreign substance that elicits the production of antibodies.

Antisense molecule A single-stranded nucleic acid that is complementary to a gene and, therefore, to the messenger RNA transcribed from that gene. It blocks the expression of the gene by interfering with protein production.

Assay A method for determining the presence or quantity of a component.

Atom The smallest particle into which a chemical element can be divided and still retain the element's properties.

Autosome A chromosome that is the same in males and females of the species (as opposed to the sex chromosomes). Humans have 22 pairs of autosomes and 1 pair of sex chromosomes. Mutations in, or traits encoded by, genes on these chromosomes can be described as autosomal. For example, cystic fibrosis is an autosomal recessive trait.

Bacillus thuringiensis A naturally occurring bacterium with pesticidal properties. *B. thuringiensis* produces a protein (Bt toxin) that is toxic only to certain insect larvae that consume it.

Bacteriophage A virus that infects bacteria. Also called a phage.

Baculovirus A class of insect viruses. They are used in basic research as cloning vectors for eukaryotic cells and in commercial applications as biocontrol agents for controlling insect pests.

Beta sheet One of a small number of stable arrangements of the peptide backbone within proteins.

Bioassay A method of determining the effect of a compound by quantifying its effect on living organisms or their component parts.

Biocatalyst A biological molecule, such as an enzyme, or a one-celled biological organism that causes a chemical reaction to occur.

Biochip DNA-based microprocessors that use DNA in its role as an information storage molecule.

Biofuel Fuels, such as ethanol, that are derived from biomass.

Bioinformatics The development and use of software for the manipulation and analysis of DNA and protein sequence information.

Biolistics A method of getting DNA into cells by using small metal particles coated with DNA. These particles are fired into a cell at very high speed.

Biological control The use of one organism to control the population size of another organism.

Biological molecules Large, complex molecules, such as proteins, nucleic acids, lipids, and carbohydrates, that are produced only by living organisms. Biological molecules are often referred to as macromolecules or biopolymers.

Biomarker A biological feature that is used to determine the progress of a disease.

Bioprocess A process in which microorganisms, living cells, or their components are used to produce a desired end product.

Bioreactor A container used for bioprocessing.

Bioremediation The use of organisms, usually microorganisms, to clean up contamination.

Biosensor An electronic system that uses cells or biological molecules to detect specific substances. It consists of a biological sensing agent coupled to a microelectronic circuit.

Biosynthesis Production of a chemical by a living organism.

Biotechnology (Broad definition) The use of living organisms to solve problems and make useful products. (Modern definition) A collection of technologies that use living cells and/or biological molecules to solve problems and make useful products.

Blastocyst In mammals, a 4- to 5-day-old embryo that consists of approximately 100 to 150 undifferentiated cells.

B lymphocyte A type of immune system cell that is responsible for the production of antibodies.

Bovine somatotropin (BST) The proteinaceous growth hormone found naturally in cattle that is also known as bovine growth hormone (BGH). BST is chemically very similar to human growth hormone and is used by some dairy farmers to increase the growth rate, the protein/fat ratio, and milk production in cows.

Callus A cluster of undifferentiated plant cells that have the capacity to regenerate a whole plant in some species.

Cancer vaccines Unlike other vaccines that provide preventative protection, cancer vaccines stimulate the lymphocytes of the immune system to attack existing tumor cells.

Capsid Protein coat of a virus. Plants can be genetically engineered to be resistant to a virus by giving them the gene that encodes the capsid.

Carbohydrates Biological molecules composed of simple sugars, such as glucose. Also known as polysaccharides. Familiar examples are starch, cellulose, glycogen, and lactose.

Carboxyl group A carbon atom bonded to two oxygen atoms, one of which is also bound to a hydrogen atom. It is represented in depictions of molecules by COOH.

Carrier protein A type of transport protein, embedded in the plasma membrane, that escorts charged molecules across the membrane.

Catabolism The sum of all metabolic reactions involved in breaking down molecules.

Catalyst A substance that speeds up a chemical reaction but is not itself changed during the reaction.

cDNA library A collection of genetic clones that contains all of the cDNA derived from a source organism. The cDNA is "housed" in the library by splicing portions of the entire complement of cDNA into suitable vectors. (See **Complementary DNA**.)

Cell culture A technique for growing cells under laboratory conditions.

Cell cycle The cyclical process of growth and cell division. In eukaryotes, it consists of four distinct stages.

Cell fusion The formation of a hybrid cell produced by the fusing of two different cells.

Cell membrane See **Plasma membrane.**

Cell therapy Treating a disease by providing healthy cells to replace dead or malfunctioning cells. Genetically engineered cells may also be used to deliver a constant supply of therapeutic compounds.

Channel protein Proteins that create pores within cell membranes to allow movement of small molecules across the membrane.

Chimera In classical mythology, an animal that was made of parts of more than one animal, such as a griffon, which had the body of a lion and the wings of an eagle. In modern biology, it refers to organisms that are composed of cells descended from more than one set of parents, such as the mice created by injecting cultured embryonic stem cells from a brown mouse into an unrelated embryo derived from two white mice.

Chitin A carbohydrate that is the principal component of fungal cell walls and insect exoskeletons.

Chromatin The DNA-protein complex found in the nucleus.

Chromosomal aberration An abnormality in chromosome number or structure.

Chromosomes Components of a cell that contain genetic information. Each chromosome contains numerous genes. Chromosomes occur in pairs: one is obtained from the mother, and the other is obtained from the father. Chromosomes of different pairs are often visibly different from each other. (See also **DNA.**)

Chromosome walking The step-by-step analysis of a long stretch of DNA by the sequential isolation of clones that carry overlapping sequences of this DNA. Used to locate unknown genes. Using a gene library and starting from a known sequence (usually the site of a restriction fragment length polymorphism), scientists isolate clones containing DNA that hybridizes to DNA probes taken from the ends of the known sequence. The ends of these clones are then used to screen the library for clones that hybridize to the ends of the first clone. This screening and isolation is repeated again and again until the unknown gene of interest is reached. If each clone

covers a long stretch of DNA, the researcher can "walk" the chromosome quickly, because each "step" is like a "giant step." Thus, researchers prefer to use clones or vectors that carry a large amount of foreign DNA. (See **Clone, Cloning, and Genetic library.**)

Clone A cell, collection of cells, or collection of individuals containing genetic material identical to that of the parent cell and of each other. Clones are produced from a single parent cell and thus show little if any variation compared with that in similar organisms produced through sexual reproduction. The word "clone" also refers to the identical pieces of DNA that a collection of cells (usually bacterial) contain.

Cloning Like the term clone, cloning also has a variety of definitions depending upon the entity being cloned. On a molecular level, cloning means isolating DNA sequences and incorporating them into plasmids or other vectors so that they can be inserted into a suitable organism (a bacterium or yeast cell) for copying. Cloning also refers to the production of genetically identical cells from a single parent cell. These genetically identical cells are referred to as a clonal population. Finally, cloning also refers to the production of genetically identical multicellular organisms.

Codon A triplet of bases that specifies an amino acid.

Colony hybridization A technique that uses hybridization to identify bacteria containing DNA that is complementary to a certain sequence.

Complementary DNA (cDNA) A single-stranded DNA that is synthesized in vitro from an RNA template by reverse transcriptase.

Computational biology Integration, analysis, and modeling of biological data.

Conjugation The transfer of genetic material from one bacterium to another through physical contact between the cells. In *Escherichia coli,* the contact occurs through a special structure called a pilus.

Cosmid A plasmid that is packaged in a phage coat. Scientists use cosmids to transfer a relatively long stretch of DNA into host organisms.

Covalent A type of chemical bond that consists of two electrons shared by two atomic nuclei.

Crossing over A natural process that occurs during meiosis in which pieces of homologous chromosomes are exchanged.

Culture To grow living organisms in prepared medium.

Culture medium A nutrient system for growing bacteria or other cells in the laboratory.

Cytoplasm The living matter within a cell, excluding the genetic material.

Cytosine A nitrogen-containing base found in DNA and RNA.

Deletion A form of chromosomal aberration in which a portion of a chromosome is lost.

Denaturation The complete unfolding of a protein or the separation of the two complementary strands of a DNA double helix.

Deoxynucleotide A compound made up of the sugar deoxyribose, phosphate, and a nitrogen-containing base. Found in DNA.

Deoxyribose The five-carbon sugar found in DNA.

DNA (deoxyribonucleic acid) The chemical molecule that is the basic genetic material found in all cells. DNA is the carrier of genetic information from one generation to the next. Because DNA is a very long, thin molecule, it is packaged into units called chromosomes. DNA belongs to a class of biological molecules called nucleic acids.

DNA chip A small piece of glass or silicon that has arrays of DNA on its surface. Its usefulness as a gene identification tool is based on DNA hybridization. (See **Microarray**.) The term DNA chip is also used to refer to a biochip, which differs significantly from DNA microarrays. (See **Biochip**.)

DNA hybridization The formation of a double-stranded nucleic acid molecule from two separate but complementary single strands. The single strands can be two DNA strands or one RNA and one DNA strand. The term also applies to a molecular technique that uses one nucleic acid strand to locate another.

DNA library A collection of cloned DNA fragments that collectively represents the genome of an organism. A complementary DNA library collectively represents the messenger RNA species that were present in the cells when the library was made.

DNA ligase An enzyme that rejoins cut pieces of DNA.

DNA polymerase An enzyme that replicates DNA. DNA polymerases synthesize new DNA complementary to a template strand. Synthesis occurs in the $5'$-to-$3'$ direction only and requires a primer.

DNA probe A relatively short single strand of DNA that is used to detect a specific sequence of nucleotides through hybridization.

DNA repair enzymes Proteins that recognize and repair abnormalities in DNA.

DNA sequence The order of nucleotide bases in the DNA molecule.

DNA vaccines Pieces of foreign DNA that are injected into an organism to trigger an immune response. The injected DNA is expressed by the host cell, producing an antigen that is recognized as foreign by both B and T lymphocytes. Unlike common vaccines, DNA vaccines contain the information for making the antigen and not the antigen itself or the whole organism.

Dominant allele An allele that is phenotypically expressed in the same way in individuals who are either homozygous or heterozygous for that allele.

Duplication A chromosomal aberration in which part of the chromosome is present in duplicate form in a cell.

Electronegativity The ability of an atom to attract electrons.

Electroporation A technique that uses an electrical current to create temporary pores in a cell membrane through which DNA can enter.

Element A substance that cannot be broken down by ordinary chemical means.

Embryonic stem (ES) cell In mammals, the cells derived from the inner cell mass of a 4- to 5-day-old blastocyst. An embryonic stem cell can give rise to any type of differentiated cell but cannot give rise to a complete organism.

Embryo transplantation A technique used in animal biotechnology. After in vitro fertilization, the zygote is cultured for a few days and then implanted into a female. The developing embryo is sometimes separated into individual cells at the four- to eight-cell stage, and each cell is implanted in a female.

Endonuclease An enzyme that cleaves a nucleic acid at nonterminal phosphodiester bonds. (See **Exonuclease** for comparison.) One class of endonucleases, the restriction endonucleases, recognize specific sequences of bases along a DNA molecule and cleave the molecule following recognition.

Endoplasmic reticulum An extensive series of internal membranes in eukaryotic cells.

Endostatin An endogenous protein that blocks the proliferation of blood vessels by inhibiting the growth of endothelial cells.

Enhancers DNA-binding sites of certain transcription activator proteins that are important for maximal transcription of associated promoters.

Enzyme   A protein that accelerates the rate of chemical reactions. Enzymes are catalysts that promote reactions repeatedly without being changed by the reactions.

Enzyme-linked immunosorbent assay (ELISA)   A technique for detecting specific proteins by using antibodies linked to enzymes.

Erythropoietin   A growth factor that stimulates the cells that give rise to red blood cells.

Escherichia coli   A bacterium commonly found in the intestinal tracts of most vertebrates. *E. coli* is used extensively in recombinant DNA research because it has been genetically well characterized.

Eukaryote   An organism whose genetic material is located within a nucleus. Yeast cells, fungi, protozoans, plants, and animals are eukaryotes.

Evolution   Changes in the gene pool of a population over time. These changes in the frequencies of certain genotypes result primarily from natural selection. Other factors that may contribute to changes in the genetic composition of a population are genetic drift and migration.

Exon   The region of a gene that determines the amino acid sequence of a protein.

Exonuclease   An enzyme that cleaves the terminal phosphodiester bond of a nucleic acid molecule, releasing a single nucleotide. Exonucleases must have access to the end of a molecule for activity; they will not cleave circular nucleic acid molecules.

Expression   The physical manifestation (protein production) of the information contained in a gene.

Expressivity   In classical genetics, the degree to which a particular genotype is expressed in the phenotype of those having that genotype.

Extremophiles   Microorganisms that live at extreme levels of pH, temperature, pressure, and salinity.

Extremeozyme   Enzymes from organisms that live under extreme conditions and, as such, can resist denaturation under conditions that typically cause other proteins to lose their tertiary structure.

Factor VIII and factor IX   Two of the many compounds involved in a cascade of events that lead to blood clotting. If one of these proteins is missing or defective, the resulting condition is a form of hemophilia. Because both can be small proteins, they can be produced in large quantities by bacteria or eukaryotic cells genetically engineered for novel-protein production.

Fatty acid   A molecule consisting of a carboxyl group and a long hydrocarbon chain.

Fermentation   In biochemistry, fermentation is the anaerobic breakdown of glucose. In biotechnology and biochemical engineering, fermentation is the process of growing microorganisms to produce various chemical or pharmaceutical compounds. Microbes are usually incubated under specific conditions in large tanks called fermentors. Fermentation is a form of bioprocess manufacturing.

Fibroblast   A type of cell found in connective tissue

Fluorescent (or fluorescence) in situ hybridization (FISH)   Detection of a specific nucleic acid sequence within its natural setting, such as intact chromosomes or thin sections of cells, by hybridizing it to a probe with a fluorescent label and visualizing the probe.

Founder effect   A type of genetic drift that occurs when a population is founded by a very small number of individuals.

Functional foods   Foods containing compounds with beneficial health effects beyond those provided by the basic nutrients, vitamins and minerals. Also called nutraceuticals.

Functional genomics   A field of research whose goal is to understand what each gene does, where it is located, and how it is regulated.

Gaucher's disease   A genetic disease caused by a problem with the enzyme that breaks down a specific lipid found in the membranes of blood cells.

Gel electrophoresis   A process for separating molecules by forcing them to migrate through a semisolid material (gel) under the influence of an electric field.

Gene   A unit of hereditary information. A gene is a section of a DNA molecule that specifies the production of a particular protein.

Gene amplification   The increase, within a cell, of the number of copies of a given gene.

Gene chip   See DNA chip and microarray.

Gene knockout   The replacement of a normal gene with a mutated form of the gene by homologous recombination. Used to study gene function.

Gene mapping   Determining the relative locations of genes on a chromosome.

Gene therapy   The addition of genetic material to an individual so that a defect or disease can be corrected. To date, human gene therapy has involved changing the genetic makeup of somatic cells only. Genetic changes to germ cells are prohibited in humans and have been restricted to animals.

Genetic code The system of nucleotide triplets in genes that encode the amino acids in proteins. All living organisms on earth use the same genetic code.

Genetic drift Change in gene (allele) frequencies due to chance occurrences.

Genetic engineering The technique of removing, modifying, or adding genes to a DNA molecule in order to change the information it contains. Also known as recombinant DNA technology. By changing this information, genetic engineering changes the type or amount of proteins an organism is capable of producing.

Genetic heterogeneity Mutations in different genotypes leading to the same phenotype.

Genetic library A collection of DNA that, taken collectively, represents all of an organism's genome. The DNA molecules are "housed" in microorganisms as recombinant DNA molecules and are copied when the microorganism replicates.

Genetic linkage See **Linkage.**

Genome The total hereditary material of a cell.

Genomics The characterization and comparison of not only the structure, but also the patterns of gene expression in entire genomes or large portions thereof.

Genotype The specific genetic makeup of an organism, as opposed to the actual characteristics of an organism. (See **Phenotype.**)

Glycolysis The series of chemical reactions through which one molecule of glucose is converted into two molecules of pyruvate.

Glycoprotein A protein molecule that is bound to a simple sugar or carbohydrate.

Golgi apparatus Membrane-bound sacs and tubules found in eukaryotic cells.

Growth factors Naturally occurring proteins that stimulate growth and reproduction of specific cell types. For example, epidermal growth factor stimulates the production and differentiation of cells in the upper skin layer. Fibroblast growth factor stimulates growth of cells in connective tissue. Growth factors are being studied as possible therapeutic compounds to be used in the treatment of diseases or injuries. For example, the two growth factors just mentioned could be useful for treating burn victims.

Growth hormones Hormones that stimulate growth in plants and animals. The growth hormones in plants bear no chemical resemblance to the growth hormones in animals. In the vertebrates, growth hormone is a protein hormone secreted by the anterior pituitary. It stimulates protein production in its target organs. Also known as somatotropin.

Guanine A nitrogen-containing base found in DNA and RNA.

Haplotype A contraction of haploid and genotype. The DNA content of one contiguous molecule, such as a mitochondrial genome, a single chromosome within a diploid organism, or a portion of a single chromosome, even down to very small regions.

Hematopoietic stem cell A type of stem cell that can differentiate into all types of blood cells.

Homologous Two chromosomes are said to be homologous if they carry alleles for the same traits. In each cell containing homologous chromosomes, each member of a homologous pair is derived from a different parent. Nonhomologous chromosomes carry genes for different traits.

Huntington's disease An autosomal dominant genetic disease in which a mutation leads to the production of a defective protein that kills nerve cells.

Hyaluronate A carbohydrate made of a chain of several thousand sugar molecules in a regular repeating sequence.

Hybridization Production of offspring, or hybrids, from genetically dissimilar parents. In selective breeding, the term usually refers to the offspring of two different species. (See also **DNA hybridization.**)

Hybridoma A type of hybrid cell produced by fusing a normal cell with a tumor cell. When lymphocytes (antibody-producing cells) are fused to the tumor cells, the resulting hybridomas produce antibodies and maintain rapid, sustained growth, producing large amounts of an antibody. Hybridomas are the source of monoclonal antibodies.

Hydrogen bonds Weak electrostatic attractions. Hydrogen bonds exist between paired bases in DNA and are important in determining protein structure.

Hydrophilic Favoring chemical associations with water molecules.

Hydrophobic Disfavoring chemical associations with water molecules.

Hypersensitive response (HR) One of the endogenous defense systems that plants use against pathogens. By causing localized plant cell death at the point of infection, the hypersensitive response denies the pathogen living host cells.

Immunoassay A technique for identifying substances that is based on the use of antibodies.

Immunotoxin A molecule that is toxic to the cell and is attached to an antibody.

Industrial sustainability Use of manufacturing processes and products that meet the current consumer demand for products without compromising the resources and energy supply of future generations.

Initiation codon The codon in messenger RNA that tells the ribosomes to start synthesizing a protein. Usually 5′ AUG 3′.

Initiation factor A protein necessary to begin translation. Initiation factors are not parts of the ribosomes and do not participate in translation once the process has begun.

Insulin A protein hormone that lowers blood glucose levels. The first commercial product derived from genetically engineered bacteria.

Interferon A protein produced naturally by the cells in our bodies. It increases the resistance of surrounding cells to attacks by viruses. One type of interferon, alpha interferon, is effective against certain types of cancer. Others may prove effective in treating autoimmune diseases.

Interleukin A protein produced naturally by our bodies to stimulate our immune systems. There are at least six kinds of interleukins.

Introns Noncoding regions within a gene. They are transcribed into RNA but are removed by splicing before protein synthesis.

Inversion A chromosomal aberration in which a section of chromosome is reversed.

In vitro Performed in a test tube or other laboratory apparatus.

In vitro selection Selection at the cellular or callus stage of individuals possessing certain traits, such as herbicide resistance.

In vivo In the living organism.

Ion An atom or molecule that has an electrical charge because it has an unequal number of protons and electrons.

Islet cells Pancreatic cells that are the source of insulin, glucagon, and somatostatin, three hormones involved in regulating glucose metabolism and absorption.

Karyotype An organized visual profile of the chromosomes of an organism. Cells are arrested in the metaphase stage of cell division, when the chromosomes are condensed. The condensed chromosomes are isolated and treated with Giemsa stain, which reveals distinctive banding patterns. Photomicrographs of the stained chromosome are then arranged in a standard format according to size, the relative position of the centromere, and other criteria, so that the number and morphology of the chromosomes can be assessed. Karyotype can also be used as a verb, meaning to create a karyotype.

Keratinocyte A predominant cell type of the epidermis, which produces the protein keratin.

Keratins The family of structural proteins that make up hair, wool, feathers, claws, hooves, and so on.

Knockout mouse A mouse in which a specific gene is inactivated by genetic engineering.

Ligation The joining of the ends of two DNA molecules.

Linkage The tendency of pairs or groups of genes to be inherited together because they occur close together on the same chromosome.

Locus (plural, loci) The chromosomal location of a gene or a physical feature, such as a microsatellite or single-nucleotide polymorphism.

Lysate The mixture of cellular components obtained after cells have been broken open.

Lysis The breaking open of cells.

Macrolesion A genetic change in which a large amount of DNA is altered by changing the total amount of DNA or changing the relative positions of genes on a chromosome. (See also **Chromosomal aberration, Duplication, Deletion, Inversion,** and **Translocation.**)

Macromutation See **Macrolesion.**

Marker A restriction fragment, microsatellite allele, single-nucleotide polymorphism, or other detectable chromosome feature.

Marker genes Genes that identify which plants, bacteria, or other organisms have been successfully transformed.

Melting temperature The temperature required to denature a DNA or protein molecule.

Messenger RNA (mRNA) The RNA molecules that carry genetic information from the chromosomes to the ribosomes.

Metabolic engineering   Changing cellular activities by manipulating the enzymatic, transport, and regulatory functions of the cell.

Metabolism   The sum of all chemical reactions occurring in a living organism. It consists of anabolism plus catabolism.

Microarray   A tiny gridded array of biologically important molecules, cells, or tissues used in basic and applied research and commercially. For example, single-stranded DNA probes arrayed in a grid are used in hybridization experiments with DNA or messenger RNA (mRNA) isolated from biological samples. Hybridization of sample DNA to arrayed DNA reveals genotypes at many loci. Hybridization of arrayed DNA to mRNA samples reveals what genes were being expressed in the sample. These microarrays allow scientists to type many genes at once or to measure the expression of every gene in a cell in a single experiment.

Microbial fermentation   The production of useful compounds, such as vitamins, antibiotics, enzymes, and certain foods. (See also **Fermentation** and **Bioprocess.**)

Microinjection   A method of delivering DNA, primarily to animal cells, by using a microscopic needle to pierce the nucleus.

Microlesion   A genetic change, sometimes called a micromutation, that involves a small amount of DNA. (See also **Point mutation.**)

Micromutation   See **Microlesion.**

Microsatellite   Any region of short repeated nucleotide sequences within a genome.

Mitosis   Nuclear division characterized by chromosome replication and subsequent creation of two identical daughter cells.

Molecular genetics   The study of the molecular structures and functions of genes.

Molecule   A particle consisting of two or more atoms held together by chemical bonds.

Monoclonal antibody   Highly specific, purified antibody that is derived from only one clone of cells and recognizes only one antigen. (See also **Hybridoma.**)

Multigenic   A multigenic, or polygenic, trait is one whose expression is governed by many genes.

Multipotent   The capacity of a cell to differentiate into many different, but not all, cell types of an organism.

Mutagen   A substance that induces mutations.

Mutant   A cell or organism that manifests new characteristics because of a change in its genetic material.

Mutation   Any change in the base sequence of a DNA molecule.

Mycorrhiza   A symbiotic association between certain fungi and the roots of vascular plants.

Nanotechnology   The study, manipulation, and manufacture of ultrasmall structures and machines made of as few as one molecule.

Natural selection   The differential rate of reproduction of certain phenotypes in a population. If those phenotypes have a genetic basis, natural selection can lead to a change in gene frequencies in a population.

Nitrogen-fixing bacteria   Bacteria that can take atmospheric nitrogen and convert it into a form that plants can use.

Noncoding DNA   DNA that does not encode any product (RNA or protein). The majority of DNA in plants and animals is noncoding.

Nonhomologous   Chromosomes are described as nonhomologous if they carry genes for different traits. Compare **Homologous.**

Northern blotting   A technique for identifying an RNA sequence by transferring it from a gel to a filter and hybridizing it to a DNA probe. It is useful for measuring gene expression.

Nuclease   An enzyme that cleaves the phosphodiester bonds of a nucleic acid molecule. (See **Endonuclease** and **Exonuclease.**)

Nucleic acid   A biological molecule composed of a long chain of nucleotides. DNA is made of thousands of four different nucleotides repeated randomly.

Nucleoside   A nucleotidelike molecule containing only a sugar and a base.

Nucleotide   A compound made up of three components: a sugar (either ribose or deoxyribose), phosphate, and a nitrogen-containing base. It is found as individual molecules (e.g., adenosine triphosphate, the "energy molecule") or as many nucleotides linked together in a chain (nucleic acid, such as DNA).

Oligonucleotide   A polymer consisting of a small number of nucleotides. Oligonucleotides can be synthesized by automated machines and so are widely used as probes and primers.

Oncogene   A gene thought to be capable of producing cancer.

Oncology   The study of tumors.

Operon   A collection of adjacent genes that are transcribed together and whose products usually have related functions.

Organelle   A distinct body in the cytoplasm of eukaryotic cells.

Origin of replication   A sequence of DNA bases that tell DNA polymerase and its helper proteins where to begin duplicating a DNA molecule.

Osmosis   The diffusion of water across a selectively permeable membrane.

Penetrance   In genetics, the proportion of individuals with a particular genotype that display the phenotype associated with that genotype.

Peptide bond   The chemical bond that links adjacent amino acids within proteins.

Phage   See **Bacteriophage.**

Pharmacogenomics   Genetic and genomic studies to determine and predict the basis for individuals' different reactions to pharmaceuticals. It is a marriage of the study of genetics, genomics, and pharmacology.

Phenotype   The observable characteristics of an organism as opposed to the set of genes it possesses (its genotype). The phenotype that an organism manifests is a result of both genetic and environmental factors. Therefore, organisms with the same genotype may display different phenotypes because of environmental factors. Conversely, organisms with the same phenotypes may have different genotypes.

Phosphate group   An important molecular grouping in biological systems that consists of a phosphorus atom bound to three oxygen atoms, two of which carry a negative charge. Phosphate groups are added and lost from larger biological molecules through processes known as phosphorylation and dephosphorylation.

Phosphodiester bonds   The chemical bonds that connect nucleotides in the backbones of DNA and RNA.

Phospholipid   A type of lipid, similar to a fat, in which a phosphate group, rather than a fatty acid, is bound to the third carbon in the glycerol molecule.

Photosynthesis   The conversion of light energy to chemical energy. In plants and some bacteria, it is the synthesis of organic molecules from carbon dioxide and water, using light energy.

Phytoremediation   The use of plants to clean up pollution.

Plasma membrane   The membrane surrounding the cytoplasm of a cell and separating its contents from the external environment.

Plasmid   A small, self-replicating piece of DNA found outside the chromosome. Plasmids are the principal tools for inserting new genetic information into microorganisms or plants.

Pleiotropy   The capacity of a gene to affect a number of different phenotypic characteristics.

Pluripotent   A cell is said to be pluripotent if it can differentiate into any cell type found in an organism.

Point mutation   A change in the DNA sequence in a single gene. Most often, this term refers to a change in a single base or a single base pair in a gene.

Polymerase chain reaction (PCR)   A method of making millions of copies of a single DNA molecule by using a heat-stable DNA polymerase.

Porcine somatotropin (PST)   The growth hormone found in pigs. (See **Growth hormones.**)

Primary structure   The linear sequence of amino acids within a protein molecule.

Primer   A single-stranded nucleic acid molecule (DNA or RNA) hybridized to a template strand in such a way that the primer's 3′ end is available to serve as the starting point for synthesis of a new DNA strand complementary to the template. Primers are required for DNA synthesis by DNA polymerase enzymes.

Probe   A single-stranded DNA or RNA molecule used to detect the presence of a complementary nucleic acid.

Prokaryotes   Organisms whose genetic material is not enclosed by a nucleus. The most common examples are bacteria.

Promoter   A special sequence of bases in DNA that is recognized by RNA polymerase enzymes. The promoter signals RNA polymerase to begin transcription of a gene.

Protein   A complex biological molecule composed of a chain of units called amino acids. Proteins have many different functions: structure (collagen), movement (actin and myosin), catalysis (enzymes), transport (hemoglobin), regulation of cellular processes (insulin), and response to stimuli (receptor proteins on the surfaces of all cells). Protein function is dependent on the protein's three-dimensional structure (tertiary structure), which depends on the linear sequence of amino acids in the protein (secondary structure). The information for making proteins is stored in the sequence of nucleotides in the DNA molecule.

Protein chip   See **Microarray.**

Proteinase   An enzyme that cleaves the peptide bonds of protein backbones.

Proteome   The sum of all the proteins expressed in a given cell at a given time.

Proteomics   A field of research devoted to discovering the structures and functions of all proteins made by a specific type of cell.

Protoplast   A plant or bacterial cell whose wall has been removed by artificial treatment.

Quaternary structure   The arrangement of multiple protein subunits in a larger complex.

Random amplification of polymorphic DNA (RAPD)   A PCR-based technique for characterizing an unknown genome that is used when nothing is known about the sequence in question. To do this, scientists use many sets of primers that are random sequences, 10 bases long. PCR products can be generated anywhere on the genome that the primers can anneal in the right orientation and close enough together. Whether a given primer yields products and of what length is a function of the sequence of the genome. The more similar two genomes are, the more amplification products they will share.

Recessive allele   An allele whose expression is masked in the heterozygous state by a dominant allele.

Recombinant DNA (rDNA)   DNA that is formed by combining DNAs from two different sources.

Recombinant DNA (rDNA) technology   The laboratory manipulation of DNA in which DNAs or fragments of DNA from different sources are cut and recombined by using enzymes. This rDNA is then inserted into a living organism. rDNA technology is usually synonymous with genetic engineering.

Recombination   The formation of new combinations of genes. Recombination occurs naturally in plants and animals during the production of sex cells (sperm, eggs, or pollen) and their subsequent joining in fertilization. In microbes, genetic material is recombined naturally during conjugation, transformation, and transduction.

Regeneration   The process of growing an entire plant from a single cell or group of cells.

Replacement gene therapy   See **Gene therapy.**

Repressors   Proteins that bind to DNA and block transcription.

Restriction endonuclease   See **Restriction enzyme.**

Restriction enzyme   An enzyme that recognizes a specific sequence of bases in a DNA molecule and cleaves the molecule at or near that sequence. The recognition sequence is called a restriction site. Different restriction enzymes recognize and cleave at different restriction sites. Also called restriction endonuclease.

Restriction fragment   A short length of DNA that results from the cleavage of a large DNA molecule by a restriction enzyme.

Restriction fragment length polymorphism (RFLP; pronounced "riflip")   A difference in restriction fragment lengths between very similar DNA molecules (such as homologous chromosomes from two different individuals). RFLPs are caused by relatively minor differences in the base sequences of the molecules. RFLP analysis is used to detect differences in DNA molecules that are, on a large scale, quite similar. Applications of RFLP analysis include DNA typing and prediction of genetic disease through DNA testing.

Restriction map   A diagram of the sites on a DNA molecule cleaved by different restriction enzymes.

Reverse transcriptase   The enzyme that uses an RNA molecule as a template for synthesizing a complementary DNA molecule.

Ribose   The five-carbon sugar found in RNA.

Ribosomes   The protein-RNA complexes that form the site of protein synthesis.

Ribozymes   RNA molecules that catalyze reactions, often the breakdown of RNA molecules. Also called catalytic RNA.

RNA (ribonucleic acid)   Like DNA, a type of nucleic acid. RNA differs from DNA in three ways: RNA nucleotides contain the sugar ribose instead of deoxyribose, RNA contains the base uracil instead of thymine, and RNA is primarily a single-stranded molecule rather than a double-stranded helix. The three major types are messenger RNA, transfer RNA, and ribosomal RNA. All are involved in the synthesis of proteins from the information contained in the DNA molecule.

RNA chip   See **Microarray.**

RNA interference (RNAi)   A process in which the presence of short double-stranded RNA (dsRNA) molecules triggers enzymes to degrade messenger RNA molecules whose base sequence matches that of the dsRNA.

RNA polymerase   The enzyme that synthesizes RNA using a DNA template.

Secondary structure Local regions of alpha helixes, beta sheets, and unstructured loops within a protein molecule.

Sex-linked inheritance A trait that is determined by a gene on a sex chromosome, most often the X chromosome. As a result, the trait shows different patterns of inheritance in males and females. In humans, the ability to discriminate color is a sex-linked trait.

Short tandem repeat (STR) A repeated sequence of one or a very few nucleotides, such as $(5'\ CA\ 3')_n$. Also called simple sequence repeat (SSR). The first short tandem repeat (or SSR) in human DNA was described in 1989 and had the sequence $(5'\ CA\ 3')_n$, where n is variable. The sequence was observed to occur at over 50,000 different locations within the human genome.

Simple sequence repeat (SSR) See **Short tandem repeat (STR).**

Single-nucleotide polymorphism (SNP) A single-nucleotide difference between two sequences of DNA; that is, different individuals have different bases at a particular site in the genome.

Somaclonal variant selection A form of plant genetic manipulation that is analogous to selective breeding at the plantlet and not the reproductive stage.

Somatotropin A synonym for growth hormone.

Southern blotting A technique for identifying a specific DNA sequence by transferring single-stranded DNA from a gel to a filter and then hybridizing the DNA with a complementary nucleic acid probe.

Splicing The process of removing introns from messenger RNA.

Stem cell A cell that divides to produce daughter cells that can either differentiate or remain as stem cells, i.e., undifferentiated.

Stop codon One of the three codons in messenger RNA that cause protein synthesis to stop.

Structural motif A simple combination of a few secondary-structure elements frequently found in protein molecules.

Subcloning Breaking a large cloned gene into smaller parts and making a new clone from each of the DNA pieces.

Substrate A chemical molecule on which an enzyme acts.

Systemic acquired resistance (SAR) One of the endogenous protection mechanisms found in plants.

In response to infection, the plant synthesizes a variety of defense-related proteins, such as the enzyme chitinase, which degrades fungal cell walls.

Systems biology The use of biological data to create predictive models of cell processes, biochemical pathways, and whole organisms.

T cells Lymphocytic cells of the immune system involved in cell-mediated immunity and interactions with B cells.

Telomeres Specialized DNA structures at the ends of chromosomes. They are segments of DNA consisting of short repeated sequences assembled into an unusual formation that includes a loop at the very end of the chromosome.

Terminator Sequence of DNA bases that tells the RNA polymerase to stop synthesizing RNA.

Tertiary structure The total three-dimensional structure of a protein.

Therapeutic cloning The use of somatic cell nuclear transfer to create cells, genetically identical to a patient's, that can be used to treat disease or repair damaged tissues.

Thymine A nitrogen-containing base found in DNA.

Tissue culture A procedure for growing or cloning cells or tissue by in vitro techniques.

Tissue engineering The synthesis of organs and tissues under laboratory conditions. It is dependent on cell culture techniques.

Tissue plasminogen activator A naturally occurring protein that dissolves blood clots; it is currently being produced for commercial use by genetically engineered bacteria. Also known as tPA.

Tobacco mosaic virus (TMV) A naturally occurring pathogen of tobacco. TMV consists only of RNA and protein. If the RNA is extracted from TMV and rubbed into a tobacco leaf, new viruses are produced. Therefore, RNA, not DNA, is the genetic material of TMV. TMV was important in research that elucidated the relationship between DNA, RNA, and protein.

Transcription The process of using a DNA template to make a complementary RNA molecule.

Transcriptional activator A protein that helps RNA polymerase begin transcription at one or more promoters.

Transduction The transfer of DNA from one bacterium to another via a bacteriophage.

Transfer RNA (tRNA) The RNA molecules that match codons and amino acids at the ribosome.

Transformation A change in the genetic structure of an organism as a result of the uptake and incorporation of foreign DNA.

Transgenic A transgenic organism is one that has been altered to contain a gene from an organism that belongs to a different species.

Transient gene therapy See **Gene therapy.**

Translation The process of using a messenger RNA template to make a protein.

Translocation A chromosomal aberration in which a segment of one chromosome breaks off and joins a nonhomologous chromosome.

Transport protein A class of proteins, embedded in cell membranes, that allow movement of charged molecules across the hydrophobic membrane. They include channel proteins, carrier proteins, and protein pumps, which move molecules against the concentration gradient.

Transposon A mobile genetic element that can move from one location in a plasmid or chromosome to another location.

Tumor necrosis factor An endogenous compound that slows cell growth and kills some tumor cells.

Tumor suppressor gene A gene that leads to tumor development when it is inactivated.

Uracil A nitrogen-containing base found in RNA.

Vector The agent used to carry new DNA into a cell. Viruses or plasmids are often used as vectors.

Virus An infectious agent composed of a single type of nucleic acid (DNA or RNA) enclosed in a coat of protein. Viruses can multiply only within living cells.

Western blotting A technique for identifying a protein by transferring it from a gel to a membrane and then probing it with a labeled antibody.

Index

Page numbers in italics indicate pages in the *Student Guide* corresponding to the preceding page numbers in this volume. Index entries from chapters 1 to 5 are listed only once, as this material is common to both volumes.

Cell transplant therapy, 24–25
Cell wall, 55, 212
Cellular cloning, 15, 516, *345*
Cellular membranes, 40, 49–51, 60, 212, 216, *185*
    fluidity of, 55
    movement of substances across, 50
    structural organization of, 50
    structure of, 54–55
Cellular processes, 58–70
    cascades and pathways, 64–68
    cell-to-cell communication, 61–62, 70–71
    chemical reactions, 63–64
    common principles governing, 62–70
    energy requirement, 62–63
    essential functions, 58–62
    growth, 58–59, 71
    maintenance of internal environment, 59–60, 71
    recombinant DNA technology and, 137
    regulation of, 68–70
    reproduction, 58–59, 71
    response to external environment, 60–61, 71
Center of origin, of crop plants, 501–502
Central dogma, 81–83, 102
Cephalosporin resistance, 315, *230*
CFTR protein, 353, 433, 436, *305*
Chain terminator, 280, 283–291, *214–217*
    as antiviral drugs, 290–291, *218–219*
Chance, 95–96, 405–406
Channel protein, 60–61
Chaperone protein, 123
Chargaff, Erwin, 81
Chemical bond, 51–52
    chemical-bond energy, 62–63
Chemical manufacturing, 6
Chemical reactions, cellular, 63–64
Chemical symbols, 52
ChemID, 464
Chicken pox, 319
Chimera, 166
Chimpanzee
    defining separate species of, 386–387
    DNA typing from hair, 386
    genome sequence of, 455
    social behavior of, 386
Chitin, 22, 55
Chloramphenicol resistance, 315, *230*
Chloride channel protein, 85–86, 134
Chloroplast, 47
Chloroplast DNA, 111
Cholera, 320, 325, *234*
Cholesterol, 54–55
    blood, 352
Chromatin, 110
Chromosomal defects, 87, 350–351, 390–392, *272–274*
Chromosomal protein, 110
Chromosome, 45–47, 86, 110–111, 195, *176*, 221, 419, *290*
    banding patterns of, 391–393, *273–274*
    chromosomal nature of inheritance, 77–78
    cytogenetics, 391–392, *273–274*
    of *E. coli*, 206, 209, *182*

mapping of, 429, *300*, *see also* Genetic mapping
    physical mapping of, 385
Chromosome painting, 260–261
Chronic myelogenous leukemia, 351
Chronic myeloid leukemia, 371, *262*
Chymotrypsin, 125, 127–130
Civic responsibility
    information sources and, 491, *337*
    science and technology and, 489–491, *335–337*
Classroom discussion, 483–484, 557
Clonal population, 91, 147
Clone, 91, 516, *345*
Cloning, 15, 244, 249, *200*, *see also* Animal cloning; Human cloning; Plant cloning
    of complex organisms, 148–150
    definition of, 516, *345*
    of DNA, 146–150, 221, 434, *303*
    DNA libraries, 147–149
    of eukaryotic gene into prokaryote, 163
    governmental policies that regulate, 584, *383*
    issues analysis, 512–522, *341–351*
    marker genes for, 146–147
    vector, *see* Vector, cloning
Cloning technology, 15–16
Clostridium botulinum, 36
CMV infection, *see* Cytomegalovirus infection
Coat color
    in dogs, 155, 336–342, *242–245*, 363, *254*, *see also* Dog hair
    in mice, 344
Codeine, 437, *306*
CODIS (Combined DNA Index System), 402–403, 409, *284*
Codominance, 342, *245*
Codon, 101, 103–105, 187, 195, *176*
Coffee, decaffeinated, 542–543, *360–361*
Coinheritance, 422, *293*
Colony, 147–148
Colony hybridization, 258–259
Comb, forming wells in agarose gel, 226–227, *190*
Combined DNA Index System (CODIS), 402–403, 409, *284*
Commercialization of product, 496
Common ancestor, 456, 459, *318*
Compartmentalization within cells, 47, 50, 69–70
Competent bacteria, 245, 249, *200*, 298
Complementary base pairing, 81, 97–99, 104–105, 140, 173, 177, *172*, 179, *see also* Hybridization analysis
    antisense technology, 198–200, *179–181*
Complementary DNA (cDNA), 142–143
    cDNA library, 147–149
    cloning eukaryotic gene into prokaryote, 163
Computational biology, 18
Conditional-lethal mutation, 321
Confidentiality, 556, 564, *372*
Conjugation, 12, 91–93
    medical implications of, 306
    transfer of antibiotic resistance in *E. coli*, 306–312, *225–227*
Conjugative plasmid, 306–312, *225–227*

Endostatin, 25
Energy, cellular requirement for, 62–63
Energy production, 6, 40
Energy source
    carbohydrates, 55–56
    lipids, 54
Enhancement therapy, 585–587, 591, *384*
Enhancer, 108–109, 124, 187
Enterococcus, vancomycin-resistant, 314, *229*
Enterohemorrhagic *Escherichia coli* (EHEC), 295–296,
    320
Enteroinvasive *Escherichia coli* (EIEC), 295–296
Enteropathogenic *Escherichia coli* (EPEC), 295–296
Enterotoxigenic *Escherichia coli* (ETEC), 294–295
Entrez, 465
Enucleation of egg, 517–518, *346–347*
Environment
    cleaning up, 6, 37–39
    development of genetic disease and, 573, 578, *377*
    impact of technology on, 482–483, 488, *334*
    influence on phenotype, 362, *253*, 573, 578, *377*
    monitoring of, 42
    organism-driven environmental change, 96
    prevention of environmental problems, 39–42
Environmental assessment, of biotechnology products,
    550–552, *368–370*
Environmental biotechnology, 37–42
Environmental Protection Agency (EPA), biotechnology
    regulation by, 525, *354*, 530–531, 541, *359*, 548,
    366
Environmental selection, for antibiotic resistance, 316–
    317, *231–232*
Enzyme(s), 7, 57, 64, 100, 195, *176*
    active site of, 64, 459, *318*
    drug-metabolizing, 437–438, *306–307*
    in food processing, 35–36
    herbicide-resistant, 524, *353*
    industrial, 6, 9, 41, 164
    in metabolic pathways, 65
    missing or malfunctioning, 68
    regulation of
        altering enzyme activity, 69–70
        altering enzyme synthesis, 70
        regulatory site of, 69
    structure of, 127–130
Enzyme engineering, 15
Enzyme replacement therapy, in Gaucher's disease, 588,
    592–594, *385–387*
EPA, *see* Environmental Protection Agency
EPEC, *see* Enteropathogenic *Escherichia coli*
Epidemiology, 386
Epidermal growth factor, 26, 125
Epistasis, 343–349, *246–249*
EPO, *see* Erythropoetin
erbB gene, 371, *262*
Erythrocytosis, benign, 132–133
Erythromycin resistance, 315, *230*
Erythropoetin (EPO), 22, 132
ES cells, *see* Embryonic stem cells
Escherichia coli
    antibiotic resistance in, 306–313, *225–228*, 321

    care and feeding of, 665
    cell size, 206
    chromosome of, 206, 209, *182*
    conjugation in, 306–312, *225–227*
    diseases associated with, 294–297, 661, *453*
    DNA extraction from, 212–217, *185–186*
    DNA of, 110
    enterohemorrhagic, 295–296, 320
    enteroinvasive, 295–296
    enteropathogenic, 295–296
    enterotoxigenic, 294–295
    finding genes for histidine biosynthesis in, 153–154
    finding specific genes in, 153–154
    genome sequence in, 207, 209, *182*
    genome size in, 113, 206–211, *182–184*
    lac operon of, 107
    long-term storage of, 665–666
    model organism for Human Genome Project, 428–
        429, *299–300*
    pathogenicity of, 294–297
    safe handling in laboratory, 295–297, 661, *453*
    strain O157:H7, 36, 296
    T4 infection of, 109
    transduction in, 325–328, *234–237*
    transformation in, 298–305, *222–224*
    translation in, 106
    trp operon of, 107–108
    uropathogenic, 296
EST, *see* Expressed sequence tag
Estrogen, 54–55, 109
ETEC, *see* Enterotoxigenic *Escherichia coli*
Ethical dilemma
    gene therapy, 585–595, *384–388*
    genetic testing, 569–581, *377–380*
Ethical principles, 555–556
    major, 556, 564, *372*
    secondary, 556, 564, *372*
Ethics, *see also* Bioethics
    effect on science and technology development, 482–
        483
    Human Genome Project, 430, *301*, 579, *378*, 582, *381*
Ethidium bromide, 231–232
Ethnic group, DNA databases, 402, 409, *284*
Ethyl alcohol, 66
Etiquette, rules for classroom discussion, 557
Etruscan culture, 414, *289*
Eubacteria, 47
Eugenics, 587
Eukaryotic cells, 47–48, 113
    cell cycle of, 59
    chromosomes of, 110
    molecular biology of, 105–106
    promoters in, 187
Eumelanin, 336, 339–340, *242–243*, 343, 346–347, *246–*
    247, 359, *250*
European corn borer, 548, *366*
European Molecular Biology Laboratory, 464
Evolution, 86–96, 390, *272*
    ancient DNA, 158–160
    definition of, 86–87
    directed protein evolution, 41

agarose concentration in gel, 230
buffer for, 231
pouring (casting) gel, 226–227, *190–191*
preservation of gels, 232
rate of migration of DNA, 230, 239, *194*
recording data, 232, 241–242, *196–197*
sample wells, 226–227, *190–191*
sequencing gels, 287, *215*
staining the gels, 227, *191*, 231–232, 241, *196*
voltage applied, 231
of hemoglobin, 362, *253*
precast gels, 233
of proteins, 150, 233, 431, *302*, 449–454, *315–317*
supplies and equipment for, 234
Gel electrophoresis chamber, 227, *191*
GenBank, 464–465
Gender determination, 410, *285*, 414, *289*
Gene(s), 101, 195, *176*
abbreviations for names of, 339, *242*
base sequence of, 86, 102
centrality of genes, 73, 83
discrete nature of, 74–77
finding specific genes, 152–156
functions of, 97–115
misunderstandings about, 84–86
nature of, 83–86
physical size of, 206, 209, *182*
relationships among genes, proteins, and observable
traits, 83–86
similar genes in related organisms, 155–156
structure of, 97–115
unit of evolutionary change, 86–96
Gene chip, *see* Microarray
Gene cloning, 244, 249, *200*
Gene expression
during development, 435, *304*
microarray analysis of, 162, 432–436, *303–305*
regulation of, 106–110, 198–203, *179–181*
Gene flow, 87
from transgenic crops, 500–506
adverse impacts of, 504–505
to conventional crop varieties, 500
defining terms and delineating the issue, 500–501
gathering information, 500
placing in context, 505–506
probability of, 501–504
to wild plants, 500
Gene gun, 146, 331
Gene identification, 570
Gene knockout, *see* Knockout mice
Gene replacement, in embryonic stem cells, 165–166
Gene therapy, 23–24
bioethics case study, 585–595, *384–388*
enhancement therapy, 585–587, 591, *384*
ethical dilemma, 585–595, *384–388*
in Gaucher's disease, 587–589, 592–594, *385–387*
germ line, 585–586, 591, *384*
growth hormone gene for healthy short children,
589–590, 594, *387*
guidelines for, 586
issues related to, 586–587

national policy on, 591, *384*
NIH guidelines for, 586, 592, *385*
replacement, 23
somatic cell, 585, 591, *384*
transient, 24
Gene transfer
horizontal, 313–318, *228–233*
in nature, 313–318, *228–233*
pathogenicity and, 296
Generalization construction, 489, *335*, 493–494, *339–340*
Genetic code, 101–103, 135, 187, 195, *176*
"second half of," 115
Genetic discrimination, 582, *381*
Genetic disease, 68, 350–372, *250–263*
adult-onset, 576–577, 579–581, *378–380*
among Anabaptists of Pennsylvania, 375–380, *264–269*
carrier status, *see* Carrier of genetic disease
diagnosis of, 21, 354
environmental influence on, 573, 578, *377*
gene identification, 570
gene therapy in, *see* Gene therapy
knockout mice as models of, 166–167
mapping of disease gene, 415–427, *290–302*
online resources, 464
severity of, 573
testing for, *see* Genetic testing
therapeutics, 23–24
transposition-related mutations, 89
Genetic distance, 386
Genetic drift, 88
Genetic engineering, 8, 12–14, 162–167, *see also*
Recombinant DNA technology
of animals, 165–167
comparison with selective breeding, 13–14
generation of genetic variation, 13
goals of, 162–163
of microorganisms, 163–164
restriction enzymes and, 221–222
Genetic heterogeneity, 84
Genetic information
ambiguity of, 573
misuse of, 574
use of, 430, *301*
Genetic linkage, 78, 419–420, *290–291*
Genetic mapping, 78, 415–427, *290–302*
ABO blood type, 420–422, *291–293*
analysis of molecular data, 424–427, *295–298*
historical note about, 422, *293*
of human disease gene, 415–427, *290–302*
Human Genome Project, 429, *300*
mapping genes to chromosome locations, 422–423,
293–294
by microarray analysis, 433
of nail-patella syndrome, 420–422, *291–293*
using short tandem repeats, 415–427, *290–302*
Genetic marker, 155–156, 419, *290*
Genetic predisposition, 571–573, 578, *377*
Genetic privacy, 430, *301*, 564, *372*, 574–575
Genetic screening, 570–571
neonatal, 571–572

Heat-stable toxin, 294–295
Height, 362–363, *253–254, see also* Short stature
Helix-loop-helix motif, 121–123, 126–129, 134
Hemoglobin, 131, 156, 188, 332, *238*, 353, 361–362, *252–253*
Hemolytic-uremic syndrome, 294–296, 320
Hemophilia, 22–23, 165, 585
Herbicidal protein, 32
Herbicide, 482, 488, *334*
Herbicide resistance, 523–524, *352–353*
Herbicide tolerance, 30–31
  in genetically engineered crops, 500, 506
Herceptin, 371, *262*
Hereditary disease, *see* Genetic disease
Herpesvirus infection, drug treatment, 290–291, *218–219*
Hershey-Chase experiment, 79–80
Heterozygote, 344, 352–353, *see also* Carrier of genetic disease
HhaI, 220, *189*, 224, *190–193*
High-fructose corn syrup, 445, *311*
HindIII, 138, 220, *189*, 224–225, *190–193*, 228, *192*, 239–240, *194*, 249, *200*
Hippocratic Oath, 554, 559, 566, *374*
Histone, 86
HIV, *see* Human immunodeficiency virus
Home pregnancy test, 5, 20, 151
Homeodomain, 124
Homeotic gene, 160, 163
Homocystinuria, 571
Homologous chromosomes, 12
Homoserine acyltransferase, 70
Homoserine kinase, 70
Homozygote, 344, 352
Honor code violation, 557–560, 565–567, *373–375*
Hooke, Robert, 45–46
Horizontal gene transfer, 313–318, *228–233*
Hormone(s), manufacture of, 8
Hormone replacement therapy, 528
Hospital mix-up of babies, 403–404, 410, *285*
Hospital-acquired infection, 314, *229*, 316, *231*
Host range, of plasmids, 306
Housekeeping genes, 204
Howard Hughes Medical Institute, 354
Human, amylase genes of, 456–457
Human chorionic gonadotropin (HCG), 151
Human cloning
  potential benefits of, 522, *351*
  risks involved in, 522, *351*
Human embryonic cells, 507–511
Human evolution, 158, 160, 455
Human genome, 113, 363, *254*, 455
  size of, 206–211
Human Genome Project, 20–21, 113, 428–431, *299–302*, 572, 578, *377*
  ethical, legal, and social issues, 430, *301*, 579, *378*, 582, *381*
  functional genomics, 430, *301*
  genetic linkage maps, 429, *300*
  goals and objectives of, 428, *299*
  information sources, 431, *302*

model organisms and, 428–429, *299–300*
  physical maps, 429, *300*
  sequencing DNA, 429–430, *300–301*
Human immunodeficiency virus (HIV), 180, 290–291, *218–219*, 544, *362*
Human molecular genetics, 350–372, *250–263*
Human remains, identification of, 399–400, *281–282*, 414, *289*
Huntington, 352
Huntington's disease (HD), 23, 352, 570, 578, *377*
  genetic testing for, 576–581, *379–380*
Hyaluronate, 22
Hybridization (breeding), transgenic plants and wild plants, 502, 504–505
Hybridization analysis (nucleic acids), 257–269, *205–209*
  comparing genotypes using, 157
  in DNA typing, 161
  DNA-DNA, 140–141
  for genome comparisons, 392, *274*
  microarray technology, 432–436, *303–305*
  probe for, *see* Probe
  with restriction analysis, 262–263, 267–269, *207–208*
  RNA-DNA, 140–141
  RNA-RNA, 202–203
  screening DNA libraries by, 257–259
  Southern, *see* Southern hybridization analysis
  wet laboratories, 264
  whole-genome, 383
Hydrocarbons, 62
Hydrogen bond
  in DNA, 97–98, 173, 257
  formation of, 118
  in proteins, 118–119
  in water, 118
Hydrogenated oils, 33
Hydrophilic molecule, 46–51, 54–55, 57, 118–119
Hydrophobic molecule, 46–51, 54–55, 57, 118–119, 173
Hypercholesterolemia, familial, 352
Hypersensitive response, in plants, 31
Hypervariable repeats, in DNA, 408, *283*
Hypothyroidism, 571

I

Identical twins, 149
"If/then" statement, 493, *339*
Imatinib mesylate, 371, *262*
Immortality, of cancer cells, 110
Immunization, risks and benefits of, 512, *341*
Immunosuppressive therapy, 25
Implantation, 507–509
In vitro fertilization (IVF), 509, 519–521, *348–350*, 571, 583, *382*
Inborn error of metabolism, 68
Inbreeding, 88, 375–380, *264–269*, 385
Independent Assortment, Principle of, 76–77, 90
Individual identity, 160–161
Individual rights, 544, *362*
Industrial nation, 488, *334*, 498

Marker gene, 303, *222, see also* Genetic marker
  for cloning DNA, 146–147
Marker-based genome comparisons, 383–384
Market opportunity, 482–483
Mass spectrometry, 135, 152
Maternal inheritance, 111
Maternity case, 410, *285*
Mating
  nonrandom, 88
  random, 88
MC1, *see* Melanocortin 1
McCarty, 298
McClintock, Barbara, 82
McKusick, Victor A., 378, *267*, 464
McLeod, 298
Medical biotechnology, 19–30
Medical device, biopolymers as, 22
Medical research tools, 29–30
Medical sleuth, 373–380, *264–269*
Medical Subject Headings, 464
MEDLINE, 464–465, 474, *326*
Melanocortin 1 (MC1), 343–349, *246–249*
Melanocortin 1 receptor (MC1R), 343–349, *246–249*
Melanocyte, 336, 340, *243*, 343, 346–347, *246–247*
Melanoma, 344
Melting temperature
  of DNA, 257–258
  of protein, 123
Membrane protein, 51, 202
Mendel, Gregor, 74–77
Mendelian principles, 76
Meningitis, 379, *268*
Mennonites, genetic diseases among, 375–380, *264–269*
Mental retardation, 360, *251*, 571
Mesoderm, 507–508
Messenger RNA (mRNA), 102, 195–196, *176–177*
  central dogma, 81–83
  enzymatic degradation of, 202–203
  making cDNA from, 142–143, 148
  microarray analysis of, 433, 435, *304*
  splicing of precursor RNA, 106, 188–189
  synthesis of, 102–104, 187–205, *176–181*
  synthesis of protein from, 103–105, 187–205, *176–181*
    prevention by antisense RNA, 198, *179*
    rate of, 109
Metabolic engineering, 16–17
Metabolic pathway, 64–68
  branching, 65–66, 70
  interruption of, 68
Metabolism, 58–59, 63–64
  metabolic problems, 68
Metamorphosis, in axolotl, 455–456
Metastasis, 367, *258*
Methicillin-resistant *Staphylococcus aureus*, 314, *229*
Methionine, biosynthesis of, 70
Methylene blue, 231, 241, *196*
Methylene chloride, 542, *360*
Microarray, 16–17, 162, 433
  DNA, 17
  features of, 432

protein, 17
small-molecule, 18
tissue, 17
whole-cell, 18
Microarray analysis, 17–18, 432–436, *303–305*
  applications of, 433
  of gene expression, 432–436, *303–305*
  of genotype, 433, 436, *305*
  identification of genes in metamorphosis, 456
  pharmacogenomic fingerprinting by, 438, *307*
  simultaneous analysis of many genes, 162
Microbial fermentation, 3, 7–8
Microbiological methods, 665–667
Microinjection of DNA, 146, 165, 330
Microlesion, 88
Micromutation, 88–89
Microorganisms, handling in laboratory, 661, *453*
Micro-RNA (miRNA), 109, 199, *180*, 204
Microsatellite, 383–384, 392–393, *274–275*, 395, *277*, 415
Miescher, Frederick, 78
Milk, "rBGH free," 35
Milkweed, 530–536, 544–549, *362–367*
Minisatellite, 408, *283*
Miscarriage, 350
Missing soldiers, 400, *282*
Mitochondria, 47
Mitochondrial DNA, 111, 157, 399–400, *281–282*, 462, *321*
  in DNA typing, 399–400, *281–282*
  of dogs, 394–395, *276–277*
  human, 399–400, *281–282*, 414, *289*
  inheritance of, 399–400, *281–282*
  of Neanderthals, 158–160
  reconstructing evolutionary relationships, 394–395, *276–277*
Mitochondrial DNA diseases, 522, *351*
Mitosis, 59
Model organisms, 155–156, 166–167
  for Human Genome Project, 428–429, *299–300*
Modular protein, 124–125
Molecular binding specificity, 49
Molecular biology, 97–136
Molecular clock, 461–462, *320–321*
Molecular cloning, 15, 516, *345*
Molecular data, use in genetic mapping, 424–427, *295–298*
Molecular genetics
  of cancer, 367–372, *258–263*
  human, 350–372, *250–263*
Molecular interactions, 46–49
  molecular binding specificity, 49
  with water, 46–49
Molecules, 51–52
  skeleton representations of, 52–53
Monarch butterflies
  Bt corn and, 527, 530–536, 544–549, *362–367*
    Cornell study, 548–549, *366–367*
    milkweed near cornfields, 534–535
    risk management, 535–536
    risk to butterfly larvae, 533–535

Product of chemical reaction, 64
Prokaryotic cells, 47–48
  cell division in, 59
  molecular biology of, 105–106
  promoters in, 187
Promoter, 103, 105, 108, 114, 187–188, 201
  alteration of promoter recognition, 109, 201
  eukaryotic, 187
  prokaryotic, 187
Propidium iodide, 260–261
Proportionality, principle of, 556, 564, *372*
"Prosecutor's fallacy," 406
Protease, HIV, 290–291, *218–219*
Protease inhibitor, 291, *219*
Protein(s), 4, 70–71
  allosteric, 69
  amino acid sequence of, 86, 101, 115–116, 158, 187,
      431, *302*, 455–462, *318–321*
    online resources, 134, 463–475, *322–327*
  analysis of, 150–152
  C terminus of, 115–116
  catabolism of, 65–67
  central dogma, 81–83
  conserved regions of, 456, 459, *318*
  denaturation of, 123, 212, 449, 452, *315*
  detection using antibodies, 150–151
  dietary, 196, *177*
  directed protein evolution, 41
  domains of, 122–125, 130, 160
  drawing of, 120–121
  effects of mutations on, 129–132
  endogenous therapeutics, 22
  as energy source, 63
  evolution of, 158, 455–462, *318–321*
  functions of, 57, 60–61, 64, 100, 115–135, 195, *176*
  gel electrophoresis of, 150, 233, 431, *302*, 449–454,
      315–317
  hydrogen bonds in, 118–119
  hydrophilic backbone of, 119
  hydrophobic core of, 119–120
  melting temperature of, 123
  membrane, 51
  metabolism of, 58
  modular, 124–125
  N terminus of, 115–116
  natural regenerative, 26
  natural selection and, 456
  nuclear magnetic resonance of, 115, 151–152
  primary structure of, 115–116
  quaternary structure of, 122–123
  relationships among genes, proteins, and observable
      traits, 83–86
  repeating units of, 53
  replacement therapies, 22–23
  ribbon drawing of, 121
  secondary structure of, 120–122
  stability of, 123–124
  structural motifs in, 121–122
  structure of, 56–57, 64, 100–101, 115–135, 195, *176*
  structure-function predictions, 132–134

  synthesis of, 67–68, 101–106, 135
  tertiary structure of, 122–123, 451
  three-dimensional shape of, 100–101, 115, 151–
      152
  X-ray crystallography of, 115, 151
Protein microarray, 17
Proteinase, 127, 212, 216, *185*
Protein engineering technology, 14–15, 41
Proteomics, 135, 152, 430, *301*
Proto-oncogene, 368–369, *259–260*
Public opinion, 482–483, 489–491, *335–337*
PubMed, 465, 468, 475, *327*
Pulp and paper industry, 6, 37–38, 41
Pumps, 60
"Pure breed," 363, *254*
Pure culture, 666–667
Purine, 173, 177, *172*
Pyrimidine, 173, 177, *172*
Pyruvic acid, 8, 66–67

Q

Quantitative PCR, 271
Quaternary structure, of proteins, 122–123
Quinolone resistance, 310, *225*, 317, *232*
Quinupristin/dalfopristin, 317–318, *232–233*
Quorum sensing, 61, 70

R

R plasmid, 316, *231*
Radiation hybrid cell line, 385
Random amplification of polymorphic DNA, 383
Random mating, 88
Rape case, 402–403, *see also* Forensic case
ras gene, 368–369, *259–260*, 371, *262*
Reading frame disruption, 354
Receptor protein, 61, 100–101, 109, 114, 132–133, 343–
      349, *246–249*, 368–369, *259–261*
Recessive gene, 76, 336, 339–340, *242–243*, 352
Recombinant DNA, 244–250, *199–201*, *see also*
      Restriction analysis
  production of, 137–140
    cutting DNA with restriction enzymes, 138
    ligation of DNA fragments, 139–140
    separating mixtures of DNA fragments, 138–140
Recombinant DNA technology, 12, 16, 135, 137–167,
      221, *see also* Genetic engineering
  cloning DNA, 146–150
  comparing genotypes and genomes, 157–162
  finding genes, 152–156
  manipulation and analysis of DNA, 137–146
  protein analysis, 150–152
  regulation in United States, 550–553, *368–371*
Recombinant plasmid, 244–250, *199–201*
Recombination, 87, 136, 390, *272*, 418–422, *290–293*
  asexual reproduction and, 91–93
  sexual reproduction and, 89–92
  as source of genetic variation, 89–93
Recombination frequency, 421–422, *292–293*

S

Saccharomyces cerevisiae, model organism for Human Genome Project, 428–429, *299–300*

Safeguards, 542, *360*

Safety
food, *see* Food safety
handling *E. coli*, 295–297
laboratory, 661–662, *453–454*

SAGE, *see* Serial analysis of gene expression

Saliva, testing for amylase activity, 441–442, 446–448, *312–314*

Salivary amylase, 440–442, 444–445, *310–311*, 449–450, 453, *316*, 465

Salmonella, 36

Sample well, 226–227, *190–191*

Sanitary infrastructure, 318, *233*

Sarcoma, 367, *258*

Satellite colony, 299

Saturated fatty acids, 53–54

Scarlet fever, 314, *229*

SCID, *see* Severe combined immunodeficiency disease

Science
civic responsibility and, 489–491, *335–337*
difference between science and technology, 487, *333*
nature of, 492–494, *338–340*
science, technology, and society, 481–494, *332–340*

Scientific literature, 491, *337*, 493, *339*
online resources, 464, 468, 474–475, *326–327*

Scientific model, 82

Scientific process, 489, *335*, 491, *337*, 493–494, *339–340*

Scientific research, *see* Research

Scientific understanding, 487–488, *333–334*

SCNT, *see* Somatic cell nuclear transfer

SDS, *see* Sodium dodecyl sulfate

"Search Gene Map," 355

Search tutorials, 465

Secondary structure, of proteins, 120–122

Segregation, Principle of, 76, 90

Selective breeding, 12–13, 74
comparison with genetic engineering, 13–14

Sequence-tagged site (STS), 430, *301*

Sequencing gel, 287, *215*

Serial analysis of gene expression (SAGE), 431, *302*

Serine proteinase, 127, 129

Severe combined immunodeficiency disease (SCID), 23–24
gene therapy for, 591–592, *384–385*

Sewage treatment, 37

Sex chromosomes, 351, 410, *285*

Sex-linked inheritance, 78

Sexual reproduction, recombination and, 89–92

Shiga toxin, 294, 296

Shiga-like toxin, 320

Shigella, antibiotic resistance in, 313, *228*

Shigella dysentery, 294–295

Shingles, 319

Shipping fever, 34

Short stature
growth hormone for, 590, 594–595, *387–388*
growth hormone gene therapy for healthy children, 589–590, 594, *387*

Short tandem repeat (STR), 382
DNA typing, 403, 408, *283*, 411, *286*
genetic mapping using, 415–427, *290–302*, 436, *305*
genome comparisons using, 384–385, 393–394, *275–276*
physical mapping of chromosome markers, 385

Shrimp, amylase genes of, 456–457

Sickle-cell anemia, 85–86, 131, 156, 353, 361–362, *252–253*, 570–571, 574, 578–579, *377–378*

Sickle-cell trait, 353, 361–362, *252–253*, 573

SIDS (sudden infant death syndrome), 379, *268*

Signaling molecule, 60–61, 64

Signaling pathway, 368–370, *259–261*

Silent mutation, 114

Simple sequence repeat (SSR), 392–393, *274–275*

Single nucleotide polymorphism (SNP), 383, 433

Single-cell organisms, 47

Single-gene disorder, 350–352

siRNA, *see* Small interfering RNA

sis gene, 371, *262*

Skin, artificial, 27

Skin cancer, 367–368, *258–259*

Slippery slope, 556, 564, *372*

SmaI, 220, *189*, 224, *190–193*, 267, *207*

Small interfering RNA (siRNA), 199, *180*, 203–204

Small-molecule microarray, 18

Smallpox vaccine, 28

Smith, E. B., 77–78

SNP, *see* Single nucleotide polymorphism

Social behavior, 386

Social skills, of students, 554

Societal issues, *see also* Bioethics
bioengineered food, 523–526, *352–355*
biotechnology's risks and benefits, 527–549, *356–367*
cloning, 512–522, *341–351*
embryonic stem cells, 507–511
gene flow from transgenic crops, 500–506
gene therapy, 585–595, *384–388*
genetic testing, 569–581, *377–380*
Human Genome Project, 430, *301*, 579, *378*, 582, *381*
rational analysis of, 495–526, *341–355*
analytical process, 498–499, 513–514, *342–343*
basic approach, 497–498
comparison of risks, 515, *344*
considering safety issues, 514–515, *343–344*
considering societal impact, 515, *344*
placing issue in context, 514, *343*
science, technology, and society, 481–494, *332–340*

Sodium dodecyl sulfate (SDS), 449, 452, *315*

Sodium-potassium pump, 60

Soil erosion, 482, 488, *334*

Soil organisms
antibiotic-producing, 315, *230*
testing for amylase activity, 441–442, 446–447, *312–313*

Somaclonal variant selection, 36

Somatic cell enhancement, 585, 591, *384*

Somatic cell gene therapy, 585, 591, *384*

Somatic cell nuclear transfer (SCNT), 510, 519–520, *348–349*, 522, *351*, 584, *383*

Southern blotting, 259–261, 267, *207*

monoclonal antibody-based, 5–6
natural products as, 21–22
pharmacogenomics, 437–438, *306–307*
plant-produced, 11
regenerative medicine, 25–28
RNAi drugs, 204
vaccines, *see* Vaccine
Thermal cycler, 271, 274–275, 278, *212*
Threonine, biosynthesis of, 70
Thymine, 57, 81, 97–98, 102–103, 172–173, 177, *172*
Thymine dimer, 368, *259*
Thyroid hormone, 456
Thyroid hormone receptor, 456
Ti plasmid, 146, 164, 329–333, *238–239*
disarmed, 330, 332, *238*
Tigecycline, 318, *233*
Tissue culture, 5
Tissue engineering, 27
Tissue microarray, 17
Tissue plasminogen activator, 22
Tobacco, genetically engineered, 23, 332, *238*
Tobacco mosaic virus, 112, 164
Tomato, genetically engineered, 36
Total-community genomics, 39
Toxic shock syndrome, 296, 314, *229*, 320
Toxicity, 528–529, *356–357*, 539–541, *357–359*
Toxicology, 540, *358*
Toxic-waste site, 37–38
Toxin, 294, 314, *229*, 320
Trait, relationships among genes, proteins, and observable traits, 83–86
Transcription, 81–82, 102–104, 187–205, *176–181*
Transcription factor, 105, 108–109, 124, 187
Transcriptional activator, *see* Transcription factor
Transcriptional regulation, 192–193
activation, 108–109, 201
alteration of promoter recognition, 109, 201
repression, 107–108, 201
Transductant, 321
Transduction, 12, 93, 113, 320
of antibiotic resistance genes, 319–328, *234–237*
in *E. coli*, 325–328, *234–237*
medical importance of, 320
natural, 320, 325, *234*
Transfer method, sterile transfer of large volumes, 664, *456*
Transfer RNA (tRNA), 102, 196, *177*
splicing of precursor molecules, 188
structure of, 104
suppressor, 321
in translation, 103–105
Transformation, 12, 92–93, 245, 249, *200*, 251, 258
to cancer cell, 368, *259*
cloning DNA, 146–147
in *E. coli*, 298–305, *222–224*
natural, 298
Transforming factor, 79
Transforming growth factor beta, 26
Transgenic animals, 13, 34
manufacture of pharmaceuticals using, 23

Transgenic plants/crops, 13, 30–31, 164–167
gene flow from, 500–506, *see also* Gene flow
manufacture of pharmaceuticals using, 23
regulation in United States, 550–553, *368–371*
regulatory approval process for, 530–531, 549, *367*
Transient gene therapy, 24
Translation, 81–82, 103–105, 187–205, *176–181*
Translational repression, 109, 201
Translocation, 87–90, 350–351, 390, *272*
balanced, 351
Transplant rejection, 25
Transport protein, 60, 100
Transposable element, 89
Transposase, 89
Transposition, 114, 316, *231*
Transposon, 83, 89, 113
carrying antibiotic resistance genes, 316, *231*
carrying pathogenicity genes, 296
discovery of, 82
of fruit fly, 154
nonreplicative, 89, 91
replicative, 89, 91
Trastuzumab, 371, *262*
Travelers' diarrhea, 294–295
Tricarboxylic acid (TCA) cycle, 66–67
Trinucleotide repeat expansion disease, 352
Triose phosphate isomerase, 121–122
Tris-borate-EDTA buffer, 231
Triton X-100, 449, 451–452, *315*
tRNA, *see* Transfer RNA
Trophoblast, 507–508
trp operator, 127
trp operon, 107–108, 201
trp repressor, 108, 126–127, 129
Truth telling, 556, 559, 564, *372*
Trypsin, 129
Tryptophan, 107, 127, 129
Tumor, 367, *258*
Tumor necrosis factor, 25
Tumor suppressor gene, 25, 351, 369–371, *260–262*
Turner's syndrome, 351
Twinning, 149
Tygacil, 318, *233*
Tyrosinase, 340, *243*, 344, 346–347, *246–247*, 359, *250*
Tyrosinase-related protein 1 (TYRP1), 336, 340, *243*, 343–344, 346–347, *246–249*
Tyrosinase-related protein 2 (TYRP2), 336, 340, *243*, 343, 346–347, *246–249*
Tyrosine, 336, 340, *243*, 343, 346–347, *246–247*, 359–360, *250–251*
TYRP, *see* Tyrosinase-related protein

U

Ultraviolet light, DNA damage caused by, 367, *258*
Underground storage tank, leakage from, 37–38
Unfertilized egg, stem cells from, 510–511
United States Department of Agriculture (USDA), biotechnology regulation by, 530–531, 541, *359*, 548, *366*
Unknown Soldier, 400, *282*